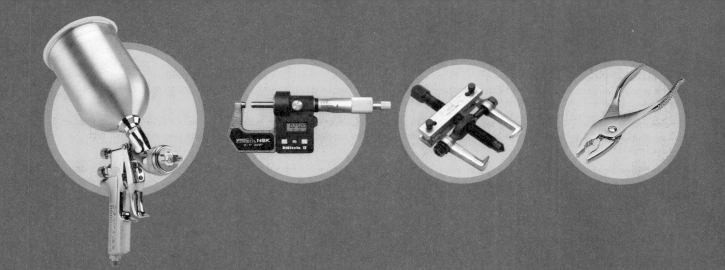

I-CAR Professional Automotive Collision Repair

2ND EDITION

JAMES E. DUFFY

DELMAR

THOMSON LEARNING

Australia • Canada • Mexico • Singapore • Spain • United Kingdom • United States

Cover Image: Vehicle Image Courtesy of General Motors Corporation, Service Operations

Delmar Staff

Business Unit Director: Alar Elken
Executive Editor: Sandy Clark
Developmental Editor: Allyson Powell
Editorial Assistant: Matthew Seeley
Executive Marketing Manager: Maura Theriault
Channel Manager: Mona Caron

Marketing Coordinator: Brian McGrath
Executive Production Manager: Mary Ellen Black
Senior Production Coordinator: Karen Smith
Project Editor: Christopher Chien
Art/Design Coordinator: Cheri Plasse
Technology Project Manager: David Porush

Library of Congress Cataloging-in-Publication Data

Duffy, James E.
 I-CAR professional automotive collision repair / James E. Duffy.- -2nd ed.
     p. cm.
   Includes index.
   ISBN 0-7668-1398-3
   1. Automobiles- -Bodies- -Maintenance and repair. I. Title: Professional automotive collision repair. II. Inter-Industry Conference on Auto Collision Repair. III. Title.
   TL255 .D84 2000
   629.2'6'0288–dc21

                    00-050856

Asia (including India):
Thomson Learning
60 Albert Street, #15-01
Albert Complex
Singapore 189969
Tel 65 336-6411
Fax 65 336-7411

Australia/New Zealand:
Nelson
102 Dodds Street
South Melbourne, Victoria 3205
Australia
Tel 61 (0)3 9685-4111
Fax 61 (0)3 9685-4199

Latin America:
Thomson Learning
Seneca 53
Colonia Polanco
11560 Mexico D. F. Mexico
Tel (525) 281-2906
Fax (525) 281-2656

Canada:
Nelson
1120 Birchmount Road
Toronto, Ontario
Canada M1K 5G4
Tel (416) 752-9100
Fax (416) 752-8102

UK/Europe/Middle East:
Thomson Learning
Berkshire House
168-173 High Holborn
London WC1V 7AA
United Kingdom
Tel 44 (0)171 497-1422
Fax 44 (0)171 497-1426

Business Press
Berkshire House
168-173 High Holborn
London WC1V 7AA
United Kingdom
Tel 44 (0)171 497-1422
Fax 44 (0)171 497-1426

Spain:
Parainfo
Calle Magallanes 25
28015 Madrid
España
Tel 34 (0)91 446-3350
Fax 34 (0)91 445-6218

Distribution Services:
ITPS
Cheriton House
North Way
Andover,
Hampshire SP10 5BE
United Kingdom
Tel 44 (0)1264 34-2960
Fax 44 (0)1264 34-2759

International Headquarters
Thomson Learning
International Division
290 Harbor Drive, 2nd Floor
Stamford, CT 06902-7477
USA
Tel (203) 969-8700
Fax (203) 969-8751

CONTENTS

SECTION 1   FUNDAMENTALS

SECTION 2

MINOR REPAIRS

SECTION 3

PREPAINTING PREPARATION

SECTION 6   OTHER OPERATIONS

SECTION 7   ESTIMATING AND ENTREPRENEURSHIP

PREFACE

ABOUT THE BOOK

Today, more than ever, automotive collision repair technicians are needed. However, to be a successful collision repair technician, the student must master a wide range of skills. The *I-CAR Professional Automotive Collision Repair, 2e* text will help students master those skills by providing them with the knowledge they need in an easy-to-understand format. The sentence structure, layout of the text, and the large number of illustrations will help students maximize their learning.

This text provides an explanation of the theory behind a successful repair. It also provides information on the latest collision repair tools, equipment, and techniques and offers important safety tips and strategies for students to use in protecting themselves and the environment. Finally, it provides the foundation needed to prepare for the ASE Certification Exam and offers insight into what it takes to become a successful, well-rounded collision repair technician.

New to This Edition

The core of this text remains the text you know and trust, but we have added the following exciting new features to increase student interest, improve comprehension, and advance learning:

- The refinishing chapters (16, 17, and 18) are now in full color.
- Over 300 new images improve visual interest, add instructional value, and update core content.
- Each chapter now includes 10 ASE-style questions to provide students with practice for the ASE Certification Exam.
- Technical terms are bolded, italicized, and defined for easy identification and mastery.
- There is an increased emphasis on safety throughout.

In addition, the content has been updated to reflect key changes in the industry. For example, new information has been added on:

- Gravity-feed spray guns, HVLP spray guns, and OBD II.
- Manual mixing scales, electronic mixing scales, and computerized service information.
- Material Safety Data Sheets (MSDS), Personal Protective Equipment (PPE), Right-to-Know Laws, material hazards, hazardous material exposure, waste disposal and recycling.
- Plastic part removal and plastic adhesive repair systems.
- Abrasive grinding discs, masking plastic, masking foam, sponge buffing pads, vehicle cleaning, and sanding guide coat.
- Electronic measuring systems, including laser measuring systems, ultrasound measuring systems, and computerized robotic arm measuring systems.
- Scan tool cartridges, scan tool controls, scan tool connection, using the scanner, switch, actuator, and snap-shot diagnostics, erasing trouble codes, and computerized estimating programs.

FEATURES OF THE TEXT

Learning how to repair collision damage on today's vehicles can be a challenge. To guide readers through this complex material, we have included a series of features that will ease the teaching and learning processes.

OBJECTIVES

Each chapter begins with a list of objectives. The objectives state the expected outcome that will result from completing a thorough study of the contents of the chapter.

DANGER

Instructors often tell us that safety is their most important concern. Danger notes appear throughout the text to alert students to important personal safety information.

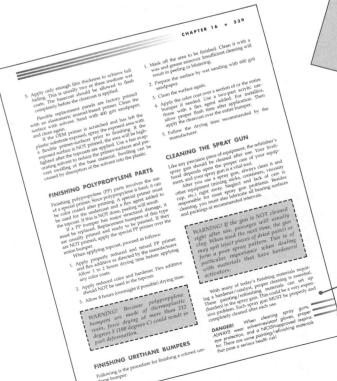

WARNINGS

Warnings provide students with important guidelines to follow to prevent damage to tools, equipment, the vehicle, or the repair.

NOTES

The notes sprinkled throughout the text emphasize key points or refer students to related information in other chapters.

TECHNICAL TERMS

Each chapter ends with a list of the terms that were introduced in the chapter. These terms are defined in the glossary and are highlighted in the text upon first use.

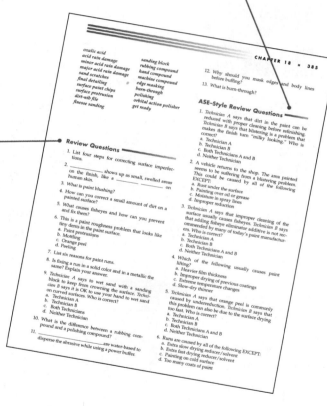

SUMMARY

Highlights and key bits of information are listed at the end of each chapter. This list can be used as a refresher and as a study tool.

ASE-STYLE REVIEW QUESTIONS

Each chapter ends with a series of ASE-style questions that will provide students with practice for the ASE Certification Exam.

REVIEW QUESTIONS

A combination of true/false, short-answer, fill-in-the-blank, multiple choice, and ASE-style questions make up the end-of-chapter review questions. Different question types are used to challenge the reader's understanding of the chapter content.

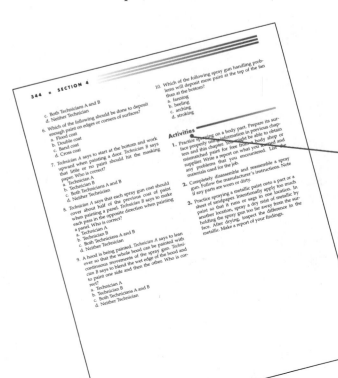

ACTIVITIES

Activities are included at the end of each chapter. These activities offer the readers the opportunity to apply what they have learned.

Delmar offers a number of excellent video and CD-ROM products on automotive refinishing and repair. For additional information, please visit our website at www.Autoed.com or call 1-800-347-7707.

SUPPLEMENTS

TECHNICIAN'S MANUAL

An all new Technician's Manual has been created for the second edition of the *I-CAR Professional Automotive Collision Repair* text. Divided into two parts, the first includes numerous Activity Sheets covering NATEF's Applied Academic Skills for Collision Repair. These sheets will give students the opportunity to develop Language Arts, Mathematics, Science, and Workplace Skills within the context of collision repair.

The second part of the Technician's Manual includes numerous Job Sheets covering NATEF's Collision Repair Task Lists. Broken up by category, these Job Sheets will give students the hands-on practice needed to master the NATEF tasks.

The Technician's Manual also includes a handy chart for students to use to track their progress toward mastery.

INSTRUCTOR'S MANUAL

The Instructor's Manual has been expanded and reformatted to meet the needs of today's busy instructor. In addition to an answer key for the core text, the Instructor's Manual now includes the following exciting resources:

- Lecture outlines to facilitate classroom instruction

- A pre-test, post-test, final exam, and a series of chapter quizzes to assess learning. (These are ready for photocopying and distribution to students.)

- A Technician's Manual answer key to make grading a snap

- Numerous transparency masters to add visual interest to classroom presentations

All of these resources will enable the instructor to spend less time preparing and more time teaching!

ABOUT I-CAR

Founded in 1979, I-CAR (Inter-Industry Conference on Auto Collision Repair) is an international, not-for-profit training organization dedicated to improving the quality, safety, and efficiency of auto collision repair for the ultimate benefit of the consumer. You may obtain additional information about this important organization by visiting page 37 of this text.

ACKNOWLEDGMENTS

The publisher would like to extend special thanks to Tom McGee of I-CAR for his support and assistance during the revision of this text. The publisher would also like to extend special thanks to Warren Barbee, Michael Jund, and Alfred Thomas for their important contributions to this second edition.

Reviewers

We would like to acknowledge and thank the following dedicated and knowledgeable educators for their comments, cricitisms, and suggestions during the review process:

Kim Helgeson, *Wyoming Technical Institute*
Lalo Hernandez, *St. Philips College (SWC)*
Daniel Ide, *Paradise Valley High School*
Michael Jund, *Scott Community College*
Thomas Mack, *Vale Tech—Remington Education Center*
Alfred M. Thomas, *Pennsylvania College of Technology*
Steven M. White, *Portland Community Colleges*

Contributing Companies

We would like to thank the following companies who provided technical information and art for this edition:

3M Automotive Trades Division; American Honda Motor Co., Inc.; American Isuzu Motors Inc.; American Suzuki; NIASE; Audi of America; Babcox Publications; Badger Air Brush Co.; BASF; Bee Line Co.; BlackHawk Collision Repair, Inc.; BMW of North America; Carborundum Abrasives North America; Car-O-Liner; Chief Automotive Systems; Cooper Tools; DaimlerChrysler Corporation; Danaher Tool Group; Dataliner AB; DeVilbiss Automotive Refinishing Products; Dorman Products; Dynabrade, Inc.; Edelmann; Equalizer Industries; Evercoat Fibreglass Co., Inc.; FMC Automotive Equipment Division; Ford Motor Company; Fred V. Vowler Co., Inc.; General Motors Corporation, Service Operations; Hein-Werner; Helicoil, Fastening Systems Division; Hopkins Manufacturing Corporation; HTP America, Inc.; Hunter Engineering Company; I-CAR; ITW Automotive Refinishing; Kansas Jack; Klein Tools, Inc.; L.S. Starrett Company; Lab Safety Supply Inc., Janseville, WI; Lincoln Electric Co.; MAACO Enterprises, Inc.; Marson Corp.; Mazda Motor of America, Inc.; McQuay-Norris; Mercedes-Benz of North America, Inc.; Mitchell International, Inc.; Mitsubishi Motor Sales of America, Inc.; Morgan Manufacturing Inc.; Munsell Color Group; Mustang Monthly Magazine; NATEF; NCG Company/Spotle Distributor; Nissan North America, Inc.; Norco Industries, Inc.; Oatey Bond Tite; OTC, a Division of SPX Corporation; Palnut Co.; PBR Industries; Perfect Circle/Dana Corp.; Porsche Cars North America, Inc.; PPG Industries, Inc.; Riverdale Body Shop; Rotary Lift/A Dover Industries Co.; Saab Cars USA, Inc.; Sartorius; Seelye, Inc.; Sellstrom Mfg.; Snap-on Tools Corporation; Stanley Tools; Subaru of America and Fuji Industries, Ltd.; Talsol Corporation; Team Blowtherm; Tech-Cor, Inc.; The Eastwood Company; TIF Instruments, Inc.; TRW Plastic Fasteners and Retainers Division; U.S. Chemicals and Plastics, Inc.; Volkswagen of America; Wedge Clamp International, Inc.

Portions of materials contained herein have been reprinted with permission of General Motors Corporation, Service Operations. The photographs of the Dodge Charger R/T and Dodge Intrepid are used with permission from DaimlerChrysler Corporation.

Section 1

Fundamentals

Chapter 1
Industry Overview and Careers

Chapter 2
Safety and Certification

Chapter 3
Vehicle Construction

Chapter 4
Hand Tools

Chapter 5
Power Tools and Equipment

Chapter 6
Basic Measurement
and Service Information

Chapter 7
Fasteners

Chapter 8
Collision Repair Hardware
and Materials

Chapter 9
Welding, Heating, and Cutting

1

Industry Overview and Careers

Objectives

After studying this chapter, you should be able to:

■ Summarize the collision repair industry.

■ List the related skills needed to become a good collision repair technician.

■ Explain the typical movement of a vehicle through a collision repair facility.

■ Summarize the major areas of repair in a body shop.

■ Describe basic procedures for repairing a collision damaged vehicle.

■ Know the meaning of being a true professional technician.

■ Identify the setup and inner workings of a typical collision repair shop and its various personnel.

Introduction

This chapter will overview the collision repair industry. It will build the "framework of knowledge" needed to fully understand later text chapters. Its content will also help you learn the basic terminology and repair methods of this field. You will, on paper, follow damaged vehicles and shop personnel through a typical process of estimating, parts ordering, structural repair, painting, detailing, and final delivery to the customer. See Figure 1–1.

Considering that we are a "nation on wheels," you have selected an excellent area of employment to study. The skills you learn will be in high demand as long as there are people driving cars and trucks.

If you are just beginning your study of collision repair, you have much to learn. Only highly skilled, knowledgeable professionals can properly repair collision damaged vehicles. Study the material in this textbook carefully and you will be on your way to a successful career in collision repair.

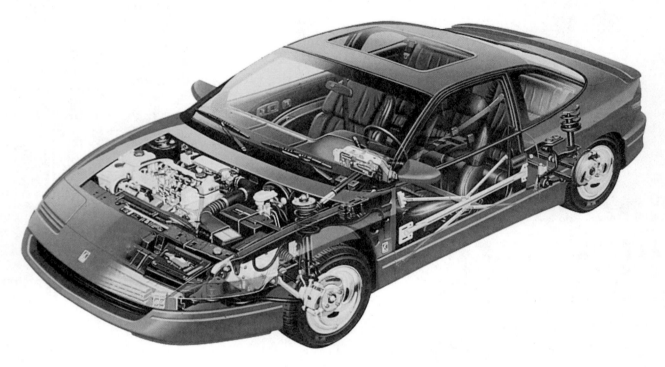

A

B

Figure 1-1 Various construction methods and designs are used in today's vehicles. To be a successful collision repair technician, you must be well educated and knowledgeable in various areas of automotive technology. (A) This vehicle uses flexible door or side panels that resist dents. It is a popular front-engine, front-wheel drive vehicle. *(Courtesy of General Motors Corporation, Service Operations)* (B) This sports car is made of high strength steel. It is a front-engine, rear-wheel drive vehicle. *(Courtesy of General Motors Corporation, Service Operations)*

HISTORY NOTE

Did you know that the first self-propelled vehicle was in a collision the very first time it was driven? The huge steam engine-powered vehicle, trying to avoid a horse-drawn carriage, crashed into a brick wall. The vehicle, the brick wall, and the driver were all "hurt." As this first crash of the first motor vehicle, on its first drive points out, there will always be collisions and a need for well-trained collision repair technicians.

COLLISION REPAIR TECHNICIAN

Collision repair technicians are skilled, knowledgeable people who know how to use specialized equipment and highly technical methods to restore collision damaged vehicles. They must possess the basic skills in the following areas:

1. **Metalworker**—A collision repair technician must be able to do all types of metalworking to properly form and shape sheet metal car bodies after a collision.

2. **Welder**—A collision repair technician must weld and cut steel, aluminum, composites, and plastics efficiently during body repairs.

3. **Auto repair technician**—A collision repair technician must remove and install mechanical systems, requiring some skill in automotive repair technology.

4. **Plumber**—A collision repair technician must work with numerous lines, hoses, and fittings during power steering, brake system, and fuel system service.

5. **Electrician**—A collision repair technician must be good at testing and repairing wiring and electrical components after damage. Finding shorts, opens, and other wiring problems is essential to today's computer-controlled vehicles.

6. **Air Conditioning Technician**—A collision repair technician may be required to work on air conditioning systems.

7. **Computer Technician**—A collision repair technician should be able to scan and repair computer problems stemming from collision damage.

8. **Painter/Refinishing Technician**—A collision repair technician should know how to restore vehicle finishes to their original condition after collision repairs are made.

As this points out, today's collision repair technician must be a highly skilled professional. Low skilled technicians can no longer survive and earn a living with modern, complex automotive technology.

OTHER PERSONNEL

The collision repair and paint technicians are the principals in any collision repair business. However, there are other jobs that must be done as well. The following describes other personnel that work in and with the collision repair shop.

The **shop owner** must be concerned with all phases of work performed. In smaller shops, the owner and shop manager are usually the same person. In large operations or dealerships, the owner might hire a shop manager. In all cases, the person in charge should understand all of the work done in the shop as well as its business operations.

The **shop supervisor** is in charge of the everyday operation of the shop. This job involves communication with all personnel who contribute to the facility's success.

The **parts manager** is in charge of ordering all parts (both new and salvaged), receiving all parts, and seeing that they are delivered to the ordering technician. Since not every collision repair shop has an actual parts manager, the task of ordering parts can fall on each employee at one time or another.

The **bookkeeper** keeps the shop's books, prepares invoices, writes checks, pays bills, makes bank deposits, checks bank statements, and takes care of tax payments. Many shops hire an outside accountant to perform these tasks.

The **office manager's** duties include various aspects of the business such as handling letters, estimates, and receipts. In many small shops, the office manager also acts as the parts manager and bookkeeper.

A **receptionist** is sometimes employed to greet customers, answer the phone, route messages, and do other tasks.

A **helper** or **apprentice** learns new jobs while assisting experienced personnel. He or she might help a technician mask a car before painting, install parts, or help clean up the work area at the end of the day.

THE PROFESSIONAL

The term "professional" refers to the attitude, work quality, and image that a business and its workers project to customers. It is interesting to note that in Europe a shop cannot open without the presence of a "Meister" or "master" craftsperson. A **professional:**

1. is customer oriented.

2. is up to date on vehicle developments.

3. keeps up with advancements in the repair industry.

4. pays attention to detail.

5. ensures that his or her work is up to specifications.

6. participates in trade associations.

NOTE! Other areas of employment in collision repair are summarized throughout this textbook.

WHAT IS COLLISION REPAIR?

Collision repair is the process of restoring a vehicle that has been damaged to its original condition by repairing both structural and cosmetic damage. Another term for collision repair is "auto body repair." See Figure 1–2.

With minor damage, this can be as simple as replacing a bumper or a grille, or fixing a small dent. With major damage, this can be a complicated process, requiring that you partially disassemble the vehicle and use frame straightening equipment, welding equipment, spraying equipment, and other specialized tools.

Many of the **mechanical systems,** including the steering, braking, and suspension components, can also be damaged in a collision. They too can require service by the collision repair technician.

Wires, sensors, and other electrical components can be damaged by the tremendous forces involved in a collision. Repair technicians must be capable of using a scan tool and test instruments to find and repair this circuit damage.

WHAT IS A COLLISION?

A *collision* is an impact that causes damage to the vehicle body and chassis. Since vehicles can weigh more than a ton, metal parts can be crushed, bent, and torn. Plastic or composite parts can be broken and deformed. The frame might even be forced out of alignment, all resulting from the tremendous force of a collision.

The direction of impact, force of impact, type of body structure, and other factors are all important. For example, damage might have resulted when two vehicles hit each other or when a vehicle hit a stationary object lying in or along the road. This kind of information can be useful to the estimator and body technician.

A collision might be as minor as a "door ding" where someone accidentally opened their door and hit another car. Or, it might be severe enough to cause a *total loss,* where repairs would be more expensive than the cost of buying another vehicle.

VEHICLE CONSTRUCTION BASICS

Vehicle construction deals with how the factory manufactured the vehicle. There are many construction variations. See Figure 1–3.

As a technician, you must understand construction differences. This information will greatly affect how you repair a vehicle. Various types of metal (steel, aluminum), plastics, and composite materials, in numerous combinations, are used in vehicles today. This makes the job of the collision repair technician very challenging. See Figure 1–4.

NOTE! Chapter 3, Vehicle Construction, will explain how a vehicle is made in more detail. Refer to this chapter for more information on this topic if needed.

Figure 1–2 During a collision, severe structural damage often occurs to the body, frame, and mechanical and electrical components in or near the impact area. *(© American Honda Motor Co., Inc.)*

Figure 1–3 Vehicle manufacturers are using more and more exotic materials in the design and construction of their vehicles. This increases the challenge to today's collision repair and refinishing technician. *(Courtesy of General Motors Corporation, Service Operations)*

Figure 1-4 When manufactured, automated equipment normally assembles and welds the body structure to help form the frame. Because of this, collision damage repair can be very challenging. *(Courtesy of General Motors Corporation, Service Operations)*

Frame and Body

The *frame* is usually a high strength metal structure used to support other parts of the vehicle. It holds the engine, transmission, suspension, and other parts in position. The frame can be separate from the body or the body can be welded together to form the frame. See Figure 1–5.

The vehicle *body* is a steel, aluminum, plastic, or composite skin forming the outside of the vehicle. The body is normally painted to give the vehicle its attractive appearance.

Frame Types. A *full frame* is a strong steel structure that often extends from the front to the rear of the vehicle. It is used on many trucks and some passenger vehicles. The frame serves as a foundation for the body and drive train.

Unibody describes a vehicle that has the body structure welded together to form an integral frame

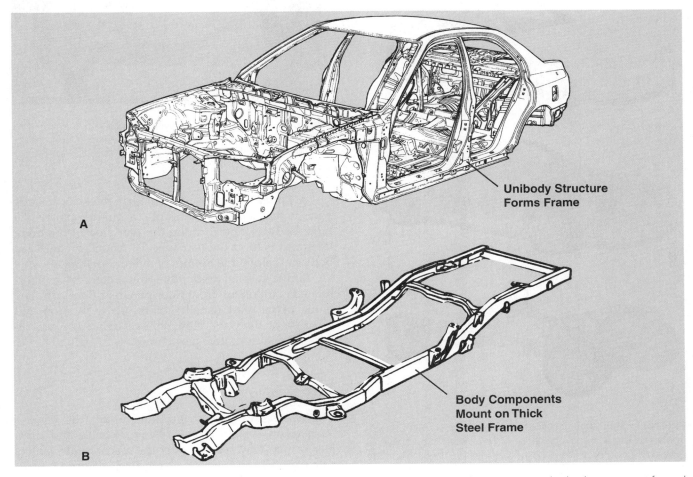

Unibody Structure Forms Frame

Body Components Mount on Thick Steel Frame

Figure 1-5 These are the two common methods of vehicle construction. (A) With unibody construction, the body structure is formed to provide a stiff, strong "frame structure." This is the most common design used on passenger cars. *(© American Honda Motor Co., Inc.)* (B) The separate frame was first used on the 1896 Oldsmobile. Because of its strength, the full frame can still be found on many sport utility vehicles (SUVs).

structure. Also called **space frame** or **unitized construction,** this is the most common type of body configuration. Unibody construction reduces weight, improves fuel economy, and has a high strength-to-weight ratio. See Figure 1–6.

With unibody construction, other body parts (fenders, doors, etc.) bolt to the unibody structure. All parts, even the roof, work together to help give the frame and body its strength. See Figure 1–7.

Parts and Components

The term **part** or **component** generally refers to the smallest units on a vehicle. An **assembly** is several parts that fit together to make up a more complex unit. For example, a car bumper assembly might be made up of several parts—bumper frame, bumper cover, bumper brackets, etc.

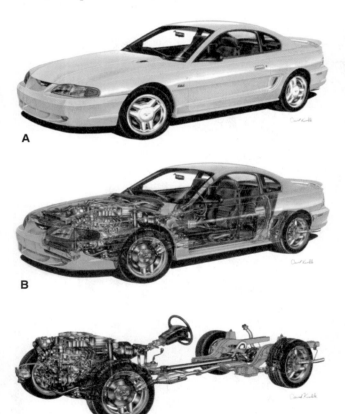

Figure 1–6 These three photos show different views of how a passenger car is made. *(Courtesy of Ford Motor Company)* (A) A vehicle body provides an attractive exterior. It can be made of steel, aluminum, and/or composite materials. (B) This body has been sectioned away to show the engine and other mechanical/electrical assemblies. (C) This view shows the powertrain for the same vehicle. It is a front-engine, rear-wheel drive configuration.

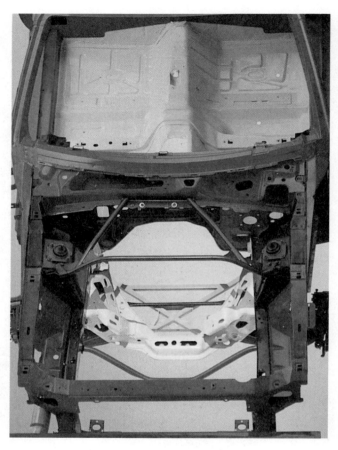

Figure 1–7 This top view of unibody construction shows how structural members are added to support the engine, suspension, and other mechanical systems. *(Courtesy of Ford Motor Company)*

A **panel** is a general term that refers to a large part of the body. For example, a quarter panel is a large body section forming the rear side of the body over and behind the rear wheels. A rocker panel is a body part along the bottom of a door opening.

Similarly, the term **pan** often refers to a floor-related component: front floor pan, rear floor pan. The name of the part usually refers to the location and purpose of the part. This makes it relatively easy to remember automotive part names.

Chassis

The **chassis** includes everything under the body—suspension system, brake system, wheels and tires, steering system. It consists of the mechanical systems that support and power the vehicle. See Figure 1–8.

NOTE! The chassis is explained further in Chapter 3, Vehicle Construction, and Chapter 24, Mechanical Repairs.

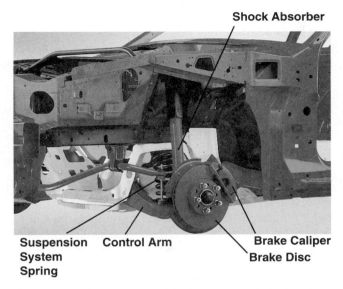

Shock Absorber

Suspension Control Arm   Brake Caliper
System   Brake Disc
Spring

Figure 1–8 Suspension and braking systems bolt directly to the body on a vehicle with unitized construction. *(Courtesy of Ford Motor Company)*

SHOP OPERATIONS

Many types of personnel are involved in repairing a collision damaged vehicle. It is important to have a general idea of the duties of all personnel associated with a collision repair facility in order to communicate and succeed on the job.

The next section of this chapter will explain the major tasks needed to fix a collision damaged vehicle. It will discuss what happens when a vehicle is sent to a repair facility after a collision, as shown in Figure 1–9.

Figure 1–9 When a damaged vehicle enters a collision repair facility, it goes through a series of preparation and repair steps.

Estimating

Estimating involves analyzing damage and calculating how much it will cost to repair the vehicle. It is critical that the quote on the repair is not too high or too low.

In most shops, a trained ***estimator*** makes an appraisal of vehicle damage and determines the parts, materials, and labor needed to repair the vehicle to its original condition. Estimators must be well-versed in how many vehicles are made. They must be good with computers, numbers, and communicating with people.

An ***estimate,*** also called a ***damage appraisal*** or ***physical damage report,*** is a written or printed form that explains what must be done to repair the vehicle. The estimate must explain which parts can be repaired and which will require replacement, summarizing all aspects of the repair and costs.

Manual estimating involves using an estimating sheet for writing out information about the vehicle, using crash estimating guides, and collision damage manuals to make the repair estimate. ***Crash estimating books*** and ***collision damage manuals*** contain vehicle identification information, the price of new parts, time needed to install the parts, refinishing or painting data, and other information. See Figure 1–10A.

A

B

Figure 1–10 Estimating is needed to find out how much the collision repair will cost. Most shops now use a computer estimating system to speed this process. *(Courtesy of Mitchell International, Inc.)* (A) Various manuals are often used by collision repair personnel. These give part prices, labor time, body specifications for the frame and unibody, and other data. (B) Computerized estimating is a fast and more efficient way of calculating repair costs. It is much more efficient than manual estimating.

Computer estimating involves using electronic **hardware** (computer, printers, hard drives, CD-ROM drives) and **software** (computer programs, CDs) to speed up the estimating process. The estimator might use a laptop to input which parts must be replaced or repaired. This saves time over writing the estimate out longhand. When this information is entered, the computer can streamline the estimating processes by automating many steps of writing the estimate. See Figure 1–10B.

When making an estimate, the estimator must make sure no damage is overlooked. For example, with a major impact, he or she must check for damage to the suspension, engine, electrical system, and interior of the vehicle. Many parts besides the body can be affected by a collision. See Figure 1–11.

Most vehicles now use an air bag. The *air bag system* uses impact sensors, an on-board computer, an inflation module, and a nylon balloon in the steering column and dash to protect the driver during a head-on collision. It can be expensive to service since the bag and all sensors may require replacement after a collision. The cost of the air bag system repair is another example of what must be considered when writing an estimate. See Figure 1–12.

The estimate is usually given to the customer, who turns the repair estimate in to the insurance company. The insurance company will then select a repair shop or give a check to the owner for completing the repair.

The *insurance adjuster/appraiser* reviews the estimates and determines which one best reflects how the vehicle should be repaired. He or she may inspect the collision damaged vehicle to make sure the repairs will be done cost effectively.

Once the owner and/or the insurance company approves the repairs, the vehicle is turned over to the shop supervisor. The supervisor and sometimes a technician will then review the estimate to determine how to proceed with the repair.

Washup

In the repair facility, the first step is washup. **Washup** involves a thorough cleaning of the vehicle before beginning work. This is needed to remove mud, dirt, wax, and water-soluble contaminants. These substances must be cleaned off before starting because they could contaminate the work area and the paint work. The vehicle must be completely dry before being moved to the metalworking area.

Bodywork

The **bodywork area** is where the vehicle is repaired in the shop. The damage can be the result of either a collision or deterioration. The repair tasks in this area of the collision repair facility are performed by collision repair technicians and their helpers. See Figure 1–13.

Depending upon the severity of the damage, the technician may have to remove and replace parts or simply use hand tools to work out small sheet metal dents.

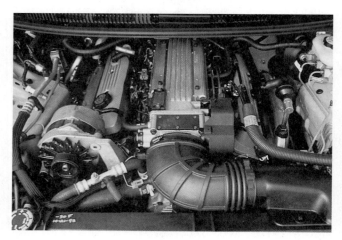

A

B

Figure 1–11 When preparing an estimate, the estimator must closely inspect for damage to all parts, body and mechanical. (A) The engine compartment can experience tremendous damage during frontal impact. The estimator must find any damage to the engine and its support systems. Wiring can also be damaged and must be accounted for in an estimate. *(Courtesy of General Motors Corporation, Service Operations)* (B) The interior can also be damaged in a collision. Air bags can deploy. The steering wheel can be damaged. Occupants can fly forward with enough force to damage the vehicle's interior. Side impacts can also push a door into the passenger compartment, damaging seats and other parts. *(Photograph by courtesy of Mercedes-Benz of North America)*

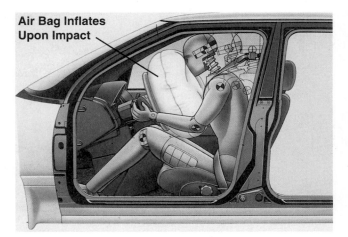

Air Bag Inflates Upon Impact

A

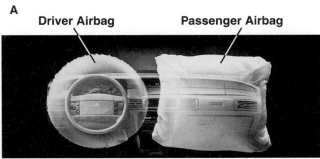

Driver Airbag   Passenger Airbag

B

Air Bag
Steering Wheel

Inflator Cartridge

C

Figure 1–12 Most vehicles now come equipped with air bags. (A) Air bags deploy on frontal impact. Sensors detect impact and signal a computer. The computer can then deploy the air bag to help protect the driver and passenger from injury. *(Courtesy of General Motors Corporation, Service Operations)* (B) When deployed, air bags help prevent injury from impact with the dash and windshield. They deflate right after impact so the driver can still see the road ahead if needed. *(Courtesy of General Motors Corporation, Service Operations)* (C) An air bag module contains a charge of pellets that produce gas pressure when electrically energized. *(Courtesy of General Motors Corporation, Service Operations)*

Minor Repairs

Minor repairs are those that require minimum time and effort. Small dents, paint scratches, and damaged

A

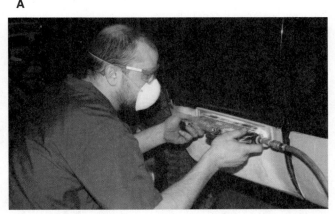

B

Figure 1–13 Minor collision repair often involves minor straightening and application of plastic body filler. (A) Here a technician is applying thin layers of plastic body filler over a straightened metal vehicle. (B) After the plastic filler hardens, power sanders are used to shape the filler to match the contour of the vehicle's body shape.

trim are typical examples. They require moderate skill and the use of hand tools, power tools, and collision repair tools.

Hand tools generally include tools used by both auto mechanics and collision repair technicians, such as wrenches, screwdrivers, and pliers. They are commonly used to remove parts, fenders, doors, and similar assemblies.

Power tools use air pressure or electrical energy to aid repairs. This classification includes air wrenches, air and electric drills, sanders, and similar tools.

Body shop tools are the most specialized tools designed for working with body parts. A few are shown in Figure 1–14. They can be used to cut metal, straighten small dents, and do similar tasks.

Panel straightening involves using various hand tools and equipment to reshape the panel back into its original contour. Dollies, body hammers, plastic filler,

A

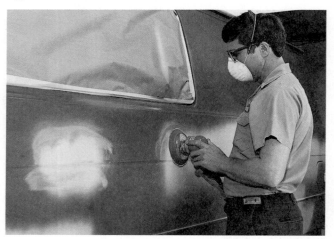

B

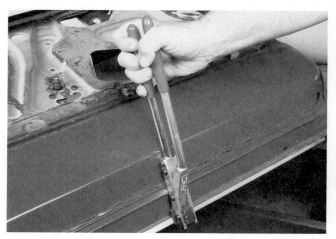

C

Figure 1–14 Many kinds of tools and equipment must be mastered by a collision repair technician. (A) A special air-supplied respirator protects the technician from paint mist and fumes generated by applying primer on a repair area. (B) Many kinds of sanders are used in collision repair. *(PPG Industries, Inc.)* (C) Special pliers are designed to fold metal over a door frame during door skin replacement.

Figure 1–15 Panel replacement sometimes is as simple as unbolting a component, like this fender. However, with welded-on panels the repair can be more time consuming. *(Chief Automotive Systems, Inc.)*

and sanders are a few of the tools and materials used to fix metal panel damage. Similar tools are needed with plastic or composite panels.

Panel replacement involves removing and installing a new panel or body part. It may be necessary to unbolt and replace a fender, door, or spoiler, for example, Figure 1–15. With quarter panels and other welded body sections, the technician has to cut off the damaged panel with power tools and then use a welder to install the new panel. This takes considerable skill.

Major Repairs

Major repairs is a general category that typically involves replacement of large body sections and frame/unibody straightening. To begin a major repair, the technician must normally remove severely damaged body parts, like fenders, bumpers, etc. Then measurements are taken at specific body points to analyze the damage. See Figure 1–16.

Measurement systems allow you to check for frame or unibody misalignment resulting from a collision. Various types of gauges and measuring devices can be used to compare known good body specifications with the actual measurements. The measurements will help determine what must be done to straighten any frame or unibody misalignment. See Figure 1–17.

Once you have measured to determine the extent and direction of frame/unibody misalignment, you can use straightening equipment to straighten the frame back into alignment.

Frame straightening equipment, also called a *frame rack,* uses a large steel framework, pulling

Figure 1–16 Major repairs typically involve replacement of large body sections and frame straightening. Before frame straightening, fenders, the hood, and other parts must often be removed.

chains, and hydraulic power to pull or push the frame back into its original position. The vehicle frame or unibody is clamped down onto the frame straightening equipment so it cannot move. Chains are then fastened to the damaged portion of the vehicle. Then, tremendous hydraulic force is applied to the chains to pull the frame or body in the direction opposite the deformation. See Figure 1–18.

Figure 1–17 A measuring system is used to check the location of specified points on the frame or unibody structure. This will tell if and how the frame is damaged so the technician can proceed with repairs. (Chief Automotive Systems, Inc.)

Figure 1–18 The frame rack or frame straightening equipment is very powerful. It will straighten a damaged unibody or frame back to its original position. It takes considerable training and skill to use frame straightening equipment. (Chief Automotive Systems, Inc.)

After straightening, more measurements are taken to determine if everything is returned to specifications. If not, frame rails and other unibody sections may have to be replaced, as shown in Figure 1–19.

Corrosion Protection

Corrosion protection involves using various methods to protect body parts from rusting. When doing repairs, you must always use recommended methods

Figure 1–19 When straightening with a rack, you may have to use heat to aid component straightening. If a unibody structural component cannot be straightened, it must be cut off and a new piece must be welded into place. (Chief Automotive Systems, Inc.)

Figure 1–20 When doing collision repair work, you must restore the manufacturer's corrosion (rust) protection materials. Here a technician is spraying a corrosion inhibitor onto a freshly repaired door. (3M Automotive Trades Division)

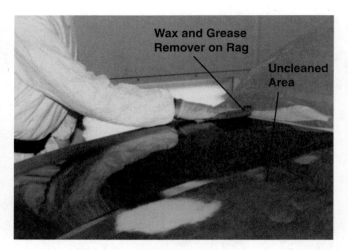

Wax and Grease Remover on Rag

Uncleaned Area

Figure 1–21 Before painting (refinishing), it is extremely important that all surfaces to be refinished are perfectly clean. Cleaning agents are available to remove wax, grease, oil, and other contaminants.

of protecting repair areas from damage. Corrosion protection is often a joint job of both the collision repair and paint technicians. Discussed later in this book, there are various methods of corrosion protection. See Figure 1–20.

Surface Preparation

Surface preparation, or "surface prep," involves inspection and treatment of the surface to prepare it for refinishing or painting. It is needed to make sure the repairs and new paint will hold up over time.

Begin with *surface inspection*, which involves looking closely at the body surface to determine its condition. This will let you decide what must be done to the surface to prepare it for new paint. Figure 1–21 shows a technician cleaning a vehicle with a cleaning agent while inspecting it for problems.

On one extreme, you might only have to sand the surface to ready it for painting. On the other extreme, if the paint is cracked and peeling, you may need to remove the paint. *Stripping* can be done by applying a chemical remover to soften and lift off the paint. Stripping can also be done using air-powered blasting equipment.

Discussed in later chapters, there are other methods of surface preparation before refinishing/painting.

Sanding uses an abrasive coated paper or plastic backing to level and smooth a body surface being repaired. Sanding with coarse, rough paper may be done to level plastic filler. Sanding with fine, smooth paper may also be done to lightly scuff the surface so the new paint will stick.

Paint Preparation

Paint preparation, or "paint prep," gets the vehicle ready for spraying or refinishing. During paint prep, you often must:

1. Remove windshield wipers, emblems, nameplates, mirrors, and other trim pieces.

2. Clean each area to be painted with an approved cleaning agent. This will remove grease, road tar, wax, and petroleum-based contaminants from the surface.

3. Machine and/or hand-sand chips and scratches.

4. Clean the interior carefully to remove all dust and dirt.

5. Double-check that all surfaces are properly cleaned before masking.

6. Closely inspect all surfaces to be painted for any remaining imperfections. Correct any surface problems you find!

NOTE! If the vehicle went through initial cleanup, some of these steps, such as interior cleanup, can be ignored.

Masking

Masking protects surfaces and parts from paint overspray. Masking protects the parts of the vehicle such as windows, trim, and lights. Masking is done by placing special tape, paper, or plastic over areas NOT to be painted.

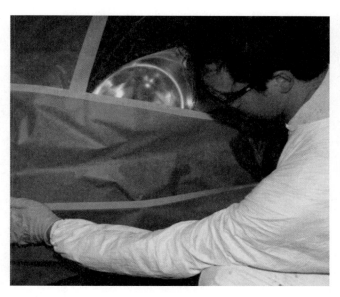

Figure 1-22 Masking involves covering any surface or component not to be painted to protect it from overspray. This technician is masking the front and left side of a vehicle so the right fender can be refinished.

Figure 1-23 When painting, the technician must wear safety gear to protect the skin and the lungs from chemicals. A spray gun is used to apply thin, even coats of primer, first, and then color. Note how the lights and tire have been masked to prevent overspray. *(PPG Industries, Inc.)*

Masking can also be done by spraying a special water-soluble material over surfaces not to be painted. The masking material can then be washed off with soap and water after painting. See Figure 1–22.

The ***masking area*** in the shop is usually equipped with masking paper and tape dispensers, tire covers, and other needed materials.

Paint overspray is an unwanted paint mist that spreads away from the surface being painted. It can stick to any uncovered or unmasked surface. Considerable time can then be wasted trying to clean off surfaces coated with a paint overspray. Masking is an important step before priming and painting.

Refinishing (Painting)

Refinishing or ***painting*** basically involves applying primer and paint over the properly prepared vehicle body. It is the most visible aspect of collision repair. Although it does not affect vehicle safety, it does affect the customer's evaluation of the repair because it is easy to see paint defects. See Figure 1–23.

Not only do collision damaged vehicles have to be painted, but older vehicles with weathered paint may also need refinishing to enhance their beauty. Sometimes the owner gets tired of looking at the same old color.

The ***paint shop area*** is where the vehicle is refinished. Here, a series of operations is performed on the vehicle to make sure the paint looks good and holds up over time.

Priming is primarily done to help smooth the body surface and help the top coats of paint adhere or stick to the body. It is done before painting. Discussed in later chapters, various types of primer and other pre-painting steps may be required.

Paint selection involves finding out what type of paint materials must be used on the specific vehicle. It must be done properly to make sure the paint matches the rest of the vehicle and to assure long paint life. The technician must determine what kind of paint is on the car and then select the right paint for the job.

Paint mixing usually involves adding solvents to give the paint the right "thickness" or viscosity for spraying. It also may involve mixing in a recommended amount of catalyst (hardener) to make the paint cure. Instructions on the container will give information about the type of materials and ratio of paint-to-solvent or thinner required. Proper paint mixing is critical to how the paint flows out of the spray gun and onto the vehicle body. Incorrect mixing makes it impossible to do a good paint job.

Spraying is the physical application of color or primer using a paint spray gun. The technician methodically moves the spray gun next to the vehicle body while spraying accurate layers of paint onto it. This requires a high degree of skill to prevent ***paint runs*** (excess paint thickness that flows down or "runs") and other painting problems.

When the vehicle moves into the spray booth, the painting technician takes over. Final cleaning is done at this point. Masking is removed after the paint has completely dried.

Custom painting involves forming various designs in the paint. Precut mask designs, airbrush work, pinstriping, and other techniques can be used with custom painting.

Drying

Drying involves using different methods to cure the fresh paint. If only partially dry when returned to the customer, the new paint can be easily damaged.

Air drying is done by simply letting the paint dry in the atmosphere.

Forced drying uses special heat lamps or other equipment to speed the paint curing process. Most shops use drying equipment to speed up drying. This is especially useful in drying today's paints. Drying equipment designs vary.

Postpainting Operations

Postpainting operations include the tasks that must be done before returning the vehicle to the customer. This would include removal of masking tape, reinstalling parts, and cleaning of the vehicle. Any paint flaws or problems must also be repaired.

Two common paint problems are **orange peel** and paint runs (sags). If the painted surface has orange peel, or a rough textured surface like the peel on an orange, you may have to wet sand the hardened paint, Figure 1–24. This will knock off the tiny bumps in the paint and smooth them out. A paint run results when too much paint is applied, too quickly. The thick wet coating of paint runs down or sags and then hardens into a large lump.

Wet sanding involves using a water-resistant, ultrafine sandpaper and water to level the paint. It can sometimes be done to fix small imperfections (orange peel, runs, etc.) in the paint. Wet sanding and buffing can also be done to a good paint job to make it even better looking.

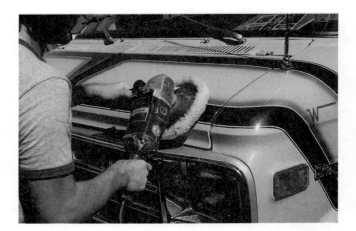

Figure 1–25 After wet sanding, or on oxidized, dull paint, buffing with compound can restore the shine. *(Badger Air Brush Co.)*

Compounding or **buffing** is done to smooth newly painted surfaces, after wet sanding for example, or to remove a thin layer of old, dull paint. See Figure 1–25. It will make the paint smooth and shiny. An air or electric buffing machine equipped with a pad is used to apply buffing compound to the finish. The abrasive action cuts off a thin layer of topcoat to brighten the color and shine the paint.

Detailing is a final cleanup and touch-up on the vehicle. It involves washing body sections, cleaning and vacuuming the interior, touching up any chips in sections not painted, and other tasks to improve vehicle cosmetics. Detailing can also involve other tasks such as polishing trim, cleaning glass, installing trim, and cleaning the tires. This will ready the vehicle for return to the customer.

Mechanical-Electrical Repairs

Mechanical repairs include tasks like replacing a damaged water pump, radiator, or engine bracket, for example. Mechanical components like these are often damaged in a collision. Many mechanical parts are easy to replace and can be done by the collision repair technician. However, other mechanical repairs may require special skills and tools. In this case, the vehicle would be sent to a professional mechanic. See Figure 1–26.

Electrical repairs include tasks like repairing severed wiring, replacing engine sensors, and scanning for computer or wiring problems. During a collision, the impact on the vehicle and the resulting metal deformation can easily crush wires and electrical components. For this reason, today's collision repair technician must have basic skills needed to work with and repair electrical/electronic components.

Figure 1–24 This technician is wet sanding to remove small surface imperfections before painting. Wet sanding can also be done to a new finish if problems are found in the paint work.

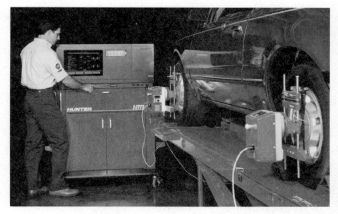

A

B

Figure 1–26 Mechanical repairs are sometimes done in larger collision repair shops. *(Hunter Engineering Company)* (A) An alignment rack has all of the equipment needed to check and correct alignment of wheels. This is often needed after collision repairs. (B) This technician is under a vehicle making adjustments to the steering and suspension systems to align the wheels.

> **NOTE!** Several chapters in this book are devoted to helping you learn to work on mechanical and electrical systems.

COLLISION REPAIR FACILITIES

There are several ways to classify a collision repair facility. A few of the most common will be discussed.

An ***independent body shop*** is one owned and operated by a private individual. The shop is NOT associated with other shops or companies.

A ***franchise facility*** is tied to a main headquarters which regulates and aids the operation of the business. The shop logo, materials used, pricing, etc. are all set by the headquarters and the franchise must follow these guidelines.

A ***dealership body shop*** is owned and managed under the guidance of a new car dealership—Ford,

Figure 1–27 Collision repair facility personnel, if properly trained, can earn a good living. If you study and learn the contents of this book, you are on your way to a solid career in collision repair. *(Photo courtesy of Team Blowtherm)*

GM, Chrysler, Toyota, etc. This type of shop often concentrates on repairs of the specific make of vehicles sold by the dealership.

A ***progression shop*** is often organized like an assembly line with specialists in each area of repair. One person might do nothing but "frame" work. Another technician might be good at "building the body" or installing parts and panels. The shop might have a wheel alignment technician, prep people, painter, and cleanup specialists. The vehicle will move from one area and specialist to the next until fully repaired.

A ***specialty shop*** concentrates on and only does specific types of repairs. For example, a collision repair facility might send a radiator with a small hole in it to a specialty radiator shop for repair with their specialized equipment.

Complete collision services means the facility might do wheel alignments, cooling system repairs, electrical system diagnosis and repair, suspension system work, and other repairs. Today, more and more collision repair shops are offering complete collision services. See Figure 1–27.

Summary

- Collision repair involves repairing structural damage after a vehicle has been in a collision. Another term for collision repair is auto body repair.

- The collision repair and paint technicians are the principals in any collision repair business. Other personnel include the shop owner, shop supervisor, parts manager, bookkeeper, office manager, receptionist, and helper/apprentice.
- A collision is an impact that causes damage to the vehicle body and chassis.
- There are several ways to classify a collision repair facility. A few of the most common are: independent, franchise, dealership, progression, specialty, and complete collision services.

Technical Terms

collision repair
   technician
shop owner
shop supervisor
parts manager
bookkeeper
office manager
helper
apprentice
professional
collision repair
collision
total loss
vehicle construction
frame
body
full frame
unibody
space frame
unitized construction
part
component
assembly
panel
pan
chassis
estimating
estimator
estimate
damage appraisal
physical damage report
manual estimating
crash estimating books
collision damage
   manuals
computer estimating
hardware

software
air bag system
insurance adjuster/
   appraiser
washup
minor repairs
hand tools
power tools
body shop tools
panel straightening
panel replacement
major repairs
measurement systems
frame straightening
   equipment
frame rack
corrosion protection
surface preparation
surface inspection
stripping
sanding
paint preparation
masking
masking area
paint overspray
refinishing
painting
paint shop area
priming
paint selection
paint mixing
spraying
paint runs
custom painting
drying
air drying
forced drying

postpainting operations
orange peel
wet sanding
compounding
detailing
mechanical repairs
electrical repairs

independent body shop
franchise facility
dealership body shop
progression shop
specialty shop
complete collision
   services

Review Questions

1. What happened to the first self-propelled vehicle?

2. _____ _____ _____ are skilled, knowledgeable people who know how to use specialized equipment and highly technical methods to restore collision damaged vehicles.

3. List and explain the basic skills often required of a repair technician.

4. The term "professional" refers to the _____, _____, and _____ that a business and its workers project to customers.

5. List six traits of a professional technician.

6. Define the term "collision repair."

7. What is a "total loss?"

8. Explain the difference between the frame and body of a car.

9. _____ describes a vehicle that has the body structure welded together to form an integral frame structure. It is also called _____ or _____ construction.

10. This term refers to several parts that fit together to make up a more complex unit.
    a. Chassis
    b. Component
    c. Assembly
    d. None of the above

11. _____ involves analyzing damage and calculating how much it will cost to repair the vehicle.

12. What is the job of an insurance adjuster/appraiser?

13. _____ _____ involves using various hand tools and equipment to bend or shape the panel back into its original contour.

14. Why would you use a measurement system?

15. _____ _____ _____,
also called a _____ _____,
uses a large steel framework, pulling chains, and
hydraulic power to pull or push the frame back
into its original position.

16. Explain two ways to strip off deteriorated paint
quickly.

17. Give two reasons for sanding a body surface.

18. List five steps that must be done during "paint
prep."

19. Basically, how and why do you mask a car?

20. _____ _____ _____
means the facility might do wheel alignments,
cooling system repairs, electrical system repairs,
suspension system work, and other repairs.

ASE–Style Review Questions

1. Which of the following jobs is NOT part of the
collision repair industry?
a. Metalwork
b. Auto repair technician
c. Finance agent
d. Welder

2. *Technician A* says collision damage is caused by an
impact on the vehicle body and/or chassis. *Technician B* says that opening a car door and accidentally hitting another vehicle is a collision. Who is
correct?
a. Technician A
b. Technician B
c. Both Technicians A and B
d. Neither Technician

3. *Technician A* says that the frame is the part of the
vehicle that supports the engine. *Technician B* says
that the frame is usually a high-strength metal
structure used to support other parts of the vehicle. Who is correct?
a. Technician A
b. Technician B
c. Both Technicians A and B
d. Neither Technician

4. *Technician A* says that unibody describes a vehicle
that has the body structure welded together to
form an integral frame structure. *Technician B* says
that space frame construction is a type of full-frame construction. Who is correct?
a. Technician A
b. Technician B
c. Both Technicians A and B
d. Neither Technician

5. *Technician A* says that a panel consists of the
mechanical systems of the vehicle. *Technician B*
says that a panel is a general term that often refers
to a large body part. Who is correct?
a. Technician A
b. Technician B
c. Both Technicians A and B
d. Neither Technician

6. *Technician A* says that an estimate is the analyzing
of damage and a calculation of the repair cost.
Technician B says that the estimate is a guide for
the repair of the vehicle. Who is correct?
a. Technician A
b. Technician B
c. Both Technicians A and B
d. Neither Technician

7. *Technician A* says that the first step in collision
repair is disassembly of the vehicle. *Technician B*
says that the first step is washing the vehicle.
Who is correct?
a. Technician A
b. Technician B
c. Both Technicians A and B
d. Neither Technician

8. *Technician A* says that major repair is a general
term that refers to repairs that involve replacement of large body parts. *Technician B* says that
corrosion protection involves using various methods to protect body parts from rusting. Who is
correct?
a. Technician A
b. Technician B
c. Both Technicians A and B
d. Neither Technician

9. *Technician A* says that masking is a job done by
the frame repair technician. *Technician B* says that
masking is a job done in the paint department.
Who is correct?
a. Technician A
b. Technician B
c. Both Technicians A and B
d. Neither Technician

10. *Technician A* says that detailing is a job done by
the office manager. *Technician B* says that detailing
is the final cleanup and touch-up to the vehicle.
Who is correct?
a. Technician A
b. Technician B
c. Both Technicians A and B
d. Neither Technician

Activities

1. Ask a technician or shop owner to visit your classroom. Ask him or her to describe the duties of their position.

2. Make a field trip to a collision repair facility. Have the shop owner or supervisor give you a guided tour of the shop. Discuss the field trip in class the next day.

3. Ask a parent, relative, or friend who has been in a collision to describe what happened during and after the collision when trying to get the vehicle repaired.

4. Go to the library and look up information on different types of insurance—full coverage, liability, no-fault. Prepare a report on this subject.

2

Safety and Certification

Objectives

After studying this chapter, you should be able to:

- List the types of dangers and accidents common to a collision repair facility.

- Explain how to avoid shop accidents.

- Outline the control measures needed when working with hazardous substances.

- Summarize hand and power tool safety.

- Describe safety practices designed to avoid fire and explosions.

- Explain the benefits of ASE certification.

- Summarize the purpose of I-CAR.

- Know the sources of professional training and certification available to collision repair facility personnel.

Introduction

This chapter will summarize the most important safety rules for a collision repair facility. It will explain the types of accidents that can happen and describe how to prevent them.

Safety involves using proper work habits to prevent personal injury and property damage. Always keep in mind that a collision repair facility can be a very dangerous place to work if safety rules are not followed. Carelessness and the lack of good work habits can be deadly.

It is up to you to learn and follow accepted safety rules. This chapter will summarize the most important dangers in a collision repair facility. Remember that you must still read all tool, equipment, and material instructions because they will give more specific details than this chapter.

Most injuries are caused by a lack of knowledge or by ignoring accepted safety rules. Smart technicians never gamble with their well-being by ignoring safety practices, Figure 2–1. A smart technician knows that shop safety does not waste time. It protects your most important investment—your health.

Figure 2–1 A collision repair facility can be a safe and rewarding place to work if everyone follows established safety rules. If anyone fails to follow these rules, it can be a dangerous place of employment. *(Riverdale Body Shop)*

You must also remember that some injuries happen instantly while others result from prolonged exposure over time. For example, if you fail to wear leather gloves when welding, you can be severely burned in a split second. However, if you fail to wear a respirator when sanding filler, you may not have symptoms of injury until you are diagnosed with lung cancer or another ailment.

WHAT CAUSES ACCIDENTS?

Accidents are unplanned events that hurt people, break tools, damage vehicles, or have other adverse effects on the business and its employees. Since a collision repair facility has so many potential sources of danger, safety must be your number one concern.

Remember! Accidents don't just happen; people cause accidents!

TYPES OF ACCIDENTS

Care must be taken to prevent several kinds of accidents including asphyxiation, chemical burns, electrocution, fires, and explosions.

Asphyxiation refers to anything that prevents normal breathing. There are many mists, gases, and fumes in a collision repair facility that can damage your lungs and affect your ability to breathe.

Chemical burns result when a corrosive chemical such as paint remover injures your skin or eyes.

Electrocution results when electricity passes through your body. Severe injury or death can result.

A *fire* is rapid oxidation of a flammable material, producing high temperatures. A burn from a fire can cause painful injuries and permanent scar tissue. There are numerous *combustibles* (paints, solvents, reducers, gasoline, dirty rags) in a collision repair facility. Any one of these can quickly cause a fire.

Explosions are air pressure waves that result from extremely rapid burning. For example, if you were to weld near a gas tank, it could explode. The fumes in the tank could ignite and the tank could blow open.

Physical injury is a general category that includes cuts, broken bones, strained backs, and similar injuries. To prevent these painful injuries, constantly think and evaluate each step. Always think about what you are doing and try to do it safely.

TEXTBOOK SAFETY

Thousands of collision repair technicians are injured or killed every year. Broken safety rules cause most of these accidents. Don't learn to respect safety rules the difficult way—by experiencing a painful injury. Learn to respect safety the easy way—by studying this book and following known safety practices.

Note that other chapters in this text give more specialized safety rules and cautions. However, even these safety rules are generic and cannot cover every challenge you will encounter in the collision repair facility. For this reason, you must always refer to manufacturer's instructions when in doubt.

Manufacturer's instructions are very detailed procedures for the specific product. They are written to guide you in the use of the exact item, whether it be a piece of equipment or a certain type of paint.

Always refer to manufacturer's instructions. They may be in the owner's instruction manual for tools and equipment. With paint and chemicals, they are normally printed on the container's label, product information sheet, or material safety data sheet (MSDS).

OVERALL SHOP SAFETY

True professionalism and shop safety begins with how a technician performs given work tasks. A professional understands the ever-present dangers in a collision repair facility and strives to avoid making dangerous mistakes.

You will be using air and electric tools, welders, cutting equipment, hydraulic frame straightening equipment, and hazardous materials. If you do not know or do not follow correct methods, you or someone else may be seriously injured. See Figure 2–2.

Figure 2–2 Make sure you have proper training before using any potentially hazardous tool or piece of equipment. Here technicians are receiving training on a new computerized measurement system used to analyze major vehicle damage. *(Chief Automotive Systems, Inc.)*

Accidents have a far-reaching effect, not only on the victim, but also on the victim's family and friends. Therefore, it is the obligation of all employees and the employer to develop a safety program. A **safety program** is a written shop policy designed to protect the health and welfare of personnel and customers. It includes rules on everything from equipment use to disposal of hazardous chemicals.

In an effort to maintain strict safety measures, customers should never be allowed into the shop's work areas.

SHOP LAYOUT AND SAFETY

The **shop layout** is the general organization or arrangement of work areas. For various safety reasons, you should fully understand the layout of your shop. This will help you learn fire exit routes, fire extinguisher locations, storage areas, and other information to make you a better worker. Study the typical collision repair facility layout in Figure 2–3.

A collision repair facility is typically divided into these work areas:

1. Metalworking area.
2. Repair stall.
3. Frame straightening equipment.
4. Front end rack (alignment rack).
5. Tool and equipment storage room.
6. Classroom area.
7. Paint preparation area.

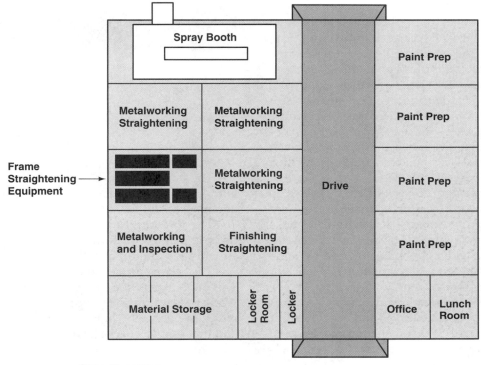

Shop Floor Plan

Figure 2–3 Collision repair shop layouts vary. Learn the layout of your shop. Memorize fire extinguisher locations and fire exit routes.

Figure 2–4 Here technicians are helping each other install a hood. Cooperation among employees is good for everyone in the facility. It increases production and safety.

8. Finishing or spray area.

9. Office.

10. Locker room.

The **metalworking area** is where parts are removed, repaired, and installed. This is the largest area in the shop where many tasks are performed. Dangers in this area are many and varied—cutting metal, welding, grinding, etc. See Figure 2–4.

A **repair stall** is a work area for one vehicle. Also termed a **bay,** it is usually marked off with painted lines on the floor. The stall has room in the front, rear, and sides for working on the vehicle. Stalls are often marked off in metalworking and prep areas.

The **frame rack** is a large piece of equipment designed to straighten damaged vehicle structures. It has powerful hydraulic rams and arms that require skill, training, and safety for use. It can exert tons of pulling force that can injure. If a frame rack chain were to come loose or break, it could fly with a tremendously destructive force, Figure 2–5.

A **front end rack** or **alignment rack** is used to measure and adjust the steering and suspension of a vehicle before and after repairs. It is needed to make sure the wheels are aligned to factory specifications. Dangers result from driving vehicles up and onto the rack and from using the small **pneumatic** (air) or **hydraulic** (oil-filled) equipment.

The **tool and equipment storage room** is an area for safely keeping specialized tools and equipment. It may have a pegboard wall or shelves for keeping tools that are not in use. Dangers result only if tools are not stored and maintained properly.

A **classroom** may be provided for lectures, demonstrations, and meetings of shop personnel.

A

B

Figure 2–5 (A) An improperly used frame rack has the potential to cause serious injury. (B) Pulling chains are used with tons of force on them. If a chain were to break or slip off the vehicle, it could fly with deadly force.

The **paint prep area** is where the vehicle is readied for painting or refinishing. The vehicle is cleaned and masked in this area.

The **finishing area** or **spray booth** is where the body is painted. It normally has a large metal enclosure to keep out dirt and circulate clean, fresh air.

The **shop office** contains business equipment. This area keeps the shop financially sound. Office workers keep track of paperwork, making sure estimates are accurate, bills are paid, parts are ordered, and that payroll checks go out.

The locker room is for cleaning up after work, for changing work clothes, and using the restroom.

As you will learn, specific rules apply to each shop area.

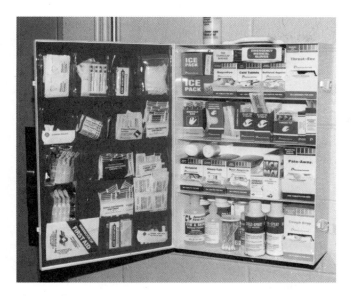

Figure 2–6 Know the location of the shop's first aid kit. The kit should be kept fully stocked with supplies to treat minor injuries.

IN CASE OF AN EMERGENCY

Have a list of *emergency telephone numbers* clearly posted next to the shop's telephone. These numbers should include a doctor, hospital, and fire and police departments.

A *first aid kit* includes many of the medical items needed to treat minor shop injuries. It will have sterile gauze, bandages, scissors, antiseptics, and other items to help treat minor cuts and burns. A fully stocked first aid kit should always be kept in a handy location, usually near the office or restroom. See Figure 2–6.

Safety signs give information that helps to improve shop safety. Signs are often posted to denote fire exits, extinguisher locations, dangerous or flammable chemicals, and other information. Safety signs should be located throughout the shop.

WORK AREA SAFETY

It is very important that the work area be kept safe. This must be a "team effort" involving everyone working in the shop.

Keep all surfaces clean, dry, and orderly. Wipe up any oil, coolant, or grease on the floor. If someone were to slip and fall, it could result in serious injuries.

Keep workbenches clean and orderly. When things are stacked on benches and tables, they can easily fall. A cluttered work area invites injury.

Hang tools up or put them away when they are not in use. Roll up idle hoses and air lines so they do not create traffic hazards.

Floor jacks, bumper jacks, jack stands, and creepers should be kept in their designated area, out of aisles and walkways.

FIRE SAFETY

Gasoline is a highly flammable petroleum or crude oil-based liquid that vaporizes and burns rapidly. Always keep gasoline and diesel fuel in approved safety cans. Never use them to wash your hands or tools.

Oily, greasy, or paint-soaked rags should be stored in an approved metal container with a lid. When soiled rags are left lying about improperly, they are prime candidates for *spontaneous combustion* (fire that starts by itself). See Figure 2–7.

Use only approved explosion-proof equipment in hazardous locations. A **UL (Underwriters Laboratories)** approved drum transfer pump along with a drum vent should be used when working with drums to transfer chemicals.

Keep all solvent containers closed when not in use. Handle all solvents and other liquids with care to avoid spillage.

Store paints, solvents, pressurized containers, and other combustible materials in approved and designated storage cabinets or rooms. Storage rooms should have adequate ventilation, Figure 2–8.

Discard or clean all empty solvent containers. Solvent fumes in the bottom of a container are prime ignition sources.

Never light matches or smoke in the spraying area! The paint mist and fumes are extremely explosive.

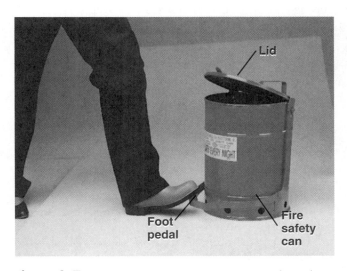

Figure 2–7 Always keep dirty rags in an approved metal can with a metal lid when they are not in use. This will help prevent a fire due to spontaneous combustion. *(Photo Courtesy of Lab Safety Supply, Inc., Janesville, WI)*

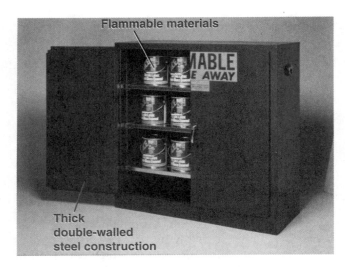

Figure 2–8 Store all flammable materials (paints, solvents, etc.) in a metal fire cabinet. Never leave them lying around the shop. (Photo Courtesy of Lab Safety Supply Inc., Janesville, WI)

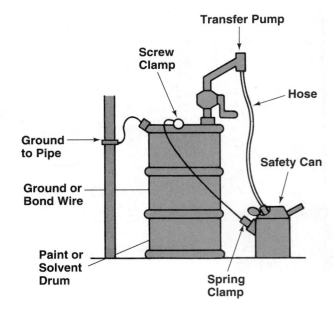

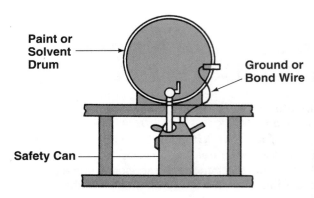

Figure 2-9 Study the proper method of transferring flammable liquids from bulk storage. Note ground wires.

Use extra caution when transferring flammable materials from bulk storage. Keep the drum grounded so that static electricity cannot cause an explosion. See Figure 2–9.

Vehicle batteries often explode! Hydrogen gas in the air near vehicle batteries can cause the battery to go off like a small bomb. Flying chunks of plastic and sulfuric acid can cause serious injury. Charge batteries only in a well-ventilated area.

Do not use torches or welding equipment at the mixing bench or in the paint area.

NEVER use automotive-type paints on household items such as toys or furniture. This could pose a dangerous health hazard to anyone who might ingest the paint.

Trash and rubbish should be removed from the shop area regularly. If it is not, serious fire and work area dangers can result.

Keep used paper towels and other paper products in a covered container separate from other trash. Empty it every day. Keep broken glass and jagged metal in separate containers. Recycle them if feasible.

Hold a rag around a fuel line fitting when disconnecting it.

You must release fuel pressure on many fuel injected vehicles before working on them because they can retain fuel pressure. Disconnect the battery before working on a fuel system.

Keep gas cylinders away from sources of heat like a furnace or room heater.

Never drop a gas cylinder. The head could break off and cause serious injury. Shut off the main gas valve on top of the tank after use—if a hose leak develops, this can prevent an explosion.

Electrical Fires

Electrical fires result when excess current causes wiring to overheat, melt, and burn. This often results when a wire or wires are short to ground while doing electrical work or after the collision cuts through wire insulation. To prevent electrical fires, always disconnect the battery when doing electrical work or when damage may have cut through wires.

Fire Extinguishers

A *fire extinguisher* is designed to quickly smother a fire. There are several types available for putting out different kinds of fires.

Know where the fire extinguishers are and what types of fires they put out, Figure 2–10.

A multi-purpose dry chemical fire extinguisher will put out ordinary combustibles, flammable liq-

GUIDE TO FIRE EXTINGUISHER SELECTION

	Class of Fire	Typical Fuel Involved	Type of Extinguisher
Class **A** Fires (green)	**For Ordinary Combustibles** Put out a class A fire by lowering its temperature or by coating the burning combustibles.	Wood Paper Cloth Rubber Plastics Rubbish Upholstery	Water*[1] Foam* Multipurpose dry chemical[4]
Class **B** Fires (red)	**For Flammable Liquids** Put out a class B fire by smothering it. Use an extinguisher that gives a blanketing, flame-interrupting effect; cover whole flaming liquid surface.	Gasoline Oil Grease Paint Lighter fluid	Foam* Carbon dixode[5] Halogenated agent[6] Standard dry chemical[2] Purple K dry chemical[3] Multipurpose dry chemical[4]
Class **C** Fires (blue)	**For Electrical Equipment** Put out a class C fire by shutting off power as quickly as possible and by always using a nonconducting extinguishing agent to prevent electric shock.	Motors Appliances Wiring Fuse boxes Switchboards	Carbon dioxide[5] Halogenated agent[6] Standard dry chemical[2] Purple K dry chemical[3] Multipurpose dry chemical[4]
Class **D** Fires (yellow)	**For Combustible Metals** Put out a class D fire of metal chips, turnings, or shavings by smothering or coating with a specially designed extinguishing agent.	Aluminum Magnesium Potassium Sodium Titanium Zirconium	Dry powder extinguishers and agents only

*Cartridge-operated water, foam, and soda-acid types of extinguishers are no longer manufactured. These extinguishers should be removed from service when they become due for their next hydrostatic pressure test.

Notes:

(1) Freezes in low temperatures unless treated with antifreeze solution, usually weighs over 20 pounds, and is heavier than any other extinguisher mentioned.

(2) Also called ordinary or regular dry chemical (sodium bicarbonate).

(3) Has the greatest initial fire-stopping power of the extinguishers mentioned for class B fires. Be sure to clean residue immediately after using the extinguisher so sprayed surfaces will not be damaged (potassium bicarbonate).

(4) The only extinguishers that fight A, B, and C classes of fires. However, they should not be used on fires in liquefied fat or oil of appreciable depth. Be sure to clean residue immediately after using the extinguisher so sprayed surfaces will not be damaged (ammonium phosphates).

(5) Use with caution in unventilated, confined spaces.

(6) May cause injury to the operator if the extinguishing agent (a gas) or the gases produced when the agent is applied to a fire is inhaled.

Figure 2-10 This chart shows four fire extinguisher classifications and the types of fires they will put out.

uids, and electrical fires. During a fire, a few seconds can be a "lifetime" for someone.

Never smoke while working on any vehicle or machine in the shop. There are too many flammables that could start a fire.

In case of a gasoline fire, do not use water. Water will spread the fire. Use a fire extinguisher to smother the flames. Never open doors or windows unless it is absolutely necessary. The draft will only make the fire worse.

A good rule is to call the fire department first and then attempt to extinguish the fire. Standing 6 to 10 feet (2 to 3 meters) from the fire, hold the extinguisher

firmly in the recommended position. Aim the nozzle at the base of the fire and use a side-to-side motion, sweeping the entire width of the fire.

Stay low to avoid inhaling the smoke. If it gets too hot or too smoky, get out. Remember, never go back into a burning building.

AVOIDING FALLS

Keep all shop floor drain covers snugly in place. Open drains have caused many toe, ankle, and leg injuries.

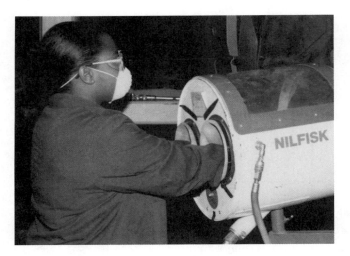

Figure 2–11 Use a vacuum system to remove asbestos dust from the vehicle. Asbestos is a powerful cancer-causing agent.

To clean up oil, use a commercial oil absorbent. Spread the absorbent on the spill. Rub it with your broom in a circular motion. Then use a dust pan and broom to pick up the absorbent.

Make sure that aisles and walkways are kept clean and wide enough for safe movement. Provide for adequate work space around all machines. Cluttered walking areas contain items waiting to cause injury. Never leave tools or a creeper lying on the floor.

AVOIDING ELECTROCUTION

Keep all water off the floor. Remember that water is a conductor of electricity. A serious shock hazard will result if a live wire happens to fall into a puddle in which a person is standing. The floor must be dry when using electric power tools.

Electrocution results when electric current passes through the human body. This can affect heart and brain functions, possibly causing serious injury or death.

Disconnect electrical power before performing any service on a machine or tool.

Some late model vehicles have heated windshields. Over 100 AC volts are sent to the windshield to quickly warm the glass for melting ice and snow. This can be enough voltage and current to cause serious injury.

AVOIDING ASPHYXIATION

Make certain the work area is well lit and *ventilated* (that it has good supply of fresh air flowing through it).

Vehicle engine exhaust produces deadly *carbon monoxide (CO)* gas, which is odorless and invisible. Connect a shop vent hose to the tailpipe of any vehicle being operated in the shop.

Check and service furnaces and water heaters in the shop at least once every six months.

Asbestos dust used in the manufacture of older brake and clutch assemblies is a cancer-causing agent. Never blow this dust into the shop. Use a vacuum system while wearing a filter mask to clean off asbestos dust safely. See Figure 2–11.

Proper ventilation is also important in areas where caustics, degreasers, undercoats, and finishes are used. Ventilation can be by means of an air-changing system, extraction floors, or central dust extraction system.

Set the parking brake when working on a vehicle with the engine running. If the vehicle has an *automatic transmission* (shifts through gears automatically using hydraulic system), place it in park.

With a *manual transmission* (must be hand shifted through each forward gear), the vehicle should be in reverse (engine off) or neutral (engine on) unless instructed otherwise for a specific service operation.

AVOIDING EYE INJURIES

Remember that your eyes are irreplaceable. They are very delicate and can be permanently damaged in a split second. Human eyes are sensitive to dust, flying particles from grinding, and mists or vapors from spraying. Such exposure could cause severe eye injury and possibly the loss of sight.

When risks to your eyes exist, wear appropriate eye or face protection. See Figure 2–12.

A good pair of *safety glasses* is suitable with only minor danger, as when blowing off dust or sanding with an air sander.

Wear a *full-face shield* when there is more danger from flying particles, as when grinding.

Wear chemical splash *goggles* when handling solvents, thinners, reducers, and other similar liquids. Because the eyes are very susceptible to irritation, approved safety goggles should also be worn when paint mists are present.

A welding helmet or welding goggles with the proper shade lens must be worn when welding. These will protect your eyes and face from flying molten metal and harmful light rays. See Figure 2–13.

AVOIDING CHEMICAL BURNS

There are many sources of chemical burns in a collision repair facility. Cleaners, some paints, refrigerants,

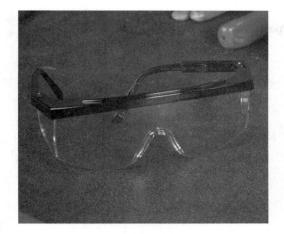

A

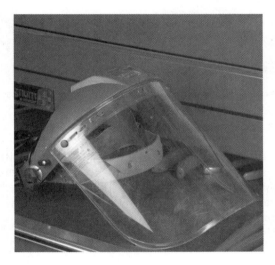

B

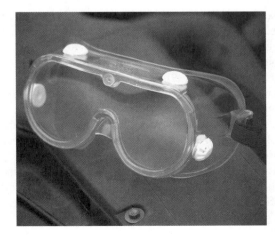

C

Figure 2-12 Note these three basic types of eye protection. (A) Safety glasses. (B) A full-face shield helps protect your eyes as well as your face. (C) Goggles help protect all around the eyes. *(Courtesy Goodson Shop Supplies)*

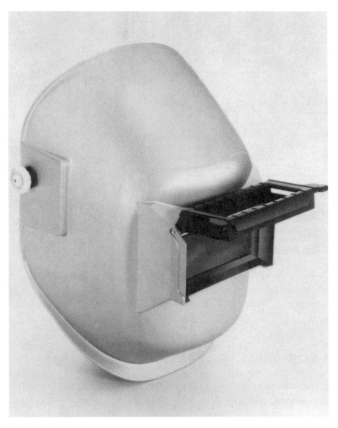

Figure 2-13 A welding helmet protects the face from molten spatter and the eyes from the blinding light of welding. The tinted glass in the helmet must be dark enough to protect the eyes from bright light. *(Courtesy of Sellstrom Mfg.)*

and solvents can cause skin and eye burns. A few rules to follow to prevent chemical burns include:

- Remove solvent-soaked clothing. The cloth will hold these chemicals against the skin, causing skin irritation or a chemical burn. Wear long sleeves for added protection.

- Wear **rubber** or **plastic gloves** to protect yourself from the harmful effects of corrosive liquids, undercoats, and finishes. Use impervious gloves when working with solvents or two-component primers and topcoats. These gloves offer special protection from the chemicals found in two-component systems.

Danger! *Chemical paint remover is very powerful. When using it, use extreme caution and protective gear, Figure 2-14.*

When washing hands, use a proper **hand cleaner.** Never use thinner as a hand cleaner. Many chemicals can be absorbed into the skin, eventually causing long-term illnesses.

Figure 2–14 Chemical paint remover will quickly soften and lift hardened paint. It will cause serious burns! (© The Eastwood Company, 1-800-345-1178, www.EastwoodCompany.com)

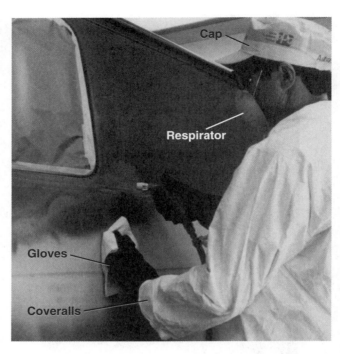

Figure 2–15 This technician is blowing off a vehicle body in the prep area to ready it for refinishing. Note goggles, coveralls, cap, and respirator. (PPG Industries, Inc.)

PERSONAL SAFETY

Personal appearance and conduct can help prevent accidents. Follow these general guidelines for personal safety.

Loose clothing, unbuttoned shirt sleeves, dangling ties, loose jewelry, and shirts hanging out are very dangerous. Instead, wear approved shop coveralls or jumpsuits. Pants should be long enough to cover the top of the shoes.

A clean *jumpsuit* or *lint-free coveralls* should be worn when in the spraying area. Dirty clothing can transfer dirt to freshly painted surfaces. See Figure 2–15.

Wear thick *leather shoes* that have nonslip soles to prevent falls and foot injuries. Good work shoes provide support and comfort for someone who is standing for a long time. Never wear athletic shoes or sandals in the shop. Toes can be broken or severed if a heavy object falls on them.

To keep hair safe and clean, wear a *cap* or hat when sanding, grinding, and doing similar jobs, Figure 2–15. Wear a protective *painter's stretch hood* in the spray booth; this is a cloth head covering with a see-through lens for protection from overspray or airborne materials. When working beneath the hood or under the vehicle, wear a head covering known as a *bump cap;* this is a thick cloth hat without a bill that protects your head from minor injuries while working. Keep clothing away from moving parts when an engine or a machine is running.

Tie long hair securely behind your head before beginning to work. If hair becomes tangled in moving parts or air tools, it can cause serious injury.

Any loose or hanging clothing, such as shirttails, ties, cuffs, or scarves, creates a risk of getting wrapped up in moving parts of the vehicle or machinery, causing serious bodily injury. Remove all jewelry before working.

Use a portable shop light when working in dark areas, such as under a vehicle. This will increase work speed, quality, and safety.

When lifting and carrying objects, bend at your knees, not your back. Do not bend your waist when lifting. Remember, heavy objects should be lifted and moved with the proper equipment for the job. Get someone to help you.

Never overreach! Maintain a balanced stance to avoid slipping and falling while working.

Proper conduct can also help prevent accidents. Horseplay is dangerous and very unprofessional.

Do not risk injury through the lack of knowledge. Use shop equipment or machinery only after receiving proper instruction.

Metalworking air hammers, the piercing noise of grinding, and the radio blaring full blast can all make it difficult to hear anything else. Some shop noise is enough to cause permanent hearing loss.

When in metalworking areas, wear *earplugs* or *earmuffs* to protect your eardrums from damaging noise levels. See Figure 2–16.

To prevent serious burns, avoid contact with hot metal parts such as the radiator, exhaust manifold, tailpipe, catalytic converter, and muffler.

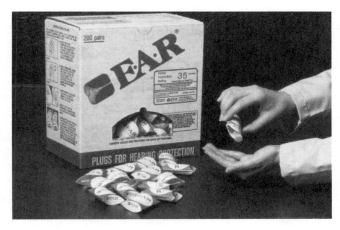

A

B

Figure 2–16 Many noises in the shop (for example, grinders and air chisels) can cause hearing loss. Wear earplugs or earmuffs when sound or decibel levels are high. (A) Earplugs *(Photo Courtesy of Lab Safety Supply Inc., Janesville, WI)* (B) Earmuffs/headphones. *(Courtesy of Sellstrom Mfg.)*

During metalworking, it is very easy to be cut by sharp, jagged metal. Use caution, Figure 2–17.

When driving a vehicle into a shop, watch out for other vehicles and people. It is best to have someone act as a guide. Leave the window open and the radio off, so that the helper's directions can be heard.

TOOL AND MACHINE SAFETY

Collision repair tools and equipment can cause serious injuries if basic safety rules are NOT observed. Some general tool rules include the following:

Keep tools clean and in working condition. Greasy or oily tools can easily slip out of your grasp, possibly causing injury.

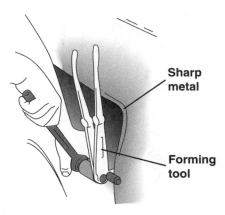

Figure 2–17 Cuts are the most common injury in a metalworking area. Sharp metal can easily lacerate skin. This technician is using a special tool to form a lip on sheet metal during corrosion repair.

Check all tools for cracks, chips, burns, broken teeth, or other dangerous conditions before using them. If tools are defective, replace them.

Be careful when using sharp or pointed tools that can cause injury. If a tool should be sharp, keep it sharp.

Do not use any hand tools for any job other than that for which they were specifically designed.

Never carry screwdrivers, punches, or other sharp items in your pockets. You could be injured or damage the vehicle.

When using an electric power tool, make sure that it is properly grounded. The third round ground prong sends electricity flow to the ground if the tool shorts out, not through your body.

Check the wiring insulation for cracks and bare wires. To avoid serious injury when using electric tools, never stand on a wet or damp floor.

Do not operate a power tool or machine without its safety guard(s) in place.

When power grinding, cutting, or sanding, or doing similar operations, always wear safety glasses.

When using power equipment on small parts, never hold the part in your hand. It could slip and you could injure your hand. Use a bench vise instead.

Before plugging in any electric tool, make sure the switch is OFF to prevent serious injury. When you are through, turn the tool off to ready it for the next person.

Use *jack stands* or *safety stands* to support a vehicle whenever you must work underneath it. Jack stands are strong steel pieces of equipment designed to support a vehicle; they have four legs to hold them steady, and an extendible shoe on which the vehicle rests. *Hydraulic jacks* are used to raise a vehicle before placing jack stands and to lower it after working. Never trust hydraulic jacks alone; they are for raising the vehicle, not holding it. Many people have

Use floor jack
to raise vehicle

A

Support on jack
or safety stands
before working

B

Figure 2–18 To avoid serious injury, take industry-recommended steps to lift and secure a vehicle. (A) Use a floor jack to raise the vehicle. The transmission should be in neutral while the vehicle is being raised so the wheels can turn. Never work under a vehicle that is supported only with a floor jack. (B) Position and lower the vehicle onto jack stands (safety stands) before working on it. After the vehicle is on the stands, place the transmission back into park or in gear and block the wheels that are on the ground.

been seriously injured when a vehicle fell on them. See Figure 2–18.

Never use a machine or tool beyond its stated capacity. If a grinding wheel specification is 2000 rpm (revolutions per minute), do not install it in a grinder with a higher operating speed. The wheel could explode if operated at a higher rpm.

Use a tool or machine only for its designed task. Never misuse or modify equipment.

Keep hands away from moving parts when the machine or tool is under power. Running vehicle engines often cause injuries. Never clear debris from a machine when it is under power, and never use your hands.

Use caution with compressed air. Pneumatic tools must be operated at the pressure recommended by their manufacturer. The downstream pressure of compressed air used for cleaning purposes must remain at a pressure level below 30 psi (2.1 kg/cm^2).

Do not use compressed air to clean clothes. Even at low pressure, compressed air can cause dirt particles to become embedded in your skin, which can result in infection. Air entering an artery could cause death. See Figure 2–19.

Store all parts and tools properly by putting them away neatly. This practice not only cuts down on injuries, it also reduces time wasted looking for a misplaced part or tool.

When working with a hydraulic press or power unit, apply hydraulic pressure in a safe manner. It is generally wise to stand to the side when operating hydraulic tools, Figure 2–20. Always wear safety glasses.

If the shop has a hydraulic lift, be sure to read the instruction manual. Check the pads to see that they are making proper contact with the frame. Then raise the vehicle about 6 inches (15 centimeters) and check to make sure it is well balanced on the lift.

Any rattling or scraping sounds mean that the vehicle is not positioned properly. If this happens, lower the lift and realign the pads. Test it again. See Figure 2–21.

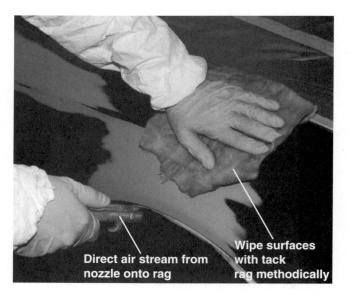

Direct air stream from
nozzle onto rag

Wipe surfaces
with tack
rag methodically

Figure 2–19 Compressed air should never be directed at your body or clothing, especially your eyes and ears. If compressed air enters your bloodstream, it could cause serious injury or death.

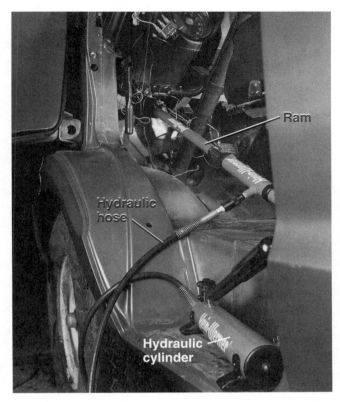

Figure 2–20 Hydraulic equipment, like this power ram, can exert tons of force. Make sure the equipment is secure and stand to one side when applying pressure. *(Hein-Werner® Corporation)*

Figure 2–21 After raising the vehicle to the desired height on the lift, engage the lift safety catch. This will prevent the lift from accidentally lowering. *(Rotary Lift/A Dover Industries Co.)*

After lifting the vehicle to full height, engage the **lift safety catch** to ensure that the lift cannot lower while you are working underneath the vehicle.

Never permit anyone—technician or customer—to remain in the vehicle while it is being lifted.

Make sure you are properly trained before using frame straightening equipment. Personal injury and major vehicle damage can result if you use improper methods.

When welding or cutting near vehicle interiors, remove seats and carpets to prevent a fire. Always have water and a fire extinguisher nearby.

PAINT SHOP SAFETY

Many paints and paint-related products present serious health hazards unless proper precautions are taken. The issue of paint shop safety has grown more important in recent years. More vehicles are now being refinished with urethane enamels, which contain harmful chemicals. The amount of exposure to chemicals has increased, and some are more toxic. These chemicals can cause nose, throat, and lung irritation. They can also cause acute respiratory prob-

lems. Symptoms can include shortness of breath, tightness in the chest, dizziness, nausea, abdominal pain, and vomiting.

RESPIRATORS

Respirators are needed in the paint shop to keep airborne materials from being inhaled. They must be used even when adequate ventilation is provided. There are three primary types of respirator available to protect technicians: the dust or particle respirator, the cartridge filter respirator, and the hood or air-supplied respirator. See Figure 2–22.

Respirator Types

A **dust** or **particle respirator** is a filter that fits over your nose and mouth to block small airborne particles. It is not designed to stop paint mist and fumes. A dust respirator is used when sanding or grinding to keep the dangerous materials out of your lungs.

Sanding operations create dust that can cause bronchial irritation and possible long-term lung damage. Protection from this health hazard is often overlooked. Just because the sanding dust does not cause immediate symptoms, does not mean that it will not cause problems when you get older. An approved dust respirator should be worn whenever sanding or grinding operations are performed. See Figure 2–23.

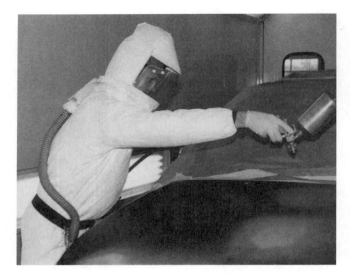

Figure 2–22 An air-supplied respirator is needed to protect your lungs from catalyzed materials or paints that use a hardener/activator.

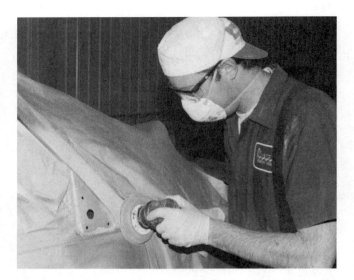

Figure 2–23 A dust respirator will trap and keep sanding debris from entering your lungs. Even though you will not feel pain at the time, long-term exposure to plastic filler dust can cause serious lung damage. This technician is also wearing safety glasses, gloves, and a bump cap.

Follow the instructions provided with the dust respirator to ensure proper fit. Bend or shape it so that air cannot leak around your face, nose, and mouth.

Cartridge filter respirators protect against vapors and spray mists of nonactivated enamels (no hardener added), lacquers, and other nonisocyanate materials. *Isocyanates* are very hazardous materials that are often used to "accelerate" (speed-up/shorten) drying times, or to "activate/harden" refinishing material. If the refinishing materials contain no isocyanates, an air-purifying cartridge filter respirator with organic vapor cartridges and prefilters can be used.

The cartridge filter respirator consists of a rubber face piece designed to conform to the face and form an airtight seal. It includes replaceable prefilters and cartridges that remove solvent and other vapors from the air. The respirator also has intake and exhaust valves, which ensure that all incoming air flows through the filters.

Cartridge filter respirators are available in several sizes. Purchase one that fits your face.

Remember that cartridge filter respirators should be used only in well-ventilated areas. They must not be used in environments containing less than 19.5 percent oxygen.

To maintain the cartridge filter respirator, keep it clean and change the prefilters and cartridges as directed by the manufacturer.

An approved *air-supplied respirator* consists of a half-mask, full face piece, hood, or helmet, to which clean, breathable air is supplied through a hose from a separate air source. It provides protection from the dangers of inhaling isocyanate paint vapors and mists, as well as from hazardous solvent vapors. See Figure 2–24.

The fresh-air-supply respirator is comfortable to wear and does not require fit testing. The fresh-air-supply respirator may include a self-contained oil-less pump to supply air to either one hood or two half-mask respirators.

The pump's air inlet must be located in a clean air area. Some shops mount the pumps on an outside wall, away from the dust and dirt generated by shop operations. If shop-compressed air is used, it must be filtered with a trap and carbon dioxide/monoxide filter. Oil, water, scale, and odor must be removed. The air supply must have a valve to match air pressure to respirator equipment specifications and an automatic control to sound an alarm or shut down the compressor in case of overheating and contamination. Overheating frequently causes carbon monoxide contamination of the air supply.

Respirator Testing and Maintenance

It is very important that an air-purifying cartridge filter respirator fits securely around your face. This will prevent contaminated air from entering your lungs. To check for respirator air leaks, a *fit test* should be performed prior to using the respirator, making both negative and positive pressure checks.

To make a *negative pressure test,* place the palms of your hands over the cartridges and inhale. A good fit will be evident if the face piece collapses onto your face.

To perform a *positive pressure test,* cover up the exhalation valve and exhale. A proper fit is evident if

Figure 2–24 When applying hardened/activated materials, you must wear a hood supplied with outside or filtered fresh air. A pump circulates the air through the hood to protect your lungs.

the face piece billows out without air escaping from the mask.

Another form of fit test consists of exposing amyl acetate (banana oil) near the seal around the face. If no odor is detected, a proper fit is evident.

Here are a few other maintenance tips for cartridge respirators:

Replace the prefilters when it becomes difficult to breathe through the respirator. Replace the cartridge(s) at proper intervals, and always at the first sign of solvent odor. Regularly check the mask to make sure it does not have any cracks or deformities. Store the respirator in an airtight container.

Note that facial hair may prevent an airtight seal, presenting a health hazard. Therefore, refinishers with facial hair should use a fresh air supply respirator.

Follow the manufacturer's instructions to ensure proper maintenance and fit.

DUSTLESS SANDING SYSTEMS

A **dustless sanding system** uses a blower or air pump to draw airborne dust into a storage container, much like a vacuum cleaner. Depending on the system, vacuum pumps, vacuum pullers, brush motors, or turbine motors can be used to provide sufficient air volume. This action pulls airborne sanding dust through holes in a special sanding pad or through a shroud that surrounds the sanding pad.

Some dustless system manufacturers claim that their machines can trap over 99 percent of the toxic dust created by sanding lead- or chrome-based automotive paints and primers.

AIR BAG SAFETY

Air bags can be very dangerous because they deploy with tremendous force. Air bags can reach speeds of over 100 miles per hour during deployment. They can break arms, hands, fingers, or cause even more serious injury if you are near the bag during accidental deployment.

When working around air bags, use caution and follow service manual directions to prevent accidental deployment. Never install or connect a new air bag module until all wiring is checked with a scan tool and repaired if necessary. A short in the wiring to an impact sensor or to the air bag module could make the new air bag deploy as soon as it is connected to its control circuit.

Refer to the vehicle's service manual to learn the precautions that must be followed when servicing an air bag. It will give the details needed to work safely. For details on air bag service, refer to Chapter 27.

MANUFACTURER'S WARNINGS

Manufacturer's warnings give important procedures for the safe use of products. Study and follow instructions and warnings given by product and equipment manufacturers. Follow them to the letter.

Most of the products used in a collision repair shop carry warning and caution information that must be read and understood by users. Likewise, all federal (including Occupational Safety and Health Administration [OSHA], Mine Safety and Health Administration [MSHA], and National Institute for Occupational Safety and Health [NIOSH]), state, and local safety regulations should not only be fully understood, but also strictly observed.

Material safety data sheets (MSDS), available from all product manufacturers, detail chemical composition and precautionary information for all products that can present a health or safety hazard.

Employers must know the general uses, protective equipment, accident or spill procedures, and other information for safe handling of hazardous material. They must provide training to employees as part of their job orientation.

The best way to protect yourself when using paint and body products is to be familiar with the correct application procedures. Follow all safety and health precautions found on MSDS and on product labels.

RIGHT-TO-KNOW LAWS

In collision repair and refinishing shops, hazardous wastes are generated. As a result, every employee is protected by Right-to-Know Laws.

Right-to-Know Laws give essential information and stipulations for safely working with hazardous materials. They began with OSHA's Hazard Communication Standard. This document was originally intended for chemical companies and manufacturers that require employees to handle potentially hazardous materials in the workshop. Since then, the majority of states have enacted their own Right-to-Know Laws. The federal courts have decided that these regulations should apply to all companies, including the auto collision repair and refinishing professions.

EMPLOYER RESPONSIBILITIES

The general intent of the law is for employers to provide their employees with a safe working place as it relates to hazardous materials. Specifically, there are three areas of employer responsibility:

All employees must be trained about their rights under the legislation, the nature of the hazardous chemicals in their workplace, the labeling of chemicals, and the information about each chemical posted on material safety data sheets (MSDS).

The shop must maintain documentation on the hazardous chemicals in the workplace. It must show proof of training programs, records of accidents and/or spill incidents. The shop must show satisfaction of employee requests for specific chemical information.

Complete **EPA (Environmental Protection Agency)** lists of hazardous wastes can be found in the Code of Federal Regulations. Materials and wastes of most concern to the technician are organic solvents: wastes that contain heavy metals, especially lead, and ignitable, corrosive, and/or toxic materials.

It should be noted that no material is considered hazardous waste until the shop is finished using it and is ready to dispose of it. For instance, a caustic cleaning solution with a heavy concentration of lead in the cleaning tank is not considered hazardous waste until it is ready to be replaced. Then it must be handled accordingly. The EPA says it is the owner's responsibility to determine whether or not the waste is hazardous, and the owner must have adequate test results to support the shop's claim.

WASTE DISPOSAL

Many shops use "full-service" haulers to test and remove hazardous waste from the property. Besides hauling the hazardous waste away, the hauler will also take care of all the paperwork, deal with the various government agencies, even advise the shop on how to recover disposal costs.

The collision repair shop is ultimately responsible for the safe disposal of hazardous wastes, even after they leave the shop. Be sure that any hauling contract is in writing.

In the event of an emergency hazardous waste spill, you or the shop owner must contact the National Response Center (1-800-424-8802) immediately. Failure to do so can result in a $10,000 fine, a year in jail, or both.

LIABILITY

You must also be aware of the liability question resulting from auto collision work. *Liability* means that you are financially responsible for damage or injuries resulting from improper repairs. There have been several lawsuits recently in which improper repairs were blamed for personal injury. Basically, if you do not do the job right, you and your shop can be held responsible or liable.

Collision repair facilities and insurance companies have the highest exposure to liability responsibility. The technician must set high standards personally and for the shop in terms of quality. Achieving these standards and employing the latest repair techniques will reduce the chance of being held liable.

TECHNICIAN LIABILITY

There are several principles of liability the collision repair technician should be familiar with:

Take "due care" to do the work thoroughly and correctly. The shop and technician(s) are liable for any faulty work. The shop is liable for failure to repair damage, which is acknowledged to exist, unless the customer specifically directs that it should not be repaired.

Warn the customer about anything that might be wrong with the vehicle after the repair. If the customer or insurance company does agree to allow you to fix the problem, put it in writing. Have the customer or insurance agent sign the work order. This will protect you and assure that the customer knows that the vehicle is not in perfect condition.

Ignorance is no defense! In the eyes of the law, the collision repair shop, technician, and insurance personnel are expected to be knowledgeable about the latest available techniques for safe, sound repair. Failure to learn and apply those techniques could result in liability.

ASE CERTIFICATION

Just as doctors, accountants, electricians, and other professionals are licensed or certified to practice their professions, collision repair and refinishing technicians can also be certified. The National Institute for Automotive Service Excellence (ASE) offers a certificate program. **ASE certification** is a testing program to help prove that you are a knowledgeable collision repair or refinishing technician.

Certification protects the general public and the professional. It assures the general public and the prospective employer that certain minimum standards of performance have been met.

ASE tests include multiple-choice questions pertaining to the service and repair of a vehicle. They do not include theoretical questions. To prepare for ASE collision repair or refinishing tests, study the material in this book carefully.

To help you prepare for the Collision Repair and/or Painting and Refinishing tests, some test questions at the end of service chapters in this text are similar to those used by ASE.

Collision repair and refinishing technicians can get certified in one or more technical areas by taking and passing written certification tests. The **National Institute for Automotive Service Excellence (ASE)** offers a voluntary certification program that is recommended by the major vehicle manufacturers in the United States. The Collision Repair and Painting and Refinishing tests contain 40 questions in various areas.

Technicians who pass the written tests are awarded a certificate and a shoulder emblem for their work clothes. See Figure 2–25.

Many employers now expect their collision repair and refinishing technicians to be certified. The certified technician is recognized as a professional by the public, employers, and peers. For this reason, the certified technician usually receives higher pay than one who is not certified.

For further information on the ASE certification program, write:

ASE
13505 Dulles Technology Drive
Herndon, VA 22071-3145
703-713-3800

Or contact ASE on the World Wide Web at: www. ase-cert.org

I-CAR

The **Inter-Industry Conference on Auto Collision Repair,** or **I-CAR,** is an advanced training organization dedicated to promoting high-quality practices in

Figure 2–25 After you pass an ASE test, you will be given a shoulder patch for your work uniform.

the collision repair industry. It is made up of vehicle manufacturers, collision repair shops, insurance companies, independent appraisers, recyclers, educators, and tool, equipment, and material manufacturers.

I-CAR has emerged as the leading advanced collision repair training organization. I-CAR classes are offered in the United States and Canada.

Some of the many courses offered to collision repair professionals by I-CAR include:

1. Collision Repair 2000.

2. Plastic Repair, Retexturing, and Refinishing.

3. Finish Matching.

4. Advanced Vehicle Systems.

5. Steering and Suspension.

Upon completion of a course, a certificate of achievement is awarded for that area, as shown in Figure 2–26. This is also an excellent way to upgrade your skills after becoming employed and to prepare for ASE tests.

For information about the I-CAR courses and when they are given, write:

I-CAR
3701 Algonquin Road
Suite 400
Rolling Meadows, IL 60008-3118

Figure 2–26 After graduation from school and when on the job, you should consider I-CAR training. It will help you keep up to date with technical advances in the collision repair field. *(I-CAR)*

or,

In the U.S., call: 1-800-ICAR-USA (1-800-422-7872) or for class schedules: 1-800-ICAR-456 (1-800-422-7456)

I-CAR Canada
10 Milner Business Court
Suite 404
Scarborough, ON M1B 3C6

or,

In Canada, call: 1-800-565-ICAR (1-800-565-4227)
In Ontario, call: 1-416-299-ICAR (1-416-299-4227)

Or contact I-CAR on the World Wide Web at:

www.i-car.com

Many equipment and finish manufacturers have training programs designed to educate their customers in the use of their products. These training programs run throughout the course of the year. In most cases, they are scheduled in major cities across the United States and Canada, or at their training facilities. Training programs vary in duration and can be tailored to individual requirements. Collision repair facilities should contact local sales representatives for more information.

DON'T BE FOOLISH!

Safety is your responsibility! Follow recommended safety practices at all times.

For example, if you believe it is inconvenient or uncomfortable to wear a respirator mask, you are not thinking! After prolonged exposure, your lungs will be damaged. You will suffer serious health problems at an older age because of your laziness and lack of wisdom. Plan a long, healthy life by protecting yourself now from work hazards!

Summary

- Care must be taken to prevent several types of accidents including asphyxiation, chemical burns, electrocution, fires, and explosions.
- A safety program is a written shop policy designed to protect the health and welfare of personnel and customers. It includes rules on everything from equipment use to disposal of hazardous chemicals.
- In case of an emergency, be sure the following are available and easily seen: emergency telephone numbers, a first aid kit, and safety signs denoting fire exits, fire extinguishers, and dangerous chemical information.
- Certification protects the general public and the technician. ASE certification assures the general public and the prospective employer that certain minimum standards of performance have been met.
- The Inter-Industry Conference on Auto Collision Repair or I-CAR is an advanced training organization dedicated to promoting high-quality practices in the collision repair industry. I-CAR classes are offered in the United States and Canada.

Technical Terms

safety	*front end rack*
accidents	*alignment rack*
asphyxiation	*pneumatic*
chemical burns	*hydraulic*
electrocution	*tool and equipment*
fire	*storage room*
combustibles	*classroom*
explosions	*paint prep area*
physical injury	*finishing area*
manufacturer's	*spray booth*
instructions	*shop office*
safety program	*emergency telephone*
shop layout	*numbers*
metalworking area	*first aid kit*
repair stall	*safety signs*
bay	*gasoline*
frame rack	*spontaneous combustion*

UL (Underwriters
   Laboratories)
electrical fires
fire extinguisher
electrocution
ventilated
carbon monoxide (CO)
asbestos dust
automatic transmission
manual transmission
safety glasses
full-face shield
goggles
rubber gloves
plastic gloves
hand cleaner
jumpsuit
lint-free coveralls
leather shoes
cap
painter's stretch hood
bump cap
earplugs
earmuffs
jack stands
safety stands
hydraulic jacks

lift safety catch
respirators
dust respirator
particle respirator
cartridge filter respirator
isocyanates
air-supplied respirator
fit test
negative pressure test
positive pressure test
dustless sanding system
manufacturer's warnings
material safety data
   sheets (MSDS)
Right-to-Know Laws
EPA (Environmental
   Protection Agency)
liability
ASE certification
ASE tests
National Institute for
   Automotive Service
   Excellence (ASE)
Inter-Industry
   Conference on Auto
   Collision Repair
I-CAR

Review Questions

1. Safety involves using proper work habits to prevent _____ and _____.

2. List six kinds of accidents that are common in a collision repair facility.

3. What are some combustible materials found in a collision repair facility?

4. Name ten collision repair facility work areas.

5. How is frame straightening equipment dangerous?

6. When soiled rags are left lying about improperly, they can cause:
   a. Spontaneous combustion.
   b. Chemical burns.
   c. Explosion.
   d. None of the above.

7. Asbestos dust is inert and does not pose a health problem. True or False?

8. Wear _____ or _____ to protect yourself from the harmful effects of corrosive liquids, undercoats, and finishes.

9. What can happen if you wear athletic shoes and drop a heavy part on your foot?

10. When using an electric power tool, make sure that it is properly _____. The _____ ground _____ sends electricity flow to the ground, not through your body if the tool shorts out.

11. Numerous people have been seriously injured when a vehicle fell on them. How could this have happened?

12. These are needed in the paint shop to keep paint mists and fumes from being inhaled.
    a. Face shield
    b. Mouth piece
    c. Respirator
    d. Shop rag

13. Technician A leaves his shirt sleeves unbuttoned in the shop. Technician B wears loose clothing instead of coveralls. Who is correct?
    a. Technician A
    b. Technician B
    c. Both A and B
    d. Neither A nor B

14. Which of the following is important to shop safety?
    a. Jack stands
    b. Attitude
    c. Approved metal containers
    d. All of the above

15. When running an engine in a vehicle with an automatic transmission, place it in _____.
    a. Neutral
    b. Reverse
    c. Park
    d. Drive

16. Which of the following statements is correct?
    a. Gasoline is an ideal cleaning solvent.
    b. Never keep more than 10 day's worth of paint outside of approved storage areas.
    c. Oily rags should be stored in approved metal containers.
    d. All of the above.

17. Which type of fire involves electrical equipment?
    a. Class A
    b. Class B
    c. Class C
    d. Class D

18. With which type of respirator can facial hair prevent an airtight seal?
    a. Cartridge filter type
    b. Air-supplied type

c. Full face type
d. Hood type

19. Why should you become ASE certified?

20. What is I-CAR?

ASE–Style Review Questions

1. A technician is going to prime a repaired panel. A catalyzed primer is going to be used. Which of the following should the technician do?
   a. Read the owner's manual for the spray gun.
   b. Read the label on the can of primer.
   c. Wear a dust respirator.
   d. Work in a well-ventilated area.

2. Which one of the following details the chemical composition and precautionary information for products that can present a health or safety hazard, and is available from all product manufacturers?
   a. MSDS
   b. DSMS
   c. Instructions
   d. Warnings

3. Which of the following is a testing program to help prove that you are a knowledgeable collision repair and refinishing technician?
   a. I-CAR
   b. NIOSH
   c. OSHA
   d. ASE

4. A technician is transferring flammable materials from bulk storage. Which of the following should be done to prevent an explosion?
   a. Keep the drum(s) grounded
   b. Pour the material slowly
   c. Pour the material quickly
   d. Ground his body

5. Which of the following should be done before removing fuel lines on a fuel injection system?
   a. Relieve fuel pressure
   b. Disconnect battery ground
   c. All of the above
   d. None of the above

6. When a battery charger is connected to a vehicle, wires start to smoke and a fuse blows. Which of the following is the most common cause?
   a. Charger set too high
   b. Shorted wires from collision
   c. Open wires from collision
   d. Charger connected backwards

7. A fire has broken out under a vehicle in the shop. *Technician A* says to open all shop doors and windows to let out the smoke. *Technician B* says to keep them closed. Who is correct?
   a. Technician A
   b. Technician B
   c. Both Technicians A and B
   d. Neither Technician

8. A technician is using electrically powered equipment. Which of the following can cause serious injury?
   a. Water on floor
   b. No ground prong
   c. Bad wire insulation
   d. All of the above

9. A technician is refinishing a vehicle with catalyzed paint. What type of respirator should be worn?
   a. Dust respirator
   b. Cartridge respirator
   c. Disposable respirator
   d. Air-supplied respirator

10. A vehicle is being raised in the air for working on the underbody. *Technician A* says to mount the vehicle on safety stands. *Technician B* says to leave the vehicle supported by the hydraulic floor jack. Who is correct?
    a. Technician A
    b. Technician B
    c. Both Technicians A and B
    d. Neither Technician

Activities

1. Ask a technician or shop owner to visit your classroom. Ask him or her to describe the duties of that position.

2. Make a field trip to a collision repair facility. Have the shop owner or supervisor give you a guided tour of the shop. Discuss the field trip in class the next day.

3. Ask a parent, relative, or friend who has been in a collision to describe what happened during and after the collision when trying to get the vehicle repaired.

4. Go to the library and look up information on different types of insurance<-->full coverage, liability, no-fault. Prepare a report on this subject.

3

Vehicle Construction

Objectives

After studying this chapter, you should be able to:

- Define the most important parts of a vehicle.

- Explain body design and frame variations.

- Compare unibody and body-over-frame construction.

- Identify the major structural parts, sections, and assemblies of body-over-frame vehicles.

- Identify the major structural parts, sections, and assemblies of unibody vehicles.

- Summarize how to classify vehicles by body, engine, and drive train configurations.

Introduction

Vehicle construction refers to how a vehicle is made. As mentioned in Chapter 1, if you know how a vehicle is "put together" or constructed, you will be prepared to repair it properly when damaged. You must be able to identify the basic parts of a vehicle and compare design variations. This will give you the vocabulary to fully grasp information in later text chapters. So study carefully! See Figure 3–1.

As you will learn, the modern motor vehicle is no longer a simple "buggy with an engine." The present-day motor vehicle is a maze of interacting mechanical–electrical systems. In fact, over 15,000 parts are used in a typical vehicle. Damage to one part can affect the operation of another seemingly unrelated part. As a result, you must understand the whole vehicle and its systems to safely and effectively repair collision damaged vehicles.

During a collision, severe damage often occurs to the vehicle's unibody or frame structure. Depending upon the type of construction, varying methods must be used to repair the damage properly. This chapter will give you essential information that will help you understand later chapters.

Figure 3–1 Various construction methods are used in today's vehicles. This chapter will help you understand how cars, trucks, and vans are "put together" so that you can repair them when they are damaged in a collision. (*Photograph courtesy of Mercedes-Benz of North America*)

Figure 3–2 Vehicles must pass motor vehicle safety crash certification tests to assure adequate passenger safety. (*BMW of North America*)

CRASH TESTING

Automobile manufacturers are challenged by having to design vehicles that are light, **aerodynamic** (have low wind resistance), and are yet strong and safe.

Computer-simulated crash testing helps to determine how well a new vehicle might survive a collision. Computer-simulated testing is used before building a prototype, or first real vehicle, to find weak or faulty structural areas before investing in mass production.

It is critical that the passenger compartment is strong enough to help prevent injury to the driver and passengers. It is also important that the front and rear sections collapse upon impact to absorb some of the energy of a collision.

Certified crash tests are done using a real vehicle and sensor-equipped dummies that show how much impact the people would suffer during a collision. Computer readings from the sensors in the dummies give feedback about each crash test for body structure evaluation, Figure 3–2.

Crush zones are built into the frame or body to collapse and absorb some of the energy of a collision. The front and rear of the vehicle collapses while the passenger compartment tends to retain its shape. This helps reduce the amount of force transmitted to the occupants, Figure 3–3.

VEHICLE CLASSIFICATIONS

Vehicle classification relates to the construction, size, shape, number of doors, type of roof, and other criteria of a motor vehicle. To communicate properly in collision repair, you must understand these basic terms.

VEHICLE CONSTRUCTION

There are three main types of vehicle construction: body-over-frame, unibody, and space frame. Each has its own unique repair challenges.

Full Frames

Body-over-frame vehicles have separate body and chassis parts bolted to the frame. The engine and other major assemblies are mounted on the frame. This type of frame consists of two side rails connected by a series of crossmembers. See Figure 3–4.

Most full frames are wide at the rear and narrow at the front. The narrow front construction enables the front wheel to make a sharper turn. A wide frame at the rear provides better support of the body.

Other characteristics of a separate frame are:

1. A full-frame vehicle is heavier.

2. High amounts of energy are absorbed by the frame during a collision.

3. Suspension and drive train parts can be quickly assembled on the frame.

4. The heavy frame is made of thick sheet metal approximately ⅛ inch (3 mm) thick.

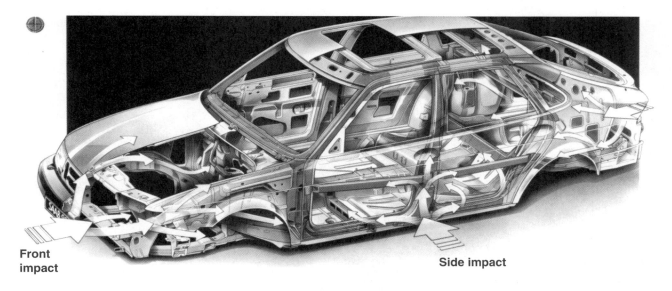

Front impact

Side impact

A

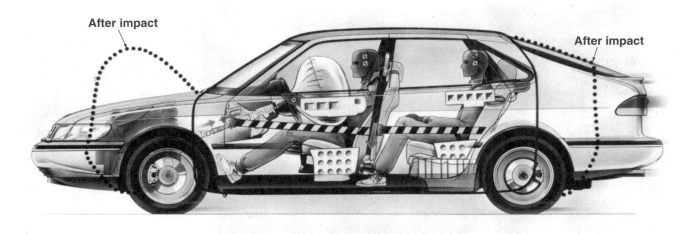

After impact

After impact

B

Figure 3–3 (A) The arrows show how impact force is spread out. (B) The front and rear of the vehicle are designed with crush zones to absorb some of the energy of an impact and protect the passenger compartment. *(Saab Cars USA, Inc.)*

The *frame rails* are the long steel members that extend along the sides of the vehicle. The *torque boxes* are the structural parts of the frame designed to allow some twisting to absorb road shock and collision impact. The *frame horn* is the very front of the frame rails where the bumper attaches.

Crossmembers are thick metal stampings that extend sideways across the frame rails. They are often used to support the engine, suspension, and other chassis parts. *Spring hangers* are sometimes formed on the frame to hold the suspension system springs. *Rubber body insulators* are used between the frame and body to reduce noise and vibration.

Although unibody construction is the trend, body-over-frame construction is still being used. Full- or partial-frame construction is used on some full-size, luxury vehicles, most full-size pickup trucks, and some small pickups.

The *frame* is an independent, separate part because it is not welded to any of the major units of the body shell. The strong side rails are normally made of U-shaped channel- or box-shaped sections. Various brackets, braces, and openings are provided for the parts of the chassis.

A *perimeter frame,* as its name implies, has the frame rail near the outside, or perimeter, of the

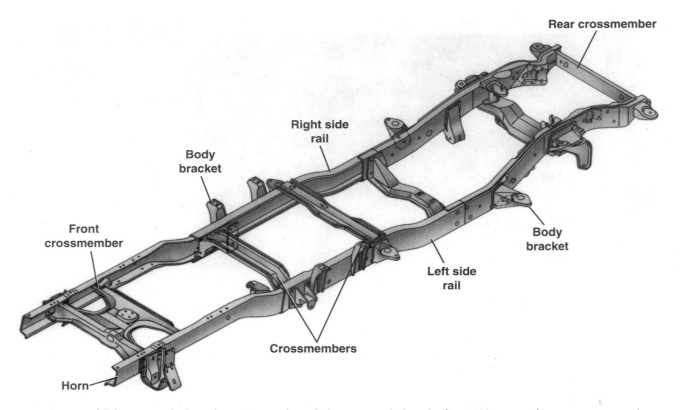

Rear crossmember

Right side
rail

Body
bracket

Front
crossmember

Body
bracket

Left side
rail

Crossmembers

Horn

Figure 3–4 A full frame is a thick steel structure. Body and chassis parts bolt to the frame. This type of construction is used in some full-size cars and many trucks. Note the frame parts. *(Courtesy of DaimlerChrysler Corporation)*

vehicle. It is the most common type of full frame. It utilizes full-length side rails with torque boxes in the four corners of the center section. A perimeter frame has good side impact strength. See Figure 3–5.

A *ladder frame* has long frame rails with a series of straight crossmembers formed in several locations. It is a seldom used modification of the perimeter frame.

A *partial frame* is a cross between a solid frame and a unibody. *Sub-frame assemblies* are used at the front and rear while the unibody supports the middle area of the vehicle. The sub-frame is used to support the suspension and drive train. See Figure 3–6.

Unibody Construction

Unibody construction uses body parts that are welded and bolted together to form an integral vehicle. No separate heavy-gauge steel frame under the body is needed. In unibody designs, heavy-gauge,

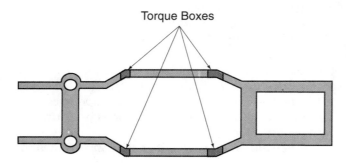

Torque Boxes

Figure 3–5 Perimeter Frame

Figure 3–6 Note how other parts attach to this frame. It provides a very strong structure for mounting other parts securely. *(American Suzuki)*

Figure 3–7 Unibody construction uses body panels to form the frame. No separate, thick steel framework is needed. This body is made of corrosion-resistant aluminum. *(© American Honda Motor Co., Inc.)*

cold-rolled steels have been replaced with lighter, thinner, high-strength steel alloys or aluminum alloy. One example is shown in Figure 3–7.

The concept of unibody construction first proved successful in the aircraft industry, where lightweight, high-strength fuselages were needed. Adapted to the auto industry, unitized construction eliminated the need for a separate, rigid frame. The strength and rigidity of the vehicle was integrated into the body shell.

The body shell is formed by welding sheet metal into a box- or egg-like configuration. Strength is achieved through shape and design of the individual parts instead of their mass and weight. The entire inner structure works together for structural integrity.

The eggshell concept is an ideal model for understanding the principle of unibody design. Even when pressing hard on an eggshell, it is comparatively difficult to destroy its thin structure. All the force applied by your finger is not concentrated in one place but is dispersed over the shell surface. In engineering, this concept is called a ***stressed hull structure.***

Space Frame

Similar to a unibody, a ***space frame*** vehicle has a metal body structure covered with an outer skin of plastic or composite panels. It is a relatively new type of vehicle construction. Quite often, the roof and quarter panels are not welded to the structure as they are with traditional unibodies. Exterior body panels are attached with mechanical fasteners or adhesives. See Figure 3–8.

After a collision, a space frame is more likely to have hidden damage because of the ability of plastic panels to hide more severe damage. Corrosion protection is also important since the plastic body panels may look good but the hidden metal frame structure may become deteriorated.

Body Structure

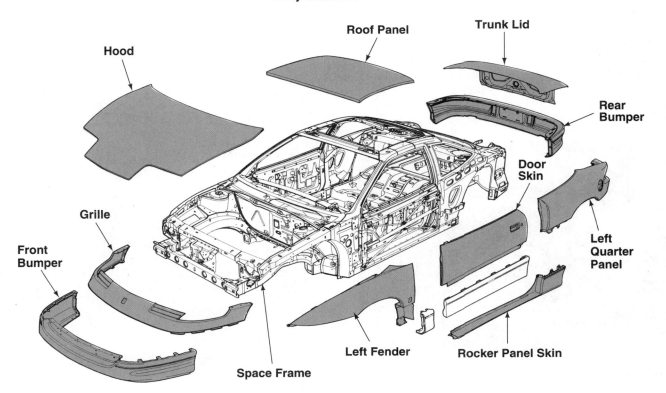

Figure 3–8 Note space frame construction. Composite (plastic) panels fasten to a metal inner body structure. Composite panels can be made flexible to resist door dings and small dents. *(Courtesy of General Motors Corporation, Service Operations)*

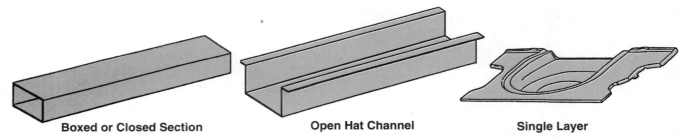

Figure 3–9 Unibody parts can be single layer, hat channel, box, or other shapes. Curves in the panels increase strength. *(I-CAR)*

Basically, unibody vehicles are made up of numerous simple parts. Single-layer stamped steel panels, channels, and boxed or closed steel sections form the unibody, Figure 3–9.

Support members are often bolted to the bottom of the unibody to hold the engine, transmission, and suspension in alignment. They are needed in high-stress areas to reduce body flex.

JOINING PARTS

Stationary parts, like the floor, roof, and quarter panels, are permanently welded or adhesive bonded into place. *Hinged parts,* like doors, hood, and deck lids, will swing up and open.

Fastened parts are held together with various fasteners (bolts, nuts, clips, etc.). Many parts, like the fenders, hood, and grille, bolt into place. These bolted-on parts also add to the strength of the vehicle, Figure 3–10.

Welded parts are permanently joined by melting the material so that it flows together and bonds when cooled. Both metal and plastic parts can be welded.

Press-fit or *snap-fit parts* use clips or an interference fit to hold parts together. This assembly method is becoming more common to reduce manufacturing costs. Grilles, bumper covers, inner door trim, and other nonstructural parts may use clips to snap into place.

Adhesive bonded parts use a high-strength epoxy or special glue to hold the parts together. Both metal and plastic parts can be adhesive-joined. *Structural adhesive* can also be used to bond parts together. See Figure 3–11.

A *composite unibody* is made of plastics and other materials, like carbon fiber, to form the vehicle. The frame is made totally of plastics or other composite materials, keeping metal parts to a minimum. This cuts weight while increasing strength, rigidity, performance, and fuel economy. Although this type of vehicle is not being mass produced, several manufacturers are experimenting with composite unibody construction.

Figure 3–10 Many parts (fenders, for example) bolt to the unibody structure. Engine and other assemblies are bolted to the body. *(Photograph provided courtesy of Mercedes-Benz North America, Inc)*

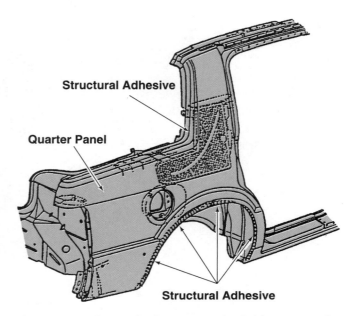

Structural Adhesive

Quarter Panel

Structural Adhesive

Figure 3–11 Structural adhesive is used to hold parts securely together and helps to make the body quieter and stronger. *(Courtesy of DaimlerChrysler Corporation)*

MAJOR BODY SECTIONS

For simplicity and to help communication in collision repair, a vehicle is commonly divided into *three body sections*—front, center, and rear. It is important to understand how these sections are constructed to properly repair them. Study Figure 3–12.

The *front section,* also called *nose section,* includes everything between the front bumper and the firewall. The bumper, grille, frame rails, front suspension parts, and usually the engine are a few of the items included in the front section of a vehicle.

The nickname "front clip" or "doghouse" is used to refer to the front body section. Front sections are often purchased in one piece from an *automotive recycler* or "salvage yard." The empty engine compartment forms the "doghouse."

The *center section* or *midsection* typically includes the body parts that form the passenger compartment. A few parts in this section include the floor pan, roof panel, cowl, doors, door pillars, glass, and related parts. A slang name for the center section is the "greenhouse" because it is surrounded by glass.

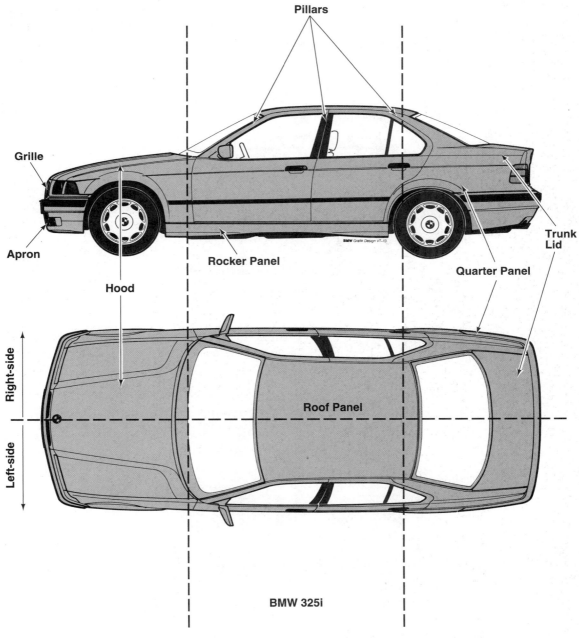

Figure 3–12 A vehicle is commonly divided into three sections for better communication. Study the external parts in each section. *(BMW of North America)*

The **rear section, tail section,** or **rear clip,** commonly consists of the rear quarter panels, trunk or rear floor pan, rear frame rails, trunk or deck lid, rear bumper, and related parts. Also called the "cathouse," it is often sectioned or cut off of a salvaged vehicle to repair severe rear impact damage.

When discussing collision repair, repair personnel often refer to these sections of the vehicle. It simplifies communication because everyone knows which parts are included in each section.

PANEL AND ASSEMBLY NOMENCLATURE

A **panel** is a stamped steel or molded plastic sheet that forms a body part. Various panels are used in a vehi-

cle. Usually, the name of the panel is self-explanatory. When panels are joined with other components, the result is called an **assembly.**

The **vehicle left side** is the steering wheel side on vehicles built for American roads. The **vehicle right side** is the passenger side or side opposite the steering wheel. Remember that vehicles built for other countries will often have their steering wheel on the right side because they are driven on the left side of the road.

Another way to determine the right and left sides of a vehicle is to stand behind the vehicle. Your right hand would be on the right side and left hand would be on the left side of the vehicle. Panels and parts are often called out as left or right side, as shown in Figure 3–13.

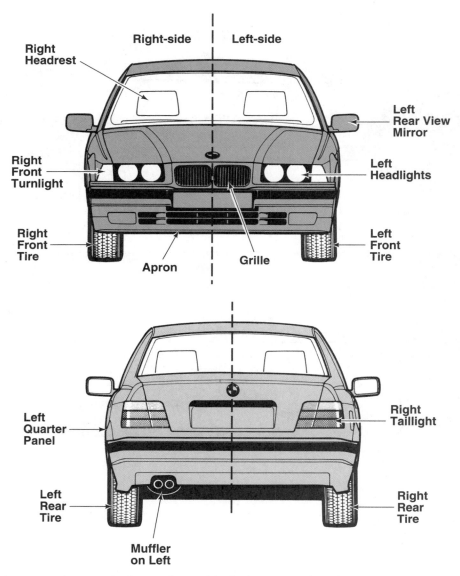

Figure 3–13 Note the names of parts at the front and rear of the body. (BMW of North America)

FRONT SECTION PARTS

The **frame rails** are the box frame members extending out near the bottom of the front section. They are usually the strongest part of a unibody.

The **cowl** is near the rear of the front section, right in front of the windshield. This includes the top cowl panel and side cowl panels.

The **front fender aprons** are inner panels that surround the wheels and tires to keep out road debris. They often bolt or weld to the frame rails and cowl.

The **shock towers** or **strut towers** are reinforced body areas for holding the upper parts of the suspension system. The coil springs, and strut or shock absorbers fit up into the shock towers. They are normally formed as part of the inner fender apron. See Figure 3–14.

The **radiator core support** is the framework around the front of the body structure for holding the cooling system radiator and related parts. It often fastens to the frame rails and inner fender aprons.

The **hood** is a hinged panel for accessing the engine compartment (front-engine vehicle) or trunk area (rear-engine vehicle). **Hood hinges,** bolted to the hood and cowl panel, allow the hood to swing open. The hood is normally made of two or more panels welded or bonded together to prevent flexing and vibration.

The **dash panel,** sometimes termed **firewall** or **front bulkhead,** is the panel dividing the front section and the center, passenger compartment section. It normally welds in place.

The **front fenders** extend from the front doors to the front bumper. They cover the front suspension and inner aprons. They normally bolt into place around their perimeter.

The **bumper assembly** bolts to the front frame horns or rails to absorb minor impacts.

CENTER SECTION PARTS

The **floor pan** is the main structural section in the bottom of the passenger compartment.

With front-wheel drive vehicles, the floor pan can be relatively flat. With rear-wheel drive vehicles, a **tunnel** is formed in the floor pan for the transmission and drive shaft. The drive shaft needs room to extend back to the rear axle assembly.

Pillars are vertical body members that hold the roof panel in place and protect in case of a rollover accident. See Figure 3–15.

The **front pillars** extend up next to the edges of the windshield. They must be strong to protect the passengers. Also termed **A-pillars,** they are steel box members that extend down from the roof panel to the main body section.

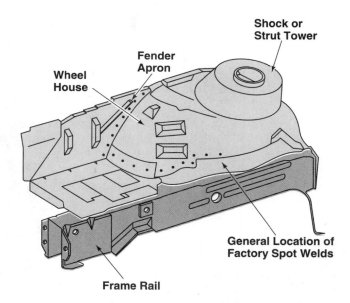

Figure 3–14 Frame rails are important structural parts in the front section of a unibody vehicle. Other major parts fasten to the rails. *(Used by permission of Tech-Cor, Inc.)*

Center pillars or **B-pillars** are the roof supports between the front and rear doors on four-door vehicles. They help strengthen the roof and provide a mounting point for the rear door hinges.

Rear pillars extend up from the quarter panels to hold the rear of the roof and rear window glass. Also called **C-pillars,** their shape can vary with body style.

Rocker panels or **door sills** are strong beams that fit at the bottom of the door openings. They normally are welded to the floor pan and to the pillars, kick panels, or quarter panels. The **kick panels** are small panels between the front pillars and rocker panels.

The **rear shelf** or **package tray** is a thin panel behind the rear seat and in front of the back glass. It often has openings for the rear speakers. The **rear bulkhead panel** separates the passenger compartment from the rear trunk area.

The **doors** are complex assemblies made up of an outer skin, inner door frame, door panel, window regulator, glass, and related parts. **Door hinges** bolt between the pillars and door frame. The **window regulator** is a gear mechanism that allows you to raise and lower the door glass. See Figure 3–16.

Side impact beams are metal bars or corrugated panels that bolt or weld inside the door assemblies to protect the passengers. They prevent the door from opening upon impact and help protect the passenger area, Figure 3–17.

The **roof panel** is a large multipiece panel that fits over the passenger compartment. It is normally welded to the pillars. Sometimes it includes a sun roof or removable top pieces, termed T-tops.

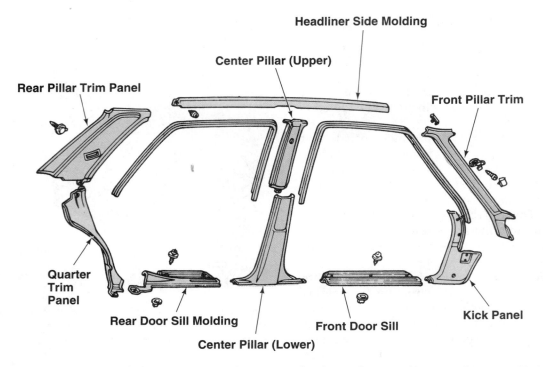

Figure 3-15 Pillars hold the roof in place and provide openings for doors. They must be extremely strong. (© American Honda Motor Co., Inc.)

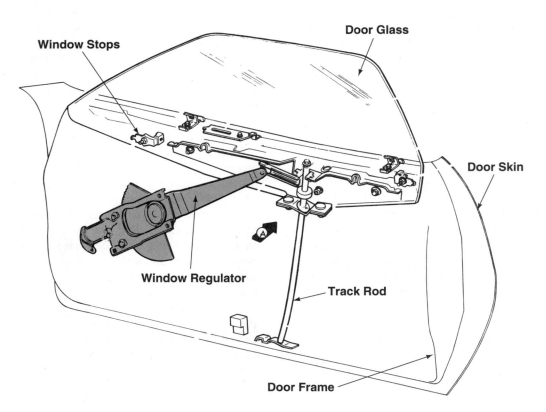

Figure 3-16 Study the basic construction of a door assembly. The regulator allows the window to go up and down. (Courtesy of DaimlerChrysler Corporation)

Door beams

Figure 3–17 Side impact beams are located inside the doors. *(Courtesy of DaimlerChrysler Corporation)*

The **dash assembly,** sometimes termed **instrument panel,** is the assembly including the soft dash pad, instrument cluster, radio, heater and AC controls, vents, and similar parts.

REAR SECTION PARTS

The **trunk floor panel** is a stamped steel part that forms the bottom of the rear storage compartment. Quite often, the spare tire fits down into this stamped panel. It often welds to the rear rails, inner wheel houses, and lower rear panel.

The **deck lid,** or **trunk lid,** is a hinged panel over the rear storage compartment, Figure 3–18. A **rear hatch** is a larger panel and glass assembly hinged for more access to the rear of the vehicle.

The **quarter panels** are the large, side body sections that extend from the side doors back to the rear bumper. They are welded in place and form a vital part of the rear body structure.

The **lower rear panel** fits between the trunk compartment and the rear bumper between the quarter panels.

Rear shock towers hold the top of the rear suspension. The **inner wheel housings** surround the rear wheels.

GASKETS AND SEALS

Various **gaskets** and **rubber seals** are used to prevent air and water leakage between body parts. Seals or **weatherstripping** are often used around doors and the rear deck lid. The rubber seal is partially compressed when the door or lid is closed to form a leakproof connection. Another example, a rubber gasket often seals the stationary glass where it fits into the body.

ANTICORROSION MATERIALS

Anticorrosion materials are used to prevent rusting of metal parts. Various types of anticorrosion materials are available. For example, **undercoating** is often a thick tar or synthetic rubber-based material sprayed onto the underbody of the vehicle. After performing repairs, you must restore all corrosion protection. See Figure 3–19.

SOUND-DEADENING MATERIALS

Sound-deadening materials are used to help quiet the passenger compartment. They are insulation materials that prevent engine and road noise from entering the passenger area.

ENGINE LOCATIONS, DRIVE LINES

There are four basic drive train designs: front-wheel drive **(FWD),** rear-wheel drive **(RWD),** rear-engine, rear-wheel drive **(RRW)** and mid-engine, rear-wheel drive **(MRD).** The vast majority of unibody vehicles on the road today are FWD with the engine in the front. These variations affect vehicle construction and repair methods.

A **longitudinal engine** mounts the crankshaft centerline front-to-rear when viewed from the top. Front-engine, rear-wheel drive vehicles use this type of engine mounting. See Figure 3–20.

A **transverse engine** mounts sideways in the engine compartment. Its crankshaft centerline extends toward the right and left of the body. Both front-engine, and rear-engine vehicles use this configuration.

A **front-engine, front-wheel drive (FWD)** vehicle has both the engine and transaxle in the front. Constant velocity (CV) axles extend out from the transaxle to power the front drive wheels. This is one of the most common configurations, Figure 3–21.

A **front-engine, rear-wheel drive (RWD)** vehicle has the engine in the front and the drive axle in the rear. The transmission is often right behind the engine and a drive shaft transfers power back to the rear axle. A few vehicles have the engine in the front and the transmission in the rear, however.

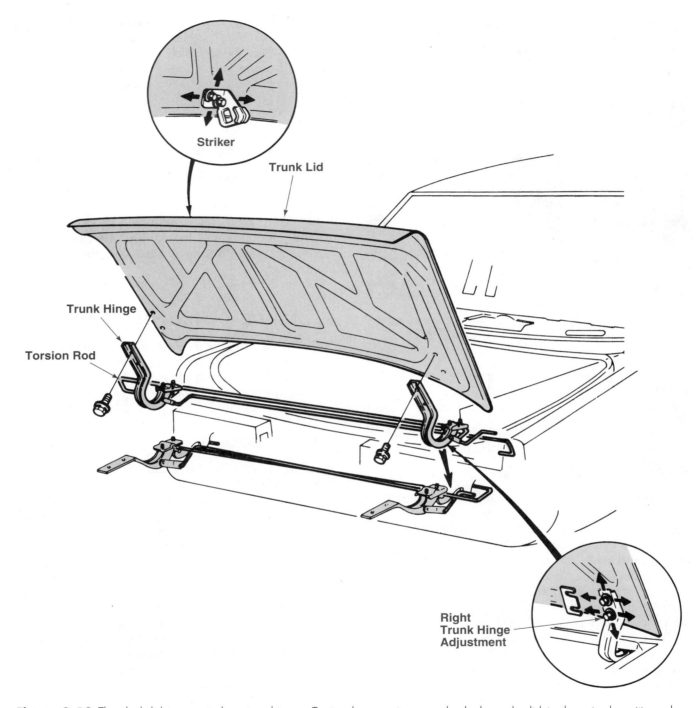

Figure 3–18 The deck lid is mounted on two hinges. Torsion bars, springs, or shocks keep the lid in the raised position when needed. The striker engages a latch to hold the lid closed. *(© American Honda Motor Co., Inc.)*

A *rear-engine, rear-wheel drive (RRD)* vehicle has the engine in the back and a transaxle transfers power to the rear drive wheels. Traction upon acceleration and cornering is good because the weight of the engine and transaxle are over the rear drive wheels, Figure 3–22.

A *mid-engine, rear-wheel drive (MRD)* vehicle has the engine located right behind the front seat, or centrally located. This helps to place the center of gravity in the middle so that the front and rear wheels hold the same amount of weight. This improves cornering ability.

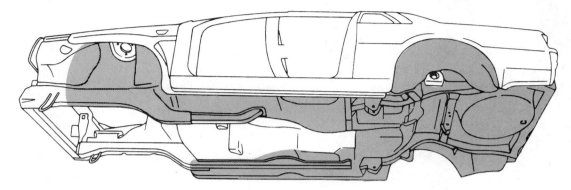

☐ **Indicates undercoating coated portions**

Figure 3–19 Anticorrosion material is often applied to underbody areas exposed to road salt, water, and other debris. This helps protect the underbody from corrosion. *(Used with permission from Nissan North America, Inc.)*

Figure 3–20 Study the underside of this front-engine, rear-wheel drive vehicle. Rear-wheel drive is still very common with large luxury cars, high-performance cars, pickup trucks, and full-size vans. Note how the front crossmember bolts to the frame. *(American Suzuki)*

All-wheel drive uses two differentials to power all four drive wheels. This is used on several makes of passenger vehicles, Figure 3–23.

Four-wheel drive systems use a transfer case to send power to two differentials and all wheels. The transfer case can be engaged and disengaged to select two- or four-wheel drive as desired. It is common on off-road vehicles.

NOTE! For more information on vehicle construction (air bags, drive lines, etc.) and related sub-

jects, refer to textbook index. For example, Chapter 24 explains the operation and repair of mechanical systems.

VEHICLE SIZES

A *compact car* is the smallest body classification. It normally uses a small, 4-cylinder engine, is very lightweight, and gets the highest gas mileage.

An *intermediate car* is medium in size. It can use a 4-, 6-, or 8-cylinder engine and has average

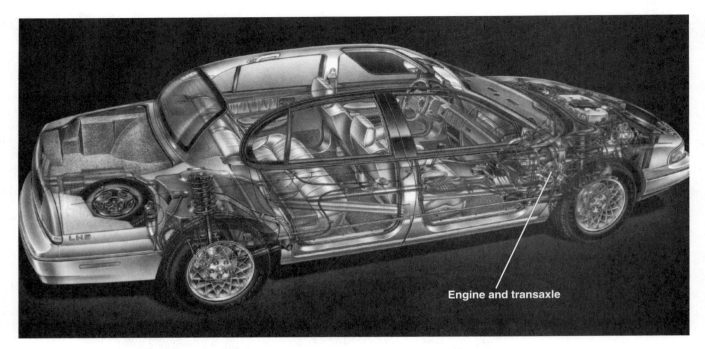

Engine and transaxle

Figure 3–21 Study this front-engine, front-wheel drive design. *(Courtesy of DaimlerChrysler Corporation)*

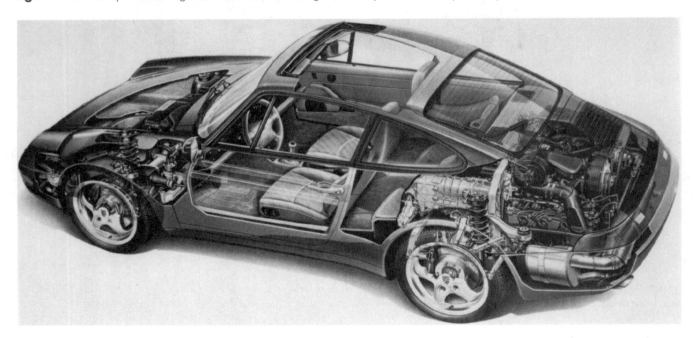

Figure 3–22 This is a high-performance, rear-engine, rear-wheel drive vehicle. The weight of the engine is over the drive wheels for good traction upon acceleration. *(Photograph used by permission from Porsche Cars North America, Inc.)*

weight and physical dimensions. It usually has unibody construction, but a few vehicles have body-over-frame construction.

A *full-size car* is the largest classification. It is large, heavy, and often uses a high performance V-8 engine. Full-size cars can have either unibody or body-over-frame construction.

ROOF DESIGNS

A *sedan* refers to a body design with a center pillar that supports the roof. Sedans come in both 2- and 4-door versions.

A *hardtop* does not have a center pillar to support the roof. The roof must be reinforced to provide

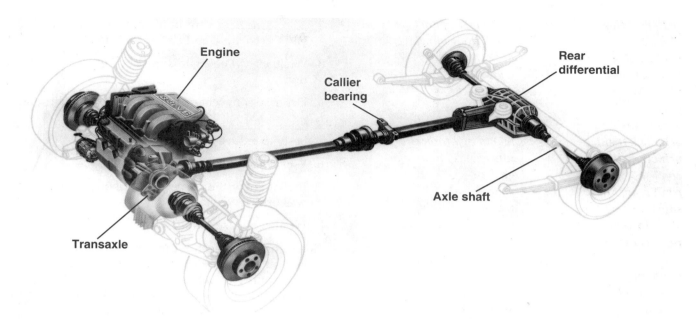

Figure 3–23 All-wheel drive uses a drive train that powers all four wheels all the time. This one is for a minivan. *(Courtesy of Daimler-Chrysler Corporation)*

enough strength. A hardtop is also available in both 2- and 4-door versions.

A **hatchback** has a large third door at the back. This design is commonly found on small compact cars so that more rear storage space is available.

A **convertible** uses a retractable canvas roof with a steel tube framework. The top folds down into an area behind the seat. Some convertibles use a removable hardtop.

A **station wagon** extends the roof straight back to the rear of the body. A rear hatch or window and **tailgate** open to allow access to the large storage area.

VANS AND TRUCKS

A **van** has a large box-shaped body to increase interior volume or space. A full-size van normally is front-engine, rear-wheel drive. A **minivan** is smaller and often uses front-engine, front-wheel drive with unibody construction.

A **pickup truck** normally has a separate cab and bed. A front-engine, rear-wheel drive is typical.

Summary

- Vehicle classification relates to the construction, size, shape, number of doors, type of roof, and other criteria of a motor vehicle.
- There are three main types of vehicle frame construction: body-over-frame, unibody, and space frame. Each has its own unique repair challenges.

- A vehicle is commonly divided into three body sections:
  1. front, or nose, section
  2. center section, or midsection
  3. rear section, tail section, or rear clip.
- There are four basic drive train designs:
  1. front-wheel drive (FWD)
  2. rear-wheel drive (RWD)
  3. rear-engine, rear-wheel drive (RRW)
  4. mid-engine, rear-wheel drive (MRD).
- The three vehicle sizes are compact, intermediate, and full size.

Technical Terms

certified crash tests
crush zones
vehicle classification
body-over-frame
frame rails
torque boxes
frame horn
crossmembers
spring hangers
rubber body insulators
frame
perimeter frame
ladder frame
partial frame
sub-frame assemblies
unibody

stressed hull structure
space frame
support members
stationary parts
hinged parts
fastened parts
welded parts
press-fit parts
snap-fit parts
adhesive-bonded parts
structural adhesive
composite unibody
three body sections
front section
nose section
automotive recycler

center section
midsection
rear section
tail section
rear clip
panel
assembly
vehicle left side
vehicle right side
cowl
front fender aprons
shock towers
strut towers
radiator core support
hood
hood hinges
dash panel
firewall
front bulkhead
front fenders
bumper assembly
floor pan
tunnel
pillars
front pillars
A-pillars
center pillars
B-pillars
rear pillars
C-pillars
rocker panels
door sills
kick panels
rear shelf
package tray
rear bulkhead panel
door hinges
window regulator
side impact beams
roof panel
dash assembly

instrument panel
trunk floor panel
deck lid
trunk lid
rear hatch
quarter panels
lower rear panel
rear shock towers
inner wheel housings
gaskets
rubber seals
weatherstripping
anticorrosion materials
undercoating
sound-deadening
  materials
longitudinal engine
transverse engine
front-engine, front-wheel
  drive (FWD)
front-engine, rear-wheel
  drive (RWD)
rear-engine, rear-wheel
  drive (RRD)
mid-engine, rear-wheel
  drive (MRD)
all-wheel drive
four-wheel drive
compact car
intermediate car
full-size car
sedan
hardtop
hatchback
convertible
station wagon
tailgate
van
minivan
pickup truck

Review Questions

1. How many parts are used in a typical vehicle?
   a. 1,000
   b. 10,000
   c. 15,000
   d. 50,000

2. _____ are built into the frame or unibody to collapse and absorb collision energy.

3. What are three main types of vehicle construction?

4. Describe some characteristics of a separate frame.

5. _____ are used between the frame and body to reduce noise and vibration.

6. What is a perimeter frame?

7. Explain how the concept of unibody construction was first developed.

8. What is a space frame?

9. A vehicle has been hit in the front. The front section must be replaced. *Technician A* says to call automotive recyclers or salvage yards to find a front clip. *Technician B* says to order new parts to assure a good repair. Who is correct?
   a. Technician A
   b. Technician B
   c. Both technicians
   d. Neither technician

10. Describe the term "doghouse."

11. How do you tell the right and left side of a vehicle?

12. This is the area at the rear of the front section, right in front of the windshield.
    a. Kick panel
    b. Crossmember
    c. Side panel
    d. Cowl

13. What is the difference between FWD and RWD vehicles regarding the floor pan?

14. This part helps strengthen the roof and provide a mounting point for the rear door hinges.
    a. A-pillar
    b. B-pillar
    c. C-pillar
    d. D-pillar

15. The _____ are the large, side body sections that extend from the side doors back to the rear bumper. They are welded in place and form a vital part of the rear body structure.

ASE–Style Review Questions

1. *Technician A* says that the "right side" of the vehicle is determined as if you were sitting in the driver seat. *Technician B* says that the "right side" of the vehicle is determined as if you were standing

in front of the vehicle looking back. Who is correct?
a. Technician A
b. Technician B
c. Both Technicians A and B
d. Neither Technician

2. *Technician A* says that the body-over-frame construction of a vehicle has separate body and chassis parts bolted to the frame. *Technician B* says that unibody construction uses body parts that are welded and bolted together to form an integral vehicle. Who is correct?
a. Technician A
b. Technician B
c. Both Technicians A and B
d. Neither Technician

3. *Technician A* says that space frame construction uses a perimeter frame to strengthen the body. *Technician B* says that space frame construction has a metal body structure covered with an outer skin of plastic or composite panels. Who is correct?
a. Technician A
b. Technician B
c. Both Technicians A and B
d. Neither Technician

4. Which of the following methods of joining parts is not used on a unibody construction vehicle?
a. Bolts
b. Welding
c. Pop rivets
d. Adhesive bonding

5. *Technician A* says that unibody parts can be single layer, hat channel, box, or other shapes. *Technician B* says that unibody parts can be flat steel, box, or open hat channel shapes. Who is correct?
a. Technician A
b. Technician B
c. Both Technicians A and B
d. Neither Technician

6. *Technician A* says that a space frame vehicle is less safe because it has plastic body parts. *Technician B* says that unibody construction uses heavier steel welded together to make it stronger because there is no real frame. Who is correct?
a. Technician A
b. Technician B
c. Both Technicians A and B
d. Neither Technician

7. *Technician A* says that a unibody-constructed vehicle is divided into four sections (front, driver, passenger, and rear). *Technician B* says that a unibody-constructed vehicle is divided into three sections (front, center, and rear). Who is correct?
a. Technician A
b. Technician B
c. Both Technicians A and B
d. Neither Technician

8. *Technician A* says that a good paint is all that is needed to protect the vehicle from corrosion following a collision repair. *Technician B* says that gaskets and seals are used to prevent air and water leakage. Who is correct?
a. Technician A
b. Technician B
c. Both Technicians A and B
d. Neither Technician

9. All-wheel drive
a. can be engaged and disengaged to select four- or two-wheel drive.
b. uses a transfer case.
c. is used on large trucks for more traction.
d. has all four wheels engaged at all times.

10. *Technician A* says that a hardtop is a body design that doesn't have a center pillar that supports the roof. *Technician B* says that a sedan is a body design that doesn't have a center pillar that supports the roof. Who is correct?
a. Technician A
b. Technician B
c. Both Technicians A and B
d. Neither Technician

Activities

1. Have a class discussion comparing the advantages and disadvantages of conventional frame versus unibody construction.

2. Examine a vehicle in the shop. See how many parts you can identify. Study how the vehicle is constructed.

3. Discuss front- and rear-wheel drive vehicles. What are the advantages and disadvantages of each?

4. Name the makes and models of vehicles that use steel, aluminum, and plastic or composite construction. Discuss each.

4

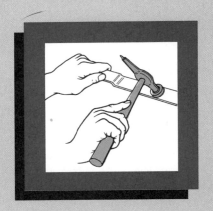

Hand Tools

Objectives

After studying this chapter, you should be able to:

- Identify general purpose hand tools.

- Explain the use of general purpose hand tools.

- Identify the most important collision repair hand tools.

- Explain the use of body shop hand tools.

- Compare the advantages and disadvantages of different tools.

- List typical safety rules for hand tools.

- Properly select the right tool for the job.

- Maintain and store tools properly.

Introduction

Professional collision repair technicians invest thousands of dollars in their tools. A good set of tools will speed repairs and improve work quality. The importance of having the right tools for the job cannot be overstated.

Remember! Tools serve to increase the abilities of your hands, arms, legs, back, eyes, and ears. They allow you to do things that would be impossible otherwise. Neither you, nor the very best technician, can do quality work using inferior tools.

As you will learn in this chapter, you must know how to use and care for your tools properly. Proper tool selection and utilization are critical to doing competent repairs. Improper use of tools is dangerous and will result in substandard repair work. Study the material in this chapter carefully and you will have a good background in collision repair hand tools.

PURCHASE QUALITY TOOLS

Never buy cheap, low-grade tools. Cheap tools slow down your work rate and efficiency because they are heavier, more clumsy, and break more easily. You usually get what you pay for. Good tools will pay for themselves in a short period of time.

A *lifetime tool guarantee* means the tool will be replaced or repaired if it ever fails or breaks. Some guarantees are for the life of the tool; some are NOT. Even though guaranteed tools are more expensive, they will save money in the long run.

TOOL STORAGE

Always store tools properly when they are not in use. This will protect them from damage or loss. It will also speed your work because you will be able to find tools more quickly. Keep related tools together in one drawer. Keep heavy tools in one drawer and small or delicate tools in another. This will keep the delicate tools from being damaged.

After use, wipe oil and grease off tools. A greasy, oily tool can easily slip from a technician's hand and cause injury. Keeping tools clean makes them safer and prolongs their life and usefulness.

Tool holders are clip racks, pouches, or trays that help you organize small tools. Tool holders allow you to carry the tool set to the job easily. This saves you from having to walk back and forth to and from your tool box.

A *tool box* stores and protects your tools. Most tool boxes are made up of a large, bottom roll-around cabinet and an upper tool chest. The *upper chest* often holds commonly used tools at eye level. The *lower cabinet* often stores heavier tools. See Figure 4–1.

■ *Danger!* NEVER open more than one tool box drawer at the same time because the box can flip over. Serious injury could result! Close each drawer before opening the next one.

GENERAL DUTY HAMMERS

Hammers are used for striking and exerting an impact on a part. Always use the right hammer for the task and use it properly. See Figure 4–2.

A *ball peen hammer* has a flat face for general striking and a round peen end for shaping sheet metal, rivet heads, or other objects.

The *brass* or *lead hammer* will make heavy blows without marring the metal surface. The soft metal head will dent and protect the part.

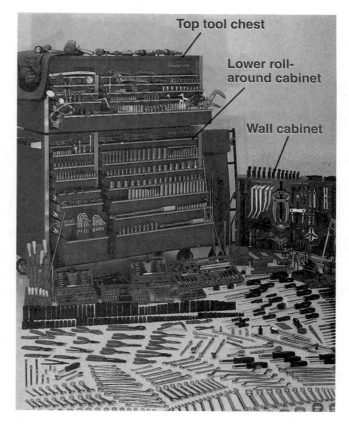

Figure 4–1 A quality tool box is the "foundation" for your set of tools. The lower roll-around cabinet often holds heavier, bulkier tools, while the upper chest is for smaller, more commonly used tools. *(Courtesy of Snap-on Tools Corporation)*

A *plastic hammer* is for making light blows where parts can be easily damaged. It is used on delicate parts.

A *dead blow hammer* has a metal face filled with lead shot (balls) to prevent rebounding. It will not bounce back up after striking. It is good for driving operations where part damage must be avoided.

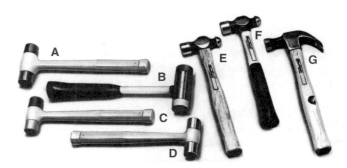

Figure 4–2 An assortment of hammers is essential. (A) Brass hammer. (B) Rubber hammer. (C) Plastic hammer. (D) Dead blow hammer. (E) Ball peen hammer. (F) Ball peen hammer with rubber handle grip. (G) Claw hammer. A claw hammer should not be used in collision repair; it is for woodworking. *(Courtesy of Snap-on Tools Corporation)*

A **rubber mallet** has a solid rubber head that is fairly heavy. It is often used to gently bump sheet metal without damaging the painted finish. It is also used to install wheel covers. Rubber mallets can be used in conjunction with a suction cup on soft "cave-in" type dents. While pulling on the cup, the technician uses the mallet to tap lightly all around the surrounding high spots. A "popping sound" occurs as the high spots drop and the low spot springs back to its original contour.

BODY HAMMERS

Body hammers are designed to work with sheet metal. They often have a point on one end and a flat head or a head of an odd shape on the other. They come in many different designs. Each style is designed for a special use. See Figure 4–3.

Picking hammers have a sharply pointed head and will remove many small dents. The pointed end is used to raise small dents from the inside. The flat end is for hammer-and-dolly work to remove high spots and ripples. Picking hammers come in a variety of shapes and sizes. Some have long picks for reaching behind body panels. Some have sharp points; others are blunted.

Bumping hammers are used to bump out large dents. They may have a round or square face. The surfaces of the faces are nearly flat and large enough that the force of the blows are spread over a large area. These hammers are used for initial straightening on damaged panels. They are also used for working inner panels and reinforced sections that require more force but not a finished appearance.

Finishing hammers are used to achieve the final sheet metal contour. The face on a finishing hammer is smaller than that of the heavier bumping hammer. The surface of the face is crowned to concentrate the force on top of the ridge or high spot.

Shrinking hammers are finishing hammers with a serrated or cross-grooved face. They are used to shrink spots that have been stretched by excessive hammering.

Danger! *When using hammers, make sure the head is tight on its handle. Injury could result if the head were to fly off. Try to hit the object squarely and hold the handle securely.*

DOLLIES AND SPOONS

A **dolly** or **dolly block** is used like a small, portable anvil. It is generally held on the back side of a panel being struck with a hammer. Together the hammer and dolly straighten high spots down and low spots up to reshape the damaged body panel. See Figure 4–4.

Body spoons are used sometimes like a hammer and at other times like a dolly. They are available in a variety of shapes and sizes to match various panel shapes. The flat surfaces of a spoon distribute the striking force over a wide area to prevent hammer dents. They are particularly useful on creases and ridges.

HOLDING TOOLS

A **vise** will secure or hold parts during hammering, cutting, drilling, and pressing operations. It normally bolts onto a workbench.

Vise caps are soft lead, wood, or plastic jaw covers that will protect a part from marring. Vise jaws are often knurled and the small teeth will damage parts if they are not covered.

WARNING! Never hammer on a vise handle when tightening it. Use only your hands to tighten a vise handle.

A **C-clamp** is a screw attached to a curved frame. It will hold objects on a work surface or drill press while you are working on them.

PUNCHES AND CHISELS

Punches and chisels are necessary in every body repair tool chest. There are several types.

Punches are used for driving and aligning operations. They come in a number of different shapes for different tasks.

A **center punch** is pointed to start a drilled hole or to mark parts. The indentation will keep a drill bit from wandering out of place.

A **drift** or **starting punch** has a fully-tapered shank. It will drive pins, shafts, and rods partially out of holes. A **pin punch** has a straight shank for use after a starting punch. It will push a shaft completely out of a hole.

An **aligning punch** is long and tapered and used to align body panels and other parts. It can be used to line up fender bolt holes and bumpers, for example.

Chisels are handy for some cutting operations. For example, a chisel can be used to shear off rivet heads or separate sheet metal parts, Figure 4–5.

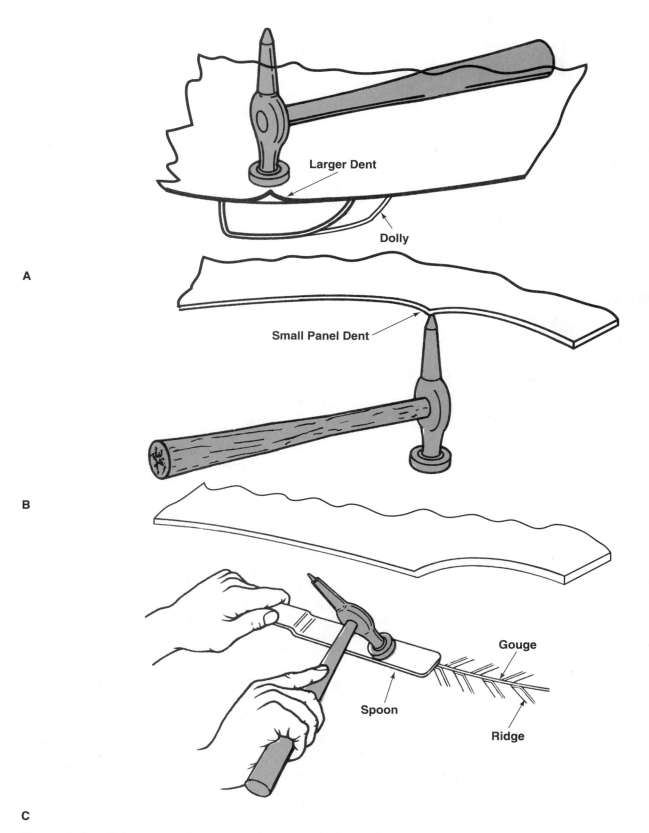

Figure 4–3 Body hammers come in various shapes. (A) The flat head of a body hammer is commonly used to tap down high areas in body panels. (B) The pointed tip on body hammer will concentrate impacts in a smaller area for working smaller dents. (C) A body hammer and a spoon are being used to lower a ridge on a damaged sheet metal component.

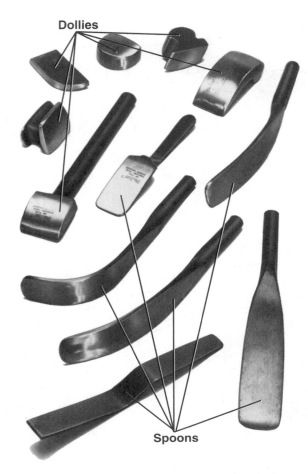

Dollies

Spoons

Figure 4–4 Dollies and spoons are essential in body repair. Dollies are often placed on the back of a panel so hammer blows can straighten the metal more effectively. Spoons are also used with hammers for less severe dents and creases. *(Courtesy of Snap-on Tools Corporation)*

Keep the ends of chisels and punches ground and shaped correctly. If the end becomes mushroomed or enlarged, grind it down. The sharp deformed end could cause hand injuries.

WRENCHES

Wrenches are used to loosen or tighten nuts and bolts. Various types of hand wrenches are shown in Figure 4–6.

Wrench size is the distance across the wrench jaws. Wrenches come in both standard (inch) and metric (millimeter) sizes. The size is stamped on the side of the wrench. The width of the jaw opening determines the size of a wrench.

For example, a ½-inch wrench has a jaw opening from face-to-face of approximately ½ inch. The actual

size is slightly larger than its normal size so that the wrench can fit around a nut or bolt head of equal size.

Most standard, or **SAE (Society Of Automotive Engineers)** wrench sets include sizes from $3/16$ to 1 inch. Metric sets usually include 6- to 19-millimeter wrenches. Smaller and larger wrenches can be purchased but are rarely used in collision repair.

It is important to remember that metric and SAE size wrenches are NOT interchangeable. For example, a $9/16$-inch wrench is $3/10$ millimeter larger than a 14-millimeter nut. If the $9/16$-inch wrench is used to turn or hold the 14-millimeter nut, the wrench will probably slip, rounding the points on the nut.

When using a wrench, PULL on the wrench—don't push. If the wrench comes off the bolt, there is less chance that you will hurt your hand. Never extend the length of a wrench with a pipe for leverage. Excess force can bend or break the wrench.

WRENCH TYPES

An *open end wrench* has three-sided jaws on both ends. This type of wrench is good if the bolt or nut is not very tight. The open jaws are weak. If a bolt is extremely tight, the open end wrench will bend or flex outward and strip the bolt head.

A *box end wrench* has closed ends that surround the bolt or nut head. It is a stronger design than an open end wrench. It is available in 12-point and 6-point types. A 6-point will grip the head better and should be used on very tight, corroded, or partially stripped fastener heads.

A *combination wrench,* as implied, has both types of ends—box end and open end. It provides the features of two wrench types.

A *tubing, flare nut,* or *line wrench* has a small split in its jaw to fit over lines and tubing. It can be slipped over fuel, brake, and power steering lines. Avoid using an open end wrench on tubing nuts because they will round off and strip easily. Tubing nuts are very soft.

An *Allen wrench* is a hex, hexagon, or six-sided wrench. It will install or remove set screws. Set screws are often used on pulleys, gears, and knobs. See Figure 4–7.

A *crescent* or *adjustable wrench* has movable jaws to fit different head sizes. It can be used when a correct size wrench is not available. For example, if a bolt is badly corroded, an adjustable wrench may be set to fit the odd head dimension.

A *pipe wrench* is another type of adjustable wrench for holding and turning round objects. Its sharp jaw teeth dig into and grasp the part. A pipe

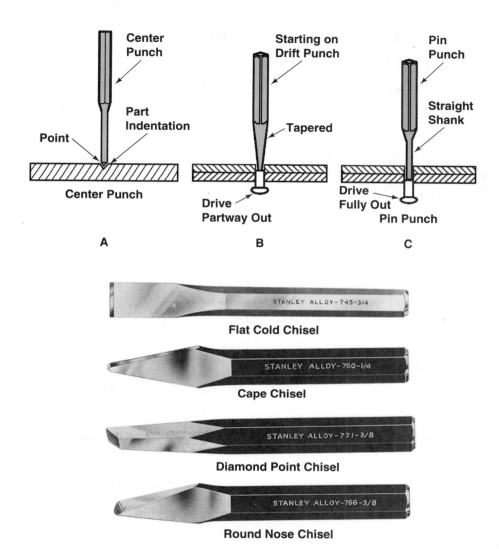

Figure 4–5 Study the use of punches and chisels. (A) The center punch will make a small indentation in a part. This will keep the drill bit from moving around. (B) The starting, or drift, punch is tapered and will start the driving operation. (C) The pin punch has straight sides and will completely drive out the part. (D) Chisels are named after their shape. Note chisel names. The cold chisel is common in collision repair. *(© Stanley Tools, New Britain, Connecticut, 06050)*

Figure 4–6 Study hand wrench names. (A) Open end wrench. (B) Open end wrench with socket on the other end. (C) Box end wrench. (D) Curved box end wrench. (E) Ratchet wrench. (F) Open end and flare nut wrench. (G) Angled head open end wrench. (H) Open end and flare nut wrench. (I) Curved box end wrench. (J) and (K) Combination wrenches. (L) Offset box end wrench. M: Long open end wrench. N: Long offset box end wrench. *(Courtesy of Snap-on Tools Corporation)*

Figure 4–7 Allen wrenches are needed to turn hex head set screws. An assortment of SAE and metric sizes are needed. *(Courtesy of Snap-on Tools Corporation)*

wrench will mar and damage part surfaces and should be used only when necessary.

A *chain wrench* is also used to grasp and turn round objects. It is used to remove stuck or damaged engine oil filters or to adjust exhaust systems.

Specialized body wrenches come in a variety of designs to meet specific demands. For example, the frame wrench is designed to bend the lip of heavy gauge frame rail.

Sockets

Sockets are cylinder-shaped, box end wrenches for rapid turning of bolts and nuts. They come in a variety of designs—deep sockets, swivel sockets, impact sockets, etc.

Deep sockets are longer for reaching over stud bolts. *Swivel sockets* have a universal joint between the drive end and socket body. *Impact sockets* are thicker and case hardened for use with an air powered impact wrench. Impact sockets are often black in color. **Conventional sockets** or nonimpact sockets, are usually chrome plated.

Socket drive size is the size of the square opening for the drive handle. Common drive sizes are ¼,

⅜, ½, and ¾ inch. Small ¼-inch drive sockets are often used when working on an instrument panel with its many small screws. Large ½-inch drive sockets would be needed on larger suspension system parts.

Sockets are also made in 4-point, 6-point, 8-point, and 12-point configurations. Again, a 6-point is the strongest and will not strip the bolt or nut head easily.

A *ratchet* has a small lever that can be moved for either loosening or tightening bolts and nuts. It is a common type of socket drive handle. One type is shown in Figure 4–8.

A *breaker bar* provides the most powerful way of turning bolts and nuts. Also called a *flex handle*, it will break loose large or extremely tight fasteners.

A *speed handle* can be rotated to quickly remove or install loose bolts and nuts.

Extensions fit between the socket and its drive handle. They allow you to reach in and install the socket when surrounded by obstructions. See Figure 4–9.

A *universal joint* allows you to reach around objects with a socket wrench and extension. It will flex, allowing you to rotate the socket from an angle. Separate universal joints and swivel sockets are available.

A *torque wrench* is used to measure tightening force. *Torque* is a measurement of twisting force. It is given in both inch-pounds and foot-pounds, as well as in metric measurements of Newton meters. See Figure 4–10.

If the fastener is tightened to a value less than its specification, the fastener can vibrate loose. If tightened too much, the threads can strip or the fastener can break. In either case, damage to other automobile parts can result. This can also endanger the driver and passengers of the vehicle.

Always torque all bolts and nuts to the *manufacturer's specifications* (factory values given in service manual). If there are no specifications listed, use the general torque specifications chart. One is given in the back of this book.

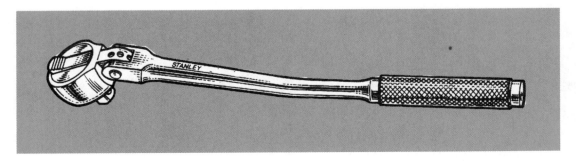

Figure 4–8 The ratchet is a common socket drive handle. This one has a swivel head for reaching and turning sockets in tight quarters. *(© Stanley Tools New Britain, Connecticut 06050)*

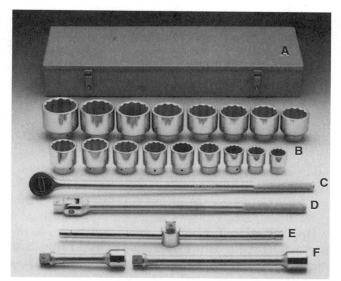

Figure 4–9 (A) Storage box. (B) Socket set. (C) Ratchet. (D) Breaker or flex bar. (E) T-handle. (F) Extensions. *(Danaher Tool Group)*

Figure 4–10 A torque wrench should be used to properly tighten critical fasteners (nuts, bolts, etc.) on steering and suspension systems. It will measure twisting force accurately. *(Courtesy of Snap-on Tools Corporation)*

PRY BARS

Pry bars are used to gain leverage for shifting heavy parts. They are hardened steel bars with various end shapes. Pry bars are often used to adjust engine belt tension. They will also shift parts to align holes. This will aid in hand starting bolt threads. See Figure 4–11.

Figure 4–11 Pry bars are helpful for aligning or shifting parts into position. They are used when adjusting engine belts, for example. *(Courtesy of Snap-on Tools Corporation)*

SCREWDRIVERS

Screwdrivers allow you to rotate screws for installation or removal. Many threaded fasteners used in the automotive industry are driven by a screwdriver. Each fastener requires a specific kind of screwdriver. A well-equipped technician will have several sizes or types of each.

When selecting a screwdriver, select a tip wide and thick enough to almost fill the screw head opening. If it is NOT the right size, screwdriver and screw damage will result. See Figure 4–12.

A *standard screwdriver* has a tip with a single flat blade for fitting into the slot in the screw. A *Phillips screwdriver* has two crossing blades for a star-shaped screw head. *Torx* and *square head*

Figure 4–12 Study the names of screwdrivers. (A) Stubby screwdrivers. (B) Standard screwdrivers. (C) Phillips screwdriver. (D) Standard screwdriver with wood handle. (E) Phillips screwdriver with wood handle. *(Courtesy of Snap-on Tools Corporation)*

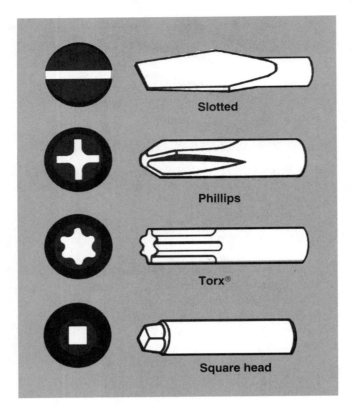

Figure 4–13 Note screwdriver and screw head names. *(Klein Tools, Inc.)*

screwdrivers have specially shaped tips, which are shown in Figure 4–13.

Stubby screwdrivers have a very short shank. They can fit into tight or restricted areas. For example, you might use them to install trim around a wheel opening. *Offset screwdrivers* have the shank at a 90-degree angle to the tip. They are also for restricted areas, like inside glove boxes.

An *impact driver* is hit with a hammer to rotate tight or stuck screws. The body of the driver can be rotated to change directions for installing or removing screws.

PLIERS

Pliers are used for working with wires, clips, and pins. They will grasp and hold parts like fingers. The collision repair technician must own several types. See Figure 4–14.

> *WARNING! Do NOT use pliers in place of a wrench. They will scar and damage parts like nuts and bolts.*

Slip-joint, or *combination, pliers* will adjust for two sizes. They are the most common design used by the collision repair technician.

Rib joint or *Channellock pliers* have several jaw settings for grasping different size objects. They will open wider than combination pliers.

Needlenose pliers have long, thin jaws for reaching in and grasping small parts. Never twist needlenose pliers or you will spring the jaws out of alignment. *Snap ring pliers* have tiny tips for fitting into circlips or snap rings. Both external and internal types are available.

Vise-Grip or *locking pliers* have jaws that lock into position on parts. The jaws can be adjusted to different widths. Then, when clamped down, they will hold the part and free your hand to do another task. Special locking pliers with extended jaws are available for welding or working with sheet metal. See Figure 4–15.

Diagonal cutting pliers have cutting jaws that will clip off wires flush with a surface. Another name for these is *side cut pliers.*

SHEET METAL TOOLS

Various tools are used by the technician to work body sheet metal. The following are some examples.

A *sheet metal gauge* is used to measure body or repair panel thickness or gauge size. It has small openings with the gauge size stamped next to it. Gauge size is a number system that denotes the thickness of sheet metal. See Figure 4–16.

To use a thickness gauge, find the opening that fits the sheet metal. The number equals sheet metal thickness. Generally, a larger gauge number means thinner sheet metal. Ten gauge sheet metal would be thicker than 20 gauge.

Tinsnips are the most common metal cutting tool. They can be used to cut straight or curved shapes in sheet metal. They are also used to trim panels to size.

Panel cutters will precisely cut sheet metal, leaving a clean, straight edge that can be easily welded. They are used to make straight or curved cutoffs in panels, often to repair corrosion.

FILES

Files are used to remove burrs and sharp edges and to do other smoothing tasks. Various shapes and coarsenesses of teeth are available, Figure 4–17.

In general, a *coarse file* works well on soft materials, such as brass, aluminum, and plastics. A *fine file* will give a smoother surface on harder materials. A fine file is normally used to smooth steel.

When using a file, always have a handle securely in place. The sharp tang, if NOT covered, can easily

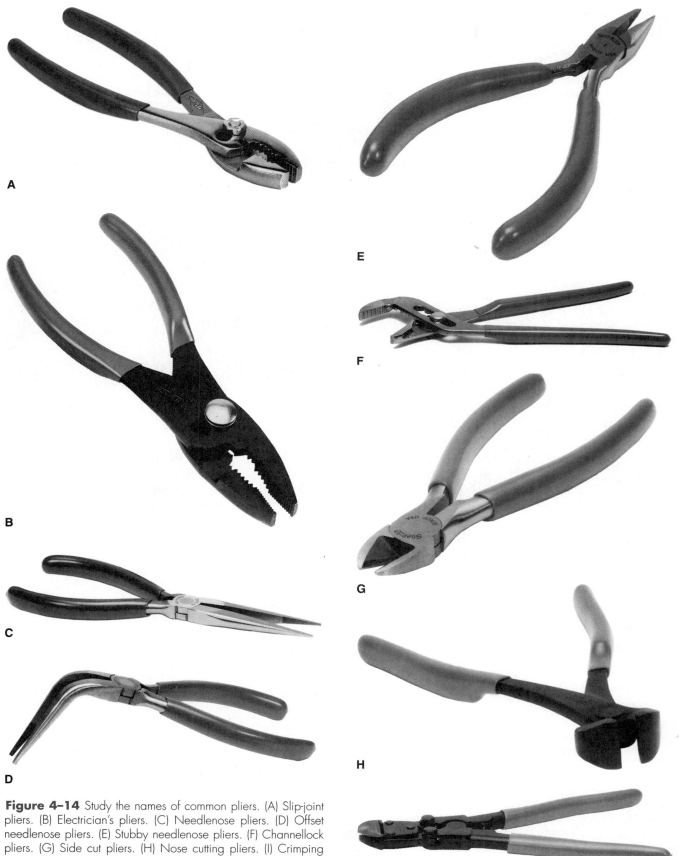

Figure 4–14 Study the names of common pliers. (A) Slip-joint pliers. (B) Electrician's pliers. (C) Needlenose pliers. (D) Offset needlenose pliers. (E) Stubby needlenose pliers. (F) Channellock pliers. (G) Side cut pliers. (H) Nose cutting pliers. (I) Crimping pliers. *(Courtesy of Snap-on Tools Corporation)*

Vise-Grip Pliers®

Sheet Metal Vise-Grips®

C–Clamp Vise-Grips®

Figure 4–15 Vise-Grip® or locking pliers are handy because they will lock and stay on a part. They will hold two parts together like a clamp.

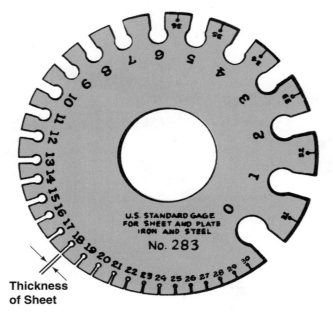

Thickness of Sheet

Figure 4–16 A standard gage or sheet metal gauge can be fitted over metal to find its thickness. The number system denotes gauge thickness.

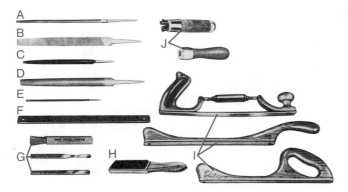

Figure 4–17 Note the names of files and file accessories. (A) Round file. (B) Flat file. (C) Diamond file. (D) Half round file. (E) Rat tail file. (F) Body file. (G) Small files. (H) File cleaning card. (I) Body file handles. (J) File handles. *(Courtesy of Snap-on Tools Corporation)*

A

B

Figure 4–18 Shown are (A) a body file in a flexible holder and (B) rigid file holders. *(Photo B Courtesy of Snap-on Tools Corporation)*

puncture your hand or wrist. Apply filing pressure only on the forward stroke. Lift the file on the back-stroke.

Never press too hard or file too rapidly. This will quickly dull the file. Do NOT hammer on a file. Never pry with a file. It is very brittle and can shatter. Clean the file with a stiff wire brush or card brush when it is clogged.

Body files have very large, open teeth for cutting plastic body filler. They often have a separate blade that fits into a handle. They are commonly used when metal finishing damaged areas. See Figure 4–18.

SAWS

A *hacksaw* is sometimes needed to cut metal parts in collision repair. Most hacksaws have an adjustable frame that accepts different blade lengths. When installing a hacksaw blade, the blade teeth should point AWAY from the handle.

A reciprocating saw has a handle with the blade extending out one end. No frame is used. After making a starting hole, a jab saw can be used to cut into a blind panel that is not accessible from the back.

Hacksaw blade coarseness is rated in teeth per inch or millimeter. Blades also come in different tooth configurations. Select the correct blade for the material to be cut.

Generally, at least two teeth must contact the material at the same time. This will help prevent one tooth from catching and being broken off, Figure 4–19.

When using a hacksaw, place one hand on the handle and the other on the frame. Push down lightly on the forward stroke and release downward pressure on the backstroke.

As a general rule, count "one thousand one, one thousand two," and so on to time each stroke. If cuts are made too fast, the blade will quickly overheat and

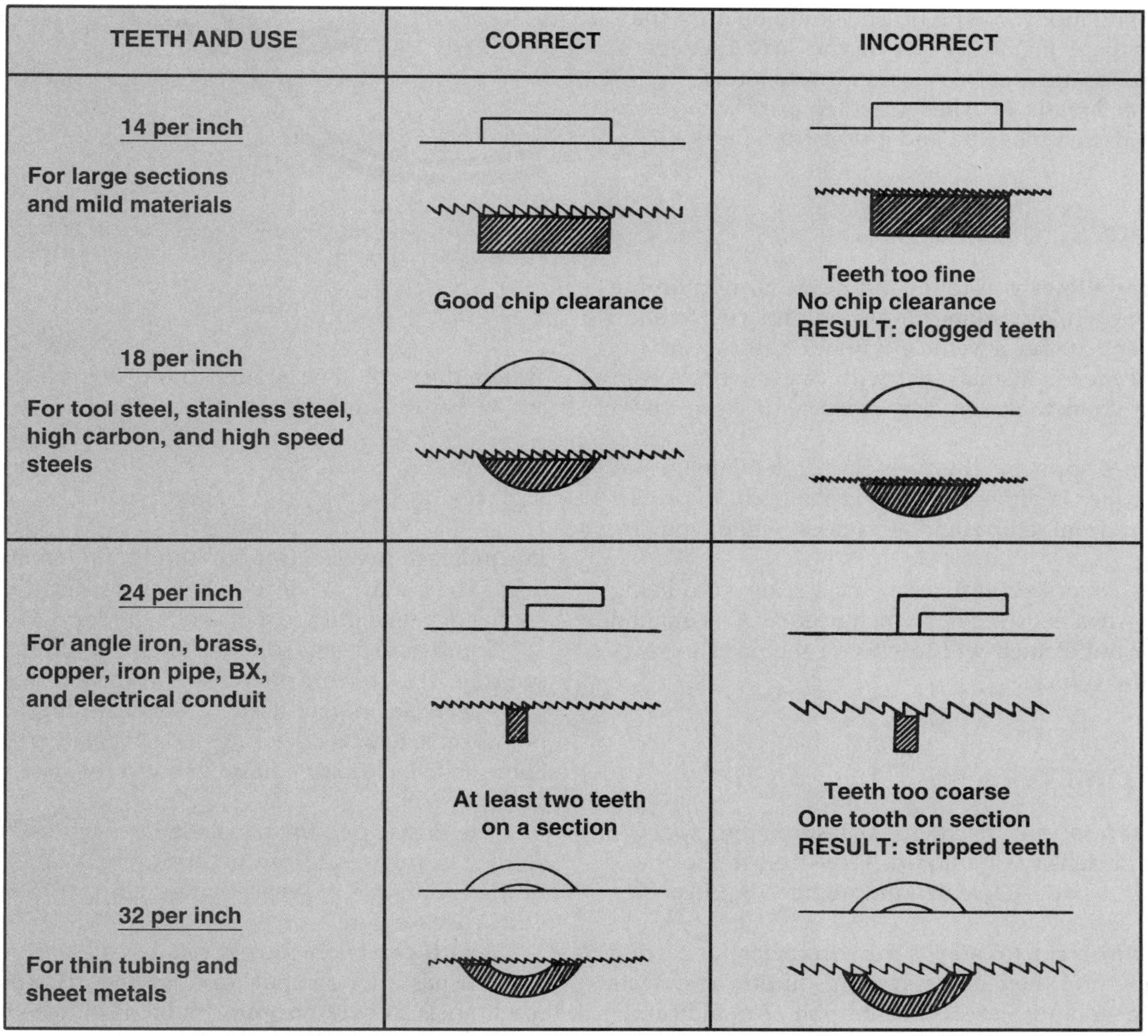

TEETH AND USE	CORRECT	INCORRECT
14 per inch **For large sections and mild materials**	**Good chip clearance**	**Teeth too fine** **No chip clearance** **RESULT: clogged teeth**
18 per inch **For tool steel, stainless steel, high carbon, and high speed steels**		
24 per inch **For angle iron, brass, copper, iron pipe, BX, and electrical conduit**	**At least two teeth on a section**	**Teeth too coarse** **One tooth on section** **RESULT: stripped teeth**
32 per inch **For thin tubing and sheet metals**		

Figure 4–19 A hacksaw is sometimes used to cut metal parts. Study this chart, which explains hacksaw blade applications.

become dull. This method of timing stroke speed works for saws and files.

CLEANING TOOLS

Collision repair technicians sometimes have to clean vehicle parts when working. For example, they may need to replace mechanical parts (suspension or steering system parts, for example), that are covered with oil or grease.

Hand scrapers will easily remove gaskets and softened paint (when paint stripping chemicals are used). They are used on flat surfaces. To prevent cuts, never scrape toward yourself. Scrapers can be made of steel for tough jobs or hard plastic to protect from gouging the surface.

Steel brushes are sometimes used to remove corrosion and dirt from parts. They are slow but suitable in some situations. A wire brush should be used sparingly on bare metal because it can leave scratches. Wire brushes are ideal for cleaning weld joints.

Softer bristle brushes are often used with cleaning solvent to remove oil and grease from parts.

CREEPERS, COVERS

A *creeper* allows you to lie and move around on the shop floor while working. It also allows you to slide around and under a vehicle without getting dirty. A *stool creeper* is a small seat with caster wheels. You can lay your tools on the creeper to keep them handy.

Covers protect the vehicle from damage. A *fender cover* is often placed over the body to protect the paint from chips and scratches while you are working.

Seat covers protect the vehicle seats from being stained. Always use covers when needed. One stained seat or paint scratch will teach you the need for protecting the vehicle.

HYDROMETERS

Hydrometers are used to measure the *specific gravity* or density of a liquid. They are used to check antifreeze, battery acid, and other liquids. See Figure 4–20.

An *antifreeze tester* is a hydrometer for testing the strength of the cooling system solution. It detects the water-to-antifreeze concentration. An indicating arrow or the number of floating balls indicates the freeze-up protection of the antifreeze. If the cooling

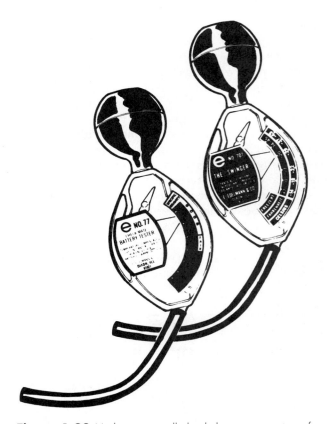

Figure 4–20 Hydrometers will check the concentration of antifreeze solution, battery acid, or other liquids. *(Edelmann)*

system does not have enough antifreeze, major engine and radiator damage can result during cold weather.

FLUID HANDLING TOOLS

Funnels are needed to pour fluids into small openings. They may be needed when adding oil to an engine or brake fluid to a master cylinder.

Spouts are designed to install into a can for handy pouring. They too are often used when adding oil.

A *grease gun* is used to lubricate high friction points on a vehicle's steering and suspension systems. Some collision repair equipment also requires lubrication.

The grease gun tip is installed over grease fittings so that heavy grease can be injected into the part or fitting. Replaceable **grease cartridges** are often used with a grease gun.

An *oil can* is often used to lubricate parts and air tools. It has a long spout for applying oil to hard-to-reach areas. A *suction gun* can be used to remove or extract liquid. By pulling on the handle, a fluid can be pulled up into the gun.

TIRE TOOLS

A *tire pressure gauge* will accurately measure tire air pressure when pressed over the tire valve stem. Various designs are available. The gauge will read pressure in psi (pounds per square inch) or kilopascals. You should always check tire air pressures before releasing a vehicle to the customer. Low air pressure is common and can be dangerous. Recommended tire pressure is printed on the sidewall of the tire.

A *tread depth gauge* will quickly measure tire wear. The gauge is placed over the most worn area of the tire tread. A small pin on the gauge is pushed down into the tread. The gauge reading equals tire tread depth. The tire should be replaced or recommended for replacement if the tread is worn thinner than about $\frac{1}{8}$ of an inch (3 mm).

PULLERS, PICKS

Wheel pullers are used to remove steering wheels, engine pulleys, and similar pressed-on parts. They are often bolted or mounted on the part, then a large screw is tightened to force the part off. See Figure 4–21.

A *slide hammer* is another type of puller that uses a hammering action to remove parts. Various ends can be installed on a slide hammer, as is shown in Figure 4–22.

A *dent puller* is a slide hammer that will accept various pulling tips: a threaded tip, a hook tip, a cutting tip, or a suction cup. With the appropriate tip

installed into the damaged panel, the slide hammer is briskly slid along the steel shaft and struck against its handle. This produces a powerful pulling force to help remove dents in sheet metal panels.

Pull rods have a handle and curved end for light pulling of dents. A small dent or crease can be pulled up with a single pull rod, but three or four may be used simultaneously to pull up larger dents. A body hammer can also be used with a pull rod. The high crown of a dent can be bumped down, while the low

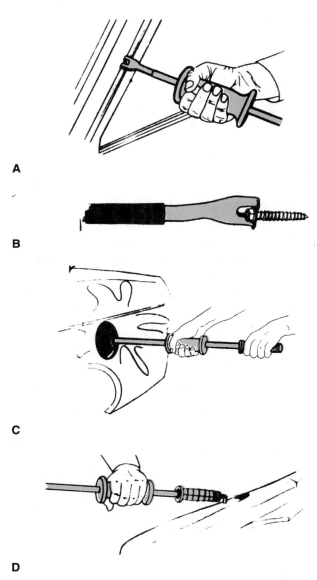

A

B

C

D

Figure 4–22 Note the uses of a slide hammer puller. (A) A slide hammer with a hook attachment will grasp and pull damaged surfaces. (B) A screw threaded into the part is common for pulling small dents. (C) A suction cup attachment will let you "pop out" or remove some minor dents. (D) A cutter attachment can also be mounted on a slide hammer.

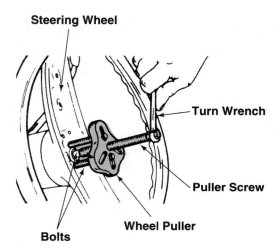

Steering Wheel

Turn Wrench

Puller Screw

Wheel Puller

Bolts

Figure 4–21 Wheel pullers will remove pressed-on parts, like a steering wheel. Wear eye protection! *(OTC, a Division of SPX Corporation)*

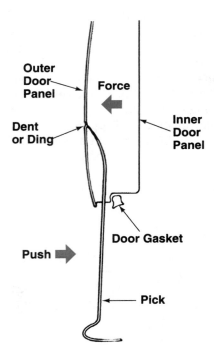

Outer
Door
Panel

Force

Inner
Door
Panel

Dent
or Ding

Door Gasket

Push

Pick

Figure 4–23 A pry pick is often used during paintless dent removal. The pick will reach behind hidden or blind panels to carefully straighten small dents and door dings.

spot is pulled up. Simultaneous bumping and pulling returns the panel to its original shape with less danger of stretching the metal.

Picks are used to reach into confined spaces for removing tiny dents or dings. They are made of strong case-hardened, spring steel. The pick is used to push up low spots. Picks vary in length and shape; most have a U-shaped handle. See Figure 4–23.

Picks are commonly used to straighten small dents in doors, quarter panels, and other sealed body sections. Picks are often preferred to slide hammers and pull rods because they do not require drilling holes in the sheet metal. They are used during *paintless dent removal* (straightening small body dings or dents without painting the panel).

Note! Straightening tools and techniques are discussed more in other text chapters. Refer to the index for additional information.

MISCELLANEOUS COLLISION REPAIR TOOLS

A variety of other miscellaneous hand tools will be useful from time to time. Many are inexpensive and have several uses. A few are very expensive but may be provided by the shop.

A common *utility knife* with a retractable blade is handy for cutting or trimming. Most come with extra blades stored in the handle. Remember that they will cause deep skin lacerations (cuts).

A *windshield knife* is designed to cut through the sealant used on some windshields. It has a sharp cutter blade and two handles. You hold one handle to guide the cutter and pull on the other handle to force the knife through the sealant. An electric heated windshield knife can be used to cut the adhesive. Air or pneumatic windshield cutters are also available.

Trim pad tools are designed to reach behind interior panels to pop out and remove clips. They often have a flat fork end that will surround the clip and push on the trim panel to prevent damage. They are often needed to remove inner door trim panels. See Figure 4–24.

A *door hinge spring compressor* is used to remove and install the small spring used on some door hinges. To use this tool, the screw is tightened to compress the spring. The spring is then shortened so that it can be installed or removed.

A *door striker wrench* is a special socket designed to fit over or inside door striker posts. It will allow you to turn and loosen the post for adjustment or replacement.

A *door alignment tool* can be used to slightly bend and adjust sprung doors and hatches. It has steel knurled blocks that fit next to the door hinge. Depending upon the placement of the tool, when the door is partially closed you can quickly realign a door. One is pictured in Figure 4–25.

A *locksmith tool*, also termed a "slim jim," is used to open locked doors on vehicles. If you do not have a key or the keys are locked in the vehicle, this type of tool will often trip and unlock the door for entry into the passenger compartment. Although designs vary, one type simply slips down between the door glass and door frame to move the lock mechanism.

Suction cup tools have a large synthetic rubber cup for sticking onto parts. They are often used to hold window glass when holding or moving the glass. They can also be used to straighten large surface area dents in panels. See Figure 4–26.

Dispensers or *cartridge guns* are used to apply adhesive or sealer to body panels. They are often spring loaded. When you press the trigger, spring tension acts on the cartridge to force the material out of its tip.

A *smoke gun* can be used to find air leaks around doors and windows. With the vehicle's blower on, one technician blows smoke around all potential leakage points. Another technician stands outside to find smoke coming out of the passenger compartment, locating the leak.

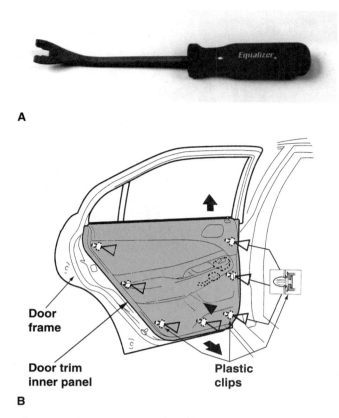

A

B

Door frame

Door trim inner panel

Plastic clips

Figure 4–24 (A) A trim pad tool has a slender, forked end for reaching behind panels to remove snap-on type clips. *(Equalizer Industries)* (B) A service manual will often illustrate clip locations so you know where to insert the trim pad tool to pry the clips out. This service manual drawing shows how to remove the clips from an inner trim panel. *(© American Honda Motor Co., Inc.)*

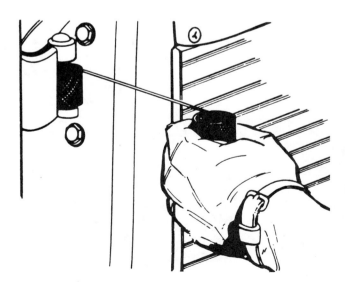

Figure 4–25 A door alignment tool is positioned next to the door hinge. By trying to push the door closed, you can realign the door jamb, thereby adjusting the front door edge with the rear portion of the front fender. *(Morgan Mfg. Inc.)*

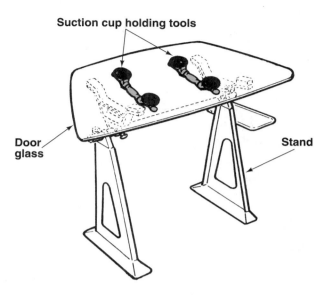

Suction cup holding tools

Door glass

Stand

Figure 4–26 Suction cup holding tools are often used to hold a windshield during installation. *(Saab Cars USA, Inc.)*

A **body filler dispenser** is used to squeeze the desired amount of plastic filler and hardener onto the mixing board. The material is forced out of the machine by the turn of a handle. This helps keep debris out of the filler. One type is shown in Figure 4–27.

Spreaders or *spatulas* are used to apply filler to low spots in body panels. They are usually made of plastic or sometimes of rubber. They come in various widths for different size dents or low spots.

A *masking machine* is designed to feed out masking paper while applying masking tape to one edge of the paper, Figure 4–28. It speeds the work because the tape is already attached to the paper. You can quickly apply the masking paper to the body with a masking machine. See Figure 4–29.

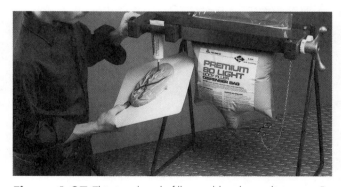

Figure 4–27 This is a handy filler and hardener dispenser. By turning the handles, you can feed out the right amount of materials. The machine keeps the filler perfectly clean and free of dust and debris. *(Oatey Bond Tite)*

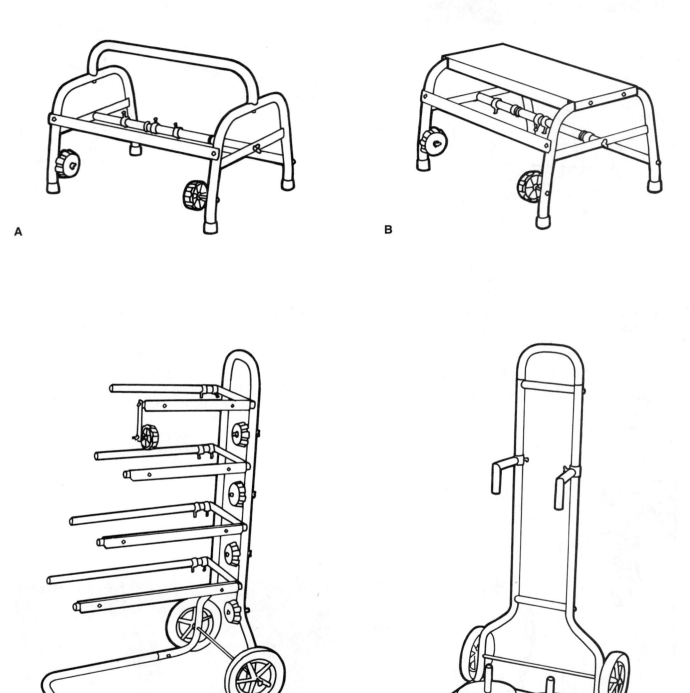

A

B

C

D

Figure 4–28 Masking machines save time. (A) A mobile masking machine has a handle for carrying it to the job. (B) This masking machine has step or seat for convenience. (C) This masking machine has several arms for different sizes of paper and tape. (D) This machine is for vinyl or plastic masking materials. *(Oatey Bond Tite)*

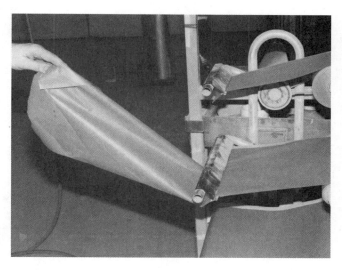

Figure 4–29 You can quickly apply the masking paper to the body with a masking machine.

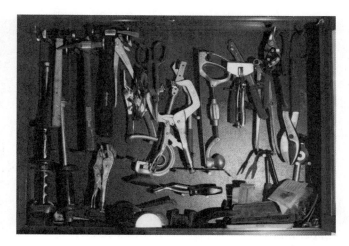

Figure 4–31 Many other more specialized hand tools are explained in this textbook. Refer to the index or specific chapters to gain more information on these specialized hand tools.

Figure 4-30 A hood/trunk tool is a telescoping rod that can be used to prop open the hood or trunk lid.

Pinstripe brushes are used to apply paint in small, controlled areas. Small paint brushes can also be used to touch up chips along the lower body panel that were not painted. Pinstripe brushes come in different widths and lengths.

A ***hood/trunk tool*** is a telescoping rod that can be used to prop open the hood or trunk lid. It has rubber tips that contact the vehicle and panel. The tool will extend out and lock at varying lengths to help

hold these panels in place while you are removing or installing fasteners. See Figure 4–30.

> *WARNING! Always ask another technician to help you remove and install hoods and trunk lids. They are very heavy and clumsy. Help may be needed to prevent damage to vehicle parts during installation and removal.*

Note that many other more specialized hand tools are explained in this textbook. See Figure 4–31 and refer to the index or specific chapters to gain more information on these specialized hand tools.

Summary

- Purchase quality tools with lifetime guarantees. Quality tools will pay for themselves in a short period of time.
- Proper storage will protect tools and speed your work when they are easy to locate.
- Some of the more widely used tools in collision repair work include general duty hammers, body hammers, dollies and spoons, holding tools, punches and chisels, wrenches, pry bars, pliers, sheet metal tools, files, saws, cleaning tools, creepers, covers, hydrometers, fluid handling tools, tire tools, pullers, picks, and knives.

Technical Terms

lifetime tool guarantee
tool holders
tool box
upper chest
lower cabinet
hammers
ball peen hammer
brass hammer
lead hammer
plastic hammer
dead blow hammer
rubber mallet
body hammers
picking hammers
bumping hammers
finishing hammers
shrinking hammers
dolly
dolly block
body spoons
vise
vise caps
C-clamp
punches
center punch
drift
starting punch
pin punch
aligning punch
chisels
wrenches
wrench size
SAE (Society of
   Automotive Engineers)
open end wrench
box end wrench
combination wrench
tubing wrench
flare nut wrench
line wrench
Allen wrench
crescent wrench
adjustable wrench
pipe wrench
chain wrench
sockets
deep sockets
swivel sockets
impact sockets
socket drive size

ratchet
breaker bar
flex handle
speed handle
extensions
universal joint
torque wrench
torque
manufacturer's
   specifications
screwdrivers
standard screwdriver
Phillips screwdriver
Torx® screwdriver
square head screwdriver
stubby screwdriver
offset screwdriver
impact driver
pliers
slip-joint pliers
combination pliers
rib joint pliers
Channellock® pliers
needlenose pliers
snap ring pliers
Vise-Grip® pliers
locking pliers
diagonal cutting pliers
side cut pliers
sheet metal gauge
tinsnips
panel cutters
files
coarse file
fine file
body file
hacksaw
hand scraper
steel brushes
softer bristle brushes
creeper
stool creeper
cover
fender cover
seat cover
hydrometer
specific gravity
antifreeze tester
funnel
spout

grease gun
oil can
suction gun
tire pressure gauge
tread depth gauge
wheel puller
slide hammer
dent puller
pull rod
pick
paintless dent removal
utility knife
windshield knife
trim pad tools

door hinge spring
   compressor
door striker wrench
door alignment tool
locksmith tool
suction cup tool
cartridge gun
smoke gun
spreader
spatula
masking machine
pinstripe brush
hood/trunk tool

Review Questions

1. A good set of tools will speed _____,
   and improve _____.

2. What is a tool guarantee? Why would you want
   one?

3. Which tools would you put in the top chest of a
   tool box and which in the lower roll-around cabinet?

4. This type of hammer is often used to install wheel
   covers.
   a. Lead hammer
   b. Plastic hammer
   c. Dead blow hammer
   d. Rubber mallet

5. _____ are used to achieve the final contour. The face on a _____ is smaller
   than that of the heavier bumping hammer.

6. How are dollies or dolly blocks used?

7. _____ are used sometimes like a hammer and sometimes like a dolly.

8. Never hammer on a vise handle when tightening
   it. True or false?

9. This type of punch is pointed to start a drilled
   hole or mark parts.
   a. Drift punch
   b. Starting punch
   c. Center punch
   d. Aligning punch

10. What is wrench size?

11. A car has damage to the front suspension and a
    brake line has to be removed. *Technician A* says to
    use vise grips first to make sure the fitting comes

loose. *Technician B* says to use a combination wrench to protect the fitting nut from damage. Who is correct?
   a. Technician A
   b. Technician B
   c. Neither A nor B
   d. Both A and B

12. List the most common socket drive sizes.

13. This tool provides the most powerful way of turning bolts and nuts.
   a. Ratchet
   b. Flex handle
   c. Speed handle
   d. Impact driver

14. Never twist on _____ pliers or you will spring the jaws out of alignment.

15. Generally, a larger gauge number means a thinner sheet metal. Ten gauge sheet metal would be thicker than 20 gauge. True or false?

16. Explain when you would use a coarse file instead of a fine file.

17. What are body files?

18. A _____ is a small seat with caster wheels.

19. Explain the use of a slide hammer dent puller.

20. _____ have a handle and curved end for light pulling of dents.

ASE–Style Review Questions

1. A technician is doing metal work on a panel. There are some ⅛-inch diameter high spots remaining. Which hammer should be used to straighten these small raised areas?
   a. Brass hammer
   b. Bumping hammer
   c. Shrinking hammer
   d. Picking hammer

2. *Technician A* is using a dolly to back up metal straightening with a finishing hammer. *Technician B* is using a dolly as a hammer to raise a low spot from behind. Who is correct?
   a. Technician A
   b. Technician B
   c. Both Technicians A and B
   d. Neither Technician

3. Which of the following is sometimes used as a dolly and sometimes as a hammer?
   a. Sanding block
   b. Pick
   c. Spoon
   d. Slide hammer

4. *Technician A* says to use soft vise caps when clamping a vehicle's control arm in a vise. *Technician B* says that you need the vise jaws to cut into the part for secure clamping. Who is correct?
   a. Technician A
   b. Technician B
   c. Both Technicians A and B
   d. Neither Technician

5. A bolt head is rusted and must be removed. *Technician A* is going to use locking pliers to try to remove the bolt. *Technician B* says to use a six-point wrench first. Who is correct?
   a. Technician A
   b. Technician B
   c. Both Technicians A and B
   d. Neither Technician

6. Which of the following is the best tool to use on a brake line fitting?
   a. Adjustable wrench
   b. Open end wrench
   c. Closed end wrench
   d. Tubing wrench

7. A vehicle's lug nuts have been overtightened. Which of the following tools should be used to loosen the lug nuts?
   a. Torque wrench
   b. Ratchet
   c. Locking pliers
   d. Breaker bar

8. A rusted bolt head has rounded off while being loosened. Which of the following tools should be used to remove it?
   a. Locking pliers
   b. Slip joint pliers
   c. Breaker bar
   d. Speed handle

9. *Technician A* is using a hydrometer to measure the specific gravity of a vehicle's antifreeze solution. *Technician B* is using a hydrometer to check the viscosity of motor oil. Who is correct?
   a. Technician A
   b. Technician B
   c. Both Technicians A and B
   d. Neither Technician

10. A slide hammer is being used to straighten a dent. *Technician A* says a hooked attachment will grip and pull exposed surfaces. *Technician B* says to weld rods or pins onto the panel for tool attachment. Who is correct?

a. Technician A
b. Technician B
c. Both Technicians A and B
d. Neither Technician

Activities

1. Tour your shop. Note the location of all hand tools. Prepare a report on your findings.

2. Visit an outside collision repair shop. Talk with other technicians about the tools they use. Discuss with the class anything you learned about new tools or techniques.

3. Order catalogs from tool manufacturers. Compare the quality, price, and warranty of tools offered by each manufacturer.

5

Power Tools and Equipment

Objectives

After studying this chapter, you should be able to:

■ Identify power tools found in a collision repair facility.

■ Explain the purpose of each type of power tool.

■ Summarize how to safely use power tools.

■ Identify the typical types of equipment used in collision repair.

■ Describe how to use collision repair equipment.

■ Explain safety precautions for using shop equipment.

■ Select the right power tool or piece of equipment for the job.

■ Explain low emissions spray equipment and regulations.

■ Explain the operation of spray booths and drying rooms.

■ Summarize atomization and how it relates to spray gun operation.

■ Identify the various types of spray guns and explain how each type operates.

Introduction

From the time of the blacksmith to today's technician, tools have served to increase the abilities of human hands, arms, legs, backs, eyes, and ears. They allow us to do things that would be impossible otherwise.

Neither you nor the best technician can do quality work using inferior tools. Professional collision repair technicians invest thousands of dollars in their tools, and for good reason. A good set of tools will speed repairs while improving work quality. The importance of having the right tools for the job cannot be overstated.

Power tools and equipment use electrical energy, compressed air, or hydraulic power. Drills, sanders, cut-off tools, scanners, air compressors, spray guns, frame straightening equipment, alignment racks, and some measurement systems all use some form of power to do their work. For you to work efficiently, a sound grasp of the technology and use of power tools and equipment is essential.

This chapter will overview the safe use of basic collision repair power tools and equipment. Understanding when and how to use each is critical to your success as a collision repair technician. Later chapters will discuss the specific application of each tool in more detail.

Danger! Power tools and equipment can be very dangerous if NOT used correctly. Always follow the instructions given in the owner's manual for the specific tool or piece of equipment. The information in this chapter is general and cannot cover all tool variations. If in doubt about any tool or piece of equipment, ask your instructor or shop supervisor for a demonstration.

Danger! The use of air-driven or electric power tools requires safety glasses, goggles, or a face shield. Also, never wear loose clothing that could get caught in a tool. Wear leather gloves when grinding, welding, or doing other tasks that could injure your hands.

Refer to Chapter 2 for more information on power tool and equipment safety.

AIR SUPPLY SYSTEM

The shop's *air supply system* provides clean, dry air pressure for numerous tools and equipment. It has an air compressor, metal lines, rubber hoses, quick disconnect fittings, a pressure regulator, filters, and other parts.

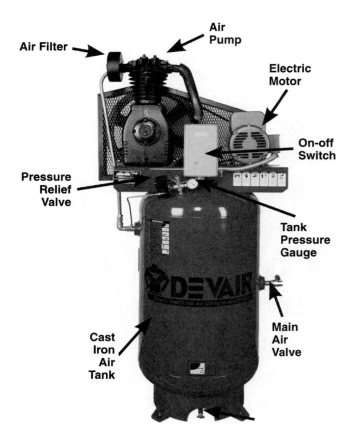

Figure 5–1 Note major parts of an air compressor, the "heart" of your air tools and equipment. *(Photo courtesy of Team Blowtherm)*

AIR COMPRESSOR

An *air compressor* is made up of an electric motor, air pump, and large air storage tank. The motor spins the air pump which works like a small, reciprocating piston engine. Piston action in the air pump pushes the air into a large, thick steel storage tank. The "heart" of any paint shop is its air compressor. See Figure 5–1.

A *compressor air filter* prevents airborne dust and dirt from entering and damaging the air pump's moving parts. The filter must be cleaned or changed at recommended intervals.

A *compressor circuit breaker* protects the electric motor from overheating damage. The circuit breaker will open to stop current flow to the motor if motor temperature and current become too high. When a circuit breaker trips, a button on the side of the motor must be pressed to reset the breaker.

A *compressor drain valve* on the bottom of the tank allows you to drain out water. The compression of the air tends to make moisture condense out of the air. This moisture must be drained out periodically to prevent it from entering the air lines.

An *automatic drain valve* periodically opens a solenoid-operated valve on the bottom of the air com-

pressor to remove moisture from the storage tank. This saves you from having to manually open the drain valve every day. The system is commonly found on large industrial air compressors.

A *compressor pressure sensor* is an electrical device that cuts current to the motor when normal system pressure is reached. It normally mounts in a small metal box at the air outlet line from the tank.

A *compressor pressure relief valve* prevents too much pressure from entering the tank. It is a spring-loaded valve that opens at a predetermined pressure, usually around 90 to 120 psi (63 to 84 kg/cm2).

An *air tank shut-off valve* is a hand valve that isolates the tank pressure from shop line pressure. It should be closed at night or when the compressor is not going to be used. If it is NOT closed, the compressor would run all night if a hose leaked or ruptured.

Compressor oil plugs are provided for filling and changing air pump oil. The oil level in the compressor should be checked regularly and the oil changed periodically. Normally use single weight, nondetergent oil or the oil recommended by the manufacturer.

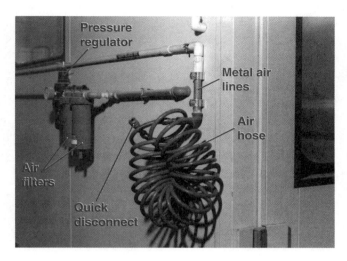

Figure 5–2 Fixed metal air lines carry air pressure to various work areas in the shop. Quick disconnect fittings allow the technician to attach high-pressure hoses to the metal air lines. *(Courtesy of Snap-on Tools Corporation)*

AIR LINES

Shop air lines are thick steel pipes that feed out from the compressor tank to several locations in the collision repair shop. Flexible synthetic rubber ***air hoses*** connect the metal pipes to the air tools and equipment.

Air hose sizes are ¼ inch (6 mm), ⁵⁄₁₆ inch (8 mm) and ³⁄₈ inch (10 mm) inside diameters (ID). There is less restriction using larger hoses. The preferred diameter is ⁵⁄₁₆ inch (8 mm) and it is used in many shops, Figure 5–2.

Air couplings allow you to quickly install and remove air hoses and air tools. Couplings should also have a large enough inside diameter to prevent air flow restrictions.

By sliding the outer fitting sleeve back and pushing, you can connect or disconnect a tool and a hose. By sliding the fitting sleeve back and pulling, you can quickly disconnect tools.

AIR PRESSURE REGULATOR AND FILTER

An ***air pressure regulator*** is used to precisely control the amount of pressure fed to air tools and equipment. A regulator is commonly used to reduce air pressure sent to paint spray guns. However, pressure regulators may NOT be used between the compressor and many air tools, like impact guns. Those tools require the power of full line pressure. See Figure 5–3.

The air pressure regulator has a thumb screw, spring-loaded valve, and pressure gauge. By turning the thumb screw while watching the gauge, you can

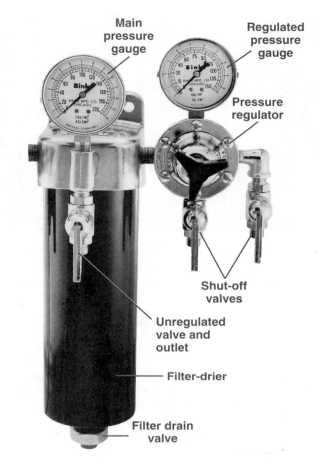

Figure 5–3 The pressure regulator can be set to match the recommended pressure for a specific tool, such as a spray gun. A filter-drier is needed before the gun to keep impurities out of both the gun and the paint work. *(Photo courtesy of ITW Automotive Refinishing, 1-800-445-3988)*

set pressure to the desired level. The regulator pressure gauge often reads in pounds per square inch (psi). Some also read in metric values.

An ***air line filter-drier*** is used in the air line to remove moisture and debris from the air flowing through the air lines. It is designed to trap and hold water and oil that passes out of the compressor air pump. If any of these materials were to enter your paint spray gun, they would ruin the paint job.

An ***air filter drain valve*** allows you to remove trapped water and oil from the filter. It should be opened every evening to purge this unwanted material.

AIR TOOLS

Air tools include hand-held spray guns, impact guns, and other tools used in the collision repair shop. It is critical that you know how to identify and use them safely. There are various types of air tools. They run cool and are very dependable.

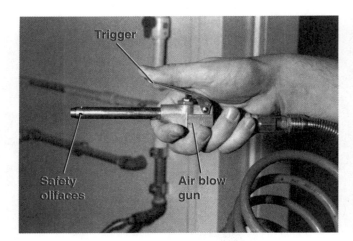

Figure 5–4 This is an OSHA-approved blow gun (blower). Note the small air holes in the tip; these can prevent air pressure buildup and injury if the nozzle contacts part of your body. *(Courtesy of Snap-on Tools Corporation)*

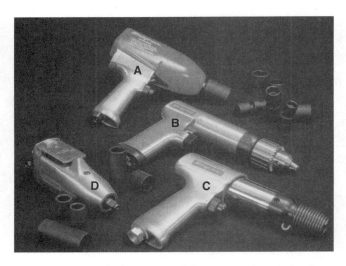

Figure 5–5 Study the names of these air tools. (A) ½-inch air impact wrench. (B) Air drill. (C) Air hammer. (D) ⅜-inch impact driver. *(Courtesy of Snap-on Tools Corporation)*

Air Blow Gun

An **air blow gun** is used to blow dust and dirt off the vehicle and its parts. It has a trigger that can be pressed to release a strong stream, or blast, of air. A blow gun is used to remove debris along trim pieces, bumpers, and other enclosed areas.

An **OSHA-approved blow gun** has pressure-relief holes in the nozzle tip. This helps prevent injury if the blow gun is accidentally pressed against your body. Non-OSHA blow guns do not have these holes and are more dangerous. Only use an OSHA-approved blow gun! See Figure 5–4.

Danger! *Never direct a blow gun air blast at yourself or others. If air enters your bloodstream, serious injury or death could result.*

Impact Wrenches

An **impact wrench** is a portable, hand-held, reversible air tool for rapid turning of bolts and nuts. When triggered, the output shaft spins freely at more than 2,000 rpm (revolutions per minute). The socket snaps over the square drive head and shaft. See Figure 5–5.

When the impact wrench meets resistance, a small spring-loaded hammer strikes an anvil attached to the drive shaft. Thus, each impact moves the socket around a little until torque equilibrium is reached, the fastener breaks, or the trigger is released. This produces a powerful rotating force to loosen or tighten fasteners.

For all their torquing power, air impact wrenches have practically no recoil. Holding one is rather easy,

but this can be misleading. The stream of air blowing through the wrench is very strong. Therefore, the wrench should be held tightly.

When using an air impact wrench, it is important that only **impact sockets** and **impact adapters** (usually black) are used. Other types of sockets and adapters (chrome plated) might shatter and fly off, endangering everyone in the immediate area.

Danger! *Never use a conventional universal joint on an impact gun. The spinning joint can fly apart and explode. Use only special impact-type swivel joints with an impact wrench.*

A **½-inch impact wrench** has a ½-inch drive head and is shaped like a pistol with a hand grip hanging down. It can tighten fasteners to over 100 foot-pounds of torque. Remember that this is enough to snap off or stretch most bolts. It is commonly used on bolt and nut head sizes over about ⁹∕₁₆ inch (14 mm). A ½-inch impact wrench is frequently used to service wheel lug nuts.

A **⅜-inch impact wrench** has a drive head of a smaller size for working with fastener heads under ⁹∕₁₆ inch (14 mm). It is lighter and more handy for working with small bolts and nuts on body panels.

The **air ratchet** wrench, like the hand ratchet, has a special ability to work in hard-to-reach places. Its ⅜-inch angle drive reaches in and loosens or tightens where other hand or power wrenches cannot fit. An air ratchet is handy for working behind panels, in tight quarters. The air ratchet wrench looks like an ordinary ratchet. It has a larger hand grip that contains the air vane motor and drive mechanism.

WARNING! *Remember that there is no reliable adjustment with an air ratchet or impact wrench. Where accurate torque is required, use a torque wrench! The air regulator on air-powered wrenches can be employed to adjust torque only to an approximate tightness. Adjust the impact to assure you do NOT overtighten. Then, use the hand torque wrench to tighten the bolt or nut to specs.*

Air Hammer

An *air hammer* uses back-and-forth hammer blows to drive a cutter or driver into the work piece. It works like a conventional hammer and chisel or punch but much quicker. Different shaped tools can be installed in the air hammer. For cutting, flat, forked, and curved cutters are useful. See Figure 5–5.

An *air chisel* is similar, but it is smaller and is often equipped with a cold chisel type of cutter.

When using an air hammer or chisel, make sure the tool is held against the work before pulling the trigger. This will prevent the cutter from possibly flying out of the tool. Avoid running into hidden hardware, frame rails, and wiring with cutting tools. Keep the cutters sharp at all times.

Drills

Drills are used to make accurately sized holes in metal and plastic parts. Both air and electric drills are available.

Air drills use shop air pressure to spin a drill bit. They can be adjusted to any speed and are more commonly used than electric drills. They are usually available in ¼-, ⅜-, and ½-inch sizes. Air drills are smaller and lighter than electric drills. This compactness makes them a great deal easier to use for most tasks.

Drill bits of different sizes fit into the chuck on drills for making holes in parts. As shown in Figure 5–6, they come in various sizes. The size is usually stamped on the upper part of each bit.

A *key* is used to tighten the chuck. The *chuck* has movable jaws that close down and hold the bit. The key has a small gear that turns a gear on the chuck.

Figure 5–6 A good set of drill bits is a must. They should be sharp and ready to use. *(Courtesy of Snap-on Tools Corporation)*

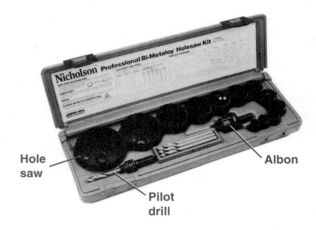

Hole saw

Albon

Pilot drill

Figure 5–7 Hole saws will quickly and accurately make large diameter holes. *(Cooper Tools)*

Never leave a key in a drill when not tightening. Also, unhook the air hose when installing a bit. Otherwise, injury could result. This also applies to a large drill press.

Hole saws are special cutters for making large holes in body panels. They fit into a drill like a drill bit. See Figure 5–7.

A *spot weld drill* is specially designed for removing spot welded body panels, Figure 5–8. It has a clamp-type head and lever arm for accurately drilling through spot welds. The cutter will not deviate from the weld center during cutting.

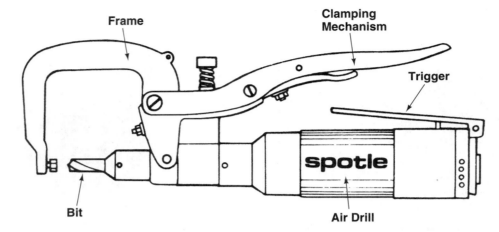

Figure 5–8 Spot weld drills have a special fixture mounted over the drill. This helps hold the drill over the spot weld for accurate removal. *(NCG Company/Spotle Distributor)*

Air Cutters

An *air nibbler* is used like tinsnips to cut sheet metal. It has an air-powered snipping blade that moves up and down to cut straight or curved shapes in sheet metal panels. Several are in Figure 5–9.

An *air cutoff tool* uses a small abrasive wheel to rapidly cut or grind metal. It is handy for cutting when a confinement prevents the use of a nibbler. See Figure 5–9.

An *air saw* uses a reciprocating (back-and-forth) action to move a hacksaw-type blade for cutting parts. It works just like a hacksaw, but will cut much more quickly.

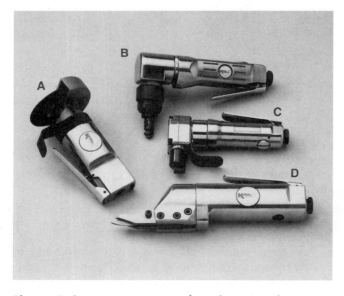

Figure 5–9 Note various types of metal cutting tools. (A) Disc-type cutoff tool. (B) Air nibbler. (C) Smaller air nibbler. (D) Air snips *(Courtesy of Snap-on Tools Corporation)*

Air Sanders

An *air sander* uses an abrasive action to smooth and shape body surfaces. Different coarseness sandpapers can be attached to the pad on the sander. **Coarser sandpaper** removes material more quickly. **Fine sandpaper** produces a smoother surface finish. Air sanders are one of the most commonly used air tools in collision repair.

A *circular sander* simply spins its pad and paper around and around in a circular motion. This type of sander is seldom used in collision repair. See Figure 5–10.

A *safety trigger* is designed to help prevent you from turning the power tool on by accident. The on-off trigger has a spring-loaded arm that locks the trigger off and keeps it from being pushed down. You must pull back on the safety lever and then push down on the trigger to energize the tool. Many potentially dangerous power tools are now equipped with a safety trigger.

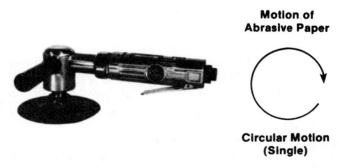

Motion of Abrasive Paper

Circular Motion (Single)

Figure 5–10 A circular motion is not commonly used for sanding operations in collision repair. The sanding disc would leave deep sanding marks. A circular motion is used only in grinding and polishing.

**Motion of
Abrasive Paper**

**Circular Motion
(Double)**

Figure 5–11 An orbital, or dual-action, sander is the most commonly used in collision repair. It produces a smoother sanded surface.

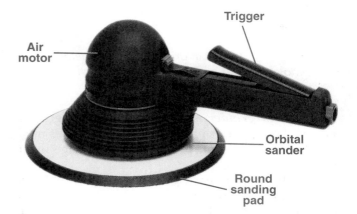

Air
motor

Trigger

Orbital
sander

Round
sanding
pad

A

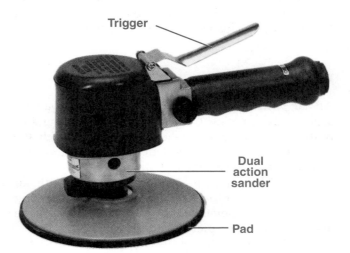

Trigger

Dual
action
sander

Pad

B

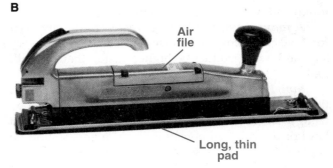

Air
file

Long, thin
pad

C

Figure 5–12 Study the different types of air sanders *(Courtesy of Snap-on Tools Corporation)* (A) The disc-type orbital sander is the most common design used. It is small and easy to handle. Its rounded sandpaper will let you work curved surfaces. (B) This dual-action sander is good for featheredging repaired areas. (C) An air file uses long, retangular sheets of sandpaper. It is good for sanding large panels.

An *oscillating sander* spins a sanding pad that is mounted to a concentric shaft that also has a counterweight on it. These sanders are used because they sand very smoothly and sand faster than some other sanders.

An *orbital* or *dual-action sander* moves in two directions at the same time. This produces a much smoother surface finish. A dual-action, or DA, sander is used to featheredge a repair area. It is the workhorse of body technicians. One is pictured in Figure 5–11.

The *sander pad* is a soft synthetic rubber mounting surface for the sandpaper. It normally screws into the sander head.

Sander pads can be designed to use self-stick sandpaper (adhesive already on the sandpaper) or a Velcro system (tiny hooks and loops formed on the sander pad and sandpaper). The Velcro system is gaining popularity because the sandpaper runs cooler, fine grits do not clog as easily, you can change sandpaper more quickly, and you can reuse the sandpaper if desired.

Danger! Never allow an air sander to free-wheel, or spin, without the sandpaper being in contact with the work surface! Due to the high rotation speeds that would result, the sandpaper could fly off the pad and cause serious injury.

An *air file* is a long, thin air sander for working large surfaces on panels, Figure 5–12. It is handy when you must true or sand a large repair area. It will plane down filler so that a large area is smooth. An air file is often used for rough shaping operations.

GRINDERS

Grinders are used for fast removal of material. They are often used to smooth metal joints after welding and to remove paint and primer. They come in various sizes and shapes.

The most commonly used portable air grinder in collision repair and refinishing shops is the *disc-type grinder.* It is operated like the single-action disc sander. An air grinder should be used carefully. It can quickly thin down and cut through body panels, causing major problems.

> ▪ *Danger!* If an air grinder will cut metal, it can
> ▪ certainly cut a technician. Wear leather gloves
> ▪ and a full face shield when grinding.

A **bench grinder** is a stationary, electric grinder mounted on a workbench or pedestal stand. It is used to sharpen chisels and punches, and to shape other tool tips.

When using a bench grinder, keep the tool rest close to the stone or wire wheel. Otherwise, the tool or part can be pulled into the grinder. This can force your hand into the wheel or brush, causing a painful injury. Also, keep the shield in place over the top of the stone or brush and wear eye protection.

POLISHERS

An **air polisher** is used to smooth and shine painted surfaces by spinning a soft buffing pad. Polishing or buffing compound is applied to the paint with the polisher pad. This removes minor imperfections to increase **paint gloss** or shine.

A polishing pad is a cotton, synthetic-cloth, or foam rubber cover that fits over the polisher's backing plate or arbor. Sometimes the pad and backing plate are **integral, or one-piece.**

A **cloth or cotton polishing pad** is often used when heavy polishing is needed. The pad is very tough and will cut paint smoothly and quickly.

A **foam rubber polishing pad** is commonly used for final polishing to remove swirl marks. It is very soft and will produce a higher luster than a cloth pad.

Swirl marks are tiny unwanted, circular marks that can be seen after buffing. They are often caused by using a course polishing compound or a pad that is too hard or dirty. When polishing, swirl marks should be kept to a minimum. Use a foam pad for final polishing.

A **pad cleaning tool** is a metal star wheel and handle that will clear dried polishing compound out of the cloth or cotton pad. The wheel is held onto the spinning pad to clean out and soften the pad material. It will help prevent swirl marks in the paint.

PNEUMATIC TOOL MAINTENANCE

Air tools need little upkeep. However, you will have problems if basic maintenance is not performed. For instance, moisture gathers in the air lines and is blown into tools during use. If a tool is stored with water in it, rust will form and the tool will wear out quickly.

Air tool lubrication involves injecting a few drops of oil into the air inlet of all air tools to prevent corrosion and friction damage. To avoid contamination of vehicle's surfaces, use only special air tool oil designed for collision repair air tools. It is formulated to prevent some surface contamination that can result from oil spraying from the air outlet on the tool. Motor oil and transmission fluid should NOT be used to lubricate air tools because they will contaminate the repair surface.

Put a small amount of oil into the air inlet or into special oil holes on the tool BEFORE AND AFTER use. This will prevent rapid wear and rusting of the vane motor and other parts of the tool. Run the tool after adding the oil. Wipe off excess oil on the tool to keep it off body parts.

An **in-line oiler** is an attachment that will automatically meter oil into air lines for air tools. It can be used on lines used for air tools but NOT for spray guns.

Remember, do NOT use in-line oilers in the paint area. Never over-oil sanders, grinders, and other air tools. You could contaminate the vehicle's surface with oil.

All air tools have recommended air pressures. If the tool is overworked, it will wear out sooner. If something is wrong with the tool, fix it. If the defect is NOT repaired, a chain reaction may occur and the other parts will become defective as well. The air compressor, in turn, will be overworked and could put out contaminated air.

Never forget! Your tools and equipment cannot function correctly without good care. To do high quality collision repair and refinishing/painting work, maintain your tools and equipment.

BLASTERS

Blasters are air-powered tools for forcing sand, plastic beads, or another material onto surfaces for paint removal. For example, they are handy when trying to remove surface corrosion from body panels. They will blast out all of the corrosion without further thinning the panel.

With today's thin-gauge, high-strength steel, plastic media blasting is often recommended over grinding to clean out corrosion pockets. Grinding thins the metal and makes it weaker.

A **hand blaster** is a small, portable tool for blasting parts and panels on the vehicle. Abrasive is installed in the blaster cup or container. Airflow through the tool pulls a metered amount of abrasive out of the tool and forces it against the work piece. This will quickly remove paint, primer, corrosion, and scale to take the surface down to bare metal.

A **cabinet blaster** is a stationary enclosure equipped with a sand or bead blast tool. A window and rubber gloves in the cabinet allow you to blast parts without being exposed to the abrasive dust. It is often used on small parts removed from a vehicle.

SPRAY GUNS

Spray guns are used to apply sealer, primer, color, and clear materials to the vehicle. Spray guns must atomize the liquid so that it flows onto the vehicle surface smoothly and evenly. The term **atomize** means to break the liquid into very tiny droplets or a fine mist. This will help the material go down smoothly. This requires sufficient air pressure and volume at the spray gun. See Figure 5–13.

If you plan on being a good painter, you must know your painting equipment inside and out.

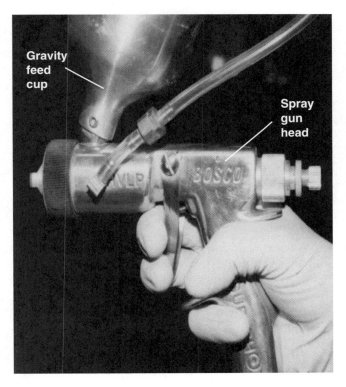

Figure 5–13 The gravity-feed spray gun is the most common type used for painting. This spray gun also uses cup pressure to feed liquid through the spray gun. This is an HVLP gun, which means it is a high-volume, low-pressure gun. It is designed to deposit more of the atomized material onto the surface being sprayed while reducing overspray.

Spray Gun Parts

To use, service, and troubleshoot a spray gun properly, you must understand the operation of its major parts. See Figure 5–14 as each spray gun part is discussed.

The **gun body** holds the parts that meter air and liquid. The body holds the spray pattern adjustment valve, fluid control valve, air cap, fluid tip, trigger, and related parts.

The **spray gun cup** often fits onto the bottom or on top of the body to hold the material to be sprayed. The cup fits against a rubber seal to prevent leakage. The seal is mounted in a lid on the gun body.

The spray gun's **fluid control valve** can be turned to adjust the amount of paint or other material emitted. It consists of a thumb screw or knob, needle valve, and spring. Turning the knob affects how far the trigger pulls the needle valve open. The **fluid needle valve** seats in the fluid tip to prevent flow or it can be pulled back to allow flow.

The spray gun's **air control valve** controls how much air flows out of the air cap side jets. It has an **air needle** that can be slid back and forth to open or close the air valve.

The **spray gun trigger** can be pulled to open both the fluid and air valves. It uses lever action to pull back on the needle valves.

The **spray gun air cap** works with the air valve to control the spray pattern of the paint. It screws over the front of the gun head.

The **spray pattern** is the shape of the atomized spray when it hits the body. With little airflow out of the side jets on the cap, you would have a very round, concentrated spray pattern. As you adjust air flow up, the cap jets narrow and better atomize the fluid flowing out of the gun.

For more information on this and other related subjects, refer to the text index.

Gravity-Feed Spray Guns

A **gravity feed spray gun** mounts the cup on top of the spray gun head so that the pull of gravity helps the material flow out of the cup. A gravity-feed design helps eliminate the problem of spitting. All of the liquid will flow down into the pickup hole formed in the bottom of the cup, Figure 5–13. No pickup tube is needed. The gravity-feed spray gun is a very common design with today's high spray-efficiency requirements.

Siphon Spray Guns

Siphon spray guns use airflow through the gun head to form a suction that pulls paint into the airstream. One is shown in Figure 5–15.

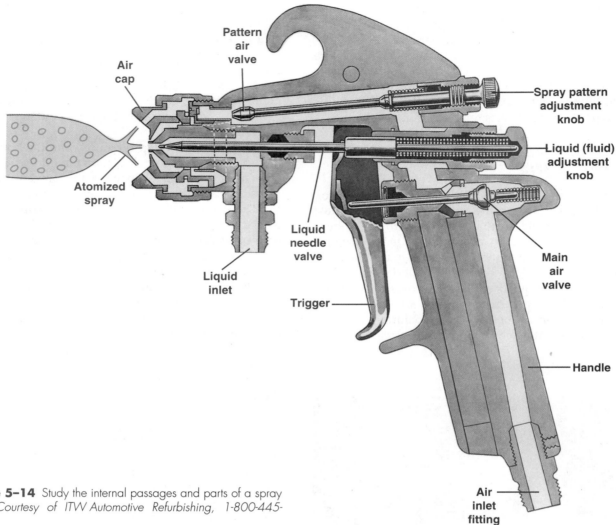

Figure 5–14 Study the internal passages and parts of a spray gun. *(Courtesy of ITW Automotive Refurbishing, 1-800-445-3988)*

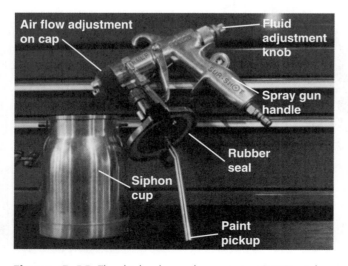

Figure 5–15 The high-volume, low-pressure (HVLP) siphon spray gun is a high-efficiency design. Airflow through the head generates a vacuum, which pulls paint up and out of the cup. It is not as common as the gravity-feed type.

An ***air vent hole*** and hose on the siphon spray gun allow atmospheric pressure to enter the cup. This vent can become clogged with dry primer or paint. If the vent is plugged, paint will not flow out of the gun.

An exploded view of a siphon-type spray gun is in Figure 5–16. Study the relationship of the parts carefully.

Mix Nozzle Types

An **internal mix nozzle** combines the air and liquid inside or under the cap. An **external mix nozzle** combines the air and liquid outside of the gun. The external mix is more common in collision repair. See Figure 5–17.

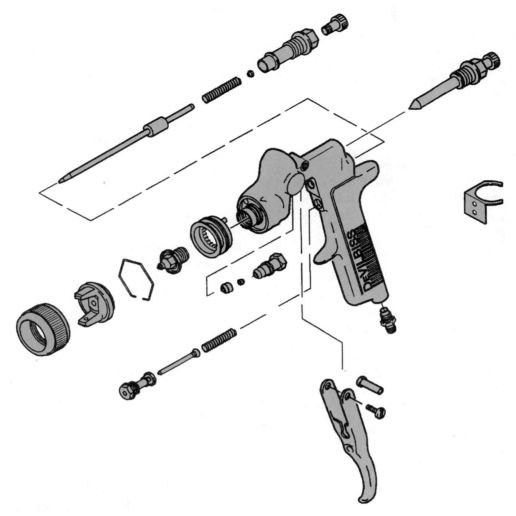

Figure 5–16 This exploded view shows every part of a HVLP high-efficiency spray gun. Study the part names and locations carefully. This type of illustration is handy when ordering new parts or when doing major service on a spray gun. *(Courtesy of ITW Automotive Refinishing, 1-800-445-3988)*

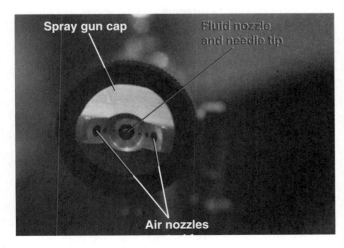

Figure 5–17 This close-up of an external mix spray gun nozzle shows the air jet openings and the center fluid opening. The volume of air leaving the jets can be adjusted to control atomization and the shape of the spray pattern.

Pressure-Feed Spray Guns

Pressure-pot or *pressure-cup spray guns* use air pressure inside the paint cup or tank to force the material out of the gun. The major external parts of this type of gun are in Figure 5–18.

Pressure-pot guns provide possible advantages over siphon-cup guns. They allow more paint speed through a smaller nozzle. Smaller paint streams atomize better.

A pressure pot, by having a remote cup, makes the gun lighter and easier to handle. It also permits spraying with the gun horizontal for painting under flared parts without danger of "spitting." Also, with the cup or tank away from the vehicle, the problem of paint dripping from the gun cup vent is eliminated.

Pressure-tank spray guns use a much larger storage container for paint materials. They hold enough paint for a complete paint job. This saves time. You are sure the paint will match throughout the whole job. This might help with spraying hard-to-match metallic or pearl paints, for example.

Remember that pressure cups have seals that must be kept clean and regularly inspected for damage. A loss of cup pressure affects the delivery of fluid to the spray gun.

> *WARNING! Always adjust line pressure to specifications to prevent damage or rupture of the cup or tank. The specifications will be given by the manufacturer.*

Pressure cups also hold pressure AFTER being disconnected from the air source. This can be messy if you open the lid and paint sprays all over the shop. Make sure you release the cup pressure before opening the lid.

Pressure cups also require some bleed-down time if the cup is initially over-pressurized. For example, if a cup is pressurized to 10 psi (68.95 kPa) and the painter desires only 6 psi (41.37 kPa), the cup maintains that 10 psi (68.95 kPa) pressure momentarily when the trigger is first pulled back. This could force out too much paint and cause problems.

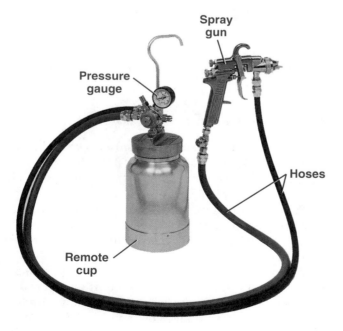

Figure 5–18 The pressure-pot or pressure-cup gun is handy and easy to use. Study the parts of this pressure-cup gun. Note how the hoses feed from the cup to the gun. High-efficiency pressure-pot spray guns are available. *(Photo courtesy of ITW Automotive Refinishing, 1-800-445-3988)*

Bleeder and Nonbleeder Guns

Bleeder spray guns allow air to pass through the gun at all times, even when the trigger is NOT pulled back. They do not have an air control valve. The trigger only controls fluid needle movement. Bleeder guns are generally used with heated air sources.

Nonbleeder spray guns release air only when the trigger is pulled back. They have an air control valve. The trigger controls both fluid needle movement and the air control valve.

Touch-Up Spray Guns

Touch-up spray guns are very small and are ideal for painting small repair areas. Often called a "door jamb gun," they have a tiny cup for holding a small amount of material. They operate like a conventional spray gun.

VOC Emissions

Volatile organic compounds (VOCs), are harmful substances produced by many painting materials. Some geographic regions require the use of high transfer efficiency spray equipment to cut VOC emissions. These laws are designed to reduce the amount of air pollutants entering the earth's atmosphere.

High-Efficiency Spray Guns

High-efficiency spray guns are designed to reduce VOC emissions by applying more of the material to the vehicle and allowing less into the air. Check local ordinances to find out if you must use these types of guns.

High-efficiency spray guns are called *HVLP* for high volume, low pressure, or *LVLP* for low volume, low pressure. They use 10 psi (68.95 kPa) or less for reducing overspray and emissions. HVLP spray guns are popular for the following reasons:

1. They use less material and thus reduce material costs.

2. They produce less unwanted overspray so less dry spray settles on unpainted surfaces of the vehicle.

3. They use less air pressure.

4. Spray suits and lenses last longer.

5. Booth air filters last longer.

6. Booth inner surfaces stay cleaner.

7. They help protect our environment.

Transfer efficiency describes the percentage of spray material a spray gun is capable of depositing on the surface. Transfer efficiency for spray guns must be at least 65 percent to meet regulations in some geographic locations. This means that at least 65 percent of the material sprayed must remain on the vehicle.

High transfer efficiency means more paint applied will remain on the surface. There is less spray material wasted as overspray. High transfer efficiency cuts VOC emissions and protects our atmosphere.

Transfer efficiency ratings are affected by the type of gun, gun adjustment, air pressure, spray pattern, and reduction of spray material.

Actual spray transfer efficiency will vary depending on a painter's skill and technique. As you become more familiar with the techniques used with HVLP equipment, transfer efficiency and skill will increase.

With *turbine spray systems,* the unit produces its own pressure and does not use shop air. This is an advantage if the shop air compressor is already near capacity. Another advantage is that turbine air is dry and free of oil.

Heated air produced by turbines can help reduce the paint's viscosity, which improves atomization. It may also improve drying times of spray materials in colder climates.

Some possible disadvantages are that the turbine's electric motor is NOT explosion proof and cannot be placed in the spray booth. The heated air also evaporates the solvents in atomized material faster than conventional air sources. This may cause a drier spray and poor appearance if proper reducer, techniques, and spraying conditions are not used.

Air conversion units act as in-line regulators to convert high pressure main line air pressure to low pressure, high volume air. They use existing shop air from the compressor.

Air conversion units allow adjustment of air pressures up to 10 psi (68.95 kPa). They can be wall mounted or worn by the painter. Air conversion units are available in both electrically heated and unheated models.

Step-down guns internally restrict the inlet pressure of 40–100 psi (275.8–689.5 kPa) to 10 psi (68.95 kPa) at the air cap. Step-down spray guns do not require the use of a turbine or air conversion unit. They use existing shop air from the compressor. A special orifice, venturi, or regulator mounted on the gun cuts pressure.

Electrostatic spray guns electrically charge the paint particles at the gun to attract them to the vehicle body. They operate on the theory that opposites attract. The charged paint particles are attracted to the grounded metal body of the vehicle. VOC regulations also list electrostatic spray equipment as complying to

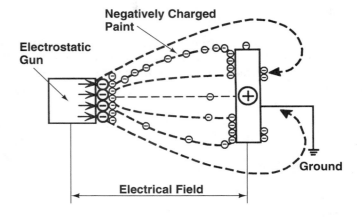

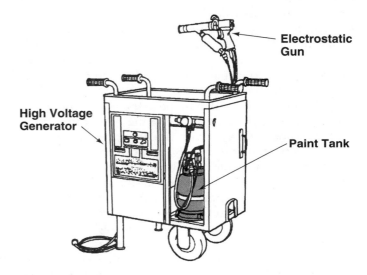

Figure 5–19 Electrostatic spray equipment uses static electrical charges to pull paint onto metal surfaces.

the minimum 65 percent transfer efficiency requirement. Its operation is shown in Figure 5–19.

Electrostatic spray guns are used for spraying finishes in automotive assembly plants and fleet use. They have higher transfer efficiency than other types of spray equipment when spraying complex shaped parts. For example, bicycle frames can be sprayed with higher transfer efficiency because the material is attracted to all sides of the part.

Electrostatic spray guns do NOT help attract the paint to nonmetal plastic parts. To spray plastics, an electrically conductive *primer* must be applied. This primer cannot be applied using electrostatic spray equipment. Maintenance and repairs are somewhat expensive. Also, the orientation of metallic flakes may be difficult to manage.

Before purchasing spray equipment, test the equipment in the shop. This will help you select the right equipment to reduce air pollution and still produce high quality painting/refinishing in minimum time.

Specialized Spray Guns

Specialized spray guns are designed to apply various materials. They can be used to spray rubberized chip resistant materials, anticorrosion materials, and other substances.

SPRAY BOOTHS

A **spray booth** is designed to provide a clean, safe, well-lit enclosure for painting and refinishing. It isolates the painting operation from dirt and dust to keep this debris out of the paint. It also confines the volatile fumes created by spraying and removes them from the shop area. See Figure 5–20.

Spray booths are designed to conform to federal, state, and local safety codes. In some areas, automatically operated fire extinguishers are required because of the highly explosive nature of refinishing materials, Figure 5–21.

A workbench is often in the paint mixing room for reducing the paint and for filling the spray gun. Paint storage should always be outside the spray booth.

An **air makeup** or **air replacement system** is important because air must be exhausted from spray booths continuously during the spraying process.

In winter, the spray area can become cold and uncomfortable. Finish problems can arise from spraying cold materials on cold vehicles in cold air. To prevent this problem, an air makeup system will often provide even temperatures and clean filtered air.

Some shops employ an independent air replacement system designed for the spray booth. This provides clean, dry, filtered air from the outside, and even heats the air in colder weather. Replacement air can be delivered to the general shop area or directly into the booth for a completely closed system.

The **downdraft spray booth** forces air from the ceiling down through exhaust vents in the floor. The downward flow of air from the ceiling to the floor pit creates an envelope of air passing by the surface of the vehicle.

By taking clean, heated air and directing it downward, the downdraft prevents contamination and overspray from settling on the freshly painted surface of the vehicle. This air movement also helps to remove toxic vapors from the breathing zone of the painter, providing a safer working environment.

A **side draft** or **crossflow spray booth** moves air sideways over the vehicle. An air inlet in one wall pushes fresh air into the booth. A vent on the opposite wall removes the booth air.

Spray booth lighting is very important. Colors appear different under different types of light. For instance, incandescent light tends to highlight reds.

Figure 5–20 The spray booth is a clean room for refinishing operations (painting, priming, etc.). The booth circulates filtered air over the vehicle to keep dust and dirt out of the finish. *(Riverdale Body Shop)*

This makes it very important to utilize **color correcting lights** in the spray booth to make color show up properly. These lights will provide a perfectly neutral light that makes accurate color matching possible. If you are going to paint only a fender for example, you want the freshly painted fender to match the color of the rest of the vehicle.

Figure 5–21 A spray booth is a large metal-and-glass enclosure with special air-handling equipment. *(Photo courtesy of Team Blowtherm)*

PAINT DRYING EQUIPMENT

A dust-free **paint drying room** will speed up drying/curing, produce cleaner work, and increase the volume of refinishing work. Some drying rooms have forced air or permanent infrared units for the force drying of paint and primers. These oven-like units can speed up the drying time by as much as 75 percent. Force drying of fillers, primers, and sealers will reduce waiting between operations. Drying rooms can also be used for fast drying spot and panel repairs.

Infrared drying equipment uses special bulbs to generate infrared light for fast drying of materials. They can be designed into booth systems. They range from portable panels for partial or sectional drying to large heating units capable of moving automatically along a track next to the vehicle for overall drying. Look at Figure 5–22.

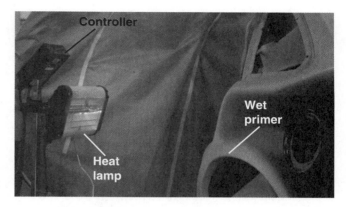

Figure 5–22 Infrared light is being used to speed curing of primer. Heat lamps can cause painful burns, so use caution when working around these units.

> *WARNING! When using a drying room, caution must be taken NOT to overheat and destroy the finish. A* surface thermometer *measures panel temperatures to prevent overheating damage. It also lets you know when the surface is hot enough for complete drying. Two types are shown in Figure 5–23.*

MEASURING SYSTEMS

A **measuring system** is used to gauge and check the amount of frame and body damage on the vehicle. It is a special machine for comparing known good measurements with those on the vehicle being repaired.

In some designs, the measuring system mounts over the vehicle. Brackets and rods are positioned so that pointers aim at specific reference points on the body. Other systems use laser light to denote properly located reference points. If any reference point does NOT line up with a measuring system, the frame or unibody must be straightened or parts must be replaced.

STRAIGHTENING SYSTEMS

Hydraulic equipment uses a confined fluid or oil to develop the pressure necessary for operation. The pressure is achieved manually (pumping on a handle or lever) or by a small motor (either air or electrically driven) that drives a pump.

A **straightening system,** sometimes called a **frame rack,** is a hydraulic machine for straightening a damaged frame or unibody structure back into proper alignment, Figure 5–24. The vehicle can be anchored to the shop floor or rack using clamps and chains. Then, pulling chains are attached to the bent or damaged areas of the vehicle. One type of frame straightening equipment is shown in Figure 5–25.

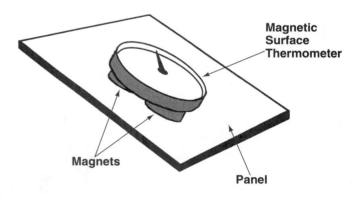

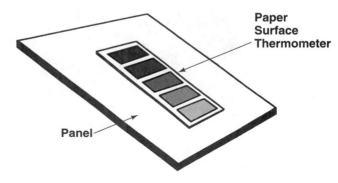

Figure 5–23 Special thermometers should be used with drying equipment. *(I-CAR)*

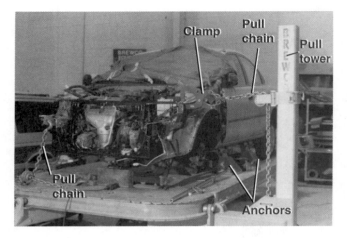

Figure 5-24 A damaged unibody vehicle is being pulled back into alignment (straightened). Note the pulling chains and the directions of the pulling forces.

Pulling posts are vertical uprights on the frame machine that allow the chains to pull in the needed directions to straighten the damage. See Figure 5–26.

There are two basic types of frame/panel straighteners on the market: portable and stationary.

Frame straightening accessories are the many chains, special tools, and other parts needed. They will vary with the equipment manufacturer. To perform the many different straightening operations involving pushing, pulling, or holding a vehicle, a large assortment of adapters is included in a typical kit. See Figure 5–27.

Body jacks, also known as porta-powers, can be used with frame/panel straighteners or by themselves. A body jack has a small hand-operated pump that operates a hydraulic cylinder. Pumping the handle extends the cylinder ram to produce a powerful pushing action.

A *body dolly* is a set of caster wheels that is clamped to the vehicle. It allows you to move the vehicle around the shop during repairs. See Figure 5–28.

LIFTS

The traditional stationary in-ground lift was usually found only in service stations, muffler shops, transmissions shops, and tire dealers. Today, collision

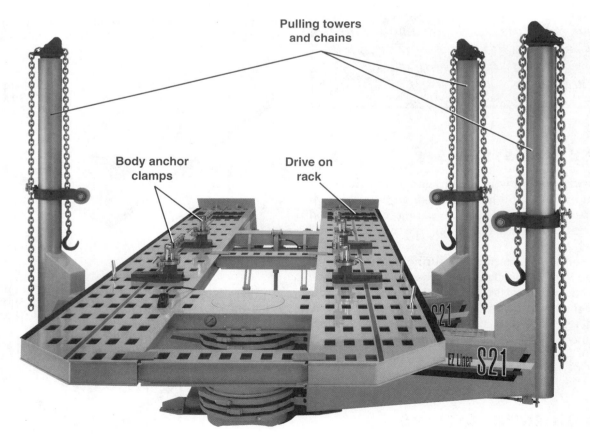

Figure 5-25 Note the major parts of a frame/unibody straightening machine. Anchors hold the vehicle down onto the rack while it is being pulled. The towers allow you to position the pulling chains at different heights and positions to remove frame or unibody damage. *(Chief Automotive Systems, Inc.)*

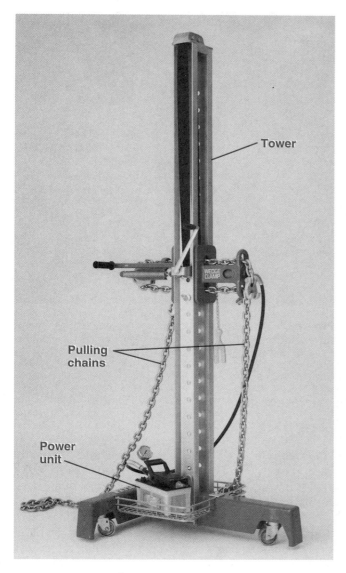

Figure 5-26 A pulling tower can either mount on the rack or be a separate unit. In either case, the tower must be anchored properly for safe pulling. *(Wedge Clamp International, Inc.)*

Figure 5-27 Many accessories are provided with a frame rack. You need to refer to the equipment manual to learn all about each of them. *(Chief Automotive Systems, Inc.)*

Figure 5-28 This technician has a vehicle on a body dolly for moving it around the shop. Hold-down clamps from the frame rack remain on the unibody structure. *(Wedge Clamp International, Inc.)*

repair shops use lifts because it is easier to evaluate underbody damage and make repairs.

As shown throughout this book, there are several ways to get a vehicle off the ground. Four-post and two-post hoists allow total movement under the vehicle. However, they take up more than a work stall in length and width. Side post hoists are great for estimating, but some access to the sides of the vehicle is impaired. Old center post hoists make some areas under the vehicle hard to reach but take up less space.

Raising a vehicle on a lift or a hoist requires special care. Adapters and hoist plates must be positioned correctly on lifts to prevent damage to the underbody of the vehicle. The exhaust system, tie-rods, and shock absorbers could be damaged if the adapters and hoist plates are incorrectly placed.

For more information on lifts, refer to the text index.

ELECTRIC TOOLS

Electric power tools plug into shop wall *AC* (alternating current) outlets. Drill presses, heat guns, welders, and vacuum cleaners are all vital electric-only tools. In addition, electric tools such as drills, polishers, and sanders perform the same tasks as their pneumatic counterparts.

Most electric power tools are built with an external grounding system for safety. A wire runs from the motor housing, through the power cord, to the third prong on the power plug. When connected to a grounded, three-hole electrical outlet, the grounding wire will carry any current that leaks past the electrical insulation of the tool. It will carry this current away from you and into the ground of the shop's wiring. See Figure 5–29.

Some electric power tools are self-insulated and do NOT require grounding. These tools have only two prongs since they have a nonconducting housing.

Never use an adapter plug that eliminates and disconnects the third ground prong or you could be electrocuted.

BATTERY CHARGER

A **battery charger** converts 120 volts AC into 13 to 15 volts **DC** (direct current) for recharging drained batteries. One is in Figure 5–30.

To use a battery charger, connect red to positive and black to negative. The red lead on the charger goes to the positive terminal of the battery. The black lead goes to ground or the negative battery terminal.

After connecting the charger, adjust its settings as needed (12 volt battery, fast or slow charge, etc.).

> **DANGER!** A battery explosion could result if charger instructions are not followed. If you connect a charger to a battery while the charger is running, any small spark could ignite explosive gases that are around the battery.

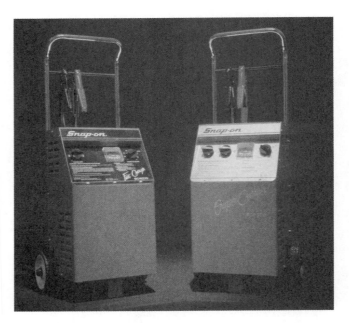

Figure 5–30 A battery charger is often needed for "dead" batteries. Connect the charger to the battery before turning it on. Set the controls so as not to overcharge the battery. *(Courtesy of Snap-on Tools Corporation)*

MULTIMETER

A **multimeter** is a voltmeter, ammeter, and ohmmeter combined into one electrical tester. Also called a VOM (volt-ohm-milliammeter), a multimeter is often used for many types of electrical tests. See Figure 5–31A.

An **analog meter** uses a needle that swings right and left for making measurements. This is an older design that is being replaced by digital meters. A **digital meter** uses a number display to show readings.

> **WARNING!** Never use a **low imped-ance** *(low resistance) meter to test modern electronic circuits. The low resistance and high current through the meter can damage and ruin delicate computer and electronic circuits. Use* **only** *high impedance (high internal resistance) meters.*

An **inductive meter** clips around the outside insulation of wiring to take electrical readings. It is time saving when making current (amp) measurements, Figure 5–31B.

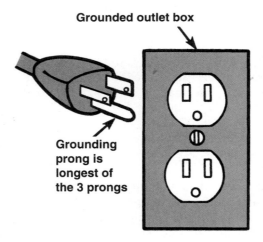

Grounded outlet box

Grounding prong is longest of the 3 prongs

Figure 5–29 Never remove a ground prong from electrical cords. That prong prevents any short in a power tool from causing an electric shock.

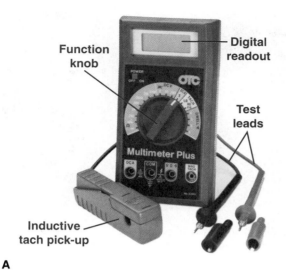

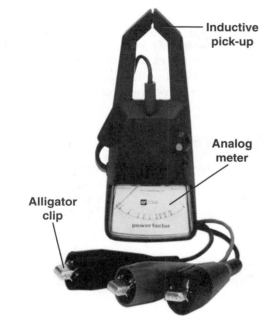

Figure 5–31 Electrical test instruments are used by collision repair technicians. Wiring is often damaged in a collision and the technician may have to find and repair minor wiring troubles. (A) A multimeter will measure voltage, current, and resistance. This is a high-impedance digital meter acceptable for use on delicate computer circuits. *(OTC, a Division of SPX Corporation)* (B) An inductive or clip-on meter will make rapid current measurements. You do not have to unhook wires to put the meter in series with the circuit. *(TIF Instruments, Inc.)*

SCAN TOOLS

Scanners or **scan tools** are used to diagnose or troubleshoot vehicle computer system and wiring problems. A scanner connects to the vehicle's diagnostic connector or **ALDL** (assembly line diagnostic link).

The scanner's electronic circuits can then communicate with the vehicle's on-board computer to help find problems.

A **scanner cartridge** is installed into the scanner for the specific make and model of vehicle being tested. The cartridge holds the information needed for that vehicle.

When there is a circuit problem, the vehicle's computer will illuminate a **malfunction indicator light** in the dash and store a trouble code number in its memory. The **trouble code number** denotes the specific problem circuit with the abnormal electrical operating value.

Modern scan tools will convert this trouble code number into a short word description of the potential problem. Most will also guide the technician through a short list of instructions for troubleshooting the specific problem. The scan tool might say that a circuit has high resistance and you should do ohmmeter tests to check for open wiring. Collision damage often severs wiring and produces trouble codes. See Figure 5-32.

Many late model vehicles have standardized **On-Board Diagnostics II** (OBD II) that monitors the operation of hundreds of circuits and components. The location of the OBD II diagnostic connector should be visible from under the left center of the dash.

Early diagnostic connectors came in various configurations. An adapter was often needed to make the scanner cable fit the vehicle's connector. Adapters are labeled for easy location for each make of vehicle. With OBD II, the standardized connector is a 16-pin

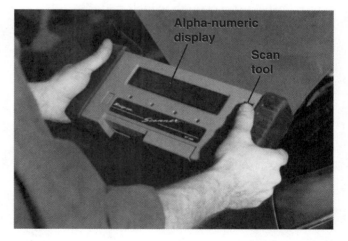

Figure 5–32 A scan tool will help you quickly find electrical/electronic circuit problems. It will indicate which circuit is not operating normally and even give help finding the exact source of the trouble. Collision damage often cuts wires or damages sensors. The collision repair technician must be able to find and repair this type of damage efficiently. *(Courtesy of Snap-on Tools Corporation)*

connector. For more detailed information on scan tool use, refer to Chapter 26.

ENGINE ANALYZER

An *engine analyzer* is a large group of test instruments mounted in a roll-around cabinet. It can be used to find various kinds of trouble in the ignition, fuel injection, charging, and other electrical systems.

PERSONAL COMPUTER

The *personal computer,* or *PC,* is now an important tool for doing various collision repair shop tasks. It is used to keep track of business transactions, complete damage reports, and even perform electrical tests on vehicles, Figure 5–33. A computer can find and manipulate huge amounts of data about all aspects of the shop operation. It is finding wider, more varied uses. See Figure 5–34.

Portable or **laptop computers** are small handheld units often used during estimating. They have small keyboards for inputting data.

SOLDERING GUN

A *soldering gun* is used to heat and melt solder for making electrical repairs. It has a transformer that converts AC into DC. DC is sent through the gun tip which heats the tip enough to melt solder.

Solder is a mix of lead and tin for joining electrical wires and components. **Rosin core solder** is designed for doing electrical repairs. The *rosin* serves as a flux to make the melted solder adhere to the parts being joined.

Figure 5–33 This shop manager is using a personal computer to write a damage estimate for a customer. *(Riverdale Body Shop)*

Figure 5–34 The personal computer is gaining importance as a tool in collision repair. It will help automate damage estimating and many other business-related tasks. *(Courtesy of Mitchell International, Inc.)*

Acid core solder is used for doing nonelectrical repairs, as on radiators. It should NOT be used for electrical repairs because the acid can corrode connections and affect their electrical resistance.

PLASTIC WELDER

A *plastic welder* is used to heat and melt plastic and composites for repairing or joining parts made of these materials. It works like a soldering gun but operates at a lower temperature. One is shown in Figure 5–35.

AIR CONDITIONING TOOLS

Because air conditioning parts are often damaged in frontal collisions, you should understand the use of basic air conditioning tools and equipment.

Air conditioning gauges are used to measure operating pressures in a vehicle's air conditioning system.

The A/C gauges are connected to *service ports* (small test fittings) in the system. The engine is then started and the A/C is set to maximum cooling. Gauge pressures are compared to known good readings. If pressures are too high or low, refer to troubleshooting charts to help find the problem, Figure 5–36.

A *leak detector* is a tool for finding refrigerant leaks in lines, hoses, and other parts of an air conditioning system. The tester is turned on and moved around possible leakage points. Since air conditioning refrigerant is heavier than air, the probe should be moved below possible leakage points. The tester will emit an audible tone or noise if a leak is present. This

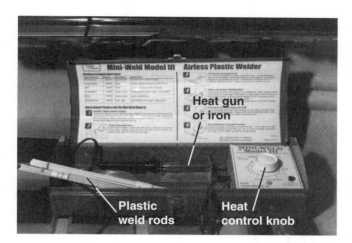

Figure 5–35 This is a plastic welder. Plastic filler rods and heat from the tool can be used to repair minor splits or cracks in plastic parts.

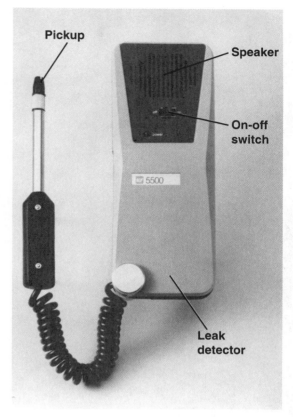

Figure 5–37 An electronic leak detector will make an audible sound or change in sound pitch if a refrigerant leak is present. *(TIF Instruments, Inc.)*

helps you find and fix refrigerant leaks more quickly. See Figure 5–37.

A *charging station* is a machine for automatically recharging an air conditioning system with refrigerant. It has pressure gauges, a tank of refrigerant, and electronic controls for metering the right amount of refrigerant into the system. See Figure 5–38.

A *recovery station* is used to capture the old, used refrigerant so that it does not enter and pollute the atmosphere. It consists of a large metal tank and vacuum pump. The recovery station pulls the refrigerant out of the vehicle and forces it into the storage tank. The station will then filter, clean, and dry the used refrigerant so that it can be recycled. See Figure 5–39.

A *recovery-charging station* can recover, evacuate, recycle, and charge a vehicle's air conditioning system. It combines the features of a recovery station and a charging station in one roll-around cabinet.

Note! For more information on power tools and equipment, refer to the textbook index. This will let you find additional information.

OTHER EQUIPMENT

Many other types of equipment are used in collision repair shops. You should have a basic understanding of their purpose.

A *tire changer* is a machine for quickly dismounting and mounting tires on and off wheels. It uses *pneumatic* (air) pressure to easily force the tire bead off or onto the wheel rim. One is pictured in Figure 5–40.

A *wheel balancer* is used to locate heavy areas on a wheel and tire assembly. It spins the wheel and tire and measures vibration. It will denote where *wheel weights* should be positioned on the wheel to equalize the weight and prevent vibration.

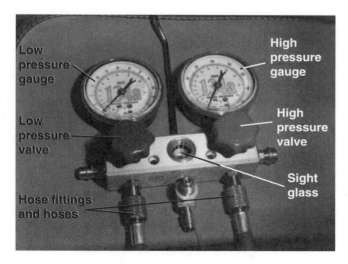

Figure 5–36 A gauge and manifold set allows you to read air conditioning system pressures to make sure the major parts are working normally. This gauge is set for new R-134a-filled systems and it reads slightly higher pressures than R-12 gauges.

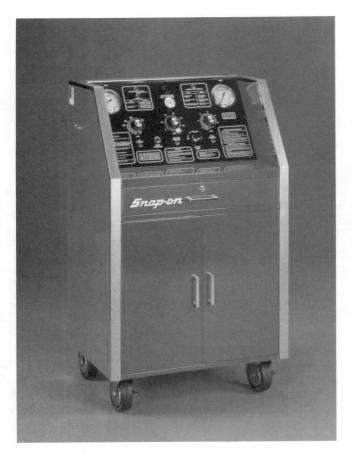

Figure 5–38 A charging station has gauges, a refrigerant tank, timing, scale, and other parts for automatic refilling of air conditioning systems. *(Courtesy of Snap-on Tools Corporation)*

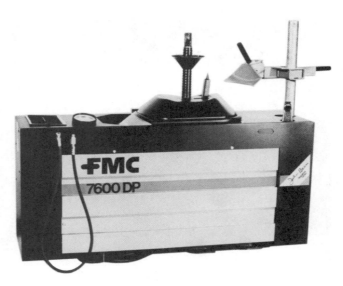

Figure 5–40 A tire changer will remove tires from wheels quickly and easily. *(FMC Automotive Equipment Division)*

A *wheel alignment machine* is used to measure wheel alignment angles so they can be adjusted. After collision repairs, the wheels of the vehicle often need realignment. The wheel alignment machine will show actual wheel angles so that they can be adjusted back to specifications, Figure 5–41.

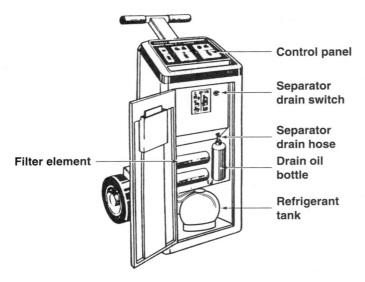

Control panel

Separator drain switch

Separator drain hose

Drain oil bottle

Filter element

Refrigerant tank

Figure 5–39 A recovery station is used to capture refrigerant from vehicles. Refrigerant damages the environment if released into the air. *(Courtesy of Snap-on Tools Corporation)*

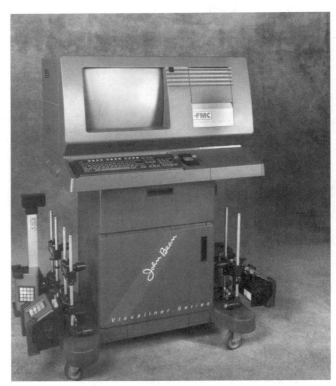

Figure 5–41 Wheel alignment machines vary. They measure the angle of each wheel so that wheel alignment angles can be adjusted after repairs. *(FMC Automotive Equipment Division)*

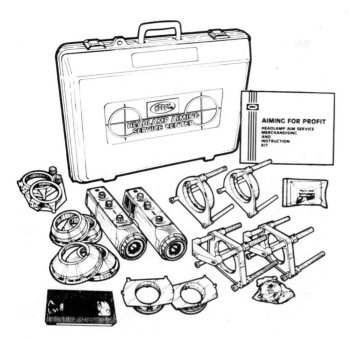

Figure 5–42 Headlight aiming equipment will let the technician quickly adjust the direction of a vehicle's headlights. Some use float-type bubbles, others use electronic indicators. *(Hopkins Manufacturing Corporation)*

Headlight aimers are used to adjust the direction of the vehicle headlights. Shown in Figure 5–42, they normally mount over the headlights. Leveling bulbs in the aimers will let you know if each bulb is aiming too high, too low, or correctly. By turning headlight aiming screws on the headlights, you can readjust the beams to aim correctly in front of the vehicle.

Summary

- Power tools and equipment use electrical energy, compressed air, or hydraulic power for operation.
- The use of air-driven or electric power tools requires safety glasses or a face shield. For hazardous operations, wear both.
- Some of the power tools used in collision repair are air compressors, air pressure regulators and filters, air blow guns, impact wrenches, air hammers, drills, air cutters, air sanders, grinders, polishers, blasters, and spray guns.
- Pneumatic tools should be lubricated BEFORE AND AFTER use to prevent wear and rusting of the vane motor and other parts of the tools.
- Volatile organic compounds (VOCs) are harmful substances produced by many painting and refinishing materials. Some geographic regions require the use of high transfer efficiency spray equipment to cut VOC emissions.

- The personal computer (PC) has become an important tool for accomplishing various collision repair tasks. It is used to keep track of business transactions, complete damage reports, and even perform electrical tests on vehicles.

Technical Terms

air supply system
air compressor
compressor air filter
compressor circuit
  breaker
compressor drain valve
automatic drain valve
compressor pressure
  sensor
compressor pressure
  relief valve
air tank shut-off valve
compressor oil plugs
shop air lines
air hoses
air hose sizes
air couplings
air pressure regulator
air line filter-drier
air filter drain valve
air tools
air blow gun
OSHA-approved blow
  gun
impact wrench
impact socket
impact adapter
½-inch impact wrench
⅜-inch impact wrench
air ratchet
air hammer
air chisel
drill
air drill
drill bit
key
chuck
hole saw
spot weld drill
air nibbler
air cutoff tool
air saw
air sander
circular sander

safety trigger
oscillating sander
orbital sander
dual-action sander
sander pad
air file
grinder
disc-type grinder
bench grinder
air polisher
paint gloss
cloth or cotton polishing
  pad
foam rubber polishing
  pad
swirl marks
pad cleaning tool
air tool lubrication
in-line oiler
blaster
hand blaster
cabinet blaster
spray gun
atomize
gun body
spray gun cup
fluid control valve
fluid needle valve
air control valve
spray gun air cap
spray pattern
gravity-feed spray gun
siphon spray gun
air vent hole
pressure-pot spray gun
pressure-cup spray gun
pressure-tank spray gun
bleeder spray gun
nonbleeder spray gun
touch-up spray gun
volatile organic
  compounds (VOCs)
high-efficiency spray gun
HVLP

LVLP
transfer efficiency
high transfer efficiency
turbine spray system
air conversion unit
step-down gun
electrostatic spray gun
primer
spray booth
air makeup system
air replacement system
downdraft spray booth
side draft spray booth
crossflow spray booth
color correcting lights
paint drying room
infrared drying
  equipment
surface thermometer
measuring system
hydraulic equipment
straightening system
frame rack
pulling post
frame straightening
  accessories
body jack
body dolly
AC
battery charger
DC
multimeter
analog meter
digital meter

low impedance
high impedance
inductive meter
scanner
scan tool
ALDL
scanner cartridge
malfunction indicator
  light
trouble code number
On-Board Diagnostics II
engine analyzer
personal computer
PC
soldering gun
solder
rosin core solder
rosin
acid core solder
plastic welder
air conditioning gauge
service port
leak detector
charging station
recovery station
recovery-charging
  station
tire changer
pneumatic
wheel balancer
wheel weights
wheel alignment machine
headlight aimers

Review Questions ▬▬▬

1. List the major parts in a shop's air supply system.

2. What is a shop air compressor?

3. This valve must be opened every night to remove moisture from the compressor's storage tank.
   a. Pressure relief valve
   b. Drain valve
   c. Bypass valve
   d. Oil valve

4. How do you use air hose couplings?

5. What is the purpose of an air pressure regulator?

6. Why is air tool lubrication important and how do you do it?

7. An _____ is a portable, hand-held, reversible air-powered wrench for rapid turning of bolts and nuts.

8. It is OK to use conventional, chrome plated sockets on impact wrenches. True or False, and why?

9. Describe some common uses for a ½-inch impact wrench.

10. A key is used to tighten the _____. The _____ has movable jaws that close down and hold the drill bit.

11. This type of sander moves in two directions as it moves, producing a much smoother surface finish.
    a. Circular
    b. Reciprocating
    c. File
    d. Orbital

12. A spray gun must _____ the liquid, that is, break the liquid into fine mist of droplets.

13. List and describe the major parts of a typical paint spray gun.

14. A siphon cup type spray gun is not working properly. Paint is not spraying out of the gun when the trigger is pressed. *Technician A* says that there may be too much solvent in the cup. *Technician B* says to check for a clogged vent hole. Who is correct?
    a. Technician A
    b. Technician B
    c. Neither technician
    d. Both technicians

15. What are some potential advantages of a remote pot or tank type spray gun?

ASE-Style Review Questions ▬▬▬

1. Which of the following is the most common maintenance task on an air compressor?
   a. Change oil
   b. Change filters
   c. Open tank drain
   d. Adjust belt

2. For safety, normal shop air pressure should not exceed:
   a. 50 psi
   b. 120 psi
   c. 150 psi
   d. 175 psi

3. *Technician A* says you have to unscrew quick disconnect fittings. *Technician B* says to pull back on the fitting sleeve. Who is correct?

a. Technician A
b. Technician B
c. Both Technicians A and B
d. Neither Technician

4. Small amounts of water come out of the spray gun during refinishing. Which of the following could be the source of the problem?
   a. High humidity
   b. Too much air pressure
   c. Contaminated filter-drier
   d. Contaminated respirator air supply

5. *Technician A* says it is not safe to blow yourself off with an air nozzle. *Technician B* says that serious injury can result if air enters your bloodstream. Who is correct?
   a. Technician A
   b. Technician B
   c. Both Technicians A and B
   d. Neither Technician

6. Large bolts must be tightened to secure a truck bed to the frame. Which drive size should be used?
   a. ⅛ inch
   b. ¼ inch
   c. ⅜ inch
   d. ½ inch

7. *Technician A* says that a foam pad will produce fewer swirl marks than a cotton pad. *Technician B* says that a finer compound will also reduce swirl marks. Who is correct?
   a. Technician A
   b. Technician B
   c. Both Technicians A and B
   d. Neither Technician

8. *Technician A* says to always store the key in the drill press chuck. *Technician B* says that this practice is dangerous. Who is correct?

a. Technician A
b. Technician B
c. Both Technicians A and B
d. Neither Technician

9. Which type of air sander should be used for initial removal of plastic filler to help quickly level the surface?
   a. Orbital sander
   b. Circular sander
   c. Belt sander
   d. Oscillating sander

10. Which sandpaper system allows you to change sandpaper more quickly and also allows you to reuse sandpaper?
    a. Velcro system
    b. Adhesive system
    c. Self-stick system
    d. Epoxy system

Activities

1. Tour your shop and note the location of all power tools and equipment. Discuss with the class any questions you might have on power tool or equipment use.

2. Visit an outside shop. Watch technicians as they use power tools and equipment. Have a class discussion about how technicians used the tools.

3. Disassemble and reassemble a shop spray gun. Note the condition of all parts. Report on the condition of the gun.

4. Inspect your shop air compressor. Find the tank drain valve, shut-off valve, and other parts.

6

Basic Measurement and Service Information

Objectives

After studying this chapter, you should be able to:

- Explain the many types of measurements needed in collision repair.

- Make accurate linear, angle, pressure, volume, and other measurements.

- Compare SAE and metric measuring systems.

- Identify and use basic measuring tools common to collision repair.

- Use conversion charts.

- Properly mix paints and other materials.

- Summarize how to measure with a paint mixing cup.

- Make paint thickness measurements.

- Use printed and computerized service information.

Introduction

Measurements are number values that help control processes in collision repair. Measurements are needed to evaluate structural damage, correct that damage, mix paint, determine paint thickness, adjust a spray gun, and perform numerous other tasks. If you cannot make accurate measurements, you will not be a successful collision repair technician. See Figure 6-1.

Remember that vehicles are precision manufactured using the latest technology. This includes the use of robotic assembly and laser measurement technology. This creates tighter tolerances than ever before. In fact, accurate measurement is one of the most important aspects of collision repair.

Vehicle manufacturers give *specifications,* or measurements for numerous body dimensions and mechanical parts. In the course of your work you will have to refer to and understand these factory specifications. This chapter will help prepare you for these factory specifications.

This chapter will summarize the basic types of measurements you will use in collision repair. It will also describe how to use various forms of published service information that are needed to complete repairs properly.

CUSTOMARY	CONVERSION	METRIC	CUSTOMARY	CONVERSION	METRIC
Multiply	by	to get equivalent number of:	Multiply	by	to get equivalent number of:
LENGTH			**ACCELERATION**		
Inch	25.4	millimeters (mm)	Foot/sec2	0.304 8	meter/sec2 (m/s2)
Foot	0.304 8	meters (m)	Inch/sec2	0.025 4	meter/sec2
Yard	0.914 4	meters	**TORQUE**		
Mile	1.609	kilometers (km)	Pound-inch	0.112 98	newton-meters (N·m)
AREA			Pound-foot	1.355 8	newton-meters
Inch2	645.2	millimeters2 (mm2)	**POWER**		
	6.45	centimeters2 (cm2)	Horsepower	0.746	Kilowatts (kW)
Foot2	0.092 9	meters2 (m2)	**PRESSURE OR STRESS**		
Yard2	0.836 1	meters2	Inches of water	0.249 1	kilopascals (kPa)
VOLUME			Pounds/sq. in.	6.895	Kilopascals
Inch3	16 387.	mm3	**ENERGY OR WORK**		
	16.387	cm3	BTU	1055.	Joules (J)
	0.016 4	liters (l)	Foot-pound	1.355 8	joules
Quart	0.946 4	liters	Kilowatt-hour	3 600 000.	joules (J = one W's)
Gallon	3.785 4	liters		or 3.6 x 10⁶	
Yard3	0.764 6	meters3 (m3)	**LIGHT**		
MASS			Foot candle	1.076 4	lumens/meter2 (lm/m2)
Pound	0.453 6	kilograms (kg)	**FUEL PERFORMANCE**		
Ton	907.18	kilograms (kg)	Miles/gal	0.425 1	kilometers/liter (km/l)
Ton	0.907	tonne (t)	Gal/mile	2.352 7	liter/kilometer (l/km)
FORCE			**VELOCITY**		
Kilogram	9.807	newtons (N)	Miles/hour	1.609 3	Kilometers/hr. (km/h)
Ounce	0.278 0	newtons			
Pound	4.448	newtons			
TEMPERATURE					
Degree Fahrenheit	(†°F-32) ÷ 1.8	degree Celsius (C)			

°F: -40 0 32 40 80 98.6 120 160 200 212
°C: -40 -20 0 20 37 40 60 80 100

Figure 6-1 This chart gives conversion factors for SAE and metric systems. Read through to compare the values for each measurement. *(Courtesy of General Motors Corporation, Service Operations)*

SAE AND METRIC SYSTEMS

The *Society of Automotive Engineers measuring system (SAE),* also called the English, U.S., customary, or conventional system, was first developed using human body parts as the basis for measurements. The length of the human arm was used to standardize the yard and the human foot devised the foot. It is primarily used in the United States, but NOT in other countries. This makes it important for you to understand both systems. Specifications are usually given in both SAE and metric values.

The SAE system uses fractions and decimals for giving number values. *Fractions* are acceptable. They divide the inch into thirty-seconds, sixteenths, and larger parts of an inch. *Decimals* are used when high precision is important. They can be used to divide an inch into tenths, hundredths, one-thousandths, ten-thousandths, and more divisions.

The *metric measuring system,* also called the *scientific international (SI) system,* uses a power of ten as its base. It is a simpler system than our conventional system. This is because multiples of metric units are related to each other by the factor **ten.** Every metric unit can be multiplied or divided by a factor of ten to get larger units (multiples) or smaller units (submultiples). There is less chance for a math error when using the metric system.

The United States passed the **Metric Conversion Act,** which committed to a voluntary and gradual transition from its customary system to the metric system. To date, only a few industries—including automotive—have made any movement toward adopting the metric system. Shop and repair manuals give values in both SAE and metric numbers.

CONVERSION CHARTS

Conversion charts are handy for changing from one measuring system to another or from one value to another.

An **SAE-metric conversion chart** allows you to change numbers to the other system. It gives multipliers that can be used to convert from SAE to metric or from metric to SAE values. For example, if a value is given in metric and you want to measure with a conventionally marked measuring tool, you would use a conversion chart.

A **decimal conversion chart** allows you to quickly change from fractions, to decimals, to millimeters. This is often used for various tasks, such as selecting drill bits. One is shown in Figure 6-2.

LINEAR MEASUREMENTS

Linear measurements are straight line measurements of distance. They are commonly used when evaluating major structural damage after a collision. There are many types of tools used for linear measurements.

SCALES

A **scale** or **ruler** is the most basic tool for linear measurement. It has an accuracy of approximately ¼ in or 0.5 mm. You may wish to study how to read a conventional rule in Figure 6-3.

An **SAE rule** often has markings in fractions of an inch (½, ¼, ⅛, ¹⁄₁₆) or in decimal parts of an inch (0.10,

FRACTION	DECIMAL	MILLIMETERS
1/64	.01563	.3969
1/32	.03125	.7938
	.03937	1.0000
3/64	.04688	1.1906
1/16	.06250	1.5875
5/64	.07813	1.9844
	.07874	2.0000
3/32	.09375	2.3813
7/64	.10938	2.7781
	.11811	3.0000
1/8	.12500	3.1750
9/64	.14063	3.5719
5/32	.15625	3.9688
	.15748	4.0000
11/64	.17188	4.3656
3/16	.18750	4.7625
	.19685	5.0000
13/64	.20313	5.1594
7/32	.21875	5.5563
15/64	.23438	5.9531
	.23622	6.0000
1/4	.25000	6.3500
17/64	.26563	6.7469
	.27559	7.0000
9/32	.28125	7.1438
19/64	.29688	7.5406
5/16	.31250	7.9375
	.31496	8.0000
21/64	.32813	8.3344
11/32	.34375	8.7313
	.35433	9.0000
23/64	.35938	9.1281
3/8	.37500	9.5250
25/64	.39063	9.9219
	.39370	10.0000
13/32	.40625	10.3188
27/64	.42188	10.7156
	.43307	11.0000
7/16	.43750	11.1125
29/64	.45313	11.5094
15/32	.46875	11.9063
	.47244	12.0000
31/64	.48438	12.3031
1/2	.50000	12.7000

FRACTION	DECIMAL	MILLIMETERS
	.51181	13.0000
33/64	.51563	13.0969
17/32	.53125	13.4938
35/64	.54688	13.8906
	.55118	14.0000
9/16	.56250	14.2875
37/64	.57813	14.6844
	.59055	15.0000
19/32	.59375	15.0813
39/64	.60938	15.4781
5/8	.62500	15.8750
	.62992	16.0000
41/64	.64063	16.2719
21/32	.65625	16.6688
	.66929	17.0000
43/64	.67188	17.0656
11/16	.68750	17.4625
45/64	.70313	17.8594
	.70866	18.0000
23/32	.71875	18.2563
47/64	.73438	18.6531
	.74803	19.0000
3/4	.75000	19.0500
49/64	.76563	19.4469
25/32	.78125	19.8438
	.78740	20.0000
51/64	.79688	20.2406
13/16	.81250	20.6375
	.82677	21.0000
53/64	.82813	21.0344
27/32	.84375	21.4313
55/64	.85938	21.8281
	.86614	22.0000
7/8	.87500	22.2250
57/64	.89063	22.6219
	.90551	23.0000
29/32	.90625	23.0188
59/64	.92188	23.4156
15/16	.93750	23.8125
	.94488	24.0000
61/64	.95313	24.2094
31/32	.96875	24.6063
	.98425	25.0000
63/64	.98438	25.0031
1	1.00000	25.4000

Figure 6-2 A decimal conversion chart will let you quickly change fractions to decimals or to metric equivalents. *(Illustration Courtesy of Perfect Circle/Dana Corp.)*

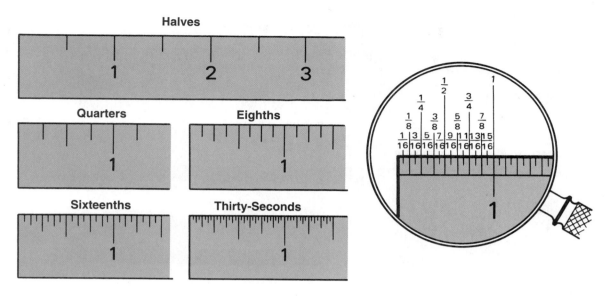

Figure 6-3 Review the divisions of a ruler. Most read in sixteenths or sixty-fourths of an inch.

0.20, 0.30, 0.40). A **metric rule** or **meter stick** is marked in millimeters and centimeters. The numbered lines usually equal ten millimeters (one centimeter). In Figure 6-4, see how to read one.

Figure 6-5 compares fractional, metric, and decimal inch scales.

Parallax error results when you read a rule or scale from an angle, instead of looking straight down. Viewing at an angle causes you to read the wrong line on the scale. Always look straight down when reading a rule.

A **pocket scale** or rule is very small (typically 6 in or 152 mm long). It will clip into your shirt pocket and can be handy for numerous small measurements. A **yardstick** or **meter stick** may also be used for some larger linear measurements.

A **tape rule** or **tape measure** will extend out for making very long measurements. A tape measure is commonly used to make large distance measurements during body damage evaluation.

Since many body measurements are made at holes in the body, the tip of the tape rule must be small or ground to a point. This will prevent measurement errors.

A **tram gauge** is a special body dimension measuring tool, Figure 6-6. It is usually a lightweight

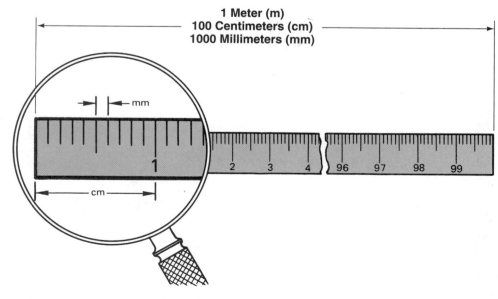

Figure 6-4 The small gradations on a metric rule equal millimeters. Numbered lines equal centimeters.

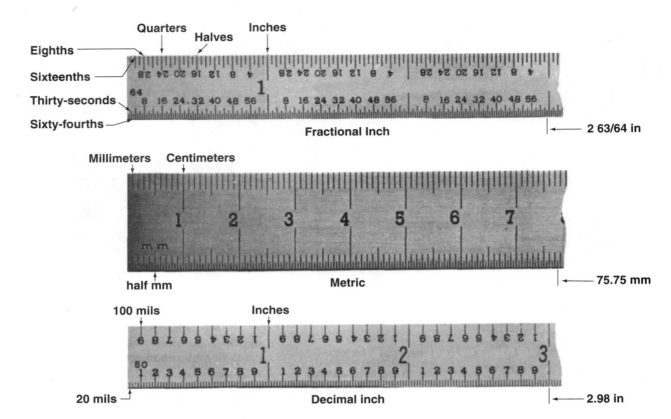

Figure 6-5 Compare the values on fractional inch, metric, and decimal inch scales.

frame with pointers. The pointers can be aligned with body dimension reference points to determine the direction and amount of body misalignment damage.

NOTE! More information on the use of tram gauges is given in Chapter 20, Measuring Vehicle Damage.

DIVIDERS AND CALIPERS

Dividers have straight, sharp tips for taking measurements or marking parts for cutting. In collision repair work, dividers are sometimes used for layout or marking cut lines. They will scribe circles and lines on sheet metal and plastic. They will also transfer and

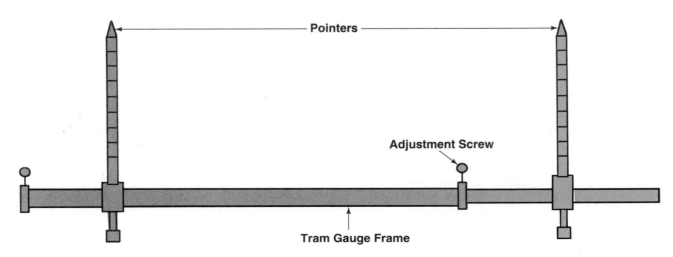

Figure 6-6 A tram gauge is a specialized measuring tool used in collision repair. It will quickly measure between body reference points to see if frame or unibody straightening is required.

Types of Calipers

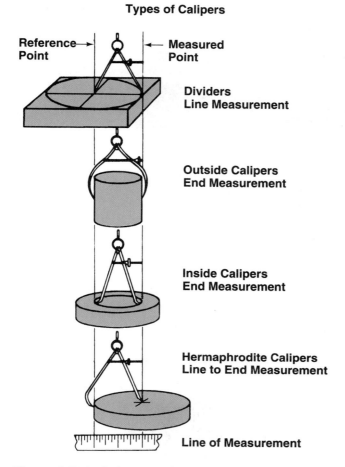

Figure 6-7 Study the types of measurements made with different kinds of calipers.

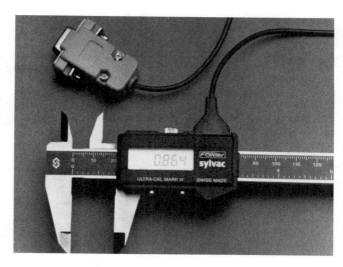

Figure 6-8 Digital sliding calipers are easy to use and read. They make outside, inside, and depth measurements. (*Courtesy of Fred V. Fowler Co., Inc.*)

make surface measurements. Dividers are sometimes used when fabricating repair pieces for corrosion repair. See Figure 6-7.

Outside calipers will make rough external measurements when ¼ in (0.40 mm) is accurate enough. The calipers are placed over the outside of parts. They are adjusted to just touch the parts. When placed next to a rule, part size can be determined. See Figure 6-7.

Inside calipers are designed for measuring inside of parts. They are used the same way. They are also accurate to about ¼ in (0.40 mm). See Figure 6-7.

Hermaphrodite calipers have one straight tip and one curved tip. They will measure cylindrical shapes easily.

Sliding calipers will measure inside, outside, and depth dimensions with a high degree of accuracy. Most sliding calipers measure to accuracy of 0.001 in (0.025 mm). It is a very convenient tool, especially digital readout, electronic calipers, when high accuracy is necessary. See Figure 6-8.

MICROMETERS

Micrometers are sometimes used for measuring mechanical parts when great precision is important. Often called a "mike," it can easily measure one-thousandths of an inch (0.001 in) or one-hundredths of a millimeter (0.01 mm).

You might use one to measure the thickness of a brake system rotor. If worn too thin, this would tell you to replace the rotor. The basic parts of an outside micrometer are shown in Figure 6-9.

To learn to read a micrometer, See Figure 6-10. It details the steps for reading an SAE micrometer. The procedure for reading a metric micrometer is in Figure 6-11.

A digital readout micrometer gives readings as a decimal in a number window on the tool. It is much faster than having to use several steps to calculate the measurement.

When using a micrometer, hold the frame in the palm of your hand. Rotate the thimble with your thumb and finger. Adjust the tool down until it drags lightly on the part. Do NOT tighten it. Then, read the micrometer.

Check the accuracy of a micrometer if dropped or after extended use. Measure a known-size feeler gauge and check the micrometer's reading or give it to your tool salesperson for an accuracy check.

DIAL INDICATORS

A **dial indicator** is often used to measure part movement and out-of-round in thousandths of an inch

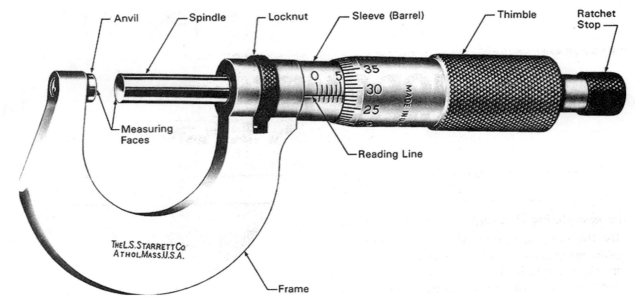

Figure 6-9 Note the parts of an outside micrometer. *(Courtesy of the L. S. Starrett Company)*

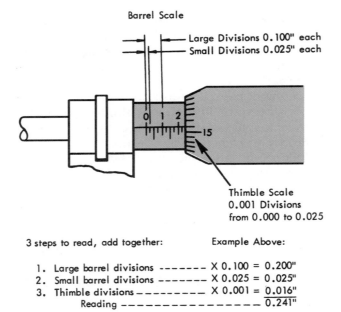

3 steps to read, add together: Example Above:

1. Large barrel divisions ------- X 0.100 = 0.200"
2. Small barrel divisions ------- X 0.025 = 0.025"
3. Thimble divisions ---------- X 0.001 = 0.016"
       Reading ----------------- 0.241"

Figure 6-10 Study the steps for reading an SAE micrometer. Basically, you must add the whole divisions showing on the barrel scale with the thousandths showing on the thimble scale.

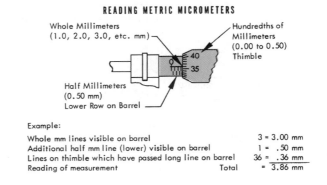

Figure 6-11 Study the steps for reading a metric micrometer.

Example:

Whole mm lines visible on barrel	3 = 3.00 mm
Additional half mm line (lower) visible on barrel	1 = .50 mm
Lines on thimble which have passed long line on barrel	36 = .36 mm
Reading of measurement Total	= 3.86 mm

(hundredths of a mm). For example, the indicator might be used to check for wheel damage. It is mounted against the wheel or tire. When the tire is rotated, the dial indicator will show how much the wheel is bent. This would let you know that the wheel should be replaced. See Figure 6-12.

To use a dial indicator, mount the tool base so it will not move. A magnetic base will stick to metal parts or a metal plate placed on the floor. Position the tool arm and indicator stem against the part. Turn the outside of the dial to zero the needle. Move or rotate the part and take your reading.

FEELER GAUGES

Feeler gauges measure small clearances inside parts. Blade thickness is given on each blade in thousandths of an inch (0.001, 0.010, 0.020) or in hundredths of a millimeter (0.01, 0.07, 0.10).

Flat feeler gauges are for measuring between parallel surfaces. ***Wire feeler gauges*** are round and are for measuring slightly larger distances between nonparallel or curved surfaces. See Figure 6-13, page 112.

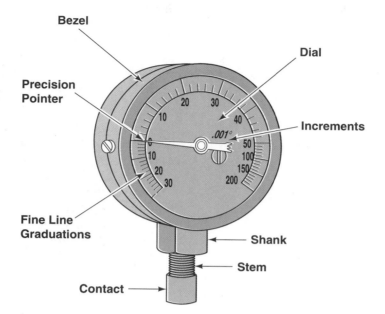

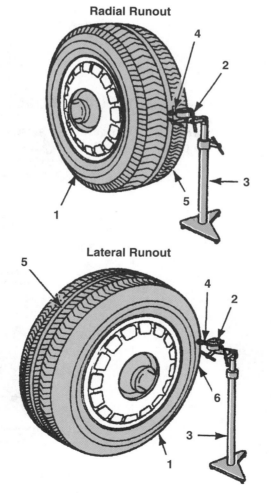

1. Tire and Wheel Asm. Mounted on Car or Balance Machine

2. Dial Indicator

3. Stand

4. Roller Wheel

5. Run Tape Around Center Tread of Tire. Measure Radial Runout of Taped Surface.

6. Measure Lateral Runout at Point Where Tread Ends. Position Dial Indicator Perpendicular To Tire at That Point.

Figure 6-12 A dial indicator is commonly used for measuring part runout or movement. (A) Note the parts of a dial indicator. *(Danaher Tool Group)* (B) Mount the indicator stand next to the part (in this case, a tire). Position the stem of the indicator against the tire and zero the needle. Slowly turn the tire to measure for out-of-round, which might mean a bad tire or a bent wheel.

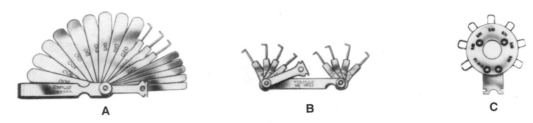

Figure 6-13 Feeler gauges are used to measure the clearance between parts. (A) A set of flat feeler gauges. (B) Wire feeler gauges. (C) Spark plug gap feeler gauges. *(Courtesy of Snap-on Tools Company)*

PRESSURE AND FLOW MEASUREMENTS

You will have to make pressure and flow measurements when working. Pressures are important for air tools, spray guns, and other equipment. Adequate airflow is important when working in a paint spray booth, for example.

A *pressure gauge* reads in pounds per square inch (psi), kilograms per square centimeter (kg/cm^2), or kilopascals (kPa). A few pressure gauges also show or measure vacuum. Note the two scales given on the pressure gauge in Figure 6-14.

Air pressure gauges are used on the shop's air compressor, on pressure regulators, or at the spray gun. Tire pressure gauges are also needed to check for proper tire inflation. Hydraulic pressure gauges can be found on hydraulic presses and other similar equipment.

Remember! Excessive pressure can be dangerous or damage parts. Low air pressure may keep the tool from working properly.

A *vacuum gauge* reads vacuum, negative pressure, or "suction." It reads in inches of mercury (in Hg) or kilograms per square centimeter (kg/cm^2).

A *flow meter* measures the movement of air, gas, or liquid past a given point. A *manometer* or airflow meter, is designed for use in a paint booth. Airflow meters can register in feet per minute or meters per minute. Liquid flow meters often register in gallons per hour or liters per hour. See Figure 6-15.

ANGLE MEASUREMENTS

Angle measurements divides a circle into 360 parts, called **degrees.** You will learn to read angles in degrees when doing wheel alignment, for example. This will be explained in later chapters.

TEMPERATURE MEASUREMENTS

Temperature is usually measured in degrees Fahrenheit (F) or Celsius (C). In the collision repair shop, temperatures are important in the drying of various materials such as primers, paints, and adhesives.

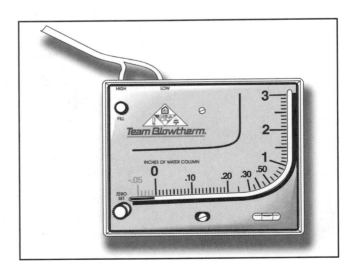

Figure 6-15 Manometers are used to measure the airflow through air circulation systems in spray booths and paint mixing rooms. Adequate airflow is needed to prevent a buildup of paint mist or fumes in the air. *(Photo courtesy of Team Blowtherm)*

Figure 6-14 Pressure gauges are common measuring tools in collision repair. Most have two scales to show both SAE and metric values. *(Photo courtesy of ITW Automotive Refinishing, 1-800-445-3988)*

Thermometers are used to measure temperature. For example, paint manufacturers often give different mixing values and recommend different thinners and reducers for different **ambient** (outside or room) temperatures.

A *mercury thermometer* encloses liquid mercury in a hollow glass rod to indicate temperature. Any change in temperature makes the mercury expand or contract. To read a thermometer, view the level of mercury next to the scale.

A *dial thermometer* is similar, but it has a needle that shows temperature. This type is often used when checking the output of an air conditioning system. It is inserted into the center vent with the system and blower on high. If the reading is too warm (above specifications), the system is not cooling properly.

Surface temperature thermometers are available in several different designs—magnetic thermometers, paper thermometers, digital thermometers.

Before applying any type of material during collision repairs, the vehicle and the material to be used, must be at room temperature. A vehicle's surface temperature must be brought to room temperature prior to applying fillers and paint materials. Failure to do so may result in formation of condensation on the surface.

Magnetic surface thermometers will stick to metal body panels for measuring temperatures during paint drying or curing. When using infrared heaters after painting, keep in mind that magnetic surface thermometers have more mass than the panel they are attached to. This may result in a lower temperature reading than the panel.

Disposable **paper thermometers** will show the highest temperature a surface reaches, but not necessarily the current temperature of the panel. They work well with infrared heaters.

A **digital thermometer** is an electronic instrument for measuring temperature. It has a digital readout that will show temperature in Fahrenheit or Celsius. Special surface types are often used in collision repair.

Most digital thermometers usually have several different probe tips. A round, hollow tip is for air temperature measurements. A flat, blunt tip is for surface temperature measurements. A tip with a strand of wire is for checking temperatures of fluids.

PAINT-RELATED MEASUREMENTS

Mixing scales are used by paint suppliers and technicians to weigh the various ingredients when mixing paint materials. Scales will precisely weigh out each ingredient. Using paint manufacturers formulations, mixing scales are used to add different pigments or other materials to make the paint the desired color.

Manual mixing scales simply weigh each paint ingredient as it is poured into the mixing cup on the scale. You must look up the paint formula to see how much of each material is needed and carefully pour out that amount while watching the scale. Most manual scales have been replaced by computerized scales.

Electronic mixing scales help automate paint mixing. After the technician programs in the amount of paint needed for the area to be refinished, a computerized scale will state how much of each material to pour into the mixing cup. It will state how much pigment, tint, flake, and binder to use by weight. As you pour each ingredient into the cup sitting on the scale, the scale will go to zero. Once zeroed, you have added the correct amount of that ingredient. The computerized scale will then prompt you to add a specific amount of the next ingredient. This will allow you to more quickly produce the correct mix of ingredients for the specific color paint. See Figure 6-16.

When mixing and using paint and solvents or other additives, you must measure and mix their contents accurately. This is essential to doing good paint and body work. You must be able to properly mix reducers or thinners, hardeners, and other additives into the paint. If you do NOT, serious paint problems will result.

A *graduated pail* is often used to measure liquid materials when mixing. It has measurement lines on it like a kitchen measuring cup. Liquid is poured into the pail until it is next to the scale for the amount needed.

Mixing instructions are normally given on the material's label. This might be a percentage or parts of one ingredient compared to the other.

A *percentage reduction* means that each material must be added in certain proportions or parts. For instance, if a color requires 50 percent reduction, this means that 1 part reducer (solvent) must be mixed with 2 parts of color.

Mixing by parts means that for a specific volume of paint or other material, a specific amount of another material must be added. If you are mixing a gallon of color in a spray gun pressure tank for example, and directions call for 25 percent reduction, you would add 1 quart of reducer. There are 4 quarts in a gallon and you want 1 part or 25 percent reducer for each 4 parts of color.

Proportional numbers denote the amount of each material needed. The first number is usually the parts of paint needed. The second number is usually the solvent (or reducer). A third number might be used to denote the amount of hardener or other additives required.

For example, the number 2:1:1 might mean add 1 part solvent and 1 part hardener to 2 parts of

A

B

Figure 6-16 (A) Electronic scales are used by paint supply stores and in-shop mixing rooms to mix ingredients in color and other materials. *(Sartorious)* (B) A computerized scale gives a readout showing how much of each component is needed to make a certain color.

color. For a half gallon of color, you would add a quart of solvent and a quart of catalyst. This can vary, so always refer to the exact directions on the materials.

A **mixing chart** converts a percentage into how many parts of each material must be mixed. One is given in Figure 6-17. Study the percentages and parts of each material that must be mixed.

PAINT MIXING STICKS

Graduated **paint mixing sticks** have conversion scales that allow you to easily convert ingredient percentages into part proportions. They are used by painters to help mix colors, solvents, catalysts, and other additives right before spraying. See Figure 6-18.

Paint mixing sticks should NOT be confused with **paint stirring sticks** (wooden sticks) for mixing the contents after they are poured into the spray gun cup or a container.

VISCOSITY CUP MEASUREMENT

A **viscosity cup** is used to measure the thickness or fluidity of the mixed refinishing materials; paints (color), primers, sealers, or even clears. It is a small cup attached to a handle, Figure 6-19.

To use a viscosity cup, dip it into the mixed paint until submerged. Lift the cup out and hold it over the paint container. As soon as the cup is lifted out, start timing how long it takes for the cup to empty. The paint will leak out of a small specific size **orifice** (hole) in the bottom of the cup.

Use the second hand on your watch or a stop watch to time draining. When the fluid stream breaks into drops, note how much time was needed. This equals the viscosity in seconds.

The paint manufacturer will give a recommended viscosity value in **viscosity cup seconds.** It will vary between 17 and 30 seconds depending upon the type of material (color, primer, sealer, or clear) and type of cup used. Refer to the paint specifications for an exact value.

If the material drains too quickly out of the viscosity cup, you have added too much solvent. More material would be needed. If the cup drains too slowly, you have not added enough solvent. Remix until the material passes the viscosity cup test.

If the paint (or primer, clear, etc.) is TOO THICK, your paint will develop orange peel or a rough film. If it is TOO THIN, excess solvent can cause the paint to have poor hiding and other problems.

PAINT MATERIAL THICKNESS MEASUREMENT

Paint thickness is measured in **mils** or thousandths of an inch (hundredths of a millimeter). Original OEM (original equipment manufacture) finishes are typi-

Reduction Percentage		Reduction Proportions	Paint (color)		Solvent
20%	=	5 parts paint / 1 part solvent		20%	
25%	=	4 parts paint / 1 part solvent		25%	
33%	=	3 parts paint / 1 part solvent		33%	
50%	=	2 parts paint / 1 part solvent		50%	
75%	=	4 parts paint / 3 parts solvent		75%	
100%	=	1 part paint / 1 part solvent		100%	
125%	=	4 parts paint / 5 parts solvent		125%	
150%	=	2 parts paint / 3 parts solvent		150%	
200%	=	1 part paint / 2 parts solvent		200%	
250%	=	2 parts paint / 5 parts solvent		250%	

Figure 6-17 This chart shows a range of possible mixing percentages and converts each percentage into parts of each material. Study the paint and solvent parts needed for each percentage.

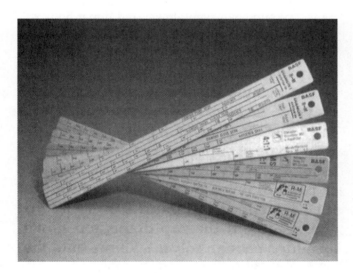

Figure 6-18 Mixing sticks have scales on them for mixing different materials—colors, solvents, and hardeners. They are a handy reference tool for measuring out needed materials. *(BASF)*

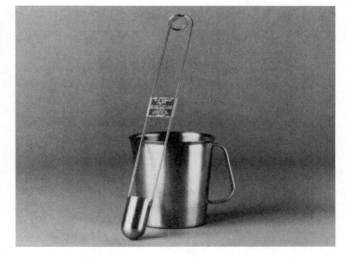

Figure 6-19 A viscosity cup will measure the consistency or fluidity (thickness) of spray materials. A mixed paint specification in seconds is often given. If you mix the paint properly, the amount of time needed for the sample to drain out of the small cup should equal the specified time. *(PPG Industries, Inc.)*

cally about 2 to 6 mils thick. With basecoat/clearcoats, the basecoat is approximately 1 to 2 mils thick. The clearcoat is about 2 to 4 mils thick. This is approximately the thickness of a piece of typing paper.

If a panel has been repainted, paint thickness will increase. If too much paint is already on the vehicle, it may have to be removed prior to refinishing.

Paint/material buildup should be limited to no more than 12 mils. The OEM finish and one refinish usually equal just under 12 mils. Exceeding this thickness could cause cracking in the new finish. Chemical stripping or blasting would be needed to remove the old paint buildup.

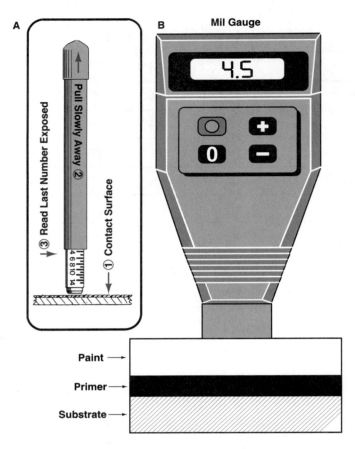

Figure 6-20 Paint thickness, or mil gauges will tell you if the paint already on the vehicle is too thick for repainting without stripping or paint removal. (A) Place the mil gauge against the panel. The magnet will be attracted to a steel panel. Slowly pull the tool away from the panel while reading the gauge, which shows paint thickness. *(Talsol Corporation)* (B) An electronic mil gauge will give an almost instant readout of paint thickness. *(I-CAR)*

A **mil gauge** can be used to measure the thickness of the paint on the vehicle. This can be done before refinishing, after refinishing, and during other finishing operations. See Figure 6-20.

There are three types of mil thickness gauges. One type, known as a **pencil mil gauge,** measures paint/refinishing materials which calibrated magnet and spring setup. The magnet is placed against painted steel components and then slowly pulled away. The tool contains a graduated scale that is exposed as the magnet sticks to the panel. The last number exposed on the scale before the magnet detaches from the panel is the paint film/material thickness in mils.

The **electronic mil gauge** shows mill thickness with a digital readout. Some electronic mil gauges can also measure mil thickness on nonmagnetic materials such as aluminum and composites.

VEHICLE IDENTIFICATION NUMBER (VIN) NUMBERS

A good technician must have a complete understanding of commonly used terms that identify parts and assemblies of a vehicle. If the technician is NOT familiar with this language, it is difficult to order parts and read a repair order.

The **Vehicle Identification Number (VIN) plate** is used to accurately identify the body style, model year, engine, and other data about the vehicle. For years, the VIN plate has been riveted to the upper left corner of the instrument panel, visible through the windshield. See Figure 6-21.

Prior to 1981 and on foreign vehicles, check the service manual for the location of the VIN, vehicle certification label, or body plate. Service manuals and collision estimating guides also contain all of the necessary VIN number decoding information.

Become familiar with each vehicle manufacturer's method of vehicle identification and the specific information it contains. Remember that you must obtain all of the information possible about the vehicle being worked on.

The **body ID number,** or **service part number** gives information about how the vehicle is equipped. It will give paint codes or numbers for ordering the right type and color paint; lower and upper body colors if the vehicle has two-tone paint. The body ID number will also give trim information. This number will be on the body ID plate on the door, console lid, or elsewhere on the body.

COLLISION REPAIR PUBLICATIONS

Various manuals or publications are used by a collision repair shop. It is important that you understand the purpose and use of each.

All automobile manufacturers publish yearly **service manuals** that describe the construction and repair of their vehicle makes and models. These manuals give important details on repair procedures and parts. Also called shop or repair manuals, they give instructions, specifications, and illustrations for their specific cars and trucks. Service manuals have information on mechanical as well as body repair.

The **contents page** of a service manual lists the broad categories in the manual and gives their page numbers. Each **service manual section** then concentrates on describing that area of repair.

Service manual abbreviations represent technical terms or words and save space. Each manufacturer uses slightly different abbreviations.

Aftermarket repair manuals are published by publishing companies (Mitchell Manuals, Motor Man-

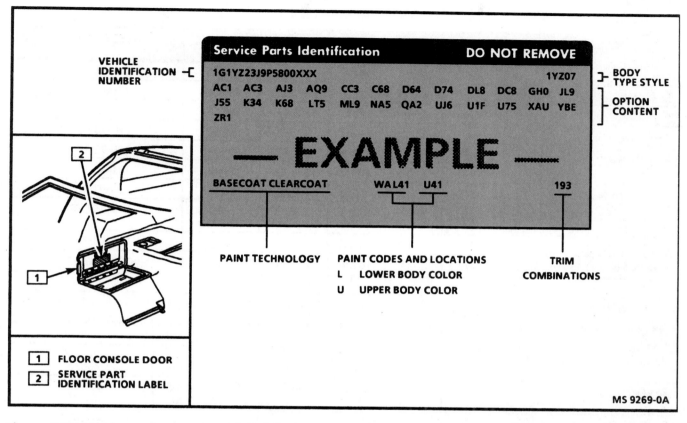

VEHICLE IDENTIFICATION NUMBER

Service Parts Identification — DO NOT REMOVE

1G1YZ23J9P5800XXX — 1YZ07 — BODY TYPE STYLE

AC1 AC3 AJ3 AQ9 CC3 C68 D64 D74 DL8 DC8 GH0 JL9
J55 K34 K68 LT5 ML9 NA5 QA2 UJ6 U1F U75 XAU YBE
ZR1 — OPTION CONTENT

— EXAMPLE —

BASECOAT CLEARCOAT — WAL41 U41 — 193

PAINT TECHNOLOGY

PAINT CODES AND LOCATIONS
L   LOWER BODY COLOR
U   UPPER BODY COLOR

TRIM COMBINATIONS

1   FLOOR CONSOLE DOOR
2   SERVICE PART IDENTIFICATION LABEL

MS 9269-0A

Figure 6-21 Vehicle manufacturers place labels or ID plates to give information about how the vehicle is equipped. You will need to refer to a service manual to locate and explain the number codes. *(Courtesy of General Motors Corporation, Service Operations)*

uals, Chilton Manuals) rather than the manufacturer. They can give enough of the information needed for most repairs.

Repair charts give diagrams that guide you through logical steps for making repairs. They can vary in content and design. Most use arrows and icons (graphic symbols) that represent repair steps.

Diagnosis charts or *troubleshooting charts* give logical steps for finding the source of problems. Mechanical, body, electrical, and other types of troubleshooting charts are provided in service manuals. They give the most common sources of problems for the symptoms being experienced.

Paint reference charts in service manuals give comparable paints manufactured by different companies. This will help you match the color of the new paint with the paint already on the vehicle. See Figure 6-22.

Collision estimating manuals or guides give information for calculating the cost of repairs. They have part numbers, prices, section illustrations, and other data to help the estimator. Discussed in later chapters, electronic or computer-based estimating guides are also available.

A *vehicle dimension manual* gives unibody and frame measurements of undamaged vehicles. Dimen-

sions are given for every make and model car and truck. These known good dimensions can be compared to actual measurements taken off of a damaged vehicle. This will let you know how badly the vehicle is damaged and what must be done to repair it. See Figure 6-23.

It is important to follow the directions in a dimension manual. See Figure 6-24. Note how it gives measurements for a door opening. You would measure across the door opening on the vehicle. If any measurements are not correct, you could determine how the impact affected the door opening. Frame straightening and/or portable hydraulic equipment could then be used to push or pull the opening back into alignment.

As with service manuals, various symbols are used in body dimension manuals. The most typical are in Figure 6-25.

COMPUTERIZED SERVICE INFORMATION

Computerized service information places service manuals, dimension manuals, estimating manuals, and other data on compact discs. This allows a

PAINT REPAIR PRODUCTS

COLOR NAME	CHRY. CODE	PPG	BASF	DuPONT	S-W ACME M-S	AKZO/SIKKENS
Black Cherry P.C.	FM9	4043	18221	B8822	37297	CHA88:FM9
Radiant Red Met. C.C.	LRF	4447	22116	B9230	45860	CHA92:LRF
Lt. Driftwood S.G.	MFA	4569	22110	B9263	46579	CHA92:MFA
Char-Gold S.G.	MJ8	4677	23044	B9327	46947	CHA93:MJ8
Emerald Green P.C.	PGF	4639	23042	B9328	46976	CHA92:PGF
Teal P.C.	LP5	4445	21094	B9232	45858	CHA92:LP5
Med. Water Blue S.G.	KBF	4270	21075	B9135	44042	CHA91:KBF
Black C.C.	DX8	9700	15214	99	34858 90-5950	CHA85:DX8
Ascot Gray C.C.	PAB	35560	24070	B9457	48759	CHA94:PAB
Bright White C.C.	GW7	4037	18238	B8833	37298	CHA88:GW7

COLOR NAME	CHRY. CODE	PPG/UCV	BASF	DuPONT	S-W ACME M-S
Agate	AZ	9856/2-1461	22135	C9208	45994
Agate/Med. Quartz (LAZ/HD5)	AD	9856/2-1461 34618/1346	22135 19133	C9208 C8904	45994/ 40075
Agate/Med. Driftwood (LAZ/MF6)	AF	9856/2-1461 27468/2-1502	22135 23061	C9208 C9301	45994/ 47481
Dk. Slate/Slate Blue (MBR/KB7)	BB	35586/2-1501 17125/2-1426	23060 21095	C9302 C9116	47457/ 44568
Dk. Driftwood/Med. Driftwood (LF8/MF6)	FF	27243/2-1463 27468/2-1502	22137 23061	C9247 C9301	45999 47481

LOWER FASCIA	CHRY. CODE	PPG	BASF	DuPONT	S-W ACME M-S	AKZO/SIKKENS
Med. Quartz (HD2)*	HD2	4321	20214	B8948	41562	CHA89:HD2

*All lower fascias that are not body color are medium quartz (HD2).

Figure 6-22 Service manuals will sometimes give charts for purchasing recommended repair materials. Simply use the body ID number for the paint color. Then find the paint number for that color in the chart. *(Courtesy of DaimlerChrysler)*

Figure 6-23 Vehicle dimension manuals give structural measurements. If the measurements on the damaged vehicle do not equal the specifications in the manual, unibody or frame straightening will be needed on a frame rack. *(Courtesy of Mitchell International, Inc.)*

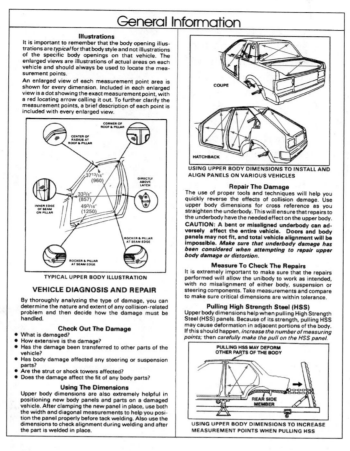

Figure 6-24 Note how measurements are given across specific points of the door opening. In a collision, the door opening can be damaged or forced out of alignment. Hydraulic equipment can then be used to realign (straighten) the door opening. *(Courtesy of Mitchell International, Inc.)*

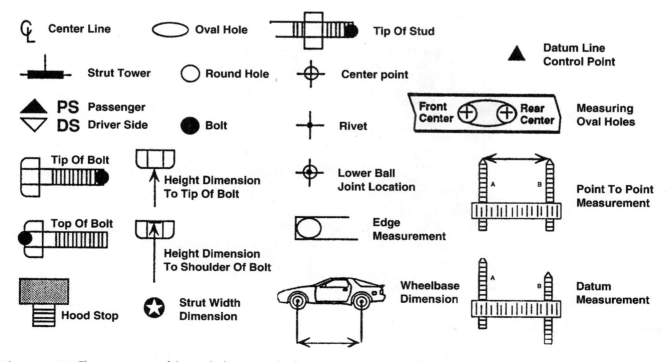

Figure 6-25 These are some of the symbols given in body repair manuals. Study them! *(Hein Werner® Corporation)*

personal computer to be used to more quickly look up and print the desired information. See Figure 6-26.

Most shops now have their service information on computer. This allows more efficient handling of shop operations. Estimating, parts ordering, bookkeeping, and the whole shop operation can be more closely monitored and controlled using computers.

For example, if a technician needs the dimensions for a specific vehicle being straightened on a frame rack, he or she can quickly pull up this information on computer because all of the data about the vehicle (VIN, year, make, model, etc.) have already been entered into the computer system. A printout of the vehicle dimensions can be made and taken out to the repair area. See Figure 6-27.

Figure 6-26 Service information in modern collision repair facilities is now on a computer. The technician or office personnel can retrieve repair instructions, dimension manuals, estimating programs, shop rate manuals, and other data more quickly from the computer than by looking them up in printed books.

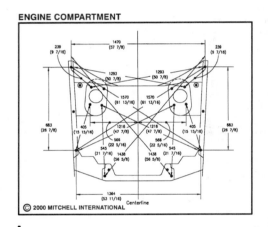

A

B

C

Figure 6-27 (A) This computer screen is displaying a page from a dimensions manual that would be needed to straighten frame or unibody damage on a frame rack. *(Courtesy of Mitchell International, Inc.)* (B) This screen shows an exploded view of a fender assembly, which might be needed when writing an estimate of damage. *(Courtesy of Mitchell International, Inc.)* (C) Here the computer is being used to write an estimate while looking at electronic photos of the damaged vehicle. *(Courtesy of Mitchell International, Inc.)*

Figure 6-28 Color matching manuals contain information that will help you match the original finish. *(BASF)*

COLOR MATCHING MANUALS

Color matching manuals contain information needed for finishing panels so that the repair has the same appearance as the finish. They have paint code information, color chips, blending and tinting data, tinting procedures, and other information. See Figure 6-28.

Summary

- Vehicle manufacturers give specifications, or measurements, for numerous body dimensions and mechanical parts. In the course of your work you will have to refer to and understand these factory specifications.

- The SAE measuring system is used primarily in the United States, while the metric system, also called the scientific international (SI) system, is used worldwide.

- Linear measurements are straight line measurements of distance. They are commonly used when evaluating major structural damage after a collision.

- Angle measurements are measured in degrees. Angle measurements are used in wheel alignment.

- Pressure measurements are measured in pounds per square inch (psi), kilograms per square centimeter (kg/cm^2), or kilopascals (kPa).

- Temperature is usually measured in degrees Fahrenheit (°F) or Celsius (°C). In the collision repair shop, temperatures are important in the drying of various materials such as primers, paints, and adhesives.

- The Vehicle Identification Number (VIN) is used to accurately identify the body style, model year, engine, and other data pertaining to the vehicle.

Technical Terms

measurements
specifications
Society of Automotive
    Engineers (SAE)
    measuring system
fractions
decimals
metric measuring system
scientific international
    (SI) system
conversion chart
SAE metric conversion
    chart
decimal conversion chart
linear measurement
scale
ruler
SAE rule
metric rule
meter stick
parallax error
pocket scale
yardstick
tape rule
tape measure
tram gauge
dividers
outside calipers
inside calipers
hermaphrodite calipers
sliding calipers
micrometer
dial indicator
feeler gauge
flat feeler gauge
wire feeler gauge
pressure gauge
vacuum gauge
flow meter
manometer
angle measurements
thermometer
ambient

surface temperature
    thermometer
mixing scales
manual mixing scales
electronic mixing scales
graduated pail
mixing instructions
percentage reduction
mixing by parts
proportional numbers
mixing chart
paint mixing stick
paint stirring stick
viscosity cup
orifice
viscosity cup seconds
paint thickness
mils
mil gauge
pencil mil gauge
electronic mil gauge
Vehicle Identification
    Number (VIN) plate
body ID number
service part number
service manuals
contents page
service manual
    abbreviations
aftermarket repair
    manuals
repair charts
diagnosis charts
troubleshooting charts
paint reference charts
collision estimating
    manuals
vehicle dimension
    manual
computerized service
    information
color matching
    manuals

Review Questions

1. _____ are number values that help control processes in collision repair.

2. Explain how you might use conversion charts.

3. How do you avoid parallax error when reading a scale?

4. In auto body, _____ will mark circles and lines on sheet metal and plastic.

5. Sliding calipers will measure _____, _____, and _____ dimensions to an accuracy of _____ in _____.

6. Explain how you use an outside micrometer.

7. This tool is often used to measure part movement and out-of-round in thousandths of an inch (hundredths of a mm).
   a. Outside micrometer
   b. Inside micrometer
   c. Feeler gauge
   d. Dial indicator

8. Describe the three types of tips available for digital thermometers.

9. If a paint requires 50 percent reduction, how would you mix the paint and its solvent?

10. When mixing refinishing materials, what might the numbers 2:1:1 mean?

11. A _____ is used to measure the actual consistency or fluidity of the mixed materials, usually paint.

12. Paint/material buildup should be limited to no more than this amount.
   a. 12 mils
   b. 2 mils
   c. 50 mils
   d. 100 mils

13. What can happen if the paint on the vehicle is too thick?

14. The _____ is used to accurately identify the body style, model year, engine, and other data about the vehicle.

15. This type of manual gives information for calculating the cost of collision repairs.
   a. Service manual
   b. Dimensions manual
   c. Estimating manual
   d. Paint code manual

ASE-Style Review Questions

1. Which of the following types of error results when you read a rule or scale from an angle?
   a. Compound
   b. Parallax
   c. Conversion
   d. Specification

2. *Technician A* is measuring between two holes to check for unibody damage using a tape measure with a small pointed tip. *Technician B* says to use a micrometer for more accurate measurement of the holes. Who is correct?
   a. Technician A
   b. Technician B
   c. Both Technicians A and B
   d. Neither Technician

3. Which of the following tools can easily measure one-thousandths of an inch (0.001 in) or one-hundredths of a millimeter (0.01 mm)?
   a. Dividers
   b. Tape measure
   c. Pocket rule
   d. Micrometer

4. Which of the following tools is often used to measure part movement and out-of-round in thousandths of an inch (hundredths of a millimeter)?
   a. Dividers
   b. Dial indicator
   c. Micrometer
   d. Tram gauge

5. Paint reduction calls for using 1 quart of reducer for each gallon of color. What reduction would this be?
   a. 125 percent
   b. 75 percent
   c. 25 percent
   d. 5 percent

6. Paint mixing instructions give the number 2:1:1. What does this mean?
   a. 2 parts color, 1 part reducer, 1 part hardener
   b. 2 parts hardener, 1 part color, 1 part reducer
   c. 2 parts reducer, 1 part color, 1 part hardener
   d. 2 hours, 1 minute, 1 second curing time

7. *Technician A* is using a mixing stick to measure out color, reducer, and hardener. *Technician B* is using electronic or computerized scales. Who is correct?
   a. Technician A
   b. Technician B

c. Both Technicians A and B
d. Neither Technician

8. *Technician A* says that original OEM paints are typically about 6 to 20 mils thick. *Technician B* says that the basecoat is approximately 10 to 12 mils thick. Who is correct?
   a. Technician A
   b. Technician B
   c. Both Technicians A and B
   d. Neither Technician

9. Which of the following would be used to verify the actual amount of frame or unibody damage on a vehicle?
   a. Dimensions manual
   b. Estimating manual
   c. Service manual
   d. Owner's manual

10. Paint drains too quickly out of a viscosity cup. This means you have added too much:
    a. Color
    b. Catalyst
    c. Reducer
    d. Toner

Activities

1. Find the VIN on several vehicles. Write down the location of the VIN for several cars. Look up the VIN data in a service manual.

2. Use a paint thickness gauge to check for excess paint layers on several vehicles. Did different panels have more paint than others? How many vehicles had paint thicker than normal? Could you see any paint cracking or other surface problems when the paint was too thick?

3. Use a micrometer to measure the blades of a feeler gauge. Did your measurements equal the actual blade thicknesses?

Fasteners

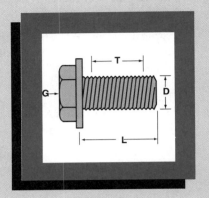

Objectives

After studying this chapter, you should be able to:

- Identify the various fasteners used in vehicle construction.

- Remove and install bolts and nuts properly.

- Explain when specific fasteners are used in vehicle construction.

- Explain bolt and nut torque values.

- Summarize the use of chemical fasteners.

- Identify hose clamps.

Introduction

Fasteners make up the thousands of bolts, nuts, screws, clips, and adhesives that literally hold a vehicle together.

As a collision repair technician, you will constantly use fasteners when removing and installing body parts. This makes it important for you to be able to identify and use fasteners properly. This chapter will summarize the proper utilization and selection of bolts, nuts, screws, clips, adhesives, epoxies, and other fasteners found in a collision repair facility.

Remember that each fastener is engineered for a specific application. Always replace fasteners with exactly the same type that was removed from the *original equipment manufacture (OEM)* assembly. Never try to "re-engineer" the vehicle. Keep in mind that using an incorrect fastener or a fastener of inferior quality can result in failure and possible injury to the vehicle occupants.

BOLTS

A **bolt** is a shaft with a head on one end and threads on the other. A **cap screw** is a term that describes a high strength bolt. Bolts and cap screws are usually named after the body part they hold: fender bolt, hood hinge bolt, etc. Their shape and head drive configuration also helps name them.

Bolt Terminology

To work with bolts properly, you must understand basic bolt terminology. See Figure 7–1.

The **bolt head** is used to tighten, or torque the bolt. A socket or wrench fits over the head, which enables the bolt to be tightened or loosened. Some metric and USC/SAE sockets are very close in size. It is very important to use the correct wrench or socket when tightening or loosening nuts and bolts. The improper wrench or socket could strip or round off the nut or bolt you are working on. This could damage your tools, or even cause you an injury if the tool slips off.

Bolt length is measured from the end of the threads to the bottom of the bolt head. It is NOT the total length including the bolt head.

Bolt diameter, sometimes termed **bolt size,** is measured around the outside of the threads. For example, a ½ inch bolt has a thread diameter of ½ inch, while its head or wrench size would be ¾ inch.

Bolt head size is the distance measured across the flats of the bolt head. In USC, head size is given in fractions, just like wrench size. A few common sizes are ⁷⁄₁₆, ½, and ⁹⁄₁₆ inch. In the metric system, 8-, 10-, 13-millimeter head sizes are typical. Common USC and metric bolt head sizes are given in Figure 7–2.

Bolt **thread pitch** is a measurement of thread coarseness. Bolts and nuts can have coarse, fine, and metric threads. Bolt threads can be measured with a **thread pitch gauge.** One is shown in Figure 7–3.

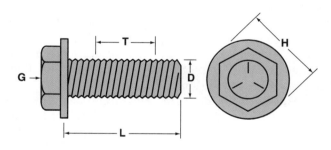

H - Head
G - Grade Marking (Bolt Strength)
L - Length (Inches)
T - Thread Pitch (Thread/Inch)
D - Nominal Diameter (Inches)

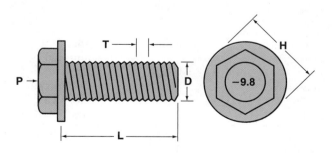

H - Head
P - Property Class (Bolt Strength)
L - Length (Millimeters)
T - Thread Pitch (Millimeters)
D - Nominal Diameter (Millimeters

Figure 7–1 Bolt measurements are needed when working. Study each dimension of both USC and metric bolts.

Common English USC/SAE Head Sizes	Common Metric Head Sizes
Wrench Size (inches)*	Wrench Size (millimeters)*
3/8	9
7/16	10
1/2	11
9/16	12
5/8	13
11/16	14
3/4	15
13/16	16
7/8	17
15/16	18
1	19
1-1/16	20
1-1/8	21
1-3/16	22
1-1/4	23
1-5/16	24
1-3/8	26
7/16	27
1-1/2	29
	30
	32

* The wrench sizes given in this chart are not equivalents, but are standard head sizes found in both inches and millimeters.

Figure 7-2 These are common bolt head and wrench sizes. Never use a USC wrench on metric bolts or vice versa. This will damage the bolt head.

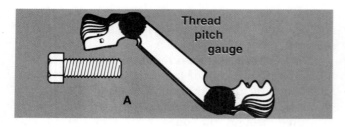
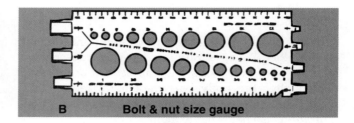

Figure 7–3 Gauges can be used to tell thread, bolt, or nut sizes. (A) Thread pitch gauge. The thread pitch gauge is fit against threads. Threads that match the gauge equal the pitch number printed on the gauge. (B) Bolt and nut size gauge. Bolts and nuts can be fitted into a bolt and nut gauge to quickly tell their sizes.

The two common metric threads are coarse and fine and can be identified by the letters SI (System International or International System of Units) and ISO (International Standards Organization).

> *WARNING! Do NOT accidentally interchange thread types or damage will result. It is easy to mistake metric threads for USC/SAE threads. If the two are forced together, either the bold or part threads will be ruined.*

Bolts and nuts are also available in right- and left-hand threads. **Right-hand threads** must be turned clockwise to tighten. Less common **left-hand threads** must be rotated in a counterclockwise direction to tighten the fastener. Left-hand threads may be denoted by notches or the letter "L" stamped on them.

Bolt Strengths or Grades

Bolt strength indicates the amount of torque or tightening force that should be applied. Bolts are made from different materials having various degrees of hardness. Softer metal or harder metal can be used to make bolts. Bolts are made with different hardnesses and strengths for use in different situations.

Bolt grade markings are lines or numbers on the top of the head to identify bolt hardness and strength. The hardness or strength of metric bolts is indicated by using a property class indicator on the head of the bolt.

Bolt strength markings are given as lines, Figure 7–4. The number of lines on the head of the bolt is related to its strength. As the number of lines increases so does the strength.

Metric bolt strength markings are given as numbers. The higher the number is, the stronger the bolt. These markings apply to both bolts and nuts, Figure 7–5.

Tensile strength is the amount of pressure per square inch the bolt can withstand just before break-

SAE GRADE MARKINGS	⬡	⬡ (3 lines)	⬡ (4 lines)	⬡ (5 lines)	⬡ (6 lines)
DEFINITION	No lines: unmarked indeterminate quality SAE Grades 0-1-2	3 Lines: common commercial quality Automotive & AN Bolts SAE Grade 5	4 Lines: medium commercial quality Automotive & AN Bolts SAE Grade 6	5 Lines: rarely used SAE Grade 7	6 Lines: best commercial quality NAS & Aircraft Screws SAE Grade 8
MATERIAL	Low Carbon Steel	Med. Carbon Steel Tempered	Med. Carbon Steel Quenched & Tempered	Med. Carbon Alloy Steel	Med. Carbon Alloy Steel Quenched & Tempered
TENSILE STRENGTH	65,000 psi	120,000 psi	140,000 psi	140,000 psi	150,000 psi

Figure 7–4 Bolt tensile strength is denoted on the head of a bolt. Slash marks (US) or numbers (metric) are used. Always replace a bolt with one of equal or higher rating to prevent failure.

Grade	Identification	Class	Identification
Hex Nut Grade 5	3 Dots	Hex Nut Property Class 9	Arabic 9
Hex Nut Grade 8	6 Dots	Hex Nut Property Class 10	Arabic 10
Increasing dots represent increasing strength		Can also have blue finish or paint dab on hex flat. Increasing numbers represent increasing strength.	

Figure 7–5 Quality nut strengths are also denoted. More dots or a higher number means more strength.

ing when being pulled apart. The harder or stronger the bolt, the greater the tensile strength.

Never replace a high grade bolt with a bolt having a lower grade marking—the weaker bolt could break! In the steering or suspension, this could seriously endanger the passengers of the vehicle.

Torque

Torque is a measurement of the turning force applied when tightening a fastener. It is critical that bolts and nuts are tightened, or torqued properly. Over-tightening will stretch and possibly break the bolt. Under-tightening could allow the bolt or nut to loosen and fall out.

Torque specifications are tightening values for the specific bolt or nut. They are given by the manu-

facturer. Discussed in the tool chapter, a torque wrench must be used to measure torque values.

If you cannot find the factory torque specification for a bolt, you can use a *general bolt torque chart.* It will give a general torque value for the size and grade of bolt. One is given in Figure 7–6. Normally, the bolt threads should be lubricated to get accurate results. Refer to the service or repair manual to see if the threads should be lubricated or dry.

A *tightening sequence,* or *torque pattern,* assures that parts secured by several bolts are clamped down evenly. Generally, tighten fasteners in a crisscross pattern. This will pull the part down evenly, preventing warpage. It is commonly recommended on wheels, as shown in Figure 7–7.

Basically, tighten the fastener in steps. Begin at approximately half-torque, then continue to three-fourths torque, and then full torque at least twice.

Metric Standard						SAE Standard / Foot Pounds					
Grade of Bolt	5D	.8G	10K	12K		Grade of Bolt	SAE 1&2	SAE 5	SAE 6	SAE 8	
Min. Tensile Strength	71,160 PSI	113,800 PSI	142,200 PSI	170,679 PSI		Min. Ten Strength	64,000 PSI	105,000 PSI	133,000 PSI	150,000 PSI	
Grade Markings on Head	5D	8G	10K	12K	Size of Socket on Wrench Opening	Markings on Head	●	◆	✚	✳	Size of Socket or Wrench Opening
Metric		Foot Pounds				Metric	U.S. Standard	Foot Pounds			U.S. Regular
Bolt Dia.	U.S. Dec Equiv.					Bolt Head	Bolt Dia.				Bolt Head	Nut
6 mm	.2362	5	G	8	10	10 mm	1/4	5	7	10	10.5	3/8	7/16
8 mm	.3150	10	16	22	27	14 mm	5/16	9	14	19	22	1/2	9/16
10 mm	.3937	19	31	40	49	17 mm	3/8	15	25	34	37	9/16	5/8
12 mm	.4720	34	54	70	86	19 mm	7/16	24	40	55	60	5/8	3/4
14 mm	.5512	55	89	117	137	22 mm	1/2	37	60	85	92	3/4	13/16
16 mm	.6299	83	132	175	208	24 mm	9/16	53	88	120	132	7/8	7/8
18 mm	.709	111	182	236	283	27 mm	5/8	74	120	167	180	15/16	1.
22 mm	.8661	182	284	394	464	32 mm	3/4	120	200	280	296	1-1/8	1-1/8

Figure 7–6 This is a general bolt torque chart. It gives different values for each bolt tensile strength rating. These values apply to dry torque unless otherwise specified.

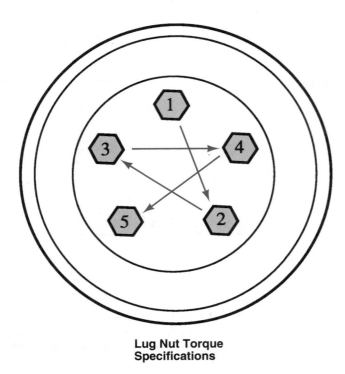

**Lug Nut Torque
Specifications**

Figure 7-7 When tightening several bolts that hold one part, a wheel for example, always use a crisscross pattern. This will prevent part warpage and damage.

Be careful when tightening bolts and nuts with air wrenches. It is easy to stretch or break a bolt in an instant. The air wrench can spin the bolt or nut so fast that it can hammer the fastener past its yield point. This can strip threads or snap off the bolt.

When torque is not critical, do NOT use the air impact wrench to run the nut full speed onto the bolt. Instead, run it up slowly until it contacts the work. Then mark the socket and watch how far it turns. Smaller air-powered speed wrenches do NOT produce the severe force of impact wrenches and are much safer to use.

NUTS

A **nut** uses internal (inside) threads and an odd shaped head that often fits a wrench. When tightened onto a bolt, a strong clamping force holds the parts together. Many different nuts are used by the automotive industry. Several are shown in Figure 7-8.

Castellated or **slotted nuts** are grooved on top so that a safety wire or cotter pin can be installed into a hole in the bolt. This helps prevent the nut from working loose. For example, castellated nuts are used with the studs that hold wheel bearings in position. Slotted nuts are also used on steering and suspension parts for safety.

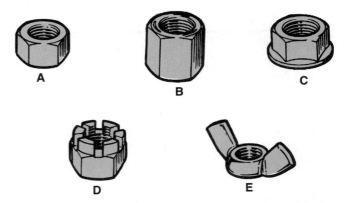

Figure 7-8 Nuts also come in various designs. Memorize their names. (A) Hex nut. (B) High or deep nut. (C) Flange nut. (D) Castle or slotted nut. (E) Wing nut. *(Dorman Products)*

Self-locking nuts produce a friction or force fit when threaded onto a bolt or stud. The top of the nut can be crimped inward. Some have a plastic insert that produces a friction fit to keep the nut from loosening. Sometimes, locking nuts must be replaced after removal. Front-wheel drive spindles sometimes use self-locking nuts.

Jam nuts are thin nuts used to help hold larger, conventional nuts in place. The jam nut is tightened down against the other nut to prevent its loosening.

Wing nuts have two extended arms for turning the nut by hand. They are used when a part must be removed frequently for service or maintenance. Air cleaners sometimes use wing nuts. See Figure 7-8.

Acorn nuts are closed on one end for appearance and to keep water and debris off the threads. They can be used when they are visible and looks are important.

Special types of nuts are used to hold specific parts onto the vehicle. Sometimes a washer is formed onto the nut. Termed **body nuts,** the flange on the nut helps distribute the clamping force of the thin body panel or trim piece to prevent warpage. See Figure 7-9.

Several applications for special nuts are shown in Figure 7-10.

THREAD REPAIR

A collision repair technician must frequently repair damaged threads. A **tap** is a tool for cutting inside threads in holes. A tap is used to repair damaged threaded holes. It is rotated down into the threaded hole to recut the threads. A **die** cuts threads on the outside of bolts or studs. It too is rotated over the threads to clean them up.

Special **t-handles** fit over the tap or die for turning. You must hold the tool perfectly square to cut

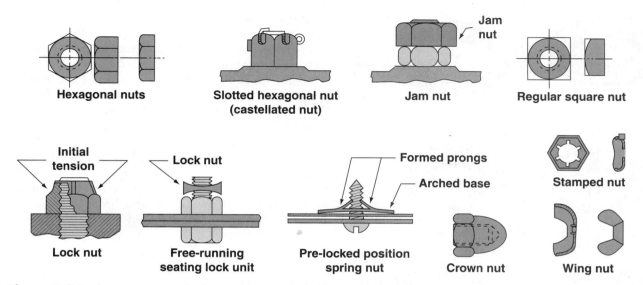

Hexagonal nuts

Slotted hexagonal nut (castellated nut)

Jam nut

Regular square nut

Initial tension

Lock nut

Lock nut

Free-running seating lock unit

Formed prongs

Arched base

Pre-locked position spring nut

Stamped nut

Crown nut

Wing nut

Figure 7–9 Body nuts are specially designed for specific holding applications.

good threads. Oil the threads. Then, rotate the tap or die about one-half of a turn and then back it out about one-fourth turn. This will clean metal shavings out and prevent tool breakage.

A **Helicoil** or **thread insert** can be used to repair badly damaged internal threads. Basically, to use a thread insert, drill the hole oversize. Then tap the hole with new threads. Use the special tool to

Lock Nut Keeps Ordinary Hex Nut Tight on Clamp Holding Cable.

Self-threading Nuts, Washer Type, Are Used on Tail Light Assembly.

Removable Pushnut® Fasteners are Slipped Over Integral Molded Studs to Secure Speaker.

Lock Nut Supports Expansion Shell That in Turn Prevents Loosening of Bolt.

Figure 7–10 Note some typical uses for jam nuts. (*Courtesy of Palnut*)

rotate the insert into the freshly threaded hole. This will allow you to use the original size bolt. See Figure 7–11.

WASHERS

Washers are used under bolts, nuts, and other parts. They prevent damage to the surfaces of parts and provide better holding power. Several types are illustrated in Figure 7–12.

Flat washers are used to increase the clamping surface area. They prevent the smaller bolt head from pulling through the sheet metal or plastic.

A **wave washer** adds a spring action to keep parts from rattling and loosening.

Figure 7–11 Study the basic thread repair procedure using a thread repair insert. (A) Drill the hole out one size larger, or as needed. (B) Run the right size tape down through the hole to cut new threads. (C) Use the special tool to screw the insert down into the threaded hole. (D) The original size bolt can now be used in the hole again. (*Helicoil, Fastening Systems Division*)

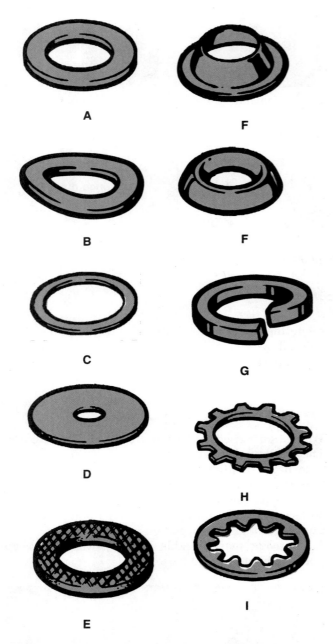

A

B

C

D

E

F

F

G

H

I

Figure 7–12 Know the washer types. (A) Plain flat washer. (B) Wave or spring washer. (C) Spacer washer. (D) Fender washer. (E) Fiber washer. (F) Finishing washer. (G) Split lock washer. H: External lock washer. (I) Internal lock washer. *(Dorman Products)*

Body or **fender washers** have a very large outside diameter for the size hole in them. They provide better holding power on thin metal and plastic parts.

Copper or **brass washers** are used to prevent fluid leakage, as on brake line fittings. **Spacer washers** come in specific thicknesses to allow for adjustment of parts. **Fiber washers** will prevent vibration or leakage but cannot be tightened to a great extent.

Finishing washers have a curved shape for a pleasing appearance. They are used on interior pieces.

Split lock washers are used under nuts to prevent loosening by vibration. The ends of these spring-hardened washers dig into both the nut and the work to prevent rotation.

Shakeproof or teeth lock washers have teeth or bent lugs that grip both the work and the nut. Several designs, shapes, and sizes are available. An external type has teeth on the outside and an internal type has teeth around the inside.

A "rule of thumb" about lock washers: if the part did NOT come with one, do NOT add one. Lock washers are extremely hard and tend to break under severe pressure. If the bolt or nut has a lock washer, use one. The manufacturer would not use one if it did not have a purpose.

SCREWS

Screws are often used to hold nonstructural parts on a vehicle. Trim pieces, interior panels, etc. are often secured by screws. See Figure 7–13.

Machine screws are threaded their full length and are relatively weak. They come in various configurations and will accept a nut.

Set screws frequently have an internal drive head for an Allen wrench. They are used to hold parts onto shafts.

Sheet metal screws have pointed or tapered tips. They thread into sheet metal for light holding tasks.

Self-tapping screws have a pointed tip that helps cut new threads in parts.

Trim screws have a washer attached to them. This improves appearance and helps keep the trim from shifting.

Headlight aiming screws have a special plastic adapter mounted on them. The adapter fits into the headlight assembly. Different design variations are needed for different makes and models of vehicles.

NONTHREADED FASTENERS

Nonthreaded fasteners, as implied, do not use threads. They include keys, snap rings, pins, clips, adhesives, etc.

Various **keys** and **pins** are used by equipment manufacturers to retain parts in alignment. It is important to be able to identify these keys and pins, order replacements, and replace them.

Square keys and **Woodruff keys** are used to prevent hand wheels, gears, cams, and pulleys from turning on their shafts. These keys are strong enough to carry heavy loads if they are fitted and seated properly. See Figure 7–14.

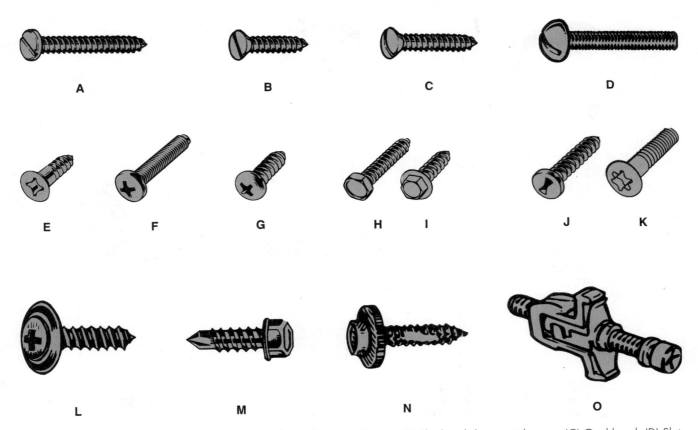

Figure 7-13 Study common screw names. (A) Pan head sheet metal screw. (B) Flat head sheet metal screw. (C) Oval head. (D) Slotted machine screw. (E) Phillips head screw. (F) Phillips machine screw. (G) Oval head sheet metal screw. H: Hex or nutdriver screw. (I) Hex screw with flange or integral washer. (J) Clutch head. (K) Torx® head. (L) Trim screw. (M) Self-taping screw. (N) Body screw. (O) Headlight aiming screw. *(Dorman Products)*

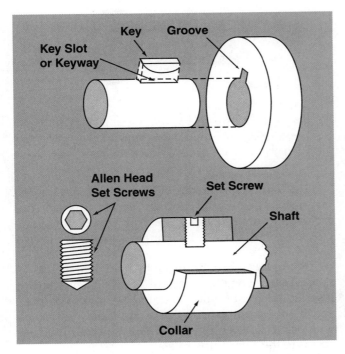

Figure 7-14 Keys and set screws are both used to align parts on shafts. (A) Key and keyway. (B) Set screw application.

Round **taper pins** have a larger diameter on one end than the other. They are used to locate and position matching parts. They can also be used to secure small pulleys and gears to shafts. See Figure 7–15.

Dowel pins have the same general diameter their full length. They are used to position and align the parts of an assembly. One end of a dowel pin is chamfered, and it is usually 0.001 to 0.002 inch (0.025 to 0.05 mm) greater in diameter than the size of its hole. When replacing a dowel pin, be sure that it is the same size as the old one.

Cotter pins help prevent bolts and nuts from loosening or they fit through pins to hold parts together. They are also used as stops and holders on shafts and rods. All cotter pins are used for safety and should NEVER be reused.

The cotter pin should fit into the hole with very little side play. If it is too long, cut off the extra length of cotter pin. Bend it over in a smooth curve. Sharply angled bends invite breakage. Bend the ends with needle nose pliers. Final bending of the prongs can be done with a soft-faced mallet.

Snap rings are nonthreaded fasteners that install into a groove machined into a part. They are used to

Figure 7–15 Know the nonthreaded fastener names. (A) Internal snap or retaining ring. (B) External snap ring. (C) E-clip or snap ring. (D) Cotter pin. (E) Clevis pin. (F) Hitch pin. (G) Split rollpin. (H): Taper pin. (I) Straight dowel pin. (J) Linkage clip. *(Dorman Products)*

hold parts on shafts. See Figure 7–15. Special snap ring pliers are designed to flex and install or remove snap rings. They have special tips that will hold the snap ring. See Figure 7–16.

> **DANGER!** *Wear safety glasses when removing or installing snap rings. Being constructed of spring steel, they can shoot out with great force.*

Body clips are specially shaped retainers for holding trim and other body pieces requiring little strength. The clip often fits into the back of the trim piece and through the body panel.

Push-in clips are usually made of plastic and they force fit into holes in body panels. Push-in clips are used to hold interior door trim panels, for example. They install easily, but can be difficult to remove, Figure 7–17.

Pop rivets can be used to hold two pieces of sheet metal together, Figure 7–18. They can be inserted into a blind hole through two pieces of metal and then drawn up with a riveting tool or gun. This will lock the pieces together. There is no need to have access to the back of the rivets.

Pop rivets should NOT be used in areas that are subject to excessive vibration or for structural pan-

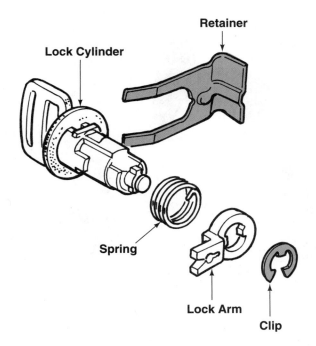

Figure 7-16 Note special retainer and E-clip used on this lock cylinder. (© American Honda Motor Co., Inc.)

els. The rivets can work loose and weaken the repair.

REPLACING FASTENERS

When replacing fasteners, observe the following precautions:

1. Always use the same number of fasteners.

2. Use the same diameter, length, pitch, and type of fasteners.

3. Observe the OEM recommendations given in the service manual for tightening sequence, tightening steps (increments), and torque values.

4. Always replace a used cotter pin.

5. Replace stretched fasteners or fasteners with any signs of damage.

6. Use the correct washers and pins as specified by the OEM.

7. Always replace "one-time" fasteners. They can be found in suspension and steering assemblies.

HOSE CLAMPS

Hose clamps are used to hold radiator hoses, heater hoses, and other hoses onto their fittings. See Figure 7-19.

A **spring hose clamp** is made of spring steel with barbs on each end. Squeezing the ends opens and expands the clamp. Special hose clamp pliers should be used to remove or install round wire type clamps. The pliers have a deep groove that will keep the clamp from slipping out of the jaws. Conventional pliers will work fine on spring strap clamps.

Worm hose clamps use a screw that engages a slotted band. Turning the screw reduces or enlarges clamp diameter. This is the most common replacement type of hose clamp.

ADHESIVES

Adhesives provide an alternate means of bonding parts together. The two types of adhesives most often used are epoxy and trim adhesive.

Epoxy is a two-part bonding agent that dries harder than adhesive. It comes in two separate containers, usually tubes. One contains the epoxy resin and the other contains a hardener. Epoxy does NOT shrink when it hardens and is waterproof and heat-resistant at moderate temperatures.

Read the instructions for the proper quantity to use. If both resin and hardener are NOT in proper proportion, the bond might fail. Some epoxy tubes automatically dispense the correct amount of resin and hardener.

Once mixed, the epoxy remains in a workable condition for only a brief time. Therefore, try to mix only as much as is required and use it as quickly as possible. Clamp the work while the glue cures, which can take several hours. Do NOT apply epoxy in low temperatures (below 50°F or 10°C), because it will NOT harden. Once an epoxy is applied, it is difficult to remove it from a surface.

Trim adhesive is used to install various trim pieces (letters, molding, emblems) onto the body surface. Trim adhesive dries to a pliable rubber-like consistency. It will bond plastics, metal, rubber, and most other materials to painted surfaces.

Make sure you are using the recommended type adhesive for the job, because performance characteristics vary. Read the label directions and refer to the vehicle's service manual for specific information on the type adhesive that will work best.

Make sure the surfaces on the part and body are properly cleaned. Mark the desired part alignment with masking tape if needed.

Apply a moderate amount of adhesive to the part. Spread the adhesive out evenly. Avoid applying too much adhesive near the outer perimeter of the part because it will squeeze out when the part is applied. When excessive adhesive squeezes out from under the part, extra clean up will be required. See Figure 7-20.

Figure 7-17 Here are a few of the special plastic retainers available. These types are often used in vehicle interiors. They quickly press into a hole. To remove them, you must carefully pry next to the retainer with a flat, forked trim tool. *(TRW Plastic Fasteners and Retainers Division, Westminster, MA and Roseville, MI)*

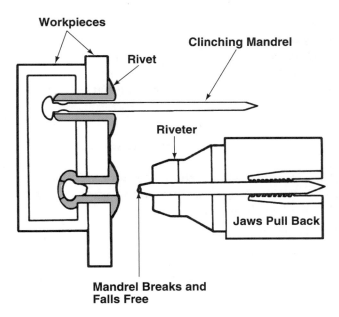

Figure 7–18 Pop rivets provide a quick way of holding sheet metal parts or panels. They are handy when the back of a panel is not accessible.

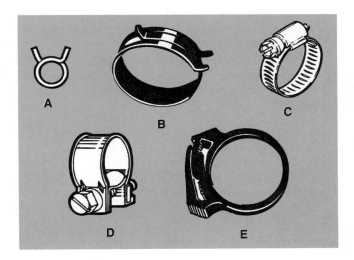

Figure 7–19 Note hose clamp types. (A) Wire spring hose clamp. (B) Wire strap hose clamp. (C) Worm hose clamp. (D) Screw-nut hose clamp. (E) Plastic hose clamp. *(Dorman Products)*

Carefully move the part straight into place on the body without smearing the adhesive. Press the part down tightly to compress the adhesive against the two surfaces. See Figure 7–21. If needed, use masking tape to hold the part in place as the adhesive dries, as shown in Figure 7–22.

NOTE! Adhesives are discussed in more detail in other chapters. Refer to the text index if needed.

Figure 7–20 When reinstalling a trim piece like this one, make sure you are using the factory recommended type of adhesive. Apply the adhesive evenly over the full surface of the part.

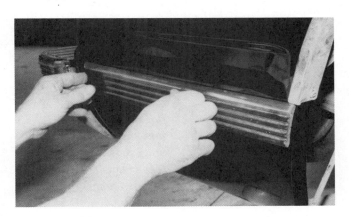

Figure 7–21 Move the trim component straight onto the body surface without smearing the adhesive. This trim piece can be aligned with the body line. However, masking tape can be used to mark alignment if needed.

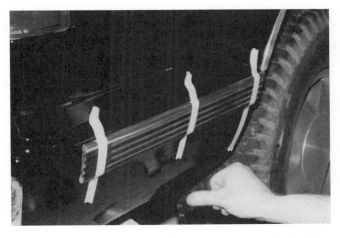

Figure 7–22 Press the trim component against the body to assure good bonding. Masking tape can be used to hold the part or trim in alignment until the adhesive dries.

Summary

- Fasteners make up the thousands of bolts, nuts, screws, clips, and adhesives that literally hold a vehicle together.
- Always replace fasteners with exactly the same type that was removed from the original equipment manufacturer (OEM) assembly. An incorrect fastener, or one of inferior quality, can result in failure and possible injury to the vehicle occupants.
- Bolt strength indicates the amount of torque or tightening force that should be applied.
- Non-threaded fasteners, as implied, do not use threads. Examples of non-threaded fasteners are keys, snap rings, pins, clips, and adhesives.

Technical Terms

fasteners	*die*
bolt	*t-handles*
cap screw	*Helicoil*
bolt head	*thread insert*
bolt length	*washer*
bolt diameter	*flat washer*
bolt size	*wave washer*
bolt head size	*body washer*
thread pitch	*fender washer*
thread pitch gauge	*copper washer*
right-hand threads	*brass washer*
left-hand threads	*spacer washer*
bolt strength	*fiber washer*
bolt grade markings	*finishing washer*
bolt strength markings	*split lock washer*
metric bolt strength markings	*shakeproof lock washer*
	teeth lock washer
tensile strength	*screw*
torque	*machine screw*
torque specifications	*set screw*
general bolt torque chart	*sheet metal screw*
tightening sequence	*self-tapping screw*
torque pattern	*trim screw*
nut	*headlight aiming screw*
castellated nut	*nonthreaded fastener*
slotted nut	*key*
self-locking nut	*pin*
jam nut	*square key*
wing nut	*Woodruff key*
acorn nut	*taper pin*
body nut	*dowel pin*
tap	*cotter pin*

snap ring	*spring hose clamp*
body clip	*worm hose clamp*
push-in clip	*adhesive*
pop rivet	*epoxy*
hose clamp	*trim adhesive*

Review Questions

1. _____ include the thousands of bolts, nuts, screws, clips, and adhesives that literally hold a vehicle together.

2. Define the term *cap screw.*

3. What is bolt thread pitch and how is it measured?

4. If you turn right-hand threads clockwise, what will happen?

5. Which of these bolts is strongest?
   a. Three head markings
   b. Two head markings
   c. No head markings
   d. One head marking

6. When installing a wheel on a vehicle, no service manual can be found for getting a factory torque specification.
   Technician A says to use an impact wrench on medium setting. *Technician B* says to use a general torque chart and a torque wrench. Who is correct?
   a. Technician A
   b. Technician B
   c. Both Technicians
   d. Neither Technician

7. What general sequence should be used when tightening a series of bolts or nuts?

8. How do you use a thread repair insert?

9. All cotter pins are used for safety and should NEVER be reused. True or False?

10. Body clips are specially shaped retainers for holding _____ and other body pieces requiring little strength.

ASE-Style Review Questions

1. Torque is a measurement of:
   a. Driving force
   b. Lifting force
   c. Turning force
   d. Pulling force

2. A technician removes a nut with three dots on it. *Technician A* says that is has the strength of grade

five. *Technician B* says that it has strength of grade three. Who is correct?
a. Technician A
b. Technician B
c. Both Technicians A and B
d. Neither Technician

3. Which of the following types of washers is used to prevent loosening by vibration?
a. Flat
b. Fender
c. Finish
d. Split lock

4. When replacing a fastener, *Technician A* says that the same number of fasteners should always be used. *Technician B* says that stretched fasteners or fasteners with any signs of damage should be replaced. Who is correct?
a. Technician A
b. Technician B
c. Both Technicians A and B
d. Neither Technician

5. Which of the following is NOT a hose clamp type?
a. Orbital
b. Wire spring
c. Screw-nut
d. Worm

6. *Technician A* says that "one-time" fasteners must always be replaced following removal. *Technician B* says that cotter pins are to be used only once. Who is correct?
a. Technician A
b. Technician B
c. Both Technicians A and B
d. Neither Technician

7. *Technician A* says that an SAE is measured in millimeters. *Technician B* says that an SAE bolt is measured in inches. Who is correct?
a. Technician A
b. Technician B
c. Both Technicians A and B
d. Neither Technician

8. *Technician A* says that tensile strength is the amount of pressure per square inch that a bolt can withstand before breaking. *Technician B* says that tensile strength is the tightening value of the specific bolt or nut. Who is correct?
a. Technician A
b. Technician B
c. Both Technicians A and B
d. Neither Technician

9. *Technician A* says that a die is a tool that is used for cutting inside threads. *Technician B* says that a tap is a tool that is used for cutting inside threads. Who is correct?
a. Technician A
b. Technician B
c. Both Technicians A and B
d. Neither Technician

10. *Technician A* says that screws are often used to hold nonstructural parts on a vehicle. *Technician B* says that machine screws are threaded their full length and are relatively weak. Who is correct?
a. Technician A
b. Technician B
c. Both Technicians A and B
d. Neither Technician

Activities

1. Inspect a car or truck. Write down the various types of fasteners you can locate. Create a chart listing their names and applications.

2. Read the directions on a few types of body adhesives. Write a short report on their use.

3. Visit a body supply house or hardware store. Inspect the various types of fasteners available.

8

Collision Repair
Hardware and Materials

Objectives

After studying this chapter, you should be able to:

- Select the right materials for the job.

- Explain the basic purpose of common materials.

- Compare the use of similar shop materials.

- Summarize when to use different kinds of filler.

- Know how to select the right type of primer and paint.

- Understand the importance of using a complete paint system.

Introduction

Collision repair materials include more than just refinishing or paint materials. They include the various fillers, primers, sealers, adhesives, sandpapers, and other compounds common to collision repair shops. It is critical that you understand their selection and use. This chapter will summarize the purpose of these basic types of materials. Later chapters will explain how to use these materials in more detail.

When consumers look at a vehicle's finish, they often only see a shiny, bright color. They seldom understand all of the technology involved in producing that long-lasting, tough, durable, high-gloss finish. There is hidden technology under the surface of the paint.

A professional collision repair and refinishing technician comprehends all of the "chemistry" and skill needed to do high quality repair. This chapter will introduce you to the materials needed to do competent repairs.

NOTE! This chapter gives you only a working knowledge of common body materials. It does not explain how to apply or use them. Later chapters will give you this information.

REFINISHING MATERIALS

A vehicle body is protected by a complete finishing system. All parts of the system work together to protect the vehicle from ultraviolet (UV) radiation, weathering, pollutants, and corrosion.

Refinishing materials is a general term referring to the products used to repaint the vehicle. Refinishing material chemistry has changed drastically. New paints last longer but require more skill and safety measures for proper application.

The *substrate* is the steel, aluminum, plastic, and composite materials used in the vehicle's construction. It will affect the selection of refinishing materials.

The vehicle's finish or paint performs two basic functions—to beautify and to protect. Can you imagine what a vehicle would look like without paint? For a day or so, it would be the drab, steel gray of bare sheet metal. Then, as corrosion covered the body, it would turn an ugly, reddish brown. Finally, this degeneration would continue until the body was coated with rust.

The term *paint* generally refers to the visible topcoat. The most elementary painting system consists of a compatible primer and final topcoat over the substrate. This process can vary considerably and is usually more complex, as you will learn later.

UNDERCOATS AND TOPCOATS

A basic finish consists of several coats of two or more different materials. The most basic finish consists of:

1. Undercoat or primer coat.

2. Topcoat (colorcoat or basecoat/clearcoat).

The *primer* or *undercoat* has to improve adhesion of the topcoat. It is often the first coat applied. Paint alone will not stick, or adhere, as well as a primer. If you apply a topcoat to bare substrate, the paint will peel, flake off, or look rough. This is why you must "sandwich" a primer undercoat between the substrate and the topcoat. Undercoats also prevent any chemicals from bleeding through and showing in the topcoats of paint.

The term *topcoat* or *colorcoat* refers to the paint applied over the undercoat. It is usually several light coats of one or more paints. The topcoat is the "glam-our coat," because it features the eye-catching color, color effects, and gloss.

Basecoat-clearcoat paint systems use a colorcoat applied over the primer with a second layer of clearcoat over the colorcoat. This is the most common paint system used today. The clear paint brings out the richness of the underlying color and also protects the color.

PAINT TYPES

The general types of paint include:

1. Enamel/urethane

2. Waterbase/waterborne paint

As you will learn, there are variations within these categories. It is important that you know what type of finishes manufacturers use because there are slightly different methods required for refinishing them. Refer to Figure 8–1.

Most *enamel* finishes used in the collision repair industry toady are catalyzed or urethane enamels. Once applied, these materials, which can be color, clear, or primers, "dry" in a two-stage process (see Figure 8–1). First, some of the solvents used to thin or reduce the material must evaporate; second, a chemical reaction occurs within the material and causes it to harden or "cure". This chemical change causes enamel finishes to dry/cure with a gloss that does not require rubbing or polishing.

Since enamels generally dry more slowly, there is more chance for dirt and dust to stick in the finish. While there is generally a slight amount of surface roughness in an enamel film, too much will cause a lower gloss.

Two-stage paints consist of two distinct layers of paint: basecoat and clearcoat. *Basecoat-clearcoat enamel* is now the most common system used to repaint cars and trucks. First, a layer of color is applied over the undercoat of primer or sealer. Next, a coat of clear is sprayed over the color basecoat.

Acrylic enamel and *acrylic urethane enamel* are two specific types of enamel paint. Both are commonly used in the industry. Acrylic urethanes are slightly harder than plain acrylics. Each is available in a variety of colors.

Water-based/waterborne paints, as implied, use water to carry the pigment. They dry through evaporation of the water. Some manufacturers are starting to use water-based paints on new vehicles. This is to help satisfy stricter emission regulations in some geographic areas. The basecoat of color is water-based. Then, an enamel topcoat is applied over the water-based paint.

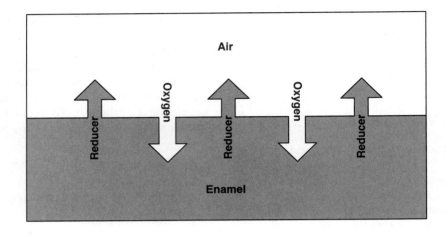

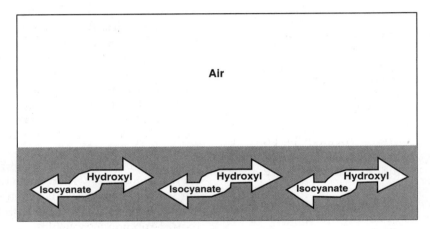

Figure 8–1 Note how paints dry or cure. *(PPG Industries, Inc.)* (A) Enamels dry by evaporation of reducer or solvent first, then by oxidation. Resin reacts with oxygen in the air for drying. Heat speeds this action. (B) Urethane and polyurethane enamels cure by molecular cross-linking.

Water-based/waterborne primers have been used for years as a fix for lifting problems. They serve as an excellent barrier coat when there are paint incompatibility problems.

ORIGINAL EQUIPMENT MANUFACTURE (OEM) FINISHES

The **OEM finishes** (factory paint jobs) on today's vehicles are either "thermo-setting" acrylic enamel (paint is furnace-hardened at the factory), new high-solids basecoat-clearcoat enamels, or water-based, low-emission paints. Common enamel finishes are baked in huge ovens to shorten the drying times and cure the paint. This is done before installing the interior and other nonmetal parts.

Vehicle manufacturers use several different types of finish materials, coating processes, and application processes. Each type of finish requires different planning and repair steps.

The most common types of OEM coating processes include:

1. Single-stage
2. Two-stage (basecoat/clearcoat)
3. Three-stage paint (tri-coat)
4. Multi-stage

These will be explained fully in later chapters of this book.

CONTENTS OF PAINT

Paint's chemical content includes the following:

1. Pigments (colors)
2. Binders
3. Solvents
4. Additives

Each of these ingredients has a specific function within the paint formula.

Pigments

The *pigments* are fine powders that impart color, opacity, durability, and other characteristics to the primer or paint. They are a nonvolatile film-forming ingredient. The main purpose of the pigment is to hide everything under the paint.

Pigment particles size and shape are also important. Pigment particle size affects hiding ability and appearance. In addition, pigment shape affects strength. Pigment particles may be nearly spherical, rod-, or plate-like. Rod-shaped particles, for example, reinforce color film as do "iron bars" in concrete.

Medium-size reflective pigment particles, such as mica, are added to *paints* to create a pearl effect. As you will learn later, **pearlescent colors** are now common and are sometimes difficult to match when repainting.

Large reflective pigment flakes are added to **metallic color.** The size, shape, color, and material in the flakes can vary. Often called "metal flakes," the flakes can be made of tiny but visible bits of metal or polyester. When light strikes the flakes, it is reflected at different angles like tiny "glittering stars" inside the paint.

If this is new to you, start looking at vehicle colors more closely. See if you can see the difference between a solid, a pearl, and a metallic color.

Solids are the nonliquid contents of the paint or primer. **High-solids materials** are needed to reduce air pollution or emissions when painting.

Binders

The **binder** is the ingredient in a color that holds the pigment particles together. It is the "backbone" or "film-former" of the paint. The binder helps the color stick to the surface. Various materials are used in the binder. The binder determines the type of paint—single-stage or base-clear.

The binder is generally made of a natural resin (such as rosin), drying oils (linseed or cottonseed), or a manufactured plastic. The binder dictates the type of paint because it contains the drying mechanism.

Binder is usually modified with plasticizers and catalysts. They improve such properties as durability, adhesion, corrosion resistance, mar-resistance, and flexibility.

Solvents

The **solvent/reducer** is the liquid solution that carries the pigment and binder so it can be sprayed. **Reduc-**

ers are composed of one or several chemicals. They provide a transfer medium. Solvents are the volatile part of a paint. They are used to reduce (thin) a paint for spraying. Solvents give the paint its flow characteristics. They evaporate as the paint dries.

Most solvents are made from crude oil or petroleum. However, water-based paints are increasing in use to meet strict pollution regulations. The solvent reduces, or thins, the binder and transfers the pigment and binder through the spray gun to the surface being painted.

Remember! Thinning and reducing are needed to give the color the right "thickness" or **viscosity** to flow (spray) out smoothly onto the surface.

Some water-based paints come **premixed** (ready to spray) and they are NOT normally reduced. If required or directed, distilled water can be added to make a thinner, more liquid solution.

When using waterborne materials, the water used for equipment cleaning must be handled correctly.

1. It contains hazardous materials and should be disposed of as hazardous waste.

2. It must NOT be poured down the drain for disposal.

3. It must NOT be combined with other waste solvents such as reducers or thinners. Keep a separate container for storing hazardous water wastes.

Due to clean air regulations, some solvents are no longer being used. To meet clean air regulations, traditional solvents are being replaced by water or other solvents. Check all federal, state, and local ordinances.

Additives

Additives are ingredients added to modify the performance and characteristics of the paint (color). Additives are used to:

1. Speed up or slow down the drying process.

2. Lower the freezing point of a paint.

3. Prevent the paint from foaming when shaken.

4. Control settling of metallic and pigments.

5. Make the paint more flexible when dry.

DRYING AND CURING

Drying is the process of changing a coat of paint (color) or other material (clear) from a liquid to a solid state. Drying is due to evaporation of solvent,

Curing

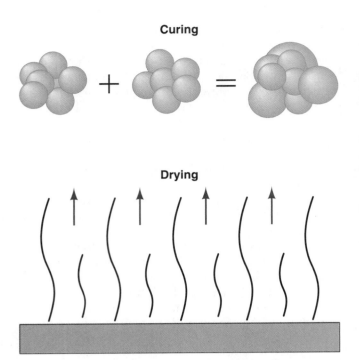

Drying

Figure 8–2 Drying is a process in which solvents evaporate from paint. Curing is a chemical reaction that changes the molecular makeup of the paint. *(I-CAR)*

chemical reaction of the binding medium, or a combination of these causes. See Figure 8–2.

The term drying is used with material that evaporates its solvent to harden. The term curing refers to a chemical action in the paint or other material itself that causes hardening.

Flash time is the first stage of drying when some of the solvents evaporate, which dulls the surface from a high gloss to a normal gloss.

A *retarder* is a slow-evaporating solvent or reducer used to retard, or slow, the drying process. A slow-drying solvent/reducer is often used in very warm weather. If a paint/clear dries too quickly, problems can result.

An *accelerator* is a fast-evaporating solvent or reducer for speeding the drying time. It is needed in very cold weather to make the paint or clear dry in a reasonable amount of time.

A general rule to follow in selecting the proper solvent or reducer is: the faster the shop drying conditions, the slower-drying the solvent or reducer should be. In hot, dry weather, use a slow-drying solvent or reducer. In cold, wet weather, use a fast-drying solvent or reducer.

A *catalyst* is a substance that causes a chemical reaction. When mixed with another substance, it speeds the reaction but does not change itself. Catalysts are used with many types of materials—paints, clears, primers, putties, fillers, fiberglass, plastics, Figure 8–3.

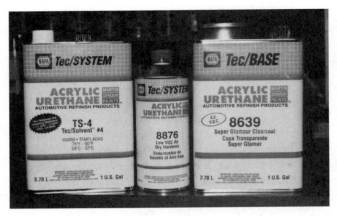

Figure 8–3 A catalyst, or hardener, is added to primers, colors, and clears to make them cure.

A *catalyst* or *hardener* is an additive used to make paint materials cure. The hardener speeds curing and makes the paint more durable.

The hardener is added to the paint right before it is sprayed. When an enamel catalyst is used, the paint can be wet sanded and compounded (polished) the next day. If you make a mistake (paint run, dirt in paint, etc.), you can fix the problem after the short curing time. The hardener will make the color or clear cure in just a few hours. Also, the vehicle can be released to the customer sooner with less chance of paint damage.

DANGER! *Isocyanate resin is a principal ingredient in some urethane catalysts and hardeners. Because this ingredient has toxic effects, you must always wear an NIOSH-approved respirator system!*

PRIMERS AND SEALERS

Primers come in many variations—primer, primer-sealer, primer-surfacer, primer-filler, etc. It is important to understand the functions of subcoating or undercoat materials. You must follow the manufacturer's instructions. Deviation from these directions will result in unsatisfactory work.

A plain primer is a thin undercoat designed to provide good adhesion for the topcoat. Primers can be used when the surface is very smooth and there is no potential problem with bleeding. If properly applied, some primers do not require sanding. See Figure 8–4.

Primers are usually two-component products because they provide better adhesion and corrosion resistance.

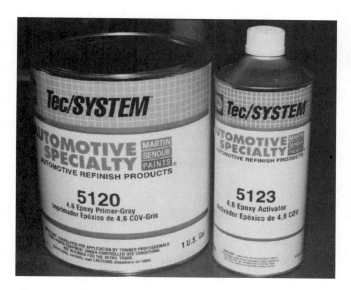

Figure 8–4 Primers are needed for the topcoat to bond to the substrate. There are many variations.

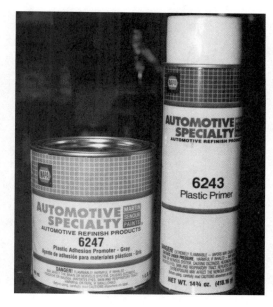

Figure 8–5 This primer-sealer is formulated for plastic parts.

Self-Etching Primer

A *self-etching primer* has acid in it to prepare bare metal so that the primer will adhere properly. A self-etching primer is often used when you have sanded a large area down to bare metal. The self-etching primer will "bite" into the metal to bond securely. This will help prevent lifting and peeling. Some primer-sealers and primer-surfacers have etching materials in them.

Epoxy Primers

An *epoxy primer* is a two-part primer that cures fast and hard. Some material manufacturers recommend epoxy primer prior to the application of body fillers. Using an epoxy primer greatly increases body filler adhesion and corrosion resistance over bare metal. Epoxy primers most closely duplicate the OEM primers used for corrosion protection.

Body fillers, once mixed with the appropriate catalyst, start a chemical reaction. This reaction in turn causes heat. The heat on bare metal will tend to cause condensation that may corrode the metal. Eventually, the plastic body filler will crack and loosen, leaving a corroded area. An epoxy primer protects against moisture entrapment caused by condensation. It can make your repair last longer.

SEALERS

Bleeding or *bleedthrough* is a problem in which colors in the undercoat or old paint chemically seep into the new topcoats. This can discolor the new color.

A *sealer* is a mid-coat between the topcoat (color) and the primer or old finish to prevent bleeding. Seal-

ers differ from primer-sealers in that they cannot be used as a primer. Sealers are sprayed over a primer or primer-surfacer, or a sanded finish. Sealers do not normally need sanding, but some are sandable.

Sealers are sometimes used when a sharp color difference is visible after sanding. They are also used to prevent sand scratch swelling problems.

A *primer-sealer* is an undercoat that improves adhesion of the topcoat and also seals old painted surfaces that have been sanded. It will solve two potential problems (adhesion and bleed) with one application. See Figure 8–5.

Primer-sealers provide the same protection as primers—adhesion and corrosion resistance. But they also have the ability to seal over a sanded old finish to provide uniform color holdout.

PRIMER-SURFACERS

A *primer-surfacer* is a high-solids primer that fills small imperfections and usually must be sanded. It is often used after a filler to help smooth the surface. Primer-surfacers are used to build up and level feather-edged areas or rough surfaces and to provide a smooth base for topcoats. See Figures 8–6 and 8–7.

A good primer-surfacer should have the following characteristics:

1. Adhesion
2. Corrosion resistance
3. Buildup
4. Sanding ease
5. Color holdout
6. Quick drying speed

Figure 8–6 Sealer is needed to prevent undercoats from bleeding through into the new color. New formulas combine sealer into primer-surfacer to save steps. (Oatey Bond Tite)

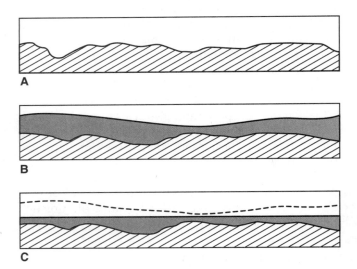

Figure 8–7 Primer-surfacer is the "workhorse" of the collision repair industry. (PPG Industries, Inc.) (A) This magnified cross-section shows that the surface is slightly rough. This might be due to sanding or the texture of plastic filler. (B) Primer-surfacer has been sprayed over surface. It has a high-solids content that flows and fills indentations. (C) Sanding primer-surfacer will quickly level and smooth the surface. This readies the surface for topcoats.

Strong adhesion is the first prerequisite of a primer. All automotive topcoat colors require the use of a primer or primer-surfacer as the first coat over bare substrate. A good primer-surfacer should be ready to sand in as short a period as 30 minutes.

A **primer-filler** is a very thick form of primer-surfacer. It is sometimes used when a very pitted or rough surface must be filled and smoothed quickly. It might be used on a solid, but pitted body panel for example.

USE A COMPLETE SYSTEM!

Remember! Always use a complete refinishing system. A **refinishing system** means all materials (primers, catalysts, reducers, colors, and clears) are compatible and manufactured by the same company. They are designed to work properly with each other. If you mix materials from different manufacturers, you can run into problems. The chemical contents of the different systems may not work well together. See Figure 8–8.

OTHER PAINT MATERIALS

A wax and grease remover is a fast-drying solvent often used to clean a vehicle. It will remove wax, oil, grease, and other debris that could contaminate and ruin the paint job.

A **flattener** is an agent added to paint to lower gloss or shine. It can be added to any color gloss paint to make it a semi-gloss or flat (dull) color. For example, some factory trim is painted semi-gloss or flat black. A flattening agent would be used in this instance. Flatteners can also be used where reflection off a high-gloss paint could affect driver visibility.

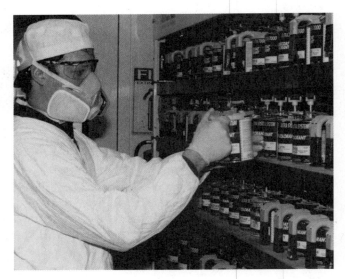

Figure 8–8 When refinishing, always use a complete refinishing system produced by one manufacturer. You will then be sure that all chemicals used in the products are compatible and will work together without problems. This in-shop system stirs cans of material so they are ready for mixing into desired colors. Even though the technician is working in a mixing room with filtered, circulated air, he is still wearing a respirator for added protection.

A *flex agent* is an additive that allows primers and colors to flex or bend without cracking. It is commonly added to materials being applied to plastic bumper covers, for example. Also called an *elastomer,* it is a manufactured compound with flexible and elastic properties that can be added to primers and paints.

Antichip coating, also called *gravel guard, Chip Guard,* or *vinyl coating,* is a rubberized material used along a vehicle's lower panels. It is designed to be flexible or rubbery to resist chips from rocks and other debris flying up off the tires. Antichip coatings are usually applied with a special spray gun, like the one in Figure 8–9.

Many manufacturers are using special chip-resistant coatings in areas that are exposed to stones and gravel. These coatings are generally between the E-coat primer and the topcoats. Some chip-resistant coatings are clear and can be applied over the topcoat. If a vehicle has chip-resistant coatings, they must be replaced during the refinishing process.

Rubberized undercoat is a synthetic-based rubber material applied as a corrosion-preventive layer. It can be applied using a production gun or a spray can.

A *metal conditioner* is an acid used to etch bare sheet metal before priming. It is a chemical cleaner that removes corrosion from bare metal and helps prevent further corrosion.

Remember the following about metal conditioners:

1. The acid cleans the metal.

2. It dissolves light surface rust.

3. It etches metal, improving adhesion.

4. It needs to be completely neutralized after applying.

5. It may have to be diluted, following product directions.

6. It is often followed by conversion coating.

7. Wear rubber gloves and eye protection.

A *conversion coating* is a special metal conditioner or primer used on galvanized steel, uncoated steel, and aluminum to prevent corrosion and aid adhesion. It is applied after acid etching or metal conditioning.

Corrosion is a chemical reaction of air, moisture, or corrosive materials on a metal surface. Corrosion of steel is usually referred to as *rust* or oxidation.

Paint stripper is a powerful chemical that dissolves paint for fast removal. If the paint has failed, you may have to use a chemical stripper. It is applied over the paint. After it soaks into and lifts the paint, a plastic scraper is used to remove the softened paint, Figure 8–10.

A *tack cloth* is used to remove dust and lint from the surface right before painting. It is a cheesecloth

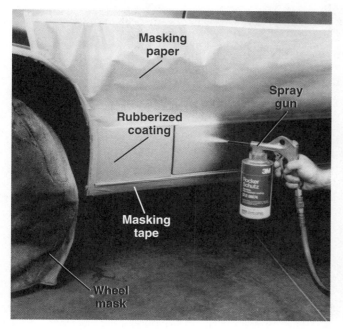

Figure 8–9 Antichip coating is a rubberized material often sprayed along lower panels of a vehicle to help resist chips. *(3M Automotive Trades Division)*

Figure 8–10 Paint stripper is a powerful chemical that will quickly dissolve paint. It is needed only if the paint is peeling, cracking, and will not take a new topcoat. Wear protective gear when using paint stripper. It will cause chemical burns. *(Evercoat Fibreglass Co., Inc.)*

treated with nondrying varnish to make it tacky. A tack cloth must be wiped gently over the surface to keep the varnish from contaminating the paint.

BODY FILLERS

A *filler* is any material used to fill (level) a damaged area. There are several types of filler. You should understand their differences.

Body filler or *plastic filler* is a heavy-bodied plastic material that cures very hard for filling small dents in metal. It is a compound of resin and plastic used to fill dents on vehicle bodies. See Figure 8–11.

Body fillers come canned and in plastic bottles. A dispenser is often used to force the filler onto a mixing board. A *mixing board* is the surface (metal or plastic) used for mixing the filler and its hardener.

Light body filler is formulated for easy sanding and fast repairs. It is used as a very thin, top coat of filler for final leveling. It can be spread thinly over large surfaces for block or air-tool sanding the panel level.

Fiberglass body filler has fiberglass material added to the plastic filler. It is used for corrosion repair or where strength is important. It can be used on both metal and fiberglass substrates. Because fiberglass-reinforced filler is very difficult to sand, it is usually used under a conventional, lightweight plastic filler. See Figure 8–12.

Short-hair fiberglass filler has tiny particles of fiberglass in it. It works and sands like a conventional filler, but is much stronger. *Long-hair fiberglass filler* has long strands of fiberglass for even more strength. It is commonly used when you must repair holes in metal or fiberglass bodies.

Cream hardeners are used to cure body fillers. They usually come in a tube. Once the hardening

Figure 8–12 Fiberglass-reinforced filler is used to improve repair strength. It is much stronger than plain plastic filler. Fiberglass-impregnated filler looks and works like regular plastic filler. It is stronger than plastic filler and is usually waterproof, which makes corrosion repairs last longer. (U.S. Chemicals and Plastics, Inc., Canton, Ohio)

Figure 8–13 Cream hardeners are added to filler to make it cure. Note that hardeners can differ and can deteriorate with age. Always use the type recommended by the filler manufacturer. (Oatey Bond Tite)

cream is mixed in, the plastic filler will heat up and harden. See Figure 8–13.

NOTE! The most common mistake is to use *too much hardener!* The plastic filler will set up or harden in a couple of minutes, before you have time to spread it on the body. Literally "tons" of plastic filler have been wasted because of hardener overuse. Using too much hardener can also cause problems with adhesion and pinholing.

Figure 8–11 Plastic filler comes in cans, buckets, and plastic bags. A dispenser will save time and keep the filler uncontaminated. (Oatey Bond Tite)

GLAZING PUTTY/FINISHING FILLERS

Glazing putty is a material made for filling small holes or sand scratches. It is similar to body-filler but it has more solid content. Putty is applied over the undercoat of primer-sealer or primer-surfacer to correct small surface imperfections. The purpose of glazing putties is to fill excessively rough surfaces or imperfections that cannot be filled with a primer-surfacer.

Spot putty is the same as glazing putty except it has even more solids. Spot putty is recommended for scratches or nicks up to 1/16 inch (1.5 mm) deep. It should NOT be used to fill large surface depressions. For larger depressions, use plastic filler or catalyzed putty.

Two-part putty comes with its own hardener for rapid curing. This is the main advantage of two-part putty. It dries much more quickly. Some two-part putties can be applied over paint to reduce sanding time.

To use two-part putty, follow label directions to mix the right amount of putty and hardener. Use a rubber spreader to work the putty into any surface imperfections. Provide for a slight buildup of material over the imperfection. After adequate curing time, sand the putty down flush with the surrounding surface.

MASKING MATERIALS

Masking materials are used to cover and protect body parts from paint overspray. *Overspray* is unwanted paint spray mist floating around the spray gun. It can stick to glass and body parts and take considerable time to clean off.

Masking paper is special paper designed to cover body parts not to be painted. It comes in a roll. When mounted on a masking machine, masking tape is automatically applied to one edge of the paper, speeding your work. See Figure 8–14. Many collision repair shops use two kinds of masking paper: green masking paper and gold masking paper.

Green masking paper is suitable for masking off primer spray but not paint spray. It is a less-expensive, more porous masking paper. If you use it to mask off for painting, the paint can bleed through and get onto the unpainted surfaces.

Gold masking paper is a nonporous paper designed for masking off paint spray. It is a more-expensive paper that will keep paint from bleeding through onto the masked surface.

Masking plastic is used just like paper to cover and protect parts from overspray. It also comes in rolls and can cover large body areas more easily than paper. Masking plastic is used to cover areas of the

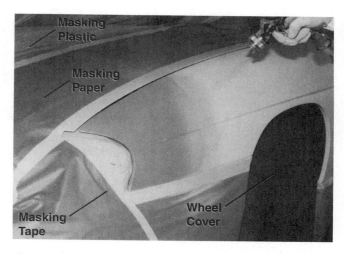

Figure 8–14 Masking paper will protect undamaged surfaces from primer and paint overspray. Masking paper has been placed around the fender for this panel repair. This will keep unwanted overspray off parts that are not to be painted: hood, door, front bumper, etc.

vehicle away from the area being painted. Plastic should not be used right next to the area being sprayed. The plastic will not absorb the paint and can cause paint to drip down onto the body surface. See Figure 8–15.

Wheel masks are preshaped plastic or cloth covers that fit over the vehicle's wheels and tires. Plastic wheel covers are disposable and should be used only once. Cloth covers are reused but should be cleaned off periodically. Preshaped plastic antenna, headlamp, and mirror covers are also available.

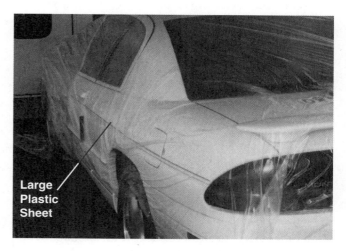

Figure 8–15 Large plastic sheets are sometimes used to mask large sections of a vehicle away from the area to be painted. Plastic can be draped over the body and held in place by a few pieces of tape. Plastic should not be used right next to a panel being painted, however. It will not absorb paint and may allow drips to fall onto the surface.

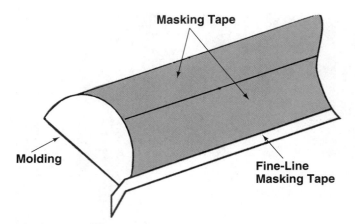

1. Apply masking tape to within 1/8" (3mm) of edge.

2. Apply masking paper.

3. Mask to edge of part with 1/4" (6mm) fine-line tape.

4. Leave a tail or handle.

5. Remove fine-line tape while paint is still wet.

Figure 8–16 Know the applications of masking tape. Note this example of how fine-line tape is used along a rubber molding to produce a good paint edge. Paper masking tape is then used to hold the paper in place next to the fine-line tape. *(I-CAR)*

Masking tape is used to hold masking paper or plastic into position. It is a high-tack, easy-to-tear tape. It comes in rolls of varying widths, ½-inch (19 mm) being the most common. See Figure 8-16.

Fine-line masking tape is a very thin, smooth surface plastic masking tape. Also termed **flush masking tape,** it can be used to produce a better *paint part edge* (edge where old paint and new paint meet). When the fine-line tape is removed, the edge of the new paint will be straighter and smoother than if conventional masking tape were used. See Figure 8–16.

Fine-line tape can help a painter mask flush-mounted parts. It may be better to mask these parts than to remove them. Door handles, side trim, mirror mounts, and moldings are a few examples. Fine-line tape is also used where two different paint colors come together, as when painting stripes or two-tone finishes. See Figure 8–17.

Duct tape is a thick tape with a plastic body. It is sometimes used to protect parts from damage when grinding or sanding. Duct tape is thicker than masking tape and provides more protection for the surface under the tape.

Masking liquid, also called *masking coating,* is usually a water-based sprayable material for keeping overspray off body parts. Some are solvent-based.

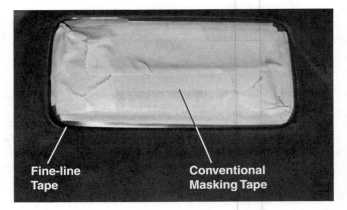

Figure 8–17 This close-up shows how fine-line tape was applied first around the door handle for very accurate masking around the edge. Then, conventional masking tape was applied over the fine-line tape. This is an alternate method preferred by some technicians. Note that it is best to remove door handles before painting, but some estimates do not allow for the extra labor time.

Masking liquid comes in a large, ready-to-spray container or drum. These materials are sprayed on and form a paint-proof coating over the vehicle.

Some masking coatings are tacky and used only during priming and some when painting. They form a film that can be applied when the vehicle enters the shop. Others dry to a hard, dull finish.

These masking coatings can be removed when the vehicle is ready to return to the owner. They wash off with soap and water. Local regulations may require that liquid masking residue be captured in a floor drain trap, and not put into the sewer or storm drain system.

ABRASIVES

An *abrasive* is any material, such as sand, crushed steel grit, aluminum oxide, silicon carbide, or crushed slag, used for cleaning, sanding, smoothing, or material removal. Many types of abrasives are used by the collision repair and refinishing technician.

Grit refers to a measure of the size of particles on sandpaper or discs. A **coarse** sandpaper would have large grit. A **fine** sandpaper would have smaller grit.

Grit Ratings

A *grit numbering system* denotes how coarse or fine the abrasive is. For example, 16 grit would be one of the coarsest and 1500 grit would be one of the finest. The grit number is printed on the back of the paper or disc.

Very coarse grit of 16 to 60 is generally used for fast material removal. It will quickly remove paint and take it down to bare metal. This grit is commonly used on grinding discs and air files paper for rapid cutting of body fillers.

A *coarse grit* of 36 to 60 is basically used for rough sanding and smoothing operations. This coarseness might be used to get the general shape of a large plastic filler area.

Medium grit of 80 to 120 is often used for sanding plastic filler high spots and for sanding paint.

Fine grit of 150 to 180 is normally used to sand bare metal and for smoothing existing paint.

Very fine grit ranges from 220 to about 2000 and is used for numerous final smoothing operations. Larger grits of 220 to 320 are for sanding primer-surfacers, and paint. Finer grits of 400 to 2000 are for color-coat sanding and sanding before polishing or buffing. Very fine grits are usually wet sandpapers to keep the paper from becoming clogged or filled with paint.

Generally, start with the finest grit that is *practical.* A courser grit will cut straighter, but will create courser scratches to fill. See Figure 8–18.

Open- and Closed-Coat Grits

Sandpaper and discs come in either open-coat or closed-coat types of grit.

With an **open coat,** the resin that bonds the grit to the paper only touches the bottom of the grit. About 50 to 70 percent of its surface is covered by grit materials. Open-coat grit will not clog as quickly.

With a **closed coat,** the resin completely covers the grit. This bonds the grit to the paper or disc more securely. About 90 percent of the surface is covered by grit materials. Closed-coat will clog faster than open-coat grit.

GRINDING DISCS

Grinding discs are round, very coarse abrasives used for initial removal of paint, plastic, and metal (weld joints). Some are very thick and do not require a backing plate. Others are thinner and require a **disc back-**

GRIT	ALUMINUM OXIDE	SILICON CARBIDE	AUTO BODY USE
Ultra Fine		1500 1200 800	Colorcoat wet sanding.
Very Fine		600	Colorcoat wet sanding before polishing.
	400 320 280 240	400 320 280 240	Sanding primer-surfacer and paint before painting. Wet or dry sanding paper.
	220	220	Dry sanding topcoat. Also available for wet sanding.
Fine	180 150	180 150	Final sanding of bare metal and smoothing paint. Dry paper.
Medium	120 100 80	120 100 80	Smoothing paint and plastic filler. Dry paper.
Coarse	60 50 40 36	60 50 40 36	Rough-sanding of filler and plastics. Dry paper.
Very Coarse	24 16	24 16	Coarse sanding or grinding to remove paint. Dry paper.

Figure 8–18 Note how sandpaper is made and classified by coarseness. Study common grits of sandpaper and their common uses. (I-CAR)

ing plate mounted on the grinder spindle. They are used for material removal operations.

Grinding disc size is measured across the outside diameter. The most common grinding disc sizes are 7 and 9 inch (175 and 225 mm). The hole in the center of the disc must also match the shaft on the grinder or sander.

SANDPAPERS

Sandpaper is a heavy paper coated with an abrasive grit. It is the most commonly used abrasive in collision repair. Sandpaper is used to remove paints and to smooth primers and fillers. There are many kinds, shapes, and grits of sandpaper. Each has its own advantage.

Sanding discs are round and are normally used on an air-powered orbital sander. They may use Velcro (hook and loop fibers) or a self-stick coating to hold the sandpaper onto the tool sanding pad. See Figures 8–19 and 8–20.

Sanding sheets are square and can be cut to fit sanding blocks. Long sheets are also available for use on air files.

Dry sandpaper is designed to be used without water. Its resin is usually an animal glue. This glue is not water resistant and will dissolve when wet, ruining the sandpaper.

Dry sandpaper is often used for coarse-to-medium grit sanding tasks, like shaping and smoothing plastic filler. One example, 80-grit dry sandpaper is often used on plastic filler. It will quickly cut the filler down.

Wet sandpaper, as implied, can be used with water for flushing away sanding debris that would otherwise clog fine grits. Wet sandpaper comes in finer grits for final smoothing operations before and

Figure 8–20 Velcro paper uses tiny hooks and loops formed into cloth fibers to hold the sandpaper on the pad. The pad and paper both have these fibers. This sandpaper can be removed and used over again. It also runs cooler and lasts longer.

after painting. Wet sandpaper is available in grits from about 220 to 2000.

Wet sandpaper is commonly used to block sand paint before compounding or buffing. Wet sanding will knock down any imperfections in the paint film. Buffing or compounding is then needed to make the paint shiny again.

When using sandpapers:

1. Follow the paint manufacturer's recommendation for use.

2. Sand, usually along the line of sight. If several grades of sandpaper are used on one area, cross sand to eliminate scratches.

3. Adjust one grade finer for hand versus machine sanding.

4. Support the abrasive with a block or pad to avoid finger marks and crowning.

5. Adjust two grades finer when wet sanding, due to faster cutting abrasives.

6. Choose one manufacturer's line so you can learn its cutting characteristics.

SCUFF PADS

Scuff pads are tough synthetic pads used to clean and lightly scratch the surface of paints and parts. Being like a sponge, they are handy for scuffing irregular surfaces, like on door jambs, around the inside of the hood and deck lids, and other obstructed areas. This will clean and lightly scuff these areas so the paint, primer, or sealer will stick. See Figure 8–21.

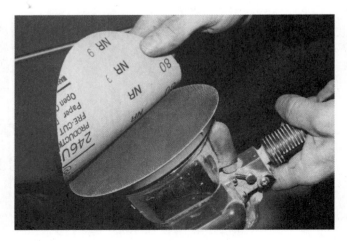

Figure 8–19 Self-stick sandpaper has a thin coating of adhesive already on the back of the sandpaper.

Figure 8–21 A scuff pad is handy for getting into tight, irregular areas, like around door jambs or frames. It will lightly scuff surfaces so new paint will bond properly without lifting or peeling.

COMPOUNDS

Compounding involves using an abrasive paste material to smooth and bring out the gloss of the applied topcoat. It can be applied by hand or with a polishing wheel on an air tool. A compound often has a fine volcanic pumice or dust-like grit in a water-soluble paste. When rubbed on a painted surface, a thin layer of paint or clear is removed. It will remove the very top layer of weathered paint, leaving a fresh new surface of paint.

A *hand compound* is designed to be applied by hand on a rag or cloth. *Machine compound* is formulated to be applied with an electric or air polisher. It will not cut as fast and will not break down with the extra friction and heat of machine application.

Rubbing compound is the coarsest type of hand compound. It will rapidly remove paint or clear, but will leave visible scratch marks. Rubbing compound is designed for hand application, not machine application. It is often used on small areas to treat imperfections in the paint surface.

Polishing compound or *machine glaze* is a fine grit compound designed for machine application. A polisher is used to carefully run the compound over the cured or dried paint or clear. Polishing compound is often used after a rubbing compound or after wet sanding. It will make the paint shiny and smooth.

Hand glazes are for final smoothing and shining of the paint. They are the last process used to produce a professional finish. They are applied by hand using a circular motion, like a wax.

Compounds come in other formulations as well. Read the label on the compound to learn about its use. See Figure 8–22.

Remember, a refinishing technician will generally use the following:

Finer polishing compound   **Coarse rubbing compound**

Figure 8–22 Compounds are used for final smoothing of painted surfaces. Some are coarser than others. Some are designed for hand application and others for machine use. Always read the label directions. *(U.S. Chemicals and Plastics, Inc., Canton, Ohio)*

Figure 8–23 Make sure the compound is the correct type. Some are designed to be applied by hand and others are for machine application.

1. Rubbing compounds to remove surface imperfections in paint.

2. Machine glazes to restore paint gloss after wet sanding. See Figure 8–23.

3. Hand glazes to remove swirl marks after machine buffing.

ADHESIVES

Adhesives are special glues designed to bond parts to one another. Various types are available.

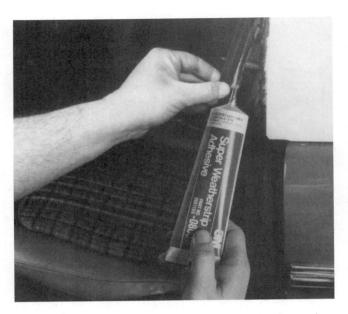

Figure 8–24 Weatherstrip adhesive is commonly used to install rubber seals around doors and trunks. Black adhesive is best on black rubber seals. *(3M Automotive Trades Division)*

Weatherstrip adhesive is designed to hold rubber seals and similar parts in place. Weatherstrip adhesive dries to a hard rubber-type consistency. This makes it ideal for holding door seals, trunk seals, and other seals onto the body. See Figure 8–24.

Plastic adhesive or *emblem adhesive* is designed to hold hard plastic and metal parts. It is used to install various types of emblems and trim pieces onto painted surfaces. See Figure 8–25.

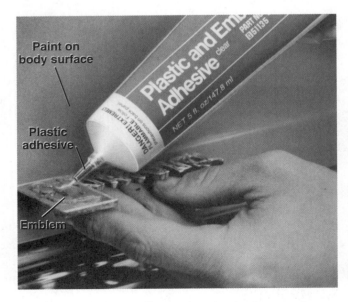

Figure 8–25 Plastic adhesive dries harder than weatherstrip adhesive. It is often used on trim pieces that bond directly to paint. *(3M Automotive Trades Division)*

Figure 8–26 Release agent will dissolve adhesive so that you can remove glued-on parts easily. *(3M Automotive Trades Division)*

An *adhesive release agent* is a chemical that dissolves most types of adhesives. It is used when you want to remove a glued part without damaging it. The agent is sprayed onto the adhesive. This softens the adhesive so the part can be lifted off easily. See Figure 8–26.

EPOXIES

An *epoxy* is a two-part glue used to hold various parts together. The two ingredients are mixed together in equal parts. This makes the mixture cure through a chemical reaction. Always use the type of epoxy suggested by the vehicle manufacturer. See Figure 8–27.

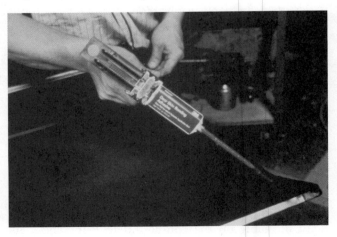

Figure 8–27 Numerous types of two-part epoxies are used to bond parts. Make sure you use the type recommended by the vehicle manufacturer. *(3M Automotive Trades Division)*

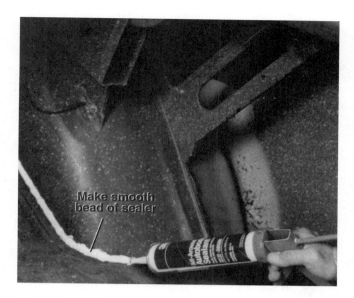

Make smooth bead of sealer

Figure 8–28 Various types of sealers are used to prevent leakage between body parts. This technician is using a caulking gun to seal a newly installed trunk floor panel. *(3M Automotive Trades Division)*

SEALERS

Sealers are used to prevent water and air leaks between parts. They are flexible, which prevents cracking. Sealers come in several variations, Figure 8–28.

Seam sealers are designed to make a leakproof joint between body panels. They are often needed where two panels butt or overlap each other. Seam sealers come in different forms and each is applied differently. Read the directions.

Tube sealers are applied directly from the tube or by using a caulking gun. They squirt out like toothpaste and cure in a few hours.

Apply primer before applying seam sealer. Seam sealers are paintable, but may need to be reprimed if product directions specify. Silicone sealers are NOT paintable and should NOT be used in auto body repair. Follow instructions on the product for finishing sealers.

Ribbon sealers come in strip form and are applied by hand. They are a thick sealer that must be worked onto the parts with your fingers.

NOTE! The proper use or application of these materials will be explained later in this book. Refer to the index if you need more information now.

HAZARDOUS MATERIALS

A *hazardous material* is any substance that can harm people or the environment. Product labels give important procedures for safe use of products to protect you and the environment. Follow all warnings given by product manufacturers. Most of the products used in a collision repair facility carry warnings and caution information that must be read and understood by all users. *Material safety data sheets* (MSDS), available from all product manufacturers, detail chemical composition and precautionary information for all products that can present a health or safety hazard. Employers must know the general uses, protective equipment, accident or spill procedures, and other information for safe handling of hazardous materials. Training about hazardous materials must be given to employees as part of their job orientation. The best way to protect yourself when using paint and body materials is to be familiar with the correct application procedures. Follow all safety and health precautions found on MSDS and on product labels.

Personal protective equipment (PPE) symbols denote what protective gear should be worn when working with hazardous materials. Symbols for rubber gloves, face shields, goggles, respirators, rubber boots, and other equipment may be given.

Right-to-Know Laws give essential information and stipulations for safely working with hazardous materials. The federal courts have decided that these regulations should apply to all companies, including the collision repair and refinishing professions. The general intent of the law is for employers to provide their employees with a safe working place as it relates to hazardous materials.

All employees must be trained about their rights under the legislation, the nature of the hazardous chemicals in their workplace, the labeling of chemicals, and the information about each chemical posted on material safety data sheets. The shop must maintain documentation on the hazardous chemicals in the workplace. It must show proof of training programs and records of accidents and/or spill incidents.

Types of Material Hazards

Irritants are materials that can affect your lungs, skin, and eyes. They can be found in solvents, reducers, polishes, plastic fillers, adhesives, and other materials. They must be life threatening if they affect your lower respiratory system and if you have prolonged exposure to them.

Toxins are poisonous substances that can be divided in several categories. **Neurotoxins** affect your nervous system; they can be found in adhesives and thinners. **Liver toxins** damage your liver; they can be found in reducers and paints. **Reproductive**

toxins can cause birth defects; they are found in gasoline and urethanes. **Blood toxins** can damage your red blood cells; they can be found in enamel clearcoats.

Corrosives can burn your skin and eyes. They are alkalines or acids. Examples are paint strippers, battery electrolyte, and some degreasers.

Carcinogens are substances that can cause cancer. They can be in some air conditioning refrigerants, older brake and clutch dust, and in some plastic fillers.

Allergens can cause allergic reactions. They can be found in adhesives, hardeners, and other substances.

Hazardous Material Exposure

There are several ways that you can be exposed to these harmful materials. You can breath them into your lungs. You can ingest them into your stomach by not washing your hands before eating or drinking in the shop (contaminants settle into food or drink). Absorption results when the hazardous material is absorbed into your skin, mucous membranes, or eyes. Hazardous materials can also enter your body through injection into an open wound or through a skin rash.

To protect yourself from hazardous materials, make sure you wear the recommended protective equipment. If you fail to wear the right kind of protective equipment, you are endangering your most valuable resource, your health!

WASTE DISPOSAL AND RECYCLING

Many shops use "full-service" haulers to test and remove hazardous waste from the property. Besides hauling the hazardous waste away, the hauler will also take care of all the paperwork, deal with the various government agencies, and even advise the shop on how to recover disposal costs.

Some shops are now *recycling* materials (like paint solvent) so that it can be reused to clean spraying equipment. A solvent recycling machine is used to remove impurities from the solvent. These impurities are filtered out and stored in a plastic bag for proper disposal. See Figure 8–29.

NOTE! For more information on hazardous materials, volatile organic compounds, pollution prevention, regulations, laws, liability, regulating bodies, and personal protection, refer to the index. Other chapters discuss these topics where they apply.

Figure 8–29 This recycling unit is purifying solvent used to clean spray guns and mixing containers. The unit will remove pigments and impurities so solvent can be used over again. The plastic bag contains impurities that must be disposed of properly. *(PBR Industries)*

Summary

- Collision repair materials include the various fillers, primers, sealers, adhesives, sandpapers, and other compounds common to collision repair facilities.
- The substrate is the metal, aluminum, plastic, or composite material used in the vehicle's construction. It will affect the selection of refinishing materials.
- There are two general types of paint: enamel and water-base (waterborne).
- A filler is any material used to fill (level) a damaged area.
- Masking materials, such as masking plastic, masking tape, and duct tape are used to cover and protect body parts from paint overspray.
- Sealers are used to prevent water and air leaks between parts. They are applied in the form of seam sealers, tube sealers, and ribbon sealers.

Technical Terms

refinishing materials   *paint*
substrate   *primer*

undercoat
topcoat
colorcoat
basecoat-clearcoat
enamel
two-stage paint
basecoat-clearcoat
    enamel
water-based/waterborne
    paints
OEM finishes
pigments
metallic color
solids
high-solids materials
binder
solvent/reducer
viscosity
premixed
additives
drying
flash time
retarder
accelerator
catalyst
hardener
isocyanate resin
self-etching primer
epoxy primer
bleeding
bleedthrough
sealer
primer-sealer
primer-surfacer
primer-filler
refinishing system
flattener
flex agent
elastomer
antichip coating
gravel guard
Chip Guard
vinyl coating
rubberized undercoat
metal conditioner
conversion coating
corrosion
rust
paint stripper
tack cloth
body filler

plastic filler
mixing board
light body filler
fiberglass body filler
short-hair fiberglass
    filler
long-hair fiberglass
    filler
cream hardeners
glazing putty
spot putty
two-part putty
overspray
masking paper
green masking paper
gold masking paper
masking plastic
wheel masks
masking tape
fine-line masking
    tape
paint part edge
duct tape
masking liquid
masking coating
abrasive
grit
grit numbering system
very course grit
course grit
medium grit
fine grit
very fine grit
open coat
closed coat
grinding disc
disc backing plate
sandpaper
sanding disc
sanding sheet
dry sandpaper
wet sandpaper
scuff pad
compounding
hand compound
machine compound
rubbing compound
polishing compound
machine glaze
hand glazes
adhesives

weatherstrip adhesive
plastic adhesive
emblem adhesive
adhesive release
    agent
epoxy
sealers
seam sealers
tube sealers
ribbon sealers
hazardous material

material safety data
    sheets
personal protective
    equipment
right-to-know laws
irritants
toxins
corrosives
carcinogens
allergens
recycling

Review Questions

1. A _____ is the steel, aluminum, composite, or plastic material used in the vehicle's construction.

2. What are two functions of a vehicle's paint or finish?

3. Define the terms "undercoat" and "topcoat."

4. This refers to a factory paint job.
   a. Lacquer
   b. Enamel
   c. OEM paint
   d. ASE paint

5. Name the four ingredients in a color.

6. The term _____ refers to a paint material that evaporates its solvent to harden. The term _____ refers to a chemical action in the paint or other material itself that causes hardening.

7. What is a catalyst?

8. This is an undercoat that improves adhesion of the topcoat and seals painted surfaces that have been sanded.
   a. Primer
   b. Primer–sealer
   c. Primer–filler
   d. Primer–surfacer

9. A _____ is a high-solids primer that fills small imperfections and usually must be sanded. It is often used after a filler to help smooth the surface.

10. What is corrosion?

11. What is masking liquid or masking coating?

12. In detail, explain the sandpaper grit numbering system and how it is used.

ASE–Style Review Questions

1. *Technician A* says that an undercoat is another name for primer. *Technician B* says that an undercoat is "sandwiched" between the substrate and the topcoat. Who is correct?
   a. Technician A
   b. Technician B
   c. Both Technicians A and B
   d. Neither Technician

2. Which of the following is NOT a paint type?
   a. Enamel
   b. Urethane
   c. Water-base
   d. Bauxite

3. *Technician A* says that OEM stands for Original Equipment Manufacturer. *Technician B* says that manufacturers use several different types of finish materials, coating processes, and application processes. Who is correct?
   a. Technician A
   b. Technician B
   c. Both Technicians A and B
   d. Neither Technician

4. Which of the following is NOT contained in paint?
   a. Solvents
   b. Binders
   c. Graphite
   d. Pigment

5. *Technician A* says that a primer-sealer will fill small imperfections. *Technician B* says that a primer-filler will fill small imperfections. Who is correct?
   a. Technician A
   b. Technician B
   c. Both Technicians A and B
   d. Neither Technician

6. *Technician A* says that it doesn't matter if you use different manufacturers' products when painting a vehicle. *Technician B* says that the same manufacturer's products should be used through the complete refinish process. Who is correct?
   a. Technician A
   b. Technician B
   c. Both Technicians A and B
   d. Neither Technician

7. *Technician A* says that green masking paper is porous and may let paint bleed through. *Technician B* says that gold paper should be used for masking if spraying paint. Who is correct?
   a. Technician A
   b. Technician B
   c. Both Technicians A and B
   d. Neither Technician

8. Which of the following is the finest grit sandpaper?
   a. 230
   b. 24
   c. 500
   d. 1500

9. *Technician A* says that rubbing compound is courser then polishing compound. *Technician B* says that following the use of polishing compound, a hand glaze is the last process used to produce a professional finish. Who is correct?
   a. Technician A
   b. Technician B
   c. Both Technicians A and B
   d. Neither Technician

10. *Technician A* says that epoxies are two-part glues used to hold various parts together. *Technician B* says the seam sealers are designed to make a leakproof joint between body panels. Who is correct?
    a. Technician A
    b. Technician B
    c. Both Technicians A and B
    d. Neither Technician

Activities

1. Visit a "paint supply" house. Read the labels on several different kinds of products. Write a report on the hazards of using different body shop materials.

2. Check out different types of sandpaper. Inspect their grits and determine whether they are open- or closed-coat. Mount different sandpapers on a piece of cardboard and explain each.

3. Apply different types of adhesive and epoxy to a piece of cardboard. Study how long it takes each to dry or cure. Write a report on your findings.

Welding, Heating, and Cutting

Objectives

After studying this chapter, you should be able to:

- Describe when to use and when NOT to use certain welding processes for collision repair.

- List safety precautions taken during welding.

- Name the parts of a MIG welder.

- Summarize how to set up a MIG welder.

- Describe differences between MIG electrode wires.

- Determine if a given 115-volt welder is powerful enough for the application.

- Explain the variables for making a quality MIG weld.

- Know the right techniques for MIG welding.

- Describe the various types of MIG welds and joints.

- Explain the resistance spot welding process.

- Know how to visually inspect and destructive test MIG welds and resistance spot welds.

- Explain the differences in welding aluminum compared to steel.

- Describe plasma arc cutting.

Introduction

This chapter will introduce welding methods common to the collision repair industry. It will define technical terms and compare different types of welding found in a collision repair shop.

With major collision repair work, many of the panels on a vehicle must be replaced and welded into place. As you will learn, this requires considerable skill and care. The structural integrity of the vehicle is dependent upon how well you weld and install panels.

> **NOTE!** If not already trained, you may want to consider taking a MIG welding course in school. Ask your instructor or guidance counselor for more information on welding courses in your school or area. Welding is an essential skill if you plan on becoming a "master auto body technician."

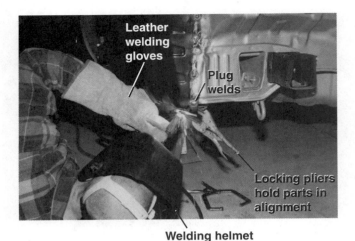

Figure 9–1 Welding is a vital aspect of collision repair. Proper training is needed to make quality welds safely. (Courtesy of Larry Maupin)

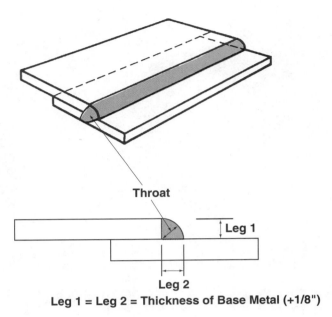

Leg 1 = Leg 2 = Thickness of Base Metal (+1/8")

Figure 9–2 Weld legs should be equal in size and slightly larger than the base metal. (I-CAR)

WHAT IS WELDING?

As defined by the American Welding Society, a **weld** is formed when separate pieces of material are fused together through the application of heat. The heat must be high enough to cause the softening or melting of the pieces to be joined. Once the pieces are heated, pressure may or may not be used to force them together. See Figure 9–1.

The **base material** is the material to be welded, usually metal or plastic. Both can be welded in the collision repair industry.

In many instances, **filler material** from a wire or rod is added to the weld joint. The filler material makes the weld joint thicker and stronger. This chapter will explain the welding of metals. Plastic welding is covered in Chapter 11.

WELD TERMINOLOGY

The **weld root** is the part of the joint where the wire electrode is directed. The **weld face** is the exposed surface of the weld on the side where you welded.

Visible **weld penetration** is indicated by the height of the exposed surface of the weld on the back side. Full weld penetration is needed to assure maximum weld strength.

A **burn mark** on the back of a weld is an indication of good weld penetration. **Burn-through** results from penetrating too much into the lower base metal, which burns a hole through the back side of the metal.

Fillet weld parts include the following. The **weld legs** are the width and height of the weld bead. Note

Figure 9–2. The **weld throat** refers to the depth of the triangular cross section of the weld.

Joint fit-up refers to holding workpieces tightly together, in alignment, to prepare for welding. It is critical to the replacement of body parts!

WELDING CATEGORIES

The three types of heat joining in collision repair are fusion welding, pressure welding, and adhesion bonding.

Fusion welding is joining different pieces of metal together by melting and fusing them into each other. The pieces of metal are heated to their melting point, joined together (usually with a filler rod), and allowed to cool. Molecules of the metals combine. MIG, TIG, stick arc, and oxyacetylene welding are all fusion processes. Refer to Figure 9–3.

There are two general types of gas metal arc welding methods: **metal inert gas (MIG)** and **tungsten inert gas (TIG).**

MIG welding is a wire-feed fusion welding process commonly used in collision repair. MIG stands for "metal inert gas," even though some shielding gases used may be active. It is the accepted industry name for **gas metal arc welding (GMAW).** MIG is sometimes called "wire-feed" because wire is automatically fed into the weld. MIG welding is the only approved method of structural welding on unibody vehicles.

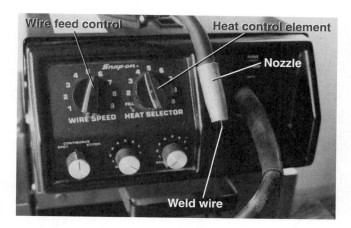

Figure 9–3 MIG stands for metal inert gas. The machine automatically feeds a small wire electrode out of the center of the gun. MIG is the recommended method for repairing collision damage.

Today's steels demand a technique that leaves a narrow heat-affected zone. Oxyacetylene and stick arc welding create too large a heat-affected zone, which will weaken the steel. Also, unibody vehicle designs require 100 percent fusion for most repairs. See Figure 9–4.

TIG welding is generally used in engine rebuilding shops to repair cracks in aluminum cylinder heads and in reconstructing combustion chambers. It has very limited use in collision repair, yet is growing.

The once popular fusion welding methods of shielded arc welding or oxyacetylene welding have been replaced in unibody repair work by MIG welding. The reason is simply that the new steel alloys used in today's vehicles cannot be welded properly with the two previous processes.

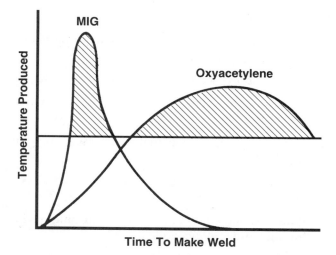

Figure 9–4 This graph compares time and temperature for oxyacetylene and MIG processes. Note how MIG welding is much hotter but takes less time, so the heat spreads less over the panel.

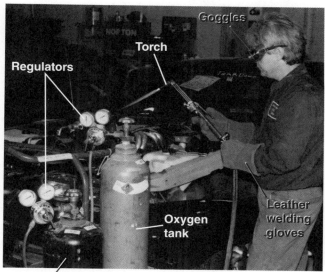

Figure 9–5 An oxyacetylene outfit can be used to cut metals, but it is no longer acceptable for structural welds on vehicles. An oxyacetylene torch is often used to heat parts for cleaning before welding or cutting.

Even though there is full fusion, stick arc welding is NOT recommended for repairs on **high-strength steels (HSS)**. The high heat used with stick arc welding weakens them. Stick arc welding can be used on body-over-frame vehicles and low carbon steel frames, however.

Oxyacetylene welding is a form of fusion welding in which oxygen and acetylene are used in combination. The process is simple. The two are mixed in a chamber, ignited at the torch tip, and used to join a welding filler rod and base metal, Figure 9–5.

Adhesion bonding uses oxyacetylene to melt a filler metal onto the work piece for joining. ***Brazing*** is a form of adhesion bonding. It weakens HSS by applying too much heat, and is NOT recommended for collision repair.

Oxyacetylene and shielded arc (stick) welding are also NOT recommended for most collision repair. They create too much heat for use with high-strength steels. However, they can still be used to repair some thick metals.

Squeeze-type resistance spot welding uses electric current through the base metal to form a small, round weld between the base metals. It is the only accepted form of pressure welding for collision repair. Resistance spot welding focuses the heat onto a small area. A series of pressure spot welds are often used to secure replacement panels onto the body structure of wrecked cars or trucks. This process is discussed later in this chapter.

METAL INERT GAS (MIG) WELDING

The advantages of MIG welding over the other fusion methods used in collision repair are numerous. Vehicle manufacturers now recommend that it be used not only for HSS, but for all structural collision repair.

The advantages of MIG welding are:

1. MIG produces 100 percent fusion in the base metals. This means MIG welds can be ground down flush with the surface (for cosmetic reasons) without loss of strength.

2. Low current can be used to MIG weld thin metals. This can prevent heat damage (warpage) that can cause strength loss.

3. The arc is smooth and the weld puddle small. This ensures maximum metal deposit with minimum spatter.

4. Gaps can be welded by making several spot welds on top of each other—immediately. There is no slag to remove, as with "stick arc" welding.

5. Most steels can be MIG welded with one common type of weld wire.

6. Metals of different thicknesses can be MIG welded together.

7. The technician can control the temperature on the weld and the time the weld takes.

8. Vertical and/or overhead welding is possible because the metal is molten for a very short time.

9. MIG welding is safer to use when making repairs under the vehicle next to brake lines, fuel lines, and fuel tanks because the wire is not energized until the gun trigger is depressed. Again, heat is concentrated at the weld joint.

10. With MIG welding, there is minimum waste of welding consumables.

11. Some aluminum body components can be MIG welded.

MIG Welding Process

MIG welding uses a small diameter welding wire electrode that is automatically fed into the weld joint. The welding machine feeds the roll or wire out to and through the welding gun. A short arc is generated between the base metal and the wire. The resulting heat from the arc melts the welding filler wire and joins the metals. Look at Figure 9–6.

During the welding process, either an inert or active **shielding gas** protects the weld from the

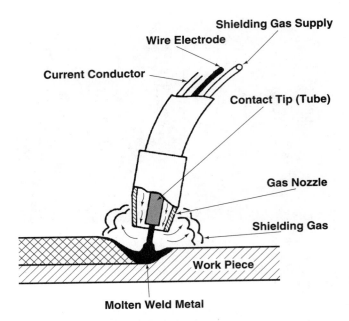

Figure 9–6 Note MIG welder components and their relationships. (Lincoln Electric Co.)

atmosphere and prevents oxidation of the base metal. The type of inert or active gas used depends on the base metal to be welded.

Several types of shielding gases and shielding gas mixtures are used with the MIG process. Argon and helium are primarily used for welding **nonferrous** (noniron) metals, such as aluminum, and can also be used in combination. Carbon dioxide and argon–carbon dioxide mixtures are other popular shielding gases for welding steel.

MIG Welder Parts

Figure 9–7 shows a MIG welder. The major parts of a MIG welder are:

1. Power supply—converts wall outlet AC to DC.

2. Welding gun—feeds wire and gas into arc when trigger is pulled.

3. Weld cable—large wires that connect welding gun to vehicle.

4. Electrode wire—small welding wire fed down through center of gun.

5. Wire feeder—mechanism for pushing wire through gun and into weld arc.

6. Shielding gas—inert gas that surrounds and protects molten metal in weld.

7. Regulator—pressure control device for setting gas pressure.

Figure 9–7 Know the parts of a MIG welder.

8. Weld clamp—ground cable for connecting welder to vehicle.

A **welder amperage rating** gives the maximum current flow through the weld joint. The thickness of the metal being welded determines the amperage needed. Generally, one amp of welding current is needed for every 0.001 inch (0.025 mm) of metal thickness.

A minimum of 90 amps is needed for collision repair, because working with two to three layers of 0.030 inch (0.8 mm) metal is common. When welding on coated steel, the next higher heat (voltage) and wire speed (amperage) settings are used. The type of shielding gas used also affects the amperage required.

Duty cycle is how many minutes the welder can safely operate at a given amperage level continuously over 10 minutes. The duty cycle rating is a percentage. A machine with 40% duty cycle can run 4 out of 10 minutes continuously (at rated amperage). A machine with 60% duty cycle can run 6 out of 10 minutes continuously (at rated amperage).

MIG Welding Gun

The **MIG welding gun,** also called a **torch,** delivers wire, current, and shielding gas to the weld site. It consists of a neck and handle, an on/off trigger, contact tip (or tube), and nozzle, Figure 9–8.

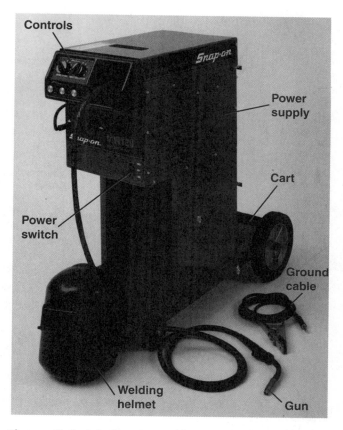

Figure 9–8 A high-quality welder is needed to do quality welds. *(Courtesy of Snap-on Tools Company)*

The **MIG contact tip,** or **tube,** transfers current to the welding wire as the wire travels through. It is usually made of copper or copper alloy. Its inside diameter must match the diameter of the electrode wire used. This is usually stamped on the tip in millimeters. The tip needs to be changed when the inside hole becomes enlarged with use, reducing current transfer efficiency.

The **MIG nozzle** protects the contact tip and directs the shielding gas flow. It must be kept clean. A dirty nozzle can cause holes (porosity) in the weld.

Antispatter compound spray may help keep spatter from sticking to the nozzle. Spatter buildup still has to be cleaned off occasionally. Spattering tends to stick to the nozzle, especially when welding steel, causing porosity. If spatter builds up, it can create a "bridge" between the contact tip and nozzle, and short out the welder.

MIG nozzle types include regular—tapered at the end—and a straight-type primarily for aluminum.

MIG Shielding Gas

MIG shielding gas protects the weld area from oxygen, nitrogen, and hydrogen, which cause porosity in

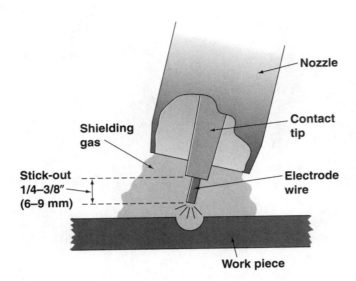

Figure 9–9 Shielding gas protects the weld from airborne contaminants. Stick-out is the distance the wire electrode protrudes from the tip.

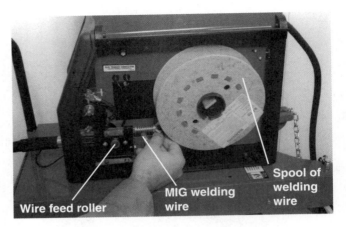

Figure 9–10 MIG welding wire is on a roll that is fed out to the gun. Set up the wire feed mechanism as directed in the operating manual.

the weld. The gas also blows dirt and particles away. It is contained in a highly pressurized cylinder. See Figure 9–9.

The shielding gas cylinder has a flow meter/regulator that meters the gas out of the tank, Figure 9–7. It must be matched to the type of shielding gas used. The flow rate is measured in cubic feet per hour (cfh). The typical flow rate for shielding gases used in collision repair is from 25–30 cfh. This may be increased to 35 cfh if there is a breeze blowing through the work area. There are also preset flow meters that cannot be adjusted.

There are two types of shielding gas:

1. **Active gas** which combines with the weld to contribute to weld quality.

2. **Inert gas** which protects the weld but does NOT combine with the weld.

Shielding gas can be pure or mixed in various combinations. If the primary gas is active, the welding process is called **metal active gas (MAG).** When nearly equal active and inert combinations are used, the process is MIG-MAG. MIG is the accepted collision industry name for any of these processes.

Types of shielding gas include: argon (inert), helium (inert), oxygen (active), and CO_2, or carbon dioxide, (active). The right shielding gas depends on the metal being welded. A gas blend of 75% argon and 25% CO_2 for general quality welds on HSS and mild steel is often recommended. Pure CO_2 is NOT recommended for welding on HSS.

MIG Wire

MIG electrode wire comes in rolls that mount inside the MIG welder, Figure 9–10. Rollers grasp and force the wire out to the gun whenever the gun trigger is pulled.

The MIG wire is identified by an **American Welding Society (AWS)** code. This code reads AWS ER70S-6 and is explained in Figure 9–11. Choose the electrode wire made of the same type of metal being welded. Electrode wire comes in different diameters. Smaller diameters are best for thin-gauge steel because less heat is required to melt them.

Common diameters for collision repair include 0.023 inch (0.6 mm), 0.030 inch (0.8 mm), and 0.035 inch (0.9 mm). Different welding machines will have different recommendations. Generally, 0.023 inch (0.6 mm) can be used on all thicknesses and is a general recommendation for most collision repair. Use 0.030 inch (0.8 mm) wire only on 22 gauge and heavier thicknesses. Use 0.035 inch (0.9 mm) only on 18 gauge and heavier thicknesses.

MIG flux core wire has its own flux contained in a tubular electrode and does NOT require a shielding gas. As with stick welding, the flux forms slag that must be chipped off. Flux core electrode wire is NOT recommended for most collision repair work.

Recently, MIG welders have been made available that run on low voltage. While low-voltage MIG welders do produce more weld spatter, they are used by smaller shops as a means of cutting down on expenses.

MIG Welding Variables

Many welders used for collision repair are 220 volt. There are also welders available that use a standard

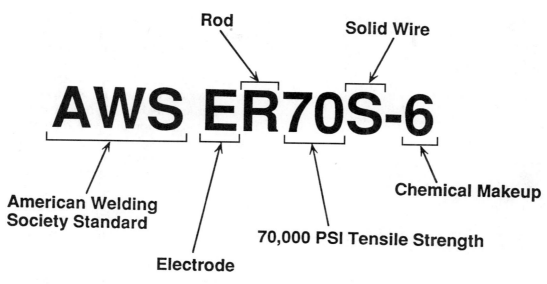

Figure 9–11 Note what each part of the AWS number code means. (I-CAR)

Welding Variables to Change	Desired Changes							
	Penetration		Deposition Rate		Bead Size		Bead Width	
	Increase	Decrease	Increase	Decrease	Increase	Decrease	Increase	Decrease
Current and Wire Feed Speed	Increase	Decrease	Increase	Decrease	Increase	Decrease	No effect	No effect
Voltage	Decrease	Increase	No effect	No effect	Increase	Decrease	Increase	Decrease
Travel Speed	Little effect	Little effect	No effect	No effect	Decrease	Increase	Increase	Decrease
Stickout	Decrease	Increase	Increase	Decrease	Increase	Decrease	Decrease	Increase
Wire Diameter	Decrease	Increase	Decrease	Increase	No effect	No effect	No effect	No effect
Shielding Gas Percent CO_2	Increase	Decrease	No effect	No effect	No effect	No effect	Increase	Decrease
Gun Angle	Backhand to 25°	Forehand	No effect	No effect	No effect	No effect	Backhand	Forehand

Figure 9–12 Study how to change variables to get better welding results.

110-volt outlet. The higher voltage welders are more expensive but usually provide more adjustments.

The operator has control over several variables when MIG welding. These include:

1. Heat (voltage)
2. Wire speed (amperage)
3. Travel speed and direction
4. Electrode stick-out (tip distance from weld)
5. Gun angle
6. Polarity
7. Electrode diameter
8. Shielding gas type and flow rate.

The quality of the finished weld will depend on all of these variables, Figure 9–12. Weld quality will also depend upon weld position, surface condition, part alignment, and how well the workpieces fit

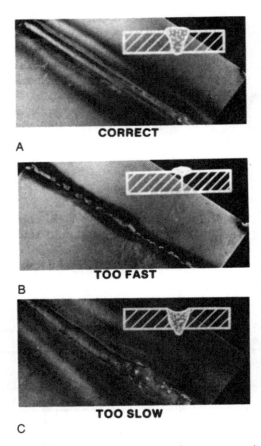

Figure 9–13 Welding speed, or how fast you move the welding gun, affects weld bead. (A) A correct weld has full penetration and proper width and height. This weld would be very strong. (B) Moving the gun too quickly would cause poor penetration. A bead would form only near the top of the base metals. This weld would break easily. (C) If you move the gun too slowly, too much heat will melt the bead down into the joint too much.

together (joint fit-up). The skill of the operator is the final controlling factor of weld quality.

The **heat setting,** or **voltage,** determines the length of the arc. The more voltage, the longer the arc. The longer the arc, the wider and flatter the weld, since the weld wire melts off in larger drops. Too long an arc for the wire diameter used results in spattering. Too short an arc for the wire diameter results in a pulsing sound.

Setting the wire speed when MIG welding also sets the amount of amperage applied to the weld. Changing wire speed and amperage mainly affects weld bead, penetration, height, and width.

To get a steady arc with a steady sound and correct penetration, both the heat (voltage) and wire speed (amperage) must be matched to each other. Set the heat, then the wire speed.

Travel speed is how fast you move the welding gun across the joint. The slower the travel speed, the deeper the weld penetration, the wider the weld, and the more bead height. Look at Figure 9–13.

If weld penetration or width is NOT right, changing travel speed is a second consideration. Changing voltage and wire speed are first considerations. If the height of the weld bead is not right, adjust travel speed first.

Travel direction refers to whether to push or pull the welding gun along the joint. **Push welding** means you aim or angle the gun ahead of the weld puddle. **Pull welding** means you aim or angle the gun back toward the weld puddle. This is shown in Figure 9–14.

Electric **wire stick-out** is the length of unmelted wire that protrudes from the end of the MIG welding gun's contact tip during welding, regardless of whether the tip sticks out of the nozzle or is recessed. Also called **tip-to-work distance** or **contact tip height,** wire stick-out should be kept between ¼ and ⅜

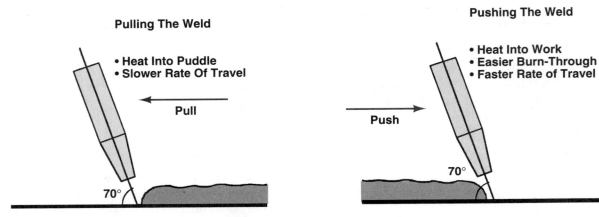

Figure 9–14 When pulling the welding gun, tilt the gun in the direction of the weld. When push welding, tilt the gun away from the direction of the weld. An angle of about 70 degrees should be formed with either method. (I-CAR)

inch (6–9 mm) for 0.023-inch (0.6 mm) wire. Wire stick-out should also be longer for larger diameter wires because of the higher heat (voltage) setting used.

Support the welding gun with the free hand, whenever possible, to control stick-out. Too much electrode stick-out lowers weld penetration. The wire becomes preheated, and melts before penetrating deep enough into the weld.

Not enough electrode stick-out increases weld penetration, because the wire is too hot when contacting the workpiece.

Gun angle consists of two different angles:

1. The angle of the gun to the workpiece.

2. The angle of direction of travel.

The angle of the gun to the workpiece varies, depending on the type of joint. For butt joints, hold the gun at 90 degrees. For "T" joints, hold the gun at 45 degrees. For lap joints, hold the gun at between 60 and 75 degrees. The thicker the workpiece, the greater the angle. The angle of direction of travel is always about 70 degrees from the workpiece, regardless of the type of joint. Look at Figure 9–15.

Even though the welder is plugged into an AC outlet, AC is changed to DC inside the welder. The polarity of the DC power source helps determine how deep the penetration will be into the workpiece.

Straight polarity, or "DC straight," is when the electrode is negative. It is also used when the workpiece, through the work clamp, is positive. In DC straight, most of the heat is on the workpiece.

Straight polarity can be used when welding very thin metal (24 gauge and lighter), to prevent burning through the metal. Straight polarity is also recommended for all positions when welding with flux cored wire.

Reverse polarity, or "DC reverse," is when the electrode is positive and the workpiece is negative. In DC reverse, most of the heat is at the arc. Reverse polarity is commonly used for collision repair, because it provides the best fusion, leaving less weld on the surface to grind off. It also has a more stable arc.

You can change polarity with a toggle switch if the welder has one. You can also reverse cable power supply hookups.

MIG Weld Types

The types of MIG welds include tack, continuous, butt, fillet, and plug welds. You should understand each, Figure 9–16.

A MIG *tack weld* is a short bead used for setup of a permanent weld. Tack welds help joint fit-up. They

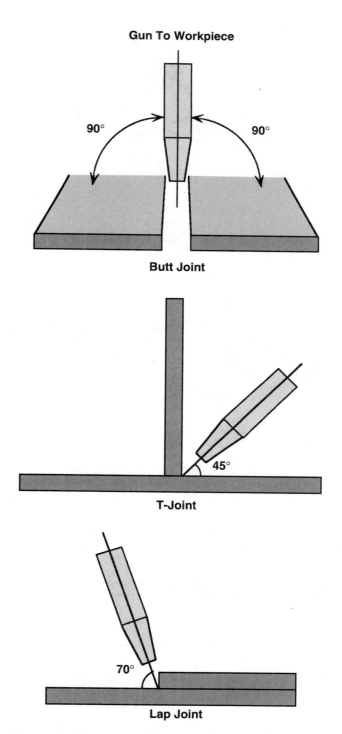

Figure 9–15 Typical welding gun angles for different joints. *(I-CAR)*

are usually placed about every 2 inches (51 mm) along a joint. Tack welds are ground flat if the permanent weld will be on the same side. This is done so there will be no break in the continuous weld.

It is possible to make spot welds with a MIG welder using a spot weld nozzle. A MIG *spot weld* fuses the top piece to the bottom piece without

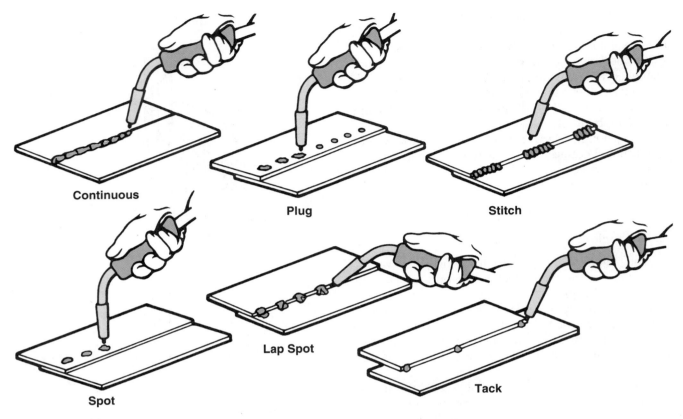

Figure 9–16 Study these six MIG welding techniques.

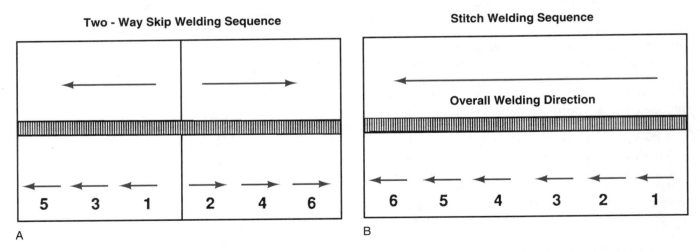

Figure 9–17 Compare stitch and skip welding methods. (A) Two-way skip welding involves running short beads in the sequence shown. Start in the middle and work your way outward following the numbers. (B) Stitch welding is often programmed into a MIG welder. The welder will automatically cycle on and off to allow cooling time between short beads. *(I-CAR)*

pre-drilling a hole. This type of weld is made by holding the nozzle at 90 degrees directly on the base metal and triggering the gun. It is hard to keep consistent quality and strength with MIG spot welds. Plug welds help assure even penetration and strength.

A **continuous weld** is a single weld bead along a joint. It could be a single pass weld. It could also be a series of short welds connected together to make one weld.

Skip welding produces a continuous weld by making short welds at different locations to prevent overheating, Figure 9–17A. All of the small welds are finally joined to produce the continuous weld. This prevents heat buildup in one location on the part.

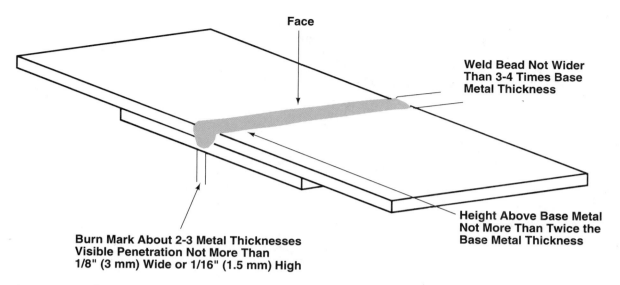

Face

Weld Bead Not Wider Than 3-4 Times Base Metal Thickness

Height Above Base Metal Not More Than Twice the Base Metal Thickness

Burn Mark About 2-3 Metal Thicknesses Visible Penetration Not More Than 1/8" (3 mm) Wide or 1/16" (1.5 mm) High

Figure 9–18 If possible, a backing piece should be placed behind butt welds as shown. This will simplify welding and increase weld strength. *(I-CAR)*

Stitch welding is a continuous weld in one location, but with short pauses to prevent overheating. The operator can pause the machine manually to allow cooling. Some welders can also be programmed to pause automatically to prevent overheating, Figure 9–17B.

MIG *butt welds* are formed by fitting two edges of adjacent panels together and welding along the mating edges. They are often used to make two joints when sectioning frame rails, rocker panels, and door pillars. For butt welds, keep a gap between the two pieces the thickness of one piece. This helps weld penetration and prevents expansion and contraction problems. Also, hold the gun at 90 degrees to the joint.

An insert, or *backing strip,* made of the same metal as the base metal can be placed behind the weld. The backing helps proper fit-up and supports the weld. See Figure 9–18.

A MIG *fillet weld* is a weld joining two surfaces with their edges or faces at about right angles to each other. The joint may be a lap joint or T-joint. Refer back to Figure 9–15.

MIG *lap joints* are welds made on overlapping surfaces. They are commonly used on front and rear rails and floor pans. When welding a lap joint, the gun angle is kept between 60 and 75 degrees. If the lower piece is thicker than the top piece, angle more at the lower piece. Look at Figure 9–19.

A MIG *plug weld* is made through a hole drilled in the top pieces, Figure 9–20. It is started by drilling or punching holes in the top piece or pieces. The hole is filled with a weld nugget. Drill or punch 3/16- to 3/8-inch (5–9 mm) holes in top piece or pieces.

Start around the edge of the hole, then fill in the hole. A 5/16-inch (8 mm) hole works well for most colli-

sion repair. A 3/16-inch (5 mm) hole is better with very thin metals (24 gauge and lighter). A 3/8-inch (9.5 mm) hole is better with heavier metals (14 gauge and heavier), Figure 9–21 on page 169 and Figure 9–22 on page 170.

When plug welds are used to join three or more panels together, holes are punched or drilled in every piece except the bottom piece. The holes are made progressively smaller from the top down. This is done to get better fusion of each layer to the adjacent ones.

Typical hole size combinations are as follows. With three layers of metal, use 5/16-inch (8 mm) and 3/8-inch (9.5 mm) holes. With four layers, use 1/4-inch (6 mm), 5/16-inch (8 mm), and 3/8-inch (9.5 mm) diameter holes.

Where MIG plug welds are used to replace factory spot welds:

1. Follow manufacturer's recommendations for number, size, and location of plug welds.

2. If this information is not available, duplicate the number, size, and location of original factory welds.

WELDING POSITIONS

Weld position refers to the orientation of the work pieces. .

In *flat welding* the pieces are parallel to the bench or shop floor. Flat welding is the easiest because gravity pulls the puddle down into the joint. Flat is used whenever possible. It provides the best control. The weld bead does not flow away from the weld area.

In *horizontal welding* the pieces are turned sideways. Gravity tends to pull the puddle into the

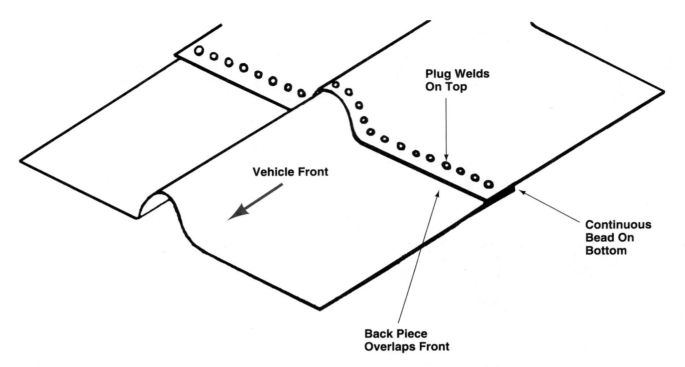

Plug Welds On Top

Vehicle Front

Continuous Bead On Bottom

Back Piece Overlaps Front

Figure 9-19 Sometimes, it is best to use a combination of weld types. With this lap joint of a floor pan, plug welds have been used along the top of the panel. Continuous weld was placed along the bottom. *(I-CAR)*

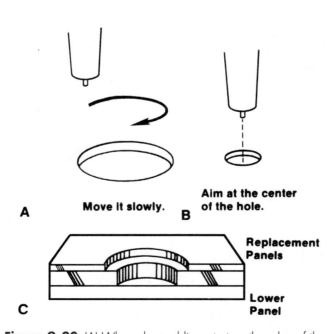

A Move it slowly.

B Aim at the center of the hole.

C Replacement Panels

Lower Panel

Figure 9-20 (A) When plug welding, start on the edge of the hole and circle inward until the hole is filled with a weld puddle. (B) If the plug weld is very small, you can often hold the gun stationary. (C) The top layers of metal must be drilled. With three layers, drill the top layer slightly larger than the second. Do not drill the bottom layer.

bottom piece. The gun can be pulled to the right or left. Pulling directs the heat at the puddle just laid, helping control penetration. When horizontal welding, drop the gun angle about 5 degrees and tip slightly in the direction of travel.

In *vertical welding* the pieces are turned upright. Gravity tends to pull the puddle down the joint. Welding in a vertical position works best when starting from the top and pulling the gun downward. The gun can be tipped downward slightly. A vertical up directs heat into the weld puddle, increasing penetration. Too much heat in the weld puddle makes it fluid, allowing gravity to pull it downward.

In *overhead welding* the piece is turned upside down. This is the most difficult type of weld position. Gravity tries to pull the puddle off the pieces and into the welder tip. Welding in an overhead position can be done by pulling the gun in either direction. Overhead welding takes the most skill because it is usually not as easy to control stick-out and gun angle. The size of the weld bead must be carefully controlled. Too large a weld bead can cause the molten metal to fall and clog the nozzle.

In collision repair, the welding position is usually dictated by the location of the weld on the vehicle. Welding parameters can be affected by the welding position. MIG welding can be done in any position, depending on the location of the base metal.

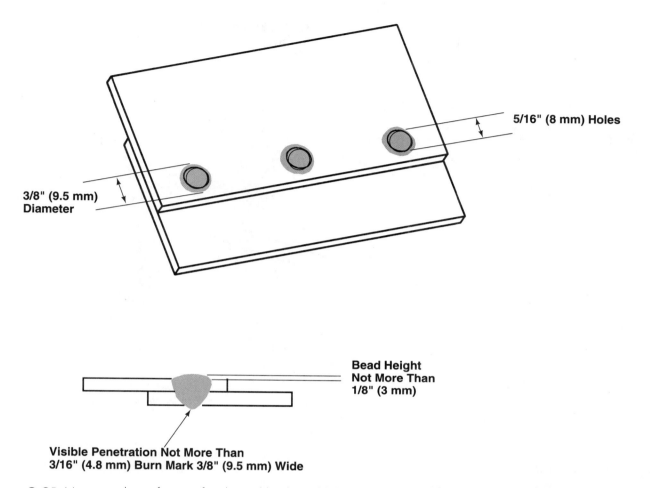

5/16" (8 mm) Holes

3/8" (9.5 mm) Diameter

Bead Height Not More Than 1/8" (3 mm)

Visible Penetration Not More Than 3/16" (4.8 mm) Burn Mark 3/8" (9.5 mm) Wide

Figure 9–21 Note typical specifications for plug welds. The weld diameter should be slightly larger than the hole in the top piece. Also note bead height and penetration. (I-CAR)

Always make test welds on same type and thickness scrap metal in positions to be used before attempting to weld on the vehicle.

WELDING SAFETY

Use common sense while welding. The most important welding safety rules include:

1. Welding voltages and amperages can kill. Keep cables and connections in good shape.

2. Do NOT place the machine on a wet floor, or stand on a wet floor when welding. Keep the shop floor dry.

3. Arc rays are ultraviolet and can cause burns. Sparks or pieces of metal can also shoot out from the weld. Wear a welding helmet with face shield and filter plate.

A *welding filter lens,* sometimes called filter plate, is a shaded glass welding helmet insert for protecting your eyes from ultraviolet burns. They

are graded with numbers from 4 to 12. The higher the number, the darker the filter. The AWS recommends grade 9 or 10 for MIG welding.

Note that there are *"self-darkening" filter lenses* available which instantly turn dark when the arc is struck. There is no need to move the face shield up and down. These filter plates work well, but have one disadvantage in that their viewing window is somewhat small.

4. Wear a heavy shirt with long sleeves. Fasten the top button. Wear pants without cuffs and welding gloves. A leather apron provides good protection.

5. Breathing welding fumes can cause respiratory problems, especially when welding zinc-coated steel. Have adequate ventilation. Wear a respirator when welding.

6. Never have matches or a butane lighter in pockets. They can ignite or explode from the ultraviolet rays of welding.

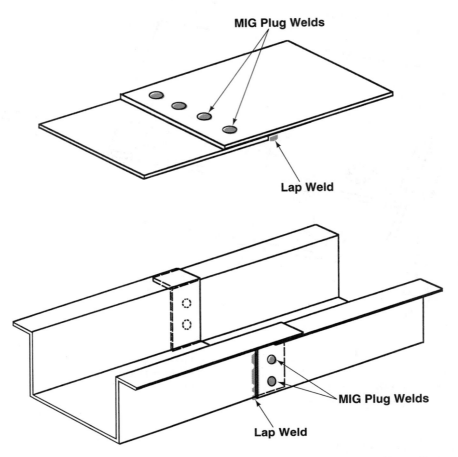

Figure 9–22 Combination plug and lap welds are often recommended when sectioning structural members. *(I-CAR)*

7. To protect the glass and other areas from sparks, use a welding blanket.

8. Protect computers and other electronic parts while welding. They can be easily damaged. Do NOT allow cables to pass near computers or sensors. Remove the computer from the vehicle if welding within 12 inches (305 mm) of the computer.

9. Keep the current path short by placing the work clamp close to the weld location. Keep the power supply as far away from the vehicle as possible.

WELDING SURFACE PREPARATION

To prepare the surface for welding, you must first remove paint, undercoating, rust, dirt, oil, and grease. Use a plastic woven pad, grinder, sander, or wire brush. Avoid removing galvanized coatings. Apply weld-through primer to all bare metal mating surfaces.

The surfaces to be welded must be bare, clean metal. If they are NOT, contaminants will mix with the weld puddle and may result in a weak, defective weld. Be sure to use a weld-through primer on ungalvanized steel or where zinc has been removed.

HOLDING PARTS

Locking jaw pliers (Vise-Grip®), C-clamps, sheet metal screws, and special clamps are all necessary tools for welding. Clamping panels together correctly will require close attention to detail. See Figure 9–23.

Clamping both sides of a panel is not always possible. In these cases, a simple technique using self-tapping sheet metal screws or rivets can be employed.

When using the plug weld technique, the panels are held in place temporarily by setting a screw in every other hole. The empty holes are then plug welded. After the original holes are plug welded, the screws are removed and the remaining holes are then plug welded.

Fixtures can also be used in some cases to hold panels in proper alignment. Fixtures alone, however, should NOT be depended upon to maintain tight clamping force at the welded joint. Some additional clamping will be required to make sure that the panels are tight together and NOT just held in proper alignment.

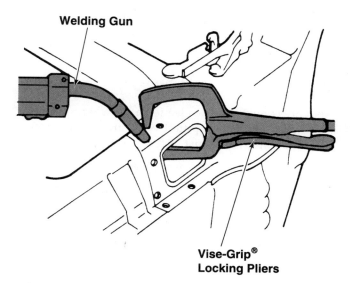

Figure 9–23 Fit-up involves making sure all areas of the panel fit together tightly and are in proper locations. Clamping pliers are used to hold parts in contact while they are being welded. Try to position them close to the area to be welded.

WELD-THROUGH PRIMER

Weld-through primer is used to add corrosion protection to weld zones. This primer must be applied to clean surfaces. Most weld-through primers have poor adhesion qualities. Do NOT overuse them. Always follow directions closely, Figure 9–24.

Weld-through primer can be applied to bare metal mating surfaces where the coating was removed during repair. After welding, remove the excess primer, because it has poor adhesion qualities.

Be sure to select a weld-through primer and NOT just a galvanized spray coating, which may interfere with welding.

After mechanical cleaning, the next step is to chemically clean the metal. Start as soon as possible

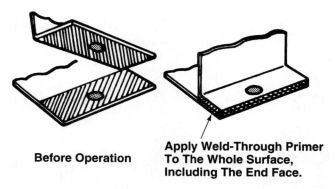

Before Operation

Apply Weld-Through Primer To The Whole Surface, Including The End Face.

Figure 9–24 Weld-through primer should be applied to bare metal before welding. It can be applied with a spray can or brush. Use only a light coat. This will restore proper corrosion protection to bare metal.

after finish sanding, to avoid the formation of light surface corrosion.

USING THE MIG WELDER

Before operating the MIG welder, be sure to:

1. Read the owner's manual carefully. Remember that when in doubt, follow directions.

2. Use the proper power source, ideally on a separate circuit.

3. Use proper wire and shielding gas.

4. Set all drive rollers, cable liners, and tensions to manufacturer's specifications.

5. Set the welder to the proper voltage and speed suggested for the thickness of the metal.

6. Adjust the shielding gas flow rate to the recommended settings.

7. Cover upholstery and glass to prevent spark damage. Sparks can be lessened by making sure the welding area is clean, reducing the voltage setting, and making sure the wire stick-out distance is correct.

8. Allow the weld to cool naturally. Do NOT use water or compressed air to cool the weld. Forced cooling can cause cracking. If the weld can be touched, it is cool enough.

Follow the recommendations for welder maintenance provided in the owner's manual.

MIG BURNBACK

Many MIG welders have a burnback setting to control the amount of time that the electrode is energized after releasing the trigger. ***Burnback*** occurs when the electrode wire melting rate is faster than the wire speed. The electrode wire melts back inside the contact tip and melts to the tip. When this happens, the wire cannot feed and the tip will need replacing.

Adjust the burnback setting while doing test welds. Always perform test welds on a piece of scrap similar to the piece being worked on. There also might be a mode setting for switching the welder to stitch, spot, or continuous mode.

Settings are different based on the type of metal being welded. The recommended settings on the welder will usually be for uncoated steel. Coated steel, compared to uncoated steel, requires heat (voltage) and wire speed (amperage) settings for next higher gauge metal. Also, use a slightly slower travel speed and more vertical gun angle.

WIRE SPEED

MIG *wire speed* is how fast the rollers feed the wire into the weld puddle. An even, high-pitched buzzing sound indicates the correct wire-to-heat ratio producing a temperature in the 9,000°F (4,982°C) range. A steady, reflected light is desirable; it will start to fade in intensity as the arc is shortened and wire speed is increased.

If the wire speed is too slow, a hissing and plopping sound will be heard as the wire melts away from the puddle and deposits the molten glob. There will be a much brighter reflected light.

Too much wire speed will choke the arc. More wire is being deposited than the head and puddle can absorb. The result is the wire melts into tiny balls of molten metal that fly away from the weld, creating a strobe light arc effect.

When welding overhead, the danger of having too large a puddle and balls is obvious. The balls are pulled down by gravity into the contact tip or into the gas nozzle. Therefore, overhead welding should always be done at a higher wire speed, with the arc and balls kept small and close together. Pressing the gas nozzle against the work will keep the wire in the puddle.

SHIELDING GAS FLOW RATE

MIG *gas flow rate* is a measurement of how fast gas flows over the weld puddle. Precise gas flow is essential for a good weld. If the flow of gas is too high, it will flow in eddies and reduce the shielding effect. If there is NOT enough gas, the shielding effect will also be reduced. Adjustment is made in accordance with the distance between the nozzle and the base metal, the welding current, travel speed, and welding environment (nearby air currents). The standard flow rate is 25 cubic feet per hour (0.75 cubic meters per hour).

HEAT BUILDUP PREVENTION

As mentioned previously, too much heat when welding or heating distorts and weakens the metal. Always make sure you do not allow excess heat to transfer into any area of a panel. Use stitch or skip welding methods described earlier.

Stitch and skip welding will prevent costly and time-consuming panel warpage. Another method of preventing heat buildup is heat sink compound.

Heat sink compound is a paste that can be applied to parts to absorb heat and prevent warpage. It comes in a can and can be applied and reused. The heat sink compound is sticky and can be placed on the

panel next to the weld. Heat will flow into the compound and out of the metal to prevent heat damage.

CLEAN AFTER WELDING

Weld areas must also be cleaned after welding. Corrosion will form first in pitted weld areas. Proper surface preparation is a must. A few rules:

1. Thoroughly clean the weld area.

2. Grind cosmetic areas flush with the surface. Use caution NOT to thin the base metal or remove zinc needlessly.

3. Use compressed air to blow the area clean.

RESISTANCE SPOT WELDING

Squeeze-type resistance spot welding has two electrode arms that apply pressure and electric current to the metals being joined. No filler metal or shielding agent is needed. This type is recommended by some European and Japanese vehicle manufacturers. It can be used to replace factory spot welds on some panels and components.

Squeeze-type resistance spot welding is primarily used for cosmetic panels when they are accessible from both sides. It is used for some structural parts like pillars and rocker panels only when recommended by the vehicle manufacturer. This process cannot be used on aluminum.

Resistance spot welding is not new, but there are some recent equipment developments. Modern spot welders for the shop are generally easier to use and more effective for making strong welds. Refer to Figure 9–25.

Figure 9–25 This technician is using a resistance spot welder to install a roof panel.

Squeeze-type resistance spot welding is a type of pressure welding. It is the main process used in the factory. Over 99% of factory welds are resistance spot welds. Factory spot welds are usually done by robots. Extremely high current and pressures are used.

Squeeze-type resistance spot welders for collision repair are less powerful, but use the same process. Two or three sheets of metal are pressed between electrode tips. Low voltage, high current passes through the material. Resistance of the metal to current flow heats and melts the metal, forming a weld as the molten metal cools. Spot weld strength depends on the amount of pressure, current, and weld time, as well as on the fit-up of the metal layers in the joint.

Resistance spot welding is fast, usually taking less than a second per weld. There is little heat effect so it will NOT warp metal. Zinc coating will vaporize only at the point between electrode tips. The surrounding zinc flows into the spot to keep corrosion protection.

A few drawbacks to resistance spot welding should be considered. It requires access to both sides of the panel being welded. Aligning and sizing of tips is critical. A hand-held unit is heavier and more awkward than a MIG gun. It is NOT recommended for welding panels with a combined thickness of over ⅛ inch (3 mm).

Many manufacturers recommend both MIG plug welds and resistance spot welding for replacing factory spot welds. They recommend resistance spot welding for repairs on body panels and some structural areas, including radiator core supports, pillars, rocker panels, exposed pinch weld areas on window and door openings and roofs.

At least one vehicle manufacturer recommends removing the galvanized coating from outside the joint before resistance spot welding. Unless otherwise directed, do NOT remove the galvanized coating. This removes corrosion protection.

Resistance Spot Welder Settings

Basic resistance spot welding parts include:

1. *Transformer with controls*—converts AC into regulated DC.

2. *Arm sets*—hold welding tips and conduct current to tips.

3. *Electrode tips*—allow current flow through spot weld.

Three separate transformer controls are needed for quality welds, including a pressure multiplier, current adjuster, and a timer. These three settings depend on each other for good resistance spot welds. There are also computer-controlled spot welders available that automatically adjust settings depending on the thickness of material dialed in.

On conventional-type spot welders, consider each setting separately.

The *pressure multiplier* usually applies welder arm pressure in a range from 100 pounds (45 kg) for two thicknesses of 25 gauge metal, to about 265 pounds (120 kg) and higher for three thicknesses or thicker metals. This allows clamping pressure to be adjusted for metal thickness. The pressure multiplier can be spring or pneumatically assisted.

Spot welders for collision repair must be able to apply a maximum of at least 250 pounds (113 kg). If there is NOT enough active and follow-up pressure, metal will be thrown out of the weld. If the spot welder relies on operator grip for needed pressure, it cannot be used for collision repair.

Note that pressures at the tips will be less than the settings indicate when using longer arm sets. Generally, doubling the length of the arms reduces pressure by ⅓ or more, Figure 9–26.

The pressure multiplier is NOT used to clamp the pieces to be welded together. Workpieces need to be fit-up tightly with clamps, screws, or other fasteners before welding. Remember that even the slightest gap results in poor welds or no weld.

The *current adjuster* is usually adjustable by percentage of peak current. Generally, the thicker the metal, the higher current required. Zinc-coated steel needs a current setting for the next thickest metal.

For collision repair, the welder needs a peak capacity of at least 7,500 amps measured at the electrodes (NOT at the transformer). The manual should state that the current capacity is measured at the electrodes. Current set too high is another cause for metal throw out.

The *spot weld timer* adjusts the time that the current passes through the joint, usually in ⅟₃₀-, ½-, and 1-second intervals (2, 30, and 60 cycles). It may be automatically set if the welder is computer equipped. The weld timer determines weld size. Generally, the longer the time, the larger the weld diameter.

Excessive spot weld time may cause the electrode tips to mushroom, resulting in no focus of current and a weak weld. A short spot weld time will result in poor penetration and a weak weld. Choose the shortest that will do the job. Longer arm sets need more current and pressure. Straight 5 inch (126 mm) length is used most often.

Electrodes for most collision repair on two to three thicknesses of 22–24 gauge steel have ⅜-inch (9 mm) diameter tips. The electrode diameter at the tips increases with use. This spreads the heat and pressure too much, weakening the weld. Keep electrode tips maintained. Before each extended use, check the tips for wear and size.

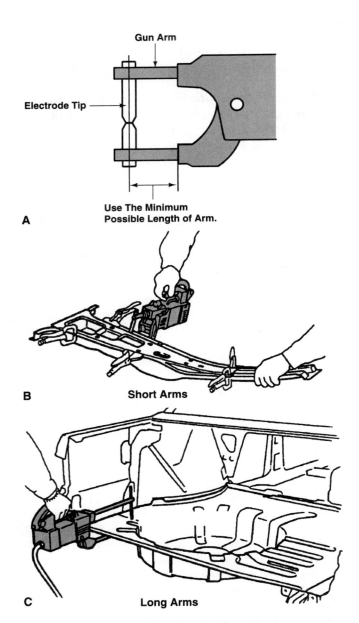

A Gun Arm / Electrode Tip / Use The Minimum Possible Length of Arm.

B Short Arms

C Long Arms

Figure 9–26 Spot weld arm length affects clamping pressure. (A) Use the shortest arm lengths that will work. (B) Short arms will reach many weld areas. (Used with permission from Nissan North America, Inc.) (C) Longer arm length might be needed on very large panels. (Courtesy of Nissan North America)

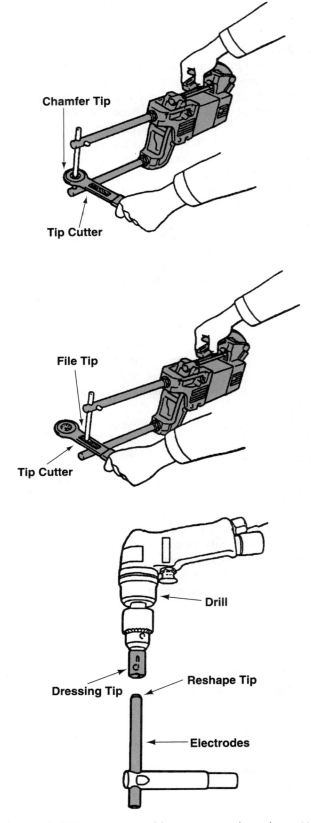

Figure 9–27 Keep spot welder tips in good condition. Use tip cutter and dresser to keep tip size correct. (Reprinted with permission)

Use a hand-held electrode shaper to maintain spot weld tips. A shaper that attaches to a drill is also available. A file can be used, but is NOT as accurate. The ridges left by filing need to be polished with emery cloth or the tips will NOT make full contact with the workpiece. See Figure 9–27.

The spot welder tips must be aligned in a straight line as seen from the front and side. If the tips come together at an angle, the weld will be weak. If the tips do NOT come together at all, the weld area will be

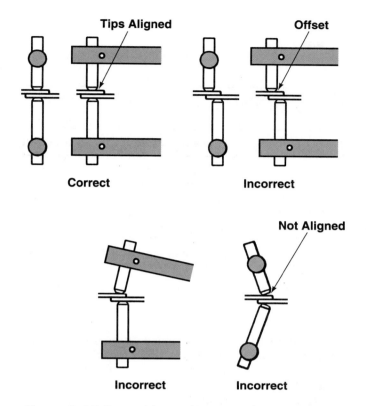

Figure 9–28 Spot welder tip alignment is also important.

offset, resulting in no weld. Always check tip alignment before welding and between groups of welds. Look at Figure 9–28.

The electrode tips also need to be tightly fastened to the arm sets. They tend to loosen during use, affecting current flow and tip alignment.

When servicing welder parts, follow the instructions in the operating manual. It will give directions for the specific make and model of welder.

Resistance Spot Welding Techniques

Check the vehicle's body repair manual to find the original factory welds and number of replacements recommended. You may also compare the number of spot welds on the original panel. Refer to Figure 9–29.

The spot welds should be in the center of the flange without riding the edge. Make sure the original spot weld locations were NOT used. Check that the pitch of the welds is a little shorter than the factory welds. Make sure the spots are evenly spaced (ensuring strength).

Body repair manuals recommend certain spacing, or *pitch,* between resistance spot welds, Figure 9–30. Pitch, and the distance from the end of the panel, determine the strength of individual spot welds. The smaller the pitch, the tighter the bond. However, if welds are too close together, current will flow to a spot already welded and may reduce weld strength. If welds are too close to the edge, current will flow to the edge of the joint, and may also weaken the weld.

Using 30 percent more welds than the original number of factory spot welds is recommended by some vehicle and spot welder manufacturers. Repair spot welders do NOT have the high amperage of factory welders. Also, OEM spot welds are larger—¼" (6 mm)—than repair spot welds, which are ⁵⁄₃₂" (4 mm). These two factors make for a stronger OEM weld in each individual nugget.

Spot weld current, time, and pressure settings are adjusted for:

1. Type of metal

2. Metal thickness

3. Length of arm sets

4. Distance electrode tips are from the metal edge.

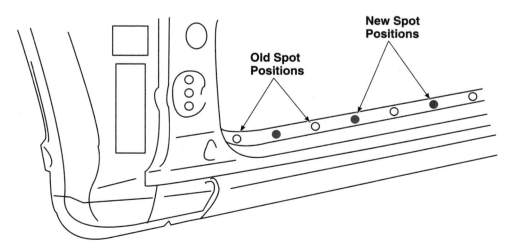

Figure 9–29 When making spot welds, do not weld over old spot welds. Use new locations to assure tight fit and good weld. *(Reprinted with permission)*

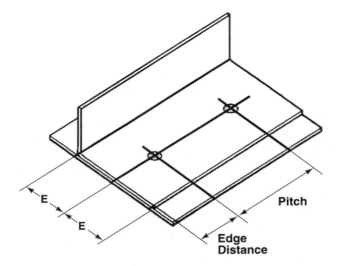

Panel Thickness mm (in)	Pitch (P) mm (in)	Edge Distance (E) mm (in)
0.6 (0.024)	11 (0.43) or more	5 (0.20) or more
0.8 (0.031)	14 (0.55) or more	5 (0.20) or more
1.0 (0.039)	18 (0.71) or more	6 (0.24) or more
1.2 (0.047)	22 (0.87) or more	7 (0.28) or more
1.6 (0.063)	29 (1.14) or more	8 (0.31) or more

Figure 9–30 Weld pitch or spacing and distance from the edge of the panel are important. Note chart recommendations for different panel thicknesses. *(Reprinted with permission)*

When welding three thicknesses of metal, set weld current and weld time for the combined thicknesses. Do NOT weld the same spot twice. This can cause metal throw out and weaken the weld.

As a starting current setting, adjust weld current 30 percent to 50 percent. Adjust timer, if not computer controlled, to about 5 or 10 cycles (½–⅙ second). Position the electrode tips and squeeze the lever. As pres-sure is applied, a built-in switch automatically turns the current on and off for a set length of time. Keep pressure on the switch momentarily after the current shuts off to fuse the weld together.

Check that there are no dents exceeding half the thickness of the panel. A shallow dent is a sign of a strong weld. There should be no visible pin holes nor a lot of spatter. Refer to Figure 9–31.

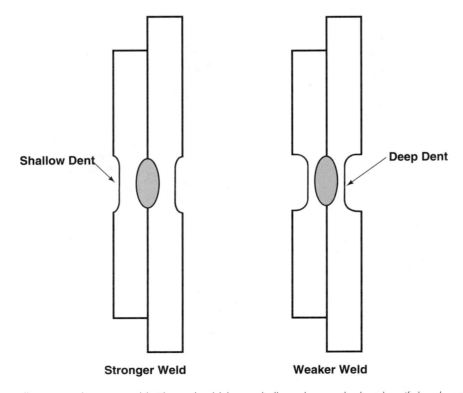

Figure 9–31 Visually inspect the spot weld. There should be a shallow dent on both sides. If the dent is too deep or not deep enough, the weld will be weak. *(I-CAR)*

NOTE! A glove should NOT catch on the surface when rubbed across it. This would indicate a lot of throw out, or too much current used when welding.

There are two basic types of weld inspection: visual inspection and destructive testing.

Visual weld inspection is the most practical method on a repaired vehicle because the weld is not taken apart. Visual weld inspection involves looking at the weld for flaws, proper size, and proper shape.

Destructive testing forces the weld apart to measure weld strength. Force is applied until the metal or weld breaks.

Basically, you have a good weld if metal tears off with weld. You have a bad weld if two pieces break cleanly at the weld with no metal tear-out.

Before making a weld on the vehicle, make a practice weld that you can visually and destructively test. It is important to duplicate the actual weld conditions when testing using the same material, metal thickness, number of pieces, and primer application.

The test weld must be made under the same conditions as the weld that will be made on the vehicle. Part of metal from one piece should tear away. If two pieces break cleanly, make another test weld after increasing weld current slightly. If the weld fails again, increase current again and make another weld. If the weld still fails, increase weld time and test again.

Pull test welds apart until the weld is good, then use the same settings for the welds on the vehicle. See Figure 9–32.

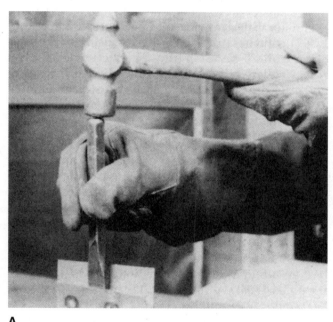

A

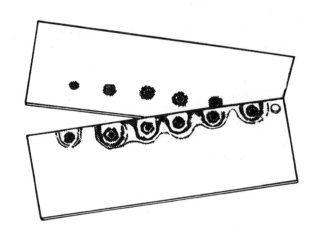

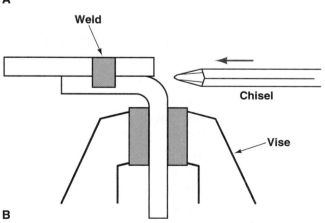

B

C

Figure 9–32 Destructive testing of welds is one way to check weld strength. It can be done on test pieces of the same thickness and material as the welds to be made on the vehicle. (A) Make test welds on scrap pieces. Then use hammer and chisel to force the weld apart. (B) Wedge the chisel down between pieces at the weld point. *(I-CAR)* (C) A bad weld will break at the weld joint. Base metal will not tear because of weld weakness.

Plug welds should have a ⁵⁄₁₆-inch (8 mm) hole. They should have a face diameter between ⅜ and ½ inch (9.5–13 mm). Plug welds should be no more than ⅛ inch (3 mm) above base metal on the face side. They should have a burn mark on the back side ⅜–½ inch (9.5–13 mm) in diameter. Plug welds should have no more than ¹⁄₁₆ inch (1.5 mm) back side visible penetration and have no burn-through.

Butt joints with a backing strip should have a face width 3 to 4 times the thickness of the thinnest base metal. A burn mark width on the back side should be about 2 to 3 times the thickness of the thinnest base metal. There should be visible penetration on the back side no more than ¹⁄₁₆ inch (1.5 mm) below base metal or more than ⅛ inch (3 mm) wide. There should NOT be any burn-through.

Fillet welds on lap joints should have equal leg sizes. Each should not be more than ⅛ inch (3 mm) thicker than the thickness of either base metal. There should be a face no more than 3 metal thicknesses wide. No visible penetration more than ¹⁄₁₆ inch (1.5 mm) below base metal should be visible. There should be no burn-through.

There should be no flaws in the weld measuring over ⅛ inch (3 mm) at their widest point. The ability to recognize weld flaws and possible causes is a good skill to develop.

WELD PROBLEMS

A welding problem causes a weak or cosmetically poor joint that reduces quality. The chart in Figure 9–33 lists weld problems and their possible causes.

1. **Weld Porosity**—holes in the weld

2. **Weld Cracks**—cracks on the top or inside the weld bead

3. **Weld Distortion**—uneven weld bead

4. **Weld Spatter**—drops of electrode on and around weld bead

5. **Weld Undercut**—groove melted along either side of weld and left unfilled

6. **Weld Overlap**—excess weld metal mounted on top and either side of weld bead

7. **Too little penetration**—weld bead sitting on top of base metal

8. **Too much penetration**—burn-through beneath lower base metal

WELDING ALUMINUM

Several vehicles now have body, frame, and chassis parts made of aluminum. Whole bodies made of aluminum are now available. As a result, the information needed for welding aluminum is growing.

Aluminum is light and relatively strong. It is naturally corrosion resistant. Aluminum forms its own corrosion barrier of aluminum oxide when exposed to air. A disadvantage of aluminum is its high cost.

There are some major differences to keep in mind when working with aluminum as opposed to steel, particularly when it comes to welding. Pure aluminum is lightweight and useful more for its ability to be formed than for its strength. When used on vehicles, it is alloyed with other elements and heat-treated for additional strength.

Concerning welding, pound for pound, aluminum is the best conductor of electricity. It conducts heat three times faster than steel. Aluminum becomes stronger, NOT brittle in extreme cold. It is also easily recyclable. Aluminum conducts heat faster than steel, and also spreads heat faster. Therefore, it requires special attention when welding.

Aluminum looks similar to magnesium which, if welded, could start a flash fire. To make sure the part is aluminum, brush the part with a stainless steel brush. Aluminum turns shiny. Magnesium turns dull gray.

DANGER! *If the part is found to be magnesium, do NOT try to weld it. It can start to burn with tremendous heat.*

When welding aluminum, be sure to protect wire harnesses and electronics from potential damage from spreading heat. Aluminum takes more voltage and amperage for the same thickness of material.

Use the following guidelines when MIG welding aluminum:

1. Use aluminum wire and 100 percent argon shielding gas.

2. Set the wire speed faster than with steel.

3. Hold the gun closer to vertical when welding aluminum. Tilt it only about 5 to 10 degrees from the vertical in the direction of the weld.

4. Use only the forward welding method with aluminum. Always push—never pull. When making a vertical weld, start at the bottom and work up.

5. Set the tension of the wire drive roller lower to prevent twisting. But do NOT lower the tension too much or the wire speed will not be constant.

6. Use about 50 percent more shielding gas with aluminum.

Defect	Defect Condition	Remarks	Probable Causes
Pores/Pits (Porosity)	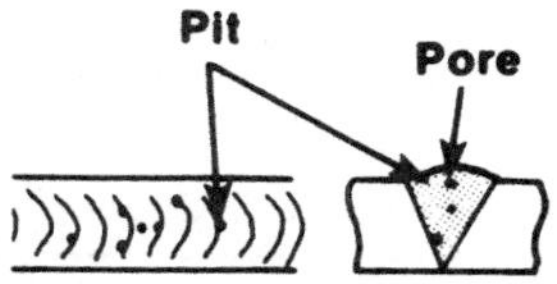 Pit Pore	Holes form when gas is trapped in the weld metal.	Rust or dirt on the base metal. Rust or moisture adhering to the wire. Improper shielding action (the nozzle is blocked or the gas flow volume is low). Weld is cooling off too fast. Arc length is too long. Wrong wire used. Gas is sealed improperly. Weld joint surface is not clean. Electrode contamination.
Undercut	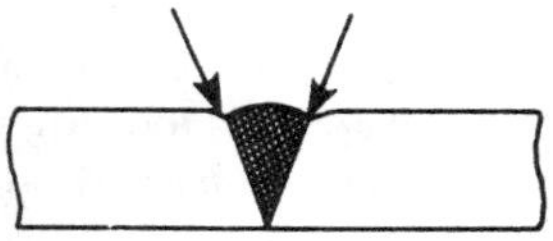	The overmelted base metal creates grooves or an indentation. The base metal's section is made smaller and, therefore, the weld zone's strength is severely lowered.	Arc length is too long. Improper gun angle. Welding speed is too fast. Current is too high. Torch feed is too fast. Torch angle is tilted.
Improper Fusion	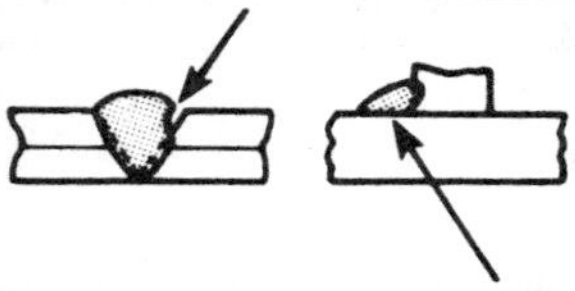	An absence of fusion between weld metal and base metal or between deposited metals.	Improper torch feed operation. Voltage is too low. Weld area is not clean. Improper gun angle.
Overlap		Apt to occur in a fillet weld rather than a butt weld. Causes stress concentration and results in premature corrosion.	Welding speed is too slow. Arc length is too short. Torch feed is too slow. Current is too low.
Insufficient Penetration	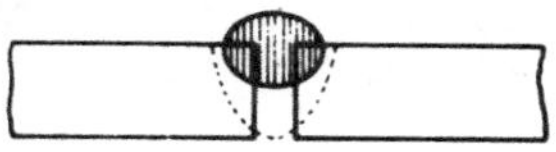	Insufficient deposition made under the panel.	Welding current is too low. Arc length is too long. The end of the wire is not aligned with the butted portion of the panels. Groove face is too small. Gun speed is too fast. Improper electrode extension.
Excess Weld Spatter	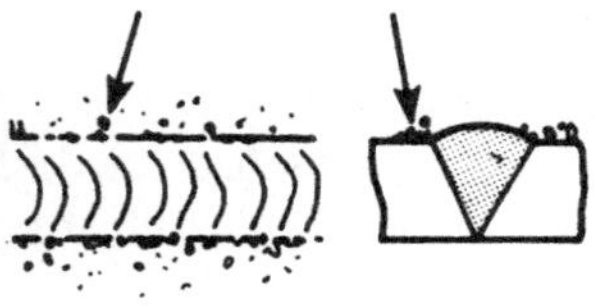	Shows up as speckles and bumps along either side of the weld bead.	Arc length is too long. Rust on the base metal. Gun angle is too severe. Worn tube or drive rolls. Wire spool too tight. Voltage is too high. Wire speed too slow.
Spatter (short throat)		Prone to occur most often in fillet welds.	Current is too high. Wrong wire used.
Waviness of Bead		Uneven bead pattern usually caused by torch mishandling.	Wavering hands. Electrode extension too long.

Figure 9–33 Study weld problems and causes carefully.

Defect	Defect Condition	Remarks	Probable Causes
Cracks		Usually occur on the top surface only.	Stains on welded surface (paint, oil, rust, etc.). Voltage is too high. Wire speed is too fast. Gun speed is too slow.
The Bead Is Not Uniform.		The weld bead is misshapen and uneven rather than streamlined and even.	The contact tip hole is worn or deformed and the wire is oscillating as it comes out of the tip. The gun is not held steady during welding.
Burn Through		Holes in the weld bead.	The welding current is too high. The gap between the metal is too wide. Gun speed is too slow. The gun-to-base metal distance is too short. Reverse polarity to straight polarity (gun negative, work positive).

Figure 9-33 *(Continued)*

7. Because there tends to be more spatter with aluminum, use an antispatter compound to control buildup at the end of the nozzle and contact tip.

8. Shop squeeze-type resistance spot welders do NOT have enough amperage for aluminum. Do NOT use "shop-type" resistance spot welders on aluminum.

9. Always use skip and stitch welding techniques to prevent heat warping. Set wire speed slightly faster. Hold the gun closer to the vertical, compared to steel.

10. Use only the push method when welding aluminum. Pushing helps take away any remaining aluminum oxide which must be cleaned off before welding. When vertical welding aluminum, always start at the bottom and work up.

11. When welding aluminum, there will be more spatter, but the spatter will NOT stick to the nozzle as with steel.

A MIG gun for welding aluminum can be either the standard equipment, or self-contained with a motor in the handle and aluminum wire spool mounted on top. The nozzle is straight, NOT tapered in at the end.

Aluminum electrode wire is classified by series according to the metal or metals the aluminum is alloyed with, and whether or not the aluminum is heat-treated. The series are set up by the Aluminum Association, not the AWS. The number does NOT indicate the strength of the electrode.

Special cleaning instructions are needed for aluminum. Remove all aluminum oxide with a stainless steel brush before welding. The metal may look clean, but aluminum oxide always needs to be brushed off. Clean the metal until it is shiny.

Never use a brush and sanding disc on aluminum after they have been used on steel. If already used on steel, iron powder will remain on the surface of the aluminum and contaminate the weld.

The typical procedure for MIG welding aluminum is as follows:

1. Clean the weld area completely, both front and back. Use wax and grease remover and a clean rag.

2. If the pieces to be welded are coated with a paint film, sand a strip about ¾ inch (19 mm) wide to the bare metal, using a disc sander and No. 80 disc. Do NOT press too hard, or the sander will heat up and peel off aluminum particles, clogging the paper.

3. Clean the metal until shiny with a stainless steel wire brush.

4. Load 0.030 (0.8 mm) aluminum wire into the welder. Trigger it to extend about 1 inch (25 mm) beyond the nozzle.

5. Set the voltage and wire speed according to the instructions supplied with the welding machine. Remember that the wire speed must be faster than for steel. Make a practice weld on scrap pieces.

6. Position the two pieces together and lay a bead along the entire joint. The distance between the contact tip and the weld should be 5/16 to 9/16 inch (8 to 15 mm).

7. If the arc is too large, turn down the voltage and increase the wire speed. The bead should be uniform on top, with even penetration on the backside.

The high heat conduction of aluminum means that the technician must protect against warpage. There are two methods for doing this: stitch welding and center out welding, explained earlier.

Parts made of aluminum are usually ½ to 2 times as thick as steel parts. When damaged, aluminum feels harder or stiffer to the touch because of work hardening.

With steel, the use of heat is avoided whenever possible to prevent reducing the strength of the metal. With aluminum, heat must be used to restore flexibility caused by work hardening. If NOT, it will crack when straightening force is applied.

Before straightening, heat is often applied to the damaged area of the aluminum. It is easy to apply too much heat since aluminum does NOT change color when heated. It also melts at a relatively low 1,220°F (660°C). Careful heat control is very important.

Heat crayons or **thermal paint** can be used to determine the temperature of the aluminum or other metal being heated. They will melt at a specific temperature and warn you to prevent overheating.

As in Figure 9–34, the crayon or paint is applied next to the aluminum area to be heated. The mark will begin to melt when the crayon's or paint's melting point is almost reached. The melting will let you know that you are about to reach the melting point of the aluminum.

Oxyacetylene Welding

Even though oxyacetylene welding is no longer used for collision repair, it can still be used for cleaning and heating on unibody vehicles. Therefore, oxyacetylene welding basics should be understood.

A typical **oxyacetylene welding outfit,** Figure 9–35, consists of the following:

1. **Cylinders**—steel tanks that hold the oxygen and the acetylene.

2. **Regulators**—diaphragm valves that reduce pressure coming from the tanks to the hoses.

3. **Hoses**—lines from the regulators to the torch.

4. **Torch**—mixes oxygen and acetylene in the proper proportions and produces a flame capable of melting steel.

Each regulator has a cylinder pressure gauge and a working pressure gauge. The gauges show the settings of the regulators. Working **oxygen pressure** ranges from 5 to 100 psi (0.35 to 7 kg/cm^2). Working **acetylene pressure** ranges from 1 to 12 psi (0.07 to 0.84 kg/cm^2).

There are two main types of torches: welding and cutting. A **welding torch** has two valves for adjusting

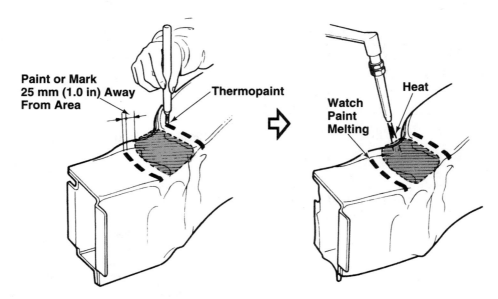

Paint or Mark 25 mm (1.0 in) Away From Area

Thermopaint

Watch Paint Melting

Heat

Figure 9–34 Thermal paint or crayon can be used to prevent overheating. Because aluminum does not change color when heated, it is easy to overheat it and blow holes in parts. Thermal paint or crayon will melt when the melting point of aluminum is almost reached. (© American Honda Motor Co., Inc.)

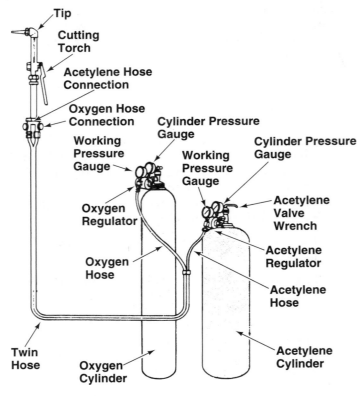

Figure 9–35 Study the parts of an oxyacetylene outfit. The tanks must be chained to the cart when in use to keep them from falling over.

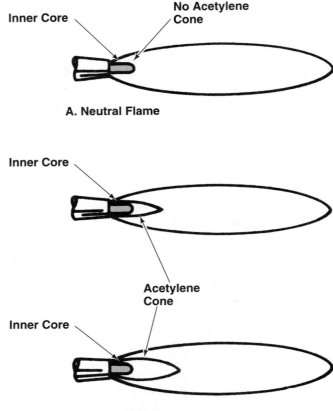

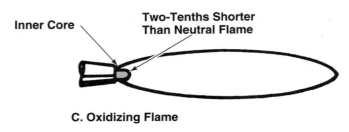

Figure 9–36 Note types of oxyacetylene flames.

gas flow. A **cutting torch** has a third oxygen valve and a lever or trigger for increased oxygen flow. The heat of the flame, combined with the blast of oxygen, will rapidly cut thick steel.

> **WARNING!** *The acetylene line pressure must never exceed 15 psi (1.05 kg/cm^2). Free acetylene has a tendency to become unstable at pressure above 15 psi (1.05 kg/cm^2) and could cause an explosion.*

All oxyacetylene welding should be done with a number 4, 5, or 6 tinted filter lens. A **spark lighter** is a necessity for producing a spark and igniting an oxyacetylene torch.

Flames and Adjustment

When acetylene and oxygen are mixed and burned, the condition of the flame varies depending on the ratio of oxygen and acetylene used.

Shown in Figure 9–36, there are three types of flames:

A **neutral flame** or standard flame is produced by mixing acetylene and oxygen in a 1 to 1 ratio. A neutral flame has a brilliant white core surrounded by a clear blue outer flame.

A **carburizing flame,** also called a reducing flame, is obtained by mixing slightly more acetylene than oxygen. Figure 9–36B shows that this flame has three parts. The core and the outer flame are the same as the neutral flame. However, between them is a light-colored acetylene cone enveloping the core. The length of the acetylene cone varies according to the amount of surplus acetylene in the gas mixture. For a carburizing flame, the oxygen-acetylene mixing ratio is about 1 to 1.4. A carburizing flame is used for heating and welding aluminum, nickel, and other alloys.

An **oxidizing flame** is obtained by mixing slightly more oxygen than acetylene. It resembles the neutral flame in appearance, but the inner core is shorter and its color is a little more violet. The outer flame is shorter and fuzzy at the end. Ordinarily, this flame oxidizes melted metal, so it is NOT used in the welding of mild steel.

Oxyacetylene Torch Flame Adjustment

When using an oxyacetylene welding torch, proceed as follows:

1. Attach the appropriate size tip to the end of the torch. Use the standard size tip for mild steel. Keep in mind that each torch manufacturer has a different system for measuring the size of the tip orifice.

2. Set the oxygen and acetylene regulators to the proper pressure.

3. Open the acetylene hand valve about half a turn and ignite the gas. Continue to open the hand valve until the black smoke disappears and a reddish yellow flame appears. Slowly open the oxygen hand valve until a blue flame with a yellowish white cone appears. Further open the oxygen hand valve until the center cone becomes sharp and well defined. This is a neutral flame and is used for welding mild steel.

4. If acetylene is added to the flame or oxygen is removed from the flame, a carburizing flame will result. If oxygen is added to the flame or acetylene is removed from the flame, an oxidizing flame will result.

Brazing

Brazing is like soldering at relatively low temperatures around 800°F (427°C). An adhesion bond is made by melting a filler metal and allowing it to spread into the pores of the workpiece. The natural flow of the filler material into the joint is called capillary attraction. In brazing, the filler metal does NOT fuse with the workpiece.

Braze welding is classified as either soft or hard brazing, depending on the temperature at which the brazing material melts. **Soft brazing** or **soldering** is done with brazing materials that melt at temperatures BELOW 900°F (468°C). **Hard brazing** is done with

brazing materials that melt at temperatures ABOVE 900°F (468°C).

Brazing is applied only for sealing purposes. This is a method of welding whereby a filler rod, whose melting point is lower than that of the base metal, is melted without melting the base metal. Brass brazing is frequently applied to automotive bodies.

The basic characteristics of brazing are:

1. The pieces of metal are joined at a relatively low temperature where the base metal does NOT melt. Therefore, there is a lower risk of distortion and stress in the base metal.

2. Brazing filler metal has excellent flow characteristics. It penetrates well into narrow gaps and it is convenient for filling gaps in body seams.

3. Because the base metal does NOT melt, it is possible to join metals that would otherwise be incompatible.

4. Since there is no penetration, it has very low strength. Brazed joints cannot resist repeated loads or impacts.

Shutting Down Oxyacetylene Equipment

Always turn off the oxyacetylene torch when it is not in use. Never lay down a lit torch. Turn the torch off when it is not being held in your hand. Do this by first closing the acetylene hand valve, then the oxygen hand valve. Shutting off the acetylene valve first will immediately extinguish the flame. If the oxygen is shut off first, the acetylene will continue to burn, throwing off a great deal of smoke and soot.

Close the main valves on the tops of the cylinders when finished. Then, crack open the torch valves to bleed the hoses of pressure. Finally, close the torch valves to ready it for the next use.

PLASMA ARC CUTTING

Plasma arc cutting creates an intensely hot air stream over a very small area that melts and removes metal. Extremely clean cuts are possible with plasma arc cutting. Because of the tight focus of the heat, there is no warpage, even when cutting thin sheet metal. See Figure 9–37.

Plasma arc cutting is replacing oxyacetylene as the best way to cut metals. It cuts damaged metal effectively and quickly but will NOT destroy the properties of the base metal. The old method of flame cutting just does not work that well anymore.

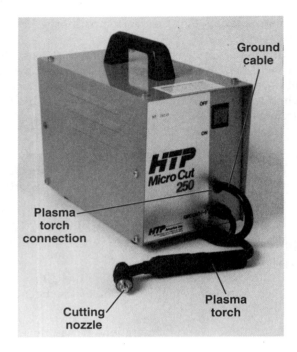

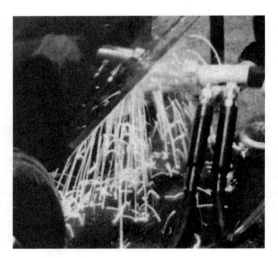

Figure 9–37 The plasma arc welder is designed to rapidly cut metal. (A) Note external parts of a plasma arc cutter. (HTP America, Inc., 1-800-USA-WELD) (B) A plasma arc cutter in use.

In plasma arc cutting, compressed air is often used for both shielding and cutting. As a shielding gas, air covers the outside area of the torch nozzle, cooling the area so the torch does not overheat.

Air also becomes the cutting gas. It swirls around the electrode as it heads toward the nozzle opening. The swirling action helps to constrict and narrow the gas. When the machine is turned on, a pilot arc is formed between the nozzle and the inner electrode. When the cutting gas reaches this pilot arc, it is superheated—up to 60,000°F (3,298°C). Look at Figure 9–38.

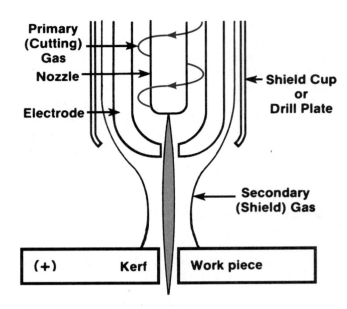

Figure 9–38 Know the plasma arc cutting process.

The gas is now so hot it ionizes and becomes capable of carrying an electrical current (ionized gas is plasma). The small, narrow opening of the nozzle accelerates the expanding plasma toward the workpiece. When the workpiece is close enough, the arc crosses the gap, with the electrical current being carried by the plasma, forming the cutting arc.

The extreme heat and force of the cutting arc melt a narrow path through the metal. This serves to dissipate the metal into gas and tiny particles. The force of the plasma literally blows away the metal particles, leaving a clean cut.

A 10- to 15-amp plasma arc cutter is generally adequate for mild steel up to 3/16 inch thick, a 30-amp unit can cut metal up to 1/4 inch thick, and a 60-amp unit will slice through metal up to 1/2 inch thick.

Controls are usually quite simple. Plasma arc cutters made specifically for thinner metals might have only an on/off switch and a ready light. More elaborate equipment can include a built-in air compressor, variable output control, on-board coolant, and other features.

On some units, a switch is provided that allows the operator to alter the current mode depending on the surface being cut. When cutting painted or rusty metal, a continuous high-frequency arc is best.

Two critical parts of the torch are the cutting nozzle and the electrode. These are the only consumables (besides air) in plasma arc cutting. If either the nozzle or the electrode is worn or damaged, the quality of the cut will be affected. They wear somewhat with each cut. However, moisture in the air supply, cutting thick materials, or poor technique will make them fail more quickly. Keep a supply of electrodes and nozzles on hand and replace them when needed.

Today's plasma arc cutters do an excellent job using clean, dry compressed air. The air can be supplied through an external or built-in air compressor or by using a cylinder of compressed air. Look for equipment designed to run on less air pressure. Cylinders of air can be expensive, while shop air is almost free. To reduce contaminants, use a regulator with a filter.

Also, check the air pressure regularly. Using the wrong pressure can reduce the quality of the cuts, damage parts, and decrease the cutting capacity of the machine.

Operating a Plasma Arc Cutter

To operate a typical plasma arc cutter, proceed as follows:

1. Connect the unit to a clean, dry source of compressed air with a minimum line pressure of 60 psi at the air connection.

2. Connect the torch and ground clamp to the unit. After plugging the machine in, connect the ground clamp to a clean metal surface on the vehicle. The clamp should be as close as possible to the area to be cut.

3. Move the cutting nozzle into contact with an electrically conductive part of the work. This must be done to satisfy the work safety circuit.

4. Hold the plasma torch so that the cutting nozzle is perpendicular to the work surface. Push the plasma torch down. This will force the cutting nozzle down until it comes in contact with the electrode. Then the plasma arc will start. Release downward force on the torch to let the cutting nozzle return to its normal position. While keeping the cutting nozzle in light contact with the work, drag the gun lightly across the work surface.

5. Move the plasma torch in the direction the metal is to be cut. The speed of the cut will depend on the thickness of the metal. If the torch is moved too fast, it will not cut all the way through. If moved too slowly, it will put too much heat into the workpiece and might also extinguish the plasma arc.

Other pointers that should be remembered when using a plasma arc cutter are:

1. When piercing materials ⅛ inch (3 mm) thick or more, angle the torch at 45 degrees until the plasma arc pierces the material. This will allow the stream of sparks to shoot off away from the gas diffuser.

   If the torch is held perpendicular to the work when piercing heavy-gauge material, the sparks will shoot back up at the gas diffuser. The molten metal will collect on the diffuser. This might plug the air holes and shorten the life of the diffuser.

DANGER! *When angling the torch, be aware that the sparks can shoot off as far as 20 feet (6 m) away. Be sure that there are no combustibles or other workers in the area.*

2. Torch cooling is important to extend the life of the electrode and nozzle. At the end of a cut, the air continues to flow for several seconds. This prevents the nozzle and electrode from overheating. Some equipment suppliers also recommend "idling" the unit for a couple of minutes after the cut is made.

3. When making long straight cuts, use a metal straightedge as a guide. Simply clamp it to the work to be cut. For elaborate cuts, make a template out of thin sheet metal or wood and guide the tip along that edge.

4. When cutting ¼-inch materials, start the cut at the edge of the material.

5. When making corrosion repairs on cosmetic panels, it is possible to piece the new metal over the rusted area and then cut the patch panel at the same time that corrosion is cut out. This process also works when replacing a quarter panel.

6. Be aware of the fact that the sparks from the arc can damage painted surfaces and can also pit glass. Use a welding blanket to protect these surfaces.

7. Make sure there is nothing behind the panel that can be damaged. Check for wiring, fuel lines, sound-deadening materials, and other objects that could cause a fire.

Summary

- A weld is formed when separate pieces of material are fused together through the application of heat. The heat must be high enough to cause the softening or melting of the pieces to be joined.

- The three types of heat joining in collision repair are fusion welding, pressure welding, and adhesion bonding.

- MIG (metal inert gas) welding is a wire-feed fusion welding process commonly used in collision repair. TIG (tungsten inert gas) is generally used in automotive engine rebuilding shops to repair cracks in aluminum cylinder heads and in reconstructing combustion chambers, but has very limited applications in collision repair.

- The major parts of a MIG welder are power supply, welding gun, weld cable, electrode wire, wire feeder, shielding gas, regulator, and wire clamp.
- The types of MIG welds are tack, continuous, butt, fillet, and plug welds.
- Plasma arc cutting creates an intensely hot air stream over a very small area, which melts and removes metal. It is replacing oxyacetylene torches as the best way to cut metals.

Technical Terms

weld
base material
filler material
weld root
weld face
weld penetration
burn mark
burn-through
weld legs
weld throat
joint fit-up
fusion welding
metal inert gas (MIG)
tungsten inert gas (TIG)
MIG welding
gas metal arc welding (GMAW)
high-strength steels (HSS)
oxyacetylene welding
adhesion bonding
brazing
squeeze-type resistance spot welding
shielding gas
nonferrous
welder amperage rating
duty cycle
MIG welding gun
torch
MIG contact tip
antispatter compound
active gas
inert gas
metal active gas (MAG)
MIG-MAG
American Welding Society (AWS)
heat setting

voltage
travel speed
push welding
pull welding
wire stick-out
tip-to-work distance
contact tip height
gun angle
straight polarity
reverse polarity
tack weld
spot weld
continuous weld
skip welding
stitch welding
butt weld
insert
backing strip
fillet weld
lap joint
plug weld
flat welding
horizontal welding
vertical welding
overhead welding
welding filter lens
filter plate
self-darkening filter lens
weld-through primer
burnback
wire speed
gas flow rate
heat sink compound
transformer with controls
arm sets
electrode tips
pressure multiplier
current adjuster

spot weld timer
pitch
visual weld inspection
destructive testing
weld porosity
weld cracks
weld distortion
weld spatter
weld undercut
weld overlap
heat crayons
thermal paint
oxyacetylene welding outfit
cylinders
regulators

hoses
torch
oxygen pressure
acetylene pressure
welding torch
cutting torch
spark lighter
neutral flame
carburizing flame
oxidizing flame
brazing
capillary attraction
soft brazing
soldering
hard brazing
plasma arc cutting

Review Questions

1. Explain the difference between a welding base material and a filler material.

2. Define the term "weld penetration." How can you tell if it is correct?

3. List ten advantages of MIG welding.

4. This term refers to metals NOT containing iron.
   a. Nonalloy
   b. Alloy
   c. Nonferrous
   d. Ferrous

5. List and explain eight parts of a MIG welder.

6. The _____ protects the contact tip and directs the shielding gas flow.

7. This type of welding gas combines with the weld to contribute to weld quality.
   a. Inert gas
   b. Active gas
   c. Biodegradable gas
   d. Butane gas

8. Explain the effects of the voltage setting on a MIG welder.

9. _____ is how fast you move the gun across the joint.

10. How does moving more slowly or quickly affect the answer in question nine?

11. Explain the difference between push and pull welding.

12. What is the difference between skip and stitch welding?

13. This is the most difficult welding position.
    a. Horizontal
    b. Vertical
    c. Flat
    d. Overhead

14. A _____ primer is used to add zinc anticorrosion protection to weld zones.

15. Body repair manuals recommend certain spacing, or _____, between resistance spot welds. Pitch, and the _____ from the _____ of the panel, determine the _____ of individual spot welds.

16. List and explain eight welding problems.

ASE–Style Review Questions

1. A technician is practicing welding on an old panel. A loud hissing and popping sound is produced. This is normally caused by:
    a. wire speed or feed too fast
    b. wire speed or feed too slow
    c. poor cable ground
    d. wrong wire size

2. When metal inert gas (MIG) welding, the wire melts into tiny balls of molten metal that fly away from the weld. A strobe light effect is produced. What could cause this welding problem?
    a. wire speed or feed too fast
    b. wire speed or feed too slow
    c. poor cable ground
    d. wrong wire size

3. Which of the following kinds of welding should always be done at a slightly higher wire speed?
    a. Horizontal
    b. Vertical
    c. Overhead
    d. Underhead

4. Technician A says that if the flow of gas is too high, it will reduce the shielding effect. Technician B says that if there is NOT enough gas, the shielding effect will be reduced. Who is correct?
    a. Technician A
    b. Technician B
    c. Both Technicians A and B
    d. Neither Technician

5. Technician A says that you should use stitch or skip welding methods to reduce heat buildup. Technician B says that heat sink compound will reduce the chance of panel warpage when welding. Who is correct?
    a. Technician A
    b. Technician B
    c. Both Technicians A and B
    d. Neither Technician

6. When welding aluminum, which of the following should you use to monitor heat buildup?
    a. Heat crayons
    b. Thermometer
    c. Pyrometer
    d. None of the above

7. Resistance spot welding is NOT recommended for welding panels with a combined thickness of over:
    a. ⅛ inch (3 mm)
    b. ¼ inch (6.35 mm)
    c. ½ inch (12.7 mm)
    d. One inch (25.4 mm)

8. Technician A says that magnesium parts should not be welded because they can begin to burn violently. Technician B says that magnesium is a metal that may start to burn during welding. Who is correct?
    a. Technician A
    b. Technician B
    c. Both Technicians A and B
    d. Neither Technician

9. Technician A says that oxyacetylene is still the most common method of welding structural panels. Technician B says that it should NOT be used for cleaning and heating on unibody vehicles. Who is correct?
    a. Technician A
    b. Technician B
    c. Both Technicians A and B
    d. Neither Technician

10. Which of the following processes often uses compressed air for both shielding and cutting?
    a. Oxyacetylene
    b. MIG
    c. TIG
    d. Plasma arc

Activities

1. Visit your guidance counselor. Ask about welding courses offered in your area. Give a report to the class about your findings.

2. Practice destructive tests of welds. Make MIG and spot welds on scrap metal. Tear them apart with a hammer and chisel. Visually inspect the welds and write a report on your findings.

3. Visit a body shop. While wearing all protective gear and a helmet with an approved lens, observe experienced collision repair welders working. Write a report on what you have learned.

Section 2

Minor Repairs

10

Straightening Fundamentals

Objectives

After reading this chapter, you should be able to:

- Describe different types of metals used in vehicle construction.

- Explain the strength ratings of metals.

- Summarize the deformation effects of impacts on steel.

- Use a hammer and dolly to straighten.

- Explain how to straighten with spoons.

- List the steps for shrinking metal.

- Prepare a surface for filler.

- Properly mix filler and hardener.

- Correctly apply filler.

- Use recommended methods for shaping filler.

- List common mistakes made when using filler and spot putty.

Introduction

This chapter will introduce you to basic metalworking (straightening) methods. It will explain how to analyze minor damage to sheet metal before showing you how to repair the damage. Good metalworking skills are critical to your success as a collision repair technician.

An untrained person can spend more time shaping and sculpting plastic filler than properly reworking the damaged metal. Not only does this waste valuable shop time, but the quality of the repair also suffers. An improperly straightened panel will have tension that can cause the filler to crack, lose adhesion, or fall off. This, of course, does nothing to build customer satisfaction.

To do quality sheet metal repairs, you must first know how to return the sheet metal to its original shape. Then, you can use a thin layer of filler to smooth the surface above the panel. This chapter will help you develop these essential skills.

SHEET METAL

There are two types of sheet metal used in automobile construction—hot-rolled and cold-rolled.

Hot-rolled sheet metal is made by rolling at temperatures exceeding 1,472°F (792°C). It has a standard manufacturing thickness range of 1/16 to 5/16 in. (1.6 to 7.9 mm). It is often used for comparatively thick parts such as frames and crossmembers.

Cold-rolled sheet metal is hot-rolled sheet metal that has been acid rinsed, cold rolled thin, then **annealed** (reheated and then cooled in a controlled manner to strengthen the metal and prevent it from becoming brittle). It has a dependable thickness accuracy, surface quality, and better workability than hot-rolled steel. Most unibodies are made from cold-rolled steel.

Low-carbon or **mild steel (MS)** has a low level of carbon and is relatively soft and easy to work. Much of the sheet metal used in vehicles today is low-carbon or mild steel. It can be safely welded, heat shrunk, and cold worked without seriously affecting its strength. Mild steel has a yield strength of up to 30,000 psi (2,100 kg/cm^2).

Because MS is easily deformed and relatively heavy, vehicle manufacturers have begun using high-strength steels in load carrying parts of the vehicle.

High-strength steel (HSS) is stronger than low-carbon or mild steel because of a heat treatment. Most new vehicles contain HSS in their structural components. It has a yield strength of up to 60,000 psi (4,200 kg/cm^2). HSS experiences an increase in stress, exceeding this yield strength, when deformed during a collision.

The same properties that give strength offer some unique challenges. When high-strength steel is deformed on impact, it is more difficult to restore than mild steel.

STEEL STRENGTH

When flat sheet steel is formed into a shape for a panel or a part, it takes on certain properties that harden it.

For example, a roof panel is relatively flat. If hit lightly in the center, the panel will usually bend and then pop back to its original shape. However, if you hit a panel with a curved shape, the panel will hardly move. Although both are the same steel, the one that has been changed the most will be stronger and more resistant to bending.

The same is true for panels whose shape has been changed during a collision. The structure of the metal in the affected areas has changed, causing the metal to

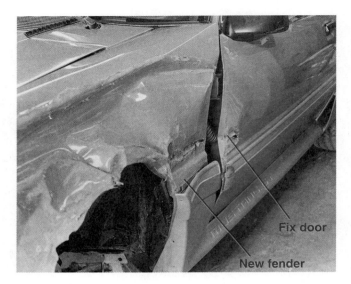

Figure 10–1 Before starting work on straightening a panel, evaluate the damage. Would it be cheaper to purchase a new or used panel?

become harder and more resistant to corrective forces. See Figure 10–1.

To repair collision damage, a technician should understand what property changes have taken place in the metal.

Deformation refers to the new, undesired bent shape the metal takes after a collision or impact. There are various ways to measure the strength of a metal. All relate to the metal's ability to resist deformation.

1. **Tensile strength** is the property of a material that resists forces applied to pull it apart. **Tension** includes both yield stress and ultimate strength. **Yield stress** is the amount of strain needed to permanently deform a test specimen. **Ultimate strength** is a measure of the load that breaks a specimen. The tensile strength of a metal can be determined by a tensile testing machine.

2. **Compressive strength** is the property of a material to resist being crushed.

3. **Shear strength** is a measure of how well a material can withstand forces acting to cut or slice it apart.

4. **Torsional strength** is the property of a material that withstands a twisting force.

Strength is expressed in pounds per square inch (psi) or kilograms per centimeter squared (kg/cm^2).

Even though most types of steel look alike, there are differences in their chemical makeup and crystalline structure. These invisible differences can affect strength and sensitivity to heat. There is a variety of high-strength steels. All have unique properties that dictate the way in which they can be repaired.

PHYSICAL STRUCTURE OF STEEL

Steel, just like all matter, is composed of atoms. These very small particles of matter are arranged to form grains that can only be seen with a microscope. Grains are formed into patterns called the grain structure.

The *grain structure* in a piece of steel determines how much it can be bent or shaped. To change the shape of flat sheet steel, you must change the shape and position of all the individual grains that are located in the area of the creases, folds, or curves.

In mild steels, the individual grains can withstand a considerable amount of change and movement before splitting or breaking. To demonstrate this, bend a piece of automotive sheet steel (part of a fender) back and forth several times. Notice that in the bend, the metal will become very hot. The heat is generated by the internal friction created as the individual grains move against each other.

EFFECT OF IMPACT FORCES

The grain pattern of a metal will determine how it reacts to force. Sheet metal's resistance to change has three properties: elastic deformation, plastic deformation, and work hardening. All of these properties are related to the yield point.

Yield point is the amount of force that a piece of metal can resist without tearing or breaking.

Elastic deformation is the ability of metal to stretch and return to its original shape. For example, take a piece of sheet metal and gently bend it to form a slight arc. When released, it will spring back to its original shape.

Spring-back is the tendency for metal to return to its original shape after deformation. It will occur in any area that is still relatively smooth. Many such areas will spring back to shape if they are released by relieving the distortion in the buckled areas.

Plastic deformation is the ability of metal to be bent or formed into different shapes. When metal is bent beyond its elastic limit, it will have a tendency to spring back. However, it will NOT spring back all the way to its original shape. This is because the grain structure has been changed.

Plasticity is important to the collision repair technician because both stretching and permanent deformation take place in various areas of most damaged panels.

WORK HARDENING

Work hardening is the limit of plastic deformation that causes the metal to become hard where it has

been bent. For example, if a welding rod is bent back and forth several times, a fold or buckle will appear at the point of the bend. The plastic deformation has been so great that the metal will be very hard and stiff at the bend.

The importance of understanding how metal stiffens, making it stronger in areas that are bent or worked, cannot be overemphasized. It is the basis of practically all collision repair.

KINDS OF DAMAGE

A part is *kinked* when:

1. It has a sharp bend of a small radius, usually more than 90 degrees over a short distance, Figure 10–2.

2. After straightening, there is a visible crack or tear in the metal or there is permanent deformation that cannot be straightened to its precollision shape without the use of excessive heat.

A part is *bent* when:

1. The change in the shape of the part between the damaged and undamaged areas is smooth and continuous.

2. Straightening the part by pulling restores its shape to precollision condition without any areas of permanent deformation.

USING BODY HAMMERS

The body hammer is often used to remove small dents in sheet metal parts. The body hammer is

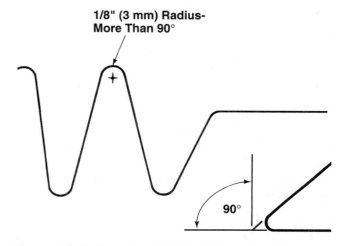

Figure 10–2 The radius of the bend determines whether you have a kink or a bend. A kink generally results when metal is folded more than 90 degrees.

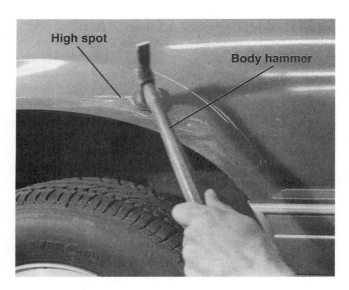

Figure 10–3 A body hammer is used to lower high spots. When using a hammer, swing at your wrist so that the hammer hits flat. If you swing with your arm and you do not hit squarely, the edge of the hammer head will form a dent.

designed to strike the sheet metal and rebound off the surface. Refer to Figure 10–3.

A **high spot** or **bump** is an area that sticks up higher than the surrounding surface. A **low spot** or **dent** is just the opposite; it is recessed below the surrounding surface. Minor low and high spots in sheet metal can often be fixed with a body hammer.

The term **raising** means to work a dent outward or away from the body. The term **lowering** means to work a high spot or bump down or into the body.

The secret of metal straightening is to hit the right spot at the right time, with the right amount of force. When using a body hammer, swing in a circular motion at your wrist. Do not swing the hammer with your whole arm and shoulder. Hit the part squarely and let the hammer rebound off the metal. Space each blow ⅜ to ½ inch (9.5 to 12.7 mm) apart until the damaged metal is level.

The face of the hammer must fit the contour of the panel. Use a flat face on flat or low-crown panels. Use a convex-shaped or high-crown face when bumping inside curves.

Heavy body hammers should be used for roughing out the damage. **Finishing,** or **dinging, hammers** should be used for final shaping. The secret to finish hammering is light taps. It is also important to hit squarely. Hitting with the edge of the hammer will put additional nicks in the metal.

STRAIGHTENING WITH DOLLIES

In the rough-out phase, a heavy steel dolly is sometimes used as an impact tool. Dollies are often used as striking tools on the back panels. Sometimes you can reach into obstructed areas with a steel dolly more easily than with a hammer. You can strike the back side of the dented panel with the dolly to raise low areas and to unroll buckles.

The contour of the dolly must fit the contour of the back side of the damaged area. This will make the blows from the dolly force the metal back into its original contour. If the wrong surface hits the panel (sharp edge of the dolly, for example), you will cause further damage to the panel.

Start out with light blows from the dolly while watching the front of the panel. Make sure you are hitting exactly where needed. Gradually increase the force of your blows to raise the damage. It is normally better to use several moderate blows than to use a few hard blows. Numerous well-placed blows with the dolly will let you better control how you work the metal back into shape.

As you hit the panel, the dolly tends to rebound slightly. This creates a secondary lifting action on the metal. You can increase rebound blows by releasing pressure as soon as the dolly hits the panel. Using a large dolly will also increase impact and rebound forces on the panel.

HAMMER-ON-DOLLY

Hammer-on-dolly is a method used to exert a powerful but concentrated smoothing force to a small area on a damaged panel. The dolly is held against the back of the damage and the hammer hits the metal right over the top of the dolly. This exerts a pinching force on the metal between the dolly and the hammer head. A small area of damaged metal is crushed and flattened between the faces of the dolly and hammer. See Figure 10–4.

Hammer-on-dolly straightening requires you to repeatedly move the point of hammer impact and dolly slightly. Each blow should overlap the next. By repeatedly moving hammer-on-dolly blows, you can steadily work out minor damage over a large area. Generally, try to work out the damage methodically. Start at the outside and gradually work toward the center of the damage. See Figure 10–5.

When learning hammer-on-dolly straightening, you may want to practice on an old scrap or discarded panel. Practice making light blows to the correct locations right over the top of the dolly. Make sure the hammer head hits the panel squarely. If you hit with the edge of the hammer head, a unwanted half-moon dent will be formed.

A proper hammer-on-dolly blow will make a slight high-pitched "ping" sound. The force of the blow goes into the panel and then into the dolly. Hit-

HAMMER-ON-DOLLY

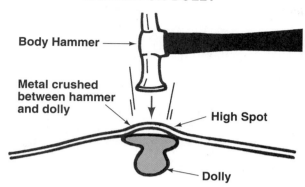

Figure 10–4 With the hammer-on-dolly method, place the dolly right behind the damage and hit the metal right over the dolly to straighten the metal in the small area between the tools.

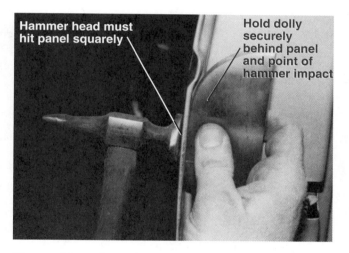

Figure 10–5 This technician is using a dolly and flat face body hammer to straighten a dent in the driver's door edge. The flat surface of the dolly is held against the back of the door while light, squarely hit blows from the hammer flatten and straighten the metal.

ting the dolly makes the pinging sound. If you accidentally miss the spot backed by the dolly, a more dull or dead sound is produced as only the panel is hit. If you miss the dolly with a hammer tap, a small unwanted dent is often produced in the panel.

With hammer-on-dolly, the shapes of the dolly and hammer head must match the desired shape of the panel. If the area to be straightened is flat, the dolly surface and hammer head must be flat. If the panel is curved, the dolly and hammer head must also be curved to match the panel's shape. When you bump or hit the damage with the hammer, the metal is flattened against the dolly and a tiny area is formed into the shape of the hammer head and dolly face. Look at Figure 10–6.

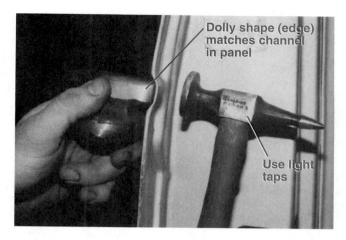

Figure 10–6 The shape or contour of the dolly and hammer head must mimic or be the same as the repair area. Note how the dolly edge was selected with same shape as the channel in the door panel. Carefully placed hammer blows have reshaped the badly damaged door skin at the channel area.

Never use a flat surface on a dolly to try to straighten a curved panel. You will damage the panel further.

Always start out with light hammer blows. A common mistake is to use excessively hard or poorly aimed hammer blows, which dent, stretch, and damage the panel. By starting light and working up to stronger blows, you can better control the movement of the metal to avoid unwanted dents. Carefully observe the results of each blow to make sure you are slowly reshaping the metal as desired. See Figure 10–7.

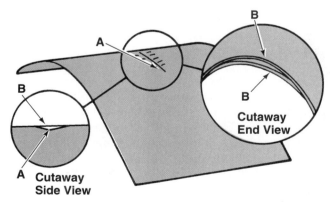

A = Metal Below Normal Level
B = Normal Level

Figure 10–7 This panel was damaged and has been roughed out. However, it is still low in the area of the buckle. Because the damage shows no evidence of a pressure area, the panel is low at point A. Point A must be raised to the original position (Point B). Begin hammering-off-dolly starting at the ends of the damage. Then, follow this up by straightening the low area up from the back side. Finally, straighten the metal with the hammer-on-dolly method. If the straightened area has stress, heat shrink it.

Hold the dolly securely against the back of the panel. Hit the area lightly so that the hammer bounces back. Light hammer-on-dolly blows are used to smooth small, shallow dents and bulges. Hard hammer-on-dolly blows can be used to stretch the metal.

To lower a bulge, place the dolly against the back side of the panel directly behind the bulge and use a hammer from the front side. There will be a slight rebound as your hammer hits the dolly. The dolly will then hit the back side of the panel. As the force of the dolly pressing against the panel is increased, the flattening action will also increase.

With hard blows using hammer-on-dolly, the metal is smashed between the hammer and dolly. This tends to crush the metal thinner and make it stretch out to fill a slightly larger surface area. All blows that are designed to stretch should be hard and accurate. Remember that an inaccurate hard blow can damage the panel.

Keep in mind that light hammer blows are for straightening, not stretching. In other words, when using the hammer-on-dolly technique for stretching, hit hard and do not miss!

Hammer-on-dolly is used only if there is access to the back side of the panel.

Remember! You can control the effects of hammer-on-dolly straightening by:

1. Using a hammer with a different shaped head.

2. Using a different shaped dolly face.

3. Altering how hard you hit the metal.

4. Changing how hard you push the dolly against the back of the panel.

HAMMER-OFF-DOLLY

Hammer-off-dolly is used to raise low spots and lower high spots simultaneously. The hammer hits the panel slightly to one side of where the dolly is being held. It is often used to rough out or shape large areas of damage during initial straightening. In this procedure, hold the dolly under the lowest area on the back of the panel. Then, hit any high area right next to the dolly with your hammer. Hammer off to one side of the dolly, not directly on top of the dolly. Look at Figure 10–8.

Generally, use the hammer and dolly to roll out the damage in reverse order than it was formed. Generally, the damage must be rolled out working toward the center. Start at the outer perimeter off the damage and work to the middle of the damage. See Figure 10–9.

For example, if the panel has a large buckle, you could use the hammer-off-dolly method. Place the

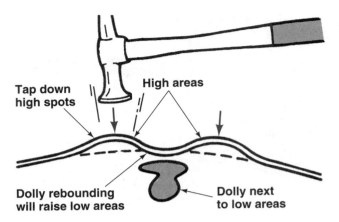

Figure 10–8 With the hammer-off-dolly method, place the dolly next to the area to be hit. Then, strike the high spots next to the dolly.

dolly on the low spot at the back of the panel. Then, hit a high spot with your hammer. This will lower the high spot and raise the low spot without stretching the metal. The hammer blow will push the high spot down and the rebound of the dolly will force the low spot up.

If the panel has a raised ridge of damage, you can also use the dolly-off method. Use a flat-faced dinging hammer to direct light to medium blows at the outer ends of the ridge. The blows from the hammer gradually force down the ends of the ridge. The dolly pressure forces the end of the channel upward. Gradually work toward the center. As the pressure is released, the metal tends to move back to its original position. The dolly can also be used as a driving tool to help work the damage. Once the area has been brought back to its basic shape, use the hammer-on-dolly method to smooth and level smaller damaged areas. You are then ready for plastic filling procedures.

Remember! You can control hammer-off-dolly straightening by:

1. Altering how hard you hit the panel. Start out with light blows and then increase their force if needed to lower high spots in the panel.

2. Changing how hard to push the dolly against the back of the panel. Pushing harder tends to increase lifting action to raise low spots.

3. Adjusting how far away the dolly is from the hammer blows. Moving the dolly father away tends to spread out the lowering-raising force to a large area on the panel. Moving the dolly next to the hammer blow tends to concentrate the lowering-raising force more.

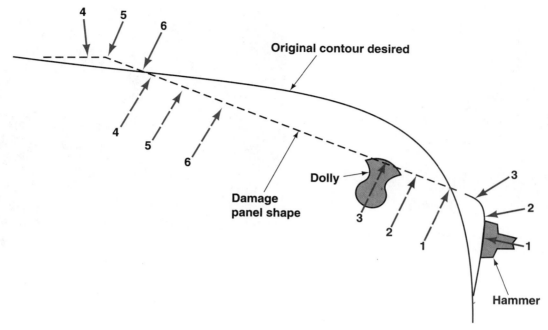

Figure 10–9 Note the steps for removing damage from a curved shape using a hammer and dolly. First, start at one end of damage, on the right. Then, work the damage on the other side. Roll out the damage toward middle. Filler would then be needed to give the area its final contour.

STRAIGHTENING WITH SPOONS

Spoons can be used to pry out damage, struck with a hammer to drive out damage, and as a dolly in hard-to-reach areas. Some are even designed to be used in place of a body hammer.

Spring hammering is a method of bumping damage with a hammer and a dinging spoon. The dinging spoon is lightweight and has a low crown. Hold the spoon firmly against a ridge or crease. Hit the spoon with a bumping hammer to work the high spot down, Figure 10–10. Always keep firm pressure on the spoon when spring hammering. It must never be allowed to bounce.

The force of the blow on the spoon is distributed over a large area of the ridge. Begin at the ends of the ridge and work toward the high point, alternating from side to side.

Spoons can be used to back up the hammer or in combination with a slapper spoon. With a long body spoon, you can often reach into restricted places. Pressure can be applied to tension areas with the spoon, while high areas are bumped down. See Figures 10–11 and 10–12.

Spoons can also be used to pry up metal or to drive out deep dents. In Figure 10–13, note how a double-end spoon is being used to pry out a dent in a door panel. The door is supported on blocks of wood to provide clearance for the panel. Use care not to stretch the metal by prying it out too much. Once the dent is roughed out, a body hammer can be used to finish the area.

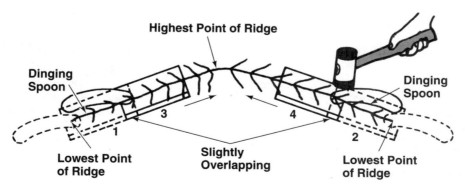

Figure 10–10 A large dinging spoon can be used to lower a ridge formed in a damaged panel. Start at the outer ends of the ridge and work your way to the center.

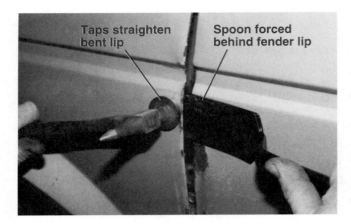

Figure 10-11 A spoon is being used as a pry bar and dolly. It has been forced between the front fender and the door to back up damage on the edge of the fender. The backs of fenders are often forced into the fronts of doors during frontal collisions.

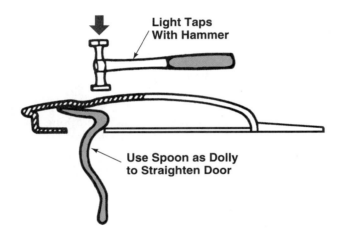

Figure 10-12 Use a curved spoon to reach in behind a damaged panel to serve as a dolly. Tap on the outside of the panel with a hammer while pressing the spoon in from the back.

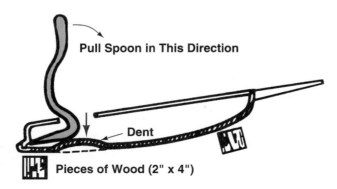

Figure 10-13 You can also use a curved spoon to pry out dents in panels, a door panel in this example. Pieces of wood can also be used to raise a door or panel off the work surface.

Figure 10-14 The pick end of a body hammer is being used to carefully lower tiny high spots remaining in the repair area of a door panel. The dolly-off method is being used to avoid hammer rebound.

STRAIGHTENING WITH PICKS

Picking hammers, pry picks, a dolly edge, and a scratch awl can be used to pick or push up metal. When picking up a small dent, it is better to use several light blows. See Figure 10-14. After an area has been raised, use a file to identify any remaining low spots.

Picks can be used to pry up metal in areas that cannot be reached with a dolly or spoon. A vehicle door is a good example, Figure 10-15. A pick can

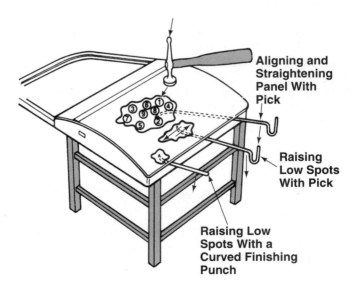

Figure 10-15 Pry picks are often used to remove small dents from doors and other hard-to-reach areas. Pry on dents with a pick while applying light taps from the hammer on the other side. Picks will fit through small openings in the side of a door or other part.

sometimes be inserted through a drainage hole. This eliminates the need to remove the inside door trim or to drill holes in the outer panel for pulling the dent.

When picks are used to remove small dents without having to repaint the panel, it is termed **paintless dent removal.** This method only works with very small dents that do not damage the finish (paint).

When prying with a pick, be careful not to stretch the metal by exerting too much pressure. Start with the point of impact or the lowest point. Slowly pry up the damaged area. On the larger areas, use a flat blade pick rather than a pointed one. Tap down pressure areas, while prying up low tension areas.

PULLING DAMAGED AREAS

Dents can be pulled out with a number of tools: suction cups, pull rods, and dent pullers. Pulling is often needed because access to the inside of many panels is blocked by reinforcements and other parts. With a puller, you can repair a simple dent in a minimal amount of time. Refer to Figure 10–16.

First, study the damage to determine the point and angle of the impact. Then, you can find where pulling force is needed to remove the damage.

Straightening with Suction Cup

A **suction cup** can be used to straighten shallow dents. Wet the area and install the cup. If hand held, pull straight out on the cup's handle. If mounted on a slide hammer, use a quick blow to pop the dent out.

A **vacuum suction cup** uses a remote power source (separate vacuum pump or air compressor air-flow) to produce negative pressure (vacuum) in the cup. This increases the pulling power because the cup will be forced against the panel tightly. Larger, deeper dents can be straightened with a vacuum suction cup.

When straightening damage and stress relieving on a frame rack, technicians sometimes punch or drill holes in old panels that are going to be replaced. The old panels are pulled and straightened just to help align adjoining parts. In this case, punching or drilling holes is acceptable. Whenever panels are going to remain on the vehicle, it is better to weld pull studs onto the damage. This technique also avoids having to weld any pull holes closed.

Straightening with Studs

A **stud spot welder** joins "pull rods" on the surface of a panel so that you do not have to drill holes. It is one way to straighten dents. Straightening with spot-welded studs avoids drilling or punching through the metal and undercoating, which can lead to corrosion. Refer to Figure 10–17.

To use a stud welding gun, remove paint in the damaged area. Install a stud into the gun. Press the gun against the dented area. Pull the trigger to weld the pin onto the surface of the panel. Weld enough pins onto the area to remove the damage. Look at Figure 10–18.

Next, attach a slide hammer to each of the pins and pull out the damage in steps, Figure 10–19. You could also use a pulling chain on a frame rack if needed, Figure 10–20. After bringing the surface of the dent almost out to the desired level, grind off the studs, Figure 10–21. This will allow you to remove the dent without drilling holes in the panel.

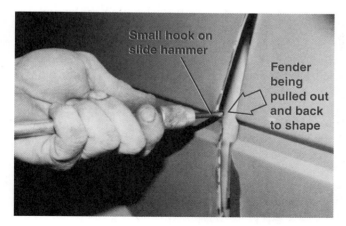

Figure 10–16 Slide hammers can be fitted with various ends, like this hook, for pulling on edges of panels. Body weight pulling against the tool is forcing a large curve of fender back into shape. It had been pushed in the door edge. Slide hammer blows have straightened the fender lip next to the channel.

Figure 10–17 A stud/nail pulling kit has a special spot welder that welds studs onto a panel. No holes must be drilled or punched for screws.

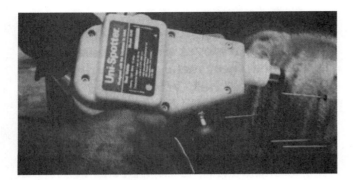

Figure 10–18 Use a stud gun to weld small metal studs onto the area to be pulled.

Figure 10–19 Attach a slide hammer to each stud. Use slide hammer blows and a body hammer to work the metal outward.

Figure 10–20 A hydraulic frame straightening chain can also be used to pull on the studs to remove the dent.

Figure 10–21 Cut off the studs when done. Then grind them flush with the surface. Use plastic filler to final shape the repair area.

SHRINKING METAL

Shrinking metal removes strain or tension on a damaged, stretched sheet metal area. During impact, the metal can be stretched. When pulled or hammered straight, the area can still have tension or strain on it. This is because the stretched metal no longer fits in the same area. The metal will tend to pop in and out when you try to finally straighten it. You would have to use heat to shrink the metal back to its original dimensions.

If a strained area is filled with plastic filler, road vibrations can cause the panel to make a popping or flapping noise. After prolonged movement of the strained area, the filler can crack or fall off. Eventually, you will be required to spend extra time correcting the work that should have been done properly in the first place.

Principles of Shrinking

A steel bar will expand (lengthen) when heated and contract (shorten) when cooled. If heated while butted against a solid object at both ends, it cannot lengthen and will bulge out in the middle or hottest area. Then when cooled, the length of the bar will decrease. This is the principle of shrinking metal.

The processes involved in shrinking a steel bar also apply to the shrinking of a warped area in a piece of sheet metal. A small spot in the center of the warped area is heated to a dull red. When the temperature rises, the heated area swells and attempts to expand outward toward the edges of the heated circle. Since the surrounding area is cool and hard, the panel cannot expand. As a result, a strong compression load is generated.

If heating continues, the stretching of the metal is centered in the dull red-hot portion, pressing it out. This causes it to thicken, thus relieving the compression load. If the red-hot area is suddenly cooled, the steel will contract and the surface area will shrink.

A variety of welding equipment can be used to heat metal for shrinking. Attachments are available for spot and MIG welding equipment to transform them into shrinking tools; the most commonly used tool is the oxyacetylene torch with a No. 1 or No. 2 tip.

Torch Shrinking

Torch shrinking uses the heat of an oxyacetylene torch to release tension in the panel. To shrink an area with a torch, heat a small spot in the bulge to a cherry red. Shrink in the highest spot of the stretched area, then in the next highest spot, until the area has been shrunk back to its proper position.

The size of the area to be heated, or the "hot spot," is determined by the amount of excess metal. The larger the area, the harder the heat is to control. An average-sized area to heat is usually about the size of a quarter. Smaller areas should be used on flat panels because they tend to warp easily. The area should never be larger than the size of the hammer face being used.

Use a neutral flame and a small tip to heat the panel. Bring the point of the cone straight down to within ⅛ inch (3 mm) of the metal. Hold the torch steady until the metal starts to turn red. Then move the torch slowly outward in a circular motion until the area is cherry red.

> *WARNING! Do not heat the metal past a cherry red. It will start to melt, and a hole may be burned through the metal. Also, remember that aluminum will not change color or turn red when heated.*

The metal usually will bulge up instead of down because the top of it is heated first. When it starts to bulge, the rest of the metal in the area follows. When the area has been heated, tap around it to drive the molecules of metal closer together. You may have to support the panel with a dolly to prevent the metal from collapsing. Push the dolly *lightly* under the metal. When the redness disappears, use a hammer and dolly to level the area around the shrink.

The metal usually will bulge up instead of down because the top of it is heated first. When it starts to bulge, the rest of the metal in the area follows. When the area has been heated, tap around it to drive the molecules of metal closer together. You may have to support the panel with a dolly to prevent the metal from collapsing. Push the dolly *lightly* under the metal. When the redness disappears, use a hammer and dolly to level the area around the shrink.

> *WARNING! Never use hard, hammer-on-dolly blows to level the area when shrinking. This will restretch the metal.*

Once the redness has disappeared and the area has been smoothed, cool the area with a wet rag or sponge. This will cause metal contraction. A slight amount of distortion could result. Straighten any warpage before heating the next spot.

Remember! It is easy to overshrink. When this occurs, the metal in the area last heated is usually collapsed or pulled flat. Sometimes the metal surrounding the heated area can even be pulled out of the proper contour. To fix overshrinking use hard, hammer-on-dolly blows to stretch the affected area.

Shrinking a Gouge

A *gouge* is caused by a focused impact that forces a sharp dent or crease into a panel. A gouge causes the metal to be stretched. Gouges must be shrunk to their original size to properly repair the damage. Simply picking up the low area would distort the panel. Filling the gouge with filler without restoring the panel's original contour will leave tension in the panel that could cause the filler to crack or pop off.

To shrink a gouge, heat the lowest point of the gouge with an oxyacetylene torch until the metal is cherry red. Use a dolly to hammer up the shrink. This will increase the tension on the soft area, forcing it to swell and return to its original position.

While the metal is still hot, hold the dolly directly under the groove and tap down the ridges on either side of the groove. This will not only drive down the ridges, but will also bump up the gouged metal.

If the gouge is long, repeat this process several times to raise the whole length of the gouge. Heat only as much of the gouge as can be worked before the metal cools.

IDENTIFYING STRETCHED METAL

Stretched metal has been forced thinner in thickness and larger in surface area by impact. When metal is severely damaged in a collision, it is often stretched in the badly buckled areas. These same areas are also sometimes stretched slightly during the straightening process. Most of the stretched metal will be found along ridges, channels, and buckles in the direct damage area. When there are stretched areas of metal, it is impossible to correctly straighten the area back to its original contour. The stretched areas can be compared to a bulge on a tire. There is no place for the area to fit within the correct panel contour.

When an area is stretched, the grains of metal are moved farther away from each other. The metal is thinned and work-hardened. Shrinking is needed to bring the molecules back to their original position and to restore the metal to its proper contour and thickness.

Before shrinking, dolly the damaged area back as close to its original shape as possible. Then you can accurately determine whether or not there is stretched metal in the damaged area. It will usually pop in and out if stretched. If it is stretched, you must shrink the metal.

Kinking Stretched Metal

Kinking involves using a hammer and dolly to create pleats, or kinks, in the stretched area to shrink the area's surface area. Instead of using heat to shrink the metal, kinking is another way to deal with stretched metal.

To kink metal, hit the stretched area with a picking hammer slightly off-dolly. This will lower the area slightly below the rest of the panel. Fill the low area with body filler. Then file and sand the area level with the rest of the panel.

FILING THE REPAIR AREA

When the area has been straightened smooth as possible, use a body file to locate any remaining high and low spots. File across the damaged area to the undamaged metal on the opposite side. This will keep the filing action on the correct plane with the good part of the panel. Look at Figure 10–22.

Push the file forward by its handle for the cutting stroke. Control file downward pressure and direction by holding the front of the file. Use as long a stroke as possible.

The scratch pattern created by the file on the metal identifies any high and low spots. You can then further work the metal into shape before using plastic filler.

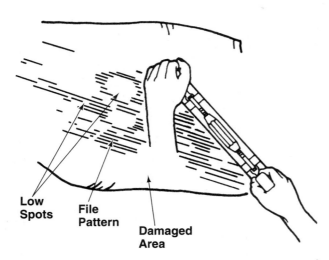

Low Spots **File Pattern** **Damaged Area**

Figure 10–22 Filing will help you locate high and low spots that need further work. Low spots will NOT have file marks.

WORKING ALUMINUM

Aluminum is used for some automotive panels, such as hoods and deck lids. The repair of aluminum requires much more care than working steel panels. Aluminum is much softer than steel, yet it is more difficult to shape once it becomes work-hardened. It also melts at a lower temperature and distorts readily when heated.

Because aluminum is soft, it responds very readily to the hammer-on-dolly method. If hammered too hard or too much, the panel will stretch. Use several light strokes rather than a few heavy blows.

Aluminum does not readily bend back to its original shape after being buckled by an impact. Therefore, it does not respond well to the hammer-off-dolly method. Be careful not to create additional damage when attempting to lower ridges with hammer and dolly blows.

Raising small dents with a picking hammer or pry bar is an excellent way to repair aluminum panels. However, do not raise the panel too far and stretch the soft aluminum.

Spring hammering with a hammer and spoon is an excellent way to unlock stress in high-pressure areas of aluminum. The spoon distributes the force of the blow over a wider area.

Because aluminum is soft, reduce hand pressure on the body file. Use a file with rounded edges to avoid scratching and gouging the metal.

Sand carefully on aluminum. A coarse grit grinding disc will quickly burn through aluminum. The heat from grinding can also warp the panel. A No. 80 grit open coat disc can be used, but sand carefully to remove only paint and primer, not the aluminum. Make two or three passes, then quench the area with a wet rag to cool the metal in the panel.

Featheredging of aluminum should be done with a dual-action sander. Use No. 80 or No. 100 grit paper and a soft, flexible backing pad.

Heat shrink stretched aluminum slowly to avoid distorting the panel. Do not heat the spot over 1,200°F (584°C). Use a thermal crayon or paint to monitor heating. Mark the part with the crayon when it is still cold. When the stated temperature has been reached, the crayon or paint mark will liquefy.

Unlike steel, aluminum does not turn red as it is heated. Instead, it turns ash gray just before it melts. Thus, a lack of caution will result in a melted panel. Also, quench or cool the heated area very slowly to avoid distortion.

USING BODY FILLER

After the damaged metal has been bumped, pulled, pried, and dinged, the application of body or plastic filler is next.

Filler	Composition	Characteristics	Application
COMPARING BODY FILLERS			
Conventional Fillers			
Lightweight fillers	Microsphere glass bubbles; fine grain talc; polyester resins	Spreads easily; nonshrinking; homogeneous; no settling	Dings, dents, and gouges in metal panels
Premium fillers	Microspheres; talc; polyester resins; special chemical additives	Sands fast and easy; spreads creamy and moist; spreads smooth without pinholes; dries tack-free; will not sag	Dings, dents, and gouges in metal panels
Fiberglass-Reinforced Fillers			
Short strand	Small fiberglass strands; polyester resins	Waterproof; stronger than regular fillers	Fills small rustouts and holes. Used with fiberglass cloth to bridge larger rustouts.
Long strand	Long fiberglass strands; polyester resins	Waterproof; stronger than short strand fiberglass fillers; bridges small holes without matte or cloth	Cracked or shattered fiberglass. Repairing rustouts, holes, and tears.
Specialty Fillers			
Aluminum filler	Aluminum flakes and powders; polyester resins	Waterproof; spreads smoothly; high level of quality and durability	Restoring classic and exotic vehicles
Finishing filler/ polyester putty	High-resin content; fine talc particles; microsphere glass bubbles	Ultra-smooth and creamy; tack-free; nonshrinking; eliminates need for air dry type glazing putty	Fills pinholes and sand scratches in metal, filler, fiberglass, and old finishes.
Sprayable filler/ polyester primer-surfacer	High-viscosity polyester resins; talc particles; liquid hardener	Virtually nonshrinking; prevents bleed-through; eliminates primer/glazing/ primer procedure	Fills file marks, sand scratches, mildly cracked or crazed paint films, and pinholes. Seals fillers and old finishes against bleed-through.

Figure 10–23 This chart shows general composition, characteristics, and applications for various types of fillers.

Body fillers, also called **plastic fillers,** are designed to cover up minor surface irregularities that remain after metal straightening. Keep in mind that the quality of the repair and finish is adversely affected by the wrong choice of filler. The chart in Figure 10–23 shows various types of filler and their uses.

NOTE! For more information on body fillers and related materials, refer back to Chapter 8.

Preparing Surface for Filler

One of the most important steps in applying plastic body fillers is surface preparation. Begin by washing the repair area with soap and water to remove dirt and grime. Then clean the area with wax and grease remover to eliminate wax, road tar, and grease. Use a cleaner that will remove the **silicones** often present in automotive waxes.

Mask any trim, parts, or adjacent panels that could be damaged by grinding, sanding, and filling. Use masking tape or duct tape to protect them.

Grind the area to remove the paint 3 to 4 inches (75 to 100 mm) around the area to be filled. Remember! Never apply body filler over paint! Apply filler only to bare metal. Filler will NOT bond properly to paint, causing problems.

Use a No. 40 grit grinding disc to remove the paint. Grinding also etches the metal to provide better

Figure 10–24 Before applying filler, grind the area to remove all paint. Coarse (36 or 40 grit) grind marks will help the filler bond to the metal. Mask trim or adjacent panels so you do not accidentally damage surfaces not being repaired. Use compressed air to blow off dust from the repair area.

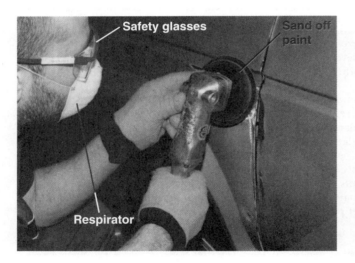

Figure 10–25 After panels have been roughed and bumped back into their original shape, remove paint in the area still requiring plastic filler. Technicians will always wear safety glasses, gloves, and a dust mask during this potentially dangerous operation.

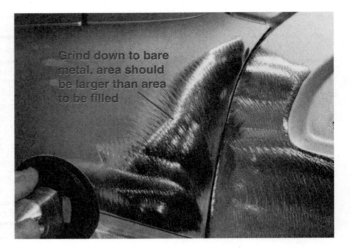

Figure 10–26 Methodically grind the repair area using a very coarse (36–40 grit) disc. Start at the top. Grind straight across the panel and then drop down so that grind marks overlap each other. You want to texture the steel so that the filler will bond to the panel securely. Do not grind in one location too long or you will thin and weaken the panel.

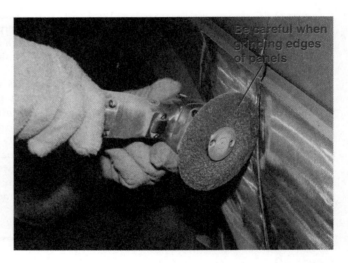

Figure 10–27 If needed, run the grinder across damaged edges to remove paint and help true the surfaces.

adhesion, Figure 10–24. Grind only enough to remove the finish. Do not grind too much or you will thin and weaken the metal. See Figures 10–25 to 10–27.

If you are applying filler over a metal patch, do not hammer down the excess weld bead. Grind it level with the surface. Hammering a weld down distorts the metal, creates stress, and increases the area to be filled.

After removing the finish from the repair area, blow away the sanding dust with compressed air and wipe the surface with a tack rag to remove any remaining dust particles.

Note that some paint and vehicle manufacturers recommend applying primer to bare metal (usually aluminum) before applying a filler. Filler may not bond as well to bare aluminum as it will to an epoxy-type primer. Refer to the paint or vehicle manufacturer's instructions.

PREPARING SURFACE FOR FILLER

Surface Preparation Problems

Surface preparation involves the many steps needed to clean, straighten, and smooth the surface before refinishing. Many problems are linked to improper surface preparation. Below is a list of rules to follow to prepare for applying filler.

1. Before applying filler, sand or grind the surface to the bare metal, making sure all corrosion, spots of paints, weld scale, etc., are removed.

2. After sanding or grinding, blow off the area with a high-pressure air gun. Wipe the area with a clean cloth to remove any fine dust or moisture that may be left on the surface.

> *WARNING! Anything that remains between the bare metal and filler will probably result in adhesion failure.*

3. Make sure no solvents are used to clean the sanded area before applying filler. Trapped vapors will result, causing pinholing and poor adhesion.

4. If body filler is to be applied to brazed seams or panel joints, thoroughly wash the area and thoroughly blow out the seam or joint with an air blow gun.

5. Avoid applying filler over seams or joints that were pop riveted. Too much movement takes place, which will cause the filler to crack.

6. For best results, the filler, shop, and parts should all be above 65°F (18°C).

7. With high humidity conditions, use a heat lamp to warm and dry the area to be prepared. This will eliminate moisture accumulation between the bare metal and filler, which will cause poor adhesion.

Mixing Filler

Mix the can of filler to a uniform and smooth consistency. It must be free of lumps and not wet on top. Fillers can be shaken on a paint shaker for several minutes.

Proper *filler mixing* is needed to prevent these problems:

1. The filler in the upper portion of the can will be too thin. This results in runs and sags when applying filler.

2. The filler can cure slowly or not at all.

3. The filler can have a gummy, soft condition when sanded. It can have a very tacky surface after curing that will clog your sandpaper.

4. The filler can cause poor featheredging. It will tear off the metal and form a small lip at its outer edge.

5. The filler can blister and lift when coated with primers and refinishing materials.

6. The filler in the bottom of the can will be too thick. When you use the bottom, the filler will be very coarse and grainy.

7. The filler will have pinholing, or small air holes will form in the filler.

8. The filler will cause poor color holdout when the area is coated with primers and refinishing materials.

Mixing the Hardener

Mixing hardener is done by squeezing the tube back and forth with your fingers to mix the material. Loosen the tube cap to release the air. Knead (squeeze) the tube thoroughly to assure a smooth, pastelike consistency. The hardener should be like toothpaste when squeezed out.

If you do NOT mix the hardener in the tube, the result can be the same problems listed for poor filler mixing.

If the hardener is kneaded thoroughly and remains thin and watery, you have *defective (bad) hardener.* It should not be used because it has broken down chemically. Hardener can spoil if frozen or if stored too long.

Mixing Filler and Hardener

Numerous problems can occur from improper *catalyzing* (mixing) of *hardener* (filler catalyst) and filler. Before catalyzing, make sure the materials (filler and hardener) to be used are compatible. They should be manufactured by the same company and be recommended for use with each other. See Figure 10–28.

The following tips will help eliminate problems relating to mixing filler and its catalyst. Open the can of filler without bending its lid. Remove the desired amount of filler. Use a clean putty knife or spreader.

Place the filler on a smooth, clean filler mixing board. A *filler mixing board* is a clean, nonporous surface for mixing filler and hardener. You can use sheet metal, glass, or hard plastic, Figure 10–28.

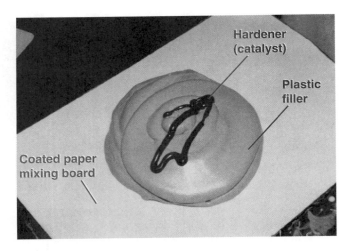

Figure 10–28 Always add the amount of hardener recommended by the manufacturer to plastic filler or glazing putty. Generally, for a baseball-size amount of plastic filler, use about a 6-inch bead of hardener, as shown.

WARNING! Cardboard should NOT be used as a filler mixing board. It is porous and contains waxes for water-proofing. These waxes will be dissolved in the mixed filler and cause poor bonding. Cardboard also absorbs some of the chemicals in the filler and hardener, changing the filler's curing quality slightly. Cardboard fibers can also stick in the filler and ruin the finish. Mixing boards with a handle are also available.

Add hardener according to the proportion indicated on the product label. Too little hardener will result in a soft, gummy filler that will not adhere properly to the metal. It will also not sand or featheredge cleanly. See Figure 10–29. Too much hardener will produce excessive gases, resulting in pinholing and hardening before you have time to apply the filler. Look at Figure 10–30.

A general rule is this: for each golf ball sized "glob" of filler, use a 1-inch (25 mm) bead of hardener. If the filler is as big as a baseball, use about 6-inch (152 mm) bead of hardener. However, always refer to the manufacturer's instructions for exact mixing directions, Figure 10–31.

Filler over-catalyzation results when you use too much hardener for the amount of filler. This must be avoided when adding hardener to the filler. In

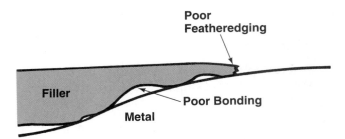

Figure 10–29 If you do not use enough hardener, the filler will not bond to the panel properly and will not featheredge or sand properly.

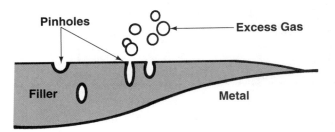

Figure 10–30 A common mistake is to use too much hardener. This will cause excess gas and pinholes in the hardened filler. The filler may also harden on your mixing board before you have time to apply it to the vehicle.

Figure 10–31 This technician is using filler and hardener in a dispenser. This helps keep the filler fresh and clean. *(Oatey Bond Tite)*

addition to paint color bleed-through and pinholing, a reverse curing action may occur, causing poor adhesion and poor sanding properties.

Filler under-catalyzation is caused by not using enough hardener in the filler. It will result in the filler not curing properly, resulting in tacky surfaces and poor adhesion properties.

With a clean putty knife or spreader, use a scraping motion (back and forth) to mix the filler and hardener together thoroughly and achieve a uniform color.

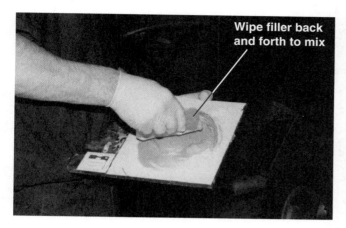

Figure 10–32 Use a perfectly clean spreader to work hardener into the filler. Do not stir in air bubbles. Wipe the spreader back and forth across the top of the material to mix it together thoroughly. This handy mixing board uses disposable sheets of coated paper so you do not have to clean your mixing board before its next use.

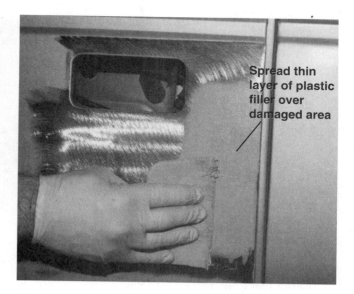

Figure 10–33 After blowing off the panel, use a clean spreader to wipe filler across the repair area. Spread the material one way (horizontally) and then the other wax (bottom to top). This will help level the filler and fill low areas on the panel. Work carefully but quickly so the filler does not start to cure.

Scrape filler off both sides of the spreader and mix it in. Every few back-and-forth strokes, scrape the filler into the center of your mixing board by circling inward. Look at Figure 10–32.

If the filler and hardener are not thoroughly mixed to a uniform color, soft spots will form in the cured filler. The result is an uneven cure, poor adhesion, lifting, and blistering.

Reinstall the cover on the can of filler right away. This will keep out dust and dirt. It will also help prevent liquids in the filler from evaporating.

Always use clean tools when removing the filler from the can and mixing the filler and hardener together. Do NOT redip the spreader or mixed filler into the can. This will cause the whole can of filler to harden with time. Hard lumps of filler might form in the can and/or applied filler. This will cause problems the next time you try to use the filler.

Use different spreaders to mix and apply the filler. A small amount of unmixed filler will always remain on the mixing spreader. If any is applied, you will have soft spots in the cured filler. The paint finish may peel.

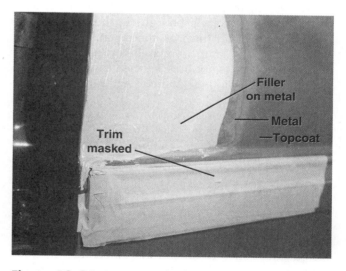

Figure 10–34 Curing time for filler will vary with shop temperature and other factors. You need to be ready to file when the filler partially hardens.

Applying the Filler

Apply the mixed filler as soon as you have finished mixing. First, apply a thin coat of filler to the repair area. Press firmly to force filler into the sand scratches and holes. This will strengthen the bond. Refer to Figure 10–33.

Work the filler patch in two directions (left to right, then top to bottom). This will greatly reduce pinholing. Buildup layers should NOT be more than ⅛ inch (3 mm) thick. See Figure 10–34.

Spread the filler approximately 3 inches (75 mm) beyond the repaired area. This will assure better adhesion, and allow you to featheredge the patch.

When this layer cures and has been sanded, apply more coats to build up the repair area to a proper contour. Allow each application to cure before sanding and applying the next coat of filler.

Build up the final layer of filler slightly above the panel surface. Make the filler slightly thicker than needed. This will allow you to sand off the waxy film that forms on the surface of the filler. You will also be able to sand the filler down smooth on an equal plane with the existing panel.

Always use a clean plastic **spreader** or **spatula** to apply filler. Chunks of old filler could fall into the new soft filler. This can cause large pockets to be gouged as you spread the filler. Also, for your final coating of filler, make sure the spreader has a smooth edge. If worn and nicked, used spreaders will not make a smooth layer of filler.

If needed, you can use your fingers to bend the spreader to match the shape of the contour. Use a smaller spreader on small repair areas. A larger spreader will fill large areas more easily.

Avoid using filler in cold temperatures. When the filler, shop, or panel are cold, the filler will NOT cure properly. It will have a tacky surface and poor sanding properties. Large pinholes could also form. Filler should be stored at room temperature (65° to 70°F or 18° to 21°C). Use a heat lamp to warm cold surfaces if needed.

In winter, moisture can form on metal surfaces when a cold vehicle is brought inside. In summer, condensation could form on the metal. Again, use a heat lamp to dry damp surfaces before applying filler. If the repair area is not first warmed to remove moisture, poor adhesion, featheredging, pinholing, and lifting could result.

Make sure all holes, gaps, joints, cracks, etc. are welded. Holes left under filler allow moisture to accumulate between the metal and filler, eventually resulting in bond failure.

Shaping Filler

Allow the filler to cure to a semi-hard consistency. This usually takes 15 to 20 minutes. Scratch the filler with your fingernail. If the scratch leaves a white mark, the filler is ready to be filed. See Figure 10–35.

Filler filing involves using a coarse "cheese grater" or body file to rough shape the semi-hard filler. You will knock off the high spots and rough edges. Since the filler is only partially hard, the body file will quickly remove excess filler. It will force the semi-soft filler through the large holes in the file face.

If you do NOT rough shape the filler with a grater, you will waste time and sandpaper. Sandpaper will become loaded quickly. It will also create unwanted dust.

To use the body file, hold it at a 30- to 40-degree angle. Pull it lightly across the semi-hard filler, Figure 10–36. Work the file in several directions. Stop filing

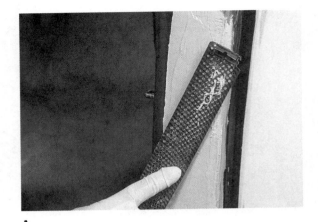

A

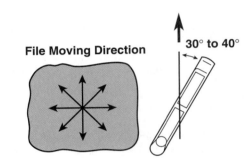

File Moving Direction   30° to 40°

B

Figure 10–35 Use a body file or "cheese grater" to rough shape filler when it is partially cured. (A) A body file will quickly remove high spots from the filler. It is faster than sanding and does not produce dust. (B) Hold the body file flat and move it in all directions to true the flat surface. Hold the file at a 30- to 40-degree angle.

when the filler is slightly above the desired level. This will be sufficient for sanding out the file marks and for feathering the edges. If the filler is undercut, additional filler must be applied. See Figure 10–37.

Applying Filler to Body Lines

Many vehicles have sharp body lines in doors, quarter panels, hoods, etc. Maintaining the sharpness of these lines when doing filler work is difficult, especially in recessed areas. The best way to get straight, clean lines is to file each plane, angle, or corner separately.

Apply masking tape along one edge. Then apply filler to the adjacent surface. Before the filler sets up, pull the tape off. This will remove the excess filler from the body line.

After the first application is dry and sanded, tape the opposite edge. Apply masking tape along the body line and over the filler. Then, coat the adjacent surface with filler. When the tape is removed and the filler sanded, the result is a straight, even line or corner.

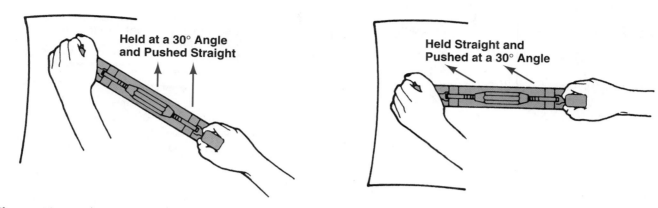

Figure 10–36 Shown are two methods for filing or board sanding of filler. Left: Push the file or sanding board sideways while holding it at a 30-degree angle. Right: Hold the tool straight and push it sideways at a 30-degree angle. Always put equal pressure on both ends of the file or sanding board.

Applying Filler to Panel Joints

Many panels have joints that are factory finished with a **seam sealer** to allow the panel to flex and move. Often, both halves of the joint suffer damage and require filling. Never make the mistake of covering the seam sealer with body filler. The filler will crack over the body flex joint.

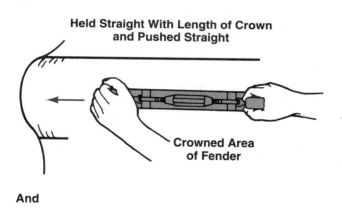

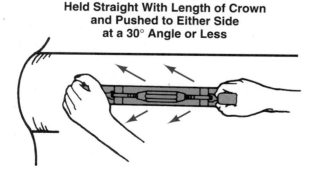

Figure 10–37 Note two methods for filing or sanding a crowned or curved surface. Top: Push straight along the crown to true the top of the crown. Bottom: Angle and push the tool off to the side to shape and true the curves on the sides of the crown. Twist the tool with your wrists to match the curved shape.

Figure 10–38 shows what happens when the inflexible filler is subjected to the twisting action of the panels under normal road conditions. A crack can develop that allows moisture to seep under the filler, which causes corrosion to form on the metal surface.

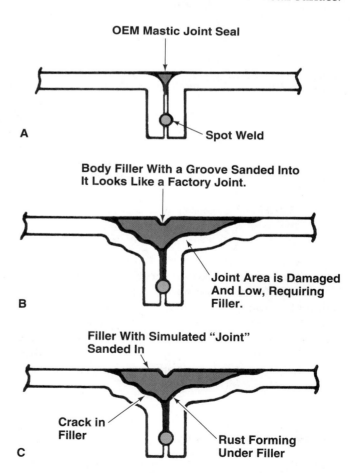

Figure 10–38 (A) Joints between two panels may be filled with sealer. (B) Never try to fill this type of joint with filler and then sand in a groove. (C) A flexible joint will crack and corrode if an improper method is used.

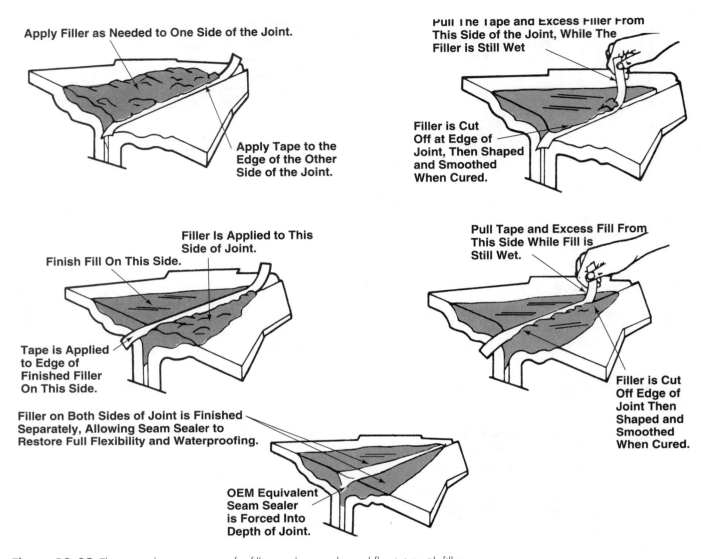

Apply Filler as Needed to One Side of the Joint.

Apply Tape to the Edge of the Other Side of the Joint.

Pull The Tape and Excess Filler From This Side of the Joint, While The Filler is Still Wet

Filler is Cut Off at Edge of Joint, Then Shaped and Smoothed When Cured.

Finish Fill On This Side.

Filler Is Applied to This Side of Joint.

Tape is Applied to Edge of Finished Filler On This Side.

Pull Tape and Excess Fill From This Side While Fill is Still Wet.

Filler is Cut Off Edge of Joint Then Shaped and Smoothed When Cured.

Filler on Both Sides of Joint is Finished Separately, Allowing Seam Sealer to Restore Full Flexibility and Waterproofing.

OEM Equivalent Seam Sealer is Forced Into Depth of Joint.

Figure 10–39 These are the proper steps for filling a damaged panel flex joint with filler.

This eventually results in the failure of the filler-to-metal bond and a weakened sheet metal joint.

The original flexibility of the joint can be preserved by taping its alternate sides. As shown in Figure 10–39, apply tape to one panel. Then apply filler to the other panel. Pull the tape up to remove excess filler. Then fill the other panel the same way. After you shape the filler on both panels, force a seam sealer into the joint.

Sanding the Filler

After filing, sand out all file marks and begin to shape the filler more accurately. Use a No. 36 or No. 40 grit disc on a sanding board or block first. An air file can also be used on large flat areas, Figure 10–40. Do not try to sand out all imperfections in the first coat of filler. Only sand the first coat to get the general shape of the repair.

A common mistake is oversanding the first layer of filler below the desired level. Two or more coats of plastic filler are normally needed to get a good, smooth surface.

After sanding the first layer of filler to shape, blow off the area and apply a second layer of plastic filler, Figure 10–41. Work the filler in two directions to fill any imperfections or holes in the repair.

After this layer cures, sand it to shape with 36 to 40 grit sandpaper. Follow this with No. 80 grit sandpaper until all large scratches are removed.

Figure 10–42 shows how several sanding methods can be used on a single repair. The technician first used an air file, then a DA (disc orbital) sander, then a sanding block, and finally a round sanding tool to progressively shape the curve on a fender.

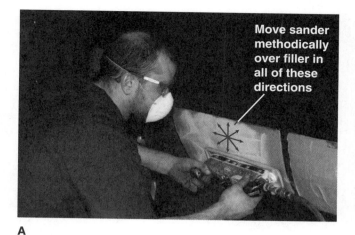

Move sander methodically over filler in all of these directions

A

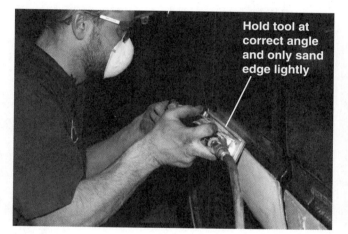

Hold tool at correct angle and only sand edge lightly

B

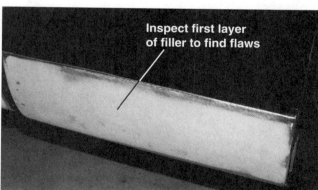

Inspect first layer of filler to find flaws

C

Figure 10–40 After filler cures fully, use coarse grit sandpaper to level the remaining high spots of plastic filler. (A) An air file works well on large surfaces like these. Move the file methodically to plane down filler to match the shape of the panel. (B) Filler edges cut very quickly. Hold the sander at a right angle and quickly slide it across the edge. Inspect the edge after each pass and sand only enough to get the basic shape of the edge with coarse paper. (C) After sanding the first layer of filler, blow it off and inspect it for surface flaws. On small repair areas, one coat of filler may be enough to fill the damaged area. This large area required a second coat of filler.

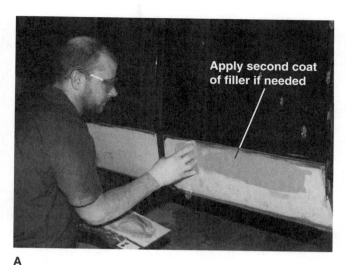

Apply second coat of filler if needed

A

Work filler down into surface imperfections

B

After spreading horizontally, spread upward while following curve of panel

C

Figure 10–41 This technician is applying a second coat of plastic filler. (A) Wipe a thin second coat of plastic filler over the first if needed. (B) Wipe the filler on in one direction and then the other to work material down into the remaining low spots. (C) The technician is wiping from the bottom to the top while matching the curve of the panel.

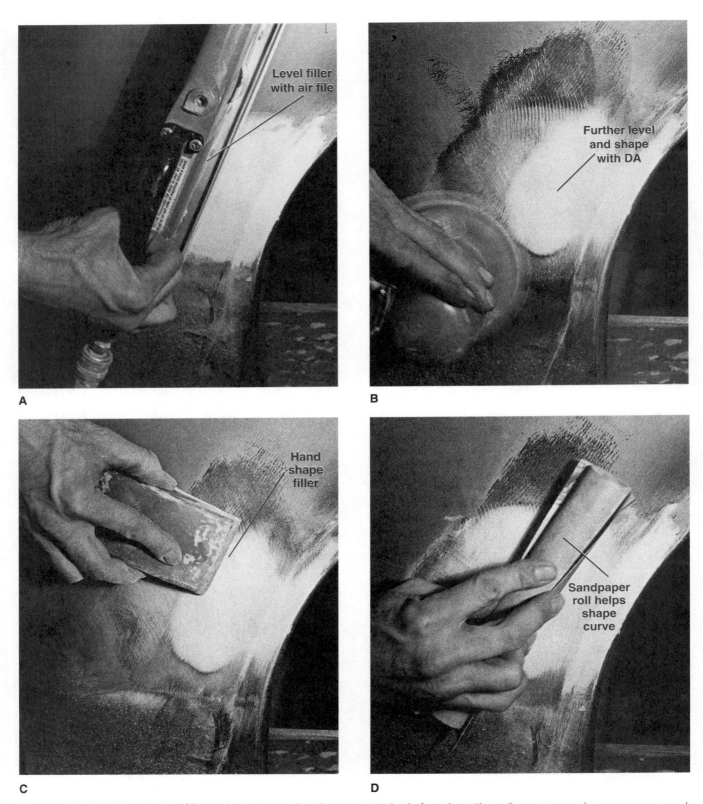

A Level filler with air file

B Further level and shape with DA

C Hand shape filler

D Sandpaper roll helps shape curve

Figure 10–42 When sanding filler, make sure you select the correct method of sanding. This will save time and improve repair quality. (A) If it is a large enough area, first sand with the air file. Its large surface area will help true large flat areas. Roll the air file to help round the contour if needed. (B) After the air file, use a disc sander to further shape the filler. Move the sander in the directions needed to featheredge and shape the filler to match the panel contour. (C) Hand block sand the area to carefully lower filler to the same plan as the metal. You want a smooth featheredge. You should not be able to feel any lip where the filler and metal meet. (D) On curves, folding sandpaper around a cardboard roll will help support the sandpaper to match the round shape being repaired.

DANGER! *Do not breathe the dust created when sanding filler. Wear the proper dust respirator to keep the plastic dust out of your lungs.*

Finally, smooth the filler with No. 180 grit sandpaper. A sanding block or air file can again be used, as well as a long sanding board. A DA sander will work fine on smaller areas.

Be careful not to oversand. **Filler oversanding** results in the filled area being below the desired level, which makes it necessary to apply more filler.

Filler undersanding leaves the filled area high or thicker than the surrounding panel. A hump would be formed at the filled area. See Figure 10–43.

After final sanding, blow with an air gun and wipe with a tack cloth. This removes any fine sanding dust that might be hiding surface pinholes. It also exposes holes lying just below the surface. These holes and remaining sand scratches must be filled.

Run your hand or straightedge over the surface to check for evenness. Do not trust "eyeballing" for accuracy. Paint does NOT hide imperfections; it highlights them. Do not be satisfied until the repaired surface is perfectly smooth.

Remember! If you can feel the slightest bump, paint will make it show up much more. The dull surface of filler and sanded paint does not visibly show surface imperfections.

Featheredging

Featheredging involves sanding the repair area until the filler and old paint blend smoothly into each other. You must use fine sandpaper, 180 grit or finer. Sand until you remove any small lip where different materials on the surface meet. Look at Figure 10–44.

Featheredging is commonly done with a DA sander. When sanding, hold the sander flat on the surface. Avoid tilting the sander to remove material more quickly. This will sand a small hole into the filler. By holding the sander flat, you will plane off the filler smoothly. Refer to Figure 10–45.

On large areas with filler, use an air file or large sanding board with 180 grit sandpaper. The large sandpaper surface area will help you quickly straighten door panels, fenders, and other large, relatively flat panels. See Figure 10–46.

Priming

Priming is done after filling to cover any bare metal as well as the filler. After using filler, primer-surfacer is often sprayed on the repair area. Since primer-sur-

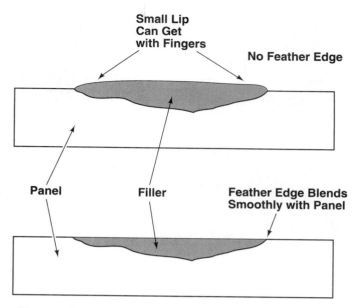

Figure 10–43 Incorrect and correct filler featheredge. (A) The filler has not been featheredged properly. You would be able to feel a small lip at the edge of the filler. (B) This filler has a good featheredge and blends smoothly into the panel.

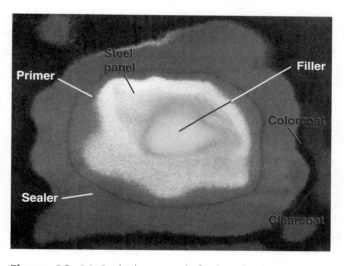

Figure 10–44 Study this properly featheredged repair area. Note the small amount of filler inside the dent in the center area. Note wide lines around the borders of primer, colorcoat, and clearcoat. If the line between different materials is sharp or thin, the area is not featheredged properly.

facer can be thick, it will help fill small sand scratches in the filler and paint. Refer to Figure 10–47.

For the best and quickest results when applying primer-surfacer, spray two or three coats with a five to fifteen minute flash time between coats. You will actually save time by following flash recommendations versus spraying coats wet on wet.

It is difficult to tell when a thick coat of primer-surfacer is truly dry. The surface will appear d

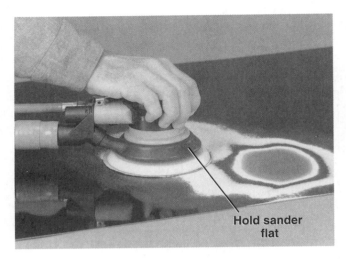

Hold sander flat

Figure 10–45 When featheredging filler, hold the sander flat so that the edge of the sandpaper does not dig into the filler. Also, pay close attention so that you do not oversand the filler. (Courtesy Dynabrade^TM, Inc.)

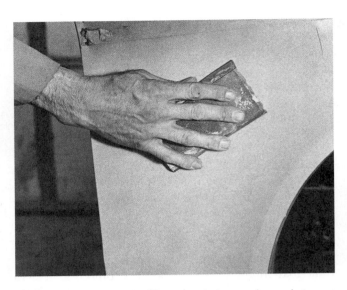

Figure 10–47 After the filler is sanded properly, apply a coat of primer or primer-surfacer. Primer will help fill small imperfections and let you further sand the repair area true.

Figure 10–46 On large areas, you should not sand through to metal in the center area of the filler. This would mean you have a metal high spot. It would show up easily as a large hump when painted.

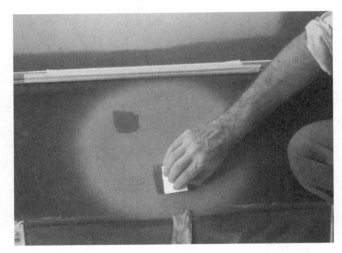

Figure 10–48 After priming, you may find small surface imperfections. These can be filled with two-component spot putty. Always use spot putty sparingly. (Marson Corp.)

while there is still a lot of solvent trapped below the surface. The lower layer of primer-surfacer is still trying to dry and shrink.

On the other extreme, thin dry coats of primer-surfacer can cause loss of adhesion, not only to the substrate, but also to the topcoat color. Always spray wet coats of primer-surfacer. Wait an hour or more before sanding the primer-surfacer.

If the primer-surfacer is sanded before all of the solvent has evaporated, the material in the scratches will continue to shrink down in the scratches. They will show up in the final finishing color topcoats as sanding scratches.

Finishing Filers (Putties)

Once the primer is dry, small pinholes and scratches can be filled with *spot putty* or *glazing putty* as shown in Figure 10–48. If you are using a one-part putty, apply it directly to your spreader. If using a polyester (two-component) putty, mix the putty and hardener according to the manufacturer's instructions.

Place a small amount of putty onto a clean rubber squeegee. Apply a thin coat over the primer. Use single strokes and a fast scraping motion. Use a minimum number of strokes when applying lacquer-based putties. They dry very fast. Repeated passes of the spreader may pull the putty away from the primer.

> WARNING! A common mistake is to use spot putty as a filler. Spot putty is NOT as strong as filler. Use spot putty only to fill small imperfections in the primer. Do NOT apply it to bare metal or painted surfaces. Most spot putties are designed to be applied over primer.

Allow the putty to dry completely before sanding smooth with No. 240 grit sandpaper. Sanding the putty too quickly results in sand scratches in the finish.

NOTE! Excessive use of glazing putties is usually an indication of a lack of skill and training. They should be used only on small pinholes and other small surface problems in the primer.

Sanding Guide Coat

A *guide coat* is often used to check for high and low spots on your repair area. A guide coat is a thin layer of a different color primer or a special powder applied to the repair area. By watching what happens to the guide coat with light sanding, you can find low and high spots. See Figure 10–49.

If the second color primer or guide coat powder does not sand off, you have found a low spot. If either sands off too quickly, you have found a high spot. Both situations require further work before painting.

Ideally, the second color primer or dry guide coat powder should all sand off at the same time. This shows that the surface is flat and ready for sealer, colorcoat, and other operations.

Final check all areas to be refinished. Look around edges carefully to find any remaining surface imperfections. You want to find any surface problems now or before painting.

Make sure you hand sand the edges on the repair panels to prepare them for painting. See Figure 10–50.

Many technicians like to wet sand repair areas as a final check of the repair. When the surface is wet, it is much easier to notice minor imperfections in the surface. Wet sanding also produces a very smooth surface. Look at Figure 10–51 and 10–52.

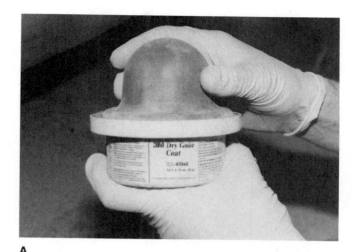

A

B

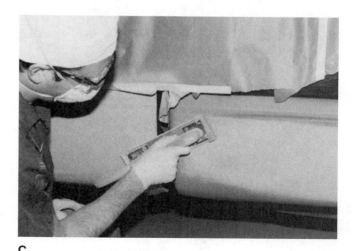

C

Figure 10–49 Dry guide coat powder can be used instead of a different color primer when checking surface shape. (A) Work the sponge applicator down into the powder. (B) Wipe the powder over the entire area of repair. (C) If the powder does not sand off in a small area, that area is still too low and would look like a dent when painted. If the powder sands off too quickly, that area is high and must be sanded level.

Figure 10–50 Carefully inspect all edges of the panel. Hand sand small chips and bits of filler and texture existing paint.

Figure 10–51 Many technicians like to wet sand the repair area before painting. The whole panel to be painted might be wet sanded, concentrating on the primed area over the filler.

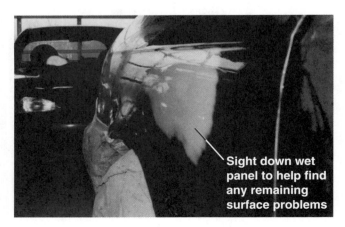

Figure 10–52 When the surface is wet from sanding, look at the repair area from every angle. Water on the surface will act like paint to help you find any small waves or imperfections remaining in the surface.

MORE INFORMATION

NOTE! The repair is now ready for final priming, sealing, and painting. This will be detailed in later chapters. Refer to the index for more information on these subjects if needed now.

Summary

- The two types of sheet metal used in auto body work are hot-rolled and cold-rolled. Hot-rolled sheet metal is used for comparatively thick parts such as frames and crossmembers. Cold-rolled sheet metal is used for most unibodies.
- The various tools used to straighten metals include body hammers, spoons, dollies, suction cups, slide hammers, and spot-welded studs.
- Shrinking metal removes strain or tension on a damaged, stretched sheet metal area.
- Aluminum is used for a variety of automotive panels such as hoods and deck lids. It is much softer than steel, yet it is more difficult to shape once it becomes work-hardened. It also melts at a lower temperature and distorts readily when heated.
- Body fillers, also called plastic fillers, are designed to fill minor surface irregularities that remain after straightening.
- Featheredging involves sanding the repair area until the filler and the finish (old paint) blend smoothly into each other.
- Priming is done after sanding body fillers to prepare the surface for refinishing.

Technical Terms

annealed	*yield point*
low-carbon steel	*elastic deformation*
mild steel (MS)	*spring-back*
high-strength steel (HSS)	*plastic deformation*
deformation	*work hardening*
tensile strength	*kinked*
tension	*bent*
yield strength	*high spot*
ultimate strength	*bump*
compressive strength	*low spot*
shear strength	*dent*
torsional strength	*raising*
grain structure	*lowering*

finishing hammers
dinging hammers
hammer-on-dolly
hammer-off-dolly
spoons
spring hammering
paintless dent removal
vacuum suction cup
stud spot welder
shrinking metal
torch shrinking
gouge
stretched metal
kinking
body fillers
plastic fillers
silicones
filler mixing

mixing hardener
defective (bad) hardener
catalyzing
hardener
filler mixing board
filler over-catalyzation
filler under-catalyzation
spreader
spatula
filler filing
seam sealer
filler oversanding
filler undersanding
featheredging
priming
spot putty
glazing putty
guide coat

Review Questions

1. What is the difference between mild steel and high-strength steel?

2. Which of these terms refers to a measure of how well a material resists a twisting force?
   a. Tensile strength
   b. Torsional strength
   c. Yield stress
   d. Compression strength

3. _____ _____ is the ability of metal to stretch and return to its original shape.

4. _____ _____ is the limit of plastic deformation that causes the metal to become hard where it has been bent.

5. When is a part kinked?

6. When is a part bent?

7. Explain the difference between a high spot and a low spot on a panel.

8. The term _____ means to work a dent outward or away from the body. The term _____ means to work a high spot or bump down or into the body.

9. Describe the difference between hammer-on- and hammer-off-dollying.

10. A vacuum suction cup uses a remote power source (separate vacuum pump or air compressor airflow) to produce negative pressure (vacuum) in the cup. True or false?

11. What is the main advantage of using a stud/nail welder over a screw-in puller?

12. When should you shrink metal?

13. How do you torch shrink metal?

14. Aluminum turns "cherry red" when heated near its melting point. True or false?

15. _____ _____, also called _____ _____, are designed to cover up minor surface irregularities that remain after metal straightening.

16. List seven rules for applying plastic filler.

17. This is done to mix a tube of hardener.
    a. Knurling
    b. Shaking
    c. Spreading
    d. Kneading

18. Why should cardboard not be used as a mixing board for filler?

19. Explain how you should file filler.

20. _____ _____ results in the filled area being below the desired level, which makes it necessary to apply more filler.

ASE–Style Review Questions

1. Which of the following types of steel is used in unibody structural components?
   a. Hot-rolled steel
   b. Mild steel
   c. High-strength steel
   d. Mild-strength steel

2. *Technician A* says that the secret of metal straightening is to hit the right spot at the right time, with the right amount of force. *Technician B* says damage should be removed in the opposite direction from how it occurred. Who is correct?
   a. Technician A
   b. Technician B
   c. Both Technicians A and B
   d. Neither Technician

3. A technician is removing a small dent from a panel with a body hammer. However, the repair is going slowly because small nicks are forming where the hammer strikes the panel. Which of the following could be the problem?
   a. Not using dolly
   b. Dolly too large
   c. Not hitting squarely with hammer
   d. Hammer has serrated face

4. *Technician A* says that the contour of the dolly must fit the contour of the backside of the damaged area. *Technician B* says the dolly contour should be the opposite of the panel to quickly drive out damage. Who is correct?
   a. Technician A
   b. Technician B
   c. Both Technicians A and B
   d. Neither Technician

5. Which of the following metalworking methods should be used to smooth small, shallow dents and bulges because it concentrates force in a small area?
   a. Hammer-off-dolly
   b. Hammer-on-dolly
   c. Hammer stretching
   d. Hammer shrinking

6. Which of the following metalworking techniques is used to raise low spots and lower high spots simultaneously?
   a. hammer-off-dolly
   b. hammer-on-dolly
   c. hammer stretching
   d. hammer shrinking

7. *Technician A* says that you should never punch or drill holes in panels. *Technician B* says that this is acceptable when the panel is going to be replaced? Who is correct?
   a. Technician A
   b. Technician B
   c. Both Technicians A and B
   d. Neither Technician

8. Filler has been applied to a panel and numerous pinholes are found in the material. Which of the following is the most common cause for this problem?

   a. Improper surface prep
   b. Undermixing
   c. Not enough hardener
   d. Too much hardener

9. Which of the following grits of sandpaper should be used when featheredging plastic filler?
   a. 36
   b. 80
   c. 180
   d. 600

10. Which of the following can be used to check for high and low spots when final sanding a repair area on a compound curve?
    a. Guide coat
    b. Straightedge
    c. Template
    d. Inspection

Activities

1. Obtain a damaged panel (fender, hood, door, or lid). Use a hammer to place a small dent in a flat area of the panel. Use the information in this chapter to grind, fill, and smooth the dent using filler.

2. Featheredge, prime, and spot putty the area filled in activity number one.

3. Read the mixing instructions on several brands of filler. Make a report on your findings. What differences in mixing hardener and filler did you find?

11

Plastic and Composite Repair

Objectives

After studying this chapter, you should be able to:

- List typical plastic and composites applications in vehicle construction.

- Identify automotive plastics through the use of international symbols (ISO codes) and by making a trial-and-error weld.

- Describe the basic differences between welding metal and welding plastic.

- Outline the basics of hot-air and airless welding.

- Repair interior and unreinforced hard plastics.

- Perform two-part adhesive repairs.

- Repair RRIM and other reinforced plastics.

Introduction

The terms *composites* and *plastics* refer to a wide range of materials synthetically compounded from crude oil, coal, natural gas, and other natural substances. Unlike metals, they do not occur in nature and must be manufactured. Plastics have become an important part of today's vehicles. Today, more and more plastic is being used in automobile manufacturing. Look at Figure 11–1.

Plastic parts include bumper covers, fender extensions, fascias, fender aprons, grille openings, stone shields, instrument panels, trim panels, fuel lines, and engine parts, Figure 11–2. Fuel saving and weight reduction programs by auto makers have made plastic parts more common. See Figure 11–3 on page 190.

This increasing use of plastics and composites has resulted in new approaches to collision repair. Many plastic parts can be repaired more economically than they can be replaced, especially if the part does not have to be removed. Cuts, cracks, gouges, tears, and punctures can be repairable. When necessary, some plastics can also be re-formed back to their original shape after distortion. Repair is also quicker than replacement. Since parts are not always available, this means less downtime for the vehicle.

Figure 11–1 Plastics and composites are now very common in modern vehicles. This car uses a composite skin over doors and many other parts. This increases impact resistance, improves corrosion resistance, and reduces vehicle weight for increased fuel economy. (Courtesy of General Motors Corporation, Service Operations)

TYPES OF AUTO PLASTICS

Two general types of plastics are used in automotive construction: thermoplastics and thermosetting plastics.

Thermoplastics can be repeatedly softened and reshaped by heating, with no change in their chemical makeup. They soften or melt when heated and harden when cooled. Thermoplastics are weldable with a plastic welder.

Thermosetting plastics undergo a chemical change by the action of heating, a catalyst, or ultraviolet light. They are hardened into a permanent shape that CANNOT be altered by reapplying heat or catalysts. Thermosets are NOT weldable, but they can be repaired with flexible parts repair materials.

The table in Figure 11–4 on page 191 shows some of the more common plastics with their full chemical name, common name, and their locations on a vehicle.

Composite plastics, or "hybrids," are blends of different plastics and other ingredients designed to achieve specific performance characteristics.

PLASTICS SAFETY

Working with plastics, fiberglass, and composites requires you to think about safety at all times. The resin and related ingredients can irritate your skin, lungs, eyes, and stomach lining. The curing agent or hardener can produce harmful vapors.

Read and understand the following safety points before using any of these types of products:

1. Read all label instructions and warnings carefully.

2. When cutting, sanding, or grinding plastics, dust control is important.

3. Wear rubber gloves when working with fiberglass resin or hardener. Long sleeves, buttoned collar, and cuffs are helpful in preventing sanding dust from getting on your skin. Disposable paint suits will keep dust away from clothes.

4. A protective skin cream should be used on any exposed areas of the body.

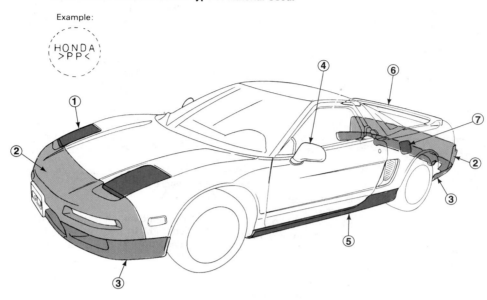

Figure 11–2 Note the types of composite parts used on this aluminum body vehicle. (© American Honda Motor Co., Inc.)

No.	Part Name	Material
①	Headlight Lid	PA6/PPE-M (Polyamide/Polyphenylene ether)
②	Front and Rear Bumpers	PBT-P (Polybutylene terephthalate)
③	Front and Rear Skirts	PP (Polypropylene)
④	Door Mirror Housing	ABS (Acrylonitrile butadiene styrene)
⑤	Side Sill Panel	PA6/PPE-M (Polyamide/Polyphenylene ether)
⑥	Trunk Lid Spoiler	UP-G (Polyster unsaturated thermoset)
⑦	Fuel Lid	PA6/PPE-M (Polyamide/Polyphenylene ether)

5. If the resin or hardener comes in contact with your skin, wash with borax soap and water or alcohol.

6. Safety glasses are always a necessity.

7. Always work in a well-ventilated area.

8. Wear a respirator to avoid inhaling sanding dust and resin vapors.

9. Some shop chemicals can react with certain plastics. Keep brake fluid, solvents, and gasoline off exposed plastics. Use only manufacturer-approved solvents to clean plastics.

10. PVC-type plastics produce a poison gas when burned. Keep them away from excess heat and flames.

PLASTIC IDENTIFICATION

There are several ways to identify an unknown plastic. One way to identify a plastic is by international symbols, or **ISO codes,** which are molded into plastic parts. Many manufacturers are using these symbols. The symbol or abbreviation is formed in an oval on the back of the part. One problem is that you usually have to remove the part to read the symbol.

If the part is NOT identified by a symbol, the **body repair manual** will give information about plastic types used on the vehicle. Body manuals often name the types of plastic used in a particular application.

A **floating test** can be used to help determine the type of plastic. Cut a piece of plastic off the part to be repaired. Drop the sliver of plastic into a container of

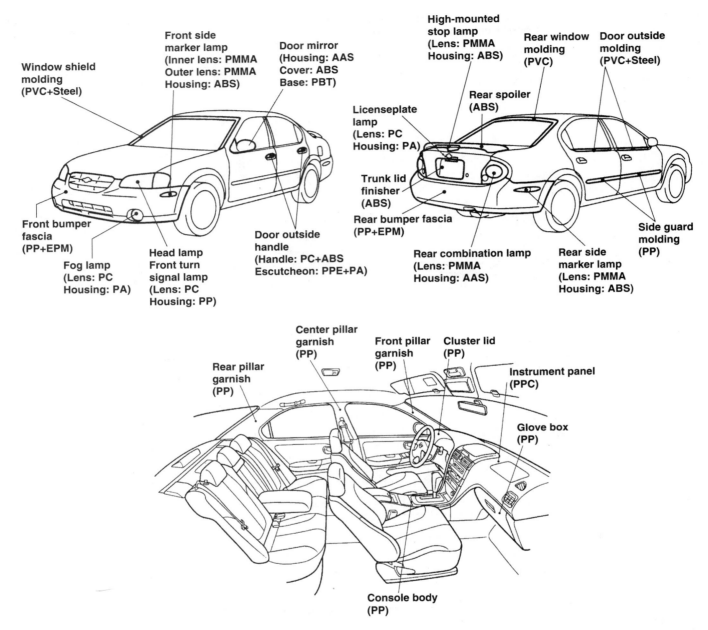

Figure 11–3 Study the location of different kinds of composites (plastics) on a typical vehicle. *(Used with permission of Nissan North America)*

water. Watch to see if the piece of plastic sinks or floats in the water.

If the plastic floats in the water, it is thermoplastic. You would then know that you can repair the part with either welding or an adhesive. If the plastic piece sinks in the water, it is thermoset plastic and must be repaired with an adhesive.

A reliable means of identifying an unknown plastic is to make a trial-and-error weld on a hidden or damaged area of the part. Try several different filler rods until one sticks. Most suppliers offer only a few types of plastic filler rods; the range of possibilities is not that great. The rods are color coded. Once you find a rod that works, the base material is identified.

Some technicians like to cut a sliver off the damaged part and use it as a welding rod. Cut the piece from a location on the part that cannot be seen after installation. Try using this piece as a welding rod. If the plastic fails to weld, you should use an appropriate adhesive to repair the plastic damage.

SYMBOL, CHEMICAL NAME, TRADE NAME, AND DESIGN APPLICATIONS OF COMMONLY USED AUTOMOTIVE PLASTICS

Symbol	Chemical Name	Trade Name	Design Applications	Thermosetting or Thermoplastic
AAS	Acrylonitrile-styrene	Acrylic Rubber	—	Thermoplastic
ABS	Acrylonitrile-butadiene-styrene	ABS, Cycolac, Abson, Kralastic, Lustran, Absafil, Dylel	Body and dash panels, grilles, headlamp doors	Thermoplastic
ABS/MAT	Hard ABS reinforced with fiberglass	—	Body panels	Thermosetting
ABS/PVC	ABS/Polyvinyl chloride	ABS Vinyl	—	Thermoplastic
EP	Epoxy	Epon, EPO, Epotuf, Araldite	Fiberglass body panels	Thermosetting
EPDM	Ethylene-propylene-diene-monomer	EPDM, Nordel	Bumper impact strips, body panels	Thermosetting
EVA	Ethylene/Vinyl Acetate	Elvax, Microthane	Soft trim parts	Thermosetting
PA	Polyamide	Nylon, Capron, Zytel, Rilsan, Minlon, Vydyne	Exterior trim panels	Thermosetting
PBT	Polybutylene	Bexloy "M"	Rocker cover moldings, fascias	Thermosetting
PC	Polycarbonate	Lexan, Merlon	Grilles, instrument panels, lenses	Thermoplastic
PPO	Polyphenylene oxide	Noryl, Olefo	Chromed plastic parts, grilles, headlamp doors, bezels, ornaments	Thermosetting
PE	Polyethylene	Dylan, Fortiflex, Marlex, Alathon, Hi-fax, Hosalen, Paxon	Inner fender panels, interior trim panels, valances, spoilers	Thermoplastic
PF	Phenol-Formaldehyde resin	Bakelite, Genal, Resinox	Ashtrays	Thermosetting
PP	Polypropylene	Profax, Olefo, Marlex, Olemer, Aydel, Dypro	Interior moldings and panels, inner fenders, radiator shrouds, bumper covers	Thermoplastic
PS	Polystyrene	Lustrex, Dylene, Styron, Fostacryl, Duraton	—	Thermoplastic
PUR	Polyurethane	Castethane, Bayflex	Bumper covers, front and rear body panels, filler panels	Thermosetting
TEO	Ethylene/Propylene	Thermoplastic, Rubber	Bumper fascia, valance panels, air dams	Thermoplastic
TPUR	Polyurethane	Pellethane, Estane, Roylar, Texin	Bumper covers, gravel deflectors, filler panels, soft bezels	Thermoplastic
PVC	Polyvinyl chloride	Geon, Vinylete, Pliovic	Interior trim, soft filler panels	Thermoplastic
RIM	"Reaction injection molded" polyurethane	—	Bumper covers	Thermosetting
RRIM	Reinforced RIM-polyurethane	—	Exterior body panels	Thermosetting
SAN	Styrene-acrylonitrite	Lustran, Tyril, Fostacryl	Interior trim panels	Thermosetting
SMC	Sheet Molded Compound	—	Body panels	Thermosetting
TPR	Thermoplastic rubber	—	Valence panels	Thermosetting
UP	Polyester	SMC, Premi-glas, BMC, Selection Vibrin-mat	Fiberglass body panels	Thermosetting

Figure 11–4 This chart lists symbols, chemical names, trade names, applications, and types of commonly used automotive plastics.

PLASTIC PART REMOVAL

There are various types of fasteners used to secure plastic parts to the vehicle. These fasteners include screws, clips, or adhesive. Refer to the vehicle's service manual for directions about part removal. See Figure 11–5.

- An assistant is helpful when removing the front bumper.
- Take care not to scratch the front bumper and body.
- Put on gloves to protect your hands.

Install in the reverse order of removal, and note these items:

- Make sure the front bumper engages the side clips and hooks of the upper beam securely.
- Replace any damaged clips.

If the plastic part has only minor damage, it can often be repaired by welding, or more commonly, by using a two-part plastic repair adhesive. By repairing the part, you save time.

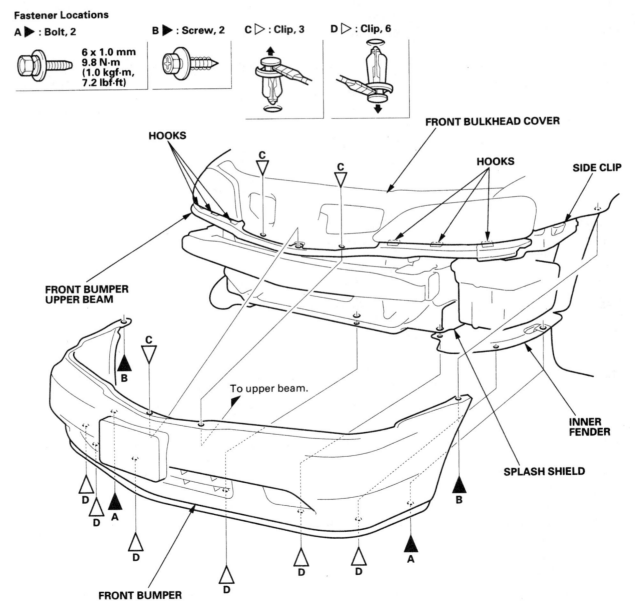

Figure 11–5 (A) Various methods are used to secure composite parts to the vehicle. Note how this front bumper is held on a unibody. (© American Honda Motor Co., Inc.) (B) Study how the rear bumper is secured to the vehicle. (© American Honda Motor Co., Inc.)

NOTE:
• An assistant is helpful when removing the rear bumper.
• Take care not to scratch the rear bumper and body.
• Put on gloves to protect your hands.

Install in the reverse order of removal, and note these items:

• Make sure the rear bumper engages the hooks of the bumper spacers and upper beam on each side securely.
• Make sure the license plate light connector is plugged in properly.
• Replace any damaged clips.

Fastener Locations

▶ : Screw, 2   A ▷ : Clip, 7   B ▷ : Clip, 2

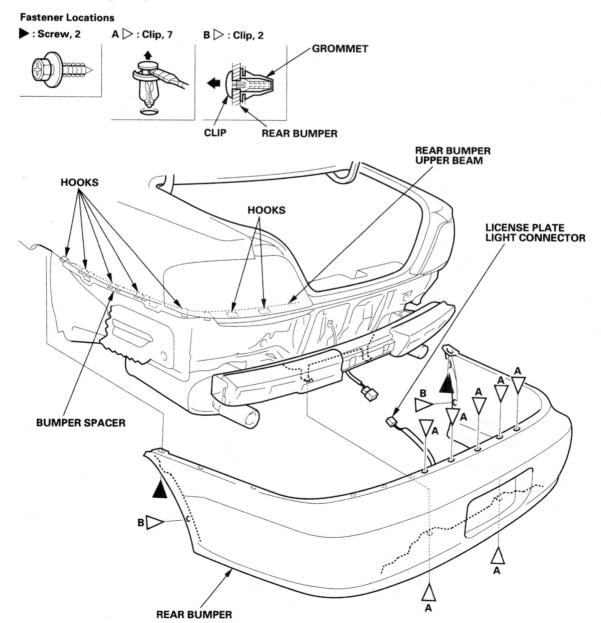

Figure **11–5** {continued}

PRINCIPLES OF PLASTIC WELDING

Plastic welding uses heat and sometimes a plastic filler rod to join or repair plastic parts. The welding of plastics is not unlike the welding of metals. Both methods use a heat source, welding rod, and similar techniques (butt joints, lap joints, etc.). Joints are prepared in much the same manner, and evaluated for strength. There are differences, however, between welding metal and welding plastics.

Plastics and composites have a wide melting range between the temperature when they soften and the temperature when they char or burn. Also, unlike metals, plastics are poor conductors of heat. This makes them difficult to heat uniformly. Plastic filler rod and plastic surface will burn before the material below the surface becomes fully softened. A plastic welder must work within a much smaller temperature range than a metal welder.

Because a plastic rod does not become completely molten, a plastic weld might appear incomplete. The outer surface of the rod becomes molten while the inner core remains semi-solid. The plastic rod will not

look completely melted, except for molten flow on either side of the bead. Even though a strong and permanent bond has been formed, you may think you have a weak, cold-joint weld.

When plastic welding, force the rod into the joint to create a good bond. When heat is taken away, the rod reverts to its original form. A combination of heat and pressure is used when welding plastic. Normally, apply pressure on the welding rod with one hand, while applying heat with the tool. Use a constant fanning motion with hot air from the welding torch.

NOTE! Too much pressure on the rod tends to stretch the bead. Too much heat will char, melt, or distort the plastic. Competent plastic welding takes practice.

HOT-AIR PLASTIC WELDING

Hot-air plastic welding uses a tool with an electric heating element to produce hot air (450° to 650°F or 230° to 340°C), which blows through a nozzle and

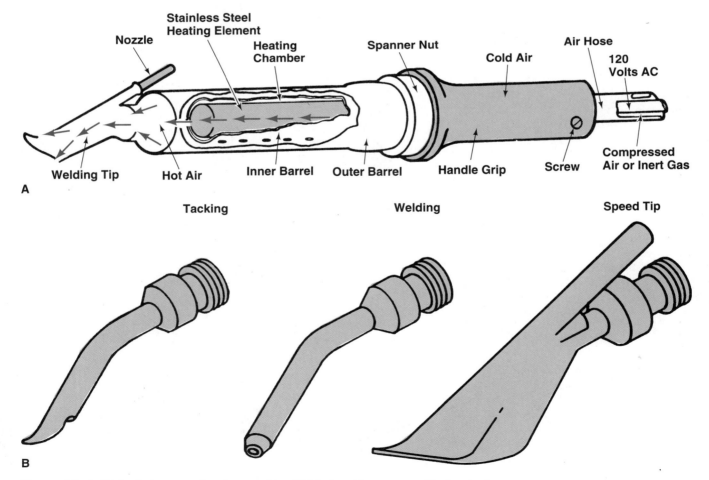

Figure 11–6 (A) Study the parts of a plastic welder. (B) Note welder tip types. *(Seelye, Inc.)*

onto the plastic. The air supply comes from either the shop's air compressor or a self-contained portable compressor that comes with the welding unit.

Most hot-air welders use a tip working pressure of around 3 psi (0.21 kg/cm^2). Air pressure regulators reduce the air pressure first to around 50 psi, and then to the working pressure of about 3 psi (0.21 kg/cm^2). A typical hot-air welder is illustrated in Figure 11–6.

The hot-air torch is used in conjunction with the welding rod, which is normally $\frac{3}{16}$ inch (5 mm) in diameter. The plastic welding rod must be made of the same material as the plastic being repaired. This will ensure proper strength, hardness, and flexibility of the repair. See Figure 11–7.

One of the problems with hot-air welding is that the plastic welding rod is often thicker than the panel to be welded. This can cause the panel to overheat before the rod has melted. Using a smaller diameter rod with the hot-air welder can often correct such warpage problems.

Three types of welding tips are available for use with most hot-air plastic welding torches. They are:

Tacking tips—shaped to tack weld broken sections of plastic together before welding. If necessary, tack welds can be easily pulled apart for realigning.

Round tips—used to make short welds, to weld small holes, to weld in hard-to-reach places, and to weld sharp corners, Figure 11–6.

Speed tips—hold, feed, and automatically preheat the plastic welding rod. This design feeds the rod into the base material, thus allowing for faster welding speeds. They are used for long, fairly straight welds, Figure 11–8.

Some hot-air welder manufacturers have developed specialized welding tips and rods to meet spe-

cific needs. Check the product catalog for more information.

Using Hot-Air Plastic Welders

No two hot-air plastic welders work exactly alike. For specific instructions, always refer to the owner's manual and other material provided by the hot-air welder manufacturer.

Some manufacturers advise against using their welder on plastic thinner than $\frac{1}{8}$ inch (3.2 mm) because of distortion. It is sometimes acceptable to weld thin plastics if they are supported from underneath while welding.

To set up a typical hot-air welder, proceed as follows:

1. Close the pressure regulator valve by turning it counterclockwise until loose. This will prevent possible damage to the gauge from a sudden surge of air pressure.

2. Connect the air pressure regulator to a supply of either compressed air or inert gas. If inert gas is used, a pressure-reducing valve is needed.

3. Turn on the air supply. Starting pressure depends upon the wattage of the heating element. Check the operating manual for specifications.

4. Connect the welder to a common 120-volt AC outlet. Use a three-prong grounded plug.

5. Allow the welder to warm up at the recommended air pressure. Air or inert gas must flow through the welder at all times, from warm-up to cool-down. This will prevent burnout of the heating element and damage to the welder.

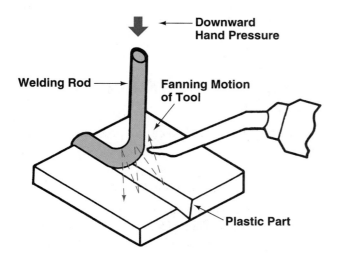

Figure 11–7 When welding plastic, you must use the proper combination of heat and pressure. Hand pressure is needed to force the plastic rod down into the weld joint. Fan the welder over the joint to heat the rod and base material properly.

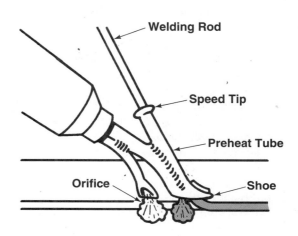

Figure 11–8 Note the construction of a high-speed plastic welding tip.

6. Select the proper tip. While wearing work gloves, insert the tip into the torch with pliers to avoid touching the barrel while it is hot.

7. After the tip has been installed, the temperature will increase slightly due to back pressure. Allow 2 to 3 minutes for the tip to reach operating temperature.

8. Check the temperature by holding a thermometer ¼ inch (6 mm) from the hot-air end of the torch. For most thermoplastics, the temperature should be in the 450° to 650°F (230° to 340°C) range. Information supplied with the welder usually includes a chart of welding temperatures.

9. If the temperature is too high to weld the material, increase the air pressure slightly until the temperature goes down. If the temperature is too low, decrease the air pressure slightly until the temperature rises. When increasing and decreasing the air pressure, allow at least 1 to 3 minutes for the temperature to stabilize.

10. Remember, the element can become overheated by too little air pressure. When decreasing air pressure, never allow the round nut that holds the barrel to the handle of the welder to become too hot to touch. This indicates overheating.

11. A partially clogged dirt screen in the regulator can also cause overheating. Watch for these symptoms.

12. When you have finished welding, disconnect the electrical cord. However, allow the air to flow through the welder for a few minutes or until the barrel is cool to the touch. Then disconnect the air supply.

AIRLESS PLASTIC WELDING

Airless plastic welding uses an electric heating element to melt a smaller ⅛ inch (3 mm) diameter rod with no external air supply. It has become very popular. Airless welding with a smaller rod helps eliminate two troublesome problems: panel warpage and excess rod buildup.

When setting up an airless welder, set the temperature dial at the appropriate setting. This will depend upon the specific plastic being worked on. It will normally take about 3 minutes for the welder to fully warm up.

Make sure the rod is the same material as the damaged plastic or the weld will be unsuccessful. Many airless welder manufacturers provide rod application charts. When the correct rod has been chosen, it is good practice to run a small piece through the welder to clean out the tip before beginning.

PLASTIC WELDING METHODS

The basic methods for hot-air and airless welding are very similar. To make a good plastic weld with either procedure, keep the following factors in mind:

1. Plastic welding rods are frequently color coded to indicate their material. Unfortunately, the **rod color coding** is not uniform among manufacturers. It is important to use the reference information provided. If the rod is not compatible with the base material, the weld will NOT hold.

2. Too much heat will char, melt, or distort the plastic. Too little heat will not provide weld penetration between the base material and the rod.

3. Too much pressure stretches and distorts the weld.

4. The angle between the rod and base material must be correct. If it is too shallow, a proper weld will NOT be achieved.

5. Use the correct welding speed. If the torch movement is too fast, it will NOT permit a good weld. If the tool is moved too slowly, it can char the plastic.

The basic repair sequence is generally the same for both processes. That is,

1. Prepare the damaged area.

2. Align the damaged area.

3. Make the weld.

4. Allow it to cool.

5. Sand. If the repair area has pinholes or voids, bevel the edges of the defective area. Add another weld bead. Then, re-sand.

6. Apply a flexible filler and topcoat.

HOT-AIR PLASTIC WELDING PROCEDURE

The typical hot-air plastic welding procedure is as follows:

1. Set the welder to the proper temperature (if a temperature adjustment is provided).

2. Wash and clean the part with soap and water. Allow it to dry. Then clean the part with a plastic cleaner. Do NOT use conventional prep solvents or wax and grease removers. To remove silicone materials, use a conventional cleaner first, making sure to completely remove all residue.

3. V-groove the damaged area.

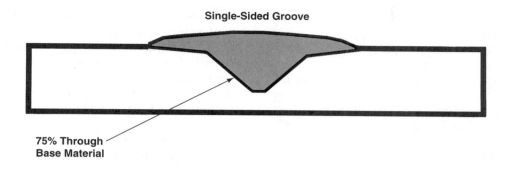

Single-Sided Groove

75% Through
Base Material

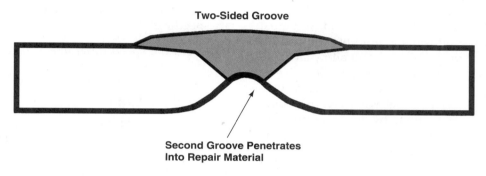

Two-Sided Groove

Second Groove Penetrates
Into Repair Material

Figure 11–9 With a single-sided weld, the weld must penetrate about 75 percent of the base material. With a two-sided weld, the second weld should fuse with the weld on the other side. *(I-Car)*

4. Bevel the part ¼ inch (6 mm) on each side of the damaged area.

5. Tack weld or tape the brake line with aluminum body tape.

6. Select the welding rod and welding tip best suited to the type of plastic and the damage.

7. Make the weld. The plastic weld should penetrate 75 percent through the base material, Figure 11–9. Allow it to cool and cure for about 30 minutes.

8. Grind or sand the weld to the proper contour and shape.

When welding plastic, single- or double-V butt welds produce the strongest joints. When using a round or V-shaped welding rod, prepare the area by slowly grinding, sanding, or shaving the adjoining surfaces to produce a single- or double-V. Wipe any dust or shavings from the joint with a clean, dry rag. Do not use cleaning solvents because they can soften the plastic edges and cause poor welds.

PLASTIC TACK WELDING

On long tears where backup is difficult, small **tack welds** can be made to hold the two sides in place

before doing the permanent weld. For larger areas, a patch can be made from a piece of plastic and tacked in place.

To tack weld, proceed as follows:

1. Hold the damaged area in alignment with clamps or aluminum body tape.

2. Using a tacking welding tip, fuse the two sides to form a thin hinge weld along the root of the crack. This is especially useful for long cracks because it allows for easy adjustment and alignment of the edges.

3. Start tacking by drawing the point of the welding tip along the joint. Press the tip in firmly, making sure to contact both sides of the crack. Draw the tip smoothly and evenly along the line of the crack. No welding rod is used when tacking.

4. The point of the tip will fuse both sides in a thin line at the root of the crack. The fused parts will hold the sides in alignment. Then, you can fuse the entire length of the crack.

WELDING A PLASTIC V-GROOVE

Prepare the rod for welding by cutting the end at approximately a 60-degree angle. When starting a weld, hold the nozzle about ¼ to ½ inch (6 to 12 mm) above and parallel to the base material. Hold the rod at a right angle to the work. Position the cut end of the rod at the beginning of the weld.

Direct the hot air from the tip alternately at the rod and the base material. Concentrate the heat more on the rod when starting to weld. Keep the rod lined up and press it into the seam. Light pressure is sufficient.

Once the rod begins to stick to the plastic, start to move the torch and use the heat to control the flow. After you start your weld, direct more of the heat into the base material. Be careful not to melt or char the plastic. As the welding continues, a small bead should form along the entire weld joint.

As you plastic weld and use up the rod, you must regrip the rod. Unless it is done carefully, you will lift the rod away from the weld. This will allow air to be trapped under the weld, weakening it. To prevent this, you must develop the skill of continuously applying pressure on the rod while repositioning your fingers. This can be done by applying pressure with your third and fourth fingers while moving the thumb and first finger up the rod.

At the end of a weld, maintain pressure on the rod as the heat is removed. Hold the rod still for a few seconds to make sure it has cooled and does not pull loose. Then carefully cut the rod with a sharp knife or clippers.

Do NOT attempt to pull the rod from the joint after making the weld. About 15 minutes cooling time is needed for rigid plastic and 30 minutes for thermoplastics.

Smooth the weld area by grinding with No. 36 grit sandpaper. Excess plastic can be removed with a sharp knife before grinding. When grinding, do NOT overheat the weld area because it will soften. Use water to cool the plastic while grinding, if needed.

After rough grinding, check the weld visually for defects. Any voids or cracks are unacceptable. Bending should NOT produce any cracks. A good plastic weld is as strong as the part itself!

The weld area can be finish sanded using 220 grit sandpaper followed by a 320 grit. Either a belt or orbital sander may be used, plus hand sanding as required. If refinishing is to be done, follow the procedure designed specifically for plastics.

PLASTIC SPEED WELDING

Plastic speed welding uses a specially designed tip to produce a more uniform weld and at a high rate of speed. You must preheat both the rod and base material. The rod is preheated as it passes through a tube in the speed tip. The base material is preheated by a

stream of hot air passing through a vent in the tip.

A pointed shoe on the end of the tip applies pressure to the rod. You do not have to apply pressure to the rod. The shoe smoothes out the rod, creating a more uniform appearance in the finished weld. On panel work, speed welding is commonly used.

With speed plastic welding, the conventional two-hand method is replaced by a faster and more uniform one-hand operation. Once started, the rod is fed automatically into the preheat tube as the welding torch is pulled along the joint. Speed tips are designed to provide the constant balance of heat and pressure. The average welding speed is about 40 inches per minute (1,000 mm per minute).

Following are some techniques essential for quality speed welding.

1. Hold the speed torch like a dagger. Bring the tip over the starting point a full 3 inches (75 mm) from the base material. You do not want the hot air to affect the part.

2. Cut the welding rod at a 60-degree angle. Insert it into the preheat tube. Immediately place the pointed shoe end of the tip on the base material at the starting point.

3. Hold the torch perpendicular to the base material. Push the rod through until it stops against the base material at the starting point. If necessary, lift the torch slightly to allow the rod to pass under the shoe.

4. Keep a slight pressure on the rod and only the weight of the torch on the shoe. Then pull the torch slowly toward you to start the speed weld.

   Caution! Once the speed weld is started, do NOT stop. If you must pause for any reason, pull the speed tip off the rod immediately. If this is NOT done, the rod will melt into the preheat tube.

5. In the first inch or two (25–50 mm) of travel, push it into the preheat tube with slight pressure.

6. Once started, swing the torch to a 45-degree angle. The rod will now feed without a need for pressure. As the torch moves along, inspect the quality of the weld.

Controlling Speed Welding Rate

The angle between the torch and base material determines the *speed welding rate.* For this reason, hold the torch at a 90-degree angle when starting the weld. Hold the torch at a 45-degree angle after you begin welding.

If the welding rate is too fast, bring the torch back to the 90-degree angle temporarily to slow it down. Then gradually move to the desired angle for proper welding speed.

Remember, once started, speed welding must be maintained at a fairly constant rate. The torch cannot be held still. To stop welding before the rod is used up, bring the torch back past the 90-degree angle and cut off the rod at the end of the shoe.

A good speed weld in a V-groove will have a slightly higher crown and more uniformity than the normal hand weld. It should appear smooth and shiny, with a slight bead on each side. For best results and faster welding speed, clean the shoe on the speed tip with a wire brush to remove any residue that might create drag on the rod.

AIRLESS MELT-FLOW PLASTIC WELDING

Melt-flow plastic welding is the most commonly used airless welding method. It can be utilized for both single-sided and two-sided repairs. Refer back to Figure 11–9.

A typical melt-flow procedure is as follows:

1. With the welding rod in the preheat tube, place the flat shoe part of the tip in the V-groove.

2. Hold it in place until the rod begins to melt and flow out around the shoe.

3. A small amount of force is needed to feed the rod through the preheat tube. The rod will NOT feed itself. Care should be used not to feed it too fast.

4. Move the shoe slowly. Crisscross the groove until it is filled with melted plastic.

5. Work the melted plastic well into the base material, especially toward the top of the V-groove.

6. Complete a weld length of about 1 inch (25 mm) at a time. This will allow smoothing of the weld before the plastic cools.

PLASTIC STITCH-TAMP WELDING

Plastic stitch-tamp welding involves melt-flow fusion followed by using pointed end of the welding tip to help bond the plastic rod and base plastic together. It is primarily used on hard plastics, like ABS and nylon, to ensure a good base-rod mix.

After completing the weld using the melt-flow procedure, remove the rod. Turn the shoe over and slowly move the pointed end of the tip into the weld area to bond the rod and base material together.

Stitch-tamp the entire length of the weld. After stitch-tamping, use the flat shoe part of the tip to smooth out the weld area.

SINGLE-SIDED PLASTIC WELDS

Single-sided plastic welds are used when the part cannot be removed from the vehicle. To make a single-sided weld, proceed as follows:

1. Set the temperature dial on the welder for the plastic being welded. Allow it to warm up to the proper temperature.

2. Clean the part by washing with soap and water, followed by a good plastic cleaner.

3. Align the break using aluminum body tape.

4. V-groove the damaged area 75 percent of the way through the base material. Angle or bevel back the torn edges of the damage at least ¼ inch (6 mm) on each side of the damaged area. Use a die grinder or similar tool.

5. Clean the preheat tube, and insert the rod. Begin the weld by placing the shoe over the V-groove and feeding the rod through. Move the tip slowly for good melt-in and heat penetration.

6. When the entire V-groove has been filled, turn the shoe over and use the tip to stitch-tamp the rod and base material together into a good mix along the length of the weld.

7. Resmooth the weld area using the flat shoe part of the tip, again working slowly. Then cool with a damp sponge or cloth.

8. Shape the excess weld buildup to a smooth contour, using a razor blade and/or abrasive paper.

TWO-SIDED PLASTIC WELDS

A **two-sided plastic weld** is the strongest type of weld because you weld both sides of the part. When making a two-sided weld, be sure to do the following:

1. Allow the welder to heat up. Then clean the preheat tube.

2. Clean the part with soap and water and plastic cleaner.

3. Align the front of the break with aluminum body tape, smoothing it out with a stiff squeegee or spreader.

4. V-groove 50 percent of the way through the back side of the panel.

5. Weld the back side of the panel using the melt-flow method. Move slowly enough to achieve good melt-in.

6. When finished, smooth the weld with the shoe.

7. Quick-cool the weld with a damp sponge or cloth.

8. Remove the tape from the front of the piece. V-groove deep enough that the first weld is penetrated by the second V-groove.

9. Weld the seam, filling the groove completely.

10. Use a razor blade or slow speed grinder to reshape the contour.

REPAIRING VINYL DENTS

Vinyl is a soft, flexible, thin plastic material often applied over a foam filler. Vinyl over foam construction is commonly used on interior parts for safety. Common vinyl parts are the dash pads, armrests, inner door trim, seat covers, and exterior roof covering. Dash pads or padded instrument panels are expensive and time consuming to replace. Therefore, they are perfect candidates for repair, Figure 11–10.

Most **dash pads** are made of vinyl-clad urethane foam to protect people during a collision. Surface dents in foam dash pads, armrests, and other padded interior parts are common in collision repair. These dents can often be repaired by applying heat as follows:

1. Soak the dent with a damp sponge or cloth for about half a minute. Leave the dented area moist.

2. Using a heat gun, heat the area around the dent. Hold the gun 10 to 12 inches (250 to 300 mm) from the surface. Keep it moving in a circular motion at all times, working from the outside inward.

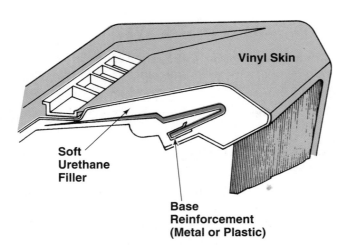

Vinyl Skin

Soft Urethane Filler

Base Reinforcement (Metal or Plastic)

Figure 11–10 Study the construction of a typical dash pad.

3. Heat the area to around 130°F (54°C). Do NOT overheat the vinyl or it will blister. Keep heating it until the area is too hot to touch. If available, use a digital thermometer to meter the surface temperature.

4. Wearing gloves, massage the pad. Force the material toward the center of the dent. The area may have to be reheated and massaged more than once. In some cases, heat alone may repair the damage.

5. When the dent has been removed, cool the area quickly with a damp sponge or cloth.

6. Apply vinyl treatment or preservative to the part.

WELDING VINYL-FOAM CUTS

To plastic weld a cut in vinyl-foam parts, a dash pad for example, proceed in the following manner:

1. Set up the welder for the proper temperature for welding urethane.

2. Wash the pad with soap and water, dry, and clean with plastic cleaner.

3. If the damaged area is brittle, warm it with a heat gun. If there are curled or jagged edges, cut them away.

4. V-groove at least ¼ inch (6 mm) into the foam padding. Bevel the edges as much as possible. Rough up the area for about ¼ inch (6 mm) around the V-groove.

5. Turn the welder so the shoe is facing up. Feed the rod slowly through the welder. Start the weld at the bottom of the groove. Completely fill it with melted plastic until it is flush with the surface.

6. Smooth out the excess rod buildup. Feather it out over the beveled edges for at least ¼ inch (6 mm) on each side of the groove.

7. Cool the weld and grind away any remaining excess rod buildup. Rough up the vinyl for about 2 inches (50 mm) beyond the weld on each side for good filler adhesion.

8. Use a flexible filler material designed for use on vinyl to get the desired contour. Allow the filler to cure.

9. Contour sand to remove the tacky glaze. If any filler is accidentally sanded through, apply a skim coat and re-sand.

10. Spray on a vinyl re-texture material to match the grain of the existing vinyl. Blend it out to a break

or contour line on the cover. This will help hide the repair.

11. Mask and paint the entire part the correct color vinyl paint. For more information on vinyl painting, refer to the text index.

HEAT RESHAPING PLASTIC PARTS

Many bent, stretched, or deformed plastic parts, such as flexible bumper covers and vinyl-clad foam interior parts, can often be straightened with heat. This is because of *"plastic memory,"* which means the piece wants to keep or return to its original molded shape. If it is bent or deformed slightly, it will return to its original shape if heat is applied.

To reshape a distorted bumper cover, use the following procedure:

1. Thoroughly wash the cover with soap and water.

2. Clean with plastic cleaner. Make sure to remove all road tar, oil, grease, and undercoating.

3. Dampen the repair area with a water-soaked rag or sponge.

4. Apply heat directly to the distorted area. Use a concentrated heat source, such as a heat lamp or high-temperature heat gun. When the opposite side of the cover becomes uncomfortable to the touch, it has been heated enough.

5. Use a paint paddle, squeegee, or wood block to help reshape the piece if necessary.

6. Quick-cool the area by applying cold water with a sponge or rag.

> *WARNING! Do not overheat textured vinyl or you will damage the vinyl surface.*

PLASTIC BUMPER TAB REPLACEMENT

When a bumper cover has been torn away from its mounting screws, the mounting tabs will often be broken or torn away. Mounting tabs must be repaired with a two-sided weld to provide enough weld strength.

If the material is a thermoplastic, either the hot-air or airless method may be used. If the piece is a thermosetting plastic, airless welding must be used. If welded properly, the repaired piece will be as strong as the original, undamaged part.

The following is the procedure for rebuilding a mounting tab that has been torn off:

1. Begin by cleaning the piece as described before.

2. Bevel back the torn edges of the mounting tab at least ¼ inch (6 mm) on both sides.

3. Rough up the plastic and wipe dust free.

4. Use aluminum body tape to build a form in the shape of the missing tab. Turn the tape edges up to form the thickness of the new tab.

5. Set the temperature dial on the welder for the type of plastic being repaired. Allow the unit to warm up.

6. Begin the weld. Push the rod slowly through the preheat tube. Slightly overfill the form, working the melted plastic into the base material.

7. Smooth and shape the weld. Quick-cool the weld area.

8. Remove the tape. V-groove along the tear line on the other side about halfway through the piece.

9. Weld the groove and quick-cool. Finish the weld to the desired contour using a slow-speed grinder with a No. 60 or 80 grit disc.

PLASTIC ADHESIVE REPAIR SYSTEMS

A plastic adhesive repair system is a high-strength glue or epoxy used to repair damage to plastic parts. The minor damage to the part is cleaned, ground out, and sanded. Then the adhesive is applied to the damaged area. A spreader is used to work the adhesive down into the damage. The repair is then sanded to shape it, much like plastic body filler.

Adhesive repair of plastics is often preferable to plastic welding, especially on more severe damage. If applied properly, adhesives produce a stronger repair that takes little more time than welding. Refer to the vehicle service manual and material instructions to make sure you are using the right kind of adhesive and correct repair procedures.

Adhesive repair systems are of two types: cyanoacrylate (CA) and two-part. Two-part is the most commonly used.

Cyanoacrylates (CAs) are one-part, fast curing adhesives used to help repair rigid and flexible plastics. They are used as a filler or to tack parts together before applying the final repair material. CAs are sometimes known as "super glues." They can be a valuable tool for the repair of plastic parts. CAs set up very quickly.

Although one-part, an activating agent can be used to accelerate the bonding process of a CA. Care must be used NOT to apply too much activating agent. If too much is used, the product foams causing a weaker bond.

CAs do NOT work equally well on all plastics. There is no hard and fast rule. When CAs are used, be sure to use products from reliable suppliers and follow the manufacturer's guidelines for using them.

Two-part adhesive systems consist of a base resin and a hardener (catalyst). The resin comes in one container and the hardener in another. When mixed, the adhesive cures into a plastic material similar to the base material in the part. Two-part adhesive systems are an acceptable alternative to welding for many plastic repairs.

Not all plastics can be welded, while adhesives can be used in all but a few instances. If adhesive repair is chosen, you must first identify the type of plastic. A good way to do this is through a plastic flexibility test.

To do a plastic flexibility test, use your hands to flex and bend the part and compare it to samples of plastic. Compare the flexibilities. Use the repair material that most closely matches the characteristics of the part's base material.

USE THE CORRECT ADHESIVE!

WARNING! When working with an adhesive system, use the manufacturer's categories to decide on a repair product and procedure. There are many plastic and repair material variations. The vehicle manufacturer's repair manual is the most accurate source of information. The service manual will recommend products and procedures for the exact type of plastic in the part.

It is important to keep in mind that there are differences between manufacturers' repair materials. When using plastic repair adhesives, remember that:

1. Mixing product lines is NOT acceptable. Choose a product line and use it for the entire repair.

2. Most product lines have two or more adhesives designed for different types of plastic.

3. The product line usually includes an adhesion promoter, a filler product, and a flexible coating agent. Use each as directed.

4. Some product lines are formulated for a specific base material. For instance, one manufacturer offers individual products for use with each type of plastic (TPO, urethanes, or Xenoy, for example), regardless of plastic flexibility.

A product line might use a single flexible filler for all plastics, or there might be two or more flexible fillers designed for different types of plastic.

TWO-PART PLASTIC ADHESIVE REPAIRS

Regardless of which manufacturer's products are used, two-part adhesive repairs share the common preparation steps:

1. Clean the part. First, use soap and water and then a plastic cleaner. See Figure 11–11.

2. Make sure both the part and the repair material are at room temperature for proper curing and adhesion.

3. Mix the two parts of the adhesive thoroughly and in the proper proportions.

4. Apply the material within the time guidelines given in the product literature. Use heat if indicated by the manufacturer.

5. Follow the cure time guidelines given in the product literature. Regulated heat can speed curing.

6. Support the part adequately during the cure time to ensure that the damaged area does NOT move before the adhesive cures. This would weaken the repair.

7. Follow the product literature for guidelines on when to reinforce a repair.

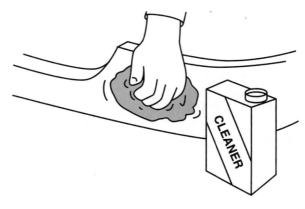

Figure 11–11 First clean the area to be repaired with soap and water. Then wipe the area with silicone and wax remover to clean the surfaces.

An **adhesion promoter** is a chemical that treats the surface of the plastic so the repair material will bond properly. Some plastics (TPO, PP, and E/P, for example) require an adhesion promoter. There is a simple test to perform that indicates whether or not the plastic will require an adhesion promoter. Lightly sand a hidden spot on the piece, using a high-speed grinder and No. 36 grit paper.

If the material gives off dust, it can be repaired with a standard structural adhesive system. If the material melts and smears or has a greasy or waxy look, then you must use an adhesion promoter. Many plastic fillers and adhesives contain an adhesion promoter. Check their labels.

In Figure 11–12, read the step-by-step procedure for repairing a flexible plastic part with two-part adhesive.

Here is a typical way to use a two-part epoxy adhesive to repair a flexible bumper cover:

1. Clean the entire cover with soap and water. Wipe or blow-dry. Then clean the surface with a good plastic cleaner.

2. V-groove the damaged area. Then grind about a 1½ inch (38 mm) taper around the damage for good adhesion and repair strength.

3. Use a sander with No. 180 grit paper to feather-edge the paint around the damaged area. Then blow off the dust. Depending on the extent of the damage, the back side might need reinforcement. To do this, use steps 4 through 6.

4. To reinforce the repair area, sand and clean the back side of the cover with plastic cleaner. Then, if needed, apply a coat of adhesion promoter.

5. Dispense equal amounts of both parts of the flexible epoxy adhesive. Mix them to a uniform color. Apply the material to a piece of fiberglass cloth using a plastic squeegee.

6. Attach the plastic-saturated cloth to the back side of the bumper cover. Fill in the weave with additional adhesive material.

7. With the back side reinforcement in place, Figure 11–13, apply a coat of adhesion promoter to the sanded repair area on the front side. Let the adhesion promoter dry completely.

8. Fill in the area with adhesive material. Shape the adhesive with your spreader to match the shape of the part. Allow it to cure properly.

9. Rough grind the repair area with No. 80 grit paper, then sand with 180 grit, followed by smoother 240 grit.

10. If additional adhesive material is needed to fill in a low spot or pinholes, be sure to apply a coat of adhesion promoter again.

TWO-PART ADHESIVE DASH REPAIR

Some two-part adhesive products are made for dash pad repair. A typical repair procedure is as follows:

1. Thoroughly clean the part.

2. Sand or grind away the broken or loose vinyl covering to get to the foam beneath. A V-groove need not be used.

3. Apply the adhesive material according to the manufacturer's directions. Make sure the recommended cure times are followed.

4. After curing, sand or grind to contour. Apply a skim coat of adhesive if necessary. Allow the adhesive to cure. Then block sand with medium grit paper, followed by fine paper.

5. Apply a sealer. Then spray re-texture to as much area as needed to hide the repair.

6. Refinish the whole part as per paint system recommendations.

REINFORCED PLASTIC REPAIRS

Reinforced plastic—including **reinforced reaction injection molded polyurethane (RRIM)**—parts are being used in many unibody vehicles. They provide a durable plastic skin over a steel unibody. The table in Figure 11–14 provides an overview of reinforced plastic repair materials.

The damage that generally occurs in reinforced plastic panels includes:

1. One-sided damage, such as a scratch or gouge.

2. Punctures and fractures.

3. Panel separation, where the panel pulls away from the metal space frame.

4. Severe damage, which requires full or partial panel replacement.

5. Minor bends and distortions of the space frame, which can be repaired by pulling and straightening.

6. Severe kinks and bends to the space frame, which require replacement of that piece along factory seams or by sectioning.

Remember! Combinations of these types of damage often occur on a single vehicle. Depending on the

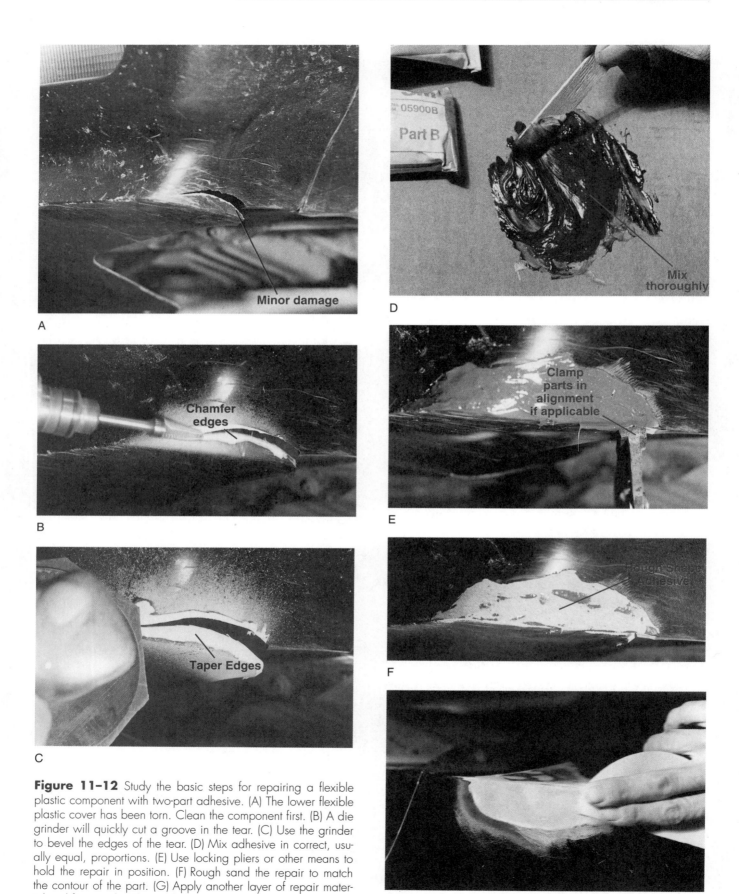

Figure 11–12 Study the basic steps for repairing a flexible plastic component with two-part adhesive. (A) The lower flexible plastic cover has been torn. Clean the component first. (B) A die grinder will quickly cut a groove in the tear. (C) Use the grinder to bevel the edges of the tear. (D) Mix adhesive in correct, usually equal, proportions. (E) Use locking pliers or other means to hold the repair in position. (F) Rough sand the repair to match the contour of the part. (G) Apply another layer of repair material and fine sand until smooth.

Figure 11–13 On larger holes, a backup patch is needed to support the repair material. The patch or backup is bonded to the back of the part. Then, repair material is applied to the front to finish the repair.

location and amount of damage, there are four different types of repairs. These are:

1. Single-sided repair.
2. Two-sided repair.
3. Panel sectioning.
4. Full panel replacement.

To select a repair method, a thorough examination of the vehicle is required. Examine all affected reinforced plastic panels. First, check the entire panel for signs of damage. Also check all panel seams for adhesive bond failure. Examine the back of the panel to determine the extent of the damage.

REINFORCED PLASTIC ADHESIVES

Many of the materials that are used for reinforced plastic repair are two-part adhesive products. Two-part adhesive means a base material and a hardener must be mixed to cure the adhesive. Each must be mixed together in the proper ratio. Both parts must be thoroughly mixed together before use.

Work life or **open time** is the time when it is still possible to disturb the adhesive and still have the adhesive set up for a good bond. This work life time will be provided by the manufacturer. The cure time of some adhesives used in reinforced plastic repair can be shortened with the application of heat. Temperature and humidity can affect work life and cure time.

After mixing, remember that each product has a work life or open time. If you move or disturb the adhesive as it starts to harden, you will adversely affect its durability.

REINFORCED PLASTIC FILLERS AND GLASS CLOTH

Two filler products are specifically formulated for use on reinforced plastic. They are cosmetic filler and structural filler.

Cosmetic filler is typically a two-part epoxy or polyester filler used to cover up minor imperfections.

REINFORCED PLASTIC REPAIR MATERIAL SELECTION CHART

Type of Repair	Applicable Repair Product				
	Panel Adhesive	Patching Adhesive	Structural Filler	Cosmetic Filler	Glass Fiber Reinforcement
Panel Replacement	X				
Panel Sectioning	X		X_1	X_1	X
One-Sided Repairs				X_1	
Two-Sided Repairs	X_2	X_2	X	X	X

Notes: 1. Some panel adhesives can also be used as structural and cosmetic fillers, depending on sanding characteristics.
2. Panel adhesives can also be used as patching adhesives, but not vice versa.

Figure 11–14 Study this reinforced plastic repair chart. It gives recommended repairs and materials.

Do NOT use fillers designed for sheet metal on reinforced plastic.

Structural filler is used to fill the larger gaps in the panel structure while maintaining strength. Structural fillers add to the structural rigidity of the part.

All two-part products will shrink to some degree. The use of heat will help to speed the drying time and will eliminate some of the shrinkage.

Check with the product manufacturer for temperatures and dry times. If the product is NOT properly heated to a full cure, shrinkage will occur as the product cures with time. The "rule of thumb" is to heat the material to a surface temperature higher than any temperatures that the vehicle will be subject to when it is on the road. If it is a black vehicle sitting in the sun in mid-summer, this could be about 170°F (77°C) or more.

Check with the product manufacturer for recommendations for heat curing. Generally, 200° to 250°F (93° to 121°C) for 20 to 40 minutes should do it. Remember that at lower temperatures, the product will have to be heated longer. Also, if there is high humidity, the cure process will take longer.

There are several different types of glass cloth available. Rovings and mattings are NOT appropriate for reinforced plastic repair. Choose unidirectional cloth, woven glass cloth, or nylon screening. The cloth weave must be loose enough to allow the adhesive to fully saturate the cloth, leaving no air space around the weave. See Figure 11–15.

REINFORCED PLASTIC, SINGLE-SIDED REPAIRS

Single-sided damage is surface damage that does NOT penetrate or fracture the rear of the panel.

Fiberglass mat   **Fiberglass cloth**

Figure 11–15 Fiberglass mat or roving has random layers of strands. Fiberglass cloth is woven and is recommended for repairs of reinforced plastics.

Damage might pass all the way through a panel, but no pieces of the panel have broken away. If the break is clean and all of the reinforcing fibers have stayed in place, then a single-sided repair would be adequate.

For a single-sided repair, you must bevel deep to penetrate the fibers in the panel. The broken fibers must come into contact with the adhesive.

The following is a typical single-sided repair procedure for reinforced plastic:

1. Clean the repair area with soap and water.

2. Clean again using mild wax and grease remover.

3. Remove any paint from the surrounding area by sanding with No. 80 grit sandpaper.

4. Scuff sand the area surrounding the damage.

5. Bevel the damage to provide an adequate area for bonding.

6. Mix two-part filler according to the manufacturer's instructions.

7. Apply the filler and cure as recommended.

Once the filler has been sanded, apply additional coats as required and re-sand. The product manufacturer will provide grit recommendations.

TWO-SIDED REPAIRS OF REINFORCED PLASTIC

A two-sided repair is normally needed on damage that passes all of the way through the panel. This would include damage to the reinforcing fibers.

A **backing strip** or **backing patch** is bonded to the rear of the repair area to restore the reinforced plastic's strength. The patch also forms a foundation for forming the exterior surface to match the original contour of the panel.

To make a two-sided repair in a reinforced plastic panel, proceed as follows:

1. Clean the surface surrounding the damage with a good wax and grease remover. Use a No. 36 grinding disc to remove all paint and primer at least 3 inches (75 mm) beyond the repair area.

2. Grind, file, or use a hacksaw to remove all cracked or splintered material away from the hole on both the inside and outside of the repair area.

3. Remove any dirt, sound deadener, and the like from the inner surface of the repair area. Clean with reducer, lacquer thinner, or a similar solvent.

4. Scuff around the hole with No. 80 grit paper to provide a good bonding surface.

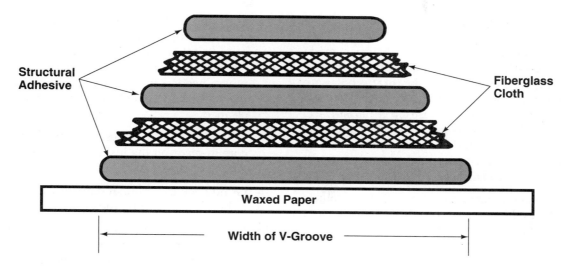

Figure 11–16 To make a filler patch, use layers of structural adhesive with fiberglass cloth as shown. Waxed paper will allow you to pick up and apply the patch to the part. The paper can then be pulled off. (I-CAR)

5. Bevel the inside and outside edge of the repair area about 30 degrees to permit better patch adhesion.

6. Clean the repair area thoroughly.

7. Cut several pieces of fiberglass cloth large enough to cover the hole and the scuffed area. The exact number of pieces will depend on the thickness of the original panel.

8. Prepare a mixture of resin and hardener. Follow the label recommendations.

9. Using a small paintbrush, saturate at least two layers of the fiberglass cloth with the activated resin mix, Figure 11–16.

10. Apply the material to the inside or back surface of the repair area. Make sure the cloth fully contacts the scuffed area surrounding the hole.

11. Saturate three more layers of cloth with the mix. Apply it to the outside surface. These layers must also contact the inner layers and the scuffed outside repair area.

12. With all of the layers of cloth in place, form a saucer-like depression in them. This is needed to increase the depth of the repair material. Use a squeegee to work out any air bubbles.

13. Clean all tools with a lacquer thinner immediately after use.

14. Let the saturated cloth patch become tacky. An infrared heat lamp can be used to speed up the process. If one is used, keep it 12 to 15 inches (300 to 375 mm) away from the surface. Do NOT overheat the repair area because too much heat will cause distortion.

15. With No. 50 grit paper, disc sand the patch slightly below the contour of the panel.

16. Prepare more resin and hardener mix. Use a plastic spreader to fill the depression in the repair area. You need a sufficient layer of material for grinding down smooth and flush.

17. Allow the patch to harden. Again, a heat lamp can be used to speed the curing process.

18. When the patch is fully hardened, sand the excess material down to the basic contour. Use No. 80 grit paper and a sanding block. Finish sand with No. 120 or finer grit paper.

There are several ways to hold the patch in place from the front side of the panel. One method is to use a pull rod. Drill a hole in the middle of the patch. Insert the end of the rod. Position the patch, and pull it snug with the pull rod. Apply heat using a heat gun on the front, or have someone else apply the heat to the back of the patch.

This same two-sided repair can be made by attaching sheet metal to the back side of the panel with sheet metal screws. Sand the sheet metal and both sides of the part to provide good adhesion. Before fastening the sheet metal, apply resin and hardener mix to both sides of the rim of the hole. Follow the procedures described earlier for the remainder of the repair.

When the inner side of the hole is NOT accessible, apply a fiberglass patch to the outer side only. After the usual cleaning and sanding operations, apply several additional layers of fiberglass cloth to the outer side of the hole. Before it dries, make a saucer-like depression in the cloth to provide greater depth for the repair material.

USING MOLDED CORES

A **molded core** is a curved body repair part made by applying plastic repair material over a part and then removing the cured material. Naturally, holes are much more difficult to repair in a curved portion of a reinforced plastic panel than those on a flat surface. Basically, the only solution (short of purchasing a new panel section) is to use the molded core method of replacement. This is often the quickest and cheapest way to repair a curved surface.

1. Locate an undamaged panel on another vehicle that matches the damaged one. It will be used as a model or pattern. A new or used vehicle can be used since it will not be harmed.

2. On the model vehicle, mask off an area slightly larger than the damaged area. Apply additional masking paper and tape to the surrounding area, especially on the low side of the panel. This will prevent any resin from getting on the finish.

3. Coat the area with paste floor wax. Leave a wet coat of wax all over the surface. A piece of waxed paper can be substituted for the coat of wax. Make sure the waxed paper is taped firmly in place.

4. Cut several pieces of thin fiberglass mat in sizes large than the area to be repaired.

5. Mix the fiberglass resin and hardener following the label instructions.

6. Starting from one corner of the mold area, place pieces of fiberglass mat on the waxed area so each edge overlaps the next one; use just one layer of matting.

7. Apply the resin/hardener to the matting with a paintbrush. Force the mixture into the curved surfaces and around corners with the tips of the bristles.

8. Use the smaller pieces of matting along the edges and on difficult curves. Additional resin/hardener can be applied if needed, brushing in one direction only to force the material into the indentations. In all cases, use only one layer of matting.

9. After matting has been applied to the entire waxed area, allow the molded core to cure a minimum of one hour.

10. Once the molded core has hardened, gently work the piece loose from the model vehicle. The core should be an exact reproduction of this section of the panel.

11. Remove the wax or waxed paper from the model vehicle. Then polish this section of the panel.

12. Since the molded core is generally a little larger than the original panel, place it under the damaged panel and align. If necessary, trim down the edges of the core and the damaged panel slightly where needed for better alignment. The edges of the damaged panel and core must also be cleaned.

13. Using fiberglass adhesive, cement the molded core in place on the inside of the panel. Allow the core and panel to cure.

14. Grind back the original damaged edges to a taper or bevel, maintaining the desired contour.

15. Lay a fiberglass mat, soaked in resin/hardener, on the taper or bevel and over the entire core. Once the mat has hardened, level it with a coat of fiberglass filler. Then prepare it for painting.

In some instances it might not be possible to place the core on the inside of the damaged panel. In this case, the damaged portion must be cut out to the exact size of the core. After the panel has been trimmed and its edges beveled, tabs must be installed to support the core from the inside. These tabs can be made from pieces of the panel or from fiberglass strips saturated in resin/hardener.

After cleaning and sanding the inside sections, attach the tabs to the inside edge of the panel and bond with fiberglass adhesive. Clamping pliers can be used to hold the tabs in place. Taper the edge of the opening and place the core on the tabs. Fasten the core to the tabs with fiberglass adhesive. Grind down any high spots so that layers of fiberglass mat can be added.

Place the saturated mats over the core, extending about 1½ to 2 inches (38 to 50 mm) beyond the damaged area in all directions. Work each layer with a spatula or squeegee to remove all air pockets. Additional resin/hardener can be added with a paintbrush to secure the layers. Allow sufficient curing time. Then sand the surface level. For a smooth surface, use fiberglass filler to finish the job.

REPAIRING RRIM

Reinforced reaction injection molded (RRIM) polyurethane is a two-part polyurethane composite plastic. Part A is the isocyanate. Part B contains the reinforced fibers, resins, and a catalyst. The two parts are first mixed in a special mixing chamber, then injected into a mold. RRIM parts are becoming more common in fenders and bumper covers.

Since RRIM is a thermosetting plastic, heat (100° to 140°F or 37° to 59°C) is applied to the mold to cure the material. The molded product is made to be stiff yet flexible. It can absorb minor impacts without damage. This makes RRIM an ideal material for exposed areas.

Gouges and punctures can be repaired using a structural adhesive. If the damage is a puncture that extends through the panel, a backing patch is required. To make a typical backing patch repair, proceed as follows:

1. Clean the damaged area thoroughly using the plastic cleaner recommended by the manufacturer and a clean cloth. Wipe dry.

2. Remove any paint film in and around the damage with an orbital sander and No. 180 grit disc.

3. Using a No. 50 grit disc, enlarge the damaged area, tapering out the damage for about 1 inch (25 mm). Wipe or blow away any loose particles.

4. Clean the back side of the damaged area with the plastic cleaner.

5. Use a No. 50 grit disc to scuff sand the area. Extend the area to about 1½ inches (38 mm) beyond the damage. Align the front of the panel with body tape, if necessary.

6. Cut a piece of fiberglass cloth to cover the damaged area and the part of the panel that has been scuff sanded.

7. Mix the adhesive according to the manufacturer's recommendations. Apply a layer of adhesive to the back side of the panel about ⅛ inch (3 mm) thick.

8. Place the fiberglass patch into position on the adhesive. Cover it with a sheet of waxed paper. Use a roller to force the adhesive into the fibers of the patch.

9. Remove the waxed paper and add another layer of adhesive. Work out the adhesive to just beyond the edges of the patch. Allow the adhesive to cure following the manufacturer's recommendation.

10. Now move to the front side of the panel. Apply a layer of adhesive, completely covering the damaged area. Build it up to slightly higher than the surrounding contour. Allow it to cure.

11. Apply heat to help speed the cure of the patch.

12. Contour the adhesive to the adjoining surface by block sanding using No. 220 grit paper.

13. Finish by feathering with an orbital sander and a No. 320 disc.

14. Follow the recommendations given in Chapter 16 for priming and painting.

SPRAYING VINYL PAINTS

Vinyl repair paints are usually ready for spraying as packaged. Since application properties cannot be controlled with thinners or other additives, air pressure is an important factor.

When applying vinyl paints by siphon feed guns, the normal air pressure range is between 40 and 50 psi (2.8 and 3.5 kg/cm^2) at the gun. With pressure systems, the air pressure is reduced to 30 to 40 psi (2.1 to 2.8 kg/cm^2). Overspray can be controlled by decreasing the air pressure.

There is no retardant for vinyl paints. If blushing occurs, allow the initial coat to set up and reapply the color in a much lighter coat.

Vinyl and soft ABS plastics should be thoroughly cleansed with vinyl cleaner and allowed to dry. Then treat the surface with vinyl prep. **Vinyl prep** is a chemical solution that opens the pores of the vinyl material. Wipe off the vinyl prep right after it is applied. The surfaces are then ready for color and/or clearcoating.

Summary

- Plastics and composites are terms that refer to a wide range of materials synthetically compounded from crude oil, coal, natural gas, and other substances. Bumpers, fender extensions, fascias, fender aprons, grille openings, stone shields, instrument panels, trim panels, fuel lines, and engine parts can be constructed of these materials.

- Two general types of plastics are used in automotive construction: thermoplastics and thermosetting plastics. Thermoplastics can be repeatedly softened and reshaped by heating with no change in their chemical makeup; thermosetting plastics undergo a chemical change and are hardened into a permanent shape that CANNOT be altered by reapplying heat or catalysts.

- One way to identify a plastic is by international symbols, or ISO codes, which are molded into plastic parts. If there is no ISO code, refer to the body repair manual for information.

- Plastic welding uses heat and sometimes a plastic filler rod to join or repair plastic parts. The types of plastic welding include hot-air and airless.

Technical Terms

composites	thermosetting plastics
plastics	composite plastics
thermoplastics	international symbols

ISO codes
body repair manual
floating test
plastic welding
hot-air plastic welding
airless plastic welding
rod color code
tack weld
plastic speed welding
speed welding rate
melt-flow plastic
   welding
plastic stitch-tamp
   welding
single-sided plastic weld
two-sided plastic weld
vinyl
plastic memory

plastic adhesive repair
   system
cyanoacrylates (CAs)
two-part adhesive
   system
adhesion promoter
work life
open time
cosmetic filler
structural filler
single-sided damage
backing strip
backing patch
molded core
reinforced reaction
   injection molded
   (RRIM) polyurethane
vinyl prep

Review Questions

1. Define the term "plastics."

2. In speed welding, the hot-air torch is held at a
   _____-degree angle to the base material
   to start the weld.
   a. 45
   b. 90
   c. 30
   d. 120

3. Explain the difference between a thermoplastic
   and a thermosetting plastic.

4. When Technician A grinds the base material, it
   melts and smears, so an adhesion promoter is
   used to make the repair. Under the same circumstances, Technician B says that an adhesion promoter is NOT needed. Who is correct?
   a. Technician A
   b. Technician B
   c. Both A and B
   d. Neither A nor B

5. A _____ _____ is a curved
   body repair part made by applying plastic repair
   material over a part and then removing the cured
   material.

6. The _____ and _____
   _____ are used to help the factory hold
   panels in place while the adhesive cures.

7. What is a backing strip or patch?

8. Which of the following is the correct repair
   method for RRIM?

a. Adhesive
b. Welding
c. Both a and b
d. Neither a nor b

ASE–Style Review Questions

1. *Technician A* says to identify a plastic by international symbols, or ISO codes, molded into plastic
   parts. *Technician B* says that the body repair manual will give information about plastic types used
   on the vehicle. Who is correct?
   a. Technician A
   b. Technician B
   c. Both Technicians A and B
   d. Neither Technician

2. *Technician A* says that one of the ways to identify
   plastic is by international symbols, which are
   molded into plastic parts. *Technician B* says that all
   plastic parts are repaired in the same manner and
   identification is not necessary. Who is correct?
   a. Technician A
   b. Technician B
   c. Both Technicians A and B
   d. Neither Technician

3. When plastic welding, *Technician A* forces the rod
   into the joint with slight pressure. *Technician B*
   uses test welds to help select the correct plastic
   welding rod. Who is correct?
   a. Technician A
   b. Technician B
   c. Both Technicians A and B
   d. Neither Technician

4. Which of the following is the preferred method of
   repairing plastic damage?
   a. Hot-air welding
   b. Airless welding
   c. Reshaping
   d. Adhesive repair

5. *Technician A* says that not all plastics can be
   welded, while adhesives can be used in all but a
   few instances. *Technician B* says plastic welding
   can be used on all plastic with better success.
   Who is correct?
   a. Technician A
   b. Technician B
   c. Both Technicians A and B
   d. Neither Technician

6. *Technician A* says that if the plastic material GIVES
   OFF DUST when ground, it can be repaired with
   a standard structural adhesive system. *Technician
   B* states that if the material MELTS and smears or

has a greasy or waxy look when ground, then you must use an adhesion promoter. Who is correct?
a. Technician A
b. Technician B
c. Both Technicians A and B
d. Neither Technician

7. Which of the following types of repair can be used when reinforced plastic damage does NOT break away large pieces of the panel?
a. Single-sided repair
b. Double-sided repair
c. Spot putty repair
d. Epoxy repair

8. When performing a two-sided repair of reinforced plastic, which of the following should be used on the rear of the panel?
a. Heat lamp
b. Duct tape
c. Baking patch
d. Nothing

9. *Technician A* says when using the single-side weld method, the weld should penetrate 75% of the base material. *Technician B* says that when using the two-sided weld method, a second weld should fuse with weld for the other side. Who is correct?
a. Technician A
b. Technician B
c. Both Technicians A and B
d. Neither Technician

10. When repairing reinforced plastic parts, *Technician A* says that full panel replacement is always needed. *Technician B* says that complete inspection of the panel is necessary to find all the damage. Who is correct?
a. Technician A
b. Technician B
c. Both Technicians A and B
d. Neither Technician

Activities

1. Inspect plastic body parts. Try to find an identification code on the back of each part. Make a report on the type of plastic used for different parts.

2. Make a one-side repair on a plastic part. Summarize the repair in a report. Describe the type of plastic, damage, and steps for repairing the part.

3. Visit a body shop. Talk to technicians about new methods of repairing plastics. Report your findings to the class.

12

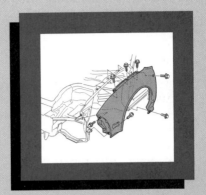

Replacing Hoods, Bumpers, Fenders, Grilles, and Lids

Objectives

After studying this chapter, you should be able to:

■ Remove and install fenders.

■ List the various methods for adjusting mechanically fastened panels.

■ Perform hood-to-hinge, hood height, and hood latch adjustments.

■ Remove, install, and adjust deck lids.

■ Remove, install, and adjust bumpers.

■ Replace grilles and other bolt-on body parts.

Introduction

A collision damaged vehicle can require a variety of repair operations. Repair steps will depend on the nature and location of the damage. Panels with minor damage can often be straightened and filled with plastic filler. Minor bulges, dents, and creases can be fixed using the techniques discussed in earlier chapters. However, quite often the damage is too great and parts must be replaced.

This chapter covers replacement procedures for hoods, fenders, bumpers, deck lids, and similar bolt-on parts. As Figure 12–1 shows, many major parts bolt on to the vehicle.

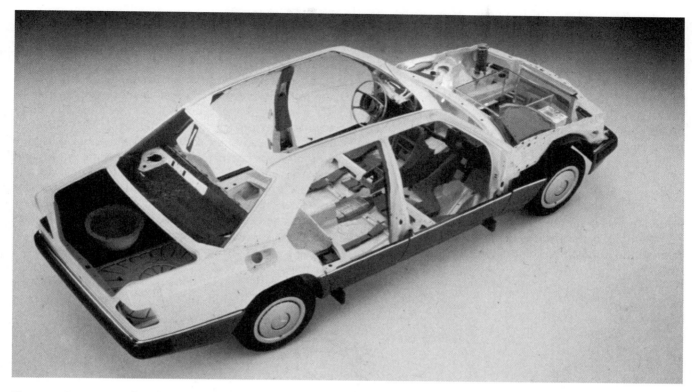

Figure 12-1 Even with unibody designs, vehicles still have many bolt-on parts and panels. It is important that you know how to properly remove and replace them. *(Courtesy of Mercedes-Benz of N.A., Inc.)*

You must know how to properly remove, install, and adjust these parts. As explained in this chapter, some parts can be removed, installed, and aligned without too much difficulty. However, when replacing several adjoining parts (front end parts, for example), you must use specific procedures to get all parts to align properly with the rest of the vehicle.

HOW ARE PARTS FASTENED?

The methods of fastening parts to cars and trucks has changed in the past few years. Many parts that were held with bolts and screws in the past now snap into place. Plastic retainers now hold these parts onto the vehicle. This is done to save time during vehicle manufacturing.

Fastener variations can make repair more challenging, Figure 12–2. Parts can be held by screws, bolts, nuts, metal or plastic clips, adhesives, and other methods. To efficiently replace parts, you must carefully study part construction. Inspect parts closely to find out how they are held on the vehicle. This will let you make logical decisions about which parts to remove first, second, and so on, and the methods needed.

Keep this in mind! On-the-job experience is the only way to become competent and fast at body part *removal and replacement (R and R)*. Sometimes

you must remove one part at a time. In other instances, it is better to remove several parts as an assembly. This chapter will give you the background information to make this learning process easier.

REFER TO ESTIMATE

When starting work, refer to the estimate to get guidance on where to begin. The estimator will have determined which parts need repair and which should be replaced. Use this information and shop manuals to remove and replace parts efficiently.

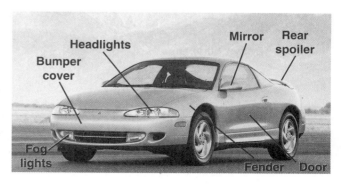

Figure 12–2 Note some of the parts that bolt onto this unibody vehicle. *(Courtesy of Mitsubishi Motor Sales of America, Inc.)*

The estimate is an important reference tool for doing repairs. It must be followed. The estimator has determined which parts must be repaired.

The estimate is also used to order new parts. You may want to make sure all ordered parts have arrived. Compare new parts on hand with the parts list. If anything is missing, have the parts person order. This will save time and prevent your stall from being tied up waiting for parts.

WHERE TO START

Generally, start removing large, external parts first. For example, if the front end has damage, you must remove the hood first. This will give you more room to access rear fender bolts. This will also allow more light into the front for finding and removing hidden bolts in the frontal area. Use this kind of logic to remove parts efficiently.

If in doubt about how to remove a part, refer to the vehicle's service manual. Factory service manuals normally have a body repair section. The **_body repair section_** of the manual explains and illustrates how parts are serviced. The manual will give step-by-step instructions for the specific make and model vehicle. It will give bolt locations, torque values, removal sequences, and other important information.

As one example of a repair variation, look at Figure 12–3. The rear quarter panels on this vehicle bolt in place. Most vehicles have welded quarter panels.

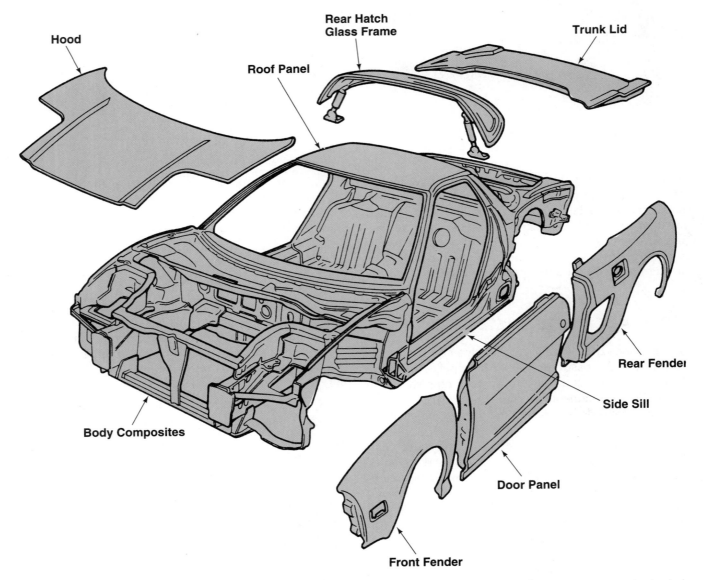

Figure 12–3 Hood, deck lid, fenders, and doors commonly bolt onto a unibody. A unique design, the rear quarter panels also bolt on in this rear-engine, high performance, aluminum bodied vehicle. *(© American Honda Motor Co., Inc.)*

This one also has aluminum body panels that must be repaired differently than steel.

■ **DANGER!** *When removing parts, especially damaged ones, be careful not to cut yourself on sheet metal edges. Torn sheet metal can be very sharp and can cause serious injury!*

HOOD REMOVAL AND REPLACEMENT

The **hood** provides an external cover over the front of the vehicle. It is one of the largest, heaviest panels on a vehicle. In a front-engine vehicle, it provides access to the engine compartment. With a rear-engine car, it serves as a trunk lid.

Before removing the hood, analyze part conditions. Open and close the hood. Check for binding and bent hinges. If applicable, inspect hood alignment with the fenders and cowl. This will help you determine what must be done during repairs.

To remove a hood, first disconnect any wires and hoses. Wires often connect to an underhood light. Hoses might run to the hood for the windshield washer system. Refer to Figure 12–4.

Next, remove the hood hinge bolts. If the hood is not badly damaged and will be reused, mark hood hinge alignment. To mark the hood, make alignment marks around the sides of the hood hinge where it

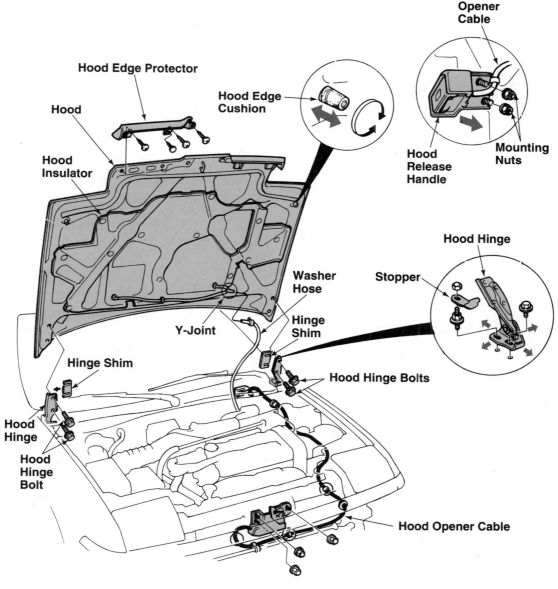

Figure 12–4 Study the parts of a typical hood assembly and cable hood release mechanism. (© American Honda Motor Co., Inc.)

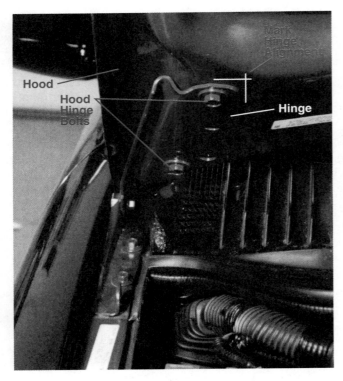

Figure 12–5 If the hood is not heavily damaged and shifted out of alignment, mark the hood next to the hinges. This will save time when initially installing the hood after other repairs.

contacts the hood. You may also want to mark the hinge where it mounts on the body. You can then use these marks to rough adjust the hinges and hood during reinstallation. See Figure 12–5.

To prevent part damage, have someone help you hold the hood. Place your shoulder under the hood while holding the bottom edge of the hood with one hand. This will keep the hood from sliding down and hitting the cowl or fenders. Use your shoulder to support the weight of the hood. With your free hand, remove the hood bolts. Your helper should do the same. Do not let the weight of the hood rest on the bolts as you loosen them.

WARNING! Do not make the mistake of removing the hood bolts without you and someone else holding the hood securely. It is easy for the hood to slip and scratch the fenders or cowl. If these parts were not originally damaged, you will have to fix them on your own time.

Note the location of any **hood hinge spacers** that help adjust the hood. If there is no major damage, you may need to reinstall the spacers in the same locations. Place the hood out of the way, where it will not fall over. Spacers are shown in Figure 12–4.

Hood Hinge Removal and Replacement

Hood hinges allow the hood to open and close while staying in alignment. They must hold considerable weight while keeping the hood open. They are often damaged in a frontal impact.

If they are badly bent, you will have to replace the hood hinges. If the hood is equipped with large coil springs, you may also have to install the old springs on the new hinges. A **hood hinge spring tool** should be used to stretch the spring off and on. It is a hooked tool that will easily pry the end of the spring off of and onto its mount.

DANGER!   Hood hinges are very strong. Wear eye protection. Also keep your fingers away from the spring as it is stretched. Your fingers could be severely pinched and cut by this spring.

If necessary, unbolt the hinges from the inner fender panels. Again, mark their alignment if the panels are not damaged. Install the new hinges. Snug the bolts down; do not tighten them. You will have to adjust the hood and hinges later.

Hood Adjustment

Install the hood in reverse order of its removal. Again, have someone help you hold the hood while installing its bolts. Snug down but do not tighten the hood bolts. You will need to adjust the hood before tightening the bolts fully. A misaligned hood is shown in Figure 12–6.

WARNING! After installing the hood, close the hood slowly. If it is not centered, it could hit and dent the fenders. It may be a good idea to place tape over the fender edges to protect them.

Hood adjustments are made at the hinges, at the adjustable stops, and at the hood latch. You can adjust the hood up or down and forward or rearward. This

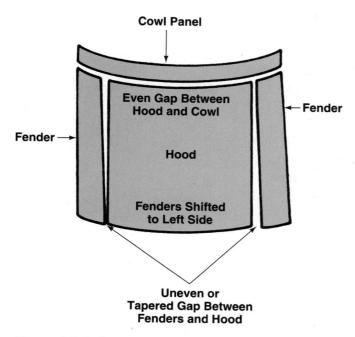

Figure 12–6 The gap around the hood should be the same. The problems may be due to fenders being out of alignment.

allows you to align the hood with the fenders and cowl vertically and horizontally. The holes in the hinges are slotted. This allows the hinges to be raised or lowered at the cowl or fender and the hood to be moved forward or rearward at the hinges, Figure 12–7.

Hood hinge adjustments control the general position of the hood in the fenders and the rear hood height. By loosening the hood-to-hinge bolts, you can move the front end of the hood right or left. You can also slide the hood to the front or back. Tighten them down when the hood is centered in the opening. You want an equal gap around the hood's perimeter. There should be enough of a gap at the back edge to clear the cowl panel. Refer to Figure 12–8A.

By partially loosening the hinge-to-body bolts, you can raise or lower the rear height of the hood. Do not loosen the bolts too much or the weight of the hood will push the hinge all the way down. Tap on the hinges with a mallet to shift them as needed. The back of the hood should be level with the fenders and cowl when fully closed.

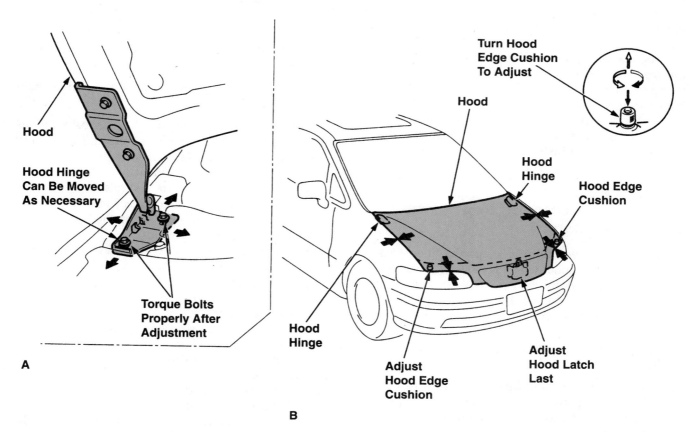

Figure 12–7 (A) When adjusting hood alignment, loosen the hinge-to-body bolts slightly. This will let you shift the hinge mounting in the elongated holes right-left and fore-aft. To avoid paint damage, slowly lower the hood while checking its alignment. Tighten the hinge bolts when the hood is aligned with the cowl and fenders. (B) Hood edge cushions must be adjusted so the height of the hood is even with the fenders. You may want to remove the latch during hood adjustment. *(Reprinted with permission by American Isuzu Motors Inc.)*

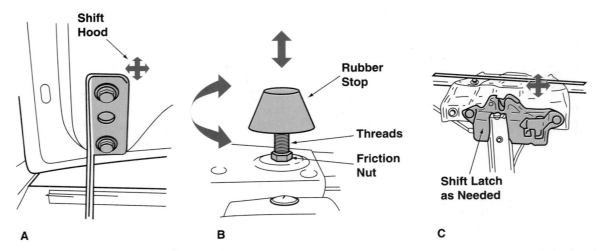

Figure 12–8 Study the basic methods of adjusting a hood. *(Courtesy of DaimlerChrysler Corporation)* (A) With the hood-to-hinge bolts loose, shift the hood as needed to position it on the body. Align the hood straight on the body and obtain the correct clearance at the back of the hood. (B) Turn the bumpers up or down to make the front of the hood level with the fenders and other parts. (C) Close the hood slowly. It should not shift sideways when the striker engages the latch. Center the latch in the striker and adjust the latch up or down to pull the hood lightly down against the bumpers.

Generally, adjust the hood hinges so that the hood is centered in the fenders. Adjust it to have the proper gap around its perimeter. A **gap** or **clearance** is the distance measured between two adjacent parts, hood-to-fender gap, for example. Then, adjust the hinges to raise or lower the back of the hood.

The **hood stop adjustment** controls the height of the front of the hood. It is usually made of rubber bumpers mounted on threaded studs. By loosening the locknut, you can rotate to raise or lower the stops and hood. Adjust the hood stops so that the hood is even with the front of the fenders and fascia. Tighten the locknuts after adjustment. See Figure 12–8B.

Hood Latch Removal, Replacement, and Adjustment

The **hood latch mechanism** keeps the hood closed and releases the hood when activated. Some hood latches are opened by moving a lever behind the grille. Most have a cable release that runs into the passenger compartment. All have slots for adjustment, Figure 12–8C.

A **cable hood release,** as shown in Figure 12–9, consists of:

1. The **hood release handle** can be pulled to slide a cable running out to the hood latch. It is normally mounted on the lower left side of the passenger compartment.

2. The **hood release cable** is a steel cable that slides inside a plastic housing. One end fastens to the release handle and the other to the hood latch.

3. The **hood latch** has metal arms that grasp and hold the hood striker. The spring-loaded arms lock over the hood striker when the hood is closed. When the cable release is pulled, the arms release the striker so the hood can open.

4. The **hood striker** bolts to the hood and engages the hood latch when closed.

Before removing a hood latch, scribe mark its location if necessary. Remove its bolts and disconnect any cable. Slots in the latch mount provide up and down and side to side adjustment.

Hood latch adjustment controls how well the hood striker engages the latch mechanism. Basically, with the hood centered and set at the right height, adjust the latch for proper closing.

Slowly lower the hood while watching to see if the striker centers itself in the latch. The hood should not pivot right or left when it engages the latch. If the hood moves to one side when closed, shift the latch right or left as needed.

The latch should also produce a slight downward compression of the rubber stops. This keeps the hood from bouncing up and down. If you must slam the hood down to engage the latch, raise the latch. If the hood will move up and down when latched, lower the latch. See Figure 12–10.

After adjustment, tighten the latch bolts. Make sure the latch releases the hood properly. Always check the vehicle's service manual for specific hood adjustment procedures.

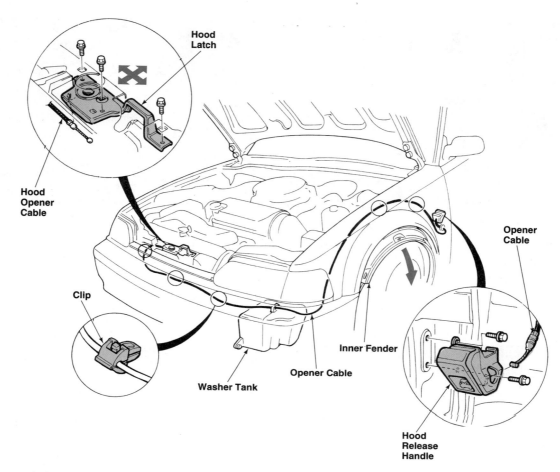

Figure 12–9 Study the parts and adjustment of the hood release mechanism. (© American Honda Motor Co., Inc.)

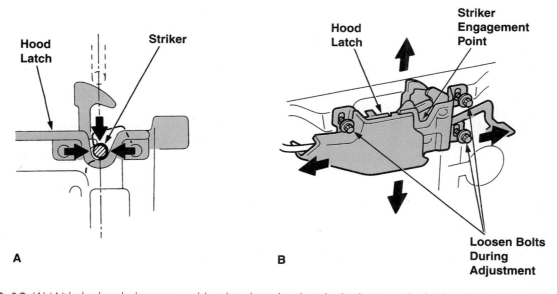

Figure 12–10 (A) With the hood alignment and height adjusted, adjust the latch. Move the latch until the striker is centered in the latch as shown. (B) The latch should be positioned so that it places a slight downward pull on the hood when it is fully engaged with the striker. (Reprinted with permission by American Isuzu Motors Inc.)

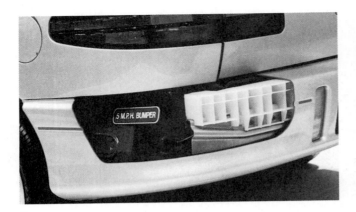

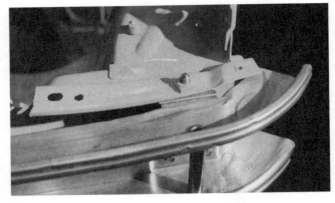

A   **B**

Figure 12–11 Note the construction of modern bumper assemblies. (A) This cutaway of a rear bumper shows plastic honeycomb that backs up the outer skin. *(Courtesy of General Motors Corporation, Service Operations)* (B) A metal, often aluminum, inner bumper holds and supports the outer skin of the bumper.

BUMPER REMOVAL, REPLACEMENT, AND ADJUSTMENT

New vehicle **bumpers** are designed to withstand minor impact without damage. They protect other parts from light impact with sheer mass and bulk.

Bumper designs are light, yet strong. Many use an outer covering of flexible plastic with a heavy steel or aluminum inner bumper. Some have a plastic honeycomb or foam structure behind their flexible cover. Many have a large one-piece cover over the lower front half of the vehicle nose. This is shown in Figure 12–11.

To remove a bumper, first disconnect the wiring going to any lights. Procedures vary so you may need to refer to a service manual. Some bolts can be hidden behind parking lights, inner fender panels, etc., Figure 12–12.

Bumpers can be heavy and clumsy. Before removing the last mounting bolts, support the bumper. Get a helper or use a floor jack. If the bumpers are to be repaired or reused, place a block of wood or piece of thick foam rubber on the jack saddle. Raise the jack to support the weight of the bumper. Remove the bolts. Then, you can work the bumper and jack away from the vehicle. These general procedures also apply to the rear bumper, Figure 12–13 on page 222.

Bumper shock absorbers are used to absorb some of the impact of a collision and reduce damage. They will compress inward to help prevent bumper and other part damage during a low speed impact. When bumper shock absorbers are used, bolts or nuts on the shocks often hold the bumper in place.

With major front end damage, sometimes it is best to remove the large assemblies, like the front clip or bumper-spoiler assembly. This will let you gain access to hard-to-reach parts on the assembly more easily. This may allow you to service damaged bumper, lights, brackets, and other front end parts in less time, Figure 12–14.

FENDER REMOVAL, REPLACEMENT, AND ADJUSTMENT

To remove a fender, find and remove all of the fasteners securing it to the vehicle. Also remove any wires going to fender-mounted lights. Fenders are usually bolted to the radiator core support, inner fender panels, and cowl. Bolts are often hidden behind the doors, inner fender panels, and under the vehicle. Refer to Figure 12–15.

With all of the bolts removed, carefully lift the fender off, Figure 12–16. Transfer any needed parts (trim, body clips, etc.) from the old fender over to the new fender.

Install the replacement fender in reverse order of the removal. If the doors or cowl are undamaged, place masking or duct tape over their edges. This will protect their finish from scratches when installing the fender.

When installing the fenders, hand start the fender bolts. Do not tighten them. Leave the bolts loose enough that you can adjust the fender. Shift the fender on its bolts so that it aligns with other body parts properly. Shift the fender forward or backward until

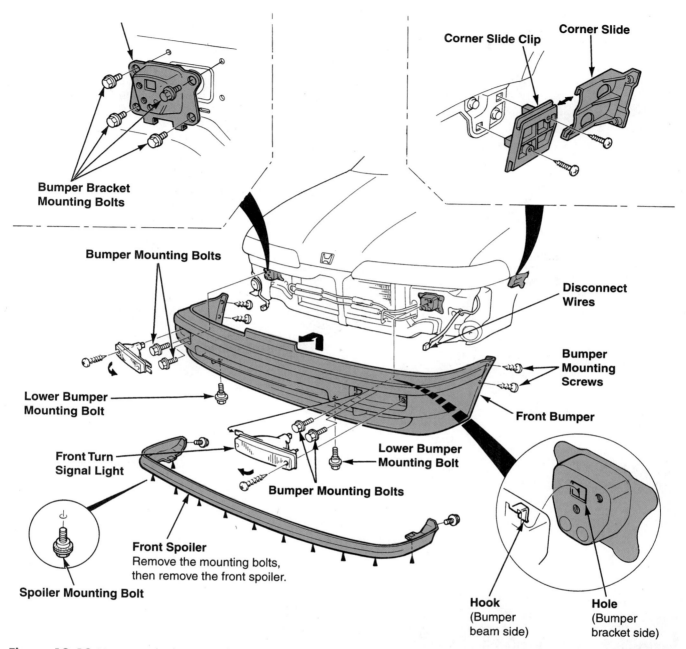

Figure 12-12 Various methods are used to secure the front bumper to a vehicle. Removal of lights and moldings will often give access to mounting hardware for removal of the bumper. Also note screws and bolts that come in from the bottom and rear of the bumper. (© American Honda Motor Co., Inc.)

the fender, door, and cowl have the correct spacing or gap. Also adjust the fender in and out so that it is flush with the door and parallel with the hood. Tighten the fender bolts after you have the fender in alignment.

Fender shimming is an adjustment method that uses spacers under the bolts that attach the fender to the cowl or inner fender panel. By changing shim

thicknesses, you can move the position of the fender for proper alignment.

Often the fender and hood adjustments must be made simultaneously to achieve a satisfactory result. The gap between fender and hood should be set to factory specifications. The front of the fender and the hood should be aligned as well. The result will be even spacing all around the fender.

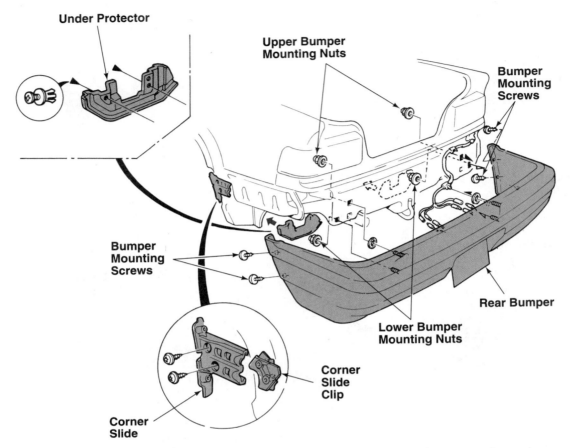

Figure 12–13 Replacement of the rear bumper is similar to that of the front bumper. Always remember to disconnect all wires so they are not damaged when you lift the bumper off. (© American Honda Motor Co., Inc.)

Figure 12–14 Sometimes it is efficient to remove the bumper and related parts as an assembly. This makes it easier to transfer parts and access parts under the assembly.

GRILLE REMOVAL AND REPLACEMENT

Grilles are often held in place with small screws, rivits, and clips. You might have to remove a cover to access grille fasteners, Figure 12–17. An air ratchet is handy for reaching down and unscrewing grille bolts. When installing a grille, make sure all clips are undamaged and installed. Since most grilles are plastic, be careful not to overtighten any bolts or screws. You could crack the grille. Select the correct rivet size.

Some grilles can be adjusted. They have slotted or oversized holes in them, Figure 12–18. By leaving the bolts loose, you can shift and align the grille with other parts. Once aligned, tighten the grille fasteners slowly.

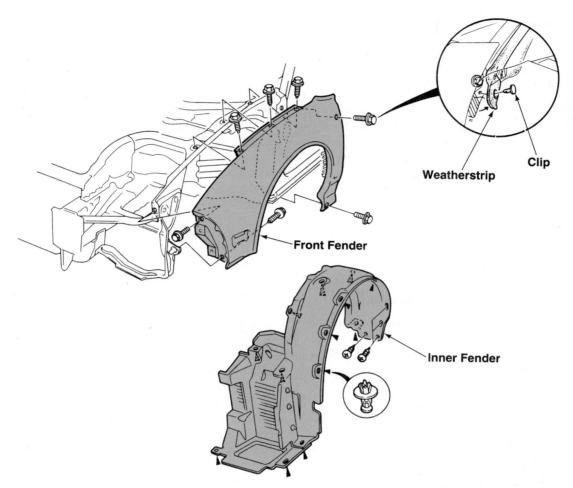

Clip

Weatherstrip

Front Fender

Inner Fender

Figure 12–15 Fender bolts are located around the outside of the fender. Some are hidden behind the inner fender panel or at the rear behind the front of the door around the cowl.(© American Honda Motor Co., Inc.)

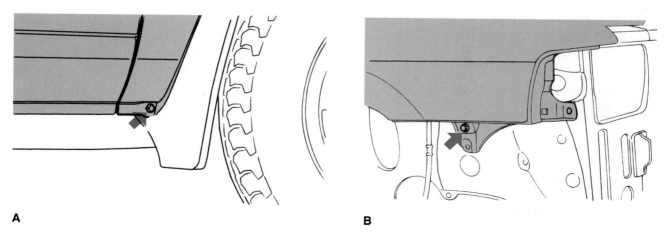

A

B

Figure 12–16 Note the basic methods for R & R of a fender. (Courtesy of DaimlerChrysler Corporation) (A) Bolts are often located at the rear bottom of the fender. (B) The bumper had to be removed first before these fender bolts could be removed.

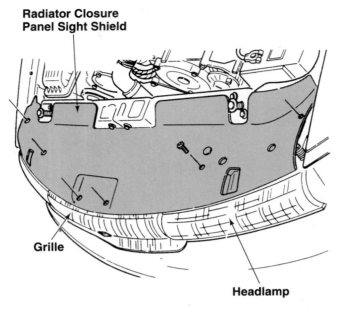

Figure 12–17 When servicing the grille and other front end parts, you often have to remove a cover panel over the front of the vehicle. Small screws normally hold this cover. *(Courtesy of DaimlerChrysler Corporation)*

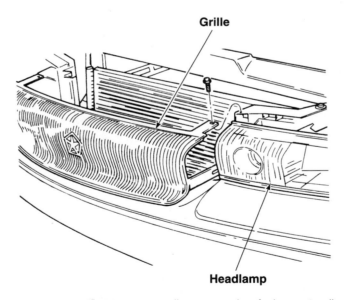

Figure 12–18 Most new grilles are made of plastic. Small clips and screws hold the grille in place. *(Courtesy of Daimler-Chrysler Corporation)*

DECK LID REMOVAL, REPLACEMENT, AND ADJUSTMENT

The deck lid is very similar to the hood in construction. Two hinges connect the deck lid to the rear body panel. The trailing edge is secured by a locking latch.

Deck lid or hatch door removal and replacement is similar to a hood. See Figure 12–19. The deck lid must be evenly spaced between the adjacent panels. Slotted holes in the hinges and/or caged plates in the deck lid allow it to be moved. To adjust the deck lid forward or rearward, slightly loosen the bolts on both hinges. Close and adjust the deck lid as required. Then raise the lid and tighten the bolts.

Weatherstripping is a rubber seal that prevents leakage at the joint between the movable part (lid, hatch, door) and the body. To prevent air and water leaks, the deck lid must contact the weatherstripping evenly when closed. The latch must be adjusted so that it holds the lid or hatch closed against the weatherstripping.

Deck lids and hatches usually do not have exterior or interior door handle mechanisms. They operate with a key (or instrument panel switch on powered units) and lock mechanism.

Lock cylinders contain a tumbler mechanism that engages the key so that you can turn the key and disengage the latch. When you insert your key into a door or lid, it engages the lock cylinder. The lock cylinder then transfers motion to the latch.

The lock cylinder on door and deck lids is usually held on the panel with a retainer. A sealing gasket may also seal out water and dust. To remove a lock cylinder, pry the retainer sideways. This will free the lock cylinder. Sometimes, lock cylinders are held by small screws. If necessary, lubricate a lock cylinder with dry powdered graphite.

Lid struts are usually spring-loaded, gas filled units that hold the lid open and cause it to close more slowly. The struts often engage small ball sockets mounted on the body. The ends of the struts snap-fit over the ball sockets. Bad lid struts will no longer support the weight of the lid. They should be replaced.

Lid torsion rods are spring steel rods used to help lift the weight of the lid. They extend horizontally across the body and engage a stationary bracket. Some torsion rod brackets have adjustment slots. You can change tension on the torsion rods by moving them in these slots. Refer to Figure 12–20.

PANEL ALIGNMENT

After installing all new body panels, you must check overall panel alignment. Make sure that clearance between panels is equal. As shown in Figure 12–21, the gap around all panels must be within specifications. Also check that all panel surfaces are even with each other. Take the time to double-check all panels to assure good alignment. This is a sign of a professional technician.

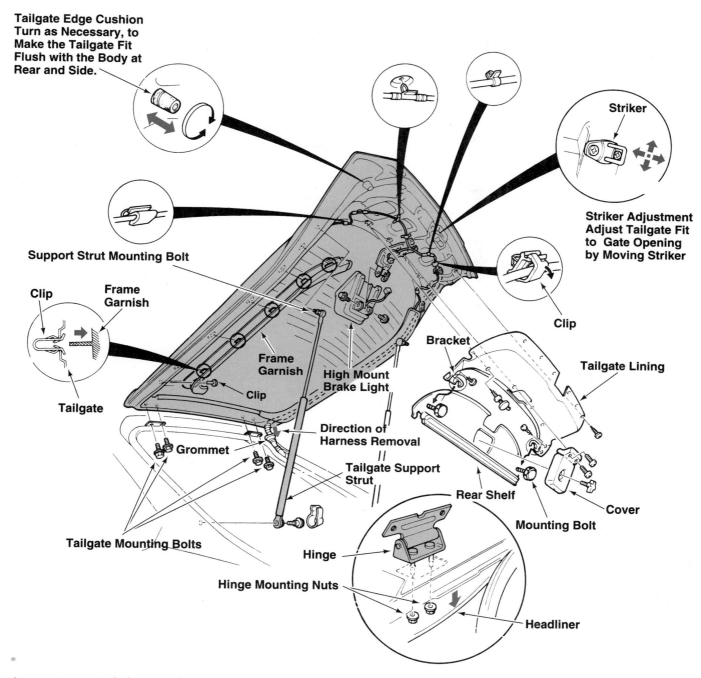

Tailgate Edge Cushion Turn as Necessary, to Make the Tailgate Fit Flush with the Body at Rear and Side.

Striker

Striker Adjustment Adjust Tailgate Fit to Gate Opening by Moving Striker

Clip

Support Strut Mounting Bolt

Clip

Frame Garnish

Bracket

Tailgate Lining

Frame Garnish

High Mount Brake Light

Tailgate

Clip

Direction of Harness Removal

Grommet

Tailgate Support Strut

Rear Shelf

Mounting Bolt

Cover

Tailgate Mounting Bolts

Hinge

Hinge Mounting Nuts

Headliner

Figure 12–19 Study the parts of a rear hatch lid. (© American Honda Motor Co., Inc.)

TRUCK BED REMOVAL AND REPLACEMENT

Truck beds are usually bolted to the frame. To remove the bed, simply remove the bolts that extend up through brackets on the frame. Keep track of bolt lengths and rubber mounting cushion locations. They must be reinstalled in their original locations. See Figure 12–22.

Truck tailgates mount on two hinges. A steel cable often limits how far down the tailgate will open, Figure 12–23. Latches in the sides of the tailgate engage strikers on the body. The handle on the outside of the tailgate moves small linkage rods that run out to the latches.

The tailgate hinges can be adjusted as described earlier for other parts. Generally, leave the hinge bolts loose while shifting the tailgate as needed. Then, tighten the hinge bolts and recheck adjustment.

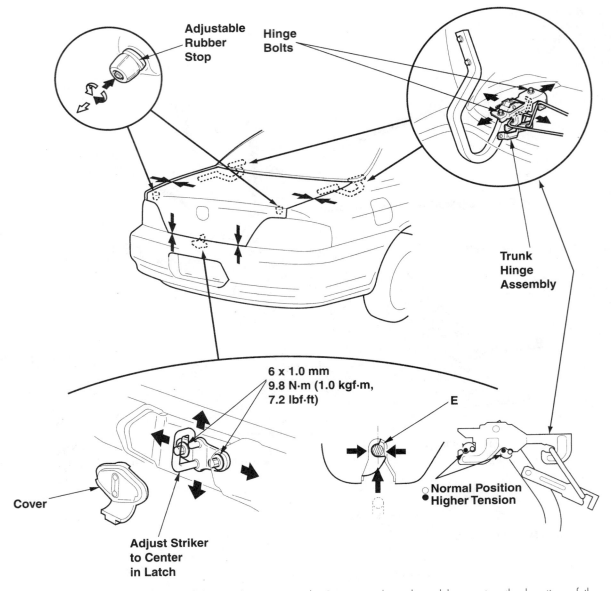

Figure 12-20 Deck lids can also be held open by torsion rods. Some can be adjusted by moving the location of the rod in a mount. (© American Honda Motor Co., Inc.)

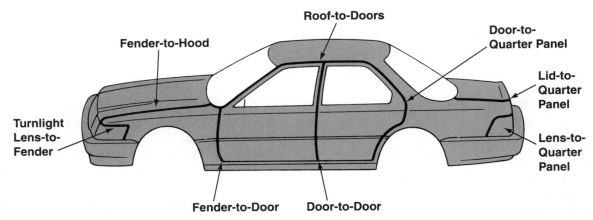

Figure 12-21 After installing panels, double-check their fit. You want even gaps or clearances among all parts. (Reprinted with permission)

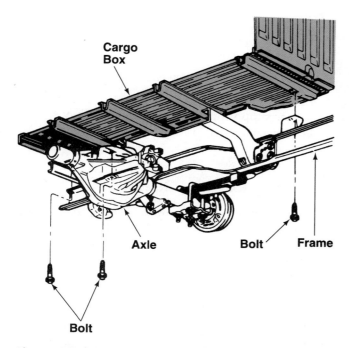

Figure 12–22 Truck beds normally bolt to the full frame. After removing all bolts, wires, and other parts, the bed can be lifted off the frame. *(Courtesy of DaimlerChrysler Corporation)*

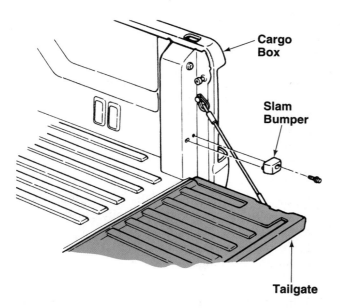

Figure 12–23 A truck tailgate is serviced like other hinged parts. It must align properly when closed and engage strikers properly. *(Courtesy of DaimlerChrysler Corporation)*

WARNING! Truck tailgates are surprisingly heavy. Ask someone to help you when removing or installing one.

INSTALLING SOUND-DEADENING PADS

Sound-deadening pads are often bonded to the inside surface of trunk cavities and doors to reduce noise, vibration, and harshness. Sound-deadening material is made of a plastic or asphalt-based material. It prevents the thin sheet metal panels from acting like large metal sounding drums to quiet the passenger compartment. The original factory material is bonded and, sometimes, heat formed to the surface.

DANGER! *Sound-deadening material is a combustible and must be removed from any area to be welded or flame cut.*

During collision repair, sound-deadening pads must be replaced to match precollision performance. After repair, the area must be properly refinished to provide corrosion protection before the sound-deadening material is applied.

Some pads are available as factory service parts. Be sure to check the factory body repair manual for details. If a factory pad is not available, the sound-deadening qualities must be restored with a suitable replacement product. These should be available from several sources.

One sound-deadening product is available with an adhesive on the back. Cut it to size and shape. Peel off the backing, and apply. The sheets can be heated with a heat gun to make them flexible to conform to the contour of the sheet metal.

Summary

- When starting work, refer to the estimate to get guidance on where to begin. The estimator will have determined which parts need repair and which should be replaced. Use this information to begin removal and replacement (R and R).

- Hood hinges allow the hood to open and close while staying in alignment. They must hold considerable weight while keeping the hood open. They are often damaged in a frontal impact.

- A cable hood release consists of the hood release handle, hood release cable, hood latch, and the hood striker.

- Bumpers today are designed to be light, yet strong. Many use an outer covering of flexible plastic with a heavy steel or aluminum inner bumper. Some have a plastic honeycomb structure behind their flexible cover. The trend is also to a large one-piece cover over the lower front half of the vehicle nose.

- Fender shimming is an adjustment method that uses spacers under the bolts that attach the fender to the cowl or inner fender panel. By changing shim thicknesses, you can move the position of the fender for proper alignment.

- After installing all new body panels, you must check overall panel alignment. Make sure that clearance between panels is equal and that the gap around all parts is within specifications.

Technical Terms

removal and repair (R and R)

body repair section

hood hinge spacers

hood hinges

hood hinge spring tool

hood adjustments

gap

clearance

hood stop adjustment

hood latch mechanism

cable hood release

hood release handle

hood release cable

hood latch

hood striker

hood latch adjustment

bumpers

bumper shock absorbers

fender shimming

weatherstripping

lock cylinders

lid struts

lid torsion rods

sound-deadening pads

Review Questions

1. List five ways parts can be held together.

2. A vehicle has a damaged fender. The estimate says to "R and R" the fender. This means you should:
   a. remove and repair the fender.
   b. remove and replace the fender.
   c. replace and realign fender.
   d. None of the above

3. Factory service manuals often have a _____ _____ section that explains and illustrates how body parts are serviced.

4. How and why might you want to mark a hood before removal?

5. Where are hood adjustments made?

6. This adjustment controls the height of the front of the hood.
   a. Hinge-to-body adjustment
   b. Hinge-to-hood adjustment
   c. Hood stop adjustment
   d. None of the above

7. Describe the parts of a cable hood release.

8. Fender _____ is an adjustment method that uses spacers under the bolts that attach the fender to the cowl or inner fender panel.

9. What is the purpose of weatherstripping?

10. Bad lid _____ _____ will no longer support the weight of the lid. They should be replaced.

ASE–Style Review Questions

1. *Technician A* says the most quarter panels are welded in place. *Technician B* says that some quarter panels are held in place with fasteners. Who is correct?
   a. Technician A
   b. Technician B
   c. Both Technicians A and B
   d. Neither Technician

2. A vehicle has minor front end damage. Before removing the hood, *Technician A* says to open and close the hood to analyze part conditions. *Technician B* says to mark hood hinge alignment. Who is correct?
   a. Technician A
   b. Technician B
   c. Both Technicians A and B
   d. Neither Technician

3. Which of the following fasteners should be removed for hood replacement if hood hinges and fender aprons are not damaged?
   a. hinge-to-apron
   b. hinge-to-cowl
   c. hinge-to-hood
   d. hinge-to-firewall

4. During hood adjustment, the hood pivots sideways when closed. This is normally caused by:
   a. Damaged hood hinges
   b. Damaged hinge springs
   c. Misaligned hinges
   d. Misaligned latch

5. When installing a new fender, *Technician A* says to tighten the fender bolts and adjust the hood to the fender. *Technician B* says to leave the fender bolts loose so the fender can be adjusted to the hood. Who is correct?
   a. Technician A
   b. Technician B
   c. Both Technicians A and B
   d. Neither Technician

6. *Technician A* says that sound-deadening material is combustible and must be removed from any

area to be welded or flame cut. *Technician B* says that these materials are fireproof. Who is correct?
a. Technician A
b. Technician B
c. Both Technicians A and B
d. Neither Technician

7. A vehicle that has suffered a hard hit to the front requires fender apron replacement and straightening of the cowl. Which of the following operations would not be necessary?
a. Get help to remove hood
b. Mark hood hinge alignment
c. Read dimensions manual on vehicle
d. Pull aprons before removal

8. A rear trunk lid has torsion bars. *Technician A* says you must always replace the torsion bars to change opening tension. *Technician B* says that sometimes torsion bar tension can be adjusted. Who is correct?
a. Technician A
b. Technician B
c. Both Technicians A and B
d. Neither Technician

9. *Technician A* says that the hood should not be adjusted to compress the rubber hood stops. *Technician B* says to close the hood briskly after replacement to seat it in the opening. Who is correct?
a. Technician A
b. Technician B
c. Both Technicians A and B
d. Neither Technician

10. *Technician A* says that fender and hood adjustments must be made simultaneously to achieve a satisfactory result. *Technician B* says to adjust the fenders first and then the hood. Who is correct?
a. Technician A
b. Technician B
c. Both Technicians A and B
d. Neither Technician

Activities

1. Inspect several cars. Determine if their bolt-on panels are adjusted properly. Is the gap between panels or components equal all the way around all panels? Make a report of your findings.

2. Using a service manual, summarize the procedures for removing a hood and fender from three different makes of cars. Write a report on how these procedures vary.

3. After getting instructor permission, loosen the bolts on a hood latch and hood. Move the hood and latch out of alignment. Adjust the hood and latch.

13

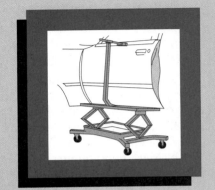

Servicing Doors, Glass, and Leaks

Objectives

After studying this chapter, you should be able to:

- R and R, and adjust a door.

- Replace both welded and adhesive-bonded door skins.

- Replace an SMC door skin.

- R and R, and adjust a door regulator.

- R and I (remove and install) a windshield.

- R and I other stationary glass.

- Find and fix air and water leaks.

- Describe gasket, full cutout, and partial cutout glass replacement procedures.

- Describe the basics of vent window and tailgate door glass service.

Introduction

Doors are the most used—and abused—parts of a vehicle. They are opened and closed thousands and thousands of times over the life of a vehicle. They must do this while still being strong enough to stay closed and protect the driver and passengers from injury during a collision. Doors must also seal out water and wind noise to keep the interior dry and quiet. Look at Figure 13–1.

Glass also plays an important role in the safety and appearance of a vehicle. Glass adds to structural integrity and visibility. Windshields, door glass, stationary glass, and related parts must all be serviced properly to keep the vehicle safe to drive.

This chapter will give you important information for working on these vital parts. It will give you the background needed to use service manual instructions and specifications effectively.

Figure 13-1 The phantom view of this vehicle shows intrusion beams in doors and large glass areas. Modern vehicles are much stronger and lighter than in the past to protect passengers from injury during a collision. *(Saab Cars USA, Inc.)*

DOORS

Vehicle doors allow entry into and exit from the passenger compartment. They are designed to be strong assemblies that provide easy access while being dependable structural units of the vehicle. It is important that you understand door construction and service because doors are commonly damaged in collisions.

DOOR CONSTRUCTION

There are two basic door designs: framed doors and hardtop doors. *Framed doors* surround the sides and top of the door glass with a metal frame, helping keep the window glass aligned. The door frame seals against the door opening. *Hardtop doors* have the glass extending up out of the door without a frame around it. The glass itself must seal against the weatherstripping in the door opening.

Illustrated in Figure 13–2, the basic parts of a door include the:

1. *Door frame*—the main steel frame of the door. Other parts (hinges, glass, handle, etc.) mount on the door frame.

2. *Door skin*—the outer panel over the door frame. It can be made of steel, aluminum, composite, or plastic.

3. *Door glass*—must allow good visibility out of the door.

4. *Door glass channel*—serves as a guide for the glass to move up and down. It is a U-shaped channel lined with a low friction material, felt for example.

5. *Door regulator*—a gear and arm mechanism for moving the glass. When you turn the window handle or press the window button, the regulator moves the glass up or down.

6. *Door latch*—engages the *door striker* on the vehicle body to hold the door closed.

7. *Inner and outer door handles*—use linkage rods to transfer motion to the door latch, allowing you to activate the latch to open the door.

8. *Door trim panel*—an attractive cover over the inner door frame. Various parts (inner handle, window buttons, speakers) can mount in the inner trim panel.

9. A plastic or paper *door dust cover* fits between the inner trim panel and door frame to keep out wind noise.

10. *Door weatherstripping* fits around the door or door opening to seal the door-to-body joint. When the door is closed, the weatherstripping is partially compressed to prevent air and water leaks.

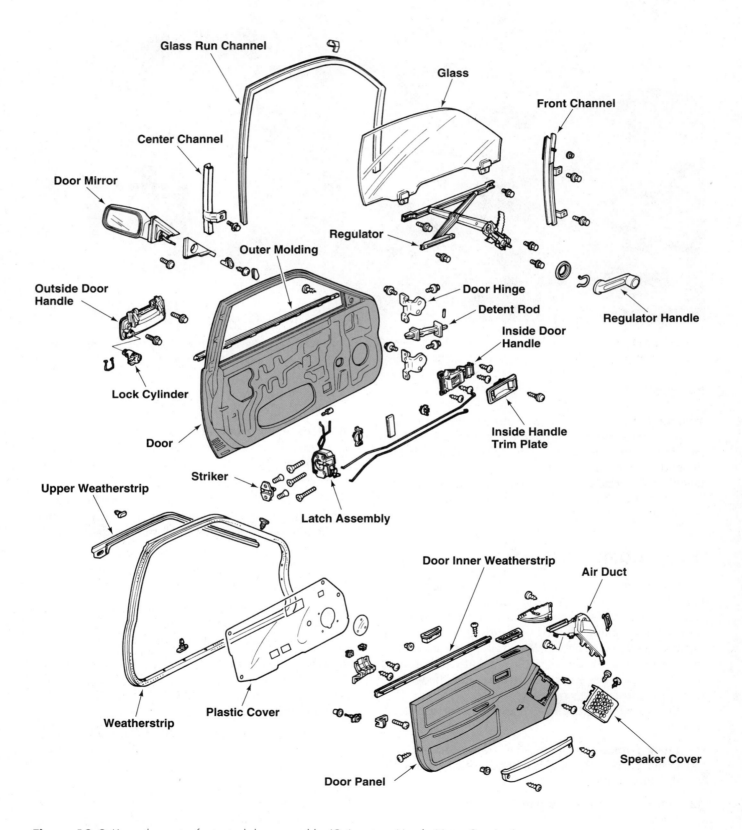

Figure 13–2 Know the parts of a typical door assembly. *(© American Honda Motor Co., Inc.)*

11. **Rearview mirror** can often mount on the outside of the door frame. A remote mirror knob may provide mirror adjustment on the inner trim panel.

DOOR ADJUSTMENTS

Doors must be accurately adjusted so that they close easily, do not rattle, and do not leak. The door hinges must be adjusted to hold the door in the center of its opening when closed. The door striker must be adjusted to engage the latch smoothly. This section will describe various door adjustments.

The door must first be adjusted to be centered in its opening. All gaps around the door must be equal and within specifications. The outer door panel must also be flush with the front fender, pillars, rocker panel, and rear door or quarter panel. After the door is adjusted in its opening, the door striker is installed and adjusted to engage the door latch properly.

Adjusting Door Hinges. Doors must fit their openings and align with the adjacent body panels.

When the doors on a four-door sedan need adjusting, start at the rear door. Since the quarter panel cannot be moved, the rear door must be adjusted to it. Once the rear door is adjusted, the front door can then be adjusted to fit the rear door. Then, the front fender can be adjusted to fit the front door.

On hardtop models, the windows can be adjusted to fit the weatherstripping. Hardtop windows are usually adjusted starting with the front edge of the glass and working toward the back. The front window is then adjusted to it. The rear door window is adjusted to the front window rear edge and the opening for the rear door assembly.

Doors are attached to the body with hinges. The bolted hinge can be adjusted forward, rearward, up, and down easily. The use of shims behind the hinge also allows the hinge to be moved as desired.

To adjust a door in its opening, follow these steps:

1. Determine which hinge bolts must be loosened to move the door in the desired direction.

2. Use a jack and wooden block to support the weight of the door. The wooden block will prevent the jack saddle from chipping the paint on the door. See Figure 13–3.

> WARNING! *Do not slam the door after adjusting it. You could damage the latch or striker if they are no longer in alignment.*

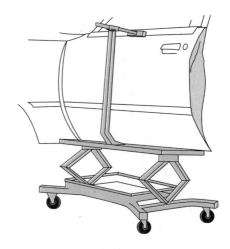

Figure 13–3 Use a door stand to adjust the door.

3. Loosen the hinge-to-body bolts just enough to permit movement of the door.

4. Move the door as needed to align it. Tighten the hinge bolts and check the fit to be sure there is no binding or interference with the adjacent panel.

5. Repeat the operation until the desired fit is obtained. Then check the striker pin alignment for proper door closing.

6. Loosen the door striker and move it up or down as needed. If the door tends to move up, move the striker down. If the door tends to drop when engaging the striker, shift the striker upward. See Figure 13–4.

7. The door and glass must also be checked to assure proper alignment to the roof rail and vertical weatherstripping.

When door adjustments are necessary, it may be helpful to remove the striker plate. Doing so allows the door to be centered in the opening more easily.

On some vehicles, a special wrench must be used to loosen and tighten the hinge bolts. If the hinges are to be removed, a line should be scribed around each one to mark its position. This makes reinstallation easier. Be careful not to cut the paint or corrosion will result.

You might have to loosen the fender at the rear bottom edge to reach the bolts. If the hinge pins are worn out, replace the hinges. Some hinges use bushings around the pins. When these bushings are worn, replace them. This will remove clearance in the hinges and also readjust the door.

In-and-out adjustments are also very important. The door must be aligned in and out to fit the body panels. To adjust the door in or out, loosen the hinge-to-door bolts. Shift the door as needed and then tighten the hinge bolts.

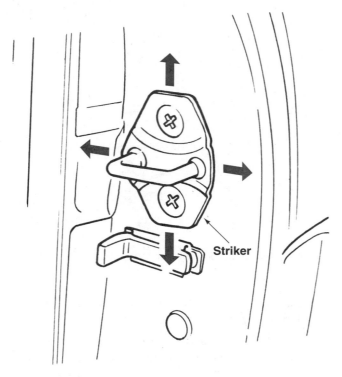

Figure 13–4 The door striker should be loosened and adjusted to hold the door tight against the weatherstripping. If the door tends to move up or down when engaging the striker, raise or lower it. (© American Honda Motor Co., Inc.)

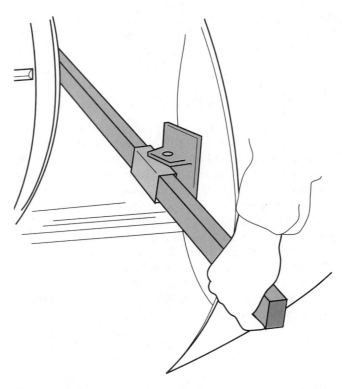

Figure 13–5 A pry bar fixture like this one can be used to do minor adjustment of a door with welded hinges.

If the top of the door is moved out, it will also move the opposite bottom corner in. If the bottom of the door is moved in on the hinge, it will move the opposite corner out.

The key is to move the door in or out equally on both hinges. Then it will only affect the front of the door because the amount of adjustment decreases toward the back. The center door post, striker pin, and lock will determine the position of the rear of the door. The front edge of the door should always be slightly in from the fender. This will help prevent wind noise from entering the gap between the door and fender.

Some hinges are welded to both the door and body. Obviously, no adjustments can be made to the welded door hinge. If only slightly out of alignment, you can spring or bend the hinge mounts to adjust welded hinges. Look at Figure 13–5.

Door Inner Trim Panel R and R

To work on parts inside the door, you must remove the inner door trim panel and related parts. Remove any screws that hold on the armrest and other trim pieces. You may have to pop out small decorative

plugs over some of the screws. Refer to the service manual if in doubt.

Remove the window crank handle. It can be held on by a screw or by a clip from behind, Figure 13–6.

With all screws out of the door inner trim panel, you will usually have to pop out a series of plastic clips. They install around the perimeter of the panel. Use a *forked trim tool* designed to remove clips in trim parts. Slide it between the door and panel. Then pry out the plastic clips without damaging the panel.

As you lift off the door inner trim panel, disconnect any wires going to the panel. Feed the wires through the panel. See Figure 13–7.

With the door panel off, peel back the paper or plastic water shield over the door. Pull slowly so you do not damage it. You now have access to the bolts and nuts securing the window regulator.

When replacing a door inner trim panel, check that all plastic clips are fully in the panel. They can slide out of position and not align with the holes in the door. Bond the paper or plastic water shield back into place, Figure 13–7B. Then, feed any wires through the panel.

Fit the panel over the top lip on the door and down into position. Starting at one end, use your hand to pop each clip into the door. Lean down and check that each clip is started in its hole. Then use your hand to pop each clip into place. Install the other parts in reverse order of removal.

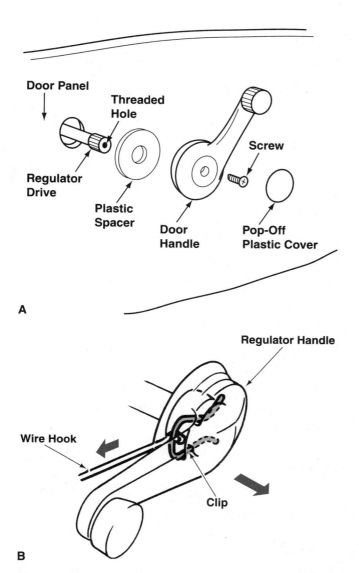

A

B

Figure 13–6 Note two common methods of holding window crank handles in place. (A) A small screw under a plastic bezel holds this window crank. (B) A spring clip holds this window crank. Hook, fork tool, or shop rag can be used to pop the clip out of its groove. (© American Honda Motor Co., Inc.)

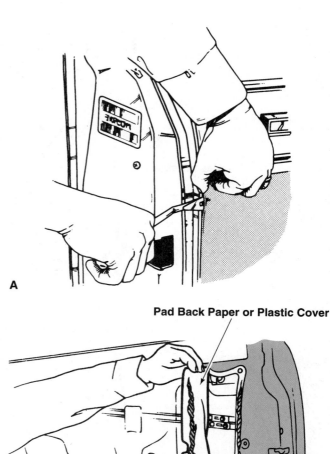

A

B

Figure 13–7 (A) A forked trim panel tool can be slid behind the inner door panel to pop out plastic clips without damage to the panel. Make sure the forked tool surrounds the plastic clip when prying. (B) To prevent tearing, slowly pull back the paper or plastic moisture shield behind the door panel. It must be reused. (Courtesy of DaimlerChrysler Corporation)

Window Glass and Regulator R and I, Adjustment

Small nuts and bolts or rivets secure the window regulator and glass guides or tracks in position. Usually, the glass bolts to the upper arms of the regulator. On a few older vehicles, the glass may be held on with special adhesive or epoxy.

To remove the glass, unbolt it from the regulator or drill out the rivets. You must then remove any parts that prevent you from sliding the glass out of the door. This can vary, so refer to the manual if needed. Look at Figure 13–8.

If the glass was broken, use a vacuum cleaner to remove all broken glass from inside the door. Install the new glass and bolt it to the regulator. Make sure you use all rubber, plastic, and metal washers.

> *WARNING! Do not slam the door before adjusting the glass. With a hardtop door, the glass can hit and break on the top of the door opening.*

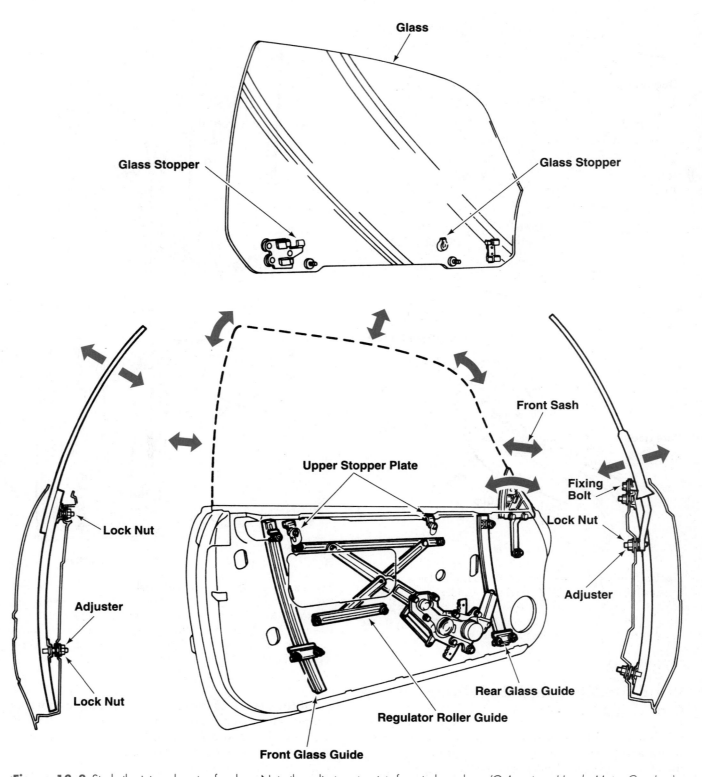

Figure 13–8 Study the internal parts of a door. Note the adjustment points for window glass. (© American Honda Motor Co., Inc.)

To adjust the glass and regulator, roll the window all the way up. Make sure it is centered in the frame or door opening. If necessary, loosen the regulator mounting fasteners. Shift the regulator and glass as needed. Turning the window crank one way or the other may help shift the regulator. When the glass is properly positioned, tighten the regulator fasteners.

Stops may also be provided to limit upward travel of the glass. Adjust the stops so they limit upward glass travel. Double-check the door glass adjustment with the door closed. Inspect where the glass contacts the inner door channel or weatherstripping. It must fully seal the door opening.

If needed, lubricate high friction points on the regulator. Lubricate the large gear and the arms where they rub together. The window glass should slide up and down easily without binding.

Door Lock and Latch R and R

Door lock assemblies usually consist of the outside door handle, linkage rods, the door lock mechanism, and the door latch. Various types of exterior door handles are available. With a push-button type handle, the button contacts the lock lever on the latch to open the door. However, most exterior door handles operate through one or more rods.

Door handles can be replaced by raising the window and removing the interior trim, panel, and water shield to gain access to the inside of the door. Exterior door handles are often held on with screws or bolts. A short screwdriver or small, ¼ inch drive socket is often needed to remove an outside door handle.

Some causes of exterior door handle problems are:

1. Worn bushings

2. Bent or incorrectly adjusted lock cylinder rods

3. No lubrication on handle, linkage, or latch

4. Worn or damaged latches

Inside door lock mechanisms are generally the pull handle type. The mechanisms are connected to the lock by one or more lock cylinder rods. Clips or bushings are used to secure the rods in place.

The door lock mechanism usually mounts through a hole in the door. A spring clip often fits over the inside of the lock to hold it in place. An arm on the lock transfers motion through a rod to the door latch.

To remove the lock, place a drop light inside the door. While looking through one of the large openings in the inside of the door, pop off the small clip holding the lock rod. Then, pry the clip off the lock. You can use a screwdriver or pliers depending upon the amount of space around the lock. Slide the lock and washer off the outside of the door.

Electric door locks often use electric solenoids to move the linkage to the door locks. When you turn the key or press a door lock button, current flows to the solenoids in the doors. The solenoids convert electrical energy into motion. The solenoid motion moves the lock linkages and latches. This locks or unlocks the doors.

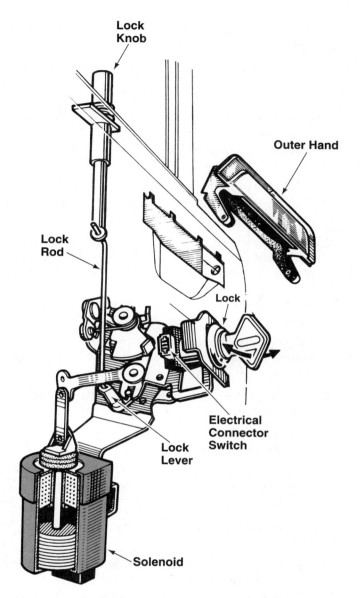

Figure 13–9 Most power door locks use a solenoid to move lock mechanisms and latches. The switch on the lock electrically energizes the solenoid. The solenoid then converts electrical energy into motion. *(Courtesy of DaimlerChrysler Corporation)*

In Figure 13–9, note the solenoid-operated door lock mechanism. For more information on electrical systems, refer to Chapter 26.

Door Skin Replacement

Like other damaged panels, a door skin can be straightened or replaced. The decision is based on the amount of door frame damage.

The door skin wraps around—and is flanged around—the door frame. The door skin is secured to the frame either by welding or with adhesives.

Typical replacement procedures are needed for both types of skins.

Replacing Welded Door Skins

A *welded door skin* has spot welds that hold the skin onto its frame. To replace a welded door skin, follow manufacturer recommended procedures and these basic steps:

1. Remove the trim panel and disconnect the negative side of the battery to isolate all power door accessories.

2. Before removing the door, check to see if the hinges are sprung. Check the alignment of the door with respect to its opening. If the door frame is bent, do not try to replace the door skin. Straighten the door frame or obtain a replacement door.

3. Inspect how the skin is fastened to the door. Determine how much interior hardware must be removed.

4. To prevent loss, place all removed parts and hardware inside the vehicle or in a parts tray.

5. If it only has minor damage, straighten and/or align the inner door frame.

6. Remove the door glass to prevent breakage or pitting while repairing the door. Place it inside the vehicle for protection.

7. Remove the door and move it to a suitable work area.

8. Apply tape to the door frame and measure the distance between the lower line of the tape and the edge of the skin. Also measure the distance between the edge of the skin and the door frame.

9. Remove the paint from the spot welds in the *skin hem* (folded door edge). Using a drill and spot weld cutter or hole saw, remove the spot welds.

10. Use a plasma arc cutter or grinder to remove the welded portion of the skin from the door frame.

11. Grind off the edge of the door skin hem flange, Figure 13–10A. Grind off just enough metal that the skin can be separated from the flange. Do not grind into the frame. Also, do not use a cutting torch or power chisel to separate the panels. The frame can be distorted or be accidentally cut.

12. Separate the reinforcing strip on the top of the skin, if applicable.

13. Using a hammer and chisel, start to separate the skin from the frame. Use a pair of tin snips to cut around any spot welds that could not be cut or ground out.

14. When the skin moves freely, remove it. Use clamping pliers to remove what remains of the hem flange. Any remaining spot welds, braze, and corrosion should be ground or blasted off.

15. With the skin removed, examine the frame for damage. Repair any remaining inner door damage at this time. If necessary, remove damage to the inner flange with a hammer and dolly.

16. Apply weld-through primer to bare metal mating surfaces. Cover other bare metal with an anti-corrosion primer or other corrosion protection material.

17. Prepare the new skin for installation. Using a drill or hole punch, make holes for plug welds. Remove the paint from the weld and braze locations with a sander. Apply weld-through primer to the bare metal mating surfaces.

18. Some door skins have a silencer pad that must be attached to the skin. To do this, clean the skin with alcohol, then heat it and the silencer pad with a heat lamp. Finally, glue the pad to the skin.

19. Apply body sealer to the back side of the new skin. Apply the sealer evenly, ⅜ inch (9 mm) from the flange, in a ⅛-inch (3 mm) thick bead.

20. Using clamping pliers, attach the new skin to the door and align it properly. Weld where required.

21. Use a hammer and dolly to flange the hem. See Figure 13–10B. Cover the dolly face with masking tape to avoid marring the skin. Bend the hem flange gradually in three steps. Be careful not to bend or throw the skin out of alignment.

22. After working the flange within 30 degrees of the frame, use a flanging tool to finish the hem flange, Figure 13–10C. Check alignment of the door in the opening.

23. Weld the plug or spot weld locations of the glass opening. Then tack weld the hem flange.

24. Drill holes into the new door skin to accommodate moldings, trim, and so forth. Make sure all edges are cut in prior to installing any parts.

25. Install all door parts. Prepare the door surfaces for refinishing.

26. Place the door on the vehicle. Align the door with the adjacent panels. Check for proper operation.

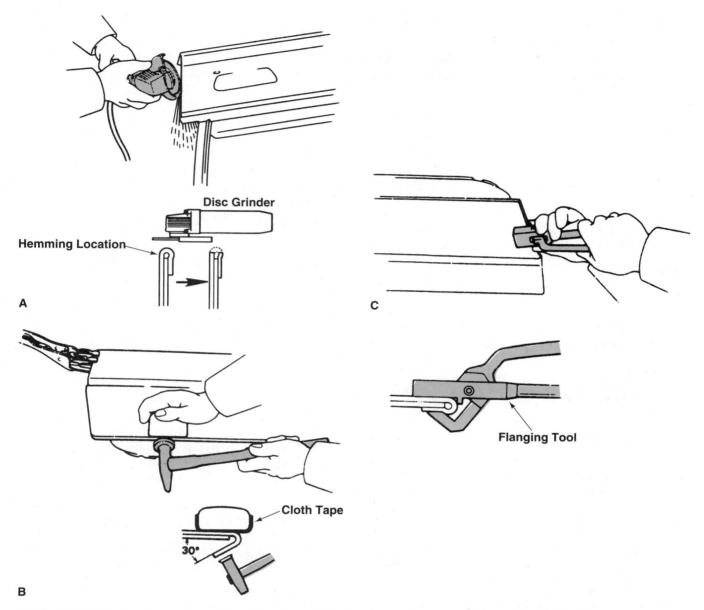

Figure 13–10 Follow the major steps for replacing a welded door skin. (A) Grind off the door skin hem flange as shown. (B) Use a hammer and dolly to straighten and fit the hem flange. (C) Use flanging pliers to force the new door skin down over the door frame.

Sealing Door Seams

Remember that all joints and seams must be protected from corrosion. Use an appropriate seam sealer to keep water and contaminants out of the door and provide long-term protection.

Allow seam sealer to dry properly before painting. Drying time depends on the type of sealer, its thickness, and the temperature and humidity in the area. Normally, the lower the temperature, the longer the necessary dry time. Also, the higher the humidity, the longer the dry time.

NOTE! Never attempt to use a seam sealer without first reading the instructions provided by the manufacturer.

Replacing Adhesive-Bonded Door Skins

An *adhesive-bonded door skin* is bonded, not welded, to the door frame. It requires different replacement methods.

1. After removing all hardware, trim material, and the glass, remove the door from the vehicle.

2. Use a grinder to grind the edge of the door. This will make it possible to safely separate and remove the damaged skin.

3. Peel off the remaining hem flange using a chisel.

4. Use sandpaper or a blaster to remove any corrosion from areas that are too tight for a grinder. Then sand or grind all areas where the door skin meets the door frame.

5. If necessary, straighten the door frame.

6. Thoroughly clean the edge of the door frame, inside and out, with a good adhesive cleaner.

7. Check the fit of the replacement door skin.

8. Prime both sides of the door frame with adhesive primer.

9. Cut the tip of the adhesive nozzle to provide a ⅛-inch (3 mm) bead of adhesive.

10. Apply the adhesive in a continuous bead. The adhesive can be applied to the inside of the door skin creases or to the door skin side of the door frame.

11. Carefully position the door skin on the frame.

12. Flange the door skin in the usual manner. Be sure to wipe away any excess adhesive along the flange. Check door alignment in the opening before the adhesive cures.

13. Seal the crimped seam with seam sealer.

14. To protect against corrosion, apply an anticorrosion compound to the inside of the door.

15. Bond or tack weld the tabs at the top edge of the door skin. Use the method that best suits the design of the door. This step may not be needed on certain vehicle makes.

16. Paint and reassemble the door.

Replacing SMC (Composite) Door Skins

Sheet molded compound or **SMC** door skin panels are similar to fiberglass. Some doors are made completely of SMC, except for steel door intrusion beams and the steel lock and hinge reinforcements.

To replace an SMC door skin, proceed as follows:

1. Cut away the center of the skin. Air shears work well because cut depth is easily controlled. If you

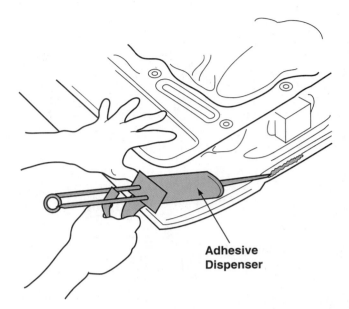

Adhesive Dispenser

Figure 13–11 Installing an adhesive-bonded door skin is similar to installing a welded-on skin. However, you must use a special two-part adhesive to bond the skin onto the frame.

use a saw, be careful not to hit or cut internal door parts.

2. To remove the remaining door skin, heat the bonding areas with a heat gun. Then apply pressure with a pry bar or chisel to remove the rest of the material. Be careful not to damage the door flange.

3. Sand the door frame flange to remove all remaining adhesive. Clean the bonding areas of the replacement skin with soap and water. Allow them to dry. Sand the bonding areas to expose the SMC fibers. Wipe dry with a clean cloth.

4. Apply a bead of two-part adhesive to the door frame flange, as shown in Figure 13–11.

5. Set the skin on the door frame and lightly clamp it. Do not squeeze too tight. You want to leave adhesive between the skin and frame. Check door alignment in the opening before the adhesive cures.

To complete the job, allow the adhesive to cure, paint the door, and then reassemble and mount it on the vehicle.

Further information on the repairs of plastic and composite panels can be found in Chapter 11.

Door Reinforcements

All the doors of unibody vehicles have inner metal reinforcements at various locations. There are some

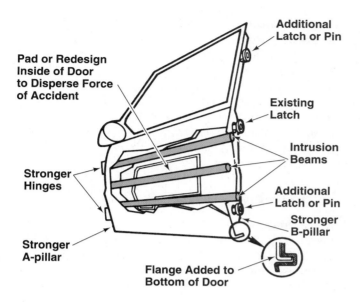

Figure 13-12 Study the inner parts relating to intrusion beams.

other door frame reinforcements, such as at the hinge locations and the door lock plate. Door intrusion beams are normally used inside side doors.

A **door intrusion beam** is welded or bolted to the metal support brackets on the door frame to increase door strength. Methods of strengthening doors are given in Figure 13-12.

GLASS

Vehicle glass allows the passengers clear visibility out of the passenger compartment and also protects them from wind, rain, and road debris. Glass is often broken during a collision and must be serviced. Passenger-side air bag deployment can break the windshield. Side air bag deployment can sometimes shatter door glass.

Windshields

Today's vehicles are built with a lot of glass that affords greater visibility for driving. Frequently, this glass is broken during a collision or by flying gravel or vandalism. Glass may also need to be removed from areas of damage before the damage can be straightened.

Glass also plays an important role in aerodynamics styling. Air drag is reduced by fitting the glass accurately to the sheet metal and eliminating trim. This approach reduces wind noise and provides a more sleek, uncluttered appearance.

Stationary glass panels in modern unibody vehicles also play an important role in the strength of the body

structure. The stationary glass is an integral part of the upper body structure. It provides lateral bracing to help prevent roof collapse during a rollover. For this reason, when stationary glass is replaced, it must be installed in the same manner as it was by the manufacturer.

If the glass is installed incorrectly, you could be putting a vehicle back on the road that is not structurally sound. It is important for you to be familiar with the various techniques used to remove and install glass properly.

Types of Glass

For years, the glass used in most vehicles was either laminated or tempered. Both are considered safety glass because of their construction. Today, other types of glass are also found in vehicles.

Laminated plate glass consists of two thin sheets of glass with a layer of clear plastic between them. It is used for windshields, Figure 13-13. When this type of glass is broken, the plastic material will tend to hold the shattered pieces in place and prevent them from causing injury. See Figure 13-14.

Tempered glass is a single piece of heat-treated glass that shatters into small pieces when broken. However, it has more impact resistance than regular glass. It is used for side and rear window glass but never for windshields.

Both laminated and tempered glass can be tinted. **Tinted glass** contains a shaded vinyl material to filter out most of the sun's glare. This is helpful in reducing eye strain and prevents fading of the interior. Some windshields are shaded to reduce the sun's glare by means of a dark tinted band or section across the top. Tinted glass is usually recommended if the vehicle is to be equipped with air conditioning.

Glass can also be tinted by adding small quantities of metal powder to the other normal ingredients of glass to give it a particular color.

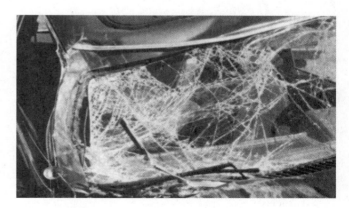

Figure 13-13 Note how lamination holds the glass together when the windshield is broken.

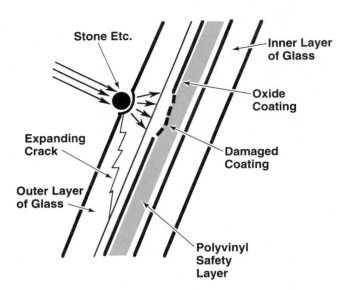

Figure 13-14 Study what happens when laminated glass is broken.

Glass can also be fitted with a defrosting circuit or antenna. **Self-defrosting glass** has a conducting grid or invisible layer that carries electric current to heat the glass. A grid type is common in rear windows. Windshields use a transparent conducting medium.

A **window antenna wire** for radio reception is placed either between the layers of laminated glass (windshield) or printed on the surface of the glass (rear window). Some glass has antenna wires and heating wires side by side.

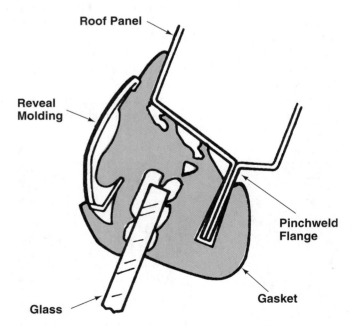

Figure 13-15 Some windshields are held in place with a gasket or rubber seal. The rubber gasket is shaped to fit around the glass and the body structure. Note this cross section of a windshield gasket.

Anti-lacerative glass has one or more additional layers of plastic affixed to the inside of the glass. This glass gives added protection against shattering and cuts during impact.

Modular or **encapsulated glass** has a plastic trim molding attached to its edge. It also fits the contours of a vehicle more closely than glass used in the past. Modular glass is the newest glass used in unibody applications.

Windshield Service

Replacement of the windshield involves two different methods: gasket installation or adhesive installation. The adhesive installation is further divided into two methods: the full cutout and the partial cutout method.

> *WARNING! When replacing glass, be aware of the following:*
> 1. *If you fail to restore the original structural integrity of a vehicle during repair, it could be violating federal law.*
> 2. *Butyl tape alone CANNOT be used to completely replace urethane because it has no structural bonding properties. It only holds the windshield in place and prevents leaks. NEVER use butyl tape ALONE to install a windshield.*

A **windshield gasket** is a rubber molding shaped to fit the windshield, body, and trim. The gasket installation method is common in older vehicles but still finds use in present day vehicles. The gasket is grooved to accept the glass, a sheet metal flange, and sometimes the exterior reveal molding, Figure 13-15. The gasket locking strip must be removed before the glass can be removed from the opening.

The **adhesive installation,** as the name implies, uses an adhesive material to secure the glass in place, Figure 13-16. The use of adhesives permits the windshield to be mounted flush with the roof panel, decreasing wind drag and noise. Rubber stops and spacers separate the glass from the metal.

Reveal molding is metal or plastic trim that fits around the vehicle's glass to cover the gap between the body and any glass adhesive. Depending upon the molding design, it can be held in place with small clips, adhesive, or a friction fit.

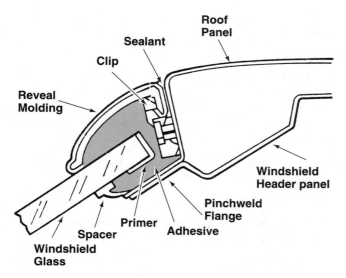

Figure 13-16 Study the installation of an adhesive-bonded windshield. Metal or plastic trim covers the edge of the glass and adhesive. Note this cross section of an adhesive-bonded windshield.

The **partial cutout method** of windshield replacement uses some of the original adhesive. The old adhesive must be in good condition and of sufficient thickness. It serves as a base for the new adhesive.

When the original adhesive is defective or requires complete removal, the **full cutout method** of windshield replacement is used. Both methods are shown in Figure 13-17.

Broken Glass

It is important to clean up pieces of broken glass from the seats, carpet, and air ducts before returning the vehicle to the customer. Vacuum the interior and blow out around the glass mounting area and vent ducts. Vacuum up as much glass as possible. Then, blow out glass from the heater and behind the dashboard with low pressure compressed air.

> **DANGER!** If you fail to clean up broken glass properly, someone could be cut and injured. Also, if you fail to clean out dash ducts, glass can blow into people's eyes when the heater or A/C blower is turned on. Wear gloves and a full face shield!

Windshield Gasket Method

Before removing the windshield, remove the interior and exterior moldings. Any garnish moldings on the interior face of the windshield are secured in place by screws or retaining clips. All of the garnish moldings

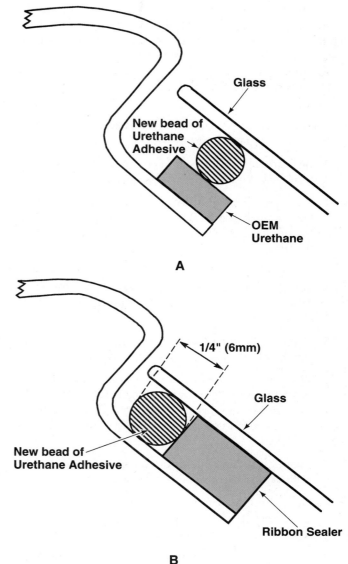

Figure 13-17 There are two ways to replace an adhesive-bonded windshield. *(I-CAR)* (A) With the partial cutout method, you can use some of the old adhesive. (B) With the full cutout method, you must remove all of the old factory adhesive. Square ribbon sealer and adhesive are used to affix the new windshield to the vehicle.

should be removed first. If necessary, the rearview mirror should be removed as well.

> *WARNING!* The instructions in this text are for general purposes. Always refer to the service manual for the exact make and model vehicle. It will give the detailed instructions and specifications needed.

On the exterior of the vehicle, remove the reveal moldings. The reveal molding is usually secured by retaining clips attached to the body opening. A projection on the clip engages the flange on the reveal molding.

To replace a windshield using a gasket, proceed as follows:

1. Place protective covers over adjacent panels. Put on safety glasses and gloves.

2. Be sure all moldings, trim, and hardware are removed. Remove the windshield wiper arms.

3. If the glass has a built-in antenna wire, disconnect the antenna leads. The wires are usually at the lower center of the windshield. Tape the leads to the glass.

4. Locate the locking strip on the outside of the gasket. Pry up the tab and remove the locking strip to open the gasket all the way around the windshield.

5. Use a putty knife to pry the gasket away from the pinchweld inside and outside of the vehicle.

6. With an assistant in the passenger compartment, push the windshield and gasket out of the body opening. If the glass was not cracked and is to be reused, exert even pressure on the glass so it will not break.

7. Clean the windshield opening with solvent to remove any dirt or residual sealer.

8. Place the glass on a cloth-covered bench or table to protect it. If the glass was removed for body repairs, leave the gasket intact.

9. If the glass was removed because it was broken, remove the gasket from the glass. Install the gasket on the replacement glass.

10. Cracks that develop in the outer edge of the glass are sometimes caused by low or high spots or poor spot welds in the pinchweld flange. Examine the body pinchweld and correct any problems.

11. Install setting blocks and spacers.

12. Carefully install the glass on the blocks to verify fit. Center the windshield. Check the gap between the glass and the pinchweld. The gap should be uniform around the entire pinchweld.

13. After the windshield is lined up, apply several strips of masking tape. Apply the tape from the glass to the vehicle body.

14. Slit each piece of tape at the end of the glass. Then lift the new windshield out and set it aside. To permanently install the windshield, you will line up the tape on the body with the tape on the glass.

15. Insert a cloth cord in the pinchweld groove of the gasket. Start at the top of the glass so the cord ends meet in the lower center of the glass. Tape the ends of the cord to the inside of the glass, Figure 13–18.

16. Squirt a soapy solution in the pinchweld groove for easier installation.

17. Apply waterproof sealer to the base of the gasket.

18. With an assistant, install the glass and gasket assembly in the body opening. Use your masking tape to align the glass. Slip the bottom groove over the pinchweld.

19. Slowly pull the cord ends so that the gasket slips over the pinchweld flange. Work the bottom section of the glass in first, then the sides, and finally the top. The glass might crack if the cord is pulled from only one end.

20. Dispense a small bead of waterproof sealer around the body side of the gasket.

21. Remove excess sealer with solvent.

22. Install the reveal and garnish moldings.

23. Check the windshield for water leaks using a low-pressure stream of water.

24. Place a soapy solution in the locking strip groove. Replace the locking strip by spreading the groove. Feed the strip into the opening. The soapy solution lubricates the groove and makes it easier to slide the strip through.

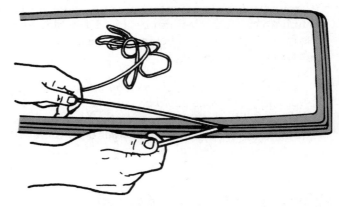

Figure 13–18 When installing windshield with a large rubber gasket, install a piece of cloth cord inside the gasket. Then, when you pull the cord, it will force the edge of the rubber gasket over the lip on the body.

Windshield Full Cutout Method

When using the full cutout method, first remove the windshield according to the following procedure:

1. Protect both the interior and the exterior of the vehicle from damage. Cover the front seat and instrument panel. Also cover the painted surfaces next to the windshield area.

2. Remove the rearview mirror. Take off the wipers, antenna, moldings, and any other parts that might get in the way. Remove the cowl vent panel if it makes it easier to reach around the edge of the windshield.

3. Using a hook tool, lift the molding at the bottom of the windshield. Once it can be securely grasped, pull it out from around the windshield. Do this carefully so the molding is not damaged.

4. With a utility knife, score the exposed urethane all the way around the outside of the windshield. Cut as close to the glass as you can. This makes it easier to insert the vibrating power knife.

5. Using a power knife, cut through the adhesive next to the glass. Cut as close to the glass as possible.

    Pneumatic windshield cutters use shop air pressure and a vibrating action to help cut the adhesive around glass. They have a variety of blade designs available for use on different windshields. Many have blades with depth stops to prevent pinchweld damage. With some designs, you may have to use two or more blades to remove some windshields.

    Electric adhesive cutters use either 12 or 120 volt power supplies to produce a vibrating action to assist in cutting the adhesive around glass. They operate similar to pneumatic windshield cutters. Some power cutters are used on the outside of the vehicle. Others are designed to be used from inside the passenger compartment.

    Piano wire (fine steel wire) can also be used to cut the adhesive, Figure 13–19. Use a 3- to 4-foot length of the lightest gauge wire that can be purchased. The cutting procedure with piano wire is a two-person job. After pushing the wire through to the interior, you and your helper must pull the wire around the windshield to cut the windshield free.

    If removing a piece of modular or encapsulated glass, especially if it is undamaged, most automakers recommend using the vibrating power knife to remove the glass instead of piano wire.

6. Carefully lift out the glass.

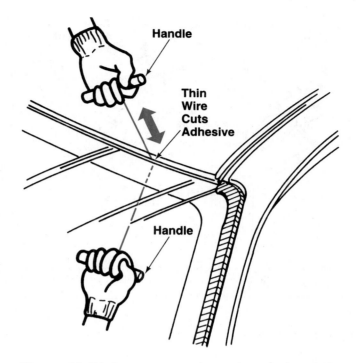

Figure 13–19 Piano wire can be used to cut through the adhesive holding a windshield. Hold the wire as close to the glass as possible. Be careful not to cut yourself. (© American Honda Motor Co., Inc.)

To prepare for the installation of the windshield, proceed in the following manner:

1. Cut out the remaining urethane adhesive from the pinchweld with a razor knife. Be careful not to cut or damage the headliner. Make sure no loose pieces of old adhesive are left. If the windshield had been replaced before with butyl tape or other unknown materials, it is especially important to remove this material thoroughly.

2. Check for corrosion in the pinchweld area. If there is corrosion, remove it with a wire brush, then sand if needed. Use a grit recommended by the adhesive manufacturer.

3. Dry set the new windshield. Check for at least ¼ inch (6 mm) of clearance around the windshield. Adjust the setting blocks as needed to get proper alignment. After the windshield is lined up, apply several strips of masking tape. Bridge from the windshield to the body. See Figure 13–20.

4. Slit the pieces of tape at the edge of the glass. Lift the windshield out and set it aside. When you set the new windshield on the adhesive, line up the tape on the vehicle with the tape on the windshield to get it positioned.

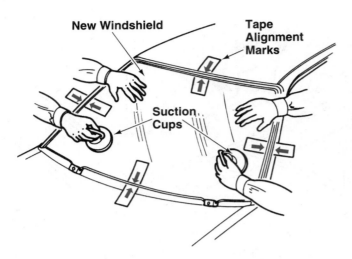

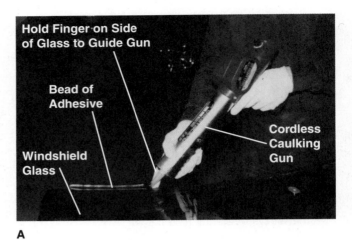

A

Figure 13–20 Ask someone to help when you are lowering a windshield straight down onto the adhesive bead. Suction cups will help hold the glass. (© American Honda Motor Co., Inc.)

5. Use a urethane adhesive cleaner on a lint-free dry cloth to wipe the pinchweld area. This removes any loose urethane. Let the cleaner dry.

6. Use a urethane primer in the pinchweld area. Do not use a liquid butyl primer or a ribbon sealer primer. Apply the primer to all bare metal in the pinchweld area.

7. Clean the windshield edge with a recommended solvent. Wipe dry with a clean, lint-free cloth.

8. Place a ½-inch (12 mm) coat of urethane primer on the inside contact surface of the windshield. Apply it all the way around. Let it dry for the required time.

9. Install square ribbon sealer to the inside edge of the pinchweld. For shallow pinchwelds, use ⁵⁄₁₆ × ⁵⁄₁₆ inch (8 × 8 mm). For deeper pinchwelds, use ⅜ × ⅛ inch (9 × 9 mm). Square ribbon sealer provides better support than round sealer. It also does not squeeze into the space needed for the urethane.

WARNING! Using an adhesive other than urethane can result in the shop and you being held liable. If the vehicle is involved in another collision and the glass reacts differently than the original installation, you and the shop could be sued.

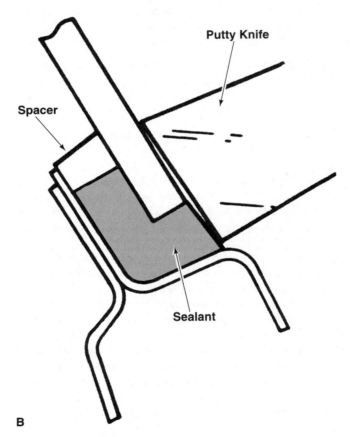

B

Figure 13–21 Apply adhesive slowly in a uniform bead of the recommended size. (A) The technician is using a dispenser to apply a continuous bead of adhesive to the flange for the windshield. (B) Use a plastic spreader or putty knife to level off and remove excess adhesive.

10. Apply a fast-curing, high-strength urethane adhesive following the manufacturer's instructions. Apply it directly behind the pinchweld, as shown in Figure 13–21. Hold the dispenser at a 45-degree angle. If too much adhesive is applied, excessive squeeze-out occurs.

WARNING! *The adhesive must be strong enough to deflect a top-mounted air bag without allowing the windshield to be blown out.*

The final step is to install the windshield. This is done as follows:

1. Set the windshield into place. Line up the pieces of masking tape on the glass and body. Press firmly so the glass makes contact with the ribbon sealer and the urethane adhesive.

2. Look for squeeze-out along the edge of the windshield. Use a paddle to spread it evenly. Smooth the surface and remove any excess adhesive. Fill all spaces between the glass and the metal.

   Remember! At least ¼-inch (6 mm) of the windshield surface must be bonded to the vehicle body.

3. Install the moldings. Clean up the area with a liquid adhesive cleaner.

4. Attach the cowl vent panel, wiper, antenna, and other parts.

The ribbon sealer will hold the windshield in place while the urethane adhesive cures. The urethane adhesive will reach its maximum strength in four or five days. Follow the manufacturer's recommendations for curing times. Curing times are affected by temperature and humidity.

Partial Cutout Method

The partial cutout method is recommended only under ideal curing conditions. Most of the steps are the same as in the full cutout method. One difference is that it is not necessary to remove all the old adhesive. The old adhesive left on the pinchweld serves as the base for the new adhesive.

If the old adhesive is not bonded securely to the vehicle body, do not use the partial cutout method. Remove all of the old adhesive and use the full cutout method. The same is true if butyl ribbon sealer or some other unknown material is discovered in the pinchweld.

Because there is no butyl ribbon sealer to hold the glass in position, allow enough time for the urethane to cure before driving the vehicle.

Inform the customer that the urethane will not achieve full strength until it is fully cured. Cure time will vary depending on temperature and humidity,

but most are fully cured in 24 to 72 hours. If possible, keep the vehicle until the windshield adhesive is fully cured.

Other Glass R and I

The removal and installation of rear and many stationary side windows follow the above methods closely. Procedures vary slightly for different vehicle makes. However, many of the operations are similar.

When working on small side windows, use duct or masking tape to secure the window while working. When inside the passenger compartment working, it is easy for the glass to fall out and break on the shop floor.

AIR AND WATER LEAKS

Air leaks will result in a complaint of wind noise in the passenger compartment. They often result from improperly adjusted doors, window glass, or damaged weatherstripping.

Water leaks will result in a complaint of water leaking into the passenger compartment or trunk. The most common causes of water leaks are poor seal around the windshield, window glass, sun roof, doors, or lids.

Improper door adjustment is the most common cause for both air and water leaks. The doors must provide a good seal between the weatherstripping and the body openings. The weatherstripping should be compressed sufficiently in the opening to prevent water, dust, and drafts from entering the vehicle. Door hinges wear and allow the doors to drop down, out of alignment.

To find body leaks, you can use several methods:

To conduct a **water hose test,** have someone squirt water onto the possible leakage areas while you watch for leaks inside the vehicle. Direct the water to flow over gaps in parts that might be leaking. Dripping water will result around any gap or leak in parts.

To conduct an **air hose leak test,** place a soapy water solution on possible leakage points and use low pressure air on the inside of the vehicle. When you find a leak, bubble will form in the soapy water.

An **air leak tape test** involves covering potential leaks with masking tape and driving the vehicle. If covering an area stops wind noise, you have found the air leak.

An **electronic leak detector** uses a signal generator tool and signal detector tool to find leaks. One person sits in the vehicle with the detector. Another person moves the generator tool around possible leakage

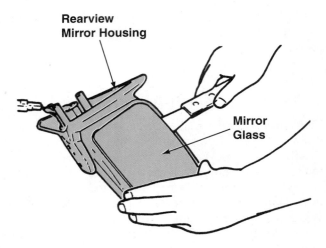

Figure 13-22 Pry off the old mirror.

Figure 13-23 When welding, cover the glass and protect it with a welding blanket. If sparks from the welding bead land on glass, they can melt or burn small indentations in glass, requiring glass replacement.

points. The detector will make an audible "beep" if the signal passes through a body opening. This shows you the location of leakage points.

After finding the leak, you must adjust the door or lid, replace the weatherstripping, or seal the glass to fix the leak. You must use common sense and the procedures given in this and other chapters to stop the air or water leakage.

REARVIEW MIRROR SERVICE

Both exterior and interior rearview mirrors require repair or replacement after collision damage, Figure 13-22.

External rearview mirrors normally bolt to the outside of the driver and passenger doors. Screws on the outside or inside of the door secure the mirrors. A gasket normally fits between the mirror and door surface.

If only the mirror is broken on external mirrors, you can sometimes replace the mirror glass. Heat the mirror with a heat gun to soften the adhesive. Then, pry off the broken pieces of mirror. Bond the new mirror into its housing with an approved adhesive.

Some interior rearview mirrors attach to the windshield with a special adhesive. You must clean the glass with a recommended cleaner. Spray on a clear primer agent. Then, apply a small bead of adhesive. Hold the mirror mounting pad stationary on the windshield for a few minutes until the adhesive cures.

PROTECT GLASS WHEN WELDING

Glass can be pitted and easily damaged when welding and grinding. If hot bits of molten metal fall on the glass, they will burn small pits into the glass. This

is an expensive, time-consuming mistake. Always cover all glass with a welding blanket or remove the glass when welding or grinding, Figure 13-23.

MORE INFORMATION

For more information on the subjects in this chapter, refer to the index. For example, heated windshields, power windows, and other related subjects are discussed in Chapter 26, Electrical Repairs.

Summary

- There are two basic door designs: framed and hardtop. Framed doors surround the sides and top of the door glass with a metal frame, helping keep the window glass aligned. The door frame seals against the door opening. Hardtop doors have the glass extending up out of the door without a frame around it. The glass itself must seal against the weatherstripping in the door opening.

- The basic parts of a door include the door frame, door skin, door glass, door glass channel, regulator, door latch, handles, door trim panel, dust cover, weatherstripping, and rearview mirror.

- Laminated plate and tempered glass are among the two most common window glasses. Laminated plate glass consists of two thin sheets of glass with a layer of clear plastic between them. It is used for windshields. Tempered glass is a single piece of heat-treated glass that shatters into small pieces when broken. It is used for side and rear windows.

- Air leaks will result from improperly adjusted doors, window glass, or damaged weatherstripping. The most common causes of water leaks are poor seal around the windshield, window glass, sun roof, doors, or lids.

Technical Terms

vehicle doors	tempered glass
framed doors	tinted glass
hardtop doors	self-defrosting glass
door frame	window antenna wire
door skin	anti-lacerative glass
door glass channel	modular glass
door regulator	encapsulated glass
door latch	windshield gasket
door striker	adhesive installation
door trim panel	reveal moldings
door dust cover	partial cutout method
door weatherstripping	full cutout method
forked trim tool	pneumatic windshield
door lock assembly	cutters
welded door skin	electric adhesive cutters
skin hem	piano wire
adhesive-bonded door	air leaks
skin	water leaks
sheet molded compound	water hose test
SMC	air hose leak test
door intrusion beam	air leak tape test
laminated plate glass	electronic leak detector

Review Questions

1. _____ are the most used—and abused—parts of a vehicle.

2. Explain the difference between a frame door and a hardtop door.

3. List and describe the 11 major parts of a door.

4. The driver's door is sagging badly and will not close easily. *Technician A* says to check for worn hinges. *Technician B* says to loosen and adjust the hinges. Who is correct?
   a. Technician A
   b. Technician B
   c. Both Technicians
   d. Neither Technician

5. This should be used when removing interior door trim panels.
   a. Pneumatic knife
   b. Electric knife
   c. Forked tool
   d. Slide hammer

6. What can happen if you slam a hardtop door after adjusting a door?

7. This can cause exterior door handle problems:
   a. Worn bushings
   b. Bent or incorrectly adjusted lock cylinder rods
   c. No lubrication on handle, linkage, or latch
   d. All of the above
   e. None of the above

8. Summarize the five major steps in replacing an SMC door skin.

9. This type of glass shatters into tiny pieces when broken.
   a. Laminated glass
   b. Tempered glass
   c. Modular glass
   d. Encapsulated glass

10. What is the difference between the partial and full cutout methods of windshield replacement?

ASE–Style Review Questions

1. *Technician A* says to adjust the door in its opening and then the striker. *Technician B* says to adjust the striker first to make sure it engages the latch. Who is correct?
   a. Technician A
   b. Technician B
   c. Both Technicians A and B
   d. Neither Technician

2. When adjusting the doors on a four-door vehicle, which of the following should be adjusted first?
   a. Front fenders
   b. Front doors
   c. Rear doors
   d. Strikers

3. *Technician A* says that door strikers never need adjustment. *Technician B* says that on some vehicles, a special wrench must be used to loosen and tighten the hinge bolts. Who is correct?
   a. Technician A
   b. Technician B
   c. Both Technicians A and B
   d. Neither Technician

4. A door on a two-door hardtop has been centered in its opening. However, the door still moves down slightly when the striker engages the latch. Which of the following should be done?
   a. Readjust door
   b. Move door hinges up
   c. Move striker down
   d. Move striker up

5. A door is going to be removed for outer skin replacement. The door frame is not damaged. Which of the following tasks should NOT be done?
   a. Remove hinge bolts
   b. Place jack under door
   c. Remove striker
   d. Mark hinges

6. *Technician A* says that you cannot adjust a door with welded door hinges. *Technician B* says that you can spring or bend the hinge mounts to adjust welded hinges. Who is correct?
   a. Technician A
   b. Technician B
   c. Both Technicians A and B
   d. Neither Technician

7. Which of the following is used to hold most door locks in doors?
   a. Adhesive
   b. Bolts
   c. Nuts
   d. Spring clips

8. A door on a vehicle is centered in its opening but the bottom is sticking out farther than the fender and rocker panel. *Technician A* says to move the upper hinge inward. *Technician B* says to move the lower hinge inward. Who is correct?
   a. Technician A
   b. Technician B
   c. Both Technicians A and B
   d. Neither Technician

9. A vehicle has an air leak somewhere around the rear door. *Technician A* says to cover the areas of potential leakage with masking tape to help locate the air leak. *Technician B* says to have an assistant use an electronic air leak detector around the back door. Who is correct?
   a. Technician A
   b. Technician B
   c. Both Technicians A and B
   d. Neither Technician

10. *Technician A* says to remove door glass during door skin replacement to gain access to the regulator. *Technician B* says that door glass removal is seldom needed during welded door skin replacement. Who is correct?
    a. Technician A
    b. Technician B
    c. Both Technicians A and B
    d. Neither Technician

Activities

1. Inspect a vehicle to check for proper alignment of doors and lids. Can you find any misaligned parts? Compare the alignment of doors and panels on different vehicles. Write a summary or make a large poster of your findings.

2. After getting permission from your instructor, adjust a hood using service manual/textbook procedures. Tape or mask any fender and hood edges that could hit together and cause damage to the vehicle finish. With the help of another student, first remove the hood, and then reinstall and adjust it properly.

3. After getting permission from your instructor, use service manual/textbook procedures to correctly remove, install, and adjust a hood latch.

Section 3

Prepainting Preparation

Chapter 14
Vehicle Surface Preparation

Chapter 15
Shop and Equipment Preparation

14

Vehicle Surface Preparation

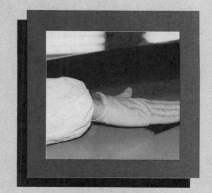

Objectives

After studying this chapter, you should be able to:

- Prepare a vehicle for painting/refinishing.

- Properly clean a vehicle using soap, water, air pressure, and a wax-grease remover.

- Evaluate the condition of the vehicle's paint.

- Describe methods for removing the damaged paint if needed.

- Properly prepare and treat bare metal surfaces.

- Correctly sand and featheredge surfaces.

- Apply an undercoat.

- Mask a vehicle properly.

Introduction

Vehicle preparation involves all of the final steps prior to painting/refinishing, including cleaning, sanding, stripping, masking, priming, and other related tasks.

Keep in mind that it is foolish to apply any kind of finish to a surface that is NOT properly prepared. The paint work will look bad or not hold up. Thorough vehicle preparation pays off in material savings, job quality, and customer satisfaction.

As you will learn, even if the original finish is in good condition, it should be lightly sanded or scuffed after washing to remove "dead film" and to block or smooth out small imperfections. If the paint surface is in poor condition, the paint should be removed down to the bare metal. In this way, a good foundation for the new finish is achieved.

Before refinishing can begin, the vehicle must be properly prepared. You must:

1. Wash the vehicle with soap and water. The underside of the vehicle should also be washed.

2. Clean the area to be refinished with wax and grease remover.

3. Correct any surface defects.

4. Mask the vehicle.

5. Tack off the vehicle using a tack rag.

6. Check that all surface preparation has been completed.

VEHICLE CLEANING

The vehicle cleaning processes should include:

1. Power washing the underbody and wheel housings.

2. Cleaning the door and trunk or hatch jambs.

3. Cleaning under the hood.

4. Using a strong wax and grease remover to remove road tars.

5. Blowing off the entire vehicle with high pressure air before and after masking.

6. Blowing off wheel covers outside of the spray booth before placing them in the spray booth.

7. Washing the repair area with a mild cleaner.

8. Tacking off the entire vehicle, including masking paper.

Initial washing involves a complete cleaning of the vehicle to remove mud, dirt, and other foreign matter. Washing should be done before bringing the vehicle into the shop. This will prevent road debris from entering the shop area. Look at Figure 14–1.

Figure 14–1 Thoroughly wash the vehicle with soap and water before starting work. All dirt and debris must be removed before starting repairs. Some shops also like to use a cleaning solvent.

Wet the whole vehicle with a water hose. Concentrate water flow onto trim pieces, around windows, and other areas that can trap and hold debris. Dirt also collects in door jambs, around the trunk and hood openings, and in wheel wells.

> *WARNING! Failure to remove foreign matter from the vehicle can allow it to blow or fall into the wet paint when spraying. This avoidable problem is very frustrating and a waste of time!*

Scrub all surfaces thoroughly with a detergent and water. A sponge works well. Wash the top first, then the front, rear, and sides. Rinse the vehicle thoroughly and let it dry completely. Remember that all adjacent panels must be as clean as the area to be refinished.

DANGER! *Never use gasoline as a cleaning solvent. A tremendous fire could result. Gasoline is also a poor wax-removing solvent and can itself deposit contaminating substances on the surface.*

> *WARNING! Do NOT use reducers for cleaning. They can be absorbed into the paint film. Blistering or lifting of the new paint can result.*

Using Wax and Grease Remover

Paint or other finish will NOT adhere to a waxy surface. For this reason, thoroughly clean all surfaces to be painted with a wax and grease solvent. Concentrate on areas where heavy wax buildup can be a problem. This would include around trim, moldings, door handles, and radio antennae.

To apply wax and grease remover, fold a clean, dry cloth. Soak it with the solvent. Then, clean the painted surface. While the surface is still wet, fold a second clean cloth and dry off the wet solvent. Work small areas and wet the surface liberally. Never attempt to clean too large an area. The solvent will dry before the surface can be wiped up.

For best results, remember to wipe off the wax and grease remover while it is still wet.

Some technicians like to use one hand to wipe on the cleaning agent and the other to dry the same area.

By using both hands and two rags, you can more easily wipe off the solvent while it is still wet and remove all surface debris.

If you fail to remove contaminants (especially tar and grease) from the surface before final sanding, this debris can be embedded in the filler or primer. Problems can then result when you are spraying on the topcoat.

Many shops like to use wax and grease remover before sanding and again right before painting. This ensures that all contaminants have been removed from the vehicle's surface before painting/refinishing.

NOTE! Always use new wiping cloths when using wax and grease remover. Laundering or washing the used rags might not move all oil, wax, or silicone residue.

To remove any last trace of moisture and dirt from seals and moldings, use compressed air at low pressure. Use a blow gun to blow out behind any area that could hold moisture.

Special attention should be paid to tar, gasoline, battery acid, coolant, and brake fluid stains. These can penetrate well beneath the surface of the paint film. These contaminants must be removed before sanding.

SURFACE EVALUATION

There are several characteristics of the finish that may affect the refinishing procedures. These finish characteristics include the type of the existing finish and the condition of existing finish.

You should make a visual inspection of the condition and appearance of the finish. During the inspection, answer these questions:

1. Has the finish been damaged because of age?

2. Has the finish been damaged because of weathering?

3. Does the finish have environmental damage?

4. Has the finish faded?

5. Will the gloss of the finish cause problems with matching?

6. How much texture or orange peel does the finish have?

7. Which conditions must be repaired prior to refinishing?

Both the customer and appraiser should be informed of any additional work that will be required prior to refinishing.

There are several factors that may affect the aging and durability of an automotive finish. The life span of a finish can be affected by its exposure to acid rain, industrial fallout, ultraviolet (UV) radiation, and hard water.

If a vehicle has been exposed to these elements, the damage may have to be repaired before refinishing. The finish may have to be removed before refinishing.

A *visual inspection* of the finish should include looking for:

1. Fading, dulling, whitening, or similar changes in gloss.

2. Cracking or checking (tiny spots are in paint film).

3. Blistering (paint film is lifting or bubbling).

4. Spotting (paint is discolored).

5. Cratering (deep cavities are formed in paint film).

6. Other paint problems.

Surface evaluation is a close inspection of the paint to determine its conditions. Once the vehicle has been cleaned, inspect the paint surface carefully.

Look for any signs of *paint film breakdown,* including checking, cracking, or blistering. Pay particular attention to the gloss level; low gloss often indicates surface irregularities.

To do a *paint adhesion check,* sand through a small area of the old finish and featheredge it. If the featheredge does NOT break or crumble, the paint may be properly adhered. If you cannot featheredge the paint, it must be removed before repainting.

Most forms of paint failure are progressive (ongoing) and cannot be stopped. In fact, trying to paint over a failing paint will usually accelerate the deterioration. If the finish is badly weathered or scarred, it is NOT suitable for refinishing.

Preexisting Damage

Preexisting damage includes paint cracking, scratches in paint, acid rain damage, and industrial fallout damage. If there is preexisting damage, the vehicle's owner should be contacted, and the situation explained. It is then the decision of the vehicle owner whether to authorize additional repairs. Some types of preexisting damage may have to be repaired before the new finish can be applied or match the existing finish.

Acid rain is caused by the release of various chemicals into the atmosphere. It results when these chemicals are absorbed by rain water and form acids. When it rains or snows these acids settle on the paint surface. They can become stronger as the water evaporates.

Acid rain damage may cause:

1. Craters to be etched into the paint film.

2. Changes in pigment colors.

3. Fading or dulling.

4. Corrosion of metallic colors.

5. Whitish spotting of clearcoats.

Acid rain damage can be identified by:

1. Using a magnifying glass.

2. Feeling the surface for depressions.

3. Visual inspection. Craters can be as small as a pinhead or as large as a quarter.

Industrial fallout is when metallic particles are released into the atmosphere from steel mills, foundries, and railroad operations. This is another source of paint damage. The metallic particles fall onto the vehicle's surface. They then corrode when exposed to moisture. The resulting damage will appear as black or brown spots or rust-colored rings on the paint surface. The particles can etch into the surface, causing a coarse or sandy surface.

Industrial fallout damage can be identified by feel or wiping a terry cloth towel over the surface. The metallic particles will catch on the cloth and can be easily seen.

Hard water spotting damage forms when rain or tap water dries on the paint surface. The water evaporates, leaving hard minerals on the paint surface. Hard water spotting:

1. Leaves a round white ring on the paint surface.

2. Is generally repairable because the mineral deposits are on top of the paint film.

UV radiation damage results from excessive exposure of the paint film to the sun's radiation. Exposure to UV radiation can cause:

1. Discoloration of the paint film.

2. Cracking of the paint film.

3. Fading or similar loss of gloss.

4. Yellowing of the paint.

When paint is badly aged or damaged, the finish should be completely removed. There are three common ways of stripping paint from metal surfaces:

1. Chemical stripping.

2. Abrasive blasting.

3. Sanding or grinding.

NOTE! For more information on surface evaluation, refer to the index. This subject is explained in several other locations, the chapter on paint problems, for example.

CHEMICAL STRIPPING

Chemical stripping uses a chemical action to dissolve and remove paint down to bare metal. A chemical paint remover is often used for stripping large areas of paint. Chemical stripping is effective in those places where a power sander cannot reach.

Before applying paint remover, mask off the area to ensure that the remover does NOT dissolve any paint NOT meant to be stripped. Use two or three thicknesses of masking paper to give adequate protection. Cover any crevices to prevent the paint remover from seeping to the bottom of a panel.

DANGER! *Paint remover (stripper) should be applied following the manufacturer's instructions. Pay attention to warnings regarding ventilation and smoking. Also wear protective clothing such as rubber gloves, long sleeve shirts, and goggles. If paint remover comes in contact with your skin or eyes, it can cause serious chemical burns.*

WARNING! Some paint removers can harm plastic filler beneath the paint. The chemical residue can be difficult to remove completely. If the chemical residue is NOT cleaned properly, it can cause paint adhesion problems.

Before applying the paint remover, slightly sand the surface of the paint to be stripped. This will help the chemical stripper penetrate and dissolve the paint more quickly.

To apply paint remover, brush on a heavy coat in one direction only, to the entire area being treated. Use a soft bristle brush. Do NOT brush the material out. Allow the paint remover to stand and soak until the finish is softened.

Although paint remover is quickly effective on most topcoats, some surfaces can prove stubborn. More than one application of remover may be needed.

Caution should be taken when removing the loosened paint coatings. Some paint removers are designed to be neutralized by water. Others are more easily removed with a squeegee or scraper. See Figure 14–2.

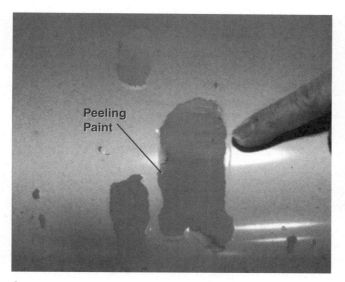

Peeling Paint

A

Brush Stripping Chemical onto Paint

B

Once Softened, Scrape Off Old Paint

C

Figure 14–2 If your evaluation finds serious problems with the vehicle's finish, you may have to strip it off. (A) This paint is peeling and would not provide a solid base for new paint. (B) Apply a thick coat of chemical paint stripper. Wear protective gear to prevent chemical burns. (C) After the paint softens, scrape it off. Then clean residue off the metal.

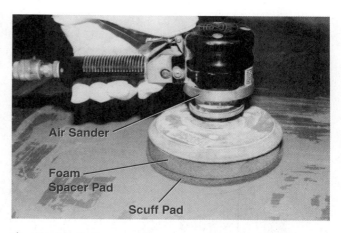

Air Sander

Foam Spacer Pad

Scuff Pad

Figure 14–3 Here a technician is using a grinder equipped with a large scuff pad. It will quickly remove softened paint without scarring the metal.

Be sure to rinse off any residue that remains by using a cleaning solvent and abrasive pad. Follow immediately by wiping with a clean rag. This rinsing operation is essential. Many paint removers contain wax. If left on the surface, the residue will prevent the paint from adhering, drying, and hardening properly.

After chemical stripping, you will sometimes have to remove small patches of remaining semi-softened paint. For fast removal, use a scuffing disc mounted on a grinder, Figure 14–3. Scuffing wheels will work well in restricted areas, Figure 14–4.

Figure 14–4 To get at hard-to-reach areas, a scuffing disc like this one will work well. (© The Eastwood Company, 1-800-345-1178, www.EastwoodCompany.com)

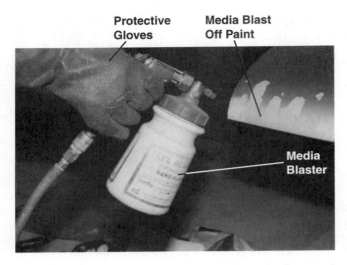

Protective Gloves **Media Blast Off Paint** **Media Blaster**

Figure 14–5 Blasting is good on small rust areas. It will remove rust from pits without thinning the metal. Blasting with soft media or plastic beads is better on larger areas.

BLASTING PAINT

Abrasive blasting involves using air pressure, a blasting gun, and an abrasive (plastic beads or other medium) to remove paint. Blasting leaves a clean, dry surface that is ideal for refinishing. Specialty shops have large blasters that can remove all of the paint from a vehicle in a few hours. Collision repair shops often have smaller blasters for removing paint from smaller areas. Look at Figure 14–5.

Blasting will quickly reveal hidden rust that can result in scaling and other problems after refinishing. In addition, blasting makes hard-to-reach areas accessible. This method also saves time when compared with sanding/grinding or chemical stripping.

Typically, blasting operations concentrate the pressure and flow of air and sand or plastic beads in a small area. On some models, the blaster can be adjusted to vary the blast volume and pattern.

> **Note!** Sandblasting can warp thin sheet steel, aluminum, plastic, and other substrates. Sand or hard media blasting is NOT recommended on large flat panels that could warp.

Plastic stripping media, however, will remove paint from almost any surface without damage. Methods of using a plastic stripping medium are similar to the conventional sandblasting methods, with the addition of these guidelines:

1. Use low (20 to 25 psi or 1.4 to 1.8 kg/cm^2) air pressure at the nozzle.

2. Hold the nozzle at a shallow (20 to 30 degrees) angle to the panel.

3. Hold the gun 18 to 24 inches (450 to 600 mm) from the panel.

4. Use a broad, sweeping back-and-forth motion.

5. Stop when the primer is reached.

Mentioned earlier in the text, some of the newer models provide dust-free or "captive" sandblasting. They contain a built-in vacuum and filtration system that vacuums up and recycles the abrasive.

REMOVING PAINT

Machine grinding is suitable for removing the finish from small flat areas and gently curved areas. Start with a No. 40 grit paper on a soft backing pad. Hold the face of the pad at a slight angle to the surface. Work back and forth evenly over the area to remove the bulk of the finish down to the metal. Follow this with a No. 80 grit paper and then No. 180 grit paper. Go over the entire area to be repaired.

> *WARNING! When using the grinder, care must be taken to prevent gouging, scarring, or heat warping the metal.*

BARE METAL TREATMENT

Proper **bare metal treatment** prepares the metal for primer. It can also inhibit erosion.

Several metal treatment systems are available that ensure a good bond. One of the best for steel is Metal Prep/conversion coating. To apply, proceed as follows:

1. Clean the metal to remove contaminants. Use a wax and grease remover. Work small areas, 2 to 3 square feet (18 to 27 sq. cm) at a time, wetting the surface liberally. While the surface is still wet, fold a second clean cloth and wipe the area dry.

2. Clean with metal conditioner. Mix the appropriate cleaner with water in a plastic bucket according to the label instructions. Apply with a cloth, sponge, or spray bottle. If rust is present, work the surface with a stiff brush or abrasive pad. Then, while the surface is still wet, wipe it dry with a clean cloth.

3. Apply conversion coating. Using the proper conversion coating for the metal (check manufacturer's recommendation), pour the conversion coating into a plastic bucket. Using an abrasive pad, brush, or spray bottle, apply the coating to the surface.

Leave the conditioner on the surface 2 to 5 minutes to create a zinc phosphate coating on the steel. If the surface dries before the rinsing, reapply. Flush the coating from the surface with cold water. Wipe dry with a clean cloth. Allow to air dry completely. Primer or primer-surfacer can then be applied.

A *wash primer* is a type of surface treatment that eliminates the need for a conversion coating. The vinyl resin in the wash primer provides corrosion resistance. The reducer in the wash primer contains phosphoric acid for strong bonding. A wash primer with phosphoric acid reducer cleans and etches the metal to aid adhesion.

Before applying a wash primer carefully read and follow the manufacturer's directions and special instructions.

Typically, metal treatment and priming are considered separate surface preparation steps. However, some new products actually combine these steps.

Self-etching primers etch the bare metal to improve paint adhesion and corrosion resistance. Self-etching primers work best on lightly sanded surfaces.

PREPARING METAL REPLACEMENT PARTS

Many manufacturers and parts suppliers prime replacement panels. The function of this *new part primer coat* is to protect the metal against corrosion. It does not necessarily provide a firm basis for a paint system. Although most primers do have this dual function, the supplier should always be consulted.

Remember that the primer on new body parts may not be intended to serve as the primer. A primer must be applied to these replacement parts or the color coat will NOT adhere properly.

Clean the new part with wax and grease remover. Then examine the part for scratches and other imperfections. Sand any imperfections until smooth but do NOT remove the coating completely. Scuff sand the entire panel. Then apply primer before painting. If in doubt about the coating, check with the manufacturer of the part for the recommended finishing procedure.

Clean e-coated replacement panels with a wax and grease remover. Wipe them with liberal amounts of solvent, changing rags frequently. Then treat the panels with a metal conditioner.

SANDING/FEATHEREDGING

One of the most important parts of surface preparation is *sanding.* Sanding prepares the surface for painting in several ways:

1. Chipped paint can be sanded to taper the sharp edges. Otherwise, the chip would show up as ridges under the new finish.

2. Cracking, peeling paint, and minor surface corrosion can be removed by sanding before applying a fresh topcoat. If they are NOT removed, these conditions will continue to deteriorate and will eventually ruin the new finish.

3. Primed areas must be sanded smooth and level.

4. The entire surface to be refinished must be scuff sanded to improve adhesion of the new finish. Scuff sanding also removes any trace of contaminants on the existing finish.

Selecting Sandpaper

Because *coated abrasives* (sandpapers) are often used to prepare the vehicle surface for painting, using the correct abrasive is critical. You must select the proper sandpaper for optimum productivity, cost efficiency, and best finish.

Open coat sandpaper is good for sanding softer materials, such as old paint, body filler, plastic, and aluminum. Open coat can prevent premature loading or clogging of the sandpaper. *Closed coat sandpaper* generally provides a finer finish and is commonly used when wet sanding.

Mentioned in earlier chapters, *grit sizes* vary from coarse to micro fine grades. Sandpaper grits are ordered by number—the lower the number, the coarser the grit. Refer to Figure 14–6.

Here are some general uses for different sandpaper grits:

1. Very coarse papers, No. 40 to 80 grit, are used to remove paint and filler.

2. Medium grit papers, No. 320, No. 360, or No. 400 grit, are used to level and sand the gloss off an old finish to be refinished.

3. Very fine and ultra fine grits are used primarily for color sanding.

4. Compounding papers, the No. 1250, No. 1500, and No. 2000 grits, are used to solve problems on newly painted surfaces.

Generally, select the finest grit paper that will do the job. Fine paper may remove material more slowly, but it will leave a smoother surface.

A common mistake is to use too coarse a sandpaper. Coarser paper will sand more quickly, but will leave deep sanding marks that can show up in the new paint job. Finer paper would have to be

	SANDPAPER GRIT SIZES			
Grit	Aluminum Oxide	Silicon Carbide	Zirconia Alumina	Primary Use for Collision Repair
Micro fine	—	2000 1500 1250	—	Used for base coat-clearcoat paint system.
Ultra fine	—	800	—	Used for color sanding.
Very fine	—	600	600	Used for color sanding. Also for sanding the paint before polishing.
	400 320 280 240	400 320 280 240	400 — 280 240	Used for sanding primer-surfacer and paint prior to painting.
	220	220	—	Used for sanding of topcoat.
Fine	180 150	180 150	180 150	Used for final sanding of bare metal and smoothing old paint.
Medium	120 100 80	120 100 80	— 100 80	Used for smoothing old paint and plastic filler.
Coarse	60 50 40 36	60 50 40 36	60 — 40 —	Used for rough sanding plastic filler.

Figure 14–6 Study uses of different sandpaper grits. You must know which coarseness of sandpaper to use for each task.

used to sand the surface down to remove these scratches.

Power Grinding and Sanding

Power grinding is done to quickly remove large amounts of paint and other materials. An air grinder is one of the fastest methods to remove material.

Power sanding normally uses an air sander to begin smoothing operations. Often, a dual-action sander is used to level surface imperfections.

DANGER! Always wear the proper dust mask when sanding. The dust can be harmful if inhaled.

When power sanding, replace the sandpaper when paint begins to cake or "ball up" on the paper. This paint buildup can scratch or gouge the surface and reduce the sanding action of the disc. Slowing down sander speed will also help prevent paint buildup and prolong sandpaper life.

Types of Sanding

There are several types of sanding that a painter must master. Most are done during the surface preparation stage, but are performed after priming.

Remember, the smoother the surface, the easier the refinishing work.

Bare metal sanding is done to smooth rough metal surfaces. If the metal work has been done properly, little sanding of bare metal should be required. But once in a while the metal is very rough from grinding or welding. In such cases, power sand it with No. 80 sandpaper to level out burrs, nibs, and deep scratches.

Paint sanding is needed when the finish is rough or in poor shape and to level and smooth primed areas. Since the primer is primarily intended to fill low spots and scratches, thorough sanding must be done to leave material in the low spots and cut away the high spots. Block sanding is highly recommended for this purpose.

Sanding Methods

Sanding can be done with a sanding block or by using power equipment. Most heavy sanding—such as removal of the old finish—is done by power sanders. Delicate final sanding operations are usually done by hand with a sanding block. Hand and power sanding can also be done wet (with water) or dry (without water). Sandpapers are available in dry or wet-or-dry types.

Block Sanding. *Block sanding* is a simple back-and-forth action with the sandpaper mounted on a blocking tool. It is often used on flat surfaces. Block sanding helps level the surface or make it flat.

Use the following general steps when block sanding:

1. Use sandpaper that fits the sanding block. If needed, cut the sheet of sandpaper to fit the block. Fit the sandpaper under the ends of the block.

2. Hold the block flat against the surface. Apply even, moderate pressure along the length of the

Figure 14–7 Dry block sanding works well when using coarser papers. Dry sanding will quickly clog finer papers. Never sand flat surfaces by hand without using a block.

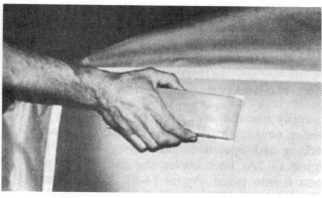

A

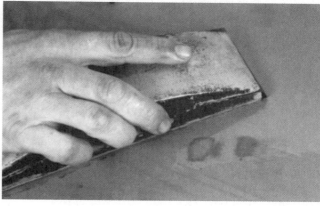

B

Figure 14–8 (A) A soft, flexible block should be used on curved panels. The soft block will flex to match the surface. (B) Note how wet sanding has been used to feather a small chip in the paint. Water will help flush away debris and allow finer sandpaper to still cut quickly.

block. Sand back and forth with long, straight strokes, Figure 14–7.

3. Do NOT sand in a circular motion. This will create sand scratches that might be visible under the finish. To achieve the best results, always sand in the same direction as the body lines on the vehicle.

4. To sand convex or concave panels, use a flexible backing pad, Figure 14–8. You can also use the side of your hand to hand sand concave (curved inward) surfaces. You can use the palm of your hand to hand sand convex (curved outward) surfaces. You must hold your hand to match the shape of the surface.

> *WARNING! Never try to sand a flat surface while holding the sandpaper in your hand. An irregular surface will result.*

5. Carefully find and sand scratches using a finer grit paper.

6. When block sanding primer or filler, make certain to sand the area until it feels smooth and level. Rub your hand or a clean cloth over the surface to check for rough spots.

Remember! If you can feel the slightest bump or dip in the surface, it will show up after painting. A surface will look smooth when sanded dull or when in primer. However, a shiny, reflective paint will exaggerate or magnify any surface roughness.

Dry and Wet Sanding. Block sanding can be done wet or dry. *Dry sanding* is often done with coarser nonwaterproof sandpaper, without using water. It is commonly done to remove material quickly. The

majority of two-component primer-surfacers can be sanded dry.

Wet sanding is done with finer waterproof sandpaper, using water to flush away sanded particles. Some primers are wet sanded. Clearcoats are sanded with sandpaper as fine as 2000 grit. Wet sanding reduces the problem of paper clogging. See Figure 14–8.

Wax and silicone can penetrate beneath the surface. Because this contamination is NOT easily detected, you might want to add wax and grease remover or soap to the water when wet sanding. Look at Figure 14–9.

Scuff Pads

Scuff padding is done with a nylon pad on hard-to-reach areas to clean and scuff the surface. Scuff padding is often done inside door jambs, around hood and trunk openings, and other restricted areas.

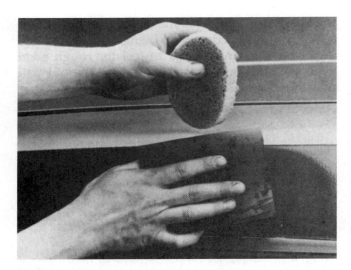

Figure 14–9 When wet sanding, dip the sponge in a bucket of water. Squeeze the water out of the sponge and over the area being sanded. *(Carborundum Abrasives North America)*

Since the scuff pad is flexible, it will scuff irregular surfaces easily.

Paint scuffing is needed to make sure the new paint bonds properly. If you paint over a smooth, hard surface, the new paint will not have anything to bond to and it will peel off. Paint scuffing can be done with fine sandpaper or a scuff pad, wet or dry.

When scuff pad sanding, you can place the pad over a sanding block on flat surfaces. You can also use your hand to hold and shape the pad on curved or restricted areas.

> **NOTE!** Scuff pad coarseness ratings can vary. Refer to manufacturer data to obtain the coarseness rating needed.

Abrasive grinding discs are used for jobs such as grinding off corrosion and removing paint. They are available in many grit sizes and in diameters of 3 to 9 inches (76 to 229 mm). The grinding disc is first assembled to the backing plate and then the disc/plate assembly is attached to the grinder.

FEATHEREDGING

If a new coat of paint is applied over a broken area of the old finish, the broken film will be very noticeable through the topcoat. The broken areas must be featheredged. **Featheredged** means the sharp edge of the broken paint film is gradually tapered down by sanding. Then the bare metal areas are filled with a primer and the entire area is sanded smooth and level.

Hand featheredging with a sanding block is usually a two-step procedure:

1. Cut down the edges of the broken areas with a coarse No. 80 grit paper and follow with No. 220 grit.

2. Complete the taper of the featheredge with either a No. 320 or No. 400 grit paper and water. This will produce a finely tapered edge and eliminate coarse paper scratches.

When **featheredging** with an air sander, use a dual action sander with a flexible backing pad. Use a No. 80 grit for the rough cut, followed by a No. 180.

Start by positioning the sanding disc flat against the work surface. Using the outer edge, or approximately 1 inch (25 mm) of the sanding disc, cut away the rough paint edges. Do NOT hold the sander at an angle greater than 10 degrees from the surface or it will cut a deep gouge in the paint.

After initially leveling the rough paint edges, lay the sander flat on the panel. Finish tapering the paint layers by moving the sander back and forth in a cross-cutting pattern. Start over the chipped area and work in an outward direction. Stop frequently and run your hand over the sanded area to feel for rough edges. When the surface feels smooth where the old paint and primer meet, featheredging is complete.

> *WARNING! A common mistake when using a DA is to tilt the sander to remove small surface imperfections. This will cause an indentation in the surface that will show up after painting. Normally, hold the sander flat on the surface to avoid this problem.*

Remember! The dust from sanding and grinding operations can often cause a major health problem. As shown in Figure 14–10, there are dust control systems available.

UNDERCOAT SYSTEMS

Proper **undercoats** of primer form the foundation for an attractive, durable topcoat of paint. If the undercoat—or combination of undercoats—is NOT correct, the topcoat appearance will suffer, possibly cracking or peeling.

Undercoats can be compared to a sandwich filler that holds two slices of bread together. The bottom slice of bread is called the substrate (or surface of the vehicle). That surface can be bare metal or plastic, or a painted or preprimed surface. The undercoat is the

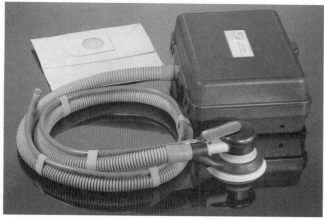

A

B

Figure 14–10 To protect your lungs from airborne dust from sanding, always work in a well-ventilated area or use a dust collection system on sanding equipment. *(Courtesy Dynabrade™, Inc.)* (A) This is a complete portable sander-dust collection system. (B) During use, sanding dust is pulled through a hose by a vacuum source and captured in a disposable bag.

sandwich filler applied to the substrate. It makes the substrate smooth and provides a bond for the topcoat. The upper slice is the topcoat, the final colorcoat that the customer sees.

There are four basic types of undercoats:

1. Primer
2. Primer-surfacer
3. Primer-sealer
4. Sealer

Most surfaces must be undercoated before refinishing for several reasons:

1. To fill scratches.

2. To provide a good base for the topcoat.

3. To promote adhesion of the topcoat to the substrate.

4. To provide corrosion resistance.

Primer-surfacers are used to provide both priming and filling in one step. *Primer-sealers* are applied to prevent solvents in the topcoat from being absorbed into the porous primer-surfacer. These three undercoats—*primer,* primer-surfacer, and primer-sealer—can be used together, singularly, or in various combinations. This will depend on the surface condition and size of the job.

Sealers are used to improve adhesion between the old and new finishes. To provide good adhesion, use a sealer over an old lacquer finish when the new finish is to be enamel. Under other conditions, a sealer is desirable but NOT absolutely necessary.

Before applying any undercoat, be sure to treat all bare metal with metal conditioner. Reduce the undercoat according to the manufacturer's instructions. Be careful to select the proper solvent for the weather conditions and mix the materials thoroughly.

Apply the first coat of undercoat and allow it to flash dry. Follow the recommendations on the label for flash time. Then apply two or three more medium wet coats for additional film buildup. Again, allow flash time between each coat. When making a spot repair, extend the additional coats several inches out from the first coat. See Figure 14–11.

Allow the undercoat to dry thoroughly. After the undercoat is dry, block sand the area until it is smooth. For best results, use No. 320 grit sandpaper. If very fine scratches still appear, another coat of undercoat might be all that is required to fill them.

Figure 14–11 Spray primer over repair areas before final sanding. Sand the primer with extra fine sandpaper to final check the surface for imperfections.

USING SPOT PUTTY

As discussed earlier in the book, during vehicle prep you may find small scratches or pits that require additional filling with glazing putty. Glazing or spot putty is used to fill small scratches and pinholes after priming.

Two-part (two-component) putties come with two ingredients that must be mixed to start the curing process. They include polyester putties (finishing fillers) and polyester primer-fillers. Both products must be mixed with hardener and can be applied to filler, bare metal, or painted surfaces.

Because today's two-component putties harden chemically, they cure quickly and can be primed and refinished without the worry of sand scratch swelling.

To apply the two-part spot putty, mix the two ingredients properly on a clean mixing board. Then use a rubber or plastic squeegee to apply the material to the small scratch or surface imperfection.

Allow the two-part putty to air dry or cure until it is hard. Refer to the instructions for drying/curing times. Test with a fingernail for hardness before sanding. If it is sanded too soon, the two-part putty will continue to shrink, leaving part of the surface imperfection unfilled. Once it hardens, dry sand the putty with No. 80 to 180 grit sandpaper. You can also wet sand the putty with No. 220 grit sandpaper.

After sanding the area, clean the surface and then re-prime. If the putty has been wet sanded, make sure to dry the surface thoroughly before applying primer.

For more information on these topics, refer to Chapters 8 and 10.

USING A GUIDE COAT

A ***guide coat*** is a very thin coat of primer or paint that assists in pointing out minor high and low spots. A guide coat is applied by spraying a very light coat of a different color material over the repair area. Sanding reveals the high spots and the low spots by contrasting the colors of the guide coat and the material under it.

A ***surface high spot*** shows up when sanding quickly cuts through the guide coat. A ***surface low spot*** shows up when sanding will NOT remove the guide coat.

When the surface is flat and ready for painting, the guide coat will sand off evenly or at the same time. Minor surface irregularities can be corrected by applying coats of primer and re-sanding. Two-component spot putty can be applied to fill deeper low spots and pits.

Light sanding or **scuffing** should be done on all areas where the old finish is in good condition. The

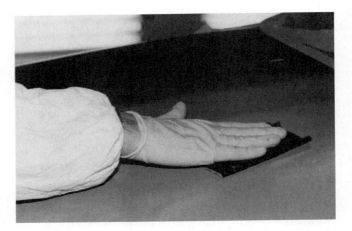

Figure 14–12 Wet sand the primer with extra fine sandpaper to check your work. Note how this technician is holding the hand and fingers flat to help sand the area flat.

purpose is to partially reduce the paint gloss to improve adhesion. Use a dual action sander, or use a sanding block. Never use a rotary disc grinder or sander.

For the basecoat/clearcoat finishes, proper surface preparation is especially critical. It is important to sand all surfaces to be refinished with No. 400 grit or finer paper, followed by 500 grit. Sanding can be done wet or dry. Wet sanding is much faster and requires less sandpaper with finer grits. Refer to Figure 14–12.

Remember! Every square inch or square centimeter to be painted must be sanded. If you fail to scuff any glossy surface thoroughly, the paint can peel off.

It is a good idea to carefully inspect the vehicle surfaces after sanding and scuffing. If you can see any shiny area on the old paint, sand or scuff that area. This will ensure that the new finish adheres properly.

MASKING

Masking keeps paint mist from contacting areas other than those to be refinished or painted. It is a very important step in the vehicle preparation process. Masking has become even more important because of two-component type paints. Once these paints dry, the overspray CANNOT be removed with a thinner or other solvent. It must be removed with a rubbing compound or by other time-consuming means.

There are several ways of masking the parts of a vehicle:

1. Masking tapes (many types and sizes). Only automotive-grade masking tapes should be used.

2. Masking paper of varying widths.

3. Plastic sheeting.

4. Special shaped cloth or plastic covers (for wheels, antenna, rearview mirrors, etc.).

5. Liquid masking material.

6. Masking foam.

> *WARNING! Traditional masking products may not work with waterborne paints because of their high water content. If using waterborne products, check with your supplier for masking papers and tapes made for use with them.*

Using Masking Paper and Tape

Masking paper is a roll of paper designed to cover parts of a vehicle. Automotive masking paper comes in various widths—from 3 to 36 inches (75–915 mm). It is heat resistant so that it can be used safely in heated booths. It also has good wet strength, freedom from loose fibers, and resistance to solvent penetration.

> *WARNING! Never use newspaper for masking a vehicle since it does NOT meet any of these requirements. Newspaper also contains printing inks that are soluble in some paint solvents. These inks can be transferred to the underlying finish or bleed into the new paint, causing staining.*

Masking plastic comes in large sheets for covering large areas of the vehicle. Masking plastic should be used away from the area to be painted. You do not want the paint spray to hit the plastic because paint can drip off the plastic and onto the paint surface.

Masking foam or ***masking rope*** is a self-stick foam rubber cord designed for quickly and easily masking behind doors, hoods, gas cap lids, and other panels to block overspray.

Masking tape is very sticky paper tape designed to cover small parts and also to hold the masking paper in place. Automotive masking tape comes in various widths—from ¼ to 2 inches (6.4 to 50 mm). Larger width tapes are used only occasionally.

Refinishing masking tape should NOT be confused with tape bought in hardware or house paint

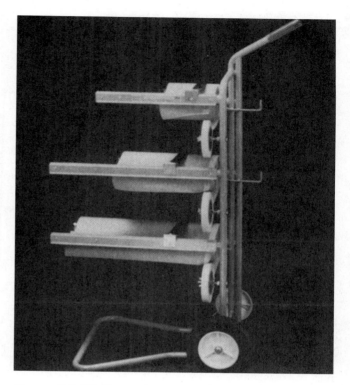

Figure 14–13 The masking paper dispenser is a time saver. This unit has different widths of paper on it.

stores for home use. This kind of tape will NOT hold up to the demanding requirements of automotive refinishing.

A ***masking paper dispenser*** applies tape to one edge of the paper as the paper is pulled out, Figure 14–13. This saves time when masking any vehicle. The average size vehicle can take two rolls of tape to be completely masked. The use of masking paper and tape dispensing equipment makes it easy to tear off the exact amount needed.

Fine line masking tape is a very thin, smooth surface plastic masking tape. Also termed **flush masking tape,** it can be used to produce a better ***paint part edge*** (edge where old paint and new paint meet). When the fine line tape is removed, the edge of the new paint will be straighter and smoother than if conventional masking tape were used. See Figure 14–14.

Masking Covers

Masking covers are specially shaped cloth or plastic covers for masking specific parts. There are several types of masking covers available. They can save time when masking.

One is the ***tire cover*** that eliminates the need for masking off the tire. Others include a body cover and a frame cover. Covers are also available in a variety of

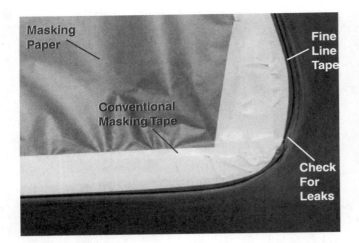

Figure 14–14 Note how a technician has applied fine line masking tape first and has then gone over it with conventional masking tape. This will let you have better control of fine line tape application for more accurate masking. Always check sharply curved masked areas for leaks. When the tape is curved, it can lift up and allow overspray to go under the tape.

sizes and shapes to mask headlights, taillights, antennas, mirrors, etc.

Masking Procedures

Before any types of masking materials are applied, the vehicle must be completely cleaned and all dust blown away. Masking tape will NOT stick to surfaces that are NOT clean and dry. It is important that the tape is pressed down firmly and adheres to the surface. Otherwise, paint will creep under it. See Figure 14–15.

In the case of a two-color job where the color break is NOT hidden by a stripe or molding, use fine line masking tape and press its edge down firmly.

There are no clear ground rules on when to mask and when to remove parts. This would include parts like trim, moldings, door handles, etc. The decision to remove or mask depends upon the design of the vehicle, and the expectations of the customer. If a part can be removed easily, this is better than trying to mask around the part. Also, if you cannot sand and clean right up to the part, the part should be removed. If it will be difficult to mask the piece, you might save time removing the part from the vehicle. Each part will require an individual decision.

Removal of trim and moldings is more often necessary when using a base/clear system than with a single stage finish. The film buildup can be greater with base/clears. This additional thickness makes the paint edge more likely to crack or chip.

It is wise to completely clean and detail the vehicle before masking and again after the refinishing job

Figure 14–15 This technician is masking the whole side of a vehicle before painting the inside of the door jamb.

is completed. This is because masking over a dirty vehicle can cause a "dirty paint job!"

If the painting environment is cold and damp, the masking tape may NOT stick to the glass or trim parts. Condensation on them can prevent the tape from sticking properly. Wipe the parts off before masking them. When masking door jambs, be sure to cover both the door lock assembly and the striker bolt. They can become filled and clogged with paint if not masked.

Although masking tape is elastic, do not stretch the tape when making a straight line, Figure 14–16. Use masking tape such as 3M Fine Line to cover tightly curved surfaces. This is especially true when masking newly applied finishes that are still soft underneath. It is also wise to avoid stretching the tape because this can increase the degree of tape marking on the finish.

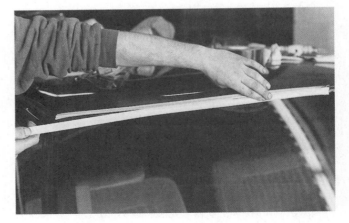

Figure 14–16 When using masking tape, pull lightly on the tape with one hand without stretching the tape. Moving the hand up or down will control direction of the tape application. The other hand is used to press the tape down securely on the surface.

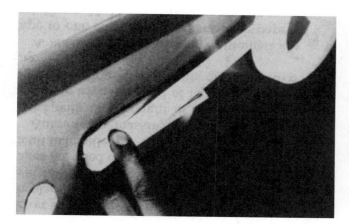

Figure 14–17 Small parts can be covered completely with masking tape. Make sure the edge of the tape does not cover the surface to be painted.

When applying masking tape, hold and peel the tape with one hand. Use your other hand to guide and secure the tape to the vehicle, Figure 14–17. This provides tight edges for good adherence. This also allows you to change directions and go around corners. See Figure 14–18.

To cut the tape easily, quickly tear upward against your thumbnail, as shown in Figure 14–19. This permits a clean cut of the tape without stretching.

WARNING! Be careful that the masking tape does NOT overlap any area to be painted. After painting, you can remove paint from parts but you cannot add missing paint to the body. Loop or overlap the inner tape edge to follow curves. The tape will stretch to conform to the curves. Difficult areas such as wheels can be masked using this process, but more often covers are used to save time.

Before masking glass areas, remove accessories such as wiper blades. The wiper shafts can be protected in the same manner as radio antennas and door handles. Use tape or special covers.

To mask glass, first apply tape along the very top and along the edges of the moldings. Then, use two pieces of masking paper to cover the glass, Figure 14–20. Overlap the tape on the paper with the tape already on the molding. The top piece of paper should overlap the bottom piece of paper. If necessary, fold and tape any pleats in the paper to prevent dust and paint seepage.

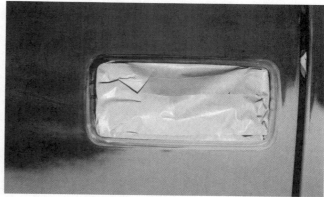

A

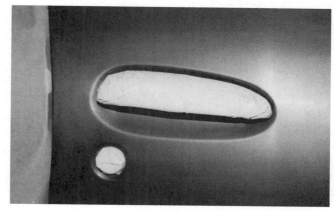

B

Figure 14–18 (A) This door handle has been removed and its hole masked closed. This is the best way to paint a door because the paint will deposit under the door handle. (B) Higher costs prevented the technician from removing this door handle and lock before painting. The handle and lock had to be carefully masked.

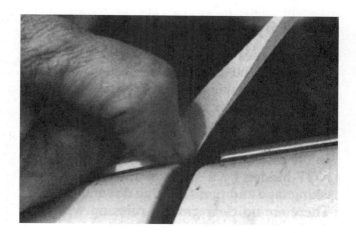

Figure 14–19 Tape can be cut or torn by holding the tape down while pulling sideways with the other hand.

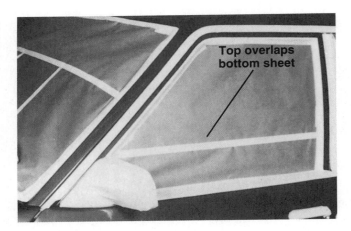

Figure 14–20 Two pieces of masking paper are often used to cover glass. Apply the lower sheet first; then the upper sheet. Tape over where the paper edges meet to prevent overspray leakage.

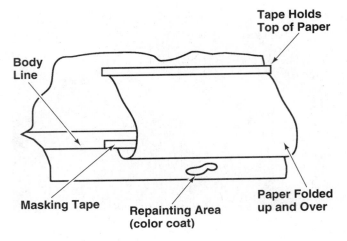

Figure 14–21 Reverse masking will help blend a repair area. Overspray will hit the folded-over paper and blend more smoothly into the old paint.

Overspray results when you do not seal the area NOT to be painted and paint/clear gets on an uncovered surface. You would have to then take the time to use solvent or compound to remove the overspray problem.

Fine line tape can be used to protect existing pinstripes from overspray. Also use fine line tape for precise color separation in two-color painting and for creating vivid, clean stripes. Its added flexibility makes painting of curved lines easier, with less reworking.

Double masking uses two layers of masking paper to prevent bleed-through or finish-dulling from solvents. It is needed when spraying horizontal surfaces (hood, trunk, etc.) next to other horizontal surfaces. Overspray will tend to soak through the adjacent masked area and onto the old paint.

Reverse masking or back taping requires you to fold the masking paper back and over the masking tape, Figure 14–21. It is often used during spot repairs to help blend the painted area and make it less noticeable. This also helps prevent bleed-through. The paper is taped on the inside and allowed to bellow slightly, which keeps it lifted a bit from the surface.

A door edge or body line next to a refinished panel can sometimes show a slight difference in an otherwise acceptable color match, particularly with metallics. This situation can be avoided by crossing the line and sanding the adjacent panel when preparing for refinishing. Then reverse mask the adjacent panel and refinish as desired.

If there is a slight difference after removing the masking paper, just paint across the line and blend in smoothly as with any spot repair. Reverse mask/back tape very carefully. Any overmasked or undermasked areas will lead to extra work after the vehicle is painted.

The most difficult area to mask can be door openings when painting the rocker panel and door pillars. You must carefully apply masking tape and paper to from a large unsupported mask to keep overspray out of the passenger compartment. Note how this was done in Figure 14–22.

Masking plastic is being applied to a vehicle in Figure 14–23. Note how paper covers the area next to the area to be painted.

Masking foam is being applied to the bottom of a hood in Figure 14–24A. This will keep overspray from going down into the gap between the hood and the fender. It will produce a soft edge of spray next to the self-stick, foam rope.

Figure 14–24B shows masking foam applied to the back of a gas cap cover. This too will quickly mask the gap between the parts to prevent a dull mist of overspray from depositing behind parts.

Figure 14–22 Note how this door opening has been masked. This is one of the most difficult masking jobs since a large unsupported opening must be masked over to protect the interior of the vehicle.

Figure 14–23 Masking plastic is used on large areas away from panels being painted. Note how paper is applied next to the fender to be painted. Plastic is being cut off and is draped over the other side of the vehicle.

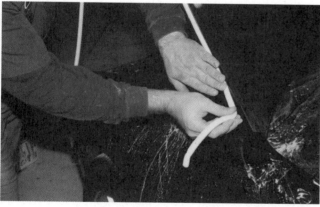

A

B

Figure 14–24 (A) A technician is applying self-stick foam rope or cord to the bottom of the hood lip. When the hood is closed, the rope will mask and prevent paint from entering the gap between the hood and fender. (B) Masking rope has been applied behind the gas cap door to keep overspray off surfaces under the door.

Liquid Masking Material

Liquid masking material seals off the vehicle to protect undamaged panels and parts from overspray. Liquid masking is used on areas where masking is necessary but difficult to apply, including wheel wells, headlights, grille, underbody chassis, and even the engine compartment.

Masking liquid, also called masking coating, is usually a water-based sprayable material for keeping overspray off body parts. Some are solvent-based. Masking liquid comes in a large, ready-to-spray container or drum. These materials are sprayed on and form a paint-proof coating over the vehicle.

Some masking coatings are tacky. They form a film that can be applied when the vehicle enters the shop. Others dry to a hard, dull finish.

Masking coatings can be removed when the vehicle is ready to return to the owner. It washes off with soap and water. Local regulations may require that liquid masking residue be captured in a floor drain trap, and not put into the sewer system.

To mask a vehicle using the liquid masking system, proceed as follows:

1. Partially mask the area to be painted by going around it with masking paper. Fold the paper over onto the area to be painted, Figure 14–25. Secure the paper with masking tape.

2. Apply the liquid masking material. Use a heavy, single overlapping coat. Apply the material to all surfaces not to be painted. This would include bumpers, grilles, doors, windshields, body panels, wheels, wheel wells, door jambs, and even the entire engine compartment, Figure 14–26.

An airless spray system is generally recommended for applying the masking material.

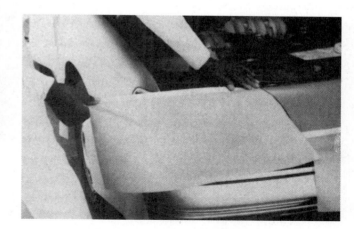

Figure 14–25 Partially mask around the area to be painted. Fold paper over the area to be painted.

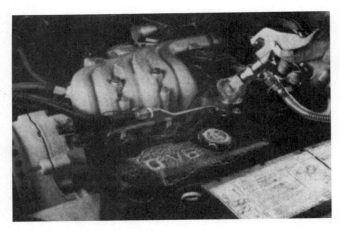

Figure 14–26 Spray liquid mask onto areas NOT to be painted.

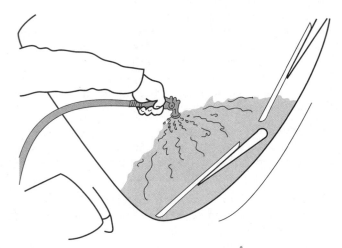

Figure 14–27 After the paint has hardened sufficiently, wash off the liquid masking material.

3. Fold the masking paper back over the liquid masking material. Wipe away any material from the area to be painted with a damp sponge. Allow the surface to dry.

4. Prepare the surface. Then apply primer and paint according to the manufacturer's instructions.

5. Allow the paint to dry, then unmask the vehicle. Liquid masking may be used in both air dry or bake conditions.

6. After the paint is cured, wash off the dried liquid masking material with a garden hose or pressure washer, Figure 14–27.

Summary

- Vehicle preparation involves all of the final steps prior to painting, including cleaning, sanding, stripping, masking, priming, and other related tasks.

- Open coat sandpaper is good for sanding softer materials, such as paint, body filler, plastic and aluminum. Closed coat sandpaper generally provides a finer finish and is commonly used in wet sanding.

- Proper undercoats of primer are the foundation for an attractive, durable topcoat of paint. If the undercoat—or combination of undercoats—is NOT correct, the topcoat appearance will suffer, possibly cracking or peeling.

- Masking keeps paint mist from contacting areas other than those to be refinished or painted. Some masking materials include masking paper and tape, masking foam, plastic sheeting, cloth, plastic covers, and liquid masking material.

Technical Terms

vehicle preparation	*wet sanding*
initial washing	*scuff padding*
surface evaluation	*paint scuffing*
paint film breakdown	*featheredged*
paint adhesion check	*featheredging*
preexisting damage	*undercoats*
acid rain	*primer*
industrial fallout	*primer-surfacer*
hard water spotting	*primer-sealer*
damage	*sealer*
UV radiation damage	*two-part (two-*
chemical stripping	*component) putty*
abrasive blasting	*guide coat*
machine grinding	*surface high spot*
bare metal treatment	*surface low spot*
wash primer	*masking*
self-etching primers	*masking paper*
new part primer coat	*masking plastic*
sanding	*masking foam*
coated abrasives	*masking rope*
open coat sandpaper	*masking tape*
closed coat sandpaper	*masking paper dispenser*
grit sizes	*fine line masking tape*
power grinding	*paint part edge*
power sanding	*masking cover*
bare metal sanding	*overspray*
paint sanding	*double masking*
block sanding	*reverse masking*
dry sanding	*liquid masking material*

Review Questions

1. *Technician A* uses a circular motion when block sanding. *Technician B* block sands in a back-and-forth direction. Who is correct?
   a. Technician A
   b. Technician B
   c. Both A and B
   d. Neither A nor B

2. Paint film breakdown will show up as _____, _____, or _____.

3. How do you do a paint adhesion check?

4. _____ are used to provide both priming and filling in one step.

5. Which of the following statements concerning masking is INCORRECT?
   a. Newspaper works well and saves money as paint masking paper.
   b. Removal of trim, as opposed to masking, is preferred in some instances.
   c. Masking tape will NOT stick to surfaces that are NOT clean and dry.
   d. Using plastic covers on tires eliminates the need for masking them.

6. List four ways of masking the parts of a vehicle.

7. Name three common ways of stripping paint from metal surfaces.

8. Define the term "abrasive blasting."

9. Why is plastic bead blasting often preferred over sandblasting?

10. _____ _____ etch the bare metal to improve paint adhesion and corrosion resistance, while providing the priming and filling properties of primer-surfacer.

11. Explain four ways that sanding prepares the surface for painting.

12. Describe four general uses for different sandpaper grits.

13. How do you use a guide coat?

ASE–Style Review Questions

1. *Technician A* says to wipe the vehicle down with wax and grease remover during vehicle prep. *Technician B* says to wipe the wax and grease remover off while it is still wet. Who is correct?
   a. Technician A
   b. Technician B
   c. Both Technicians A and B
   d. Neither Technician

2. When inspecting a vehicle before repairs, *Technician A* says to check for environmental damage. *Technician B* says to check the entire repair area for imperfections. Who is correct?
   a. Technician A
   b. Technician B
   c. Both Technicians A and B
   d. Neither Technician

3. *Technician A* says that some paint removers can damage plastic body panels. *Technician B* says that sand blasting is better than plastic media blasting on aluminum. Who is correct?
   a. Technician A
   b. Technician B
   c. Both Technicians A and B
   d. Neither Technician

4. When sanding off old paint down to the metal, which of the following grits of abrasive should be used?
   a. 24 grit
   b. 40 grit
   c. 100 grit
   d. 400 grit

5. *Technician A* says to use the finest sandpaper possible for adequate material removal. *Technician B* says that this would be too slow. Who is correct?
   a. Technician A
   b. Technician B
   c. Both Technicians A and B
   d. Neither Technician

6. Which of the following should a technician use to final sand a curved body surface?
   a. Sanding board
   b. Hard rubber block
   c. Soft, flexible sanding block
   d. Air file

7. Which of the following is often used to final check the repair area for low and high spots?
   a. Straightedge
   b. Micrometer
   c. Guide coat
   d. Color coat

8. Which sandpaper grit would be used to dry sand and featheredge a repair area?
   a. 36 or 40 grit
   b. 40 or 80 grit
   c. 80 or 120 grit
   d. 180 or 240 grit

9. A technician finds very small flaws in the sanded repair area. Which product would help level these areas the most?
   a. Primer
   b. Primer-seal
   c. Primer-filler
   d. Primer-putty

10. *Technician A* says to only use masking paper away from the area to be painted. *Technician B* says to use masking plastic right next to the area to be painted. Who is correct?
    a. Technician A
    b. Technician B

c. Both Technicians A and B
d. Neither Technician

Activities

1. Visit a body shop. Ask the owner to let you watch a worker prepare a vehicle for painting.

2. Make a chart showing the steps for preparing a vehicle for painting.

15

Shop and Equipment Preparation

Objectives

After reading this chapter, you should be able to:

- Describe the recommended maintenance program for a spray booth.

- Explain the importance of proper material atomization, viscosity, and temperature.

- Measure the viscosity of material using a Zahn cup.

- Describe the advantages of a captive spray gun system.

- Operate and maintain a spray gun.

- Complete a spray pattern test.

- Explain the use of electronic or computerized mixing scales.

- Adjust a spray gun to prepare for refinishing a vehicle.

Introduction

To complete a professional painting/refinishing job, your shop and equipment must be in perfect condition! A dirty spray booth, a poorly maintained spray gun, contaminated air supply, and other avoidable situations will all ruin your work. Sloppy shop conditions will usually result in a sloppy paint job.

This chapter summarizes the steps needed for preparing the finishing equipment and spray area for spraying the topcoat. There are a number of shop and equipment variables that affect the refinishing operation. These variables include the spraying environment, as well as the spraying equipment and their adjustments. ALL are important. You must pay close attention to these variables because they can affect the quality of your work. See Figure 15–1.

Remember! A professional painting technician will spend more time maintaining the shop and equipment than on any other single task. Spraying the vehicle takes only a very short amount of time by comparison.

NOTE! You should have studied Chapters 3, 5, 6, and 8 before beginning this chapter. These chapters give vital information for comprehending this material.

Figure 15-1 The cleanliness of your prep area, spray booth, and shop in general affects the quality of the finished product. Always keep all areas clean and organized.

PAINTING ENVIRONMENT

A painting technician must have a suitable work environment. Automotive refinishing materials must be kept free of dust and dirt during spraying and while drying. In fact, today's clearcoat finishes will magnify any dust, dirt, or other flaws in the finish.

The proper *painting environment* must address these variables:

1. Cleanliness—to keep dirt out of paint

2. Temperature/humidity—to provide proper paint curing or drying conditions

3. Light—to properly light vehicle and paint as it is applied

4. Compressed air/pressure—to send clean air, at the right pressure, to the spray gun

There are many ways of keeping dirt from becoming a problem in a finish. During body repairs, dirt can be controlled by:

1. Using a dustless, or vacuum, sanding system

2. Cleaning the vehicle before bringing it into the shop

3. Having the paint prep area separate from the body repair area

4. Using prep stations

5. Maintaining the compressed air system's filters and water traps

Before beginning a refinish operation, you must **prep** (prepare) the equipment and painting area. As diagrammed in Figure 15–2, you should:

1. Clean the spray booth.

2. Check the spray booth filters for clogging and replace them if needed.

3. Replace any burned out bulbs.

4. Drain any moisture from the compressed air system. Check filters and dryers.

PREPARING SPRAY BOOTH

Improper booth cleanliness may result in foreign matter entering the fresh finish. This can require you to sand out the debris and redo the area. You will waste

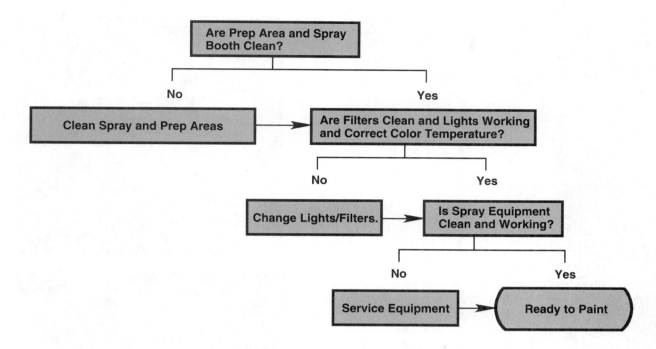

Figure 15–2 This chart shows the major steps for preparing equipment and the painting area.

time and materials if you try to paint a vehicle in a dirty environment.

The **spray booth** must provide the technician with:

1. Clean environment, Figure 15–3.

2. Constant temperature. The materials, vehicle, and airflow should be at room temperature during spraying.

Figure 15–3 Spray booth provides controlled, clean conditions for painting. *(Photo courtesy of Team Blowtherm)*

3. Daylight-corrected lighting.

4. Proper air movement for removing hazardous fumes and overspray, and for correct drying and curing of the paint film.

Before pulling the vehicle into the booth, make sure the area is perfectly clean and organized. Routine maintenance is necessary for proper spray booth performance. A regular program of cleaning, filter replacement, examining seals and lighting is important for the spray booth to operate properly.

Daily spray booth cleaning includes these tasks:

1. Remove all masking paper, empty cans, and other nonessential items. Clean up scrap, masking paper, rags, and so forth. Maintain and clean ALL equipment used in the booth.

2. Vacuum the booth, Figure 15–4. The collected material should be stored in a fireproof container. Keep the booth free of dirt and overspray at ALL times. Floors and walls should be cleaned after every job.

3. To help keep any stray dust down, wet the spray booth floor before spraying.

4. Drain oil and water filters and traps.

5. Clean the air hoses by wiping them with a damp rag or tack cloth. Clean ALL air hoses several times a day. Accumulations of dirt and dust on their outer surfaces might drop off onto the freshly painted surfaces.

Figure 15–4 Before pulling a vehicle into the spray booth, make sure the booth is perfectly clean. Remove all unneeded objects, clean surfaces, and vacuum the booth. *(Photo courtesy of Team Blowtherm)*

6. Inspect filters and replace them if necessary.

The continuous flow of air through the booth will eventually load the filters with dirt and overspray, Figure 15–5. These filters should be inspected and replaced at the recommended intervals. When they are replaced, the filters recommended by the booth manufacturer should be used, Figure 15–6. Several types of filtration systems are available. See Figure 15–7.

Figure 15–6 Filters should be checked and changed periodically. (A) These are ceiling filters in a spray booth. (B) These are floor gratings over filters.

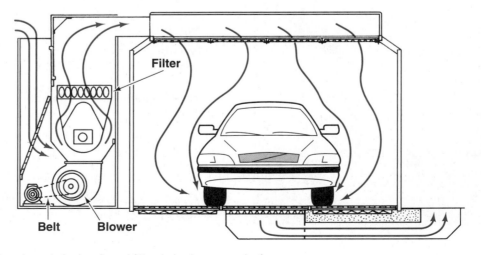

Figure 15–5 Airflow through the booth and filters helps keep any dirt from entering wet paint.

Wet or Wash Type

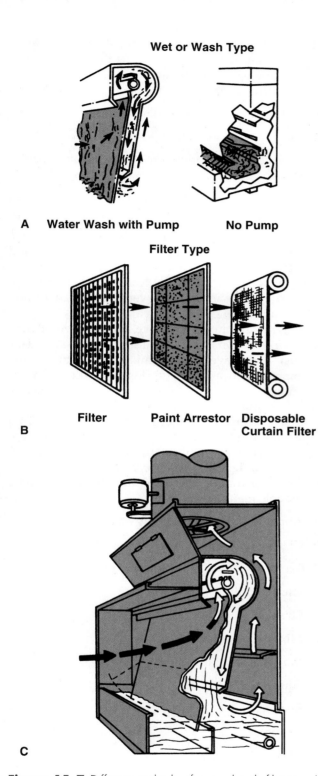

A Water Wash with Pump   No Pump

Filter Type

Filter   Paint Arrestor   Disposable Curtain Filter

B

C

Figure 15–7 Different methods of spray booth filtration: (A) Wet filtration. (B) Dry filtration. (C) A combination of water spray and centrifugal separation.

Monitor the manometer readings daily and know what a normal reading should be. The *manometer* indicates when the intake filters are overloaded. Some booths have a pressure switch that shuts off the air

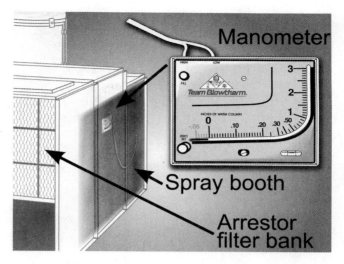

Figure 15–8 A manometer is mounted on the spray booth. It will tell you when filters are becoming clogged, requiring replacement. *(Photo courtesy of Team Blowtherm)*

supply and exhaust fan when the intake filter is clogged. Refer to Figure 15–8.

DANGER! *Clogged booth filters are a fire hazard because they could ignite under certain conditions.*

Weekly spray booth cleaning includes the following:

1. Vacuum the top of the spray booth and blow off exterior glass and walls.

2. Secure access doors and blow off interior walls and ceiling with the fan running.

3. Clean the light fixtures. Periodically check the lighting inside the booth and replace weak or burned-out bulbs. Improper lighting can lead to poor finishes.

To do a **monthly spray booth cleaning:**

1. Inspect the spray booth for air leaks and caulk and reseal when necessary.

2. Inspect and clean sprinkler heads. Replace plastic bags over sprinkler heads.

Annual spray booth cleaning includes these tasks:

1. Clean the plenum chamber and fan blade. Pay special attention to corners at the base of the chamber. Paint dust accumulation in this area creates a flammable situation.

2. Oil the plenum and fan blade.

3. Inspect the exhaust stack for paint buildup in the damper area. Excessive buildup here will prevent

Figure 15-9 After cleaning the spray booth thoroughly, you can pull the vehicle into the booth.

dampers (butterflies) from opening and closing properly.

Prep with Vehicle in Booth

After driving the vehicle into the spray booth, close the booth doors tightly, Figure 15-9. Turn on the air filtration system and allow it to run a few minutes. Then **tack rag** the entire vehicle to remove any loose dust and other debris on surfaces. This is needed before proceeding with the refinishing operation.

The booth doors must be kept tightly closed during spraying. If it becomes necessary to open a door, be sure the exhaust fan and air supply are turned off. In fact, many spray booths are equipped with door switches that automatically shut off the exhaust fan.

Personal cleanliness cannot be overemphasized. Never enter the booth wearing dirty clothes or shoes. The dirt can easily be transferred to the vehicle and into the finish. All clothing and gloves should be clean and lint-free.

MIXING AND STORAGE AREA PREP

The **paint mixing room** should provide a safe, clean, well-lit area for a painting technician to store, mix, and reduce paint materials. The mixing room should be located near the spray booth. It should also contain the mixing equipment if an intermix (in-shop mixing) system is used. The paint mixing room may also include a technician prep area. This area must be constructed to meet applicable building, fire, and electri-

cal codes. The paint mixing area must be kept clean to reduce the amount of dust and dirt that can be introduced to the paint and painter.

A **painter prep area** is a shop area that provides:

1. A clean area for a technician to dress
2. Clean storage for painting suits
3. A place to clean and store personal safety equipment
4. A good location to have emergency eye wash and first aid equipment

Dirt can enter the material as it is being mixed, so a clean, well-lit mixing area is required. Federal and local laws will dictate the type of mixing area the shop must have. In many cases, this means an enclosed space with its own ventilation system and explosion-proof lighting.

The mixing area must have adequate ventilation and fire protection (per federal, state, and local codes). It must also have proper mixing and measuring equipment such as paint shakers, paddle agitators, churning knives, paint mixing sticks, mixing cans, and other equipment.

Many refinishing materials are expendable. That is, they are used up in day-to-day shop activities. For this reason, it is very important to keep plenty of such items on hand. It is foolish to run out of a material while in the middle of a job. Have plenty of these expendable materials on hand:

1. Abrasive paper and sanding blocks
2. Clean wiping cloths or paper towels
3. Tack rags
4. Paint paddles
5. Strainers
6. Mixing buckets
7. Masking paper and tape
8. Squeegees
9. Reducers
10. Activators
11. Cleaning solvents

NOTE! For more information on materials, refer to Chapter 8.

PREPARING SPRAY EQUIPMENT

If spray equipment is NOT adjusted, handled, and cleaned properly, it will apply a defective coating to the surface. The most common problems and suggested remedies are given in the equipment service manuals. These variables can affect the appearance of the color and the outcome of the refinishing operation. When preparing the spray equipment, you must:

1. Clean the spray equipment.

2. Set up the spray equipment with the proper fluid needle, fluid nozzle, and air cap for the material to be sprayed.

3. Adjust the temperature in the spray booth if necessary.

4. Adjust the air pressure.

5. Make a test spray pattern.

Compressed Air System Preparation

The compressed air system can be a source of problems for the technician. **Air supply system problems** can introduce dirt, moisture, and oil into the air supply. From there, this contamination can get into the paint. To avoid air supply problems:

1. Check and replace oil and water filters and traps on a regular basis.

2. Drain moisture from the system daily. Draining the system in the morning allows more moisture to be removed because the air is cool and moisture has condensed.

3. Replace air hoses as necessary. Deteriorated air hoses can also introduce dirt into the system.

Air lines should slope away from the compressor. This allows moisture to collect away from the compressor. Air hose fitting should be at the top of the metal air line. This will reduce the amount of moisture getting into the air filters and hoses.

Atomization Is Important!

A thorough understanding of atomization is the key to using a spray gun correctly. **Atomization** breaks

primer, paint, or clear into a spray of tiny, uniform droplets. When properly applied to the vehicle's surface, these droplets flow together to create an even film thickness with a mirrorlike gloss. Proper atomization is essential when working with today's basecoat-clearcoat finishes.

Illustrated in Figure 15–10, atomization takes place in three basic stages.

In the first stage, the material passes through the fluid tip and is surrounded by air streaming from the annular ring. This turbulence begins the breakup of the material.

The second stage of atomization occurs when the material stream is hit with jets of air from the containment holes. These air jets keep the stream from getting out of control and aid in the breakup of the material.

In the third phase of atomization, the material is struck by jets of air from the air cap horns. These jets hit the material from opposite sides, causing it to form into a fan-shaped spray.

Basecoat-clearcoat finishes require very fine atomization. This is because the basecoat is so thin. Basecoat-clearcoat finishes will also not obtain proper hiding and coverage if they are NOT atomized properly.

In Figure 15–11, note the common types of spray guns. These were covered in detail in Chapter 5.

Spray Gun Parts Review

The principal components of a typical spray gun are illustrated in Figure 15–12.

The **spray gun air cap** directs the compressed air into the material to atomize it and form the spray pattern. Each of the orifices has a different function, Figure 15–13. The **center orifice** located at the fluid nozzle creates a vacuum for the discharge of the paint, primer, clear, etc. The **side orifices** determine the spray pattern by means of air pressure. The **auxiliary orifices** promote atomization of the paint.

Airflow through the two side orifices forms the shape of the spray pattern. When the **pattern control knob** is closed, the spray pattern is ROUND. As the

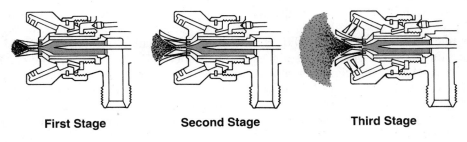

First Stage   **Second Stage**   **Third Stage**

Figure 15–10 Note the three stages of atomization shown by this cutaway view of a spray gun.

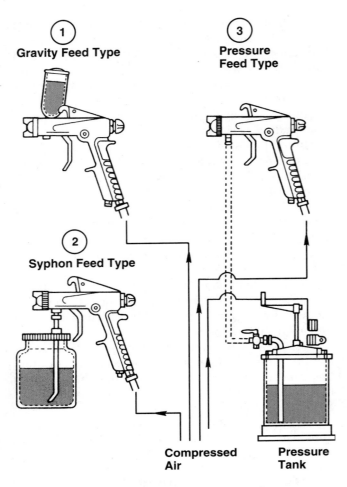

1 Gravity Feed Type

3 Pressure Feed Type

2 Syphon Feed Type

Compressed Air

Pressure Tank

Figure 15–11 Compare the four basic types of spray guns.

knob is opened, the spray becomes more OBLONG in shape.

The fluid needle and the fluid nozzle both meter and direct the flow of material from the gun into the air stream.

The *fluid nozzle* forms an internal seat for the fluid needle, allowing the *fluid needle* to open or shut off the flow of material. The amount of material that leaves the gun depends on fluid nozzle size and how far the needle is pulled back (adjusted). Fluid nozzles are available in a variety of sizes to properly handle paints of various types and viscosities.

The *fluid control knob* changes the distance the fluid needle moves away from its seat in the nozzle when the trigger is pulled. This controls flow.

Like the fluid valve, the air valve is opened by the trigger. The distance the air valve opens is regulated by a screw adjustment. When the trigger is pulled partway, the air valve opens. When it is pulled a little farther, the fluid nozzle opens.

Note how the pressure feed spraying system is hooked up.

1. Connect the regulated air hose from the air control device on the tank to the air inlet on the gun.

2. Connect the mainline air hose from the regulator to the air regulator inlet on the tank.

3. Connect the fluid hose from the fluid outlet on the tank to the fluid inlet on the gun.

Paint pressure tanks are available in sizes from two quarts to 10 gallons (2–38 liters). They are available in single, dual, and nonregulated models.

MIXING PAINT AND SOLVENT

When preparing for spraying, you must reduce the paint, primer, or clear to the proper viscosity. Directions on the can or product information sheet will explain reducing procedures.

When mixing and using paint and solvents or other additives, you must measure and mix their contents accurately. This is essential to doing high-quality paint work. You must be able to properly mix reducers, hardeners, and other additives into the paint. If you do NOT, serious paint problems will result.

Mixing instructions are normally given on the product information sheet or can. This may be a percentage or parts of one ingredient compared to the other.

A *percentage reduction* means that each material must be added in certain proportions or parts. For instance, if a paint requires 50% reduction, this means that one part reducer (solvent) must be mixed with two parts of paint.

Mixing by parts means that for a specific volume of paint or other material, a specific amount of another material must be added. If you are mixing a gallon of paint, for example, and directions call for 25% reduction, you would add one quart of reducer. There are four quarts in a gallon and you want one part or 25% reducer for each four parts of paint.

Proportional numbers denote the amount of each material needed. The first number is usually the parts of paint needed. The second number might be used to denote the amount of hardener or other additives required. The third number is usually the solvent (reducer).

For example, the number 4:1:1 might mean to add four parts of color, one part hardener, and one part solvent. For a gallon of color, you would add one quart of hardener and one quart of solvent. This can vary, so always refer to the exact directions on the paint can.

A *mixing chart* converts a percentage into how many parts of each material must be mixed. Study the percentages and parts of each material that must be mixed.

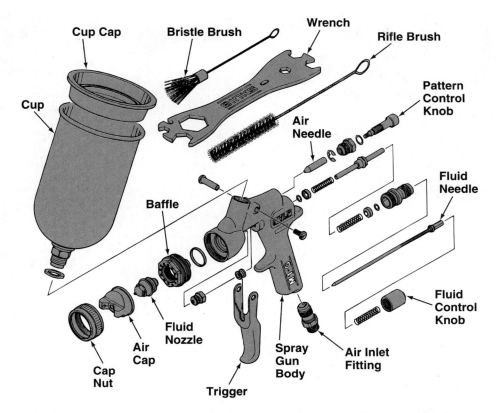

Figure 15-12 Study the basic parts of a typical gravity feed spray gun. The parts of other spray gun designs are similar. *(Courtesy of ITW Automotive Refinishing, 1-800-445-3988)*

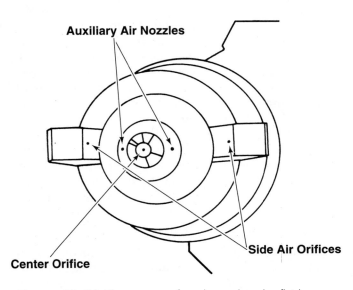

Figure 15-13 The center orifice, located at the fluid nozzle, creates a vacuum for the discharge of paint, primer, clear, etc. The side orifices determine the spray pattern by means of air pressure. Auxiliary orifices help atomize paint. *(PPG Industries, Inc.)*

Using Mixing Sticks

Graduated *mixing sticks* have scale markings that allow you to easily mix paint, primer, clear, solvent, and additives in the correct proportions. They are provided by the paint system manufacturer. Several are needed to provide mixing guides for each type of paint. Look at Figure 15–14.

Each mixing stick will have a ratio or percentage printed at the top. Measurement marks and numbers are placed along the tool for pouring out the correct quantity of each material quickly. Select the mixing stick with the correct ratio and type of material to be sprayed, Figure 15–15.

Obtain a can, pail, or container with straight sides. Tapered sides on the container would upset your measurements. It should be big enough to hold ALL paint, hardener, solvent, etc. needed for the job. A gallon or larger container saves mixing time for an overall or complete refinishing job. If you are only doing a small area and one cup will do, mix the materials in the spray gun cup. If you are using a pressure tank spray gun, measure and mix the materials in the tank.

Figure 15-14 Mixing sticks are provided by paint manufacturers for adding correct amounts of color, hardener, and solvent. They are designed to be used with their paint systems.

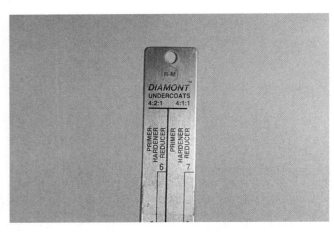

A

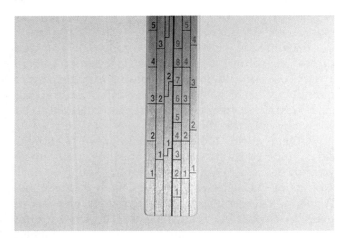

B

Figure 15-15 The top of the mixing stick will give mixing ratios. The ratio should be the same as the instructions given on the label of the material to be sprayed.

Figure 15-16 Place the correct mixing stick into a clean container for mixing the materials. The side of the container must not be tapered.

For this example, let us say that the mixing stick is for a mixing ratio of 4:1:1 for color, hardener, and solvent.

1. Place the correct paint mixing stick for the type of material into the container, Figure 15–16.

2. Pour the amount of paint or primer you might use into the bucket. Stop pouring when the paint is even with any of the numbers on the left of the stick. This might be 1 through 7, for example, depending on the quantity of paint needed.

   For more paint, you might fill to lines 6 or 7. For a spot repair, you might only need to make the material even with lines 2 or 3, for example. Make sure the paint is perfectly even with any of the numbers on the mixing stick. If color is listed on the left, the paint should be even with any number column in that column.

3. Pour in hardener until it is even with the same number (from step 2) in the next column on the stick. If the material already in the container aligns with 3, pour in hardener until it aligns with 3 but in the second column.

4. Pour in the final ingredient (usually solvent) until it aligns with the same number in the last column on the mixing stick. This is demonstrated in Figure 15–17.

After adding the correct amounts of each material, mix them thoroughly with your metal mixing stick. You can then fill your spray paint gun with properly mixed paint.

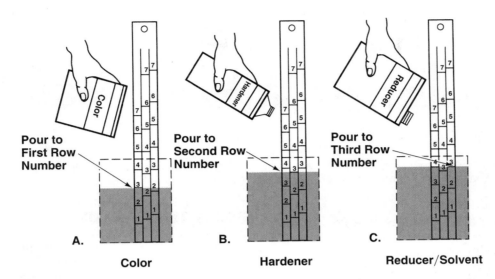

Pour to First Row Number

Pour to Second Row Number

Pour to Third Row Number

A. Color

B. Hardener

C. Reducer/Solvent

Figure 15–17 Study the basic steps for using a mixing stick. *(BASF)* (A) Depending upon information on top of the mixing stick, you usually pour in color or primer first. Pour in the amount of material needed for the job. Stop when you reach any of the numbers on the left column of the mixing stick. (B) If used, pour hardener or catalyst in next. Pour material into the container until it is even with the same number on the mixing stick, but in the next column. If the paint was even with number 3, pour in hardener to number 3 on the next, or center, column on the stick. (C) Pour in solvent until the liquid is even with 3 on the right-hand column of the stick. Pour all materials slowly so that you do not add too much. Stir the materials with a metal stick.

When filling your spray gun, always use a paint strainer over the cup. The **paint strainer** is a paper funnel-mesh strainer that keeps debris out of the spray gun. It should be used when anything is poured into your spray gun cup! Contaminants can accidentally get into new materials. Dirt and dust can also fall off the top of containers when you're pouring!

VISCOSITY

Viscosity refers to the thickness or ability to resist flow of the paint, primer, or clear. Using an incorrect viscosity paint will result in various finish defects. The material must be thoroughly mixed and properly thinned or reduced.

Viscosity can be measured by means of a **viscosity cup.** This is another very accurate method of assuring proper material thickness. The flow characteristics of liquids relate directly to the degree of internal friction. Therefore, anything that influences the internal friction (such as solvents or temperature change) will influence flow. Similarly, paint flow affects how well the paint will atomize and "flow out" on the vehicle surface.

The two types of paint viscosity measuring cups are the **Ford cup** and the **Zahn cup.** The Zahn type cup is inexpensive and more common.

The paint manufacturer will give a recommended viscosity value in **viscosity cup seconds.** It will vary

between 17 and 30 seconds depending upon the type of paint and type of cup used. Refer to the paint specifications on the can for an exact value.

If the material DRAINS TOO QUICKLY out of the viscosity cup, you have added too much solvent. More paint will be needed. If the cup DRAINS TOO SLOWLY, you have not added enough solvent. Remix until the paint passes the viscosity cup test.

If the material is TOO THICK, your paint will develop orange peel or a rough film. If paint is TOO THIN, excess solvent can cause the paint to have poor hiding and other problems.

Make sure you use the reducer designed for the shop temperature and conditions. The amount of reduction should be the same regardless of temperature.

Various finishes are manufactured to spray at specific viscosities. It is very important to follow the label directions for each type of material being mixed.

Using a Zahn Viscosity Cup

The Zahn cup is cylindrical and has an orifice at the bottom. To determine viscosity with a Zahn cup, proceed as follows:

1. Prepare the paint to be tested. Mix, strain, and reduce as directed by the manufacturer.

2. Fill the cup by submerging it in the paint, Figure 15–18.

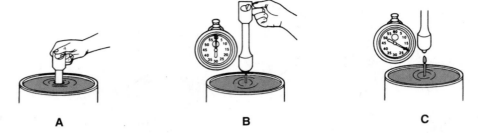

A   B   C

Figure 15–18 Using a Zahn cup. *(BASF)* (A) Dip the cup into the paint until it is full. (B) Remove the cup, and as it clears the surface of the paint, begin timing the flow of paint from the small hole in the bottom of the cup. (C) Stop the timer when the stream of paint breaks.

3. After removing the cup, release the flow of the paint and trigger the stopwatch. Keep your eyes on the flow, NOT on the watch.

4. When the solid stream of paint "breaks" (indicating air passing through orifice), stop the watch. A stopwatch is ideal for measuring paint viscosity with a viscosity cup. Most painting technicians prefer a digital stopwatch because it is easily read.

The result is expressed in seconds. Manufacturers recommend specific viscosities for their various types of paint.

Using a Graduated Mixing Cup

A *graduated mixing cup* is a clear plastic cup with paint ratio markings for mixing spray materials (Figure 15–19). Its markings are similar to the paint mixing sticks explained earlier. You must add each paint ingredient until the material is equal to the correct marking on the side of the cup. Ounce markings are also provided for adding each ingredient.

Using a Computerized Mixing Scale

A *computerized mixing scale,* or electronic scale, is programmed to automatically indicate how much of each paint ingredient to use by weight. After selecting the needed paint formula on the keypad, the computerized scale will prompt you to add a specific amount of each ingredient. As you pour each material into a cup on the scale, the scale readout will go to zero or beep when the right amount has been dispensed. When zero is reached, you have added the right amount of paint material for that paint code. The scale will then prompt you to add the next ingredient until the proper mixture is achieved.

Electronic scales are commonly used in large collision repair facilities that mix their own paint. Refer to Figure 15–20.

PAINT TEMPERATURE

The temperature at which paint is sprayed and dried has a great influence on the smoothness of the finish. This involves not only the temperature of the shop, but the temperature of the work surface as well.

You should pull the vehicle into the shop and booth long enough before spraying that its surfaces

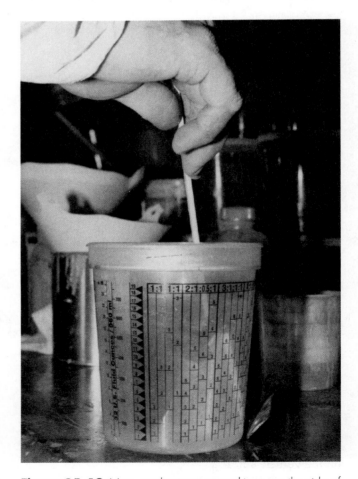

Figure 15–19 Note graduations, or markings, on the side of this mixing cup. As on mixing sticks, these markings are provided for different mixing ratios. An ounce scale is also provided.

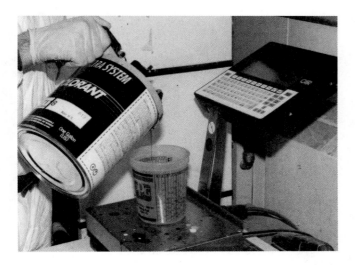

Figure 15–20 Electronic or computerized scales will automatically indicate when the right amount of material has been poured into a container. You select the correct paint code and the scale will tell you which and how much of each ingredient to add to produce the correct formula or color.

can warm up. Never spray warm paint on a cold surface or cool paint on a hot surface. This will upset the flow characteristics of the paint.

Appropriate reducers should also be used for warm and cold weather applications. For example, a **hot weather solvent** (reducer) is designed to slow solvent evaporation to prevent problems. A **cold weather solvent** is designed to speed solvent evaporation so the paint flows and dries in a reasonable amount of time.

CAPTIVE SPRAY GUN SYSTEM

Because different materials require different setups, many shops use a captive spray gun system. In a **captive spray gun system,** individual guns are set up and used for applying each type of material. Each would be clean, adjusted, and ready for use. You would have separate spray guns for each type of material, as follows:

1. For applying primers
2. For applying sealers
3. For applying basecoats
4. For applying clearcoats

Having a gun adjusted for applying each material saves time. Also, your work will be better because you will know how each gun setting will apply its own material.

Adjusting the Spray

Proper spray gun operation is critical to refinishing. Spray guns are precision engineered tools. They must be treated as such.

A **good spray pattern** will deposit an even, oval-shaped mist of paint on the surface being painted. The paint should go on smoothly in a medium to wet coat, without sagging or running.

A good pattern depends on the proper mixture of air and paint droplets. Adjusting the spray gun is much like fine-tuning an engine fuel system for the proper mixture of air and fuel.

There are three basic adjustments that will give the proper spray pattern, degree of wetness, and air pressure: air pressure, pattern control, and fluid control.

To adjust a typical paint spray gun:

1. Adjust the air pressure. If this is set at the dryer-regulator (or transformer), you must account for pressure loss through the hose and fittings. Due to friction, as air passes from the dryer-regulator to the gun, pressure will be lost. The air pressure at the dryer-regulator and at the gun varies depending on the length and diameter of the hose.

   A **gun-mounted pressure gauge** or gun-mounted gauge-regulator is the best method to measure and adjust spray gun air pressure. Install the gauge or regulator between the hose coupler and the gun. This will let you set actual pressure at the gun. Refer to Figure 15–21.

Figure 15–21 It is good idea to place a pressure gauge or gauge-regulator at the gun. This will give an actual reading at the gun without pressure loss through the air line and fittings.

The optimum spraying pressure is the lowest needed to obtain proper atomization, flow rate, and pattern width. **High spray gun pressure** results in excessive paint loss through overspray and poor flow due to high solvent evaporation before the paint reaches the surface. **Low spray gun pressure** produces poor drying characteristics due to high solvent retention, and makes the paint film prone to bubbling and sagging.

Proper spray gun air pressure varies with the kind of material sprayed and type of gun. Many low volatile organic compound (VOC) regulations require 10 psi (0.7 kg/cm^2) or less at the air cap. Always follow the spray gun manufacturer's air pressure recommendations for the type of material to be sprayed.

2. Set the size of the spray pattern using the ***pattern control knob.*** Turn the pattern control knob all the way in to create a small, round pattern. Back the knob out to produce a wide pattern.

3. Use the ***fluid control knob*** to adjust the amount of fluid leaving the spray gun. As shown in Figure 15–22, regulate the volume of paint according to the selected pattern size. Back the fluid knob out to increase the paint flow. Screw the knob inward to decrease the flow.

Spray Pattern Test

A ***spray pattern test*** checks the operation of the spray gun on a piece of paper. Before attempting to paint the vehicle, it is very important to test the spray pattern. See Figure 15–23.

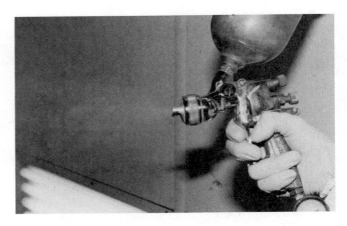

Figure 15–23 After filling the spray gun with material, always check the spray pattern and spray gun operation. You want to find and correct any problems before spraying the vehicle.

Hold the gun 2 to 6 inches (50 to 150 mm) away from the paper. Hold the gun 4 to 6 inches (100 to 150 mm) away when spraying clears. Pull the trigger all the way back and release it immediately. This burst of paint should leave a long, slender pattern on the test paper, Figure 15–24.

Figure 15–25 shows some incorrect spray patterns. These problems are usually caused by a clogged passage in the gun.

A spray pattern test that is:

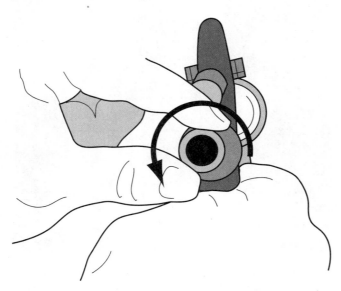

Figure 15–22 The fluid control knob on the spray gun controls the volume of color, primer, or clear leaving the gun.

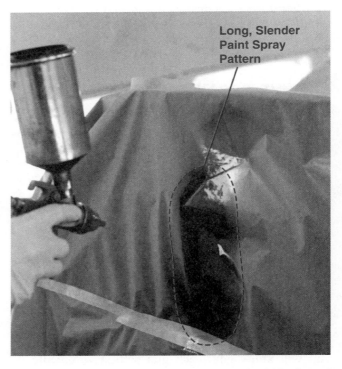

Long, Slender Paint Spray Pattern

Figure 15–24 A long slender spray pattern should be formed on the masking paper during your test.

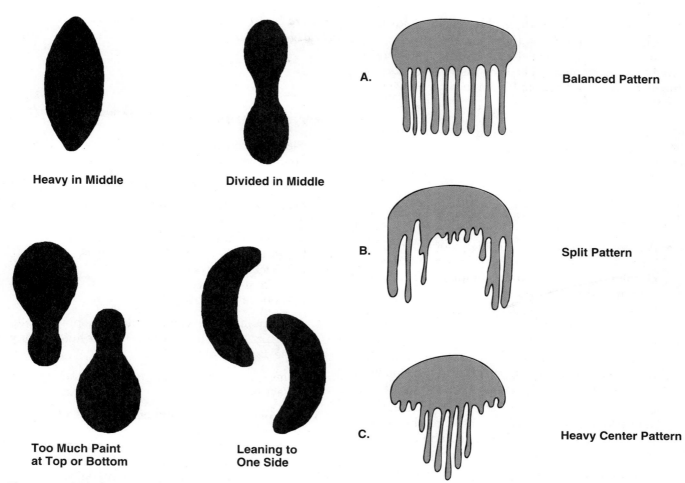

Heavy in Middle

Divided in Middle

Too Much Paint at Top or Bottom

Leaning to One Side

A. **Balanced Pattern**

B. **Split Pattern**

C. **Heavy Center Pattern**

Figure 15–25 If your test shows incorrect spray patterns like these, there is something wrong with the gun. You will need to disassemble and clean the gun or replace damaged parts. (I-CAR)

Figure 15–26 With the cap turned sideways on the gun, test spray further. Spray a heavy coat of paint onto the test paper to flood the pattern. Inspect the lengths of the runs. If all adjustments are correct, the runs will be almost equal in length.

1. Heavy in the middle could mean too little air flow.

2. Divided in the middle indicates too much air flow.

3. Too much paint at top or bottom might be caused by a restriction at the fluid needle or air cap horn.

4. Leaning to one side could mean that there is a restriction at the fluid needle or air cap horn.

5. If the pattern is heavy on one side or the top or bottom, try turning the air cap 180 degrees. If the pattern remains the same, replace the fluid needle and fluid nozzle. If the pattern rotates 180 degrees, then the problem is in the air cap horns.

Spraying primer usually requires a smaller spray pattern. Turn in the pattern control knob until the spray pattern is 6 to 8 inches (150 to 200 mm) wide. For spot repair, the pattern should be about 5 to 6 inches (125 to 150 mm) from top to bottom.

If the paint droplets are coarse and large, close the fluid control knob about one half turn or increase the air pressure 5 psi (0.35 kg/cm^2). If the spray is too fine or too dry, either open the fluid control knob about one half turn or decrease the air pressure 5 psi (0.35 kg/cm^2).

Next, test the spray pattern for uniformity of paint distribution. Loosen the air cap retaining ring and rotate the air cap so that the horns are straight up and down. In this position, you will get a horizontal spray pattern instead of a vertical one.

Spray again on your test paper. However, hold down the trigger until the paint begins to run. This is known as flooding the pattern. Inspect the lengths of the runs. If ALL adjustments are correct, the runs will be almost equal in length. See Figure 15–26A.

The uneven runs in the split pattern shown in Figure 15–26B are a result of setting the spray pattern too wide or the air pressure too low. Turn the pattern control knob in one half turn or raise the air pressure 5 pounds (0.35 kg/cm^2). Alternate between these two adjustments until the runs are even in length.

If the runs are longer in the middle than on the edges, too much material is being discharged. Turn in the fluid control knob until the runs are even in length, Figure 15–26C.

> **Danger!** *Always wear a suitable air respirator when doing any spraying.*

Summary

- A painter must have a suitable painting environment. The following variables must be addressed: cleanliness, temperature/humidity, light, and compressed air/pressure.

- Atomization breaks paint into a spray of tiny, uniform droplets. When properly applied, these droplets flow together to create an even film thickness with a mirrorlike gloss. Proper atomization is essential when working with today's basecoat-clearcoat finishes.

- Paint viscosity refers to the paint's thickness or ability to resist flow. It can be measured by means of a Ford cup or a Zahn cup. The Zahn cup is the more common, and less expensive, cup used.

- In a captive spray gun system, individual guns are set up and used for applying each type of material. This system saves time and improves work quality because each gun will already be at the correct setting when needed.

Technical Terms

painting environment	*mixing chart*
spray booth	*mixing sticks*
manometer	*paint strainer*
tack rag	*viscosity*
paint mixing room	*viscosity cup*
painter prep area	*Zahn cup*
atomization	*viscosity cup seconds*
spray gun air cap	*graduated mixing cup*
center orifice	*computerized mixing*
side orifices	*scale*
fluid nozzle	*hot weather solvent*
fluid needle	*cold weather solvent*
fluid control knob	*captive spray gun system*
mixing instructions	*good spray pattern*
percentage reduction	*gun-mounted pressure*
mixing by parts	*gauge*

pattern control knob	*spray pattern test*
fluid control knob	

Review Questions

1. Clearcoat finishes will magnify any dust, dirt, or other flaws in the finish. True or false?

2. List four functions of a spray booth.

3. Daily spray booth cleaning requires you to do ALL of the following EXCEPT:
   a. remove ALL nonessential items.
   b. vacuum the floor.
   c. check blower belt tension.
   d. all of the above.
   e. none of the above.

4. After pulling the vehicle into the spray booth, _____ the entire vehicle to remove any loose dust and other debris.

5. Name some expendable materials that should always be kept on hand.

6. To avoid air supply problems, do ALL of the following EXCEPT:
   a. check and replace oil and water filters on a regular basis.
   b. drain moisture from the system daily.
   c. replace deteriorated air hoses because they can introduce dirt into the system.
   d. ALL of the above.
   e. none of the above.

7. _____ breaks the material into a spray mist of tiny, uniform droplets.

8. A spray gun has too much paint being applied on one side during a pattern test. *Technician A* says to check for a clogged side orifice in the cap. *Technician B* says to check for proper fluid needle adjustment. Who is correct?
   a. Technician A
   b. Technician B
   c. Both Technicians A and B
   d. Neither Technician

9. If the label on a can of material says to reduce 100 percent, what would you do?

10. Paint mixing sticks have scale markings that allow you to easily mix _____, _____, and _____ in the correct proportions.

11. In your own words, how do you use paint mixing sticks?

12. If the liquid material drains too quickly out of the viscosity cup, you have added too much color and not enough reducer. True or false?

13. What is a captive spray gun system?

14. Summarize the three steps for adjusting a spray gun.

ASE–Style Review Questions

1. *Technician A* says that you will usually spend more time maintaining the shop and the equipment than on any other single task. *Technician B* says that you will spend more time painting. Who is correct?
   a. Technician A
   b. Technician B
   c. Both Technicians A and B
   d. Neither Technician

2. *Technician A* says that the spray booth should be self-cleaning because of the air flow through the system filters. *Technician B* says that booth cleaning is a daily process. Who is correct?
   a. Technician A
   b. Technician B
   c. Both Technicians A and B
   d. Neither Technician

3. *Technician A* says to use a tack rag to clean the vehicle once in the paint booth. *Technician B* says you should also blow off the vehicle. Who is correct?
   a. Technician A
   b. Technician B
   c. Both Technicians A and B
   d. Neither Technician

4. A spray gun spray pattern is too small. *Technician A* says to turn the fluid control knob in for more flow. *Technician B* says to turn the air flow control knob out for more flow. Who is correct?
   a. Technician A
   b. Technician B
   c. Both Technicians A and B
   d. Neither Technician

5. What does the mixture ratio 4:1:1 mean?
   a. Add four quarts of hardener and one quart of solvent to a gallon of paint.
   b. Add four parts of solvent and one part of hardener to a gallon of paint.
   c. Add one pint of hardener and one pint of solvent to a gallon of paint.
   d. Add one quart of hardener and one quart of solvent to a gallon of paint.

6. Which of the following should be used when filling a spray gun with paint?
   a. Strainer
   b. Funnel
   c. Mixing stick
   d. Zahn cup

7. *Technician A* says that paint temperature affects finish smoothness. *Technician B* says that solvent speed also affects finish smoothness. Who is correct?
   a. Technician A
   b. Technician B
   c. Both Technicians A and B
   d. Neither Technician

8. *Technician A* says that a pressure gauge-regulator between the spray gun and hose will help obtain more accurate gun pressure setting. *Technician B* says using the pressure gauge-regulator on the wall is better. Who is correct?
   a. Technician A
   b. Technician B
   c. Both Technicians A and B
   d. Neither Technician

9. How far away should you hold an HVLP spray gun during a pattern test?
   a. 2–4 inches
   b. 4–6 inches
   c. 6–8 inches
   d. 12–14 inches

10. To check spray pattern uniformity, *Technician A* rotates the air cap so that the horns are straight up. *Technician B* floods the pattern with excess paint. Who is correct?
    a. Technician A
    b. Technician B
    c. Both Technicians A and B
    d. Neither Technician

Activities

1. Inspect the condition of your spray booth. Is it ready for use? Make a report on its condition.

2. Make a wall chart showing how to use one type of paint mixing stick. Place callouts and steps on the chart for display in the classroom.

3. Make a spray pattern test of a spray gun. Try adjusting the gun knobs to make good and poor spray patterns.

Section 4

Refinishing

16

Painting and Refinishing Fundamentals

Objectives

After studying this chapter, you should be able to:

- Explain the difference between spot refinishing, panel refinishing, and overall refinishing.

- Properly use a spray gun.

- Describe how to spray different types of materials.

- Summarize the different kinds of spray coats.

- Outline general topcoat application procedures.

- Explain the key points to keep in mind when applying multi-stage finishes.

- Finish and/or retexture various types of automotive plastics.

- List general rules for painting/refinishing a vehicle.

Introduction

From the customer's standpoint, the topcoat or colorcoat is the most important aspect of a refinishing job. The topcoat is what the customer sees and evaluates. Most customer complaints stem from problems with the topcoat. As a painting technician, you must therefore take special pride in producing a beautiful finish. To do this, you must understand the topcoat materials and how they are applied.

> **NOTE!** Previous text chapters provided information relevant to this chapter. For more specific information on paint types, paint mixing, shop prep, and vehicle prep, refer to the index.

With today's high-solids, low-VOC paints, and high-efficiency spray guns, refinishing procedures have changed. The industry is now using high volume, low pressure (HVLP) spray equipment to reduce paint waste and emissions. They are also using paint products that have more solid content to reduce solvent waste. These changes have made painting and refinishing even more challenging. This chapter will help summarize the basics for using today's materials and equipment.

Figure 16–1 When it is ready for refinishing, pull the vehicle into a clean spray booth. Center it in the booth so you have adequate distance around the vehicle.

PREPARATION REVIEW

As detailed in Chapters 14 and 15, all straightening should be done and the vehicle should be ready for painting/refinishing. Carefully pull the vehicle into a clean paint booth and center it, Figure 16–1.

Make sure the air circulation system of the spray booth is turned on and working properly. See Figure 16-2. To prevent airborne contamination, close the booth door and allow the air circulation system to purge the booth of airborne debris. Refer to Figure 16–3.

Figure 16–2 Make sure the air circulation system in the spray booth is turned on before refinishing/painting.

Figure 16–3 Keep the doors on the spray booth closed so the air circulation system can pull any airborne dust through filters. If the doors are left open, dust will be pulled into the booth from outside work areas.

Check all masking tape and paper one last time. Make sure none of the tape has pulled up or paper has been torn. Inspect all edges and paper closely for openings that could allow overspray leaks, Figure 16–4. If needed, shake off and install wheel cover masks over the tires to protect them from overspray. Refer to Figure 16–5.

Wipe the surfaces to be painted with wax and grease remover (cleaning solvent). Use one clean rag to wipe on the wax and grease remover. Use another clean rag to wipe the surface dry, Figure 16–6.

While wearing protective gear, blow off any remaining dust and lint with an air gun. Blow off all surfaces that might hold dust or debris that could get

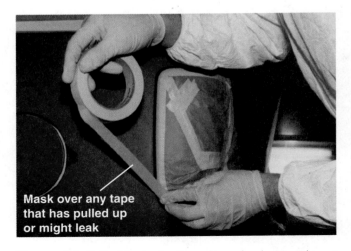

Mask over any tape that has pulled up or might leak

Figure 16–4 Double-check all masking closely. You do not want overspray to get on glass, lenses, and other surfaces not to be refinished.

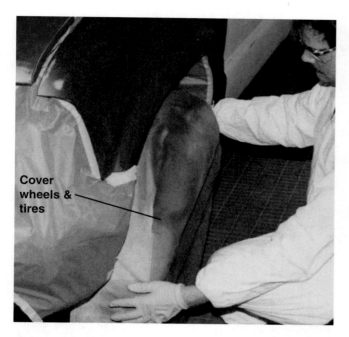

Figure 16–5 This technician is covering tires with wheel masks.

into the finish. As you blow off surfaces, wipe the vehicle down with a **tack cloth** (a rag with coating that holds dust and debris). Refer to Figure 16–7.

After wiping with a tack cloth, be careful NOT to touch the surface being refinished. Oil from your skin could contaminate the vehicle's surface.

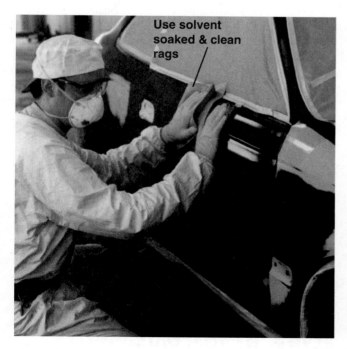

Figure 16–6 Wipe repair areas down with wax and grease remover (cleaning solvent). Use one clean rag to wipe solvent over the surface. Follow this with a clean dry rag to wipe off the wax and grease remover (cleaning solvent) while it is still wet.

Figure 16–7 Wipe repair surfaces down with tack rag while blowing with the air gun.

Make sure the material has been reduced with the proper solvent to the desired viscosity. If needed, add the right amount of catalyst/hardener. Again, refer to the directions on the materials, Figure 16–8. You may also want to use a viscosity cup to check for proper material thickness or fluidity.

Make sure the material is mixed thoroughly, Figure 16–9. A wide, flat bottom, stirring paddle or steel spatula of at least one inch (25 mm) works well. You can also use a mixing machine (shaker) if one is available. Insufficient mixing is a common reason for paint problems.

Use a strainer to filter the material as it is poured into the spray gun, Figure 16–10. This will remove any larger particles that might have entered the mixing container and would cause paint problems.

Figure 16–8 Always read refinishing material direction. The label will give mixing and application information for the specific material to be sprayed.

A

B

Figure 16–9 (A) Mix the material correctly. Here a technician is using a computerized scale to add the recommended amount of tints, dyes, metallic flakes, binders, etc. (B) Use a clean mixing stick to stir the material thoroughly.

Strainer

Spray gun cup

Figure 16–10 Filter all material being poured into your spray gun. This is a must to prevent paint problems.

Paint settling occurs when the paint sits idle and pigments, metallic flakes, or mica particles settle to the bottom. This can cause problems because the color sprayed out first will contain a high percentage of these heavier ingredients. When the spray gun cup is almost empty, the paint will be lighter and have less metallic or mica in it.

Remember! If a color with heavy pigments is allowed to stand in your spray gun for 10 to 15 minutes without being stirred, the pigment will settle. This can cause color variations when sprayed.

Check that spray gun air pressure is correct. Perform a spray test pattern on a sheet of paper. Make sure the spray gun is working properly before applying paint to the vehicle.

Check the temperature of the vehicle, the spray booth, and spray materials, Figure 16–11. They should all be equalized and within specs. Use a thermometer to check the surface temperature of the vehicle. If vehicle surface temperature is too cold or hot, painting problems will result.

REVIEW OF TOPCOATS

The type of refinishing system or paint ultimately determines the attractiveness of the color, gloss, and overall finish. The type of paint also affects other variables.

Figure 16–11 The technician is checking the surface temperature of a vehicle with a small thermometer. Vehicle surface temperatures should be within the range specified by the paint manufacturer.

Most vehicles are refinished with **acrylic urethane enamel** with a hardener or catalyst. A color basecoat is covered with a clearcoat to produce a high gloss, yet durable finish. The hardener helps the paint cure so that the vehicle can be released to the customer in less time.

Various other types of topcoats are also available: alkyd enamels, polyurethane enamels, acrylic enamels, and acrylic lacquers. They are seldom used in modern body shops. However, they can be used or requested by the customer for special applications. For this reason, you should have a basic understanding of all types of paint materials. For example, acrylic enamel will sometimes be requested for refinishing large trucks to reduce material costs. Lacquer may still be requested for custom painting of show cars and motorcycles.

The table in Figure 16–12 is a summary of the properties of topcoats used for refinishing. The table in Figure 16–13 compares the durability of various topcoat finishes.

SUMMARY OF TOPCOAT FEATURES

Nomenclature		One-Part Type			Two-Part Type	
		Alkyd Enamel	Acrylic Lacquer	Acrylic Enamel	Polyurethane	Acrylic Urethane Enamel
Spray characteristics		Excellent	Excellent	Good	Good	Good
Possible thickness per application		Fair	Fair	Good	Excellent	Excellent
Gloss	without polishing	Fair	Good	Good	Excellent	Excellent
	after polishing	Good	Good	Good	—	Good
Hardness		Good	Good	Good	Excellent	Excellent
Weather resistance (frosting, yellowing)		Fair	Fair	Good	Excellent	Excellent
Gasoline resistance		Fair	Fair	Fair	Excellent	Good
Adhesion		Good	Good	Fair	Excellent	Excellent
Pollutant resistance		Fair	Fair	Fair	Excellent	Excellent
Drying time	to touch	68°F 5-10 minutes	68°F 10 minutes	68°F 10 minutes	68°F 2-30 minutes	68°F 10-20 minutes
	for surface repair	68°F 6 hours 140°F 40 minutes	68°F 8 hours 158°F 30 minutes	68°F 8 hours 158°F 30 minutes	—	68°F 4 hours 158°F 15 minutes
	to let stand outside	68°F 24 hours 140°F 40 minutes	68°F 24 hours 158°F 40 minutes	68°F 24 hours 158°F 40 minutes	68°F 48 hours 158°F 1 hour	68°F 16 hours 158°F 30 minutes

Figure 16–12 Study the characteristics of different types of topcoats.

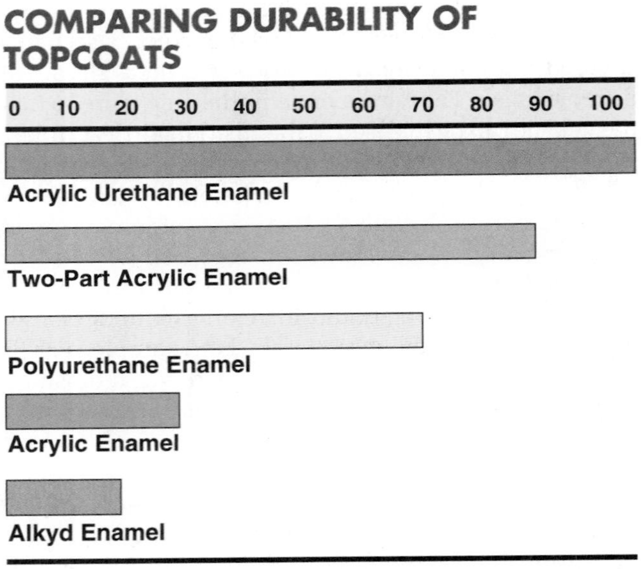

COMPARING DURABILITY OF TOPCOATS

0	10	20	30	40	50	60	70	80	90	100

Acrylic Urethane Enamel

Two-Part Acrylic Enamel

Polyurethane Enamel

Acrylic Enamel

Alkyd Enamel

Figure 16–13 Note how two-part acrylic enamel and acrylic urethane enamel are more durable than other finishes.

Before applying the topcoat, carefully read the paint manufacturer's directions that appear on the paint can. Each has specific formulations for its products. For this reason, the best source of data on how to apply a specific brand of paint is the label. Another source of good information can be found in the manufacturer's literature. Refer back to Figure 16–8.

Some of the more important label and literature data that should be checked include:

1. Viscosity recommendations

2. Air pressure recommendations

3. Use of additives, reducers, solvents, activators, and hardeners

4. Application techniques

5. Number of coats required

6. Blending and mist coat procedures

7. Polishing and compounding recommendations

8. Cleanup procedures

Make sure you are using the correct type of refinishing materials for the job. Refer to the paint code information on the body plate to find the correct paint color and type needed. The **body ID number** or **service part number** gives information about how the vehicle is equipped. It will give paint codes or numbers for ordering the right type and color paint. Lower and upper body colors will be given if it is a two-tone paint. The body ID number will also give trim information. This number will be on the body ID plate on the door, console lid, or elsewhere on the body.

Typical locations for paint codes are shown in Figure 16–14. An example of an actual body ID plate is given in Figure 16–15.

DETERMINING TYPE OF FINISH

Before planning any refinishing job, you must find out what type of paint is on the vehicle. The vehicle might have its original finish or it could have been repainted with a different type paint. Methods for finding out the type of paint on a vehicle include the following.

1. With the **solvent application method,** rub the paint with a white cloth soaked in lacquer thinner to see how easily the paint will dissolve. If the paint film dissolves and leaves a mark on the rag, it is a type of air-dried paint. If it does not dissolve, it is either an oven-dried or a two-part reaction type paint. Acrylic urethane paint film will not dissolve as easily as an air-dried paint, but sometimes the thinner will penetrate sufficiently to blur the paint gloss.

2. With a **heat application method,** wet sand an area with No. 1000 grit sandpaper to dull the paint film. Then heat the area with an infrared lamp. If a gloss returns to the dulled appearance, the paint is acrylic lacquer.

3. With the **hardness method,** you must check the general hardness of the paint. Paints do not dry to the same hardness. Generally, two-part reaction and oven-dried paints dry to a harder film than air-dried paint.

4. With an **inspection method,** inspect closely for signs of repainting. Look for masking tape–created paint lines, overspray, and other signs of repairing. If the vehicle has not been repainted, you can use the body color code identification plate to determine the type of paint on the vehicle.

SPRAY GUN APPLICATION STROKE

The **application stroke** is a side-to-side movement of the spray gun to distribute the mist properly on the vehicle. The proper stroke is critical to the vehicle finish. Practice on masking paper. Any problems with the spray pattern must be corrected before painting the vehicle. Look at Figure 16–16.

To use a spray gun, hold the gun at the proper distance from the surface, Figure 16–17. **Spray gun distance** is measured from the gun tip to the surface being painted. Typically, hold the gun 6 to 8 inches (152 to 203 mm) away for lacquer and 8 to 10 inches (203 to 254 mm) away from the surface for enamel.

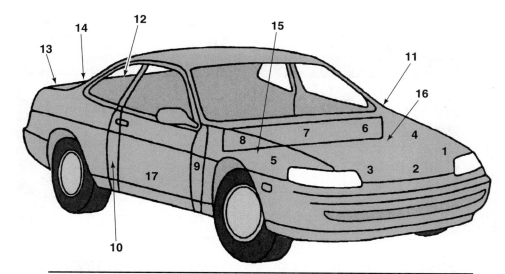

Model	Position	Model	Position
Acura	9	Honda	8,10
Alfa Romeo	4,13	Hyundai	6,7
AMC	9,10	Isuzu	2,10
Audi	12,13	Lexus	7,8
Austin Rover	17	Mazda	1,2,3,4,6,8
BMW	4,5	Mercedes	2,7,9
Chrysler	3,5,16	Mitsubishi	7
Chrysler Corp	3,5	Montero / Pickup	3
Caravan / Voyager / Ram Van	6	Cordia / Tredia	4
Chrysler Imports	1,2,4	Others	1,2,3
Colt Vista	16	Nissan	1,3,4,6,8,15,*
Conquest	7	Peugeot	2,3,4,5,8
Diahatsu	1,6,7	Porsche	9
Datsun	2	Renault	1,3,4,5,8
Dodge D50	3	Rover	1,3,4,5
Ford	10	Saab	5,6,8
Ford Motor Co	10	Subaru	2
General Motors		Suzuki	7,11
A, J and L Bodies	14	Toyota Passenger	7,8,14
E and K Bodies	12	Truck	4
B,C,H and N Bodies	13	Volkswagen	2,11
GM Imports	2,12,13,14	Volvo	6,7,8
		Yugo	12

* Under Right Front Passenger Seat

Figure 16–14 This chart shows typical locations for paint codes for major makes of vehicles. *(Courtesy of Mitchell International, Inc.)*

A **short spray distance** causes the high velocity mist to ripple the wet film. A **long spray distance** causes a greater percentage of the thinner to evaporate, resulting in orange peel or dry spray, Figure 16–18.

A slower-drying thinner will permit more variation in the distance of the gun from the surface. However, it will produce runs if the gun gets too close. Excessive spraying distance also causes more overspray.

Spray gun angle refers to whether the gun is tilted up or down or sideways. Normally, hold the gun parallel and perpendicular to the surface. Keep the gun at a right angle to the vehicle. This should be done even when spraying curves in the body. Refer to Figure 16–19 (page 330).

If you tilt the gun when spraying the sides of the vehicle, an uneven paint film will result, Figure 16–20 (page 300). On flat surfaces such as the hood or roof, the gun should be pointed almost straight down. See Figure 16–21 (page 330).

> *WARNING! When painting a hood, roof, or deck lid, the spray gun cup must NOT leak. If the seal around the cup is faulty, paint will drip out of the gun and onto the vehicle. If the drip will NOT flow out, you might have to sand and repaint the panel.*

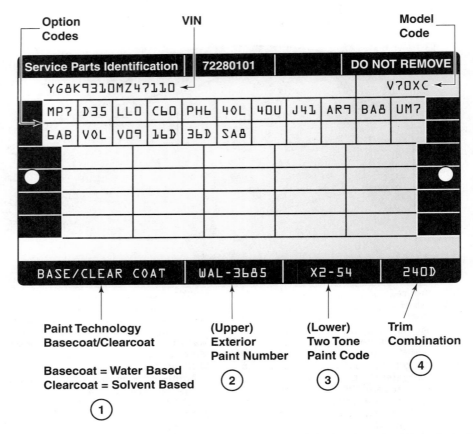

Figure 16-15 The body ID plate or service parts identification plate gives VIN and other data about the vehicle. Note the paint and trim code numbers.

Avoid fanning the gun with your wrist. Fanning the gun will deposit an uneven paint film. The paint film will be thicker right in front of your gun and thinner on the sides. The only time it is permissible to fan the gun is when blending a small repair spot. With a spot repair, you want the paint film thinner at the edges to blend out the spray. See Figure 16–22 (page 330).

Spray gun triggering involves stopping the paint spray before you stop moving the gun sideways. When you pull halfway back on the trigger, only air blows out of the nozzle. When you pull all the way back, material is atomized and sprayed out.

During the application stroke, release halfway on the trigger right before you stop moving the gun sideways. This will prevent too much material being deposited when the gun changes direction. It will also keep air moving through the nozzle to help prevent a sudden burst of paint. Release the trigger halfway at the end of each pass, then pull it back when beginning the pass in the opposite direction. Look at Figure 16–17 again.

Spray gun speed refers to how fast the gun is moved sideways while spraying. Move the gun with a steady, deliberate pass, about one foot (0.3 m) per second. The speed must be consistent or it will result

in an uneven paint film. If you move the gun too quickly, not enough paint will be deposited on the surface. If you move too slowly, too much paint will deposit on the surface. Never stop the gun while spraying, or a paint sag and run will result.

A **banding coat** is done to deposit enough paint on edges or corners of surfaces. Spray these difficult areas first, Figure 16–23 (page 331). Aim directly at the corner or edge so that half of the spray covers each side of the corner or edge. Hold the gun an inch or two (25–50 mm) closer than usual while moving the gun up or down along the edge. After all of the edges and corners have been sprayed, spray the face or front of the panel.

Generally, start spraying at the top of any upright surface, such as a door panel. The gun nozzle should be level with the top of the panel. The upper half of the spray pattern should hit and cover the masking paper. Move the gun all the way across the top of the panel. Make sure you hold the gun square with the panel and keep it the same distance from the panel. See Figure 16–24 (page 331).

Overlap strokes involve making each spray gun coat cover about half of the previous coat of paint. Make each pass in the opposite direction. Keep the

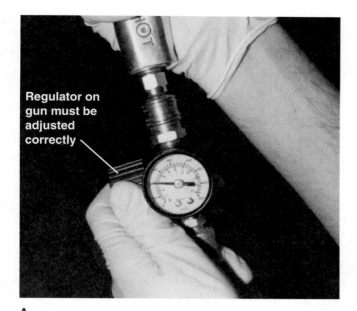

A

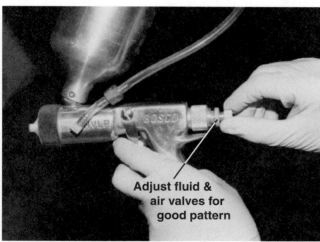

B

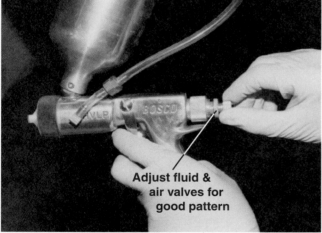

C

Figure 16–16 (A) Make sure air pressure to the spray gun is correct. Refer to recommendations from the spray gun manufacturer. (B) Adjust the spray gun before spraying the vehicle. (C) Trigger the gun to check that the spray pattern is acceptable.

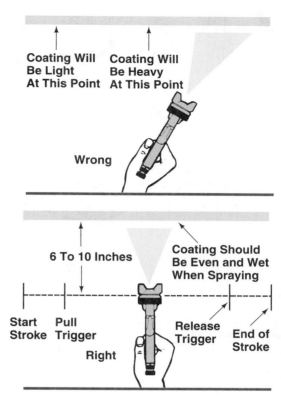

Figure 16–17 Note the right and wrong ways of using a spray gun. *(Illustration courtesy of ITW Automotive Refinishing, 1-800-445-3988)* Wrong—If you fan the gun, the paint film will be thicker in the middle and thinner, drier on the sides. Right—Like a machine, hold the gun perfectly straight and at an equal distance from the surface as you move it sideways. Release the trigger before the end of each stroke.

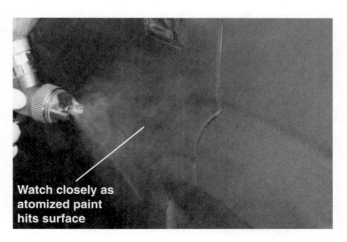

Figure 16–18 This photo shows how a spray gun atomizes paint into a fine mist that deposits smoothly on the vehicle surface.

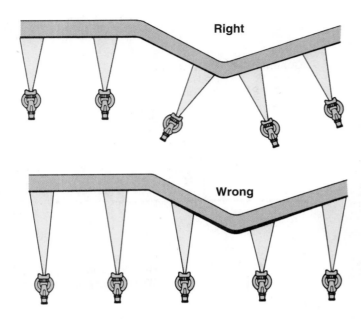

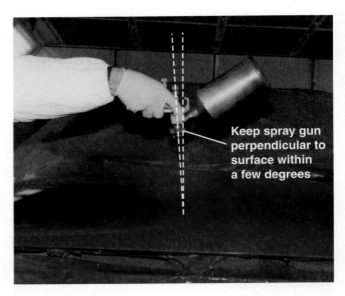

Figure 16–19 During normal painting, the spray gun should always be aimed directly at the surface and kept an equal distance from the surface. If the surface curves, redirect the spray gun to spray right at the surface. Right—The spray gun is rotated so that it stays aimed right at the surface. Note that the correct distance from the surface is also maintained. Wrong—Since the spray gun is not kept aimed at the surface and distance decreases at the curved surface, the paint would be too thick at the curve. Run or sag might result.

Figure 16–21 When spraying horizontal surfaces, such as the hood, roof, or trunk, hold the spray gun almost straight down. This will assure that the pattern applies evenly to the surface. Make sure the cup is not leaking and cannot drip on the surface.

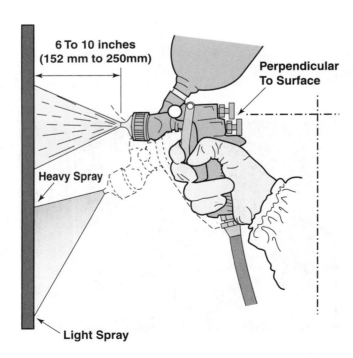

Figure 16–20 Do not let the spray gun tilt up or down. This would also result in uneven paint film.

Figure 16–22 Normally, keep the spray gun aimed straight at the surface. Do not fan the gun unless you are blending the paint out over the next panel.

nozzle level with the lower edge of the previous pass. Thus, one half of the spray pattern overlaps the previous coat. The other half of the paint pattern is applied to the unpainted area.

Always blend your paint into the "wet edge" of the previously sprayed section. If you allow the wet edge to dry or cure too much before painting the panel next to it, an unwanted dry band of paint may occur where the two panels meet. By always spraying into the wet edge where you just painted, the paint will tend to melt together and not form a dry band.

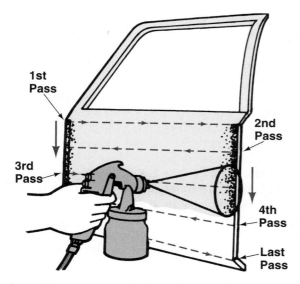

Figure 16–23 A banding coat assures that edges of the panel receive enough coverage. After making banding coats on the edges, follow the arrows to spray the rest of the door or panel.

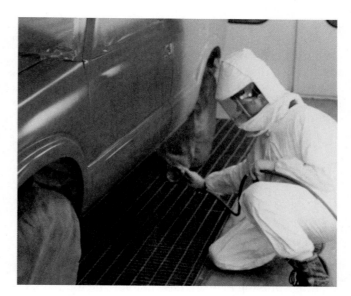

Figure 16–25 Make sure you apply enough color to the bottoms of panels. The spray gun may have to tilt upward to get under lower panels, like this rocker panel.

A

B

Figure 16–24 (A) Normally, start spraying at the top. (B) Methodically work your way down.

Proper triggering where the sections meet will avoid a double coat and the possibility of paint sags or runs.

Continue the back-and-forth passes. Trigger the gun at the end of each pass and lower each successive pass one half the top-to-bottom width of the spray pattern. Make your last pass with the lower half of the spray pattern hitting below the surface being painted. This will assure enough paint film thickness at the bottom of the panel. Refer to Figure 16–25.

For a double coat, repeat the previous procedure. Allow flash time of several minutes between coats.

Two or three single coats may be required for clear topcoats. Allow the first coat to set up or become tacky before applying additional coats. Refer to the paint label for details.

For painting very narrow surfaces, switch to a gun with a smaller spray pattern. This can also be accomplished by reducing the air pressure and spray pattern on a full-size gun.

PAINT/MATERIAL THICKNESS MEASUREMENT

Paint thickness is measured in **mils** or thousandths of an inch (hundredths of a millimeter). Original OEM paints are typically about 2 to 6 mils thick. With basecoat-clearcoats, the basecoat is approximately 1–2 mils thick. The clearcoat is about 2–4 mils thick. This is approximately the thickness of a piece of typing paper.

When you repaint a vehicle, paint thickness will increase. If too much paint is already on the vehicle, it may have to be removed prior to refinishing. Paint buildup should be limited to no more than 12 mils. The OEM finish and one refinish usually equal just under 12 mils. Exceeding this paint thickness could cause cracking in the new finish. Chemical stripping or blasting would be needed to remove the old paint buildup.

A *mil gauge* can be used to measure the thickness of the paint on the vehicle. This should be done before refinishing.

For more information on mil gauges, refer to Chapter 6.

GUN HANDLING PROBLEMS

The inexperienced painting technician is prone to making spraying errors. These include the following:

Gun heeling occurs when the gun is allowed to tilt. Because the gun is no longer perpendicular to the surface, the spray produces uneven film thickness, excessive overspray, dry spray, and orange peel.

Gun arching occurs when the gun is NOT moved parallel to the surface. At the outer edges of the stroke, the gun is farther away from the surface than at the middle of the stroke. The result is uneven film thickness, paint runs, excessive overspray, dry spray, and orange peel.

Incorrect gun stroke speed means you are moving the gun too slowly or too quickly. If the stroke is made too quickly, the paint will NOT cover the surface evenly. If the stroke is made too slowly, sags and runs will develop. Proper stroke speed is something that comes with experience. Generally, you must watch the paint go onto the surface to carefully watch for a proper coat.

Improper spray overlap means you are NOT covering half of the previous pass with the next pass. Improper overlapping results in uneven film thickness, contrasting color hues, and sags and runs. If you overlap coats too much, the paint can run. If you do NOT overlap enough, poor coverage can result.

Improper coverage means the paint film thickness is NOT uniform or it is too thin. This is caused by NOT triggering exactly over the edge of the panel or from improper spray overlap.

NOTE! Examples of these and other paint problems are given in Chapter 18.

TYPES OF SPRAY COATS

There are varying degrees of thickness for a spray coat: light, medium, or heavy. The easiest way to control thickness is by changing how fast you move the gun sideways. If you move the gun slower, a heavier, thicker coat will be applied. If you move more quickly, film thickness is decreased.

A **light coat** is usually produced by moving the spray gun a little more quickly across the surface of the vehicle. A thinner than normal coating of paint film will be deposited on the surface. A light coat is sometimes used when applying the colorcoat. The colorcoat can go on light, without a gloss. This will help you blend the new finish into the existing one with a less dramatic change of color. It will also distribute any metallic flakes more evenly. The wetter clearcoats will give the colorcoat its gloss.

A **medium wet coat** is produced by moving your spray gun at a normal speed over the surface being refinished. This is the most common coat used by the professional painter. It will produce a medium gloss with adequate coverage, and will avoid runs and sags. A medium wet coat is the most common coat recommended by paint manufacturers on colorcoats and initial clearcoats.

A **full wet coat** is done by moving the spray gun slightly slower than normal. It will deposit more paint on the surface. A full wet coat is used when applying the final layer of clearcoat or single-stage colors. It is important that the last coat of clearcoat goes on wet to produce a high gloss or shine. Spraying a full wet coat requires skill and practice to prevent runs. The wet coat makes the paint lie down smooth and shiny. It prevents a dull, textured "orange peel" type paint film.

A **mist coat**, also called *dust coat,* **drop coat,** or *tack coat,* is a very light, thin coat. The spray gun pressure is reduced and it is held a little farther from the surface and moved more quickly from side to side. The mist coat cures or dries in a short period of time to bond and form a lightly textured paint film.

Mist coats can be used and are recommended to help avoid problems (mottling and blotching) with metallic paints. For example, some paint manufacturers recommend a mist coat as the last colorcoat with some metallics. Some painters also mist coat the first layer of clearcoat on troublesome metallics (troublesome silver and gold metallics) to prevent movement of the colorcoat flakes, which could affect paint matching.

Shading coats or **blend coats** are progressive applications of paint on the boundary of spot repair areas so that a color difference is not noticeable. They are applied in two or more coats. The second and third coats are lighter and sprayed over a wider area than the first.

PAINT BLENDING

Paint blending involves tapering or fanning the new paint gradually into the old paint to provide a smooth

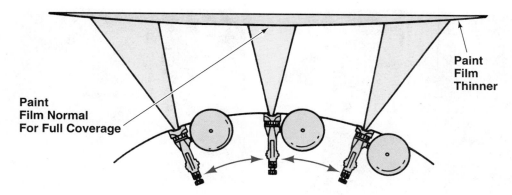

Figure 16–26 This view shows how fanning the spray gun at the wrist will blend paint thinner near the outer edges of the repair area. Work the gun from the center outward. When the paint is thinner, it will gradually blend into the existing paint to hide the repair.

transition between the two finishes. This makes any difference in the new paint color or texture less noticeable. It helps hide any slight differences between the new and existing paints. Blending is the key to a successful spot repair. Look at Figure 16-26.

To blend paint, apply the colorcoat (basecoat-clearcoat) or topcoat (single-stage paint) with a fanning motion, working from the center outward. This allows each coat to be thinner at its outer perimeter and to blend out a bit further than the previous one. Where the old and new paints meet, the existing paint will be visible through the thin layer of newly fanned paint. Any differences will be difficult to detect.

If you do NOT blend the paint, a marked contrast in paint color and texture will normally be obvious. Paint blending is commonly done when applying the colorcoat on a spot repair. The colorcoat is applied only where needed to the repair area. The undamaged colorcoat is not painted. Then, when the whole panel is sprayed with the clearcoat, the old and new colors will show through the clearcoat. Any minute difference in the colorcoats will not be noticeable and more of the vehicle will retain its original color.

An alternative method is to apply the finish in short strokes from the center outward. Again, extend each coat so that it blends out farther than the previous one.

With either method, the spray pattern should be narrowed and the fluid delivery reduced. To minimize overspray, the air pressure may also have to be reduced, depending on the material being sprayed.

REFINISHING METHODS

Refinishing methods are selected according to several variables:

1. Condition of the original finish

2. Size of the area to be finished

3. Location of the repair on the vehicle

From these variables, you must determine how much of the vehicle must be painted/refinished. The three categories of finish repair are spot repair (area of panel painted); panel repair (complete panel painted); and overall refinishing (whole vehicle painted).

SPOT REPAIRS

Spot repair involves painting an area smaller than the panel. The paint must be blended out to match the existing finish.

Spot repair generally involves:

1. Minor body repair

2. Metal conditioning

3. Application of undercoat system

4. Application of topcoat to blend into the old finish surrounding the repair

Spot repairs are recommended where a complete panel repair is either uneconomical or impractical. This might be due to the tiny size of the damage, its hidden location, or another factor. It is also commonly used when two large panels have no break line. For example, when painting a quarter panel, the paint is blended into the sail panel where the roof joins the quarter panel. Look at Figure 16–27.

Solid Color Spot Repairs

You must use experience and common sense when deciding the range of blending needed for spot repairs with solid colors. Blending with a solid color of paint is used for situations such as light damage to the fender edge.

In this case, there are two methods of blending: at the hood-to-fender gap and at the body line. When blending at a body line, you do NOT have to paint the

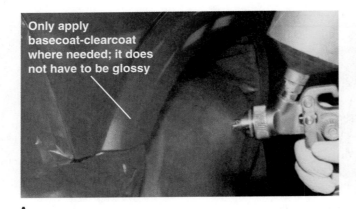

Only apply basecoat-clearcoat where needed; it does not have to be glossy

A

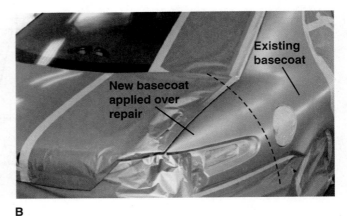

Existing basecoat

New basecoat applied over repair

B

Figure 16–27 When spot repairing a quarter panel, (A) Apply basecoat to the repaired area. (B) Apply basecoat to the area just beyond the repaired area, and then apply clearcoat to the entire quarter panel.

upper portion of the fender where paint shading shows more. This can help avoid problems with color and texture differences.

As shown in Figure 16–28, when spot repairs are made with a solid color, the blend area should be treated with rubbing compound or sanded with 1000 grit sandpaper before refinishing.

Metallic Color Spot Repairs

Matching metallic colors is complicated because you have to match the color and the density of the metallic flakes in the paint. If spot repairs are done with a metallic color, skill is required in matching the color and the metallic flakes through proper distribution of the paint. The blend area will be less noticeable if it is angled away from the body line. Refer to Figures 16–29 and 16–30.

PANEL REPAIRS

Panel repairs involve painting/refinishing a complete body part separated by a definite boundary,

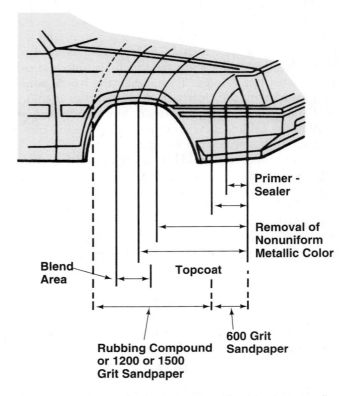

Figure 16–28 Note the basic steps for blending solid color. Solid colors are difficult to blend.

Figure 16–29 Study the basic steps for blending metallic color.

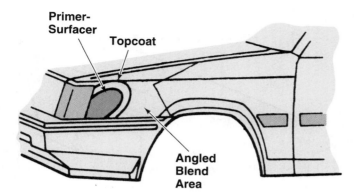

Figure 16–30 The blend area will be less noticeable if it is angled away from the body line.

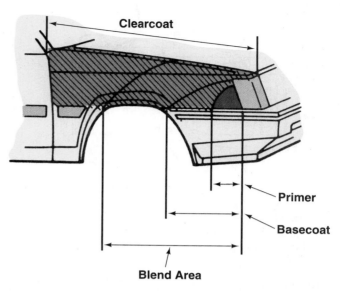

Figure 16–32 Note the panel repair with metallic basecoat-clearcoat. Basecoat is blended only over the filler area. Then, the top section of the fender or the whole fender must be sprayed with clearcoat.

such as a door or fender. Usually, you do NOT have to blend the paint with panel repairs. Blending is only needed when it is difficult to match the paint, as with a metallic color. However, blending is routinely done for such areas as between the quarter panel and roof panel.

Solid Color Panel Repairs

Panel repair with a solid color covers an entire panel (door, hood, etc.). The color match is made at the panel joints. Panel repair is more common than spot repair. It usually results in a better looking repair and can be just as fast as spot repair on most panels.

For a complete panel repair, mask off the area NOT to be painted, Figure 16–31. If a panel has damage at two different locations, the whole panel should be repaired. Blending can be done to the molding or extended below the molding.

Figure 16–31 This technician is doing a complete panel repair on the front fender.

When blending new color into old, you must use common sense to determine where to blend the paint. Try to find a body section that gives a natural break line or small area for blending.

Metallic Color Panel Repairs

New and old paint differences tend to be very noticeable with bright metallic colors. It is almost impossible to match the new metallic finish exactly with the previous one. You must extend the blending over a wide area to help hide the differences. This will make the repair less visible, Figure 16–32.

If a panel has damage at both ends, or if the whole panel is to be refinished, the blending may have to extend onto adjacent panels.

OVERALL REFINISHING

Overall refinishing involves painting/refinishing the whole vehicle. This is usually done when the surface has weathered and deteriorated. Surface prep is time consuming since the entire surface must be cleaned and sanded or scuffed. Any damage must be straightened. Also, all parts NOT to be painted (trim, glass, mirrors, etc.) must be masked or removed. There are several things you must remember when refinishing the whole vehicle.

For overall refinishing, keep a wet edge while maintaining minimal overspray on the horizontal

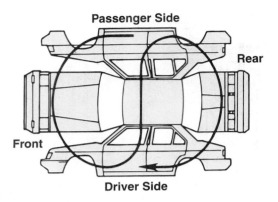

Figure 16–33 This is a typical sequence for doing an overall refinishing job when using a downdraft spray booth. Some painting technicians like to use a slightly different sequence.

surfaces. This prevents spray from settling onto dry surfaces, which would cause a gritty surface.

Avoid paint runs or sags by changing the point of overlap on each coat. **Paint overlap** refers to the area where one vertical painted area overlaps the new area being painted.

There is no universal sequence for overall refinishing of a vehicle. However, most experienced painting technicians agree that the following method is an excellent technique for nondowndraft spray booths.

Start by painting the roof, then the rear, then the driver's side, then the front, and finally the passenger side. It will produce minimum overspray on horizontal surfaces. This will also help keep a wet edge when starting a new section.

The term **wet edge** means that the area just sprayed will still be wet when starting on a new adjoining area.

When using a downdraft spray booth, overall paint sequence is different, Figure 16–33. This is because of the direction of the airflow (top to bottom). Following this procedure allows the three main horizontal surfaces to remain wet while maintaining minimum overspray.

APPLYING BASECOAT-CLEARCOAT

Basecoat-clearcoat systems pose a major challenge to painting technicians. Since more and more vehicles have basecoat-clearcoat finishes, it is very important to become familiar with them.

You must first spray the color basecoat over the repair area, Figure 16–34. The basecoat does NOT have to be sprayed where the old color is acceptable. Blend the basecoat of color out around the repair. Then, spray the whole panel with clear. This will normally make the old color and new basecoat blend and match.

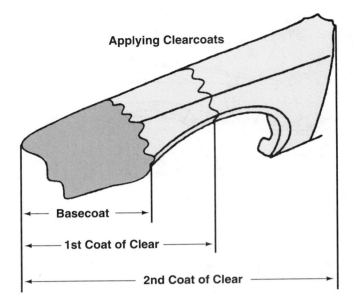

Figure 16–34 Basecoat-clearcoats are the most common finishes. When spraying the basecoat, blend the color out over the repair area. Spray color only where it is needed. Then, spray half of the panel with clear. After flash time, spray the whole fender with clear. *(I-Car)*

When estimating a basecoat-clearcoat repair, carefully examine the finish on the area adjacent to the damage. If it is chalked, dulled, or otherwise impaired, color matching might prove impossible. Ideally, such repairs should be performed as overall refinishing jobs. This approach will eliminate many problems in repairing basecoat-clearcoat finishes that are severely weathered.

When spraying, two medium coats of basecoat should be applied. The basecoat does NOT have to be glossy, and only enough should be used to achieve hiding, Figure 16–35. Two or three medium wet coats of clear should be applied next. Allow at least 15 minutes flash time between coats.

Figure 16–35 Basecoats or the colorcoats of base-clear systems do not have to be glossy or shiny. Medium coats will help new color match existing color.

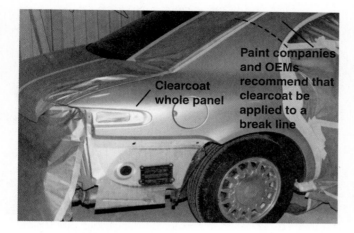

Figure 16–36 Clearcoats must be reduced properly to give a glossy finish.

Avoid sanding the basecoat. If sanding must be done because of dirt or imperfections, allow time for it to dry. Wet sanding with 800 to 1200 grit sandpaper will minimize sand scratches. The sanded area must then be given another coat of basecoat to prevent streaking and mottling.

Several tips to remember when using basecoat-clearcoat finishes include:

1. On panel repairs, spray the basecoat only where needed. You do NOT have to basecoat the whole panel if the existing color is acceptable. Use as much of the existing color as possible to help avoid a color mismatch.

2. Do NOT load clearcoats on heavily. Because they are clear, it is easy to use too much trying to increase the desired glamour effect. As a result, the thick clear will "bury" the basecoat and hide the color.

3. Do NOT use over- or under-reduced clearcoats. Contrary to some opinions, clears do NOT perform better when they are under-reduced. Reduce clearcoats according to the label instructions. See Figure 16–36.

4. Do NOT use economy reducers with clearcoats. These only weaken the performance of clears by trapping solvents. They also affect flow and leveling characteristics of the clearcoat. Use a quality reducer. Allow each coat to flash thoroughly before applying the next one.

APPLYING TRI-COAT FINISHES

Tri-coat finishes, like pearlescent finishes, require somewhat different spraying techniques because they are difficult to match. However, they have much in common with basecoat-clearcoat finishes.

Follow the recommendations furnished by the tri-coat paint manufacturer. Pay close attention when the instructions call for the use of adhesion promoters, antistatic materials, and so on.

Before painting with a tri-coat, you should make a *let-down panel.* A let-down panel helps you find the correct number of coats needed to achieve a color match with tri-coat finishes. Gun pressure, reduction, and spray technique can all affect the finished job. Therefore, the time spent spraying one or more let-down panels will be repaid by a job that satisfies the customer. A mismatch in the basecoat, mica, or clearcoats can affect the overall finish match.

NOTE! Instructions for making let-down panels and other methods of color matching are detailed in Chapter 17.

REFINISHING RULES

There are several rules you should remember when refinishing a vehicle. Ask yourself these questions and answer them before proceeding:

1. Are the vehicle surfaces straight and ready for refinishing? A common mistake is to overlook a surface problem. Since the surface is usually sanded dull, imperfections are easy to overlook. Double check metal straightening work, plastic filler, and primer before proceeding. Even overlooking a small paint chip can ruin the job.

2. Are all surfaces perfectly clean and scuffed? Finish will NOT adhere to a dirty or glossy smooth surface. If the paint peels, you will have hours of rework. Blow off and tack-rag the vehicle before refinishing.

3. Are you working in an ideal environment? Are the spray booth and vehicle the right temperature? Are the booth filters and blower working properly? Do NOT try to paint a vehicle in an open shop. Dirt will almost always settle in the finish. Work only in a spray booth!

4. Is the mixing correct? Mix materials following label directions! Are all additives mixed into the paint? If you forget to add hardener, you will NOT be able to wet sand or buff the surface for weeks.

5. Is the spray gun working properly? Test and adjust the gun spray pattern on a sheet of paper or old part. If the gun is spitting or NOT working properly, you do NOT want to try to spray the vehicle. This would result in hours of rework.

6. Does the spray gun cup leak? A leaking cup can drip and ruin the job. You must tilt the gun down when spraying the roof, hood, and trunk lid. If the cup is leaking, material might drip out and onto these surfaces.

7. Are you allowing enough flash time between coats? Flash time is the time needed for a fresh coat of finish to partially dry. Flash time is needed to prevent the material from sagging or running.

8. Are you applying the material properly? Hold the gun the right distance from the surface, typically 8 to 10 inches (203 to 254 mm). Aim the gun directly at the surface while moving it at the correct speed. Do NOT fan the gun unless you are blending the paint.

9. Is the material going on the surface properly? Closely watch as it deposits on the surface. Check for application problems—dry coat, excessive wet coat, improper spray overlap, and so on. Constant inspection of the wet coat as it hits the surface is critical to doing high-quality work. Is the lighting good enough to see the material go onto all surfaces?

10. Are you applying the correct film thickness? A common mistake is to apply too much. Remember that "thin is in." You want to apply only enough material to provide good color coverage and gloss. Use the number of coats recommended by the manufacturer.

11. Could you accidentally touch the wet surface? Keep air hoses and yourself a safe distance away. It is easy to brush up against the wet finish and damage it.

NOTE! Information relating to painting a vehicle is given in several chapters of this text. Refer to the index for added information if needed.

REMOVAL OF MASKING MATERIALS

If the finish has been force dried, remove the masking tape while the finish is still warm. If the finish is allowed to cool, the tape will be difficult to remove. It can also leave adhesive behind.

Pull the tape slowly so that it comes off evenly, Figure 16–37. Take care NOT to touch any painted areas because the paint might not be completely dry. Fingerprints or tape marks could result.

If you used liquid masking material, wash it off with soap and water. Do NOT wash the freshly painted surfaces until they are fully dry.

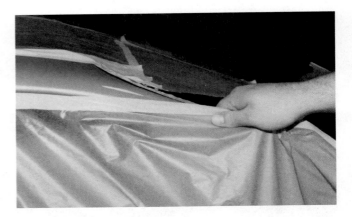

Figure 16–37 Remove masking tape slowly and pull it away from the new finish to keep it from peeling off new paint.

WARNING! Never mask a vehicle and let it sit for a prolonged period. Also, do NOT let the masking paper and tape get wet. Either will cause problems. The tape edge can roll up and allow paint to spray onto parts NOT meant to be painted. Also, the tape can stick and be difficult to remove. You might have to carefully wash off the adhesive.

FINISHING RIGID EXTERIOR PLASTIC PARTS

While most rigid exterior plastics do NOT require a primer, some paint manufacturers nevertheless recommend a primer application before the topcoat. Conversion coatings, metal conditioners, self-etching primers, and flex agents should NOT be used.

The finishing procedure for rigid plastic is as follows:

1. Before beginning, use soap and water to clean plastic parts.

2. Wash the part thoroughly with the recommended solvent.

3. Sand the area to receive the basecoat with 400 grit sandpaper. The blend area to be cleared should be sanded with 600 grit or finer sandpaper.

4. Apply color coat using the appropriate acrylic lacquer, acrylic enamel, urethane, or basecoat-clearcoat system.

5. Apply only enough film thickness to achieve full hiding. This is usually two or three medium wet coats. The basecoat should be allowed to flash completely before the clearcoat is applied.

Flexible replacement panels are factory primed with an elastomeric enamel-based primer. Clean the surface with solvent. Sand with 400 grit sandpaper, and clean again.

If the OEM primer is scratched and has left the plastic substrate exposed, spray the exposed area with a flexible primer-surfacer prior to finishing. If the exposed surface is NOT primed, the area will be highlighted after the topcoats are applied. Use a fast evaporating solvent to reduce the primer-surfacer and prevent swelling of the base material. Swelling can be caused by absorption of the solvent into the plastic.

FINISHING POLYPROPYLENE PARTS

Finishing polypropylene (PP) parts involves the use of a special primer. Since polypropylene is hard, it can be color coated after priming. A special primer must be used for the undercoat and a flex agent added to the topcoat. If this is NOT done, peeling will result.

If a PP bumper has major structural damage, it must be replaced. Replacement bumpers of this type are usually primed and ready to be painted. If they are NOT primed, apply the special PP primer over the entire bumper.

When applying topcoat, proceed as follows:

1. Apply properly reduced and mixed PP primer and flex additive as directed by the manufacturer. Allow 1 to 2 hours drying time before applying any color coats.

2. Apply reduced color and hardener. Flex additive should NOT be used in the topcoat.

3. Allow 8 hours (overnight if possible) drying time.

> WARNING! Because polypropylene bumpers are made of thermoplastic resin, force drying at more than 212 degrees F (100 degrees C) could result in part deformation.

FINISHING URETHANE BUMPERS

Following is the procedure for finishing a colored urethane bumper:

1. Mask off the area to be finished. Clean it with a wax and grease remover. Insufficient cleaning will result in peeling or blistering.

2. Prepare the surface by wet sanding with 600 grit sandpaper.

3. Clean the surface again.

4. Apply the color coat over a section of or the entire bumper if needed. Use a two-part acrylic urethane with a flex agent added. For metallics, allow proper flash time after application. Then apply the clearcoat over the entire bumper.

5. Follow the drying time recommended by the manufacturer.

CLEANING THE SPRAY GUN

Like any precision piece of equipment, the refinisher's spray gun should be cleaned after use. Your livelihood depends upon the proper care of your equipment, and your spray gun is a vital tool.

After you use a spray gun, always clean it and other equipment (mixing sticks, containers, viscosity cup, etc.) right away. Neglect and lack of care is responsible for most spray gun problems. Besides cleaning, you must also lubricate all bearing surfaces and packings at recommended intervals.

> WARNING! If the gun is NOT cleaned right after use, passages will usually clog. When used the next time, the gun may spit (eject pieces of dried paint) or form a poor spray pattern. This is of particular importance when dealing with materials that have hardeners/activators.

With many of today's finishing materials requiring a hardener/catalyst, proper cleaning is essential. These painting/refinishing materials can set up (harden) in the spray gun. This could be a very expensive problem. Each spray gun MUST be properly and completely cleaned after each use.

DANGER! When cleaning spray guns, ALWAYS wear solvent-resistant gloves, proper eye protection, and a NIOSH-approved respirator. There are some painting/refinishing materials that pose a serious health risk!

SPRAY GUN CLEANING STEPS

1. Dispose of the remaining paint or refinishing materials in accordance with federal, state, and local laws and requirements.

2. Put a small amount of cleaning solvent into the paint cup.

3. Place the lid on the paint cup or place the paint cup on the bottom of the spray gun.

4. Shake the spray gun so that the cleaning solvent rinses the inside of the paint cup. On gravity-fed spray guns, pull the trigger all the way open and let some of the cleaning solvent come out of the spray gun and into a container for proper disposal later.

5. Remove the air cap. Take a cleaning brush and use some of the solvent in the paint cup to clean the air cap, fluid tip, and needle. See Figure 16–38.

6. Reinstall the air cap.

7. Pour out the remaining (dirty) cleaning solvent into a container for proper disposal.

8. Place a small amount of fresh cleaning solvent in the paint cup.

9. While you are in the spray booth, spray this clean solvent through the spray gun.

10. Wipe the spray gun dry inside and out with a clean, dry, lint-free towel. See Figure 16–39.

From time to time, every spray gun will need a more thorough cleaning. This will involve a full disassembly. For this type of cleaning, always refer to the spray gun manufacturer's instructions for the proper procedures.

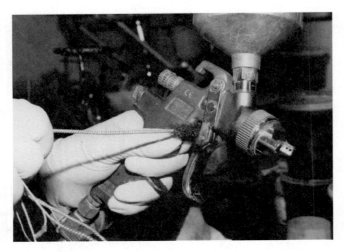

Figure 16–38 A small brush can be used to keep the outside of the spray gun clean.

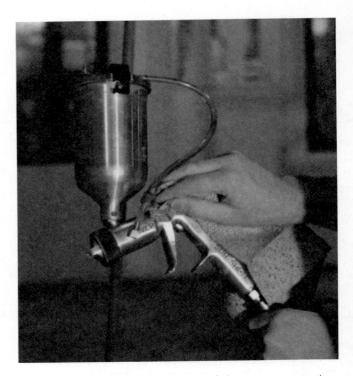

Figure 16–39 Wipe the outside of the spray gun with a clean cloth.

Many shops have "automatic," or self-contained, spray gun cleaners. Since each one of these is different, always refer to the manufacturer's instructions. This will ensure proper cleaning of all spray equipment.

> *WARNING! Never use wire or nails to clean the precision opening in a spray gun. This could enlarge or damage the openings and affect spray gun operation.*

To clean a pressure-feed spray gun, release the pressure in the cup first. Loosen the air cap. Then force the paint into the cup by triggering the spray gun. Empty the contents into a suitable container. Refill the cup with a clean, compatible solvent. The air cap can be left off. Spray the solvent out of the spray gun to wash out internal passages. If a tank is used, clean as directed by the manufacturer and reassemble.

Using a Spray Gun Washer

Areas in the United States with air pollution problems require the use of enclosed spray gun cleaning equipment. Spraying equipment (guns, cups, mixing sticks,

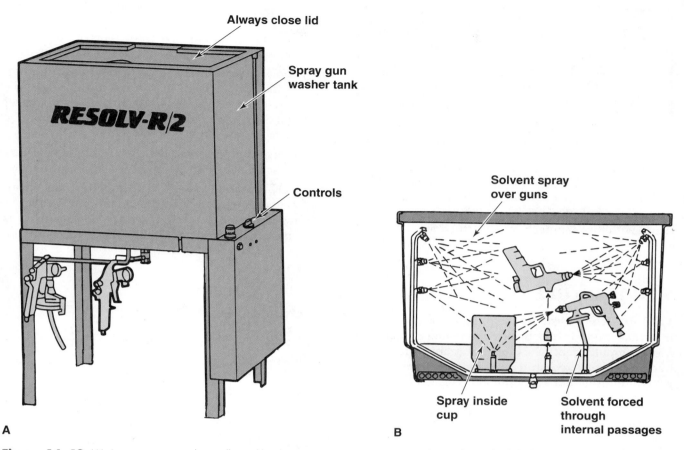

Always close lid

Spray gun washer tank

Controls

A

Solvent spray over guns

Spray inside cup

Solvent forced through internal passages

B

Figure 16–40 (A) A spray gun washer will quickly clean your spray equipment. Always keep the lid closed, even when it is not in use to keep solvent from evaporating. (B) Note how the washer circulates solvent over the outside of spray guns, inside the cup, and through internal passages. *(PBR Industries)*

and strainers) is placed in the tub of the **spray gun washer.** The spray gun washer will automatically clean them.

After the washer lid is closed, a pump circulates the solvent to clean the inside and outside of the equipment. In less than 60 seconds, the equipment is clean and ready for use. See Figure 16–40.

The spray gun washer saves time and increases safety. Compared with manual cleaning, the spray gun washer saves about ten minutes on each color change.

Check the owner's manual for complete operational details and the proper solvents to use. Always lubricate a spray gun after cleaning in a spray gun washer.

Spray Gun Lubrication

Spray gun lubrication involves placing a small amount of oil on packing and high friction points. Most manufacturers recommend daily lubrication of the parts shown in Figure 16–41.

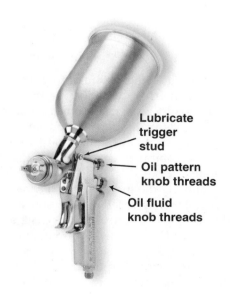

Lubricate trigger stud

Oil pattern knob threads

Oil fluid knob threads

Figure 16–41 Note the points on the spray gun that should be lubricated. Use manufacturer-recommended lubricant and do not use too much. Refer to your owner's manual for details. *(Courtesy of ITW DeVilbiss)*

> *WARNING! Do NOT overlubricate a spray gun. The excess oil could overflow into the fluid passages and mix with the paint. This could result in a defective paint job and "fish-eyes" (small dimples in paint film).*

Always examine the needle and nozzle periodically for excessive wear. Packings, springs, needles, and nozzles will have to be periodically replaced due to normal wear. This should be done only in accordance with the manufacturer's instructions.

VOC TRACKING

VOC tracking systems help monitor and record materials used to abide by environmental regulations. Most computerized VOC (volatile organic compound) tracking systems require a painting technician to enter general information about the job. This general information includes:

1. Date
2. Technician's name or initials
3. Repair order number
4. Customer and vehicle information
5. Gross vehicle weight or vehicle type
6. Size of mixing container or amount mixed

After entering all the general information, a painting technician enters information for the material to be sprayed. This includes: paint formulation number, reducers, catalysts or hardeners, and all other additives used in the mixture. As the product information is entered, the computer enters the VOC content of each material.

After applying the material, the technician enters the amount of material left in the spray gun. This allows the computer to calculate the exact amount of VOCs sprayed, NOT just the VOC total of the amount mixed. These steps are performed for each material used during the refinishing operation. The computer continues to enter the general information for each step.

By keeping records of this detail, a technician can use the computer to print the records needed to document total amount of VOCs sprayed.

Summary

- The most common types of topcoats are: polyurethane enamels and acrylic urethane enamels.
- Spray gun distance is measured from the gun tip to the surface being painted. Typically, hold the gun 6 to 8 inches (152–203 mm) away for lacquer and 8 to 10 inches (203–254 mm) away from the surface for enamel.
- When applying basecoat, two medium coats are sufficient. Use enough to achieve hiding, no more.
- VOC tracking systems help monitor and record materials used in order to abide by environmental regulations. As the product information is entered, the computer enters the VOC content of each material.

Technical Terms

tack cloth	*improper spray*
paint settling	*overlap*
body ID number	*improper coverage*
application stroke	*full wet coat*
spray gun distance	*dust coat*
short spray distance	*tack coat*
long spray distance	*shading coat*
spray gun angle	*paint blending*
spray gun triggering	*spot repair*
spray gun speed	*panel repairs*
banding coat	*overall refinishing*
overlap strokes	*paint overlap*
paint thickness	*wet edge*
mil gauge	*let-down panel*
gun heeling	*spray gun washer*
gun arching	*spray gun lubrication*
incorrect gun stroke	*VOC tracking*
speed	*systems*

Review Questions

1. From the customer's standpoint, the topcoat or colorcoat is the most important aspect of a painting/refinishing job. True or false?

2. How do you use a tack cloth?

3. What is some of the more important label information that should be checked before spraying?

4. A short spray distance causes the high velocity mist to _____ the wet paint _____.

A long spray distance causes a greater percentage of the thinner to _____, resulting in _____ _____ or _____ spray.

5. How do you properly trigger a spray gun?

6. Spray gun speed refers to how fast the spray gun is moved _____ while spraying.

7. What are overlap strokes of a spray gun?

8. What is the maximum film buildup allowable?
   a. 2 mils
   b. 5 mils
   c. 12 mils
   d. 50 mils

9. What can happen if the film is too thick after you refinish a vehicle?

10. This refers to whether a paint film is uniform.
    a. Heeling
    b. Arching
    c. Spattering
    d. Coverage

11. *Technician A* says that the first coat of color on an overall finish should always be a very wet coat. *Technician B* says that the first coat of color depends upon the paint manufacturer's directions and they usually recommend a medium wet coat. Who is correct?
    a. A only
    b. B only
    c. Both A and B
    d. Neither A nor B

12. A spot repair is being made on the front of a fender using a silver metallic base-clear material. *Technician A* says to use a mist coat to help even out the metallic basecoat when blending over the existing finish. *Technician B* says to use a full wet coat of basecoat to assure a good metallic match. Who is correct?
    a. A only
    b. B only
    c. Both A and B
    d. Neither A nor B

13. _____ _____ or _____ coats are progressive applications of paint on the boundary of spot repair areas so that a color difference is NOT noticeable.

14. Explain the difference between a spot repair and a panel repair.

15. How do you avoid runs and sags in paint during application?

16. When repairing a small area with basecoat-clearcoat, you need to spray the basecoat only over the repair area, but the whole panel should be cleared. True or false?

17. List ten rules or questions you should remember when refinishing a vehicle.

18. When manually cleaning spray paint guns, it is wise to wear _____, _____, and a _____.

ASE–Style Review Questions

1. *Technician A* says that color variations can be caused if a color with heavy pigments is allowed to stand in a spray gun for 10 to 15 minutes without being stirred. *Technician B* says that all materials being poured into a spray gun should be strained. Who is correct?
   a. Technician A
   b. Technician B
   c. Both Technicians A and B
   d. Neither Technician

2. *Technician A* says to always read the labels on paint refinishing materials. *Technician B* says that procedures for mixing paint will vary with the manufacturer. Who is correct?
   a. Technician A
   b. Technician B
   c. Both Technicians A and B
   d. Neither Technician

3. A painter is having trouble with orange peel when spraying a vehicle. Which of these would be the most common cause of this problem?
   a. Spray gun held too close to surface
   b. Spray gun held too far from surface
   c. Spray gun moved too quickly
   d. Spray gun moved too slowly

4. A paint drip has been found on a freshly painted vehicle. Which of the following is a common source of this problem?
   a. Fluid nozzle
   b. Spray gun body
   c. Cup seal
   d. Packing

5. *Technician A* says to keep the spray gun trigger held down to prevent spitting. *Technician B* says to release halfway on the trigger right before you stop moving the gun sideways. Who is correct?
   a. Technician A
   b. Technician B

c. Both Technicians A and B

d. Neither Technician

6. Which of the following should be done to deposit enough paint on edges or corners of surfaces?
   a. Flood coat
   b. Double coat
   c. Band coat
   d. Cross coat

7. *Technician A* says to start at the bottom and work upward when painting a door. *Technician B* says that little or no paint should hit the masking paper. Who is correct?
   a. Technician A
   b. Technician B
   c. Both Technicians A and B
   d. Neither Technician

8. *Technician A* says that each spray gun coat should cover about half of the previous coat of paint when painting a panel. *Technician B* says to make each pass in the opposite direction when painting a panel. Who is correct?
   a. Technician A
   b. Technician B
   c. Both Technicians A and B
   d. Neither Technician

9. A hood is being painted. *Technician A* says to lean over so that the whole hood can be painted with continuous movements of the spray gun. *Technician B* says to blend the wet edge of the hood and to paint one side and then the other. Who is correct?
   a. Technician A
   b. Technician B
   c. Both Technicians A and B
   d. Neither Technician

10. Which of the following spray gun handling problems will deposit more paint at the top of the fan than at the bottom?
    a. fanning
    b. heeling
    c. arching
    d. stroking

Activities

1. Practice spraying on a body part. Prepare its surface properly using information in previous chapters and this chapter. You might be able to obtain mismatched paint for free from a body shop or supplier. Write a report on what you learned and any problems that you encountered. List the materials used for the job.

2. Completely disassemble and reassemble a spray gun. Follow the manufacturer's instructions. Note if any parts are worn or dirty.

3. Practice spraying a metallic paint onto a part or a sheet of sandpaper. Intentionally apply too much paint so that it runs or sags in one location. In another location, spray a dry mist of metallic by holding the spray gun too far away from the surface. After drying, inspect the difference in the metallic. Make a report of your findings.

17

Color Matching

Objectives

After studying this chapter, you should be able to:

- Describe color theory and how it relates to refinishing.

- Define the terms relating to color.

- Describe the use of a computerized color matching system.

- Make let-down and spray-out test panels.

- Explain how to tint solid and metallic colors.

- Summarize the repair procedures for multi-stage finishes.

Introduction

Color matching involves the steps needed to make the new finish match the existing finish on the vehicle. Even if you use the body color code numbers and correct paint formula, the new paint may NOT be the exact same color. With today's multi-stage finishes and factory robotic painting, it can be very difficult to match colors when making spot and panel repairs.

Multi-stage paints like metallic pearl colors, for example, are very difficult to match. Even if you have a new vehicle with a fresh OEM finish, it can be a challenge to make your repair look like the color already on the vehicle. Today, there are many variables that affect the color and appearance of your paint work.

This chapter will help you develop the skills needed to match any type of paint. It will summarize color theory, color evaluation, color matching, computer analysis of paint, tinting, and other factors. See Figure 17–1.

Figure 17–1 The finish on today's vehicle is much more difficult to match than in the past. Multi-stage metallic and pearl colors can be a challenge to match. *(Courtesy of DaimlerChrysler Corporation)*

COLOR THEORY

Color is caused by how objects reflect light at different frequencies or wavelengths into our eyes. The color seen depends on the kind and amount of light waves the surface reflects. When these light waves strike the retina in your eye, they are converted into electrical impulses that the brain sees as color.

When the eye sees a colored object, that object is absorbing all of the light except for the color that it appears to be. A red ball appears red because the ball absorbs all of the colors in the light shining on it except

for the reds. In contrast, a black object absorbs almost all light, while polished nickel (chrome) absorbs none.

White light is actually a mixture of various colors of light. By passing light through a prism, light is broken down into its separate colors, called the *color spectrum.* This is the same principle of how rain breaks up light to form a rainbow. Refer to Figure 17–2.

The colors in the spectrum are easily remembered with the name "Roy G. Biv." which stands for the colors:

Red

Orange

Yellow

Green

Blue

Indigo

Violet

LIGHTING

Sunlight contains the entire visible spectrum of light. It is the standard by which other light sources are measured. Since the vehicle will be seen in sunlight,

Composition of Light

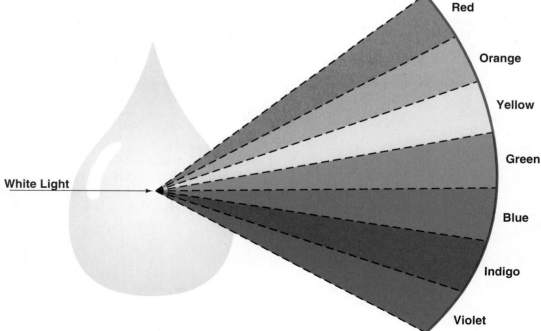

Figure 17–2 White contains all colors. When it is sent through a prism, these colors separate and can be seen. *(I-CAR)*

you should always use sunlight, or daylight-corrected lighting, when making color evaluations.

Each light source has a different mixture of colored light. When light sources are plotted on a graph, a difference can be seen in the amount of colored light each light source contains. The same color of paint will look very different under different kinds of light. This is why it is so important to check the color match in daylight or under a balanced artificial light, Figure 17–3.

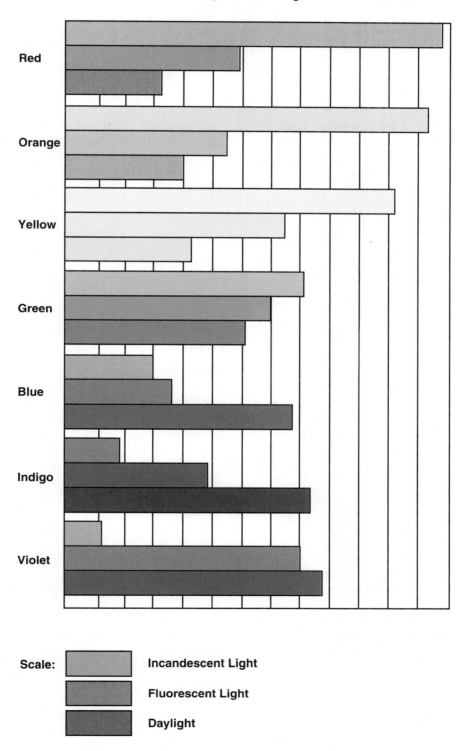

Comparison of Light

Figure 17–3 Note how the type of lighting affects how we see light. *(I-CAR)*

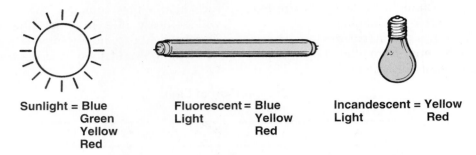

Sunlight = Blue
             Green
             Yellow
             Red

Fluorescent = Blue
Light   **Yellow**
               Red

Incandescent = Yellow
Light   **Red**

Figure 17–4 When evaluating color, use sunlight or color-corrected lighting. Colors will look different under fluorescent or incandescent lighting.

Compared to daylight, **incandescent light** has more yellows, oranges, and reds. **Fluorescent light** can have violets and reds. They should NOT be used when analyzing a color. See Figure 17–4.

The index for measuring how close a lamp in indoor lighting is to actual daylight is the **Color Rendering Index,** abbreviated **CRI.** A CRI of 100 duplicates daylight. A range of 85 to 100 is preferred for spray booth lighting.

Lamps may also have a **lumen rating** for brightness. Lamps are normally between 1,000–2,000 lumens, with a higher lumen rating producing a brighter light.

Lamps may also be rated for **light temperature** in "Kelvin." Daylight is 6,200 Kelvin. For painting, a lamp rating of 6,000–7,000 Kelvin is recommended.

What the eye sees as color is really light reflected from an object. The eye might see different shades of a color depending on the type of light source used. Look at Figure 17–5.

Always evaluate and match paint colors in daylight or while using daylight-corrected lighting. The characteristics of the light in the shop will affect how you perceive the color on a vehicle. The color of the panel does NOT change. What changes is the amount of colored light reflected from the panel. Figure 17-6 shows how a light source can affect the appearance of colors. Fluorescent and incandescent drop lights have been hung in front of a tool box. Note how the color of the red toolbox looks different. This shows why it is important to evaluate color in sunlight or under color-corrected lights.

For this reason, choose lamps that are the closest to simulating actual sunlight. The spray booth manufacturer or representative may have recommendations.

COLOR BLINDNESS

Color blindness makes it difficult for a person's eye to see colors accurately. If problems arise when you are matching certain colors, it might be wise to have your eyes checked for color vision. Nearly 10 percent of all men have trouble seeing one or more colors. Blue-greens are most often the problem. However, almost no women have this difficulty. This is one of the reasons women do the touch-up work for vehicle manufacturers. If you have a color vision problem, ask someone in the shop to help you match colors.

To do finish matching, the technician must be able to recognize colors as they actually are. It is important not only to see the color that is to be worked on, but also the overtones within that color, including the shades of darkness or lightness and the richness or fullness of the color.

DIMENSIONS OF COLOR

Many people describe color in terms of what they see. For example, you might have heard these descriptions: sky blue, ruby red, grass green, or midnight blue. These terms cause confusion when describing colors.

To minimize the confusion when painting, color should be based upon three **dimensions of color,** which are:

1. Value—lightness or darkness

2. Hue—color, cast, or tint

3. Chroma—saturation, richness, intensity, muddiness

These three dimensions are used to organize colors into a logical sequence on a color tree. The **color tree** is used to locate colors three-dimensionally when matching colors. Colors move around the color tree in a specific sequence—from blue to red to yellow to green, Figure 17–7. This sequence is easier to remember if you think of "BRYG." These are the first letters of blue, red, yellow, and green.

Value refers to the degree of lightness or darkness of the color. It is one dimension of color, Figure 17–8.

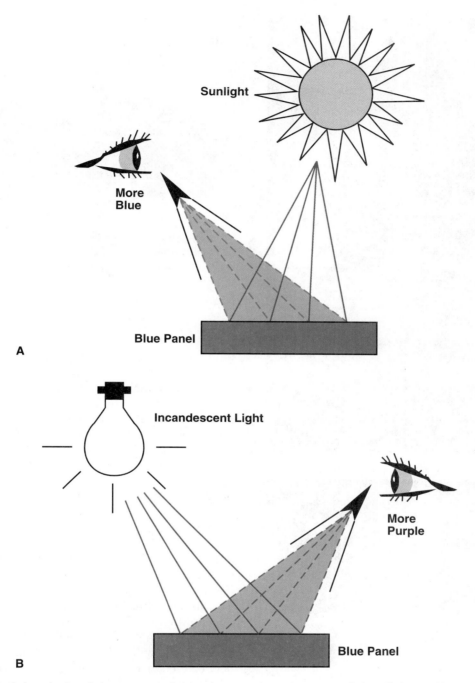

Figure 17–5 Sunlight is the best lighting to use when evaluating color. *(I-CAR)* (A) Sunlight will show a blue panel correctly. (B) Incandescent light will make the same blue panel look more purple.

When using the color tree, the value scale runs vertically through the tree. It is white at the top and black at the bottom. It is neutral gray at the center.

Hue, also called color, cast, or tint, describes what we normally think of as color. Hue is the color that is seen, moving around the outer edge of the color tree. It moves from blue to red to yellow to green. See Figure 17–8. When using the color tree, the hue scale shows color position around the color tree. It uses four main colors: blue, red, yellow, and green.

Chroma refers to the color's level of intensity, or the amount of gray (black and white) in a color. It is also called saturation, richness, intensity, or muddiness. It moves along the spokes that radiate outward from the central gray axis of the color tree. Weak, washed-out colors with the least chroma are at the core of the color tree. Highly chromatic colors that are rich, vibrant, and intense are at the outer edge. When using the color tree, chroma increases as it moves outward from the neutral gray center. It

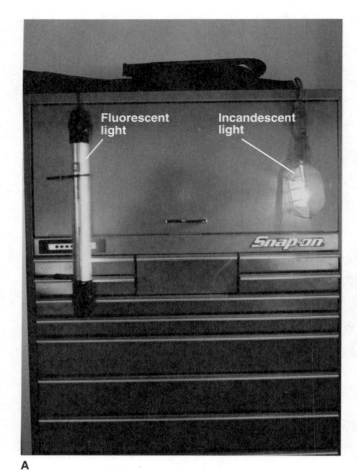

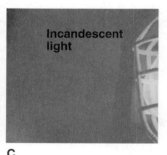

Figure 17–6 (A) Fluorescent and incandescent drop lights have been hung in front of a red toolbox. Note how they make the color of the toolbox look different. (B) This is a close-up of the color with fluorescent light. (C) This is a close-up of the color with incandescent light. *(Courtesy of Snap-on Tools Corporation)*

decreases as it moves closer to the neutral gray center, Figure 17–8.

METAMERISM

Metamerism is how different light sources affect paint pigments and metallics differently. A paint may have some blue in it which is NOT noticeable in daylight, but which becomes very evident under street

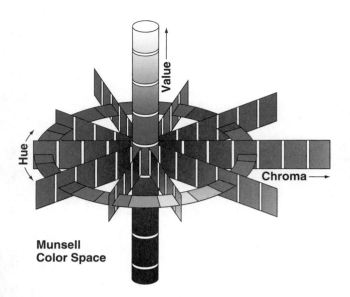

Figure 17–7 Note this graphic representation of the three dimensions of color. *(Munsell Color Group)*

lights. The problem is that the refinish and OEM color formulas are NOT made of the same pigments. This causes the pigments to look different.

Paint manufacturers formulate refinish colors to minimize the effect of metamerism. Metamerism is often a problem when a painting technician leaves the paint formula during tinting operations. That is why it is important to use only the tints called for in the formula.

PLOTTING COLOR

Plotting color is the process of identifying paint color in a graphic way based on value, hue, and chroma. This knowledge will help you become familiar with the refinishing evaluation process.

Plotting is NOT an exact science. It only has meaning when it is compared to the color on the vehicle. Plotting will help you recognize the differences between changes in value, hue, and chroma and adjust color as needed.

Evaluate colors in this order:

1. Value

2. Hue

3. Chroma

Using the plotting chart, develop a tinting plan that makes sense for all three dimensions of the color. In most cases only one dimension of a color will require adjustment.

For example, a plot for white is in Figure 17–9. Note how the vehicle color (V) is different from the

Solid Colors

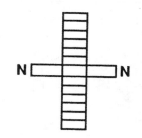

Value:
- Adjusted First.
- Adding White Increases Value.
- Adding Black/Colored Pigments Decreases Value.
- Changes in Value Affect Chroma.
- Hue Adjustments Affect Value.

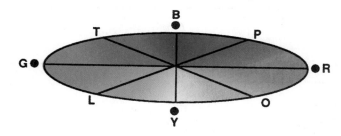

Hue:
- Adjusted Second.
- Adding Colored Pigments Moves Colors Around Color Wheel.
- Adding Colored Pigments Affects Chroma.

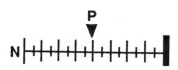

Chroma:
- Rarely Adjusted.
- Colored Pigments Affect Chroma.
- Increasing Chroma Decreases Value.
- Use Gray to Decrease Without Changing Value.

Figure 17–8 Note the major steps for evaluating color using value, hue, and chroma. (I-CAR)

paint (P). You would need to adjust the color. The value is too high. The hue is too red and the chroma is too low.

When working with colors that are between the four main colors on the color wheel, a technician should remember that:

1. Orange, bronze, and gold colors can move toward red or yellow.

2. Maroon and purple can move toward blue or red.

3. Lime can move toward green or yellow.

4. Aqua and turquoise can move toward green or blue.

When adjusting hue, a color can be moved to either side of the dominant color, or toward one of two dominant colors if it is between major colors.

Always use bluer, redder, yellower, or greener to describe how the hue is to be moved.

COLOR MATCHING

Because a variety of factors can cause color mismatch problems, it is very important to follow a step-by-step process on every job. There are acceptable variances in color that occur at both the refinishing and OEM levels.

In other words, color can fall to either side of the standard and still be acceptable. Thus, there can be a difference in the two finishes even though they are officially the same color. This can pose problems when refinishing.

Remember! Whatever the reason for color variance, you must match the only color standard that

Plotting White

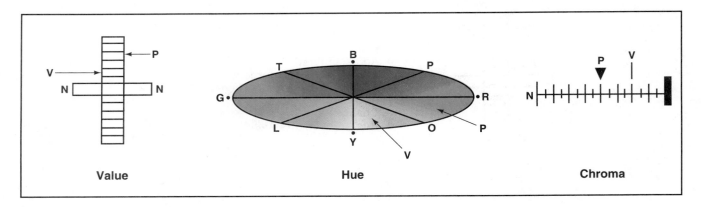

Value   Hue   Chroma

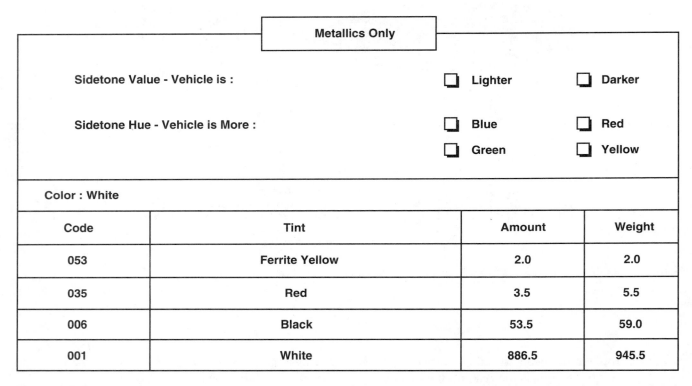

Metallics Only			
Sidetone Value - Vehicle is :		☐ Lighter	☐ Darker
Sidetone Hue - Vehicle is More :		☐ Blue	☐ Red
		☐ Green	☐ Yellow

Color : White

Code	Tint	Amount	Weight
053	Ferrite Yellow	2.0	2.0
035	Red	3.5	5.5
006	Black	53.5	59.0
001	White	886.5	945.5

Figure 17–9 When plotting color, the letter "V" stands for vehicle and "P" stands for paint. First, you would tint the paint to adjust the value. You would adjust hue second, by adding color to move the hue around the wheel. Chroma is rarely adjusted. *(I-CAR)*

really matters—the vehicle itself. For better or worse, the vehicle is the standard!

Color Directory

A **color directory** is a publication containing color chips and other paint related information for most makes and models of vehicles. Most collision repair and refinishing shops have a refinishing color directory. See Figure 17–10.

To use a color directory, locate the vehicle manufacturer's paint code. Then you can identify the color chip next to it. It is wise to compare the color chip with the vehicle color. There is always the chance that the vehicle has been repainted with a different color.

Figure 17–11 shows an actual page from a color matching manual (or color chip book). Note how it gives recommendations for changing a specific color.

The OEM has a group of sample panels, called **color standards,** for each vehicle color. These color standard panels are also provided to paint manufac-

Figure 17-10 Color manuals give instructions, color chip samples, and charts for matching colors. Color matching manuals are provided by the paint manufacturer. Color chips, since they are a photo, are not a completely accurate way of matching colors.

turers for formulation of refinish colors. These color standard panels can be used to identify acceptable color variances. For example, a blue color can vary slightly to either the green or the red side of the standard and still be acceptable.

Remember! The vehicle must be the standard because it may already have been repaired before.

PAINT FORMULAS

The *paint formula* gives the percentage or amount of each ingredient that is needed to match an OEM color. The formula will be available from the mixing system or from the local paint jobber.

A *basecoat patch* is a small area on the vehicle's surface without clear to enable the technician to check for color match with tri-coat colors. The manufacturer masks the patch before clearcoating to help you match the color more easily. The basecoat patch is sometimes located under the deck lid or hood. If the color match is correct, order the topcoat by color stock number.

Refinish suppliers supply topcoat (colors) in two ways—factory colors and custom colors. If it is a recent model or a popular color, chances are they will have the paint ready-mixed. These ready-mixed paints are called *factory packaged colors.*

If it is an older color, they might have to mix it. *Custom-mixed colors* are those colors that are mixed to order at the paint supply distributor. Custom-mixed colors can always be identified easily. The contents of the can must be written or typed on the label by the person who mixed the paint.

An *intermix system* is a full set of paint pigments and solvents that can be mixed at the collision repair shop. Most paint manufacturers have made a color mixing system available to painters. With such

Figure 17-11 Study this typical page from a color manual. It gives formulas for making the color match the vehicle more closely.

an "intermix" system, it is possible to mix thousands of colors at a savings over the cost of factory-packaged colors or jobber mixing of colors.

An in-shop mixing system also allows you to better match the color on the vehicle more easily.

SPECTROPHOTOMETER

A *spectrophotometer* is an electronic device for analyzing the color on a vehicle. It electronically reads the color frequencies in the finish to quickly and accurately find the correct formula for a color match. See Figure 17–12.

The spectrophotometer wand or box is placed on the surface (either vehicle or test panel) to be checked. Some systems require that a test panel be sprayed for a comparison. The multi-angle spectrophotometers get a reading at 45, 25 and 75 degrees. Each angle is read for several variables.

Depending upon the type of system, they will read:

1. **LAB** or lightness, red to green hue, and yellow to blue hue.

2. **LCH** or lightness, chroma, and hue.

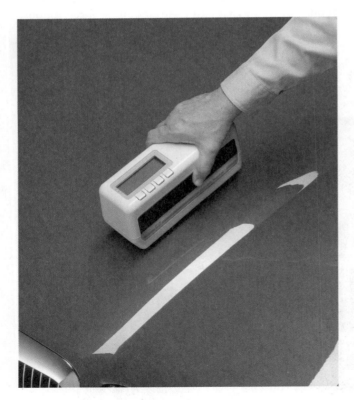

Figure 17–12 A spectrophotometer, or electronic color analyzer, uses electronic technology to "read" the actual color of the vehicle. It can then communicate with a computer system and paint formula software to mix the correct color. *(PPG Industries, Inc.)*

Most systems compare the vehicle and test panel to one another. The refinishing technician will get a reading on the relative lightness/darkness, hue, and chroma of the vehicle to the panel checked. It is still up to the technician to decide how to move the color closer to the vehicle's color. Decisions on which tint and how much will be added must still be made using human judgment.

COMPUTERIZED COLOR MATCHING SYSTEMS

Computerized color matching systems use data from the spectrophotometer to help match the vehicle's color. Many spectrophotometer systems can input color data into a computer. The computer can then use its stored data to help determine how to mix or tint the color.

Depending upon the sophistication of the system, a computerized color matching system may be able to:

1. Compare the actual color of the vehicle to a computer-stored set of color formulations.

2. Make a recommendation on which tint in the formula will move the sample panel closer to the vehicle.

3. Automatically keep a record of the mixing or tinting procedure. This will let you quickly match the paint if the vehicle returns for another repair.

COLOR VARIANCE PROGRAMS

Color variance programs compare the color of refinish paints to OEM color standards and actual painted parts and panels obtained from collision repair shops. If a particular OEM finish variation is noted often enough, a paint manufacturer may develop a **color variation formula** to match the OEM finish.

Variance chips are several samples of color used to help match color variations. One paint code may have a series of variance chips on each side of a typical color. They can be used to adjust the color formulation more precisely than by just using one chip. The chips are organized into books by the paint manufacturer. Many paint manufacturers have color variance programs. See Figure 17–13.

To use variance chips, lay the chips on the vehicle under proper lighting. Find the variance chip that best matches the color on the vehicle. Using the number for this chip, the computer will then give instructions for mixing the color to match the variance chip and vehicle. Computerized formulas will give variance to change the color as needed. See Figure 17–14.

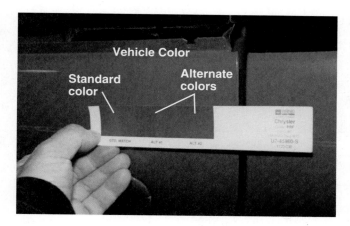

Figure 17–13 Alternate, or variance, chips are often provided so you can more closely match the formula to the actual color on the vehicle. Hold alternate or variance chips next to the vehicle and select the one closest to the vehicle's color.

SPRAY METHODS AFFECT COLOR

Although a technician cannot control the variables that can affect colors at the manufacturing plant, there are variables in the shop that can be controlled. For example, a refinishing technician can control the:

1. Agitation of the paint
2. Application techniques
3. Amount of material applied
4. Type of spray gun
5. Spray gun setup
6. Atomizing of the color

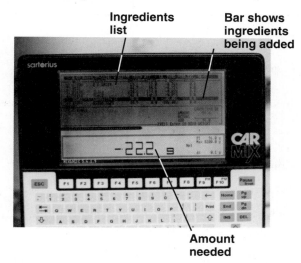

Figure 17–14 Computerized mixing systems will give amounts of each ingredient to use when matching color. Spectrometer or variance chips can be used to input the desired color formula. Electronic scales will then give a readout of each ingredient needed to mix the color properly.

7. Type and amount of solvent used for reduction
8. Identification of the correct paint code and mixing formula

All of these will help you make the new color match the existing finish. Refer to Figure 17–15.

Varying the spraying technique can affect color. That is, the application technique could cause the color to vary. Technicians who spray wet end up with a darker color than those who spray drier, especially with metallics.

MATCHING SOLID COLORS

For many years vehicles were solid colors, such as black, white, tan, blue, green, maroon, and so on. Solid colors reflect light in only one direction. Solid colors are still used on vehicles, but to a lesser degree when compared with a few years ago.

Matching solid colors is easier than matching metallic or mica paints. You only need to match the color pigment and NOT the metal or mica flakes suspended in the paint.

Oddly enough, a mismatch in a panel repair will usually show up more than a mismatch in a spot repair—even though the spot repair is smaller. That is because a panel, such as a vehicle door, has a distinct edge. And the repair, obviously, cuts off at that edge. Mismatched panels, as in the case of front and rear doors, will be right next to each other and will show a sharp contrast.

A spot repair, on the other hand, is performed by blending the repair into the surrounding area. In spot repairs, the first coat is applied to the immediate area being repaired. Subsequent coats extend beyond this area gradually. Finally, a blend coat extends beyond the color coats. Thus, if there is a slight mismatch, the blend coat and the last color coat will allow enough **show-through** of the old finish to make the color difference a gradual one.

MATCHING METALLIC FINISHES

Metallic colors contain small flakes of aluminum suspended in liquid, Figure 17–16. The position of the flakes and the thickness of the paint affect the overall color. The flakes reflect light while the color absorbs a higher amount of the light. The thicker the layer of paint, the greater the light absorption.

A **dry application of paint** makes the color appear lighter and more silver. The aluminum flakes are trapped at various angles near the surface of the color film. Light reflection is NOT uniform. The light has less paint film to travel through; little of it is

Effects Of Gun Setup

Changing Of Color Tone Spraying Condition	Color Becoming Lighter	Color Becoming Deeper	Degree Of Effect
Evaporation Speed Of Solvent	Fast	Slow	Medium
Paint Viscosity	Low	High	Large
Paint Discharge Quantity	Small	Large	Large
Spray Pattern Width	Wide	Narrow	Medium
Gun Speed	Fast	Slow	Medium
Gun Distance	Apart	Close	Medium
Air Pressure	High	Low	Large
Paint Film Thickness	Thin	Thick	Large

Note : Color Tone Will Become Vivid By Clearcoating.

Figure 17–15 This chart shows the effects of spray gun setup. Study them carefully. *(I-CAR)*

absorbed. The result is nonuniform light reflection and minimum light absorption.

A **wet application of paint** makes the color appear darker and less silver. The flakes have sufficient time to settle in the wet color. The flakes lie parallel to and deeper within the paint film. Light reflection is uniform and, because the light has to go farther into the paint film, light absorption is greater. The result is a painted surface that appears deeper and darker in color.

NOTE! Metallic colors must be stirred and mixed thoroughly before use. The pigment quickly settles below the binder. Also, the aluminum flakes settle below the pigment. If flakes stay at the bottom of the can, the paint will NOT match the same color on the vehicle being refinished.

In summary, the shades of metallic colors are controlled by:

1. Choice of solvents

2. Color reduction

3. Air pressure

4. Wetness of application

5. Spraying techniques

A good technician must know how to handle metallic colors. They are very sensitive to the solvents with which they are reduced and the air pressure with which they are applied. Metallic colors are also affected by a number of variables. This would include spraying conditions such as temperature, humidity,

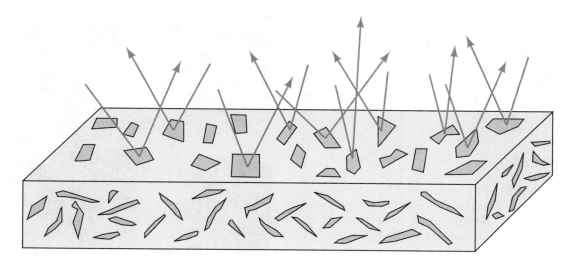

Metallic Topcoat

Figure 17–16 Metallic flakes in a color will reflect almost all light back out. The location and orientation of the flakes is critical to matching metallic finishes.

and ventilation. It also includes the spraying process, such as amount of reduction, evaporation, speed of solvents, air pressure, and type of equipment.

To darken a metallic color:

1. Use a larger fluid needle and fluid nozzle.

2. Increase fluid flow.

3. Decrease fan width.

4. Decrease air pressure.

5. Decrease travel speed.

6. Use a slower evaporating solvent.

To lighten a metallic color, do just the opposite.

MATCHING MULTI-STAGE FINISHES

Another type of glamour finish is the multi-stage finish. Shown in Figure 17–17, **multi-stage finishes** generally consist of:

1. Metal treatment (phosphate)

2. E-coat

3. Primer coats

4. Colored basecoat

5. Mica intermediate coat or pearlcoat (if applicable)

6. Two coats of clearcoat

A multi-stage system may use a fluorine clear over a urethane clearcoat. The **fluorine clearcoat** is used for the following reasons:

1. It improves resistance to UV radiation.

2. It aids protection from weathering, chalking, oxidation, and industrial fallout.

3. It improves luster of the finish.

4. It reduces required maintenance, such as waxing and polishing.

Mica and carbon graphic pigments have special requirements when it comes to color evaluation. In vehicle finishes, mica may be coated with titanium dioxide. The thickness of the titanium dioxide coating determines the colors that are reflected and allowed to pass through.

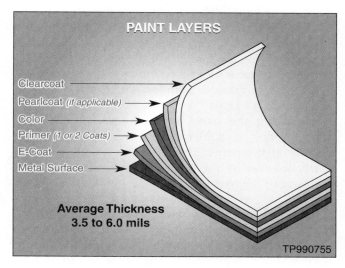

Figure 17–17 This cutaway shows the basic makeup of a multi-stage finish. (With permission from Nissan North America, Inc.)

ZONE CONCEPT

The *zone concept* divides the horizontal surfaces of the vehicle into zones defined by character lines and moldings. It requires refinishing of an entire zone or zones with basecoat, mica intermediate coats, and clearcoats.

COLOR FLOP

Flop, also termed **flip-flop,** refers to a change in color hue when viewing from head-on and then from the side. The lightness or darkness of a color varies from head-on to side-tone views. Flop occurs most often in paints containing metallic pigments, but a type of flop can also occur in mica paints. Solid colors do NOT exhibit flop.

To understand flop, think of the flakes in metallic and pearl colors as tiny mirrors. When light shines onto a mirror's surface, it is reflected back in exactly the opposite direction. Similarly, when light shines onto the aluminum flakes in a vehicle's metallic finish, it is reflected back in a straight line in the opposite direction. The brightness caused by this reflection is greatest when you are standing in the angle from which the light is reflecting. The brightness could drop off rapidly as you walk to one side or away from this angle. The direction and intensity of the light being reflected back by the flakes in the paint film is what causes the flop phenomenon. See Figure 17–18.

Head-on view compares the test panel to the vehicle straight on, or perpendicular. **Side-tone view** compares the test panel to the vehicle on a 45–60 degree angle. Compare the test panel to the vehicle on both angles with metallic finishes. The position of the metallics in the paint film can cause the color of the paint to change from head-on to side-tone.

The first approach to correcting the problem is to adjust the spraying technique to compensate for this effect. Spraying the fender a little wetter will slightly darken the appearance when looking directly into the panel. When viewed from an angle, the resulting appearance is lighter.

Spraying the panel slightly dryer reverses the effect, giving a lighter appearance when looking directly at the panel. This is because the aluminum flakes are close to the surface. The result is a darker appearance viewed at an angle, as light becomes trapped.

If spray techniques cannot correct the condition, the addition of a small amount of white will eliminate the sharp contrast from light to dark when the surface is viewed at various angles. The white acts to dull the transparency, giving a more uniform, subdued reflection through the paint film. Care should be taken

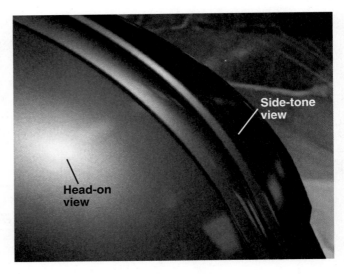

Figure 17–18 Flop occurs when a metallic or multi-stage color looks different when viewed at different angles. Note how this metallic color appears different on this curved bumper.

when adding white since the change occurs quickly. Once too much white is added, recovering the color match becomes impossible.

When confronted with an extremely difficult flop condition, the best method involves adding white plus blending the color into the adjacent panels. When blending, extend the color in stages. Spraying a little wetter will darken the paint head-on, and lighten side-tone. Spraying drier will lighten the paint head-on, and darken side-tone.

CHECKING COLOR MATCH

There are two ways most often used to check color match:

1. Spray-out panels—used with conventional paints

2. Let-down panels—used with multi-stage paints

Spray-out Panel

The *spray-out panel* checks the paint color and also shows the effects of the technician's technique on a test piece, Figure 17–19. Spray-out panels are prepared by applying the paint as near to the actual spraying conditions as possible. When done properly, a spray-out panel shows the color exactly as it will look when sprayed on a vehicle.

Before making a spray-out panel, double-check the paint code. Reduce the paint correctly. Set up the spray gun with the material that will be sprayed. Adjust the air pressure at the spray gun for the material to be applied.

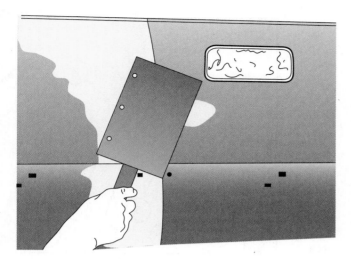

Figure 17–19 A spray-out panel tests the color match before applying the color to the vehicle.

When making a spray-out panel, apply a primer that matches the primer on the vehicle. Apply basecoat to full hiding. Allow proper flash time between coats. Apply clearcoat to half the panel. The panel should be fully dry or cured prior to evaluating color match.

When evaluating color match on basecoat-clearcoat systems, apply clear to only half of the panel. The uncleared section can be used to check the color match of the basecoat prior to applying the clear. You can refer to any noncleared patch on the vehicle.

> **NOTE!** On base-clear finishes, the basecoat can also be applied to a piece of clear plastic. After the basecoat dries, the sheet is turned over to give an idea of what the basecoat will look like with clear applied.

Let-down Panel

A *let-down panel* is used to evaluate color match on tri-coat and multi-stage paint systems, Figure 17–20. Directions for making a let-down panel are available from the manufacturer of the refinishing system used. Each coat of the mica intermediate coat will darken the appearance of the finish.

Here is how to make a let-down panel for a tri-coat finish.

1. Prepare the let-down panel with the same primers being used on the vehicle. If a primer-sealer is going to be used, apply the primer-sealer to the let-down panel also. Generally, a light color primer (or sealer) is preferred for tri-coat finishes.

2. Apply the basecoat color to hide the test sheet, using the same air pressure and spray pattern that

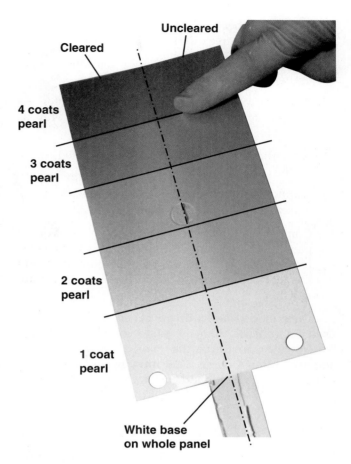

Figure 17–20 A let-down panel tests a multi-stage paint system for match before spraying the vehicle. You must mask each layer to apply different amounts of basecoat and clearcoat. You can then use the panel as a large paint chip for comparison to the vehicle's finish.

will be used on the vehicle. Duplicating the actual spray techniques when preparing the let-down panel is an important point. Make sure NOT to vary your procedures.

3. After the panel has dried, divide it into four equal sections, Figure 17–20. Next, mask off the lower three quarters of the panel, exposing the top quarter.

4. Apply one coat of mica midcoat color over the top quarter of the panel.

5. After the mica coat has flashed, remove the masking paper and move it down to the middle of the panel, exposing the top half.

6. Apply another coat of mica midcoat color over the exposed top half of the panel.

7. After this second coat has flashed, remove the masking paper and move it down to expose three quarters of the panel.

8. Apply another coat of mica midcoat color over the exposed three quarters of the panel.

9. After flashing, remove the masking paper entirely.

10. Apply a fourth coat of mica midcoat color. As always, spray the coating in the same way as would be done on the vehicle.

11. After the entire let-down panel has dried, mask off the panel lengthwise this time.

12. Apply the manufacturer's recommended number of clear coats to the exposed half.

13. Compare the different shades on the let-down panel with the paint on the vehicle. Use the same number of coats used on the matching section of the let-down panel to achieve the correct paint match.

Once made, the let-down panel can be kept and used on vehicles with the same color code. On the back of the panel note the color code, gun settings, and technician's name. One must be made for each different tri-coat color and each technician.

BLENDING CLEARCOATS

Clearcoats may be blended by using additives provided by the manufacturer. When working with base-clear finishes, remember:

1. Clearcoats are NOT perfectly clear. They will change the appearance of a color.

2. Blend the basecoat and apply clear to the entire panel.

3. You may have to step-out the clear if it must be blended.

4. You should clear the entire surface of horizontal panels.

5. Blend into the smallest area possible to help hide the repair.

FLUORINE CLEARCOAT REPAIRS

The basic steps for spot repair with a fluorine clearcoat system are shown in Figure 17-21. They include:

1. Compound or sand with 1200–1500 grit sandpaper for better adhesion.

2. Apply first coat of basecoat.

3. Apply second coat of basecoat.

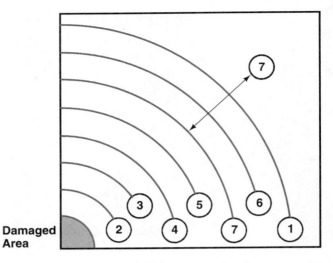

Figure 17–21 Compare the numbers on this illustration to the following: 1: Compound or sand with 1200–1500 grit sandpaper. 2: Apply the first coat of basecoat. 3: Apply the second coat of basecoat. 4: Apply the third coat of basecoat or until hiding is obtained. 5: Apply color blender if necessary. Dry at 140° F (60° C) for 20 minutes. 6: Apply 3–4 coats of fluorine clearcoat. Dry properly between coats. 7: Polish with fine compound. *(Used with permission from Nissan North America, Inc.)*

4. Apply third coat of basecoat or until hiding is obtained.

5. Apply color blender if necessary. Dry at 140°F (60° C) for 20 minutes.

6. Apply 3–4 coats of fluorine clearcoat. The area between numbers 5 and 6 in Figure 17–21 is faded out or blended if required. Dry at 60–70°F (16–21°C) for 10 minutes between coats. After applying final coat, force dry at 170°F (75°C) for 45 minutes.

7. Polish with fine compounds.

When working with tri-coat and multi-stage finishes, a technician must match the basecoat prior to applying the intermediate and clearcoats. Some vehicle manufacturers leave an area of basecoat that is NOT coated with mica or clear for comparing the basecoat spray-out to the vehicle.

TRI-COAT SPOT REPAIR

A **halo effect** is an unwanted shiny ring or halo that appears around a pearl or mica color repair. It is caused by the paint being wetter in the middle and drier near the outer edges of the repair.

Avoid a halo effect by applying the first coat of mica to the basecoat only. The more intermediate mica

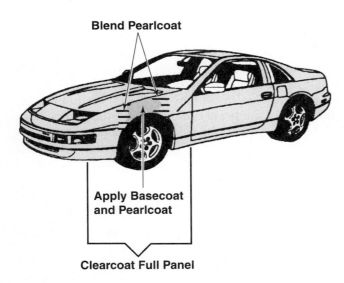

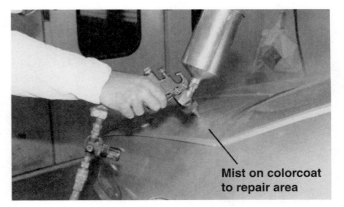

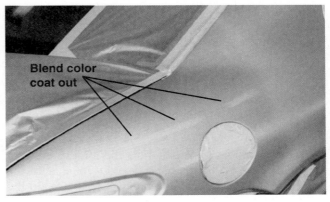

Figure 17–22 Note the coverage required for spot repairing a multi-stage finish. You must blend the pearl coat over an area larger than the repair. Then, clear the whole panel. *(Used with permission from Nissan North America, Inc.)*

coats that are applied, the darker the finish will appear. Allow a larger area in which to blend the mica intermediate mica coats. They require more room to blend than a standard basecoat. Keep the tri-coat repair area as small as possible.

This is one paint manufacturer's method for a spot or partial repair on a tri-coat system:

1. Apply primer to area over the body filler.

2. Apply adhesion promoter to all unsealed panels. Adhesion promoter should extend beyond the repair area.

3. Apply two or more coats of basecoat to areas to full hiding. Extend each coat slightly beyond the previous one, allowing the surface to dry between coats. Look at Figure 17–22.

4. Check let-down panel for total number of coats of mica intermediate coats needed to match the OEM finish. Apply the intermediate coat to repair area, extending each coat beyond the last. Allow adequate flash time between coats.

5. Apply two coats of clear over entire panel. The clear may have to be blended into the sail panel. See Figure 17–23.

TRI-COAT PANEL REPAIR

These are typical steps for a panel repair with a multi-stage finish:

1. Apply primer to area over the body filler.

Figure 17–23 (A) Apply colorcoat only where it is needed, or over the repair area. Mist metallic colors on lightly to help orient metallic flakes evenly. (B) Note how the colorcoat has been applied only to the repair area on the left of the quarter. (C) Clear the whole panel and blend the clear into the sail panel.

2. Apply adhesion promoter to a large area to be repaired.

3. Apply two or more coats of basecoat to areas to full hiding. Extend each coat slightly beyond the previous one, allowing the surface to dry between coats. See Figure 17–24.

Full Panel Repair

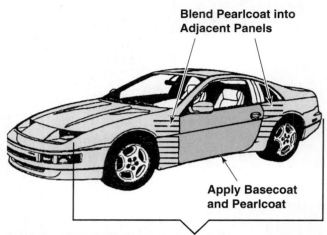

Blend Pearlcoat into Adjacent Panels

Apply Basecoat and Pearlcoat

Clearcoat Entire Side, Blending into Pillars and Sail Areas

Figure 17–24 Note this panel repair of a multi-stage finish. You will usually have to blend the pearl into adjacent panels. Then, clear the whole side of the vehicle. *(Used with permission from Nissan North America, Inc.)*

4. Check the let-down panel for the total number of mica intermediate coats needed to match the OEM finish to areas. Extend each coat beyond the previous one, with only the last coat extending into adjacent panel. Allow adequate flash time between coats.

5. Apply two coats of clear to the entire repair area, ending at panel ends. (Edge to edge of doors, quarter panels, etc.)

BLENDING MICA COATS

This is a typical mica intermediate coat blending procedure:

1. Apply mica intermediate coat to the area covered by the basecoat.

2. Apply second mica intermediate coat well beyond the edge of the first coat.

3. Apply third mica intermediate coat so it extends just beyond the edge of the first coat but within the second coat.

4. Apply fourth mica intermediate coat to just beyond the edge of the second coat.

Spot repair recommendations for tri-coats are now available. However, the zone repair is still a workable repair option, and may be required on certain vehicles.

MATCHING MICA COLORS

Many base-clear finishes contain mica pigments. Some of these finishes have proven to be especially challenging to match.

Because the finish may NOT provide full hiding, the color of the primer may show through enough to change the look of the color. If the refinishing technician applies the color to full hiding, there may be a color mismatch because the primer color is no longer visible through the topcoat.

A color effect test panel is required for base-clear finishes that contain mica. To make this test panel:

1. Apply a primer that matches the primer color on the vehicle. Allow it to dry.

2. Mask the panel into thirds.

3. Apply one coat of basecoat to the exposed section. Allow it to dry.

4. Unmask the next section, and apply another coat of basecoat.

5. Unmask the last section of the panel, and apply an additional two coats of basecoat to the entire panel. Allow it to dry.

6. Apply two coats of clear to the panel. Allow it to dry.

7. Compare the test panel with the vehicle to determine the number of basecoat coats that provide the best match.

Applying an extra coat of refinish basecoat to the vehicle may change the color to the point of NOT matching. Reds or other colors that traditionally provide poor hiding work best for making this panel.

When working with multi-stage finishes, do NOT tint the mica intermediate coat. Adjust the number of intermediate coats that are applied. Micas used in basecoat-clearcoat finishes are treated the same as metallic colors. They require special attention to side-tone angles. Mica will make the paint appear lighter and less chromatic at the head-on angle.

Mica will also reduce the metallic effect. For example, if the vehicle seems darker and more chromatic on the side-tone, add mica. If the test panel is darker and more chromatic on the side-tone, there is too much mica pigment.

Mica cannot be removed from the paint; however, this condition can be corrected by adding small amounts of both black and white. You can also add small amounts of the coarsest metallic in the formula.

Keeping good records during and after the tinting operation is important. Use the plotting chart or a separate record sheet.

WHY A COLOR MISMATCH?

Tinting should be used only as a last resort. If the color of the refinishing varies from the original, check the following possible reasons for the mismatch:

1. Has the original finish faded? Check the color on unexposed areas such as door jambs or under the deck lid or hood to determine if the finish has faded. You can restore the luster by compounding the original finish beyond the repair area. This can remove chalking and oxidation before making a color comparison.

2. Was the wrong color used? Check the vehicle manufacturer's paint code and the paint manufacturer's stock number of the color used to make sure that it is the right color.

3. Were the pigment and/or metallic flakes mixed thoroughly? Leaving pigment, metallic flake, or pearl in the bottom of the can could cause a mismatch. Be sure to agitate (mix) the paint thoroughly before and while spraying.

4. Has the amount of reducer been measured carefully? Over-reducing will lighten or desaturate a color. Remember that it is easy to add more reducer, but it cannot be taken out.

5. When using a test panel, was enough time allowed for the paint to dry? Be sure to allow proper flash and dry times for each coat. The paint usually gets darker as it dries. Dry time can be shortened by using heat lamps, heat guns, or other drying methods.

6. If using a clearcoat, remember that compounding the clear will make the paint appear darker. If testing a basecoat-clearcoat finish, color judgment cannot be made until the clear is applied to the basecoat.

TINTING

When the color does NOT match the vehicle, you must determine if:

1. Paint code was properly identified.

2. Proper paint code was used, Figure 17–25.

3. Spray-out test panel was made properly (proper mixing, application, etc.).

   Was the spray-out test panel checked:

1. Against a clean vehicle?

2. In the proper light?

3. On both face and side-tone views?

   Finally, check the paint manufacturer's variance chips for a formula that may obtain a better match. If

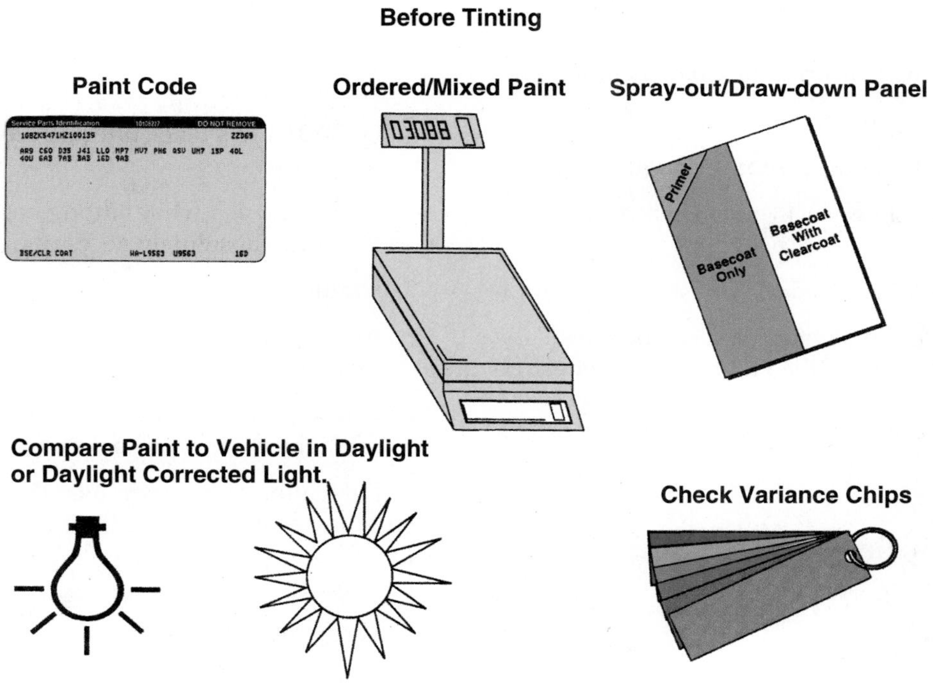

Before Tinting

Figure 17–25 Note things to check before tinting. *(I-CAR)*

there is NOT a variance formula, determine if the color is close enough to blend. If NOT, the color must be tinted to get a blendable match.

A color mismatch does NOT automatically mean tinting. There are a number of things to be checked first. Tinting should be done only to move the color close enough for blending. Do NOT try to tint to a perfect match. Blending the color out over a larger area takes care of any final color variations.

It must be remembered that the color might NOT be exactly right because all finishes gradually change color over time. Some colors fade lighter, others go darker. Yellow, for instance, fades fairly rapidly. If the yellow fades to a cream, the color will usually go lighter and whiter.

Tinting involves altering the color slightly to better match the new finish with the existing finish. Tinting may be one of the least understood tools of finish matching. There are three basic reasons for tinting:

1. To adjust color variations in shades to match the color from the manufacturer.

2. To adjust color on an aged or weathered finish.

3. To make a color for which there is no formula or for which there are no paint codes available.

Some computerized paint systems provide color variance information. Before making the decision to tint, determine if a color variance chip or formula is available. These may provide a blendable match, and reduce or eliminate the need to tint. Computerized paint systems also provide tinting information. See Figure 17–26.

Computer tinting information may include:

1. A list of tints by number and name

2. Notes on tint strength, or hiding characteristics

3. Description on how each tint affects value, hue, and chroma

4. Cautionary notes, if needed, for using each tint

This computer matching information can be printed out with a copy of the paint formula when the paint is mixed.

The printout can be referenced during the tinting operation.

Review of Tinting

Remember:

1. Tint only to blend.

2. Use only one tinting base at a time and check the color after every "hit," or adjustment.

Figure 17–26 This technician is tinting a color by adding needed color pigment.

3. Always tint within the color formula.

4. Evaluate the color match in daylight or under daylight-corrected lighting.

5. Do NOT use a black tinting base unless absolutely necessary.

6. Do NOT use a white tinting base for metallic colors unless absolutely necessary.

7. Evaluate the color match against a clean vehicle.

8. Adding small amounts of coarse metallic tinting base can darken side-tone.

9. Adding small amounts of metallic tinting base can lighten a metallic color.

Evaluating and tinting metallic colors is more difficult than evaluating and tinting solid colors. Evaluation of both metallic and solid colors should be done in this order:

1. Verify formula.

2. Plot the color on the plotting chart.

3. Compare the test panel to the vehicle.

4. Check and adjust value.

5. Check and adjust hue.

6. Check and adjust chroma.

Using kill charts and adding tints NOT in the formula will also move the color, but NOT around the outside of the color wheel. Adding tints suggested by a **kill chart** will move the paint more directly toward the gray center of the color wheel, affecting both hue and chroma.

There may be situations where this is desirable, but as a rule of thumb, do NOT use kill charts to change hue. The proper tint for changing hue is already in the formula. Adding tints which are NOT in the formula may also cause problems with metamerism.

Adding paint instead of tints will add the dominant tint and all the other tints in that paint formula. Again, use only tint bases and do NOT add paint.

Some paint manufacturers produce metallic tinting bases designed to correct a specific problem, usually having to do with changes in side-tone. Use these products according to the manufacturer's recommendations. Using the color formula and tinting guide, select the tinting base that will move the color in the right direction. Refer to Figure 17–27.

When tinting, a refinishing technician should:

1. Use only half the can of paint.

2. Use only tint bases. Never add mixed paint (color).

3. Use only tints that are in the paint formula.

4. Add tint bases in small amounts, and check the color following each addition.

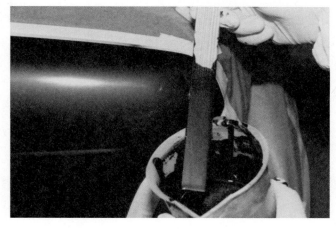

Figure 17–27 Tinting can be complex. Take your time and compare the tinted color to the color on the vehicle.

5. Keep records of each tint base and the amount added. This will be useful if more needs to be mixed.

A chart summarizing the tinting process is given in Figure 17–28 on page 366. Read through it carefully.

Summary

- The colors in the spectrum are easily remembered by the name "ROY G. BIV" which stands for the colors red, orange, yellow, green, blue, indigo, and violet.

- The same shade of paint will look very different under incandescent and fluorescent lights. Therefore, it is very important to view a color in daylight or under a balanced artificial light.

- Color should be based upon value, hue, and chroma.

- Computerized paint matching systems use data from the spectrophotometer to help match a color.

- The two ways to check color match are:
  1. Spray-out panels—used with conventional paints.
  2. Let-down panels—used with multi-stage paints.

Technical Terms

color matching	*custom-mixed colors*
color	*intermix system*
white light	*spectrophotometer*
color spectrum	*LAB*
sunlight	*LCH*
incandescent light	*computerized color*
fluorescent light	*matching system*
Color Rendering Index	*variance chips*
(CRI)	*dry application of paint*
lumen rating	*wet application of*
light temperature	*paint*
color blindness	*multi-stage finish*
color tree	*fluorine clearcoat*
value	*zone concept*
hue	*flop*
chroma	*head-on view*
metamerism	*side-tone view*
plotting color	*spray-out panel*
color directory	*let-down panel*
paint formula	*halo effect*
basecoat patch	*tinting*
factory packaged colors	*kill chart*

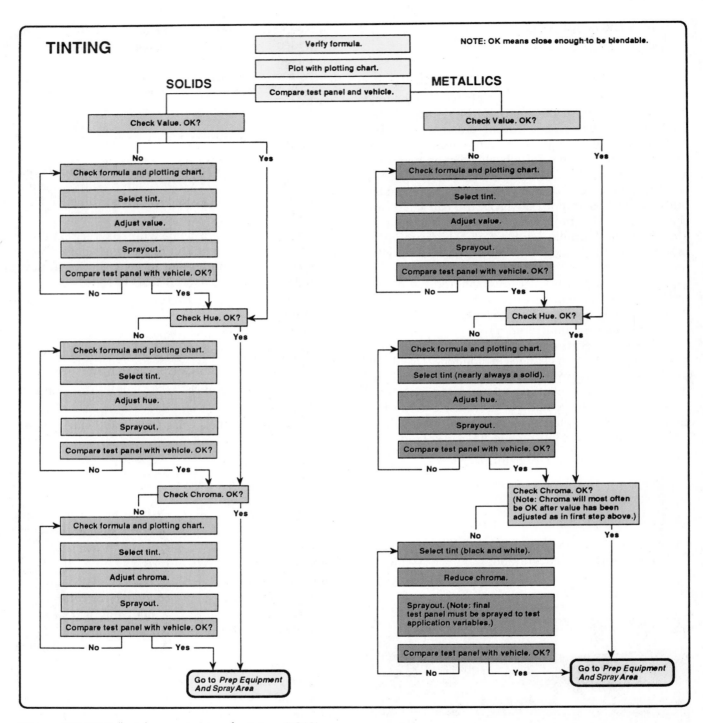

Figure 17–28 Follow these major steps for tinting. *(I-CAR)*

Review Questions

1. By passing light through a prism, light is broken down into its separate _____, called the _____.

2. When evaluating the color of a finish, it should be viewed under:
   a. sunlight.
   b. incandescent light.
   c. fluorescent light.
   d. drop light.

3. How do we see the color of a finish or any object?

4. _____ refers to the degree of lightness or darkness of the color.

5. _____, also called _____, or_____, describes what we normally think of as color.

6. _____ refers to the color's level of intensity, or the amount of gray (black and white) in a color.

7. What is a color directory?

8. A basecoat patch is a small area on the vehicle's surface without clear to enable the technician to check for color match. True or false?

9. This is a full set of paint pigments and solvents that can be mixed at the collision repair shop.
   a. Factory packaged colors
   b. Custom-mixed colors
   c. Intermix system
   d. Manufacturer system

10. What is a spectrophotometer?

11. List three tasks that a computerized color matching system may be able to do.

12. Explain variance chips in detail.

13. A dry metallic paint spray makes the color appear darker and less silver. True or false?

14. List six ways to darken a metallic color.

15. Define the term "flop."

ASE–Style Review Questions

1. Which of these paints is the most difficult to match?
   a. Single-stage white
   b. Pearl paint
   c. Black/clearcoat
   d. White/clearcoat

2. Technician A says that sunlight should be used when evaluating color. Technician B says that when painting, a lamp rating of 6000–7000 Kelvin is recommended. Who is correct?
   a. Technician A
   b. Technician B
   c. Both Technicians A and B
   d. Neither Technician

3. Which of the following is NOT a dimension of color?
   a. Value
   b. Hue
   c. Chroma
   d. Gray scale

4. Technician A says in order to use a color directory, locate the vehicle manufacturer's paint code. Then you can identify the color chip next to it. Technician B says it is wise to compare the color chip with the vehicle color. There is always the chance that the vehicle has been repainted with a different color. Who is correct?
   a. Technician A
   b. Technician B
   c. Both Technicians A and B
   d. Neither Technician

5. Technician A says that paint can always be buffed so that the original paint code can be used. Technician B says that you may have to custom mix the paint to match the weathered paint. Who is correct?
   a. Technician A
   b. Technician B
   c. Both Technicians A and B
   d. Neither Technician

6. Technician A says that a spectrophotometer is an electronic device for analyzing the color on a vehicle. Technician B says that most of these systems require that a test panel be sprayed for a comparison. Who is correct?
   a. Technician A
   b. Technician B
   c. Both Technicians A and B
   d. Neither Technician

7. Technician A says that variance chips are no longer a viable way of selecting a paint color. Technician B says that spray methods affect color. Who is correct?
   a. Technician A
   b. Technician B
   c. Both Technicians A and B
   d. Neither Technician

8. Technician A says that matching solid colors is easier than matching metallic or mica paints.

Technician B says that matching solid colors is more difficult than metallic, in which the flakes are all silver. Who is correct?

a. Technician A
b. Technician B
c. Both Technicians A and B
d. Neither Technician

9. *Technician A* says that spraying metallic colors dry makes the color appear darker and less silver. *Technician B* says that spraying metallic colors dry makes the color appear lighter and more silver. Who is correct?

a. Technician A
b. Technician B
c. Both Technicians A and B
d. Neither Technician

10. All of the following will darken a metallic color EXCEPT:

a. larger fluid nozzle
b. increased fluid flow
c. increased fan width
d. decreased air pressure

Activities

1. Obtain a color matching manual. Read its directions and use it to find a color match for a vehicle. Write a report on your findings.

2. Locate a shop with a computerized color matching system. Write a summary on how to use the equipment.

3. Make either a let-down or a spray-down panel. Make it into a display for the classroom.

18

Paint/Refinish Problems and Final Detailing

Objectives

After studying this chapter, you should be able to:

- List and explain the most common paint/refinish problems.
- Repair common finish problems.
- Use a dirt-nib file.
- Wet sand to remove minor finish problems.
- Hand and machine compound a finish.
- Properly final detail a vehicle.

Introduction

After doing the straightening work and painting/refinishing the vehicle, you must check your work quality, Figure 18–1. Ideally, the vehicle can be released to the customer after a minor cleanup. Sometimes, however, you will find small imperfections in the paint film that must be corrected. In rare occasions, you may have to solve major paint problems on existing or freshly painted surfaces.

If you do find problems after refinishing, it is very important that you discover what caused the trouble. Paint problems are usually caused by a mistake during straightening, vehicle prep, shop prep, or color mixing. You must try to analyze the problem and find out what must be done differently to prevent the same trouble from happening again. This chapter will provide you with the information needed to analyze, prevent, and correct paint problems.

The last section of the chapter summarizes what must be done to final detail the vehicle. Since *first impressions* are important, you must know how to properly clean the vehicle before releasing it to the customer.

Figure 18–1 After refinishing, you must inspect all surfaces for problems. The slightest speck of dust or any other imperfection must be found and corrected. (Courtesy of DaimlerChrysler Corporation)

SURFACE IMPERFECTIONS

The steps for correcting surface imperfections include:

1. Examining the finish for imperfections

2. Correcting the imperfection

3. Finesse finishing

4. Final detailing

There are a number of defects in the repair area that must be corrected before the vehicle is delivered to the customer. A few of these paint problems include runs, sags, orange peel, dirt or dust, and overspray.

SURFACE FINISH MEASUREMENTS

A good panel repair will have several desirable characteristics. It should not be rough, wavy, nor have surface flaws.

Surface roughness is a measurement of paint film or other surface smoothness in a limited area. If you look at any paint surface with a magnifying glass, it will have some surface roughness. Ideally, you want the finish to be as smooth as possible. Paint film roughness is normally due to improper mixing or application (spraying).

Surface waviness is a measurement of an area's general levelness or trueness. If a panel is supposed to be perfectly flat, there should be no waviness when viewed from an angle. Improper sanding or blocking of plastic filler or primer is the main reason for waviness.

The first requisite for a high-quality refinishing job is a smooth body surface. The body technician will

make extra work for the painting technician if the straightening/filling work is poor.

PAINT PROBLEMS

Paint problems include a wide range of troubles that can be found before or after painting/refinishing. You must be able to inspect and analyze problems efficiently. If you fail to find a problem before spraying, you will have extra work fixing the mistake. You must also be able to find and solve finish problems that are found after you paint/refinish a vehicle.

The following is a detailed explanation of the most common finish problems.

Acid and Alkali Spotting

Acid and **alkali spotting** cause an obvious discoloration of the surface, as in Figure 18–2. Various paint pigments react differently when in contact with acids or alkalies. The cause of acid and alkali spotting is a chemical change of pigments. This chemical change results from atmospheric contamination in the presence of moisture. This problem is found on older finishes that have been exposed to industrial pollution.

To remedy acid spotting, wash the vehicle with detergent and water. Follow this with a vinegar bath. You might try wet sanding and compounding if there is only minor spotting. If the spots have absorbed deep into the finish, you will have to sand and refinish. If contamination has reached the metal or subcoating, the spot must be sanded down to the metal before refinishing.

To prevent acid spotting, steps should be taken to minimize contact between the vehicle finish and

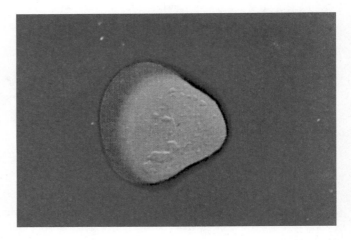

Figure 18–2 This is a close-up of acid or alkali spotting of paint. (PPG Industries, Inc.)

Figure 18-3 This paint has suffered from bleeding. *(PPG Industries, Inc.)*

Figure 18-4 Here is a blistering paint problem. *(PPG Industries, Inc.)*

the contaminated atmosphere. Advise the customer to keep the vehicle in a protected area, such as a garage, as much as possible. In addition to this, the vehicle finish should be washed frequently and vigorously flushed with cool water to remove contaminants.

Bleeding

Bleeding is a discoloration of the new paint after refinishing. It results when solvent in the new finish penetrates the materials underneath the finish. This causes dyes or colors in the existing paint, filler, or putty to dissolve, and these are then absorbed up into the new finish. Usually, reds and maroons release a dye that comes to the surface of the fresh finish to cause a bleeding problem. An example of bleeding is shown in Figure 18-3.

To remedy bleeding, remove all colorcoats and refinish. You can also allow the surface to cure. Then apply bleeder sealer and recoat.

To prevent bleeding, apply bleeder sealer over suspected bleeder colors before spraying new color.

Blistering

Blistering shows up as small, swelled areas on the finish, like a "water blister" on human skin. There will be a lack of gloss if blisters are small. You will find broken edged craters if the blisters have burst. See Figure 18-4.

Blistering is usually caused by corrosion under the surface. It is also caused by spraying over oil or grease, by moisture in spray lines, and trapped solvents. Blistering can also be due to prolonged or repeated exposure of the surface to high humidity.

To remedy blistering, sand and refinish the blistered areas. It can be prevented by thoroughly cleaning and treating metal.

Frequently drain your air lines of water. Avoid overly fast-drying solvents/reducers when temperature is high. Also allow proper drying time between coats.

Blushing

Blushing is a problem that makes the finish turn "milky." This is shown in Figure 18-5. It is caused by fast solvents/reducers in high humidity or condensation on the vehicle surface while painting.

To remedy blushing, use a slower-drying reducer or add retarder to the solvent/reducer and respray. You may also have to sand and refinish.

To prevent blushing, keep the material and the surface to be sprayed at room temperature. Select a good quality solvent/reducer. Use a retarder or reflow solvent when spraying in high humidity and warm temperatures.

Dirt in Finish

Dirt in finish is caused by foreign particles dried in the paint film. This is due to improper cleaning, not properly blowing off the vehicle, dirt on the technician, or failure to tack-rag the surface before spraying. This can also be due to a bad air regulator filter. A dirty work area or dirty spray gun can also deposit dirt in the finish. Refer to Figure 18-6.

Figure 18-5 Blushing makes the paint look "milky." *(PPG Industries, Inc.)*

Figure 18–6 Dirt in the finish can come from various sources. (PPG Industries, Inc.)

To correct light dirt in the finish, rub out the finish with polishing compound. If the dirt is deep in the finish, sand and compound to restore gloss. Metallic finishes may show mottling with this treatment and will then require additional colorcoats.

To prevent dirt in the finish, blow out all cracks and body joints. Clean with solvent and tack-rag the surface thoroughly. Be sure all of your equipment and work area is clean. Replace inlet air filters if dirty or defective. Strain out foreign matter from the material. Keep all containers closed when they are not in use to prevent contamination.

Fisheyes

Fisheyes are a problem with separation of the wet film. Small indentations will form in the wet film. Sometimes, the previous finish under the new material can be seen in these spots. An example is in Figure 18–7.

Fisheyes are usually caused by improper cleaning of the old surface. They can also be due to spraying over finishes that contain silicone.

To correct a fisheye problem, wash off the new finish while it is still wet and respray. You can also add a fisheye preventer additive to the paint and respray over the problem. This will usually make the fisheyes flow out smooth. To prevent fisheyes, clean the surface with wax and grease remover.

Mottling

Mottling is a streaking of the color, usually with metallic finishes. It is caused by excessive wetting or a

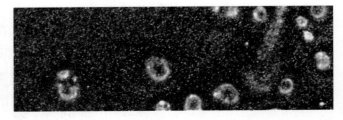

Figure 18–7 Fisheyes are small indentations in the surface film that look something like the eyes of a fish looking at you. (PPG Industries, Inc.)

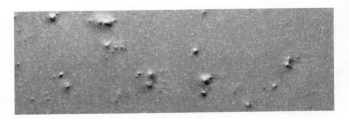

Figure 18–8 Mottling is a streaking problem common to metallic colors. (PPG Industries, Inc.)

heavier film thickness in some areas. Look at Figure 18–8.

To remedy mottling of a freshly applied finish, back the spray gun away when spraying. Also increase air pressure for the final coat. Avoid overreduction. On a dried finish, scuff down the surface and apply additional color to correct mottling.

To prevent mottling, avoid excessive wetting or heavy film buildup in local areas. Be careful not to overreduce the color.

Orange Peel

Orange peel is a paint roughness problem that looks like tiny dents in the paint surface. It resembles the skin of an orange. An example is in Figure 18–9.

The droplets formed in the spray pattern hit the surface and dry before they have time to flow out. This causes a roughness in the surface.

When this occurs with primers, excessive sanding is required to make the surface smooth enough to apply the topcoats. While there is always a slight amount of orange peel in a film, too much will cause roughness, low gloss, or both.

Orange peel is commonly caused by underreduction or an improper solvent/reducer. The paint will lack proper flow. It can also be due to the surface drying too fast and improper air pressure at the spray gun.

To fix orange peel while you are still painting, you can often thin or reduce the final coat to its maximum limit and apply the last coat wet. This may make the orange peel flow out. If not, with enamel, rub the sur-

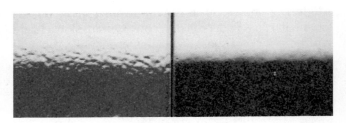

Figure 18–9 Orange peel is a rough, bumpy surface film. (PPG Industries, Inc.)

Figure 18–10 With peeling, a noticeable area of finish lifts. *(PPG Industries, Inc.)*

Figure 18–11 Pitting, or cratering, looks similar to dry spray. *(PPG Industries, Inc.)*

Figure 18–12 With a bleed-through problem, colors under the surface are visible. *(PPG Industries, Inc.)*

face with a mild polishing compound. With severe cases, you may have to sand and refinish.

To prevent orange peel, properly adjust air pressure to the spray gun. Also make sure you have properly reduced the material.

Peeling

Peeling shows up as a separation of the surface film from the subsurface. It is caused by improper surface preparation or incompatibility of one coat to another. See Figure 18–10.

To repair a peeling problem, you must remove the peeling finish completely. Prepare the metal properly and refinish with compatible materials.

To prevent peeling, thoroughly clean and treat the surface. Use recommended primers for special metals. Follow acceptable refinish practices using compatible materials.

Pin Holes or Blistering over Plastic Filler

Pin holes or blistering over plastic filler are noticed as air bubbles raising the surface film. These bubbles can then pop and cause craters when they erupt.

One cause for pin holes over plastic filler is excessive amounts of hardener. Vigorous stirring or beating of the hardener will also form bubbles in filler.

To fix this problem, sand thoroughly and recoat with a glaze coat of body filler or putty.

To prevent pin holes and blistering, mix in recommended quantities of hardener. Stir plastic filler mildly. Work out possible trapped bubbles when applying filler.

Pitting or Cratering

Pitting or **cratering** is a problem where small holes form in the paint film. It can look like dry spray or overspray. It is caused by the same things as blistering (except, in this case, the blisters have broken). Correc-

tions are also the same as for blistering. Refer to Figure 18–11.

Plastic Bleed-Through

Plastic bleed-through is a discoloration, often a yellowing, of the color due to chemicals from the plastic filler. Although it is very rare today, it may be caused by too much hardener or by applying the topcoat before the plastic filler is fully cured. See Figure 18–12.

To remedy plastic bleed-through, allow the topcoat to cure and refinish. To prevent this problem, use the right amount of hardener in plastic fillers. Allow adequate cure time before refinishing.

Corrosion under Finish

Corrosion under finish will cause the finish to peel or blister. You can also have raised spots on the surface film. This problem is caused by improper metal preparation. Broken film has allowed moisture to creep under the surrounding finish. Corrosion can also be caused by water in air lines. Look at Figure 18–13.

To remedy corrosion under the finish, seal off the entrance of moisture from the inner part of panels. Sand down to the bare metal, prepare the metal and treat it with phosphate before refinishing. You can

Figure 18–13 Corrosion under a finish is due to improper surface preparation.

also prevent a corrosion problem by applying epoxy primer directly to the metal. Locate any source of moisture and stop it.

When replacing ornaments or molding, be careful not to break the film and allow dissimilar metals to come in contact with each other. This contact can produce electrolysis that may cause a tearing away or loss of good bond with the film.

Runs

Runs occur when gravity produces a mass slippage of an overwet, thick film. The weight of the film will cause it to slide or roll down the surface. A large area of finish will flow down and form large globules. See Figure 18–14.

Runs are caused by:

1. Overreduction with low air pressure
2. Extra slow solvents/reducers
3. Spraying on a cold surface
4. Improperly cleaned surface
5. Spray gun too close or moved too slowly
6. Too many coats

To fix a run in a metallic colorcoat, you must usually allow the finish to cure. Then sand and refinish

Figure 18–14 Runs and sags are due to too much wet spray material in one place. (PPG Industries, Inc.)

the panel. If the run is only in the clearcoat and not in the metallic colorcoat, you can often wet sand to level the run without refinishing. On smaller parts or areas, you can also wash the wet paint off with solvent/reducer and start over.

With solid colors, you can also allow the finish to dry. Then, wet sand the run or sag before buffing. This will not work with metallics.

To prevent runs, use the recommended reducer at specified reduction and air pressure. Do not spray over a cold surface. Clean the surface thoroughly. Allow sufficient flash time between coats. Use proper gun spray pattern, speed, and distance from the surface.

Sag

A **sag** is a partial slipping down of the surface film created by a film that is too heavy to support itself. It appears something like a "curtain." A large area of film is pulled down by gravity.

Sags are caused by:

1. Underreduction
2. Applying successive coats without allowing dry time
3. Low air pressure (lack of atomization)
4. Spray gun too close to surface
5. Spray gun out of adjustment
6. Moving spray gun too slowly

To fix a sag, sand or wash it off and refinish. To prevent sags, use the proper solvent/reducer at recommended reduction. Adjust the air pressure and the gun for correct atomization and move it at the right speed. Keep the gun at the the right distance from the work.

Undercoat Show-through

Undercoat show-through is a problem in which the color of the undercoat is seen through the topcoat. It is caused by insufficient colorcoats or repeated compounding. Undercoat show-through can also be caused by not having uniform color under the topcoat. To fix undercoat show-through, sand and refinish. Refer to Figure 18–15.

To prevent undercoat show-through, apply good coverage of color. Avoid excessive compounding.

Water Spotting

Water spotting is a dulling of gloss in spots. It can occur as a mass of spots that appear as a large distortion

Figure 18-15 This is an undercoat show-through problem. (PPG Industries, Inc.)

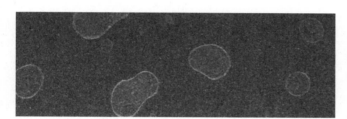

Figure 18-16 Water spotting dulls paint gloss. (PPG Industries, Inc.)

of the film. It is due to spots of water drying on a finish that is not thoroughly dry. It can be caused by washing the finish in bright sunlight. Refer to Figure 18–16.

To repair water spots, you can try wet sanding and polishing. However, you might have to sand and refinish.

To prevent water spotting, keep the fresh paint/refinish job out of the rain. Do not allow water to dry on the new finish.

Wet Spots

Wet spots are a finish problem seen as off-colored and/or slow drying spots of various sizes. It is usually due to improper cleaning. Heavy undercoats not properly dried is another reason. So is contamination with gasoline or other incompatible solvents.

To remedy wet spots, sand or wash the surface off thoroughly and refinish. To prevent this trouble, clean the surface with wax and grease remover. Allow undercoats to dry thoroughly. Use only water as a sanding lubricant.

Wrinkling

Wrinkling is a severe puckering of the film that appears like the skin of a prune (fruit). It is more common with enamel paints. There is a loss of gloss as the

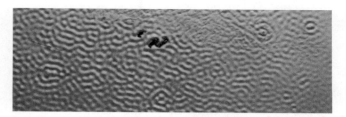

Figure 18-17 Wrinkling gives the paint a "prune" look. (PPG Industries, Inc.)

paint dries. Minute wrinkling may not be visible to the naked eye. See Figure 18–17.

Wrinkling is caused by:

1. Underreduction
2. Air pressure too low causing excessive film thickness
3. Excessive coats
4. Fast reducers creating overloading
5. Surface drying trapping solvents
6. Fresh film subjected to heat too soon

To fix wrinkling, break open the top surface by sanding and allow it to dry thoroughly. Then, remove the finish and refinish.

To prevent wrinkling, reduce and apply enamels according to directions. Do not force dry until solvents have flashed off.

Metal Dust Damage

Metal dust damage, also known as "rail dust," occurs when metallic particles from industrial fallout settle onto the paint. The particles soon begin to corrode, eating into the finish. Study Figure 18–18.

Metal dust damage can be identified on light colored finishes by the visible small brown or black specks. On darker colored surfaces, you must feel for prickly protrusion in the paint film. You can also rub a cotton towel across the surface. The iron particles will snag bits of cotton in the rag.

Metal dust damage is repaired by:

1. Blowing off the vehicle with compressed air
2. Washing and degreasing the vehicle
3. Treating with mild oxalic acid solution to dissolve the corrosion
4. Washing again to remove loose metallic particles
5. Color sanding, compounding, and glazing

Oxalic acid is also known as industrial fallout remover. It works by dissolving the corrosion holding

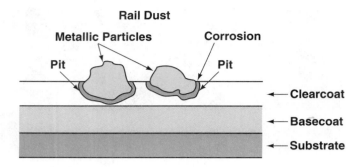

Rail Dust

Metallic Particles — Corrosion
Pit — Pit
Clearcoat
Basecoat
Substrate

Rail Dust—After Treatment

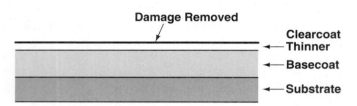

Damage Removed
Clearcoat
Thinner
Basecoat
Substrate

Figure 18–18 Metal dust fallout can etch into the finish. (A) Metal dust etched into clearcoat can be felt with a hand or rag. (B) After treatment, this damage can be repaired by wet sanding the clearcoat and compounding.

Acid Rain—Minor Damage

Not through clearcoat
Clearcoat
Basecoat
Substrate

Acid Rain—Severe Damage

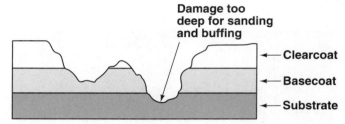

Damage too deep for sanding and buffing
Clearcoat
Basecoat
Substrate

Figure 18–19 Acid rain damage can also "eat" into the finish. (A) With minor acid rain damage, clearcoat is not penetrated. Wet sanding and buffing will usually fix the damage. (B) Major acid rain damage is too deep and the area must be refinished.

the particles to the paint. To use the oxalic solution:

1. Soak rags in the solution.

2. Lay the cloths over the affected area. Work in small areas.

3. Remove the cloths and rinse thoroughly with clean water. Remove excess water with clean towels.

4. Check the surface for remaining signs of iron particles.

5. If more particles are present, repeat the above steps.

6. If the solution does NOT dislodge all of the particles, wet sand the affected area. Then try machine compounding and polishing the area. If you do not cut through the finish or clear, this will repair the damage. The area must be refinished if it has gone through the topcoat.

DANGER! *Wear rubber gloves, respirator, and face shield when using oxalic acid. Protect yourself properly from the fumes and an accidental splash of acid.*

Acid Rain Damage

Acid rain damage varies because it is chemical damage from industrial fallout. The contents of the fallout vary with geographic locations. Various types of minerals sit on top of the surface film and damage it. Refer to Figure 18–19.

Minor acid rain damage produces shallow craters, dulling, fading, and chalking that do not eat fully through the topcoat. Minor acid rain damage can be removed by:

1. Washing the vehicle.

2. Neutralizing the acid by using a baking soda solution. The solution is a mixture of one tablespoon of baking soda per quart of water. The solution is applied with a spray bottle or by laying a saturated cloth on the area.

3. Rinsing with clean water.

4. Wet sanding with 1500 grit or finer sandpaper.

5. Compounding and glazing.

Test this in a small area on the worst acid rain damage first. If this does not restore the finish, there is major acid rain damage.

Major acid rain damage is severe chemical etching that has eaten completely through the topcoat. It may have eaten into the primer or all the way down to the body part. It requires that you strip all finish from the damaged areas. Otherwise, the acid residue will continue to eat through from beneath.

NOTE! Only horizontal surfaces (hoods, deck lids, roofs) usually suffer major acid rain damage and require stripping. The sides of the vehicle are usually unaffected and do NOT have to be stripped.

Sand Scratches

Sand scratches result from two sources:

1. Final sanding with incorrect (too coarse) sandpaper.
2. Not allowing materials (usually primer or spot putty) to dry fully before painting/refinishing.

If it is a solid color, you might be able to wet sand and machine compound the area. If the damage is in a colorcoat of a basecoat-clearcoat system, you will usually have to resand and refinish the area. See Figure 18–20.

FINAL DETAILING

The objective in ***final detailing*** is to locate and correct any defect that may cause customer complaints. Corrective steps for final detailing are:

1. Sanding and filing
2. Compounding
3. Machine glazing
4. Hand glazing

Each of these steps has its own requirements. As a general rule, finer and finer grades of products will be used for all of these steps. Progressively use finer wet sandpaper and compounds. Also, a single product line should be used throughout the repair. Manufacturer's recommendations should be followed.

Surface Paint Chips

Surface paint chips result from mechanical impact damage to the film: door dings, damage from road debris, etc. If the whole vehicle is not refinished, you should take the time to touch-up chips in the finish.

Use the material mixed for the repair. It will usually have hardener in it.

Degrease the area with wax and grease remover. If you use a small paint brush, slowly move the touch-up paint straight into each chip. If you are using a solid color, use a thicker viscosity touch-up paint to fill the chip in one application. If you have metallic

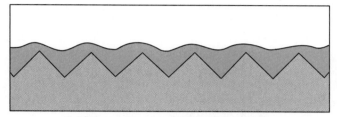

Primer-surfacer applied over sanded metal simulates the contours of the metal, with shrinkage more over deeper fills.

A

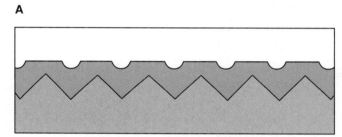

If sanded level before all solvents have evaporated, further evaporation of solvents will cause shrinkage leaving furrows over the sanding marks in the metal.

B

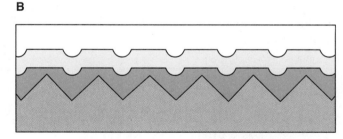

If the colorcoat is polished before all the thinner evaporates from the primer-surfacer and color, it will shrink back showing sand scratches. This condition may also be caused by using coarse rubbing compounds or too high of air pressure or too fast of solvent, when applying primer-surfacer.

C

Figure 18–20 If damage is not due to too coarse a sandpaper, sand scratches can be due to not allowing materials under the surface to dry before refinishing. (A) Primer has been applied to a rough sanded surface. (B) If sanded level before all solvents have evaporated, the primer will shrink more over sand scratches. (C) If the topcoat (color and clear) is then applied, sand scratches will appear in the finish. *(PPG Industries, Inc.)*

paint, use thinner touch-up paint and several coats to help match the color.

Allow the paint to cure sufficiently before wet sanding and polishing the chip repairs level.

Surface Protrusions

A *surface protrusion* (dirt) is a particle of paint or other debris sticking out of the film after refinishing. This problem results from a lack of cleanliness. The spray gun, spray materials, or booth was contaminated.

If the protrusion is larger, repair it with a knife or single-edge razor blade as follows.

1. Cut off the protrusion, being careful not to take off more finish than is necessary. The tip of the knife or razor blade should be pointed slightly upward.

2. Smooth the area with a finesse sanding block or 3500 grit sandpaper.

3. Blow off any particles. Finish with an extra-fine rubbing compound.

Dirt-nib Filing

Dirt-nib files can be used to remove any defect that is on or above the surface. This includes runs, sags, and dirt. A dirt-nib file will remove the protrusion with minimum damage to the surrounding film. Dirt-nib files are available commercially, or you can make your own, as shown in Figure 18–21.

To make a dirt-nib file, cut off a short piece of a vixen body file. After dressing down the sharp corners on a grinder, place a piece of 400 grit sandpaper on a flat surface and draw the piece of file back and forth over the sandpaper until the teeth become smooth. A light machine oil will speed up the operation.

DANGER! Do not try to break off a piece of the vixen body file. It can shatter!

To use a dirt-nib file, place the file lightly on the surface film. Use short, straight strokes in one direction only. Making two or three light passes will remove most defects. After filing, the area must be sanded to remove the file marks.

Finesse Sanding

After dirt-nib filing, *finesse sanding* can be used to level the protrusion with the surrounding film. Use a finesse sanding block as follows:

1. Dress the surface of the block with wet 220 grit sandpaper to make it smooth and flat.

2. Thoroughly soak the block in clear water.

3. Place the block over the protrusion and move it back and forth. If necessary, use a little water to help make the movement smoother. Refer to Figure 18–22.

4. When the protrusion has all but disappeared, blow off the loose particles and finish the job with rubbing, then polishing compound. See Figure 18–23.

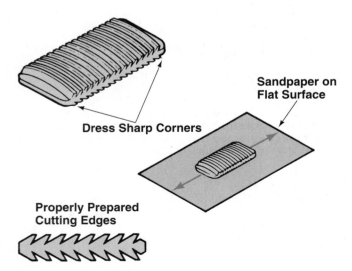

Figure 18–21 Note how to make a dirt-nib file.

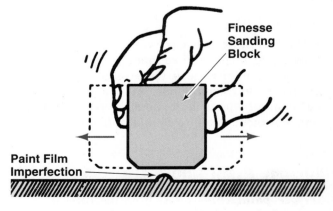

Figure 18–22 A finesse sanding block is soaked in water and then carefully rubbed over small surface imperfections to remove them. *(I-CAR)*

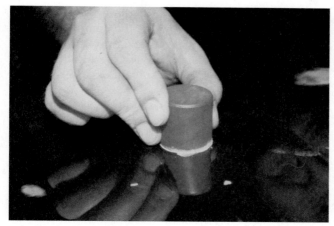

A

B

Figure 18–23 (A) Carefully rub a finesse sanding block over small imperfections in the film. (B) Rub only until the raised imperfection is sanded level. Note the dull areas that have been finesse sanded.

Wet Sanding

Wet sanding can be done to smooth the paint surface on larger areas, as when removing orange peel.

Wet sanding should normally be done with a **backing pad** or rubber **sanding block** to avoid crowning of the paint surface. A pad or block will help keep large, relatively flat surfaces level and uncrowned. On restricted and curved surfaces, you can use only your hand to wet sand. See Figure 18–24.

Sanding blocks and sandpapers are available in a variety of grit sizes. For surface repairs, use coarser wet sandpapers, 500 to 600. For finesse finishing, 1,000, 1,200, and finer grits of wet sandpaper are typically used.

Wet sand in a small circular motion. Use plenty of water to flush away paint debris. Dip the block in a bucket of water or use a hose to flow water over the

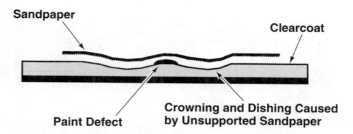

Figure 18–24 Under normal conditions, use a sanding block when wet sanding large, relatively flat surfaces. If not, crowning can result. (I-CAR)

area. A sponge can also be used to help keep the sandpaper wet. Refer to Figure 18–25.

Check the defect often when using a sanding block. You do not want to cut too deep into the finish. If you cut through the clearcoat or color, repainting will be necessary. Wash surfaces thoroughly with clean water and a sponge after wet sanding.

Rubbing Compound

Rubbing compounds generally contain coarser grit abrasives than polishes. They are used to more rapidly cut the surface film by hand. Rubbing compounds produce a low surface gloss.

Rubbing compounds are available in various cutting strengths for both hand and machine compounding. **Hand compounds** are oil based to provide lubrication. Small areas or blended areas are best done by hand compounding. On large surfaces, machine compounding is recommended.

Rubbing compounds are used:

1. To eliminate fine sand scratches around a repair area

2. To correct a gritty surface

3. To smooth and bring out some of the gloss of topcoats

4. Sometimes as a final smoothing step to remove light scratches and small dirt particles before painting/refinishing

Hand Compounding

Fold a soft, lint-free flannel cloth into a thick pad or roll it into a ball and apply a small amount of hand compound to it. Use straight, back-and-forth strokes and medium-to-hard pressure until the desired smoothness is achieved.

A

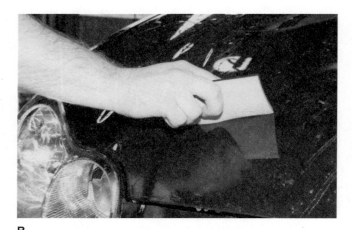

B

C

Figure 18–25 (A) Wet sanding should follow the use of a finesse sanding block. Wet the area. Hold your hand flat over the sanding block and rub in a circular motion. (B) Squeegee off the area to see if more sanding is needed. (C) A wet rag or sponge can be used to apply water to wash away sanding debris so the sandpaper will cut more quickly.

Hand compounding takes a lot of "elbow grease" and is time consuming. To keep the compounding of lacquer topcoats to a minimum, it is important to apply the finish as wet as possible (without sags or runs) by using the proper solvent/reducer for the shop temperature.

When using hand polishes or glazes, apply the glaze to the surface using a clean dry cloth. Rub the glaze thoroughly into the surface. Then wipe it dry. Glazes can fill and cover up some scratches which should be buffed out. These kinds of scratches will reappear after the vehicle has been washed a few times.

Machine Compounding

Machine compounds are often water-based to disperse the abrasive while using a power buffer. A buffing pad is rotated by an electric or air (pneumatic) polisher to force the compound over the paint surface. If done properly, this will quickly bring it to a high gloss, Figure 18–26.

Edge masking involves taping over panel edges and body lines prior to machine buffing or polishing to protect the paint from burn-through. Masking tape is applied to these surfaces to protect them.

Burn-through occurs when the pad removes too much paint on an edge, lip, or body surface. Since machine buffing will cut more quickly on these areas, always protect them with masking paper.

After the compounding is completed, remove the tape and compound the edge by hand—just enough to produce a smooth finish. Keep in mind that body lines usually retain less paint than flat surfaces and thus should get only minimum compounding.

Many finesse finishing systems recommend the use of specific buffing and polishing pads. Use a wool pad first with coarser machine compound to quickly remove the wet sanding marks. Then buff again with a softer foam pad and finer compound to bring out the paint gloss and to remove any swirl marks from the wool pad. Refer to Figure 18-27.

Apply machine compound over a small area. Then compound using a slow-speed power buffer. It is important to use a slow-speed machine to avoid static buildup and high surface temperatures. Do not push down on the buffer. Let the weight of the machine do the work. Refer to Figure 18–28 (page 382).

Because the compound has a tendency to dry out, do not try to do too large an area at one time. Always keep the machine moving to prevent cutting through or burning the topcoat. As the compound starts to dry out, lift up a little on the machine so pad speed increases. This will make the surface start to shine.

A

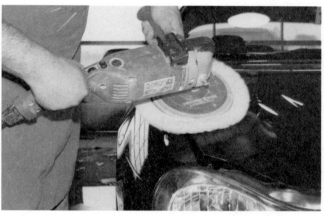

B

C

Figure 18–26 (A) Apply machine compound to a small area on the vehicle. (B) To prevent the compound from flying all over, spread it out with the buffer pad before depressing buffer trigger. Start the buffer. Do not press down on the buffer. Allow the weight of the tool to do the work. (C) Mask parts that could be damaged or sharp body lines. It is easy to cut through or burn through paint on sharp edges.

Figure 18–27 A foam pad is softer than a wool pad and will reduce swirl marks in the finish. It is often used after a wool pad. Use finer machine compound with a foam pad.

Polishing

Polishing involves using very fine compound to bring the paint surface up to full gloss. It is usually done after compounding. You can hand polish small or hard-to-reach areas. Machine polish larger areas to save time.

Instead of a circular action buffer, you should use an orbital action machine for final polishing. An *orbital action polisher* will move the polishing compound in a random manner to prevent swirl marks left from machine compounding. Swirl marks are tiny lines in the surface film from the abrasive action of the coarser compound.

Slight defects in the topcoat can be repaired by polishing. The choice of rubbing compound depends on the extent of the damage. Final polishing should always be done with an extra-fine polishing compound.

Sponge Buffing Pads

A change in automobile finishes in recent years has led to the development of an entirely new technology in the buffing and final delivery prep of collision repaired vehicles. The finishes produced by automobile manufacturers are glossier and virtually free of any texture or orange peel. The high solids finishes commonly used in the repair of these vehicles are difficult to apply without leaving a textured appearance on the surface. In order to duplicate the manufacturer's finish, it is often necessary to buff or compound the finish to remove any surface blemishes and texture normally associated with spray painting operations.

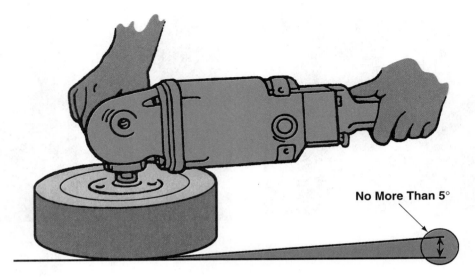

No More Than 5°

Figure 18–28 When buffing, do not press down on the buffer. Let the weight of the machine do the work. Also, do not tilt the machine more than approximately five degrees. *(I-CAR)*

Buffing with traditional wool pads often leaves an artificial-looking gloss. The subsequent development and the increased popularity of sponge pads for buffing operations has occurred for a variety of reasons. Sponge pads tend to be more user friendly and are less apt to damage the new finish than their wool counterparts. When buffing with a wool pad, both the compound and the fiber of the pad abrade and remove material from the surface. This can lead to excessive film removal, swirl marks, and even burning of the finish. When buffing with a sponge pad, the rubbing compound alone abrades the surface. Due to its softness and resilience, the sponge merely serves as the vehicle for the buffing action. This "softer" buffing action leaves the finish with a more natural-looking gloss, as it is less likely to leave swirl marks.

Sponge pads are available in various degrees or levels of aggressiveness. This is determined to a large degree by the firmness and density of the pad. They are usually available in two basic designs: One has a smooth, slightly crowned face, and the other has what is referred to as a waffle surface. Their surface configuration and their density determine their specific application, i.e., buffing or polishing. A soft waffle pad, for example, would be used for polishing and glazing operations. Conversely, a smooth denser pad would likely be used on a surface where removal of orange peel and other surface irregularities were necessary.

Irrespective of the buffing pad used, the successful buffing is achieved largely due to using the correct technique. The most common mistakes made when using the buffer are turning the pad up on edge, or "nose buffing," and using the incorrect buffer speed. The pad should be held flat to ensure the use of its entire surface, thus maximizing its efficiency and minimizing swirl marks. The correct buffer speed, also referred to as revolutions per minute (rpm), is equally important to obtaining a smooth, swirl-free finish. It is advisable for the technician to experiment with the buffer speed and adjust it to the type of pad being used to obtain the ideal finish. Each manufacturer's pads are somewhat unique and will require the user to adjust accordingly to obtain the best combination of speed and angle.

One of the major disadvantages of using sponge pads is the amount of heat they generate on the surface. Unlike the wool pad, which dissipates the heat through the fiber, the sponge pad transmits the heat back into the surface being compounded. Running the buffer at too high a speed may cause the surface to overheat, causing the rubbing compound to gum up and collect on the surface in a glue-like mass. This can readily be removed with water and a sponge, but if overdone, it can cause a blemish on the surface.

Using Buffers and Polishers. When using buffing and polishing pads:

1. Inspect, clean, or replace pads often to avoid residue buildup.

2. Use separate pads for different grades and types of products.

3. When applying the compound, apply an "X" of the product to the surface. Work it around the face of the pad before hitting the machine's trigger. This will help prevent compound from flying all over.

The buffer has an effect on the cutting action. For example, the higher the rpm, the higher the cutting rate and the lower the rpm, the lower the cutting rate. The faster the orbital buffer is moved across the panel, the slower the cutting rate. The slower the buffer is moved, the higher the cutting rate.

Excessive buffing heat can cause swirl marks, warping, discoloring, hazing, and the material can dry out too quickly. If the area is hot to the touch, there is too much heat. Cool it with water.

When using a buffer, a technician should:

1. Keep the pad flat or at about a 5 degree angle to the surface.

2. Let the weight of the machine do the work.

3. Use care around panel edges and character lines to avoid burn-through.

4. Check the repair area often and apply more product as needed.

5. Compound until the product begins to dry. Do not keep polishing on a dry surface.

Static electricity created during machine compounding causes the product to cling to the surface being repaired. Avoid static by grounding the vehicle. You might also want to add 5% rubbing alcohol to the water used to cool the surface.

When using rubbing compounds and machine glazes:

1. Use a single manufacturer's product line.

2. Follow the manufacturer's recommendations for use.

3. Use the materials sparingly.

4. Use the buffing wheel to distribute the material evenly over the area to be repaired.

5. Keep the pad flat and directly over the surface being repaired.

6. Use a slow, circular motion.

7. Use the finest product possible. Using a finer product may take a little longer initially, but it will generally require less time to complete the repair.

8. Reduce swirl marks by avoiding coarse products and worn buffing pads.

GET READY

Get ready is the last, thorough cleanup before returning the vehicle to the customer. You must do all the "little things" that make a big difference to customer satisfaction. The interior and exterior of the vehicle should be cleaner than when the customer brought it in.

Vacuum the interior of the vehicle carefully. Clean the seats, door panels, seat belts, and carpets. If dusty, clean and treat the vinyl surface with a conditioner. Be sure to remove all excess reconditioner from the seat crevices and folds. Stubborn stains should be cleaned with a recommended cleaning solution.

> *WARNING! Avoid using strong cleaning agents on the plastic parts in the dashboard. Some will dissolve and damage plastic. Also avoid using anything containing silicone.*

Carefully remove any overspray that may be on the vehicle. If it can be done without dripping on the new finish, use solvent (thinner or reducer). Clean body seams, moldings, and lights. Thoroughly clean all the glass, including windows, mirrors, and lights.

Use a brush with soap and water to clean the tires and wheels. Do not let dirty wheels spoil the appearance of an otherwise quality job. Coat them with a conditioner.

Chassis black can be used to blacken wheel openings and any other exposed undercarriage parts, since overspray often gets on these areas.

Replace wipers, moldings, and emblems that were removed before finishing. Take the time to clean off these items and be certain that everything is replaced.

As a finishing touch, clean the engine compartment. The easiest way to do this is to spray it with a heavy-duty engine cleaner. Then flush the engine compartment out with high-pressure water. A clean engine compartment usually makes a big impression on the customer.

Finally, inspect the vehicle with a careful eye for details. If a window is smeared, clean it again, Figure 18–29. If a piece of masking tape remains, remove it. If an emblem is missing, replace it before the customer asks where it is.

If the vehicle gets dirty while waiting to be picked up, wipe it down. The number one objective should always be a satisfied customer. Look at Figure 18–30.

CARING FOR NEW FINISH

A newly refinished vehicle must receive special care, as the finish can take several months to cure. Each paint manufacturer will have specific recommendations for caring for a new finish. Explain all precautions to the vehicle owner.

Figure 18-29 Make sure the vehicle is perfectly clean before releasing it to the customer. First impressions of the vehicle are very important.

To care for a new finish, you and the customer should:

1. Avoid commercial car washes and harsh cleaners for 1–3 months.

2. Hand wash using only water and a soft sponge for the first month. Dry with cotton towels only. Do not use a chamois.

3. Avoid waxing and polishing for up to three months. After that time, use a wax designed for basecoat-clearcoat finishes, as they are the least aggressive.

4. Avoid scraping ice and snow near newly refinished surfaces.

5. Flush gas, oil, or fluid spills with water as soon as possible for the first month. Do not wipe them off.

Summary

- Some of the most common paint/refinish problems include pinholes, mottling, runs, and sags.

- The steps for correcting surface imperfections include examining the finish for imperfections, correcting the imperfection, finesse finishing, and final detailing.

- The objective in final detailing is to locate and correct any defect that may cause customer complaints. The four steps are sanding and filing, compounding, machine glazing, and hand glazing.

- A newly refinished vehicle must receive special care, as the paint may take several months to cure. Know the specific manufacturer's recommendations and explain all precautions to the vehicle owner.

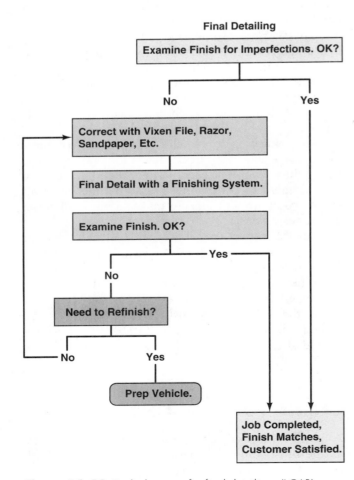

Figure 18-30 Study the steps for final detailing. (I-CAR)

Technical Terms

surface roughness
surface waviness
acid spotting
alkali spotting
bleeding
blistering
blushing
dirt in finish
fisheyes
mottling
orange peel
peeling

pin holes
pitting
cratering
plastic bleed-through
corrosion under finish
runs
sag
undercoat show-through
water spotting
wet spots
wrinkling
metal dust damage

oxalic acid
acid rain damage
minor acid rain damage
major acid rain damage
sand scratches
final detailing
surface paint chips
surface protrusion
dirt-nib file
finesse sanding

sanding block
rubbing compound
hand compound
machine compound
edge masking
burn-through
polishing
orbital action polisher
get ready

Review Questions

1. List four steps for correcting surface imperfections.

2. _____ shows up as small, swelled areas on the finish, like a _____ _____ on human skin.

3. What is paint blushing?

4. How can you correct a small amount of dirt on a painted surface?

5. What causes fisheyes and how can you prevent and fix them?

6. This is a paint roughness problem that looks like tiny dents in the paint surface.
   a. Paint protrusions
   b. Mottling
   c. Orange peel
   d. Peeling

7. List six reasons for paint runs.

8. Is fixing a run in a solid color and in a metallic the same? Explain your answer.

9. *Technician A* says to wet sand with a sanding block to keep from crowning the surface. *Technician B* says it is OK to use your hand to wet sand on curved surfaces. Who is correct?
   a. Technician A
   b. Technician B
   c. Both Technicians
   d. Neither Technician

10. What is the difference between a rubbing compound and a polishing compound?

11. _____ are water-based to disperse the abrasive while using a power buffer.

12. Why should you mask edges and body lines before buffing?

13. What is burn-through?

ASE–Style Review Questions

1. *Technician A* says that dirt in the paint can be reduced with proper cleaning before refinishing. *Technician B* says that blistering is a problem that makes the finish turn "milky looking." Who is correct?
   a. Technician A
   b. Technician B
   c. Both Technicians A and B
   d. Neither Technician

2. A vehicle returns to the shop. The area painted seems to be suffering from a blistering problem. This could be caused by all of the following EXCEPT:
   a. Rust under the surface
   b. Painting over oil or grease
   c. Moisture in spray lines
   d. Improper reduction

3. *Technician A* says that improper cleaning of the surface usually causes fisheyes. *Technician B* says that adding fisheye eliminator additive is not recommended by many of today's paint manufacturers. Who is correct?
   a. Technician A
   b. Technician B
   c. Both Technicians A and B
   d. Neither Technician

4. Which of the following usually causes paint lifting?
   a. Heavier film thickness
   b. Improper drying of previous coatings
   c. Extreme temperature changes
   d. Slow-dry thinner

5. *Technician A* says that orange peel is commonly caused by underreduction. *Technician B* says that this problem can also be due to the surface drying too fast. Who is correct?
   a. Technician A
   b. Technician B
   c. Both Technicians A and B
   d. Neither Technician

6. Runs are caused by all of the following EXCEPT:
   a. Extra slow drying reducer/solvent
   b. Extra fast drying reducer/solvent
   c. Painting on cold surface
   d. Too many coats of paint

7. *Technician A* says that a small run in a solid color can often be repaired with wet sanding and buffing. *Technician B* says that a run in a metallic color cannot be repaired this way. Who is correct?
   a. Technician A
   b. Technician B
   c. Both Technicians A and B
   d. Neither Technician

8. *Technician A* says that sand scratches result from final sanding with too coarse a grit of sandpaper. *Technician B* says it is from not allowing the primer to dry fully before painting. Who is correct?
   a. Technician A
   b. Technician B
   c. Both Technicians A and B
   d. Neither Technician

9. Which of the following would normally be used to correct a small protrusion in a paint film?
   a. Finesse sanding and buffing
   b. DA sanding and buffing
   c. Buffing only
   d. Repaint panel

10. *Technician A* says to use a cotton buffing pad before a foam pad. *Technician B* says a foam pad should be used first to eliminate swirl marks. Who is correct?
    a. Technician A
    b. Technician B
    c. Both Technicians A and B
    d. Neither Technician

Activities

1. Inspect several vehicles. How many paint problems can you find? Make a report of your findings. Is there any industrial fallout problem in your area?

2. Obtain an old painted part. Intentionally paint the part to produce various paint problems. Allow the paint to dry and try to repair some of the problems without repainting. Write a report on this activity.

3. Final detail a vehicle. Clean all interior and exterior surfaces with recommended products.

19

Decals, Custom Painting, and Trim

Objectives

After studying this chapter, you should be able to:

■ Properly remove vinyl decals and striping.

■ Prepare the surface before applying adhesive overlay material or before custom painting.

■ Install adhesive tape and decals.

■ Explain various techniques for doing custom paint work.

■ Remove, align, and install molding and emblems.

Introduction

People of all ages often want their vehicles customized or personalized. Also, factory installed pinstripes and decorative tape is now common. If you can do this kind of work, you will find it to be in high demand and very profitable.

This chapter will introduce you to basic methods for doing custom paint work, for painting stripes, and applying custom decals or pinstriping tape. The chapter will also summarize methods for working with moldings and emblems. See Figure 19–1.

Figure 19–1 If you can do custom painting or install vinyl overlays, your value will increase.

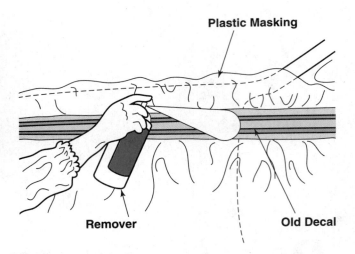

Figure 19–2 To remove vinyl stripes and decals, mask around the material. Then, spray on two coats of remover. After soaking them, you can more easily peel off vinyl materials.

OVERLAYS AND TAPE STRIPES

An *overlay* is a large, self-adhesive vinyl material applied to the finish. It provides an easy way to customize the looks of a vehicle. Overlays might be multicolor stripes, artwork, or artificial wood grain. Overlays can also be body color, mounting along the lower areas of the body to help protect the finish from stone chips.

Vinyl tape is a tough, durable decorative striping material adhered to the vehicle's finish. It has a pressure-sensitive backing that bonds securely to the finish.

Vinyl decals are generally more complex and decorative pressure-sensitive material than tape stripes. Both tape and decals can be original equipment or aftermarket installed.

When repairing collision damage, you may have to replace vinyl tape and decals. If you use established methods, this can be a rewarding task.

Overlay and Tape Stripe Removal

There are several ways to remove vinyl tape stripes and decals. These include:

1. Spray remover and scraper

2. Heat gun and scraper

3. Scuff pad (if repainting panel)

4. Rubber wheel (eraser wheel)

Vinyl overlay remover is a chemical that dissolves plastic vinyl to aid its removal. To use spray vinyl remover, mask the area around the stripe or decal with plastic, Figure 19–2. Scuff sand the decal with 220 grit sandpaper to help speed the action of the chemical. Do not cut through the vinyl or sand the paint or you may have to repaint the panel.

If original equipment or if material on the other side of the vehicle is not going to be removed, measure the location of the existing stripe or decal before removal. Write down your measurements for later reference.

Spray the entire stripe or decal with remover spray. Use smooth steady strokes so that the material is evenly coated with the chemical, Figure 19–2. After a few minutes, spray the chemical on the vinyl again. Allow the remover to soak for about 15 minutes.

DANGER! *Follow all safety precautions when using vinyl remover. It is a powerful chemical. Use it only in a well-ventilated area and wear eye and respiratory protection.*

After allowing it to soak, peel the stripe or decal away from the panel. Start in one corner and try to peel it off in one piece. If needed, use a sharp plastic squeegee to help peel off the old material.

After removal of the stripe or decal, use the squeegee to remove any remaining adhesive from the panel. Then, use a rag and *adhesive remover* to clean off the remaining vinyl glue.

On smaller pressure-sensitive overlays, a heat gun can sometimes be used to warm, soften, and remove the vinyl material. Without overheating, move the heat gun to gradually soften the stripe. As the vinyl and adhesive soften, slowly peel off the material. Then, clean the area with adhesive remover.

A scuff pad mounted on an air tool can also be used to remove narrow vinyl material, like pinstripes. It will abrade the soft plastic material without clogging. Sandpaper will quickly clog if it is run over vinyl stripes or decals.

Measure and
mark locations
for tape

A

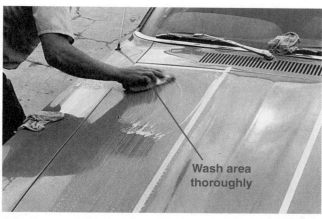

Wash area
thoroughly

B

Figure 19–3 Preparation before installing vinyl materials is important. (A) Always measure the location of existing overlays before removal. Then you can mark their locations on the vehicle for properly locating new stripes or decals. (B) Clean the area with wax and grease remover (cleaning solvent). Then wash with detergent and water. Note that masking tape is to be used as a straightedge when installing vinyl striping. *(Mustang Monthly Magazine)*

An **eraser wheel** is a rubber wheel and arbor designed to remove tape stripes without damaging the paint. The wheel is mounted in an air drill or a small grinder. When spun against a stripe, the spinning wheel will abrade the tape and adhesive off the painted surface quickly and easily.

Overlay and Stripe Application

Use a water-soluble marker and tape measure to mark where you want to install the new striping or decal. Place tiny reference marks on the finish where needed. See Figure 19–3A.

Then place masking tape on the vehicle to serve as a guide. Align the tape with the marker reference marks.

Before tape or decal application, make sure the surface of the panel is ready. It should be wet sanded and buffed if needed. Also, clean the panel with wax and grease remover. This will remove any wax and other foreign matter. Look at Figure 19–3B.

Check to be sure that the temperature of the panel is about room temperature. It should be no cooler than 60°F (16°C) or no hotter than 90°F (32°C). If brought in from outside, allow the vehicle's surface to reach room temperature.

Refer to the instructions provided with the stripe or decal material. Methods can vary.

Installing Overlays

The **overlay wet method** of installation involves using soapy water to help simplify the application of larger decals and striping. Soapy water will allow you to shift and position the material without the adhesive instantly sticking to the panel. The water will also aid removal of the backing material from the overlay.

Fill a clean bucket with water. Then add some dishwashing liquid to the water and mix. Place the decal into the soapy water. Peel off the backing material carefully so that you do not tear the vinyl, Figure 19–4A.

Position the overlay on the vehicle. Position it up against your alignment masking tape, Figure 19–4B. The soapy water will let you shift the overlay without it sticking tightly to the finish. Carefully straighten and align the material, removing any wrinkles. Working from the middle outward, use a rubber squeegee to flatten out the overlay. Make sure you have removed all air bubbles and wrinkles, Figure 19–4C. Then, use a rag to rub and adhere the overlay to the panel. As you rub toward the edges of the overlay, you will work any water out from under the vinyl. This will make the overlay stick to the finish.

Once the water is removed and almost dry, carefully peel off the carrier paper. Grasp at one edge of the carrier paper. Then, slowly peel it off the vinyl. Go slowly so that you do not lift and tear the vinyl. Refer to Figures 19–5A and B.

Next, if needed, cut off the ends of the overlay or stripe at the panel ends. Use a sharp razor blade knife to cut the material to the desired length. You may want the material to wrap around the panel edge or you may prefer it to stop right at the edge. See Figure 19–6.

Pierce any remaining air bubbles with a needle. Then, rub the overlay with a rag to work the air out from under the vinyl.

A

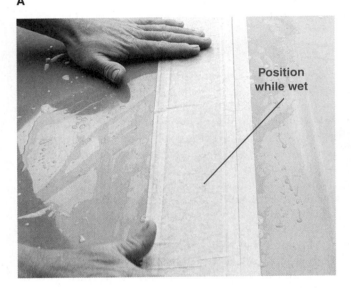

B

C

Figure 19–4 Note the basic steps for installing vinyl overlay wet. (A) Wet the vinyl overlay in soapy water and remove the backing sheet over the adhesive. (B) Position the overlay on the vehicle. Align it with your marks or masking tape. Move quickly before the water dries. (C) Use a rubber squeegee to remove wrinkles and air bubbles. Wipe outward, or away from the center of the overlay. *(Mustang Monthly Magazine)*

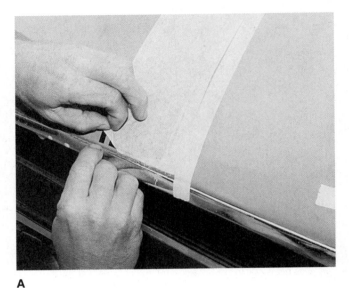

A

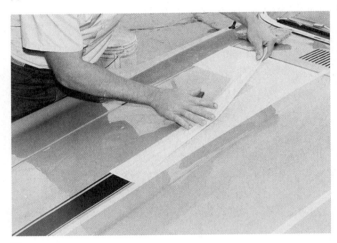

B

Figure 19–5 Note how the carrier paper is removed from the vinyl striping. (A) After the overlay is smooth and dry, peel off the carrier paper. Use your fingers to hold the ends of material down if needed. (B) Slowly remove the carrier paper by pulling it straight back. You must not tear the thin vinyl material. *(Mustang Monthly Magazine)*

Rubbing away from the overlay, double-check to be sure that all edges are adhered tightly to the finish. Rub over all edges a couple of times. The result will be a long-lasting, attractive addition to the vehicle's appearance.

Installing Vinyl Pinstripes

The **overlay dry method** of installation does not require soapy water to aid application. All surfaces must be clean and dry. Narrow pinstripe tape can be installed dry without wrinkling or bubble problems. Generally, if an overlay is narrower than about 2

Figure 19-6 With the carrier paper off, cut the ends of the striping to length. Cut them so they just start to wrap around the panel edge. *(Mustang Monthly Magazine)*

inches (50 mm), you can use the dry method of installation.

Pinstripe tape may or may not have a carrier paper. Both types are installed in about the same way. With carrier paper, you will just have to peel off the paper after adhering the tape to the vehicle's surface.

A few rules for installing pinstripe tape include:

1. Make sure the surface is perfectly clean. Use soap and water. Then, wash the area with wax and grease remover. If not, the tape will come off and your customer will return unhappy.

2. Use a grease pen or water-soluble marker to locate where you want the striping. If needed, you can also lay down masking tape as a guide. With the dry method, the tape will stick to the surface and can be difficult to reposition.

3. Position and adhere one end of the tape stripe. You might want to place masking tape over the end to keep it from pulling off.

4. Use one hand to lightly stretch the tape and the other to lightly push and tack the tape into position. You want to lightly stick the tape in position first. Then, if it is not straight, you can pull the tape up and reposition it.

5. Never stretch the tape striping too tight. If you do, the tape will distort or come off after exposure to the heat of the sun.

6. Do not overstretch the tape when rounding corners. Again, this will thin the material and cause it to lift later.

7. After using your finger to lightly tack the tape into position, stand back and check its alignment

and straightness. If it is not straight, pull the tape back off and realign it.

8. Once you are sure the tape is positioned properly, use a rubber squeegee to bond the tape securely to the finish. Start at the center of the stripe and wipe toward the ends. Do not use too much pressure near the ends or you can tear the tape.

9. If the pinstripe tape has a carrier paper, slowly pull if off. Then, check for kinks and air bubbles. Work them out. Again, rub a squeegee over the stripe to adhere it tightly to the finish.

10. Do not use your hand to try to bond pinstripe tape. It will tend to distort and wrinkle the tape. After using the squeegee, rub a soft rag over the tape to final bond it in place.

PAINTED PINSTRIPES

There are two common methods to paint on pinstripes. One uses vinyl masking tape to mask on each side of the paint stripe. The other uses a special striping tool to apply a thin line of paint over the finish. Painted-on pinstripes are sometimes requested for a custom paint job or during a restoration job. A painted stripe will generally last longer and look better than vinyl tape.

Using Pinstripe Tape

Pinstripe masking tape uses two pieces of vinyl tape held with a paper carrier. Position and apply the tape as you would pinstripe tape. Then, remove the carrier paper.

Next, use a small pinstriping brush to apply the paint in the area between the tapes. Most painters use lacquer or catalyzed enamel so the strip will dry quickly. Use long smooth brush strokes parallel with the stripe. Using a brush saves time and material over a spray gun or airbrush because you do not have to mask large areas around the stripe.

When the paint stripe has flashed, carefully remove the masking tape. Slowly pull it straight back and off the vehicle. Do not pull sideways or you could peel off some of the freshly painted stripe. Wet sand and polish the painted stripes if needed.

Using Pinstriping Tool

A ***pinstriping tool*** uses a serrated roller to deposit a painted pinstripe. The tool holds the paint in a cylinder shaped cup. The roller, when moved over the surface, deposits the paint onto the finish. Various widths of roller can be installed in the tool to change paint stripe widths.

A *magnetic guide strip* can be used to help guide the tool for making straight or slightly curved pinstripes. It is a flexible plastic material filled with magnets. When placed on a steel body part, the guide strip will stay in place without damage to the finish. A guide rod on the pinstriping tool contacts this guide when striping.

When the tool is held squarely and moved over the surface, it will make a uniform width pinstripe. If the tool is tilted, the width of the pinstripe will narrow. With practice, you will be able to make pointed or curly ends on the painted pinstripes.

Using Pinstriping Brushes

Pinstriping brushes have very long soft horse hairs on them. Special talent is needed to use them. Using a very steady hand, a freehand technique is sometimes used to pull the brush along the finish to apply painted pinstripes. Body lines or trim are often used as a guide when freehand striping.

Painting Wider Stripes

Wider painted stripes are done by spraying a different color or tint over the existing finish after masking. Before painting, wet sand the area to be painted with very fine (1000 or finer grit) wet sandpaper. This will prepare the surface for good adhesion of the new paint.

Apply fine line masking tape where the stripes are wanted. Fine line tape is needed so the stripes have a smooth edge when the tape is removed. After carefully applying the fine line tape, mask the area on each side with paper or plastic and conventional masking tape.

Spray the color for the stripes onto the unmasked area. After drying, remove the masking material. Wet sand the area and buff it to a high gloss. If the stripe is wide, use a full-size spray gun. If the stripes are smaller, use an airbrush or touch-up gun to save materials and speed cleanup.

CUSTOM PAINTING

Custom painting can involve using multiple colors, metal flake paints, multilayer masking, and special spraying techniques to produce a personalized look. Multicolor stripes, flames, murals, landscapes, names, and other artwork can be added to the finish. Com-

plex images require you to have special artistic talent. If you cannot paint an attractive image on paper, you will not be able to do it on a vehicle.

Custom painting requires considerable talent, skill, and knowledge. You need to plan the custom job carefully. This will let you determine how to mask and spray or apply each color. Custom painters are good at using airbrushes, striping tools, and masking materials.

Before custom painting, make sure the base finish is in good condition. You do not want to waste your time trying to paint over a weathered or problem finish. Wet sand and clean the area to be custom painted. Use surface preparation methods detailed in other chapters.

To produce a feathered custom stripe on a van, the area to be striped is masked and painted with the main color for the stripe. Then, an airbrush and a different color paint is misted around the outside edge of the stripe. This color is misted over the edge of the masking paper. Then, when the masking is removed, the two colors blend into each other for a custom stripe effect.

Custom masks can be made by drawing a design on thin posterboard and then cutting them out. The posterboard design is then taped onto the vehicle and spray painted. Using an airbrush and translucent paint, various attractive effects can be produced.

Card masking involves using a simple masking pattern to produce a custom paint effect. Usually, an airbrush is used to mist the paint over the edge of the masking card. The card can be moved to repeat the pattern to produce a wide range of paint effects.

Lace painting involves spraying through lace fabric to produce a custom pattern in the paint. Various lace designs can be purchased at fabric stores. The cloth pattern will allow the paint to pass through the holes in the lace but mask it in other areas.

A *marble effect* can be made by forcing crumpled plastic against a freshly painted stripe or area. When the plastic is lifted off, it will remove paint in random areas with a marble type effect.

Spider webbing is done by forcing paint through the airbrush in a very thin, fibrous type spray. Air pressure from the gun can also be used to spread and smear the wet paint to produce a varying effect.

Painted flames are a custom painting technique often used on "hot rods" or older street rods. First, fine line masking tape is used to form the outline of the flames. Then, the area around this shape is covered with masking paper or plastic.

First, the base color for the flames is applied. Then, a second translucent color is blended. A third color may be used to darken the outer edges and center area of the flames. The flames are finally wet sanded and polished when dry.

Painted lettering involves masking off letters over the finishing and spraying or brushing them on with a different color. This can be time consuming but is sometimes requested by customers.

When doing custom paint work, do not "bite off more than you can chew." Start out simple with minor complexity. As you learn to successfully do custom work, you can progress to more complex paint work. Experience is the best teacher with custom paint work.

A good idea is to practice techniques on old parts using leftover paint. Then, you can learn from your mistakes and successes without working on a customer vehicle.

MOLDINGS AND TRIM

Moldings and trim can be held on with adhesive, mechanical fasteners, or both. If in doubt, your service manual or computer database will describe how each molding or trim is held in place.

To remove molding or trim that is held on by adhesive, place masking tape next to the molding or emblem to protect the surface from scratches, Figure 19–7. Heat and warm the molding or emblem with a heat gun to soften the adhesive. Then, force a **mold-ing removal tool** or knife under the part to separate the adhesive from the body. Pull outward on the molding or emblem as you carefully work the tool behind the part.

If a quantity of adhesive remains on the vehicle, you might want to cover the spot with a sheet of plastic and heat the adhesive again. This will soften the remaining adhesive and allow you to remove it easily with a putty knife or scraper. Be careful not to overheat the surface. Heating the old adhesive to about 104–122 degrees F (40–50 degrees C) should soften it enough for easy removal.

If you are reusing the old molding or trim, clean the remaining adhesive off its mounting surface. Scrape off the bulk of the old adhesive without cutting and damaging the molding or emblem. Then, wrap the part in plastic film and heat soften the adhesive. Scrape the remainder of the adhesive off the part, Figure 19–8.

When installing an adhesive-backed molding or emblem, make sure the vehicle surface is perfectly clean. If not, the molding can come off.

If needed, use masking tape as a guide for molding installation. Carefully apply the masking tape to the vehicle in a perfectly straight line. Then, position the molding next to the masking tape before pressing it into place.

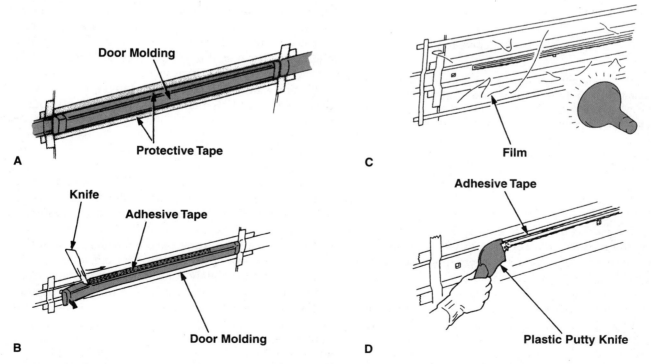

Figure 19–7 Note the steps for removing molding held on with adhesive tape. *(Reprinted with permission by American Isuzu Motors Inc.)* (A) Mask around the molding to protect the finish from damage. (B) Carefully push a knife between the molding and tape to free the molding from the body. (C) Cover the remaining adhesive with film and heat soften it with a heat gun or lamp. (D) Carefully remove the rest of the adhesive with a dull scraper to avoid paint damage.

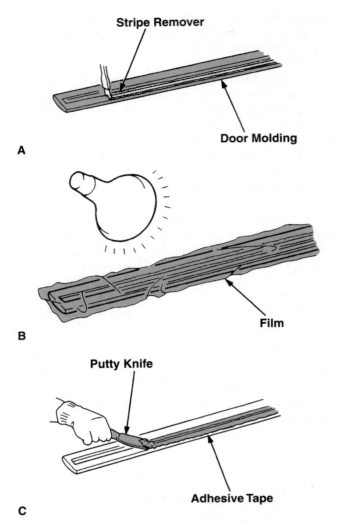

Figure 19–8 If the old molding is going to be reused, remove excess adhesive from its mounting surface. *(Reprinted with permission by American Isuzu Motors Inc.)* (A) Scrape adhesive off the molding without marring the surface. (B) Wrap it in plastic film and heat the molding to soften the remaining adhesive. (C) Scrape the remainder of the adhesive off the part.

Service publications will give measurements for locating the molding or trim properly. Use a pocket rule or measuring tape to mark where the molding or trim should be mounted. See Figure 19–9.

If you are reusing adhesive-backed molding or emblems, use aftermarket emblem adhesive. Spread a thin coating of emblem adhesive on the back of the molding or emblem. Then, move the part straight into position without smearing the adhesive. Use masking tape to hold the part in place until the adhesive dries.

When molding is held by fasteners, you will have to gain access to them. This might involve removing door panels or inner fender aprons.

Attaching Moldings with Two-Sided Tape

Moldings, ornamentation, and trim have always played a crucial role in highlighting the crowns and style lines on the automobile. Historically these have been made of metal and attached with either a metal or plastic clip or fastener. This method often necessitated drilling holes into the metal to attach the fasteners, which left an area for corrosion to gain a foothold and spread undetected. Securing the attaching hardware frequently required access to the back side of the panel. The attaching hardware also had to fit the shape and configuration of each specific molding. Many times the hardware would break during the removal of the part, making it difficult to reinstall without modifications.

The preferred method of securing moldings and trim has changed from using the mechanical fasteners described above to the use of double-sided foam tape. The use and placement of the trim accessories has taken on a more diverse and integral function as they are often shaped to conform to the crowns, recesses, and irregular shapes of the surfaces to which they are applied. The double-sided tape holds the molding closer to the surface of the vehicle, reducing the possibility of its becoming snagged and pulled loose. The weight, mass, and often irregular shapes of these moldings do not lend themselves to being attached with traditional mechanical fasteners. Two-sided foam tape has become a very popular alternative because it is available in various sizes and can be used on virtually any size, shape, or trim configuration. In addition, it is no longer necessary to drill holes into the panels, which eliminates a source of corrosion.

The removal of moldings and trim secured with the two-sided tape is best done with a tool specifically designed for this application. This method cuts through the tape, leaving a clean smooth surface that requires a minimal amount of cleanup prior to installing the replacement part. Removing the molding with this tool also allows the technician to reinstall the same part, either with the aid of a liquid adhesive designed for this particular application or by completely removing the old tape from both surfaces and using fresh tape on the back side of the part. Reinstalling the molding with new tape is the preferred method because the tape is permanently secured the instant it contacts the vehicle's surface. When installing molding with liquid glue, it must be held in place with tape until the adhesive is dry. There is also a possibility that glue will squeeze out and damage the surface next to the molding edge.

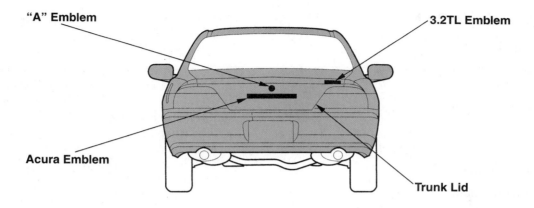

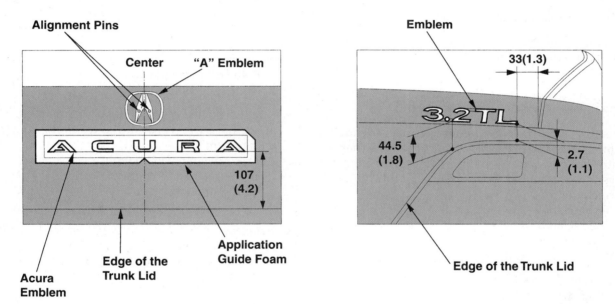

Figure 19–9 Note this service manual illustration showing how to locate emblems on this particular vehicle (© American Honda Motor Co., Inc.)

Summary

- An overlay is a self-adhesive vinyl material applied to the finish. It may be multicolor stripes, artwork, or artificial wood grain.

- There are several ways to remove vinyl tape stripes and decals, including spray remover and scraper, heat gun and scraper, and scuff pad (if repainting the panel).

- The overlay wet method of installation involves using soapy water to help simplify the application of larger decals and striping. The overlay dry method does not require water. Generally, if an overlay is narrower than 2 inches (50 mm), you can use the dry method.

- Custom painting involves using multiple colors, metal flake paints, multilayer masking, and spe-

cial spraying techniques to produce a personalized look. Stripes, flames, murals, or names may also be added.

Technical Terms

overlay
vinyl tape
vinyl decals
vinyl overlay remover
adhesive remover
eraser wheel
overlay wet method
overlay dry method
pinstriping tool
magnetic guide strip

pinstriping brushes
custom painting
custom masks
card masking
lace painting
marble effect
spider webbing
painted flames
molding removal tool

Review Questions

1. What is a vinyl overlay?

2. Why do you sand vinyl overlays before using overlay remover?

3. List ten rules for installing pinstripe tape.

4. A _____ _____ uses a serrated roller to deposit painted pinstripes.

5. Explain card masking.

6. How do you do custom lace painting?

7. How do you produce a spider webbing effect when custom painting?

8. What is a molding removal tool?

9. Custom paint work is easy if basic methods are followed. True or false?

ASE–Style Review Questions

1. Pinstripe tape must be removed from a fender before painting. *Technician A* says to use a sharp putty knife to remove the pinstripe. *Technician B* says to use a rubber wheel in an air drill. Who is correct?
   a. Technician A
   b. Technician B
   c. Both Technicians A and B
   d. Neither Technician

2. *Technician A* says to measure the location of the existing stripe or decal before removal. *Technician B* says that service information might publish stripe or decal locations. Who is correct?
   a. Technician A
   b. Technician B
   c. Both Technicians A and B
   d. Neither Technician

3. *Technician A* says that vinyl remover is not harmful. *Technician B* says that there is no need to wear eye protection when using vinyl remover. Who is correct?
   a. Technician A
   b. Technician B
   c. Both Technicians A and B
   d. Neither Technician

4. Which of the following will allow you to shift and relocate large overlays or decals?
   a. Adhesive remover
   b. Rubber wheel
   c. Solvent
   d. Soapy water

5. An air bubble is found near the middle of a large overlay or decal. *Technician A* says to poke a hole in the air bubble with a needle. *Technician B* says to use a razor blade to cut an opening in the air bubble. Who is correct?
   a. Technician A
   b. Technician B
   c. Both Technicians A and B
   d. Neither Technician

6. *Technician A* says that narrow pinstripe tape can be installed with the body surface dry. *Technician B* says that large decals or overlays can be installed with the body surface dry. Who is correct?
   a. Technician A
   b. Technician B
   c. Both Technicians A and B
   d. Neither Technician

7. Newly installed pinstripe tape is coming off when in the hot sun. Which of the following might be the problem?
   a. Tape not stretched enough
   b. Tape stretched too tight
   c. Heat gun not used
   d. Not installed straight enough

8. *Technician A* says you can use special masking tape before painting on pinstripes. *Technician B* says that you can use a special pinstripe tool or roller. Who is correct?
   a. Technician A
   b. Technician B
   c. Both Technicians A and B
   d. Neither Technician

9. When painting very wide stripes, which of the following sandpaper grits should you use to sand the area?
   a. 80 grit
   b. 120 grit
   c. 500 grit
   d. 1000 grit

10. When heating adhesive for molding removal, which of the following is the maximum temperature commonly recommended to prevent part warpage or damage?
    a. 74–80 degrees F (23–27 degrees C)
    b. 84–90 degrees F (29–32 degrees C)
    c. 94–104 degrees F (34–40 degrees C)
    d. 104–122 degrees F (40–50 degrees C)

Activities

1. Obtain a body part with an overlay on it from a salvage yard. Use proper methods to remove the vinyl overlay without damaging the finish.

2. Using a scrap part, practice doing custom paint work. Try doing lacing, flames, and other effects.

3. Attend a custom car show. Using a camera, document the types of custom paint effects found. Ask the car owners about their custom paint effects. Write a report on how some of the custom paint jobs were done.

Section 5

Major Repairs

Chapter 20
Measuring Vehicle Damage

Chapter 21
Straightening Full-Frame and Unibody
Vehicles

20

Measuring Vehicle Damage

Objectives

After studying this chapter, you should be able to:

- Distinguish between body-over-frame and unibody vehicles.
- Explain how impact forces are transmitted through both frame and unibody construction vehicles.
- Describe how to visually determine the extent of impact damage.
- List the various types and variations of body measuring tools.
- Analyze damage by measuring body dimensions.
- Analyze impact damage to mechanical parts of the vehicle.
- Explain the importance of the datum plane and centerline concepts as related to unibody repair.
- Interpret body dimension information and locate key reference points on a vehicle, using body dimension manuals.
- Discuss the use of tram bars, self-centering gauges, and strut tower gauges.
- Diagnose various types of damage, including twist, mash, sag, and side sway.
- Given a damaged vehicle and a body specification manual, locate and measure key points using a tape measure, tram bar, and self-centering gauges.

- Describe the two types of universal measuring systems.
- Explain the operation of electronic laser, ultrasonic, and robotic arm measuring systems.

Introduction

When a vehicle is in a high-speed collision, powerful impact forces can bend the frame or unibody structure. The frame, body, or unibody is designed to absorb some of the energy of the collision and protect its occupants. When a heavily damaged vehicle enters the shop, the extent of the damage must be carefully evaluated. Sometimes, measurements are needed to help the estimator calculate the costs of the repairs. See Figure 20–1.

Vehicle measurement involves using specialized tools and equipment to measure the location of reference points on the vehicle, Figure 20–2. These measurements are then compared to published dimensions from an undamaged vehicle. By comparing known good and actual measurements, you can determine the extent of damage. The difference in the two measurements indicates the direction and amount of frame or body misalignment.

After studying damage measurements, straightening equipment is used to pull the frame, body, or unibody back into alignment. Straightening procedures are explained in the next chapter.

399

Figure 20-1 An estimator and technician are carefully evaluating the general extent of damage to the vehicle on the lift. Rough measurements are being taken to determine if more precise measurements and straightening will be required. Accurate measurements cannot be made when the vehicle is on a lift. *(Rotary Lift/A Dover Industries Co.)*

A

B

Figure 20-2 A vehicle measuring system is designed to check the location of specific points on a damaged vehicle. If any points are not where they should be, the frame or unibody must be straightened. (A) This measuring system has telescoping rods and pointers that can be swung at various angles for quick measurements of body and chassis parts. *(Wedge Clamp International, Inc.)* (B) This measuring system uses rods and pointers that mount at right angles to each other. When adjusted to proper lengths, you can quickly tell if parts have been forced out of alignment by a collision. *(Car-O-Liner)*

WHY IS MEASUREMENT IMPORTANT?

There is good reason for close body structure tolerances. Steering and suspension systems, for instance, are mounted to the frame or unibody structure. Severe body or frame damage can change steering or suspension geometry or misalign mechanical parts. This can result in poor handling, vibration, and noise problems. Thus, the tolerances of critical manufacturing dimensions must be held to within a maximum value of less than ⅛ inch (3 mm).

The challenge faced by the collision repair industry is to find out which panels are out of place and in which direction and how far they have moved. Only then can a proper plan be devised to bring the vehicle back to precollision specifications. This process involves accurate measuring and monitoring of the entire vehicle.

To correctly analyze damage on a unibody vehicle, the entire structure must be considered. To do this, it is necessary to be able to take proper measurements to locate damage. It will also help to plan where to pull.

Measurement gauges are special tools used to check specific frame and body points. They allow you to quickly measure the direction and extent of vehicle damage.

Control or *reference points* are specific locations on the frame or body for making measurements. They may be holes, specific bolts, nuts, panel edges, or other locations on the vehicle. To repair a badly damaged vehicle, you must restore the reference points to their factory dimensions. This must be done while ref-

erence points in the undamaged area remain in their correct locations.

Therefore, the collision repair technician must work with the whole vehicle. This is done by measuring and recording dimensional changes. The most widely accepted method of checking body dimensions is to use the charts supplied in the body dimension manuals.

DAMAGE DIAGNOSIS

To repair a vehicle properly, you or the estimator must first accurately diagnose the collision damage.

Someone must assess the severity and extent of damage and find all parts that have been affected. Once this has been determined, a plan can be made for repair.

Damage found from an inaccurately diagnosed vehicle will be uncovered during repair. When this happens, the repair method or procedure must be changed.

Generally, physical damage is rarely missed during an inspection by a competent estimator or body technician. However, the effects of the damage on unrelated systems and damage next to the impacted part can be accidentally overlooked. A visual inspection alone is inadequate. Accident damage should be assessed by measurements with the proper tools and equipment.

The following is a basic diagnosis procedure:

1. Know the type of vehicle construction (full frame, partial frame, unibody).

2. Visually locate the point of impact.

3. Determine the direction and force of the impact. Once this is determined, check for possible damage.

4. Determine if the damage is confined to the body, or if it involves mechanical parts (wheels, suspension, engine).

5. Systematically inspect damage to the parts along the path of the impact and find the point where there is no longer any evidence of damage. For example, pillar damage can be determined by checking the door opening alignment.

6. Measure the major parts and check body height by comparing the actual measurements with the values in the body dimensions chart. Use a centering gauge to compare measurements of the height of the left and right sides of the body.

7. Check for suspension and overall body damage with the proper fixtures.

SAFETY

Before starting a vehicle damage evaluation, keep the following safety pointers in mind:

1. Once the vehicle is in the shop, check for broken glass edges and jagged metal. Edges of broken glass should be masked with tape and labeled "DANGER." Sharp jagged metal edges can be taped or covered.

2. If fluids, such as motor oil or transmission fluid, are leaking from the vehicle, wipe them up. This

may prevent someone from slipping and falling.

3. Disconnect the battery. This will eliminate the possibility of a charge igniting flammable vapors. It also protects the electrical system.

4. Make the damage diagnosis in a well-lit shop. If the damage involves functional or mechanical parts, a detailed inspection of the underbody, using a lift or a bench, is required.

NOTE! For more safety rules, refer to the index in the back of this book.

IMPACT EFFECTS

A modern vehicle is designed to withstand the shocks of normal driving and to provide protection for the occupants during a collision. Special consideration is given to designing the body so that it will collapse and absorb the maximum amount of energy in a severe collision.

The front body and rear body are, to some extent, made to deform, forming a structure that absorbs impact energy. The center passenger compartment area is designed to stay relatively unaffected. For example, during a head-on collision with a barrier at 30 mph (48 km/h), the engine compartment compacts by about 30 to 40 percent of its length, but the passenger compartment is compacted by only one to two percent of its length.

Described in earlier chapters, there are two basic types of automotive construction:

1. Body-over-frame (BOF)

2. Unibody (or monocoque)

In the body-over-frame construction, the passenger (pay-load) area is enclosed with panels of steel attached to a structural frame. The frame also supports most of the drive train and mechanical accessories. In the unibody construction, the metal body panels are welded together making a structural unit.

With BOF vehicles, collision damage can be cosmetic and/or structural. However, when a unibody vehicle is hit, the collision usually results in both cosmetic and structural damage.

Under the force of collision impacts, the frame-type vehicle and the unibody-type vehicle react quite differently. Also, damage assessment and repair techniques are different, even though the basic repair skills are similar.

When the collision damage has been identified using the proper identification and analysis procedures, anyone skilled in the mechanics of collision damage repair is capable of repairing the damage successfully.

ANALYZING COLLISION FORCES

The impact force of a collision and the extent of damage differ depending on how the collision occurred. The damage can be partly determined by understanding how the collision happened. To understand the circumstances of the collision, contact persons directly involved or eyewitnesses. It is quite possible for the person responsible for making the estimate to get a direct response from the customer. This method of damage assessment is sometimes necessary to estimate the cost of the repair.

The body technician may need to know the following items:

1. The size, shape, position, and speed of the vehicles involved in the collision

2. Speed of the vehicle at the time of the collision

3. Angle and direction of the vehicle at the time of the impact

4. The number of passengers and their positions at the time of the impact

A good body/frame technician can usually determine what actually happened to cause the damage. Certain types of collision damage often occur in a predictable pattern and sequence.

If a driver's first reaction is to turn away from the danger, the vehicle will be forced to take the hit on the side, Figure 20–3. If the driver's reaction is to slam on the brakes, the direction of impact would be frontal, Figure 20–4. A frontal collision where the point of impact is high on the vehicle could cause the cowl and roof to move rearward and the rear of the vehicle to move downward. If the point of impact is low at the front, the inertia of the body mass could cause the rear of the vehicle to distort upward, forcing the roof forward. This could leave an excessively large opening between the front upper part of the door and the roof line, Figure 20–5.

Given vehicles with similar weights traveling at the same speed, vehicle damage will vary depending on what is struck—another vehicle, a utility pole, or a wall.

If the impact is spread over a larger area (as in hitting a wall), damage will be minimal, Figure 20–6A. Conversely, the smaller the area of impact, the greater the severity of the damage, Figure 20–6B. In this example, the bumper, hood, radiator, and so forth have been severely damaged. The engine has been pushed back and the effect of the collision has extended as far as the rear suspension.

Another consideration is when one vehicle hits another while moving. If vehicle Number One in Figure 20–7 drives into the side of vehicle Number Two while Number Two is moving, the motion of the

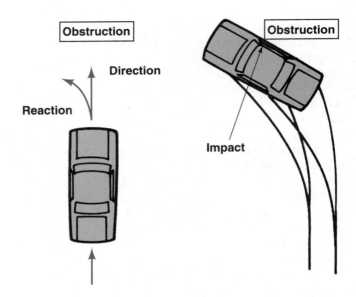

Figure 20–3 The direction of the impact affects the type of vehicle damage. If the driver reacts and turns away from an object, the hit is on the side of the vehicle, often causing side-sway damage.

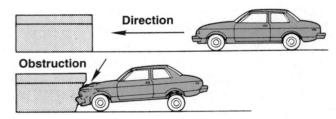

Figure 20–4 If a driver hits the brakes before a collision, this causes the nose of the car to dive downward. The impact can then cause frame or unibody sag. The front area will be forced up and the area around the cowl sags down.

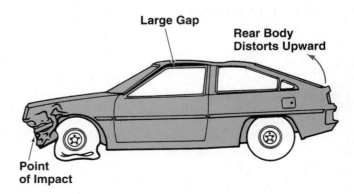

Figure 20–5 A hard frontal impact causes major or primary damage to the front end. Minor, secondary damage occurs elsewhere from the shock wave flowing through the body.

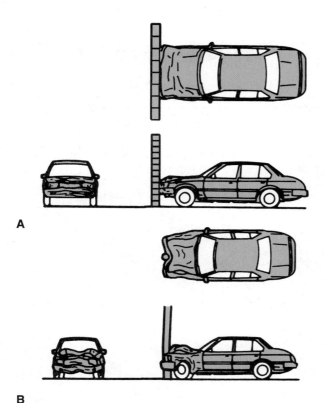

A

B

Figure 20–6 Note impact damage differences. (A) The vehicle hits a wall straight on. The large area absorbs energy, and damage is limited to front body parts. (B) Another vehicle hits a utility pole. The small area must absorb all the energy so the damage travels much deeper into the engine compartment.

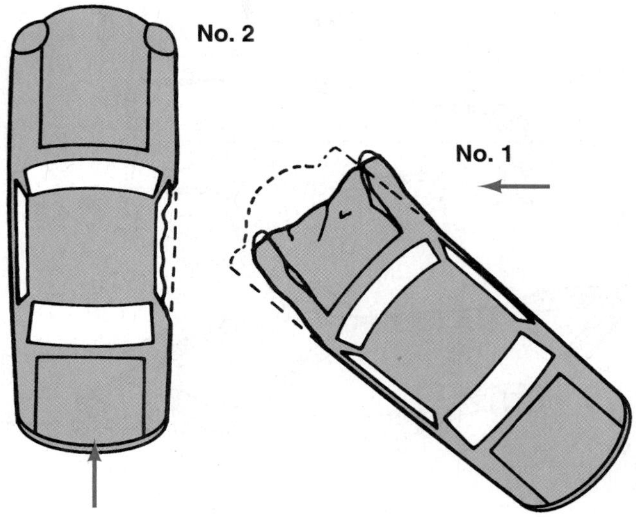

first vehicle will drive the front end of the vehicle back. However, the motion of vehicle Number Two will also "drag" that same front end to the side. There is only one collision but the damage is in two directions.

On the other hand, there might be two collisions in only one direction. This is a fairly common occurrence in highway pile-ups. For example, a vehicle might collide with another vehicle. Then, it might leave the road and hit a pole or guardrail. This results in two completely separate types of damage.

There are many other variables and possible combinations of damage types. It is important to determine what actually happened before an accurate diagnosis can be made. Get as many facts as possible and combine them with physical measurements and gauge readings to determine the exact collision repair procedure that should be taken. A little extra time spent here can save many hours in the overall repair procedure. "Think time" saves "work time."

NOTE! For more information on estimating, refer to Chapter 28.

No. 2

No. 1

Figure 20–7 A broadside collision often causes heavy frame or unibody damage. Vehicle Number One would suffer side sway, and front end damage. Vehicle Number Two would suffer major side sway damage in the center section.

TYPES OF FRAME DAMAGE

Full frame vehicle damage can be broken down into five categories:

1. **_Side sway damage_** results from collision impacts that occur from the side. It often causes side bending of the frame, Figure 20–8. Side sway usually occurs in the front or rear of the vehicle. Generally, it is possible to spot side sway damage by noting if there are buckles on the inside of one rail and buckles on the outside of the opposite side rail.

Side sway can be recognized by abnormalities such as a gap at the door on the long side and wrinkles on the short side. Look for impact damage obviously from the side. For example, hood and deck lid will not fit into proper opening.

2. **_Sag damage_** is a condition where a section of the frame is lower than normal, Figure 20–9. The structure has a swayback appearance. Sag damage generally is caused by a direct impact from the front or rear. It can occur on one side of the vehicle or on both sides, Figure 20–10.

Sag can usually be detected visually by an irregular gap between the fender and the door. The gap will be narrow at the top and wide at the bottom. Also look for a door hanging too low at the striker.

Enough sag can be present in the frame to prevent body panel alignment even though wrinkles or kinks are not visible in the frame itself.

3. **_Mash damage_** is present when any section or member of the vehicle is shorter than factory

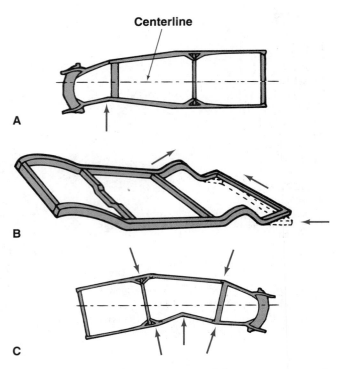

Centerline

A

B

C

Figure 20–8 Note the various types of damage. (A) Front side sway damage is present when the frame is forced to one side. (B) Rear side sway is similar but at the back of the vehicle from a rear impact. (C) Double side sway is caused by severe side impact.

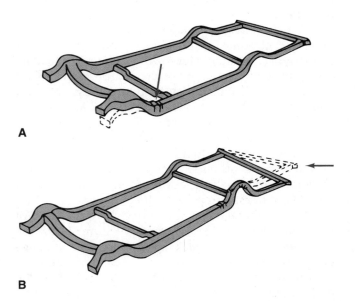

A

B

Figure 20–9 Sag causes the frame to drop lower from severe impact. (A) Front sag is often in the cowl area from the buckling of frame rails. (B) Rear sag is similar, but at the rear rails.

specifications. Mash is usually limited to forward of the cowl or rearward of the rear window. Doors could fit well and appear to be undisturbed. See Figure 20–11.

Mash is indicated by wrinkles and severe distortion in fenders, hood, and possibly frame horns or

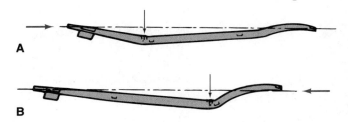

A

B

Figure 20–10 (A) Side rail sag is due to a front end collision. (B) Side rail sag is due to a rear collision.

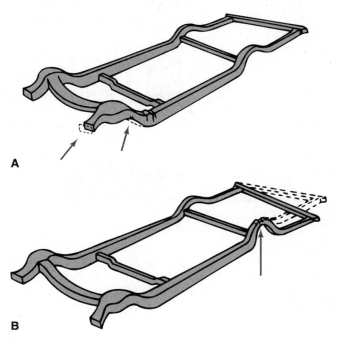

A

B

Figure 20–11 (A) Note the mash damage on the left front rail. (B) Note the mash damage on the left rear rail.

rails. The frame will rise upward at the top of the wheel arch, causing the spring housing to collapse, Figure 20–12.

With mash damage, there is very little vertical displacement of the bumper. The damage results from direct front or rear collisions.

4. **_Diamond damage_** is a condition where one side of the vehicle has been moved to the rear or front. This type of damage causes the frame to be out of square. It will form a parallelogram and is caused by a hard impact on a corner or off-center from the front or rear. Refer to Figure 20–13.

Diamond damage affects the entire frame or unibody, NOT just the side rails. Visual indications are hood and trunk lid misalignment. Buckles might appear in the quarter panel near the rear wheel housing or at the roof to quarter panel joint. Wrinkles and buckles probably will appear in the passenger compartment and/or trunk floor. There usually will be some mash and sag combined with the diamond, Figure 20–14.

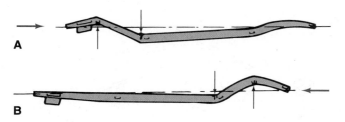

Figure 20–12 (A) Note the heavy mash and buckling damage from a front end collision. (B) This is the same damage, but from the rear.

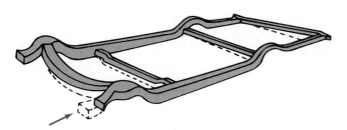

Figure 20–13 Diamond damage forces only one side of the frame rearward. The hard impact may affect both sides.

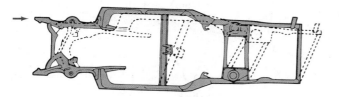

Figure 20–14 Study the diamond condition shown here.

5. **Twist damage** is a condition where one corner of the vehicle is higher than normal; the opposite corner might be lower than normal. It is another type of total frame damage. Twist can happen when a vehicle hits a curb or median strip at high speed. It is also common in rear corner impacts. Look at Figure 20–15.

A careful inspection might reveal no apparent damage to the sheet metal; however, the real damage is hidden underneath. One corner of the vehicle has been driven upward by the impact. The adjacent corner may also be twisted downward. If one corner of the vehicle is sagging close to the ground as though a spring were weak, the vehicle should be checked for twist.

Unfortunately, most accidents result in a mix of one or more of these damage problems. Side sway and sag frequently occur almost simultaneously. Also, some of these collision solutions affect crossmembers, especially the front member. In a rollover accident, for example, the front crossmember on which the motor

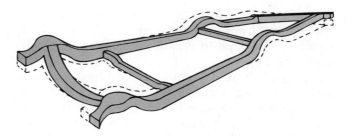

Figure 20–15 Twist damage forces one side of the frame up and the other side down.

mounts are attached will be pulled or pushed out of shape because of the engine's weight. This will result in a sag of this crossmember.

UNIBODY VEHICLE DAMAGE

The damage that occurs to a unibody vehicle as the result of an impact can best be described by using the **cone concept.** The unibody vehicle is designed to absorb a collision impact. When hit, the body folds and collapses as it absorbs the impact. As the force penetrates the structure, it is absorbed by an ever increasing area of the unibody. This characteristic spreads the force until it is completely dissipated. Visualize the point of impact as the tip of the cone.

The centerline of the cone will point in the direction of impact. The depth and spread of the cone indicate the direction and area that the collision force traveled through the unibody. The tip of the cone and point of impact is the **primary damage** area, Figure 20–16.

Since unibodies are pieces of thin sheet metal welded together, the shock of a collision is absorbed by a large portion of the body shell. The effects of the impact shock wave as it travels through the body

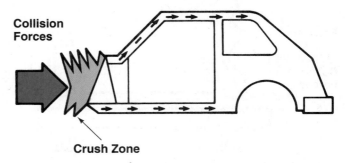

Collision Forces

Crush Zone

Figure 20–16 Unibody vehicles use the cone or egg shell principle to help absorb the energy of an impact. *(Babcox Publications)*

structure is called *secondary damage.* Generally, this damage is toward the inner structure of the unibody or toward the opposite end or side of the vehicle, Figure 20–17.

To provide some control on secondary damage and to provide a safer passenger compartment, a unibody vehicle is designed with *crush zones.* These crush zones are engineered to collapse in a predetermined fashion, Figure 20–18. The effects of the impact shock wave is reduced as it travels through and is dissipated by the body structure.

In other words, front impact shocks are absorbed by the front body and crush zones, Figure 20–19. Rear shocks are absorbed by the rear body. Side shocks will be absorbed by the rocker panel, roof side frame, center pillar, and door.

Impact damages on unibody vehicles can be described as follows.

Frontal damage results from a head-on collision with another object or vehicle. The impact of a colli-

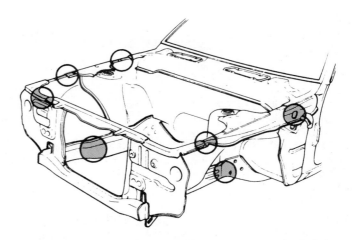

Figure 20–18 These are the crush zones on a unibody vehicle. They should be checked closely after a collision.

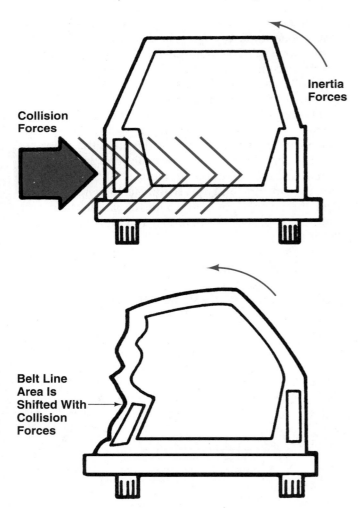

Figure 20–17 Note how the roof has shifted sideways because its inertia tried to keep it from rapidly moving with the impact. *(Babcox Publications)*

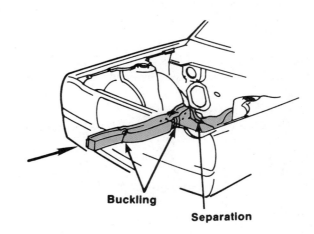

Figure 20–19 With a hard collision, unibody frame rails can buckle and even separate or tear away from other parts.

sion depends upon the vehicle's weight, speed, area of impact, and the source of impact. In the case of a minor impact, the bumper is pushed back, bending the front side members, bumper stay, front fender, cone support, radiator upper support, and hood lock brace.

If the impact is further increased, the front fender will contact the front door. The hood hinge will bend up to the cowl top. The front side members may also buckle into the front suspension crossmember, causing it to bend, Figure 20–19. If the shock is great enough, the front fender apron and front body pillar (particularly front door hinge upper area) will be bent, which will cause the front door to drop down. In addition, the front side members will buckle. The front suspension member will bend. The instrument panel and front floor pan may also bend to absorb the shock.

If a frontal impact is received at an angle, the attachment point of the front side member becomes a turning axis. Lateral as well as vertical bending occurs (sway). Since the left and right front side members are connected together through the front crossmember, the shock from the impact is sent from the point of impact to the front side member of the opposite side of the vehicle.

Rear damage occurs if the vehicle is moving backward and hits something or if it is hit by another vehicle from behind. When the impact is comparatively small, the rear bumper, the back panel, trunk lid, and floor pan will be deformed. Also, the quarter panels will bulge out.

If the impact is severe enough, the quarter panels will collapse to the base of the roof panel. On four-door vehicles, the center body pillar might bend. Impact energy is absorbed by the deformation of the above parts and by the deformation of the kick-up of the rear side member.

Side damage will cause the door, front section, center body pillar, and even the floor to deform. When the front fender or quarter panel receives a large perpendicular impact, the shock wave extends to the opposite side of the vehicle.

When the central area of the front fender receives an impact, the front wheel is pushed in. The shock wave extends from the front suspension crossmember to the front side member. If severe, the suspension parts are damaged and the front wheel alignment and wheelbase may be changed. The steering gear or rack can also be damaged by side impacts.

Top impacts can result from falling objects or from a rollover of the vehicle. This type of damage involves not only the roof panel, but also the roof side rail, quarter panels, and possibly the windows as well.

When a vehicle has rolled over and the body pillars and roof panels have been bent, the opposite ends of the pillars will be damaged as well. Depending on the manner in which the vehicle rolled over, the front or back sections of the body will be damaged, too. In such cases, the extent of the damage can be determined by the deformation around the windows and doors.

The typical **collision damage sequence** on a unibody structure happens like this:

1. In the first microseconds of impact, a shock wave attempts to shorten (mash) the structure. This causes a lateral or vertical bending in the central structure. Most of the forces that broadcast impact shock to remote areas occur at this instant. Since the structure is stiff and springy, it tends to snap back to its original shape—at least momentarily.

2. As the collision event continues, visible crushing occurs at the point of impact. Impact energy is absorbed in the deforming structure (helping protect passenger compartment). Remote areas might buckle, tear, or pull loose.

3. In some unibody vehicles, impact forces reaching the passenger compartment cause the side structure to bow out away from the passengers (never in) distorting side rails and door openings.

Widening is similar to side sway damage in BOF vehicles and is indicated by the width measurement being out of tolerance.

4. Even if the initial impact is dead center, secondary impact can introduce torsional loads that cause a general twisting of the structure. Unibody structural twisting, like twisting of a conventional vehicle frame, is usually the last collision event. It is indicated by combinations of height and width measurements being out of tolerance.

DIMENSIONAL REFERENCES

Two major dimensional references are indicated in all body dimension manuals: the datum plane and centerline.

A **datum line,** or **datum plane,** is an imaginary flat surface parallel to the underbody of the vehicle at some fixed distance from the underbody. It is the plane from which all vertical or height dimensions are taken. It is also the plane that is used to measure the vehicle during repair. The datum is normally shown on dimension charts from the vehicle's side view.

The **center plane,** or **centerline,** divides the vehicle into two equal halves: the passenger side and the driver side. The centerline is shown on dimension charts in either the bottom or top views. It can sometimes be found on some vehicles in the form of body center marks. **Body center marks** are stamped into the sheet metal in both the upper and lower body areas of the vehicle. They can save time when taking measurements.

All width or **lateral dimensions** of symmetrical vehicles are measured from the center. That is, the measurement from the centerline to a specific point on the right side will be exactly the same as the measurement from the centerline to the same point on the left side. One side of the structure would be a perfect mirror image of the other. Most vehicles are built symmetrically. But if the vehicle is NOT symmetrical (asymmetrical), the self-centering gauges will NOT align and will NOT indicate a true center reference.

It is usually necessary to think of the vehicle as a rectangle divided into three zero plane sections. The **three zero plane sections** break the vehicle into three areas—front, center, and rear. The torque box location is used as the dividing lines. This three-section principle is a result of the vehicle's design and the way it reacts during a collision.

Symmetrical means that the dimensions on the right side of the vehicle are equal to the dimensions

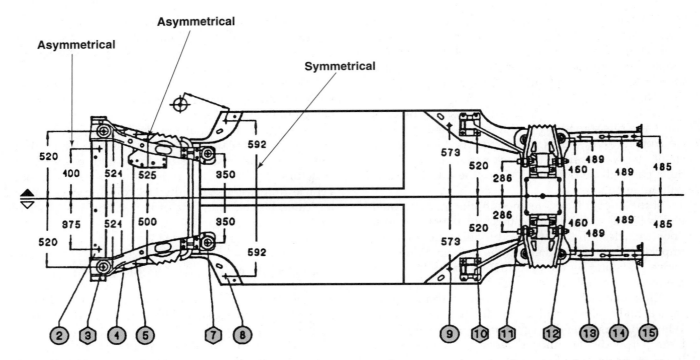

Figure 20–20 A body dimension manual will give measurements and reference points for a specific make and model vehicle. If symmetrical, measurements on both sides are the same. If asymmetrical, measurements will be different on each side. *(Hein-Werner® Corporation)*

on the left side of the vehicle. If the vehicle is ***asymmetrical,*** these dimensions are NOT the same. In such a case, use gauges that can be adjusted to compensate for the asymmetry. See Figure 20–20.

VEHICLE MEASURING BASICS

In unibody construction, each section should be checked for diagonal squareness by comparing diagonal lengths. Length and width should also be compared. The center section should be used as a base when reading structural alignment. All measurements and alignment readings should be taken relative to the center section.

Start measuring in the center or middle section. If it is NOT square, then move to the undamaged end of the vehicle to find three correctly positioned reference points.

Keep in mind that to accurately measure a vehicle, you must start with at least three dimensions you know are undamaged. The way to do this is to check the squareness of the vehicle. If the vehicle is NOT symmetrical, refer to the dimension chart for correct measurements.

Remember! The terms "control point" and "reference point" have different meanings. The **control points** used in manufacturing are NOT necessarily the same as the reference points the collision repair technician uses to measure the vehicle. **Reference**

points refer to the points, bolts, holes, and so on, used to give unibody dimensions in body specification manuals. The distance between reference points can be measured with either a tram bar or a tape measure, Figure 20–21.

A vehicle cannot be properly repaired unless all of the reference points are returned to precollision specifications. To achieve this, a collision repair technician must:

1. Measure accurately to analyze damage.

2. Measure often when pulling out damage.

3. Recheck all measurements after straightening.

TYPES OF MEASURING EQUIPMENT

Because of the importance of measuring during repairs, many kinds of special equipment have been developed by manufacturers. Each provides the capability to measure very quickly and accurately. Although a number of styles of measuring equipment can be found in collision repair shops, most of it can be divided into four basic systems:

1. Gauge measuring systems

2. Universal measuring systems

3. Dedicated fixture systems

4. Electronic measuring systems

Measurement With a Tram Bar

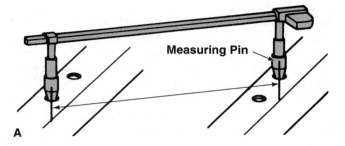

Measuring Pin

A

Measurement With a Tape Measure

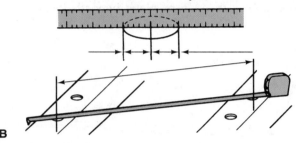

B

Figure 20–21 These are two basic methods of measuring on a vehicle. (A) Tram bar tips are inserted or touched on reference points. (B) A tape measure can also be used to measure between points.

Gauge Measuring Systems

Gauge measuring systems use sliding metal rods or bars, and adjustable pointers with ruled scales to measure body dimensions. The tape measure, tram bar, the self-centering gauge, and the strut tower gauge can be used separately or in conjunction with one another.

Tram bars are used for measurement; self-centering gauges check for misalignment. Supported by MacPherson strut tower domes, the strut tower gauge allows visual alignment of the critical reference points of unibody vehicles. The tram bar, self-centering gauge, and strut tower gauge are available as a unit or as separate vehicle diagnostic tools.

The tracking gauge is used to check alignment of the front and rear wheels. It is another popular measuring gauge. If the front and rear wheels are NOT in alignment, the vehicle will NOT handle properly.

Keeping Records. Since reference point measurements must be taken and written down several times in a repair operation, a method of tabulation must be devised. One way to accomplish this is to use a **data chart** or *tabulation chart.*

In the first column of the tabulation chart, write the location of what is being measured. Numbers or letters can be used. The second and third columns contain the manufacturer's specifications taken from the body dimension manual and actual distance as it exists on the damaged vehicle. The 1-2-3 are the readings taken at measurement step 1, measurement step 2, and so on. That is, as each pull is made, the measurements should be recorded, including those dimensions that have been corrected. This chart tells the collision repair technician at a glance how the job stands in restoring the vehicle to its precollision state.

NOTE! Straightening or pulling will be explained in detail in the next chapter.

Using Tram Bars and Tapes. A *tram bar/tram gauge* is a measuring rod with two adjustable pointers attached to it, Figure 20–21. The pointers slide along the bar's length and are adjustable for height. Since the tram bar measures one dimension at a time, each must be recorded and cross checked from two additional reference points. At least one must be a diagonal measurement.

The best areas to select for tram bar measurements are the attachment points for suspension and mechanical parts. These are critical to alignment. Throughout the pulling and straightening process, critical reference points must be measured (and recorded) repeatedly with the tram bar to monitor progress and to prevent overpulling.

The tram bar might have a scale superimposed on it. However, since almost all dimension charts list measurements in metric, use a tape measure with both English and metric scales to set up the tram bar. The tape measure can also be used to take quick measurements between reference points. Be sure the tape measure has been checked for accuracy before using it.

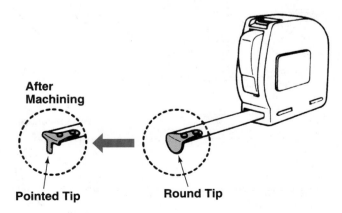

After Machining

Pointed Tip

Round Tip

Figure 20–22 Grind or file down the tip of a tape measure like this. A smaller tip will make more accurate measurements in reference holes.

NOTE! More accurate measurements can be taken if the front end of a tape measure is machined as in Figure 20–22. Its tip can then be more fully inserted into the reference point hole.

Reference point holes are frequently larger in diameter than the tram bar tip. To measure accurately with the tram bar when the holes are the same diameter, measure like edge to like edge. Study Figure 20–23.

When the holes are NOT the same size, they will usually be the same type of hole: round, square, or oblong. To find the center-to-center measurement, measure inside edge to inside edge, then outside edge to outside edge. Add the results of the two measurements and divide by 2.

For example, two round holes, one 12.5 mm in diameter, the other 38 mm in diameter, have an inside measurement of 762 mm and an outside measurement of 812 mm. The center-to-center dimension is (762 mm + 812 mm) divided by 2 = 787 mm. The 787 mm is the dimension for the tram bar. Center at the smallest hole and the other end should center at the larger hole. If it does NOT, there is damage to the unibody. Refer to Figure 20–24.

The term *point-to-point measurement* refers to the shortest distance between two points, as shown in Figure 20–25. The term *datum measurement* refers to the distance between two points when measured from the datum plane/line.

A few body dimension manuals give the measurements based on the bar's length. Here, the bar of the tram becomes the datum plane beneath the vehicle.

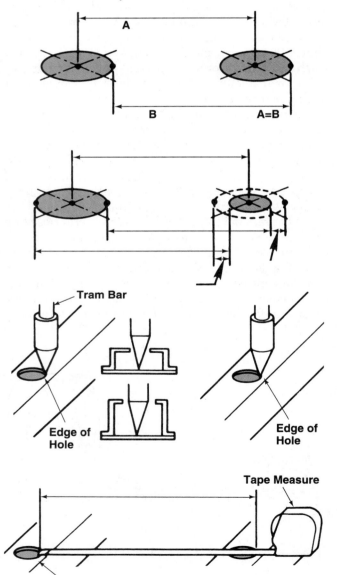

Figure 20–23 Note various methods for measuring the distance between reference points.

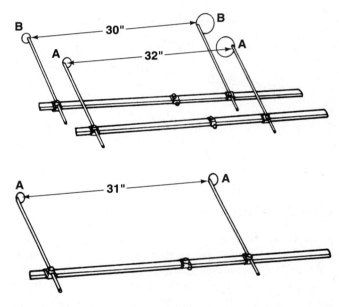

Figure 20–24 Note how to measure holes of different sizes with a tram gauge.

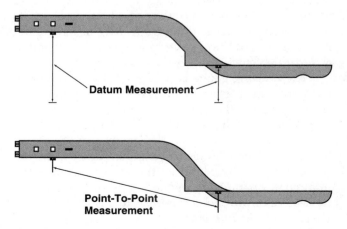

Figure 20–25 When specs are given, they can be point-to-point or datum measurements.

The pointers are set at the horizontal height dimensions specified on the dimension chart. It is important to always check the method used on the dimension chart because the two distances are NOT the same.

In using a tram bar for measuring, the vehicle dimensions are needed to accurately assess and correct the damage. If the vehicle dimensions are NOT available, use an undamaged vehicle of the same make, year, model, and body style as a source to obtain the correct factory dimensions. Frequently, if only one side of a vehicle has been damaged, it also is possible to take measurements on the undamaged side. Set the tram bar using the undamaged side. Then use it to measure the damaged side to make comparison measurements.

Upper Body Measurement. Upper body damage can also be determined by the use of a tram bar and a steel measuring tape. Their use is basically the same as when doing an underbody evaluation. Dimension charts have measurements for many upper body reference points.

Front Body Measurement. When a damaged vehicle needs the hood and front side member replaced, you should take measurements during the repair. Even if only the front, right side of the vehicle body received the impact, the left side will usually be damaged also. Therefore, the extent of deformation must be checked before pulling.

Figure 20–26 shows the typical front body reference points. They should be measured and checked against the dimension chart for the specific vehicle.

When checking front end dimensions, measure the attachment points for suspension and mechanical parts. These are critical to proper alignment. Each dimension should be checked from two additional reference points with at least one reference point being a diagonal measurement.

Note that the longer the dimension, the more accurate the measurement. A measurement from the lower cowl to the engine cradle is a better gauge than a measurement from a lower cowl to another lower cowl area. This is because the longer dimension takes in a larger area of the vehicle. The use of two or more measurements from each reference point assures greater accuracy. It also helps identify the extent and direction of any panel damage.

Body Side Panel Measurement. Body side structure damage can be evaluated by the fit and operation of the doors when opened and closed. Deformation around the door openings can cause an uneven gap and air or water leaks. Thus, accurate measurements must be taken. The tram bar is used to measure the body side panel. See Figure 20–27.

Diagonal line measurement compares reading across an opening or between four reference points to determine damage. Damage can be detected if the left-to-right symmetry of the body is used for measuring diagonal lines, as shown in Figure 20–28A. Use this measuring method if:

1. The data on the engine compartment and underbody are missing.

2. There are no data available in the body dimension chart.

3. The vehicle has been severely damaged in a rollover.

The diagonal line measurement method is NOT adequate when inspecting damage to both sides of the vehicle or in the case of twisting. This is because the left-to-right difference in the diagonal lines cannot be measured, Figure 20–28B.

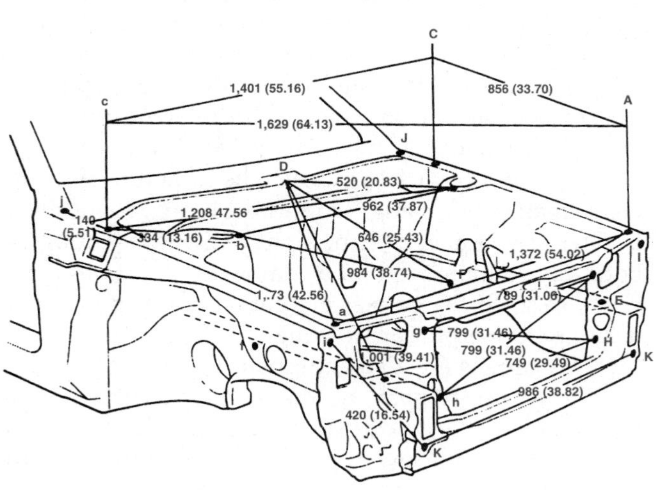

Figure 20–26 A dimensions manual will give front end body reference points like these.

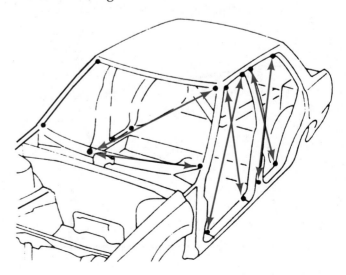

Figure 20–27 Note typical side reference points.

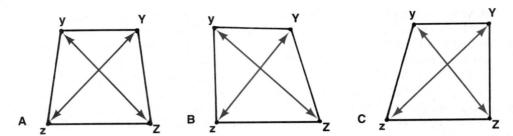

Figure 20–28 Study the use of the diagonal line measuring technique. (A) Yz equals yZ, no warping. (B) Yz is less than yZ, deflection to left. (C) Yz is greater than yZ, deflection to right.

If deformation is the same on the left and right, the left to right difference will NOT be apparent. Refer to Figure 20–28C.

In Figure 20–29, the measurement and comparison of the left and right lengths between yz and YZ will give an even better indication of damage conditions. This method should be used in conjunction with the diagonal line measurement method. This method can be applied where there are parts that are symmetrical on the left and right sides.

Rear Body Measurement. Damage to the rear body can be analyzed by the fit and action when the deck lid is opened and closed, Figure 20–30. Damage to this area can cause water leakage around the deck lid seal. Any wrinkle in the rear floor is usually due to buckling of the rear frame rail, a major problem. Thus, measure the rear body together with the underbody. In this way, the straightening work can be performed effectively.

Keep in mind that all reference points need to be checked from two or more reference points. Diagonal measurements are a good way to cross-check dimensional reference points. Correct use of self-centering gauges will also reveal the kind of misalignment that would be shown by diagonal tram bar measurements.

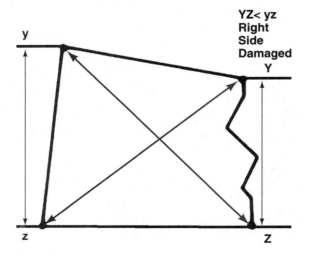

Figure 20–29 Study this diagonal line measurement example.

Note the various reference points in Figure 20–31. Refer to the body dimension manual to find where these points are located.

Self-centering Gauges

Self-centering gauges show alignment or misalignment by projecting points on the vehicle's structure into the technician's line of sight. See Figure 20–32.

Self-centering gauges are installed at various control areas on the bottom of the vehicle, Figure 20–33, page 385. They have two sliding horizontal bars that remain parallel as they move inward and outward. This action permits adjustment to any width for installation on various areas of the vehicle.

Self-centering gauges are used to establish the vehicle centerline and datum plane. Use the gauges in sets of four or more. Adjust and hang them at the locations indicated in the dimensional data manual to establish the datum plane. The body manual will also give information for adjusting them to hang on the same plane under the vehicle. Scales are provided for adjusting the gauges to hang the correct distance down under the vehicle, Figure 20–34.

Begin by setting up the *base gauges* or the two gauges hung near the center off the vehicle. Usually, the torque box area is used. Generally, these will be gauges numbered two and three. Add other gauges as needed and use the base gauges for reference.

When inspecting for damage, first hang self-centering gauges from two places where there is no visible damage, then hang two more gauges where there is obvious damage. Then, sight the gauges hung at both the undamaged and damaged locations and check to see if the gauges are parallel to each other or if the centering pins are misaligned.

The gauges are also equipped with center pins or sights, which remain in the center of the gauge regardless of the width of the horizontal bars. This allows the technician to read centerline throughout the length of the vehicle.

When sighting gauges for parallel, always stand directly in the middle, scanning with both eyes. To

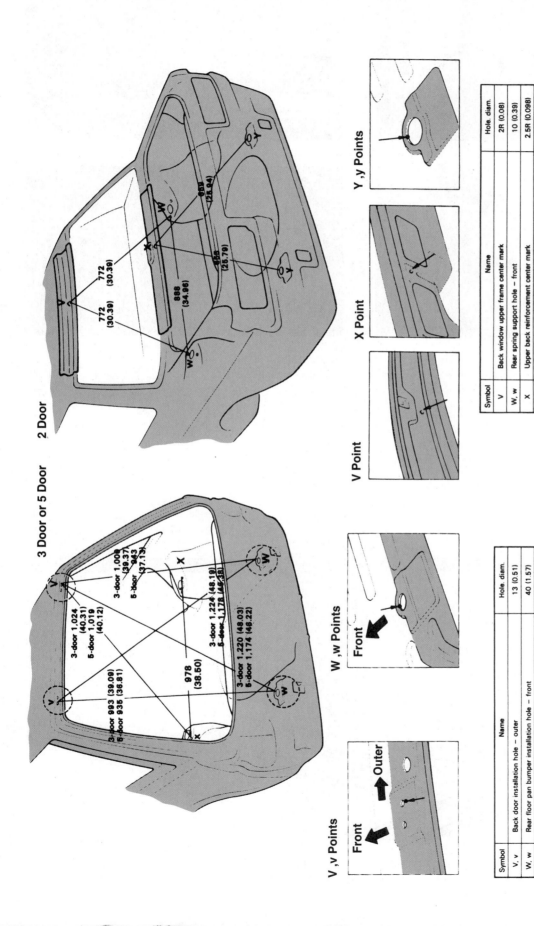

2 Door

772
(30.39)

772
(30.39)

888
(34.96)

655
(25.79)

655
(25.94)

772
(30.39)

3 Door or 5 Door

3-door 1,024 (40.31)
5-door 1,019 (40.12)

3-door 1,008 (39.37)
5-door 943 (37.13)

3-door 1,224 (48.19)
5-door 1,178 (46.38)

3-door 1,220 (48.03)
5-door 1,174 (46.22)

978
(38.50)

3-door 993 (39.09)
5-door 935 (36.81)

V ,v Points

Symbol	Name	Hole. diam.
V, v	Back door installation hole – outer	13 (0.51)
W, w	Rear floor pan bumper installation hole – front	40 (1.57)
X, x	Rear spring support hole – rear	10 (0.39)

Front

Outer

W ,w Points

Front

Symbol	Name	Hole. diam.
V	Back window upper frame center mark	2R (0.08)
W, w	Rear spring support hole – front	10 (0.39)
X	Upper back reinforcement center mark	2.5R (0.098)
Y, y	Rear floor pan bumper installation hole – front	40 (1.57)

V Point

X Point

Y ,y Points

Figure 20–30 Note rear body reference points for two body styles.

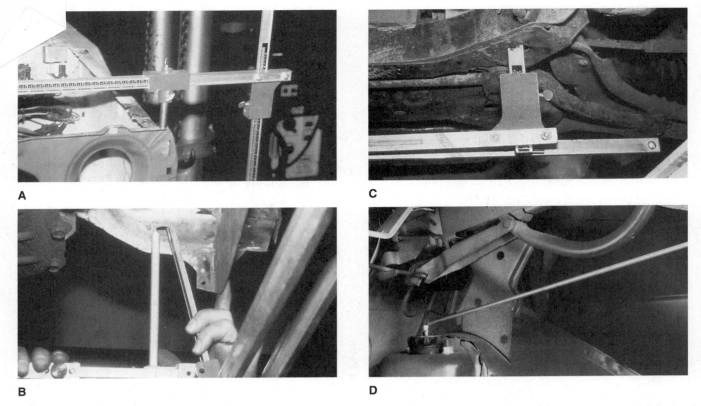

Figure 20–31 Here are several body reference point examples. *(Chief Automotive Systems, Inc.)* (A) Checking location of upper corner of core support. (B) Checking reference hole location on frame rail. (C) Checking reference hole in engine cradle. (D) Checking reference point on rear shock tower.

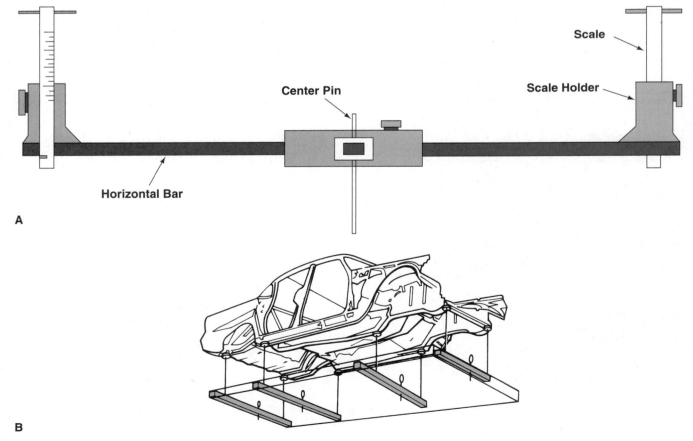

Figure 20–32 A self-centering gauge is used to check for major frame and unibody damage. (A) Note the parts of a self-centering gauge. *(I-CAR)* (B) Four self-centering gauges are often suspended under a vehicle to form a datum plane and center plane. *(Blackhawk Collision Repair, Inc., Subsidiary of Hein-Werner® Corporation)*

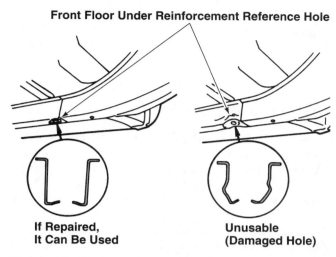

Front Floor Under Reinforcement Reference Hole

**If Repaired,
It Can Be Used**

**Unusable
(Damaged Hole)**

Figure 20–33 Do not hang gauges from damaged reference holes or points.

ensure accuracy, readings should be made at the outer edge of the self-centering gauge, NOT in the middle. See Figure 20–35.

The farther one stands from the gauges while reading, the more accurate the reading will be. Standing close changes the line of sight to the front gauges so drastically that an accurate reading is nearly impossible.

Self-centering gauges should always be set at the same height or plane. Different heights will change the angle of sight and give a false reading. Sighting from the end of the vehicle opposite the damage can make readings more accurate. With practice, you can improve the damage analysis with self-centering gauges.

The sighting of centerline pins must be done with one eye. Since the center section is the base for gauging, your line of sight must always project through the pins of the base gauges. Observing pins in other sections of the body will then reveal how much they are out of alignment.

NOTE! Never attach self-centering gauges to any movable parts such as control arms or springs.

Each self-centering gauge has two vertical scales—one on the left side, one on the right. These scales are adjusted vertically to assure that the horizontal bars accurately reflect true positions.

Once hung in specific locations, the gauges generally remain on the vehicle throughout the entire repair operation. This is true unless they interfere with straightening, anchoring, or with the tram bar.

Special centering gauges are available that can be used to check such items as body pillar damage.

The same system of alignment is employed when using gauges to check underbody damage. Some self-

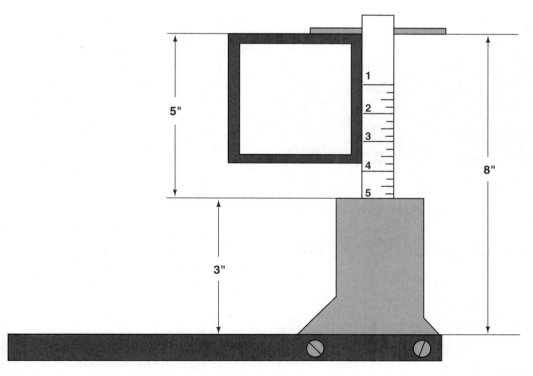

Figure 20–34 The scales on gauges should be set to match specifications in the dimensions manual. Note how the scale reading and the part of the tool must be added to adjust the gauge properly. *(I-CAR)*

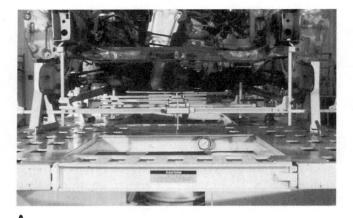

A

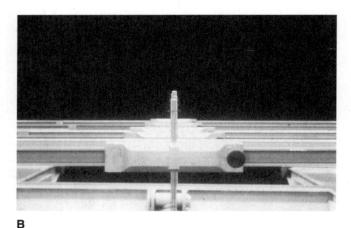

B

C

Figure 20–35 Note the basic method of using self-aligning gauges. *(Chief Automotive Systems, Inc.)* (A) After properly mounting self-centering gauges, sight down the pins to see if the vehicle is out of alignment. (B) The pins on all the gauges are in alignment, indicating that the frame or unibody is not damaged from the collision. (C) These pins and gauge rods are NOT in alignment, indicating that major frame or unibody damage has resulted.

centering gauges can be adjusted for asymmetrical vehicles as well. Again, check the dimensional data for these asymmetrical vehicles. Follow directions provided with the gauges.

Using the Datum Plane

To read for datum, all gauges must be on the same plane. After hanging all four gauges, read across the top to determine if datum is correct. If all four gauges are parallel, the vehicle is on datum. If they are NOT parallel, the vehicle is off datum. Refer back to Figure 20–35.

Since the datum line is an imaginary plane, datum heights can be raised or lowered to facilitate gauge readings. Just remember that if the datum height is changed at one gauge location, all the gauges MUST be adjusted an equal amount to maintain accuracy.

While datum readings are usually obtained from self-centering gauges, there are individual gauges available for measuring datum heights. These datum gauges are usually held in position by magnetic holders. Remember that the dimensions that allow the vehicle to be level with the road are measured from the datum plane.

Using the Centerline

To check for centerline misalignment, use the four self-centering gauges hung on the datum. To establish the true centerline, the center pin on the No. 2 gauge must be lined up with the center pin on the No. 3 gauge. Then the center pins of No. 1 and No. 4 can be read relative to the centerline of the base.

Frame or unibody damage will affect the centerline reading. If a vehicle has a diamond condition, a shortened rail or subrail, or an out-of-level condition, the centerline reading may be affected. Further inspection by gauging or measuring will be necessary.

Using Strut Tower Gauges

Supported by the shock-spring towers, the **strut tower gauge** allows visual alignment of the upper body area. It shows misalignment of the strut tower/upper body parts in relation to the vehicle's centerline plane and datum plane. See Figure 20–36.

The strut tower gauge features an upper and lower horizontal bar, each with a center pin. The upper bar is usually calibrated from the center out. Pointers are used to mount the gauge to the strut tower/upper body locations.

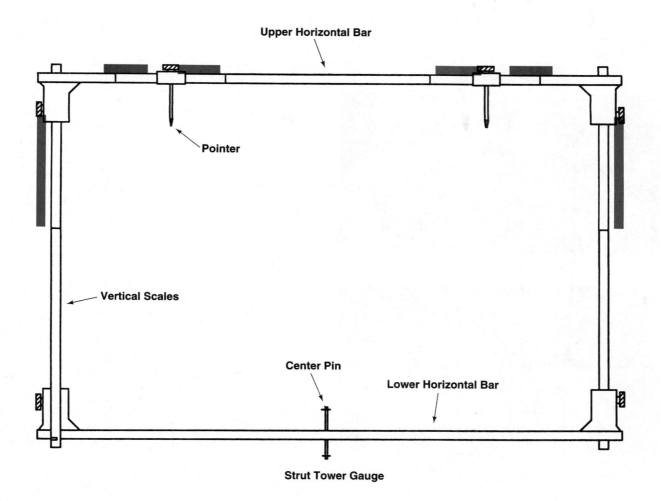

Upper Horizontal Bar

Pointer

Vertical Scales

Center Pin

Lower Horizontal Bar

Strut Tower Gauge

A

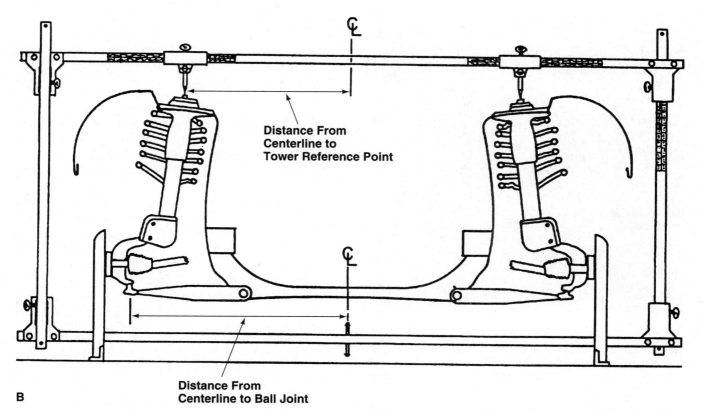

Distance From Centerline to Tower Reference Point

Distance From Centerline to Ball Joint

B

Figure 20–36 A strut tower gauge is commonly used with unibody vehicles. It is also used with other gauges. *(I-CAR)* (A) Study the parts of the strut tower gauge. (B) When the strut tower gauge is installed, you can check alignment of the top of the strut towers and also of lower parts. They should generally be at equal distances from the centerline.

A

B

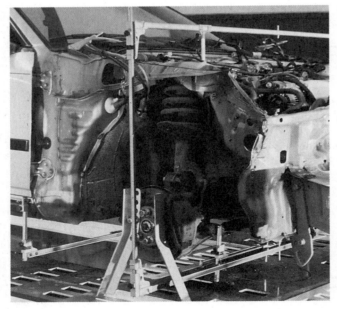

C

Figure 20–37 Note how a strut tower gauge is used. *(Chief Automotive Systems, Inc.)* (A) The strut tower gauge is adjusted and mounted using specifications from the manual. (B) Pointers on the strut tower gauge should touch exact reference points on the strut towers. If they do NOT, the towers are bent out of alignment and must be straightened. (C) The lower centering pin on the strut tower gauge can also be used with other self-centering gauges to check for damage.

Using dimension charts, the strut tower gauge is adjusted to the correct dimensions. You must use the vertical scales that link the upper and lower horizontal bars to set the lower bar at the datum plane. With the lower bar set to align properly, the upper pointers should be located at reference points on the strut towers. If they are NOT, the strut towers are damaged and pushed out of alignment. This would tell you that straightening would be needed so that the front suspension and wheels could be aligned properly. Look at Figure 20–37.

NOTE! The strut tower gauge is used most often to detect misalignment of strut towers. However, it can also be used to detect misalignment of a core support, cowl, quarter panels, and so on.

Diagnosing Damage with Gauges

To correct collision damage, pull or push the damaged area in the direction opposite to the impact. The sequence of the repair must be the reverse of the sequence in which damage happened. Therefore, the damage must be measured in the reverse sequence.

When measuring damage, keep in mind that a vehicle is similar to a building. If the foundation is NOT square and level, the rest of the structure will be uneven also. The vehicle's foundation, which is the center section of the vehicle, is measured for twist and diamond first. These two measurements will tell the collision repair technician if the foundation is square and level. The remaining measurements use the foundation as a reference.

Twist Damage. The first damage condition to look for is twist. Twist is the last damage condition to occur to the vehicle. Therefore, it should be the first measured. Twist is a condition that exists throughout the entire vehicle.

Twist damage exists when one side of the vehicle is pushed low or high on one end (front or rear) and then moves in the opposite direction (low or high) on the other end. At the same time, the opposite side would have exactly the opposite damage.

To check for twist, sight down your properly hung self-centering gauges. Twist will show up when the gauges are not parallel. If a true twist exists, the gauges would read like those shown in Figure 20–38.

Mash Damage. *Mash damage* is present when any section or frame member of the vehicle is shorter than the factory specifications. It is measured with a tram bar.

When using a tram bar on a mash damaged vehicle, be sure to make the measurement called for in the dimension chart. The amount of mash is determined

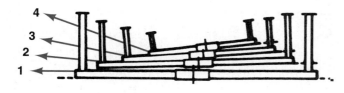

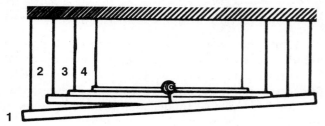

Twisting

Figure 20–38 If self-centering gauges are out of alignment as shown here, the frame or unibody is twisted.

by subtracting the actual measurement from the specified measurement. See Figure 20–39.

Sag Damage. *Sag damage* is a condition where the area of the frame or unibody is lower than normal. Sag often occurs at the cowl area or at the front crossmember. The ends of the crossmember will be closer than normal and the center will be too low.

Again, your four self-center gauges are used to check for a sag condition. Sag is indicated when one gauge is lower than the others. Look at Figure 20–40.

Side Sway (Widening) Damage. *Side sway* is present when the front, center, or rear portion of the vehicle is out of lateral alignment. Use your four self-centering gauges to check for side sway.

If the centering pin on any gauge does NOT line up horizontally with the others, side sway damage is present. If they all line up, there is no side sway deformation.

A center hit on the vehicle causes a misalignment known as double side sway. While this results from a severe impact in the center section, it affects the entire vehicle. The dimension of both front and rear sections must be checked during the pulling of double side sway damage.

Universal Measuring Systems

Universal measuring systems have the ability to measure several reference points at the same time, making the job much easier and more accurate, Figure 20–41. However, universal systems still require a degree of skill and attention to detail to operate properly. But to get the proper readings, the equipment

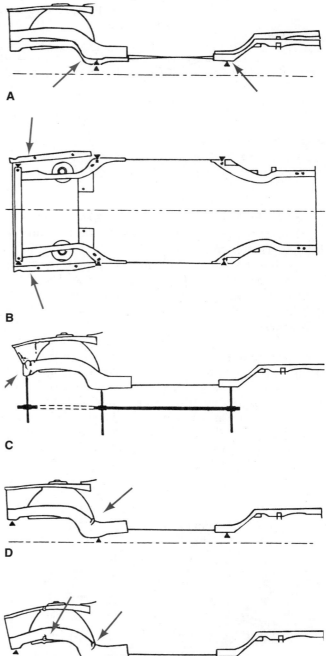

Figure 20–39 Note how to measure mash damage on unibody vehicles: (A) High front impact with secondary damage in the rear. (B) Right front corner impact. (C) Direct front impact. (D) Low front impact. (E) High front impact. (F) High rear impact.

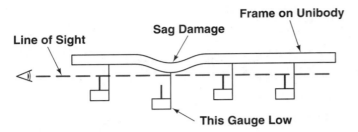

Figure 20–40 This side view of gauges shows sag. One gauge is lower than the others in the cowl area.

must be set using the manufacturer's specifications.

Universal measuring systems can be mechanical, laser, or a combination of the two.

With a universal measuring system, all the reference points can be checked by just moving around the vehicle. You can quickly determine where each reference point on the vehicle is in comparison to the measuring system.

If a reference point on the vehicle is NOT in the same position as the dimension chart says it should be, the reference point on the vehicle is wrong. When the system is set up properly, you can monitor the key points by simply looking at the pointers. If the pointers are out of position, then the vehicle is NOT dimensionally correct. Thus, that reference point is out of position and must be brought back to precollision specifications. See Figure 20–42.

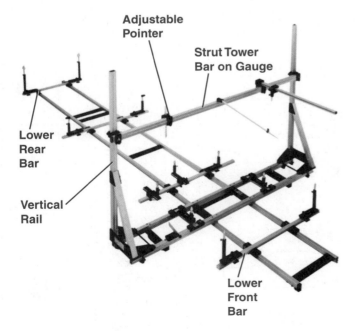

Figure 20–41 Note the major parts of this universal measuring system. (*Chief Automotive Systems, Inc.*)

> *WARNING! The pointers can be damaged during pulling/straightening. Lower the pointers, if necessary, to clear the way for the pull. This will avoid damaging or misaligning the measuring system.*

While the important vehicle dimensions can be found in body dimension manuals, most universal measuring equipment manufacturers have specific dimension charts for their equipment. These charts, one for each vehicle model manufactured, serve as guides to use before and during the repair.

The dimension chart usually illustrates two views of the vehicle underbody. Some charts also give underhood and upper body dimensions. The latter is most important with the mill and drill pad dimensions.

Remember! Many equipment manufacturer's dimension charts are intended for that specific piece of equipment only. Because of this variation between systems, it would be difficult to explain how each manufacturer's system measures a vehicle. You will need to read the equipment owner's manual for these details.

Before beginning any universal measuring operations, be sure to:

1. Remove detachable damaged body parts, both mechanical and sheet metal body panels.

2. If the damage is severe, perform rough straightening to the center section or foundation of the vehicle.

3. If the mechanical parts are left in the vehicle and an overhang condition exists, this must be compensated for.

Mechanical Measuring Systems

A typical mechanical system, Figure 20–43, consists of:

1. A bridge(s) that runs the length of the vehicle from front to back.

2. Sliding arms that mount to the bridge. They can be moved from front to back on the bridge for length measurements. They can also be moved outward or inward, for width measurements.

3. Pointers that are mounted on the arms. They can be adjusted up or down for height measurements.

4. Specific fixtures for the make and model vehicle, Figure 20–44.

Figure 20–42 This measuring system shows major damage. While pulling equipment is being used to straighten damage, measurements of progress must be taken. *(Chief Automotive Systems, Inc.)*

Figure 20–43 Note this mechanical measuring system mounted on a frame rack. *(Car-O-Liner)*

The pointer may accept special adapters to fit over a bolt head or into a reference point hole. The equipment manufacturer will provide instructions for this type of setup.

Mechanical measuring systems are designed and used according to guidelines found in the equipment manufacturer's publications. They are written for each family of body styles. Mechanical measuring systems look complicated, but they are easy to use after reading their directions carefully.

Dedicated Bench and Fixture Measuring Systems

The *dedicated bench and fixture system* acts as "go/no-go" gauge. It is a completely different type of measuring method. Instead of taking actual measurements, dedicated fixtures are used to check body or frame alignment.

The *dedicated bench* consists of a strong, flat work surface to which fixtures are attached.

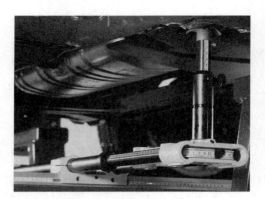

Figure 20–44 Different adapters may be provided with a measuring system. This will simplify measurements from different kinds of reference points. *(Car-O-Liner)*

Fixtures are thick metal parts that bolt between the vehicle and bench to check alignment. They are designed from the vehicle manufacturer's drawings and specifications. The fixtures physically check mountings or other key locations of the underbody.

If the fixtures fit the vehicle properly, the technician knows that the underbody, strut towers, etc. are in perfect alignment. All that is necessary is to straighten the vehicle until the reference points match the fixtures. Other underbody measurements are usually not required.

The dedicated bench and fixture measuring system requires a specific set of fixtures for each family of body styles. If the collision repair shop does not own the required fixtures, they can be rented. The bench has built-in reference positions, and the fixtures are positioned to the specific references according to instructions supplied by the manufacturer. At least three fixtures must be positioned on the bench before the vehicle is mounted. Generally these are the torque-box fixtures. However, the damage to a specific vehicle will dictate which fixtures should be used.

Figure 20–45 shows four of the more common types of fixtures. They are:

1. **Bolt-on Fixtures**—used when attachment is required to steering or suspension mountings. The studs or bolts on the vehicle are used to attach the fixtures. Depending on the damage, the fixtures can either be bolted to the vehicle first and lined up with the bench during the repair, or they can be attached to the bench first and lined up with the vehicle during repair.

2. **Pin-type Fixtures**—used most often to mate with reference point holes in the underbody. They can

also be used to mate with suspension mounting holes.

3. **Strut Fixtures**—used in the same manner as a pin-type fixture. A typical strut fixture consists of a bottom plate assembly as with the other fixtures, an adjustable shaft with a cross pin, and a bolt-on top plate.

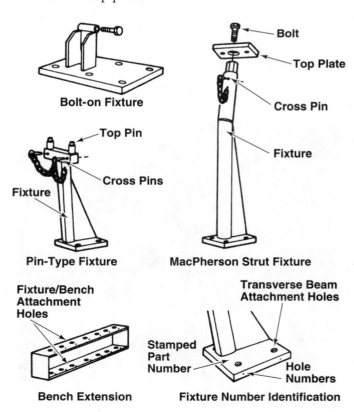

Figure 20–45 Note various fixture types.

4. **Bench Extensions**—included where the length of the vehicle requires that fixtures be positioned beyond the bench surface. The extensions are always used at the rear of the vehicle. Each extension is drilled with holes on top and on the bottom.

Most fixtures share a common attachment method to the transverse beams. The base plate for each fixture is drilled with holes and is marked with a part number and the location on the transverse beam. Stamped at the rear of each fixture base are the numbers that correspond to the numbers on the transverse beam.

Make sure these numbers are on the same side as the numbers on the transverse beam.

Once the vehicle is mounted on the bench, at least three fixtures should be set up on the undamaged area of the vehicle. Then, place as many fixtures as possible in the damaged area of the vehicle. The fixtures perform a number of functions:

1. They show where a reference point should be located.

2. They provide gauging of all the reference points at the same time. No measuring is required. If all the points line up, the steering, suspension, engine mounts, and so on are in alignment.

3. Once lined up with the fixtures, they can hold these parts in position while further straightening is done. This eliminates the "pull-measure, pull-measure" sequence required with universal measuring systems.

4. They allow accurate assembly of parts on the vehicle before actually welding those pieces together. A good example of this would be a lower rail and strut tower assembly on some unibody vehicles.

The sequence is as follows:

1. Position and hold the lower rail pieces on the fixtures.

2. Weld them together.

3. Position and hold the strut tower pieces on top of the rail.

4. Weld them in their correct position.

Electronic Measuring Systems

Electronic measuring systems use a computer, or PC, to control the operation of the measuring system. Computerized measuring systems can use:

1. Laser scanner and reflective targets

2. Ultrasound generating probes and receiver beam

3. Robotic measuring arm

Any of these system variations can input accurate measurements for computer analysis. A PC provides for fast entry and checking of vehicle dimensions against electronic specifications.

Laser Measuring Systems

The laser measuring system uses a strong beam of light, beam splitters, and targets to measure vehicle damage. It is extremely accurate when properly installed and used. See Figure 20–46.

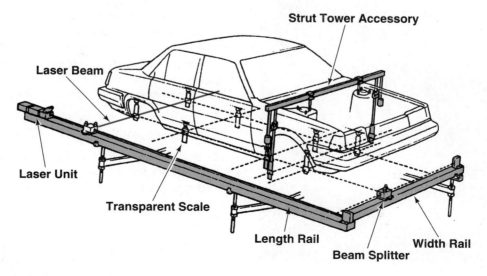

Figure 20–46 Study the basic parts of a typical laser measuring system. The laser unit directs a beam of light throughout the system. Beam splitters and scales give readings quickly and accurately.

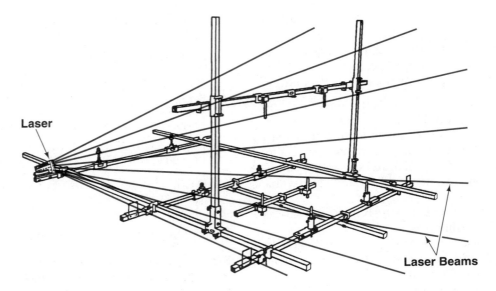

Laser

Laser Beams

Figure 20–47 Laser light moves in a straight line unless it is redirected by a reflector or splitter. *(Hein-Werner® Corporation)*

The word *laser* is an acronym for Light Amplification by Stimulated Emission of Radiation. All laser measuring systems operate in basically the same way. The laser unit is aimed at a target or measuring scale. The target or scale is either hung from or attached to the vehicle. Some systems even use parts of the vehicle as targets. Look at Figure 20–47.

There are several types of systems available. However, all laser systems work either from the centerline or from the datum plane reference. See Figure 20–48.

Measurements are taken by observing the laser beam on the targets. Some targets are clear, allowing the laser beam to pass through them. Several **clear targets** can be used with one laser light source.

Beam splitters, or *laser guides,* are capable of reflecting the laser beam to additional targets. Using combinations of transparent targets and beam splitters, it is possible to measure several dimensions on a vehicle at the same time using a single laser source.

Three-dimensional laser systems use up to three laser units to give length, width, and height coordinates anywhere on the vehicle. They will check deck lid openings, cowls, door openings, hinges, pillars, and roof lines, Figure 20–49. With such a system, a single laser unit can be used to make measurements in conjunction with measuring devices such as a tape measure or calibrated bar. See Figure 20–50.

Laser measuring systems are made up of both optical and mechanical parts. For example, in the clear target system, the optical parts are:

1. Laser power unit, which emits a safe, low-powered laser beam

2. Beam splitters that project beams at a precise right angle so height, width, and length can be measured simultaneously

3. Laser guide that reflects the laser beam at exactly 90 degrees

Other items used with laser measuring devices include calibrated bars. They attach to the vehicle or act as support devices for the laser unit itself. Scales are hung from reference points on the underside of

Figure 20–48 With some systems, one laser can be used to make all measurements. *(Hein-Werner® Corporation)*

Figure 20–49 A three-dimensional laser system can take length, width, and height measurements at one time. (Hein-Werner® Corporation)

Figure 20–51 Some laser measurement systems will also let you check and adjust wheel alignment, which saves time. (Hein-Werner® Corporation)

the vehicle. The laser beam passes through the center of the scale on the target when the measuring point is in its correct position.

Some laser measuring systems permit the technician to monitor upper body dimensions. For instance,

Figure 20–50 A laser unit can also be used in combination with other tools, like a tape measure. (Hein-Werner® Corporation)

pillar locations and window openings can be checked easily with a laser.

Some laser systems also offer an integral four-wheel alignment capability, Figure 20–51. Wheel alignment on the measuring system is beneficial. Suspension problems can be measured and corrected during frame or unibody repairs.

The accuracy of laser measurement systems depends on their proper installation. When a laser beam is projected through one target to another, the targets must be optically perfect. Otherwise, the laser beam will be deflected as it passes through the targets. The deflection will be magnified by the distance between the targets. This effect can result in serious error. Laser targets that become scratched or warped should be discarded.

When properly set up, the laser system can generally stay with the vehicle during the pulling and straightening process, like the mechanical systems. However, if the laser or mounting bars are in danger of being damaged during the pull, they must be moved.

Laser measuring systems provide direct, instantaneous dimensional readings so that the reference points in both the damaged and undamaged areas of the vehicle can be monitored continually during the pulling and straightening operation.

Laser targets are special reflective mirrors that mount on reference points of the vehicle. Targets can be hung on reference points using snap-in clips, nylon bolt clips, plugs, and magnets. This allows you to mount targets quickly and easily.

Figure 20–52 Genesis Velocity's compact scanner uses four lasers and two rotating mirrors. It weighs less than 18 lb., so it is easy for one technician to maneuver it under the vehicle single-handedly. (Chief Automotive Systems, Inc.)

Figure 20–53 This is an initial inspection report given on the monitor of an electronic laser measuring system. Note locations of targets, readings in millimeters, and locations for width, length, and height. (Chief Automotive Systems, Inc.)

The **body scanner,** or laser assembly, has two spinning lasers that strike and reflect off each target. The body scanner usually mounts on the rack under the vehicle center section. Rotating lasers send out perfectly straight light beams that hit and reflect off the targets. This allows the scanner to accurately measure the location of the targets to determine if the vehicle has frame or unibody damage. See Figure 20–52.

The lasers spin at approximately 850 rpm. Each revolution of a laser is divided into more than one million counts (divisions of a circle). The number of counts made while the laser beam travels out to a target and back to the laser hub is monitored by the computer. This allows the system to triangulate the distance of each target from the laser scanner or hub.

The distances to each target can be mathematically calculated by the computer into length and width measurements for damage analysis. This can be done simultaneously for all targets on the vehicle. The computer can compare these live measurements with

electronically stored dimensions to quickly display the amount and direction of frame or unibody damage on the computer monitor.

A graphic display of the frame or unibody is shown on the monitor in Figure 20–53. Small boxes or circles next to the drawing will give numbers that represent the amount and direction of damage in millimeters. Small arrows will point in the direction of damage.

As you straighten the damage, the monitor will display numbers that change to show how much each target and reference point has moved. When the numbers next to each target read zero (or within specs), the vehicle has been straightened and all measurements are correct. This provides constant measurement feedback as you straighten the damage.

Ultrasound Measuring Systems

Ultrasound measuring systems use many of the same principles as electronic laser systems. However, sound waves, not light beams, are used to measure

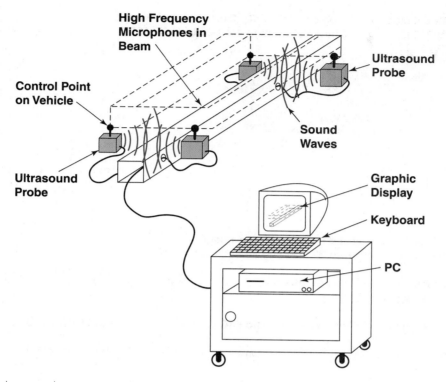

High Frequency Microphones in Beam

Ultrasound Probe

Control Point on Vehicle

Sound Waves

Ultrasound Probe

Graphic Display

Keyboard

PC

Figure 20–54 Note how an ultrasonic measuring system is similar to a laser system. High-frequency sound waves are produced by probes at reference points. Microphones on a receiver pick up these sound waves and send them to the computer, which can then calculate the locations of the probes and reference points to determine the extent and direction of damage.

vehicle reference point locations. This is illustrated in Figure 20–54.

A lightweight aluminum receiving device known as an **ultrasonic receiver beam** mounts under the vehicle center section. Instead of targets, high-frequency emitters or **ultrasound probes** that generate high-frequency sound waves are mounted on vehicle reference points and then connected to the receiver beam and the computer with leads. Each probe produces high-frequency sound waves, which are heard as clicking sounds. The receiver has high-frequency microphones that detect when each probe's sound wave returns to the beam. This allows the computer to calculate the exact location of each probe and reference point being measured.

This system also uses a graphic display of the vehicle unibody or frame to show whether each probe and reference point is within specifications. The computer can compare actual readings to its electronic dimensions manual to show the direction and extent of damage. As the vehicle is straightened, you can change to real-time display to watch as the points are pulled back to zero, which indicates perfect alignment.

A printer can be used to output a hard copy of the display screen. One example is given in Figure 20–55.

Computerized Robotic Arm Measuring Systems

A **computerized robotic arm measuring system** uses a track-mounted robot arm to measure reference points on the damaged vehicle. The robot must be moved by hand into contact with each reference point. A button is then pushed on a remote control unit to store the reading in the computer.

With the control unit connected to the computer, measurements appear immediately on the computer screen. The control unit has storage capacity for data and specifications of up to four different vehicle models. The vehicle measurements can be quickly compared with specifications. Deviations are also displayed.

If the measurements and deviations need to be documented, the control unit can be connected to a printer to provide a printout. Measurements are available either on disk or via modem from the manufacturer's main computer. All data and specifications can be stored in the shop's personal computer.

MORE INFORMATION

Manufacturers of measuring systems constantly furnish updates and bulletins on their products. Be sure

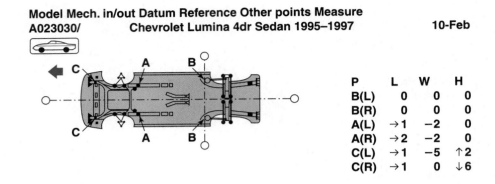

Figure 20–55 This printout shows readings from an ultrasonic measuring system. Note how points C left and C right are not within specifications.

to read and study them because they will help to make the repair procedure easier.

Remember! The best measuring system in the world will NOT help if you do NOT take the time to learn how to use it correctly. Damage cannot be repaired if it cannot be found; and damage cannot be found without precise, careful measurements.

Summary

- There are two basic types of automotive construction: body-over-frame (BOF) vehicles and unibody (or monocoque) vehicles.
- Vehicle damage is broken down into five categories: side sway, sag, mash, diamond, and twist.
- The four basic measuring equipment systems are gauge, universal, dedicated, and electronic.
- Universal measuring systems have the ability to measure several reference points at the same time, making the job easier and more accurate. Universal measuring systems can be mechanical, electronic, or a combination of the two.
- The laser measuring system uses a strong beam of light, beam splitters, and targets to measure vehicle damage. It is extremely accurate when properly installed and used.

Technical Terms

vehicle measurement
measurement gauges
control points
reference points
side sway damage
sag damage
mash damage
diamond damage
twist damage
cone concept
primary damage
secondary damage
crush zones
widening
datum line
datum plane
center plane
centerline
body center marks
lateral dimensions
symmetrical
asymmetrical
gauge measuring systems
tabulation chart

tram bar/tram gauge
point-to-point
    measurement
datum measurement
diagonal line
    measurement
self-centering gauge
base gauge
strut tower gauge
universal measuring
    system
dedicated bench and
    fixture system
dedicated bench
fixture
strut fixture
electronic measuring
    system
laser
beam splitter
laser guide
body scanner
ultrasonic measuring
    system

ultrasonic receiver beam   *computerized robotic*
ultrasound probe   *arm measuring system*

Review Questions

1. On some unibody designs, which has a fixed (nonadjustable) value?
   a. Camber
   b. Caster
   c. Toe
   d. Offset

2. The _____ is a measuring rod with two adjustable pointers attached to it.

3. From what part of a unibody vehicle are all height dimensions taken?
   a. Datum plane
   b. Center plane
   c. Centerline
   d. Reference point

4. Explain the difference between control points and reference points.

5. All unibody measurements and alignment readings should be taken relative to what section?
   a. Front
   b. Rear
   c. Center
   d. Side

6. Where are body center marks located?

7. The _____ _____, or _____, divides the vehicle into two equal halves: the passenger side and the driver side.

8. Define the term "datum plane" and explain how it is used.

9. When checking front end dimensions, *Technician A* measures from the lower cowl area to the front mount of the engine cradle. *Technician B* measures from one lower cowl area to another. Whose measurement will provide a better reading?
   a. Technician A
   b. Technician B
   c. Both technicians A and B
   d. Neither A nor B

10. Describe the four events during a typical collision damage sequence on a unibody structure.

11. What are self-centering gauges used to establish?
   a. Datum plane
   b. Vehicle centerline

   c. Mash
   d. Sag

12. Technician A stands three feet (one meter) away from a self-centering gauge when taking a reading, while Technician B stands six feet (two meters) away. Who will get the more accurate reading?
   a. Technician A
   b. Technician B
   c. Both technicians A and B
   d. Neither A nor B

13. List five safety rules to follow before starting a vehicle damage evaluation.

14. _____ _____ are special tools used to check specific frame and body points. They allow you to quickly measure the direction and extent of vehicle damage.

15. What is sag damage?

ASE–Style Review Questions

1. *Technician A* says that most computerized measuring systems will show live readouts giving the direction and extent of damage. *Technician B* says that computerized measuring systems store dimensions for the undamaged vehicle. Who is correct?
   a. Technician A
   b. Technician B
   c. Both Technicians A and B
   d. Neither Technician

2. Which of the following types of damage is a condition in which a section of the frame or unibody is lower than normal?
   a. Sway
   b. Crush
   c. Mash
   d. Sag

3. *Technician A* says that diamond damage can occur in body-over-frame construction. *Technician B* says that diamond damage may be severe on a unibody vehicle. Who is correct?
   a. Technician A
   b. Technician B
   c. Both Technicians A and B
   d. Neither Technician

4. When inspecting for damage with self-centering gauges, *Technician A* says to first hang gauges from two places where there is no visible damage. *Technician B* says to hang two more gauges where there is obvious damage. Who is correct?

a. Technician A
b. Technician B
c. Both Technicians A and B
d. Neither Technician

5. An electronic laser measuring system is being used to determine the extent of vehicle damage. *Technician A* says that the lasers rotate at several million rpm. *Technician B* says that each rotation of the laser is divided into over a million counts. Who is correct?
   a. Technician A
   b. Technician B
   c. Both Technicians A and B
   d. Neither Technician

6. An ultrasonic measuring system is being used. *Technician A* says that the probes emit a clicking sound. *Technician B* says that the receiver beam contains high-frequency microphones that measure the probe locations. Who is correct?
   a. Technician A
   b. Technician B
   c. Both Technicians A and B
   d. Neither Technician

7. An electronic measuring system shows a sideways arrow and a small 7 in a box next to the arrow and target. What should you do?
   a. Pull 7 mm in the same direction as the arrow.
   b. Pull 7 mm in the opposite direction of the arrow.
   c. Pull 7 inches in the same direction as the arrow.
   d. Pull 7 inches in the opposite direction of the arrow.

8. An electronic measuring system shows zeros for all targets except one, which shows a reading of 1. What should be done?
   a. Nothing; this is within specs.
   b. Pull target 1 mm in the direction of the arrow.
   c. Pull target 1 mm in the opposite direction of the arrow.
   d. Pull all targets until readings are correct.

9. An electronic measuring system shows a minus 5 for a width reading. What does this mean?
   a. Width is too small by 5 inches.
   b. Width is too small by 0.5 inches.
   c. Width is too small by 5 millimeters.
   d. Width is too small by 5 meters.

10. *Technician A* says that spring clips can be used to mount targets on reference points. *Technician B* says that magnets can be used to mount targets when holes are not available on a reference point. Who is correct?
    a. Technician A
    b. Technician B
    c. Both Technicians A and B
    d. Neither Technician

Activities

1. Obtain and study the operating manual for several types of measurement systems. Write a report or summary of unique methods for using each type of equipment.

2. Using a body dimensions manual, locate the data for one specific make and model vehicle. Note the locations of the reference points. Make a few practice measurements on this vehicle.

3. Visit collision repair shops. Ask the owners if you can watch a technician measuring damage on an actual vehicle. Write a report on your visit.

21

Straightening Full-Frame and Unibody Vehicles

Objectives

After studying this chapter, you should be able to:

- List the types of straightening equipment and explain their operation.

- Describe basic straightening and aligning techniques.

- Identify safety considerations for using straightening equipment.

- Plan and execute repair procedures.

- Identify signs of stress/deformation and make the necessary repairs.

- Determine if a repair or replacement can be done before, during, or after straightening.

Introduction

The major steps for repairing a heavily damaged vehicle are given in Figure 21–1. Study them!

Vehicles with major damage must often have their frame, body, or unibody structures straightened. Vehicle straightening involves using high powered hydraulic equipment, mechanical clamps, chains, and measuring systems to bring the frame, body, or unibody structure back to its original shape, Figure 21–2. At the same time, it might involve the replacement of welded panels that are damaged.

Body aligning or straightening is often thought to be a "rough and tough" physical operation. Actually, it is a relatively easy step-by-step task if proper equipment and methods are used. An important requirement when straightening is accuracy. For example, wheel alignment is directly affected by body, frame, and unibody alignment; if the vehicle is not properly straightened, the technician will not be able to align the wheels of the vehicle.

Improper straightening techniques are costly and time-consuming mistakes. Accurate vehicle alignment affects safety, repair time, repair quality, and the confidence of your customer. This chapter will summarize the most important methods for realigning a vehicle with major damage.

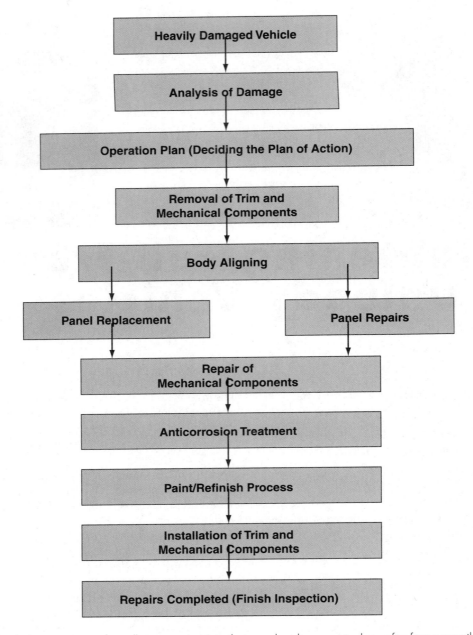

Figure 21-1 Study the major steps for collision repair. Note that panel replacement is done after frame or unibody straightening.

ALIGNMENT BASICS

The term *straightening* refers to using alignment equipment to pull the damaged metal back out to its original shape. The vehicle is secured and held stationary on the equipment. Then, clamps and chains are attached to the damaged area. When the hydraulic system is activated, the chains slowly pull out the damage. Measurements are made at unibody/frame reference points while pulling to return the vehicle to its original dimensions.

When realigning a vehicle, a pulling force or **traction** should be applied in the direction opposite the

force of the impact. This is illustrated in Figure 21–3. When determining the direction of a pull, you must set the equipment to pull perpendicular to the damage.

The *single pull method* uses only one pulling chain and it works well with damage on one part. A small bend in a part can often be straightened with a single pull.

With damage to several panels, a *multiple pull method* with several pulling directions and steps are needed, Figure 21–4. With major damage, body panels are often deformed into complex shapes with altered strengths in the damaged areas. To pull in only the

Figure 21-2 Straightening or aligning equipment is powerful enough to pull out even the worst collision damage. Knowledge of the equipment, vehicle construction, repair procedures, and common sense are needed to repair badly damaged vehicles. *(Chief Automotive Systems, Inc.)*

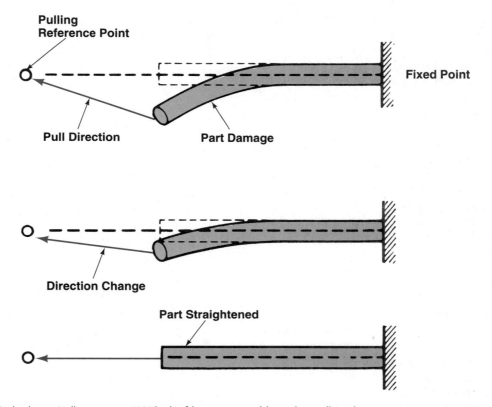

Figure 21-3 Study the basic pulling action. (A) Think of how you would need to pull on the component to return it to its original position. (B) As you pull on the component, the angle of pull might change slightly. (C) The component has been pulled back to its original shape.

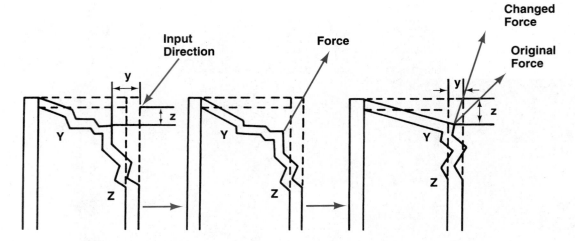

Figure 21-4 Note how to find the general pulling direction. (A) Damage input came from the right front. The damage was from directions Y and Z. (B) Basically, apply straightening equipment force in the direction opposite to the input force. (C) If metal pulls out at different rates, change the pulling direction as needed.

opposite direction would not work because of the differences in the strength and recovery rates of each panel.

Use the method that works best for the given situation. Since applying force in only one place will not always work, you often have to exert pulling force on many places at the same time. For convenience, the term "direction opposite to input" will be used to describe the effective pulling direction.

To alter the direction while pulling, divide the pulling force into two or more directions. This will allow you to change the direction of the **composite force,** that is, the force of all pulls combined. Look at Figure 21-5.

The pulling and straightening process must remove both direct and indirect damage. It must return all of the damaged metal back to precollision dimensions. To do this, the equipment must reverse the direction and sequence the damage occurred in. The damage that occurred last during the collision should be pulled out first.

STRAIGHTENING EQUIPMENT

Straightening equipment is used to apply tremendous force to move the frame or unibody structure back into alignment. Straightening equipment includes anchoring equipment, pulling equipment, and other accessories. See Figure 21-6.

The **anchoring equipment** holds the vehicle stationary while pulling and measuring. Anchoring can be done by fastening the frame or unibody of the vehicle to anchors in the shop floor or to the straight-

$$X + Y = Z$$
$$X + Y' = Z'$$
$$X' + Y + Z''$$

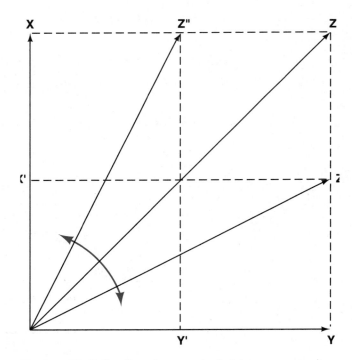

Figure 21-5 If pulling force is divided between two directions (X and Y), the composite, or total, force direction (Z) will change freely with adjustments to force in the two directions. This shows that using two chains to pull allows for better adjustment of pulling direction.

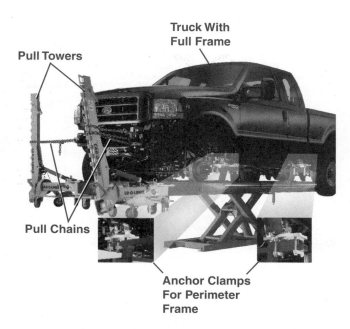

Pull Towers

Truck With Full Frame

Pull Chains

Anchor Clamps For Perimeter Frame

Figure 21–6 Although the specific instructions vary with equipment, the principles of frame or unibody straightening are the same. Note how large clamps are placed under the unibody to keep the vehicle from moving. Large towers or posts on the front of the equipment can then be used to pull out (straighten) damage. Measuring equipment is needed to monitor the pulling and straightening process. (Car-O-Liner)

ening equipment rack or bench. The objective of the anchoring system is to hold the vehicle solidly in place while pulling forces are applied. It must also distribute pulling forces throughout the vehicle.

The ***pulling equipment*** uses hydraulic power to force the body structure or frame back into position. There are many different types of pulling equipment available. Regardless of their design or operating features, each system uses the same basic pulling theory and is used in a similar manner.

Hydraulic rams use oil under pressure from a pump to produce a powerful linear motion. When you electrically activate the system, oil is forced into the ram cylinder. The ram is then pushed outward with tremendous force. This pulls on the chain attached to the vehicle (often with a powerful clamp or strong hook) to remove the damage. The rams can be mounted in or on the pulling towers, posts, or between the vehicle and anchoring system.

Pulling posts or towers are strong steel members used to hold the pulling chains and hydraulic rams. Depending upon equipment design, they can be positioned at whatever location is needed to make the pull. They push against the bench as the pull is made. This eliminates the need for separate anchoring to keep the pulling equipment from sliding under the bench as the pull is made.

DANGER! *The amount of straightening pressure required to remove damage should not be too high. If the straightening equipment is straining during the pulling process, something is wrong. If this happens, stop pulling! Release tension, and re-evaluate the setup to find the problem. If too much pressure is applied, parts or equipment can be damaged and serious injuries could result.*

You should be familiar with a variety of anchoring and pulling systems and their general operation.

In-Floor Straightening Systems

In-floor straightening systems have anchor pots or rails cemented or mounted in the shop floor. Some use ***anchor pots*** or small steel cups in various locations in the shop floor, Figure 21–7A. Others use a

A

B

Figure 21–7 These systems are called in-floor systems because they anchor the vehicle to the shop floor. (Blackhawk Collision Repair, Inc., Subsidiary of Hein-Werner® Corporation) (A) An anchor-pot system uses small steel cups embedded in the shop floor for anchoring a vehicle. (B) A rail anchoring system uses steel beams in the floor for holding a vehicle secure while straightening.

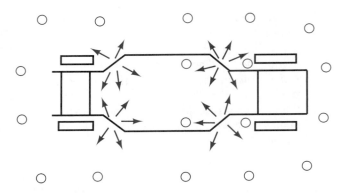

Figure 21-8 The anchoring setup must balance holding power as needed for the pulling direction and force.

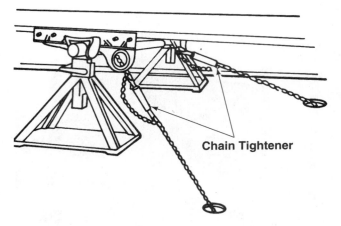

Chain Tightener

Figure 21-9 Chain tighteners or shorteners are used to remove slack from the anchoring chains.

system of steel *anchor rails* in the floor so that an infinite number of pulling-holding locations can be used, Figure 21–7B. Both systems must be balanced both in direction and force of the pull. Look at Figure 21–8.

To provide the pulling force for straightening, the anchor pot system can use hydraulic rams and pulling posts or towers. The floor grid system generally uses hydraulic rams to provide the pulling force.

An in-floor system is ideal for a small collision repair shop. After the rams and the other power accessories have been neatly stored away, the area can be used for other shop purposes. Also, in-floor systems provide single or multiple pulls, and positive anchoring without sacrificing shop space.

Anchor clamps are bolted to specific points on the vehicle (unibody pinchwelds for example) to allow the attachment of anchor chains. They distribute pulling force to prevent metal tearing.

The vehicle is usually supported using **cross tube anchor clamps** that link both sides of the vehicle. The cross tube anchor clamps are placed over the cross tube and tightened securely. Chains are then attached to the cross tube anchor clamps.

Anchor chains are attached from the floor anchors to the clamps attached to the vehicle. The anchor chains and clamps must hold the vehicle securely while straightening. **Chain tighteners** or shorteners are used to take slack out of the anchor chains. See Figure 21–9.

Rack Straightening Systems

A *rack straightening system* is generally a drive-on system with a built-in anchoring and pulling mechanism. Most racks tilt hydraulically so that vehicles can either be driven on or pulled into position with a power winch.

A *power winch* normally uses an electric motor and steel cable to provide a means of pulling the dam-

aged vehicle onto the rack. One is shown in Figure 21–10.

Rack straightening systems are generally stationary. They are capable of making multiple pulls. Pulling towers are generally attached to the rack. Depending on the system, there can be two or more towers. On some systems, the towers can be positioned 360 degrees around the rack. In other systems, the positioning is limited to the front and sides of the rack. See Figure 21–11.

Bench Straightening Systems

A *bench system* is a portable or stationary steel table for straightening vehicle damage. Some benches tilt. Others have drive-on ramps and the bench can be raised straight up or down as needed. Refer to Figure 21–12.

Winch

Figure 21-10 Most frame straightening machines tilt hydraulically so that the vehicle can be winched or driven into position. Note the winch on this rack. *(Hein-Werner® Corporation)*

A

B

Figure 21-11 Note the basic pulling directions on this rack. *(Kansas Jack, Division of Hein-Werner® Corporation)* (A) Hydraulic pulling towers and chains are positioned to force damage upward. (B) Pulling towers and chains are positioned to pull downward.

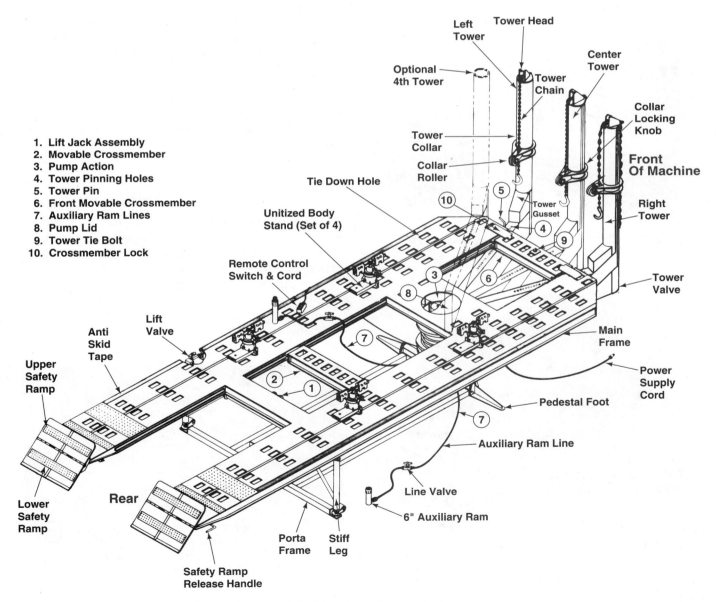

1. Lift Jack Assembly
2. Movable Crossmember
3. Pump Action
4. Tower Pinning Holes
5. Tower Pin
6. Front Movable Crossmember
7. Auxiliary Ram Lines
8. Pump Lid
9. Tower Tie Bolt
10. Crossmember Lock

Figure 21-12 Study the parts of a bench system. *(Chief Automotive Systems, Inc.)*

A **bench-rack system** is a hybrid machine that allows quick loading like a rack and other features of a bench. The table will often tilt like a rack for quick loading of the vehicle. It will also provide the accuracy and convenience of a bench once the vehicle is on the machine.

Straightening system accessories include the various pulling chains, clamps, hooks, adapters, straps, and stands needed to mount various makes and models of vehicles. Some are shown in Figure 21–13.

A **bench computer** can sometimes be provided to assist you in the straightening-measuring process. Shown in Figure 21–14, it can be wired to a main PC (personal computer) for the transfer of data to and from the office. The computer might be used to retrieve body dimension specifications and also do other functions.

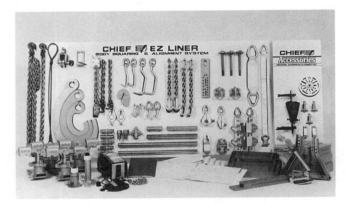

Figure 21–13 Note the various accessories used during straightening. *(Chief Automotive Systems, Inc.)*

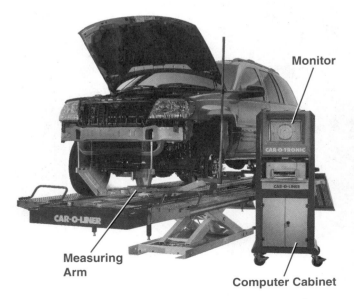

Monitor

Measuring Arm

Computer Cabinet

Figure 21–14 Some alignment equipment uses a small computer and keyboard to help speed the alignment process. *(Car-O-Liner)*

An **operating manual** and other publications are provided to give detailed information on using the exact type of straightening equipment properly. You must always refer to these materials when working. They give anchoring and pulling instructions, the accessories needed for each vehicle, and other essential information. See Figure 21–15.

> **NOTE!** It is beyond the scope of this text to detail the use of every type of body-frame alignment equipment. Always refer to the manufacturer's instructions!

Portable Pullers

A **portable puller,** sometimes called a "damage dozer," is a hydraulic ram and post mounted on caster wheels. This type of pulling of equipment can be easily rolled around the vehicle to extract damage by means of chains and clamps. It is often used to repair or pull minor damage, Figure 21–16.

Being easily moveable, this equipment can easily set the traction direction to the damage input, Figure 21–17. Many units of this type will pull in only one direction.

Other Straightening Equipment

An **engine crane** can be formed on most racks or benches to raise and remove an engine. A bar is added to the post or tower to provide a vertical pulling action for engine removal. Look at Figure 21–18.

An **engine stand** is sometimes needed to hold a power plant after removal from the vehicle. It will hold the engine on a small rollaround framework for convenience.

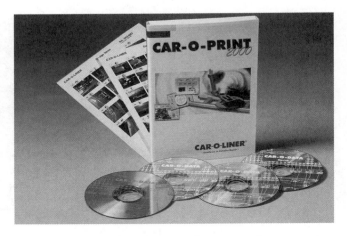

Figure 21–15 When using straightening or alignment equipment, always refer to the manufacturer's instructions. *(Car-O-Liner)*

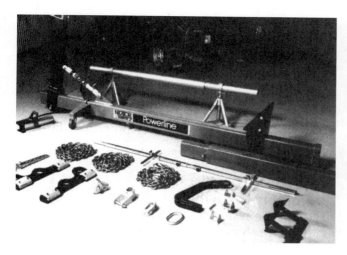

Figure 21–16 A portable frame/panel straightener unit can be moved around the shop to do minor frame straightening. (Courtesy of Dataliner AB)

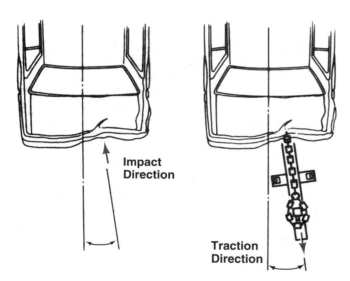

Figure 21–17 With a portable puller (dozer), you can easily set traction or pulling direction. (Used with permission from Nissan North America, Inc.)

Impact Direction

Traction Direction

Figure 21–18 This pulling tower has been set up to remove the powertrain from a vehicle. (Hein-Werner® Corporation)

Figure 21–19 A portable pulling-pushing arm anchors to the floor. (Blackhawk Collision Repair, Inc., Subsidiary of Hein-Werner® Corporation)

An *engine holder* can be used to support the engine-transmission assembly when the cradle must be unbolted or removed. It rests on the inner fenders and is adjustable in width. A chain(s) is (are) used between the holder and engine to keep it from falling down while working.

A *portable pulling-pushing arm* can be anchored next to the vehicle to remove damage, Figure 21–19. With a pulling/pushing arm, the unit can pivot completely around the end of the bench from the center position on the special flange. In the other positions, it can reach the end and one side of the bench. The unit can also be used anywhere along the side of the bench by hooking the inner clamp on the outer flange on the opposite side.

Portable hydraulic rams or *portable power units* are small piston and cylinder assemblies for removing minor damage. Shown in Figure 21–20, they are possibly the most versatile of all body aligning tools. They can be used to push or pull in restricted areas. Portable power units can spread, clamp, pull,

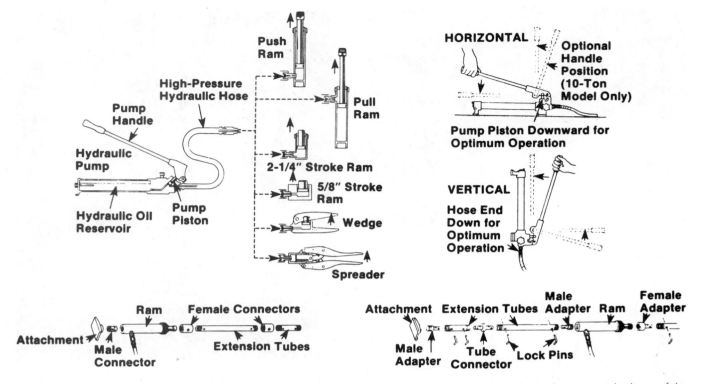

Figure 21–20 Study the parts of a portable hydraulic ram system. *(Norco Industries, Inc.)* (A) The hydraulic pump is the heart of the system. When the handle is moved up and down, hydraulic pressure flows out the hose to various attachments, like the ram. (B) When in use, the hose must not face upward or air can enter the system. (C) Note the various attachments for the ram. (D) Threaded connections and pins hold the attachments on the ram.

and stretch parts, Figure 21–21. In Figure 21–22, some actual repairs are being done.

MEASURE AS YOU PULL!

As discussed, *vehicle measurement* is done by using specialized tools and equipment to measure the location of reference points on the vehicle. These measurements are then compared to published dimensions from an undamaged vehicle. After studying damage measurements, straightening equipment can then be used to pull the frame or body back into alignment. See Figure 21–23.

Always measure as you pull! This is critical to competent work. You must monitor how the vehicle is reacting to the pulling operation. This will let you modify pulling directions and locations as needed.

The challenge faced by the collision repair industry is to find out which panels are out of alignment, and in which direction and how far they have moved. Only then can a proper plan be devised to bring the vehicle back to proper dimensions. You must accurately measure and monitor the entire vehicle while pulling. Refer to Figure 21–24.

For example, if you accidentally overpull parts, you may have to replace them. Since replacement was not initially required on the estimate, you may have to do this extra work free of charge. Both you and the shop would lose money and time. Remember! Always measure as you pull!

A vehicle cannot be properly repaired unless all of the reference points are returned to precollision dimensions. To achieve this, you must:

1. Measure accurately to analyze damage.

2. Measure often when pulling out damage, Figure 21–25.

3. Recheck all measurements after straightening.

PART REMOVAL

A general rule: Remove only the parts that prevent you from getting to the area of the vehicle being repaired. Major straightening operations can be done with major mechanical parts intact.

Depending on the construction of the vehicle and the location and degree of damage, there will be cases

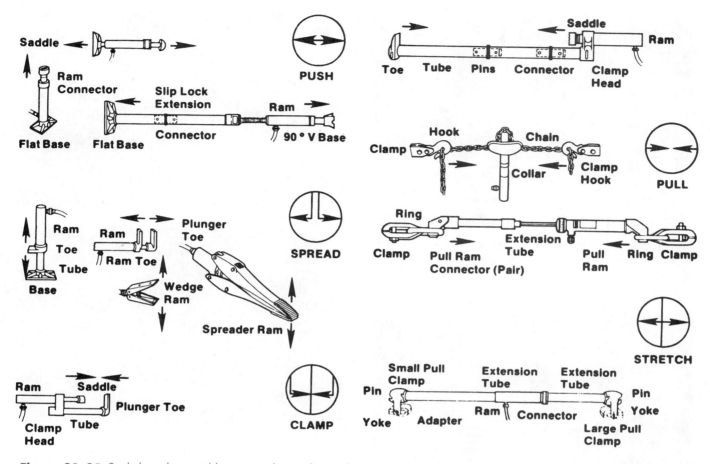

Figure 21–21 Study how the portable ram can be used to push, spread, clamp, pull, and stretch damaged parts. *(Blackhawk Collision Repair, Inc., Subsidiary of Hein-Werner® Corporation)*

Figure 21–22 Here a portable power unit, chains, and clamps are being used to pull the roof lip down so the windshield fits properly. *(Hein-Werner® Corporation)*

Figure 21–23 Measurement is a vital aspect of straightening. Measurements give you feedback about the progress of the pull. *(Chief Automotive Systems, Inc.)*

Figure 21–24 Measurement is done before pulling so the technician can develop a plan for removing the damage. *(Chief Automotive Systems, Inc.)*

Figure 21–25 Measurement is done while pulling to check the progress of straightening equipment. *(Chief Automotive Systems, Inc.)*

Figure 21–26 Before straightening, it may be necessary to remove some parts so you can properly repair the damage. *(Chief Automotive Systems, Inc.)* (A) The bumper and other parts over the unibody are commonly removed before straightening. (B) This technician is removing the suspension system so straightening can be done.

where it will be more convenient to remove parts before proceeding with the repair. Carefully analyze the vehicle and the damage to determine what must be removed. It is sometimes best to remove parts before putting the vehicle on the bench. You might have better access to the fasteners.

Take the time to carefully study the locations of the engine and transmission mounts, suspension mounts, and whether or not these parts themselves are damaged. See Figure 21–26.

PLANNING THE STRAIGHTENING

When planning the straightening process, the technician should determine the following:

1. The direction of the "pulls"

2. How to repair the damage in the reverse (first-in, last-out) sequence to which it occurred during the collision

3. Plan the straightening sequence with the "pulls" in the opposite direction from those that caused the damage

4. The correct attachment points of the pulling clamps

5. The number of "pulls" required to correct the damage

6. Which parts must be removed to make the pulls

Many times it may be best to draw out the repair plan prior to actually straightening the vehicle. This drawing should show OEM and actual dimensions, anchoring, and pulling locations.

The easiest way to determine where to straighten from is to picture the damage being removed by pulling with "your bare hands." The pulling process will work in the exact same manner.

Before attempting any repair work, determine exactly the collision procedures that should be taken. A little extra time spent on such an analysis and operational plan can save hours of work.

As a general rule, vehicle straightening is needed whenever the damage involves the suspension, steering, or powertrain mounting points. This, of course, would include situations such as a side collision where the suspension parts and their mountings are not damaged directly, but because of deformation in the center section of the vehicle's structure, the whole body is out of alignment.

Determine whether damage from a particular collision meets this rule either by eye, where there is obvious damage, or by making some general measurements with a tape measure or tram gauge. These would include diagonal measurements to check for diamond, and length measurements to check for mash. Try to understand where the damage begins and ends. Use all the dimension data available, including unibody/frame dimension books, vehicle manufacturer's manuals, or by checking against an undamaged vehicle.

STRAIGHTENING SAFETY

When using straightening equipment, inadequate attention to safety can result in part damage and serious injury.

1. Use the straightening equipment correctly according to the instruction manual.

2. Never allow unskilled or improperly trained personnel to operate straightening equipment.

3. Make sure the rocker panel pinchwelds and chassis clamp teeth are tight.

4. Always anchor the vehicle securely before making a pull. Check that the chassis clamps and anchor bolts are tightened.

5. Always use the size and grade (alloy) chain recommended for pulling and anchoring. Use only the chain supplied with the straightening equipment.

6. Drawing chains must be positively attached to the vehicle and/or anchoring locations so that they will not come off during the straightening operation. Avoid placing chains around sharp corners.

7. Before pulls are made, apply counter supports so you do not pull the vehicle off the bench or rack.

8. Never use a service jack (floor jack) for supporting the vehicle while working on or under it. Always use jack stands (safety stands) for supporting the vehicle.

9. A pull clamp can always slip and cause sheet metal tear. Prevent bodily harm and material damage by always using safety wires.

10. Never stand in line with a pulling chain or clamp. Chain breakage, clamp slippage, or sheet metal tearing could cause injury or damage. Remember, it can be dangerous to work inside the vehicle when pulls are being made.

11. Cover pulling chains with a heavy blanket. If a chain breaks, this will keep the chain from being thrown.

Before doing any straightening work, protect the vehicle and externally attached parts as follows:

1. Remove or cover interior parts (seats, instruments, carpet).

2. When welding, cover glass, seats, instrument panels, and carpet with a heat-resistant material.

3. Be careful not to scratch or damage parts that do not require painting. If the painted surface is scratched, be sure to repair that portion. Even a small flaw in the painted surface might cause corrosion and an unhappy customer.

STRAIGHTENING BASICS

Body-over-frame vehicles can usually be straightened and realigned with a series of single-direction pulls.

Remember that a unibody vehicle is designed to spread collision forces throughout the structure. Most unibody repairs demand multiple pulls, which sometimes means four or more pulling points and directions during a single straightening and alignment setup.

Remember! A single, hard pull in one direction on a unibody vehicle can tear the metal before it is straight. There simply is not enough material available in any one place to transmit sufficient force to complete a repair. Again, as in the anchoring system, the pulling force must be distributed through several attaching points.

Multiple pulls should be used whenever possible, especially when making the initial pulls. This will spread the force of the pull over a larger area to minimize tearing. They will also allow the technician to "pull" on several areas, and at different angles, at one time. Multiple pulls often make the correction of indirect damage easier. They can be made using one or more pulling posts or towers, hydraulic rams, or adjustable pulling towers on racks or benches.

Making a light "pull" first, holding the pressure, then pulling on another area, allows damage to be removed in a more controlled manner.

When pulling, DO NOT:

1. Cut away any sheet metal that could possibly be used to attach pulling clamps. The damaged sheet metal is usually the best place to attach pulling clamps.

2. Attach a "pull" to any suspension or mechanical parts; use only the mounting points.

The most important points to remember when pulling are:

1. Reverse direction and sequence.

2. Use multiple pulls whenever possible.

3. Pull, measure, pull, measure.

4. Always think safety. Be safe!

ANCHORING PROCEDURES

If there are four anchoring points used during straightening operations, the anchoring points must be able to withstand the total pulling force applied without damaging them. See Figure 21–27. Depending on the angle of the pull, the forces may not be distributed equally. The anchoring points nearest the pull may be loaded more than the others.

Multiple anchoring systems are available in many different designs. The anchoring systems are designed for use on different designs of equipment and type of vehicle to be anchored.

On body-over-frame vehicles, multiple anchoring is accomplished by using chains. The anchoring chains are attached to the frame in the four torque box areas. These are often the strongest points on the frame. The other ends of the chain are attached to the bench, rack, floor grid, or anchor pots in the floor.

On most unibody vehicles, multiple anchoring is accomplished using four **_pinchweld clamps,_** Figure 21–28. The pinchweld clamps are attached to the body at the front and rear of the rocker panel pinchwelds. The bottoms of the pinchweld clamps are fastened to the bench or rack, Figure 21–29.

Some floor grid or anchor pot systems use pinchweld clamps and cross tubes. The **cross tubes** pass through the pinchweld clamps and run from one side of the vehicle to the other. They are supported by four safety stands. One safety stand is positioned at each pinchweld clamp location. They are anchored to a floor grid or anchor pots using chains or a lock arm and wedges. This is also called **chainless anchoring.**

Some unibody vehicles do not have pinchwelds, or they are too weak to withstand pulling forces. These vehicles require the use of a different type of anchoring system that attaches to suspension or mechanical parts mounting locations. Many equipment manufacturers have special anchoring adapters available for these vehicles. If the equipment manufacturer does not have specific anchoring recommendations for the vehicle being repaired, a creative anchoring solution will be needed.

Sandwich clamps can be used to anchor the vehicle if they are installed in areas where there are double or triple layers of structural metal. If anchoring in this manner, always use caution and common sense when anchoring the vehicle. Monitor the anchoring locations closely.

Here are some common cautions and tips to be used when anchoring vehicles:

1. Do not wrap chains around suspension parts.

2. Always remove all slack from anchoring chains prior to pulling.

3. Always follow equipment manufacturer's pinchweld clamp tightening sequences.

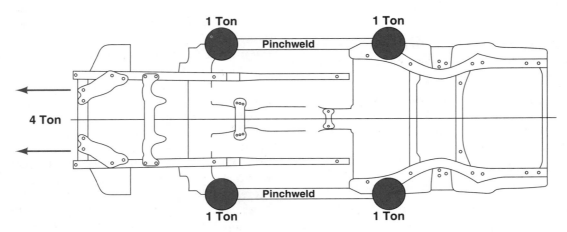

Figure 21–27 When anchoring a vehicle, remember that anchoring strength must be equal to or greater than the pulling force that will be applied.

Figure 21–28 Unibody cars are anchored using pinchweld clamps. Four are normally used in center section of vehicle. (Car-O-Liner)

Figure 21–29 The pinchweld clamps must be securely tightened to the lower underbody of the vehicle. The bottom of the clamps are fastened to the bench. (Chief Automotive Systems, Inc.)

4. Always remove all grease, undercoating, and dirt from the pinchweld flanges and the clamp jaws before installing clamps.

5. Do not use chains that are not designed for pulling and anchoring equipment, for example, do not use tow chains.

6. Make sure that fuel and brake lines will not be crushed by the pinchweld clamps, or straightening equipment.

7. Follow the equipment manufacturer's recommendations for pinchweld clamp locations. Many times the clamp locations are provided on dimension charts so that the clamps do not obstruct the measuring system.

When straightening, make sure that the anchoring force is at least as great as the straightening force being applied. Also, make sure that the straightening force does not exceed the anchoring force. If the straightening force is greater than the anchoring force, anchoring points on the vehicle will be damaged. Monitor the pinchwelds during the entire straightening process. Make sure pinchwelds are not straining, buckling, or distorting.

A full-frame vehicle can be anchored by placing a suitable plug hook in the fixture holes located on the bottom of the frame rail. Blocking should be used to keep the hook in line with the frame rail. Make an identical hookup on both sides of the vehicle, Figure 21–30.

When anchoring the vehicle in preparation for straightening, lean toward "over anchoring" or "over clamping." An extra anchor point or two takes very little time and surely cannot hurt anything.

WHEEL CLAMPS

Wheel clamps are used to load the suspension with the wheel and tires removed with the vehicle on the rack or bench. They allow you to access parts behind the wheels and tires. They also provide more accurate measurement of body alignment because the suspension is not left hanging unsupported. See Figure 21–31.

After raising the suspension with a jack and removing the wheel, bolt the wheel clamp over the lug studs, Figure 21–32. You can then lower the wheel stand onto the surface of the bench or rack.

When anchoring a vehicle, remember that the anchoring force must oppose the pulling force. A good example is a side-front pull illustrated in Figure 21–33.

ATTACHING PULLING CHAINS

Pulling chains transfer pulling force from the straightening equipment to the damaged area on the vehicle. One end of the pulling chain is fastened to the hydraulic ram, post, or tower, and the other end to the pulling clamps or adapters.

Pulling clamps bolt around or onto the vehicle so a pulling chain can be attached to the vehicle, Figure 21–34. After determining where to pull from, the pulling clamps can be attached to any point that can withstand the force of the pull. Examples of where to attach the pulling clamps are:

1. Bumper energy absorber mounts and bolt holes, Figure 21–35

2. Steering, suspension, and mechanical mounting points

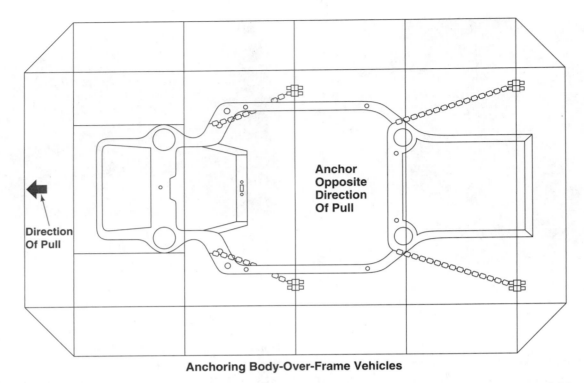

Anchoring Body-Over-Frame Vehicles

Figure 21–30 With body-over-frame/full-frame vehicles, anchoring cannot be done with pinchweld clamps. Since full frame is thick and strong, chains are secured to the frame for straightening. *(Blackhawk Collision Repair, Inc., Subsidiary of Hein-Werner® Corporation)*

3. Damaged sheet metal

4. Weld-on tabs

5. Pinchweld flanges

Do not attach a "pull" to any suspension or mechanical parts; use only the mounting points.

Body-over-frame vehicles have frames that are ⅛–¼ inch (3–6 mm) thick.

On unibody vehicles the technician must use more care when pulling. Single hard "pulls" can tear the thin metal. To prevent tearing, multiple "pulls"

should be used. If needed, use more than one clamp to assure that the metal does not tear, Figure 21–36.

Set the pulling clamp so that the line extending along the path of the pulling force passes through the middle of the teeth of the clamp. If this is not done, rotational force will act on the clamp to pull it off, further damaging the section.

Nylon pull straps provide an additional means of making pulls in difficult areas. They are sometimes used on double pull hook-ups, Figure 21–37.

Figure 21–31 Note the use of pinchweld clamps and wheel clamps. *(Chief Automotive Systems, Inc.)*

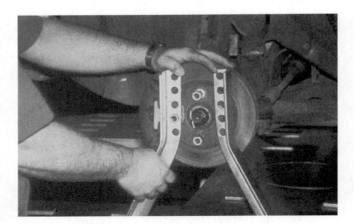

Figure 21–32 If you need to remove wheels and tires, wheel clamps are needed to apply load to the suspension system so measurements are more accurate. *(Chief Automotive Systems, Inc.)*

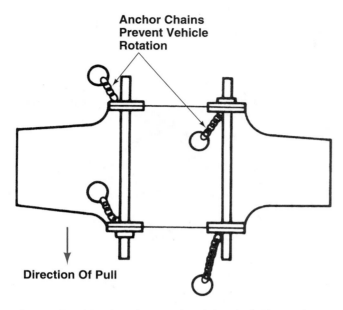

Anchor Chains Prevent Vehicle Rotation

Direction Of Pull

Figure 21-33 Note the position of the chains for anchoring a front, side pull. This will keep the vehicle from rotating. *(Blackhawk Collision Repair, Inc., Subsidiary of Hein-Werner® Corporation)*

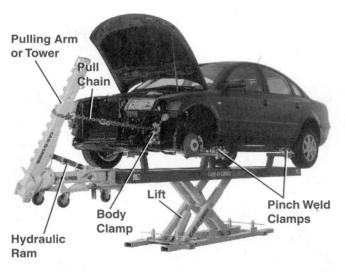

Pulling Arm or Tower

Pull Chain

Hydraulic Ram

Body Clamp

Lift

Pinch Weld Clamps

Figure 21-34 Pulling clamps bolt to the structure of the vehicle. Then, pulling chains can be attached to these clamps. *(Car-O-Liner)*

When you are straightening a bend in a frame rail, make sure the clamp is attached to the correct portion of the rail, as shown in Figure 21–38. This will help stretch and pull out the damage. With this type of damage, you might use a second pulling chain from the side to help straighten the damage. See Figure 21–39.

A **welded pull-tab** can be used if you need to pull where there is no place to attach a clamp. Illustrated in Figure 21–40, weld a small piece of steel plate onto the vehicle unibody where needed. Then,

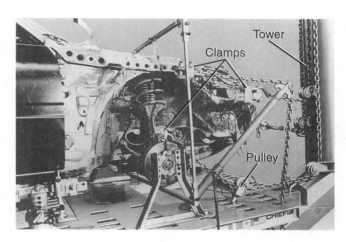

Tower

Clamps

Pulley

Figure 21-35 Note the clamps used for this multiple "pull." Also note the pulley used to produce down-pull. *(Chief Automotive Systems, Inc.)*

Figure 21-36 Additional clamps should be used when there is a chance that metal tearing will result. *(Blackhawk Collision Repair, Inc., Subsidiary of Hein-Werner® Corporation)*

Pull Strap

Figure 21-37 Nylon pull straps are handy when you cannot use normal clamping methods. *(Bee Line Co.)*

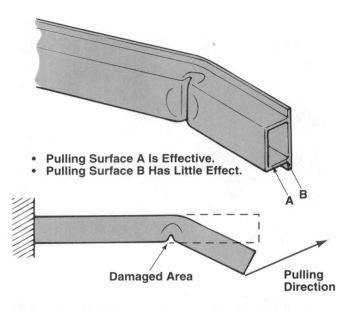

- Pulling Surface A Is Effective.
- Pulling Surface B Has Little Effect.

A B

Damaged Area

Pulling Direction

Figure 21–38 When straightening a bent box section, clamp on the side of the component that is bent. This will help stretch and straighten the component better than clamping on the other side of the box section.

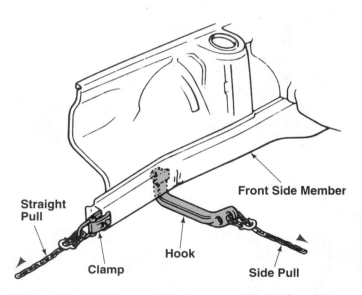

Straight Pull

Front Side Member

Clamp

Hook

Side Pull

Figure 21–39 When straightening a part like this front rail, pulling from two directions is an efficient method of straightening. Note the pulling arm on one pulling chain.

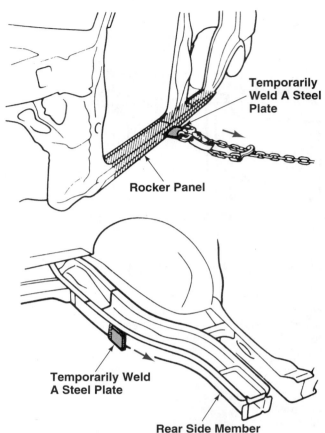

Temporarily Weld A Steel Plate

Rocker Panel

Temporarily Weld A Steel Plate

Rear Side Member

Figure 21–40 When you need to "pull" and there is no place to attach clamps, weld small metal tabs onto the body structure. Then, you can attach a clamp and chain to the metal tab.

you can bolt the pulling clamp onto the metal tab. After pulling, use an air cutoff tool to carefully remove the weld without damaging the body.

Pulling adapters are special straightening equipment accessories that allow you to pull in difficult situations. They come in various shapes and sizes for specific tasks.

Strut pulling plates are designed to bolt onto the top of strut towers for pulling. Spacers and hardened

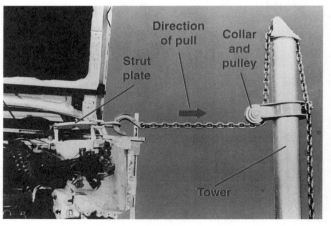

Direction of pull

Collar and pulley

Strut plate

Tower

Figure 21–41 Special adapters, like this strut tower adapter, are handy when making some "pulls." The adapter bolts to the top of the shock tower for pulling it. Note how the collar and roller and tower have been positioned for a level side "pull." *(Chief Automotive Systems, Inc.)*

bolts are used to secure the strut pulling plate to the threaded holes in the strut tower. Then, the pulling chain can be attached to the plate. See Figure 21–41.

EXECUTING A PLANNED STRAIGHTENING SEQUENCE

As mentioned earlier, the overall pulling sequences should be preplanned. However, determining the proper attachment points and pulling directions should be fine-tuned and evaluated as you work. By noting the amount of body displacement and its direction related to alignment reference points, you can adjust pulling direction and force as needed.

The progress toward alignment should be monitored with the measuring system during the "pull." Since the body (sheet metal) has elasticity, the structure will partially return to its postdamaged condition even if the body is pulled back to the prescribed dimensions. Therefore, estimate the amount of return in advance and make allowance for it during the straightening operations.

The straightening procedure or sequence simply consists of solving a variety of small problems rolled into one. Find your first problem, and begin working on it. Then, move to the next problem and so on.

Because of the power of the rams, the metal will begin moving as soon as the chain slack is taken up. There is no need to worry whether or not the ram has the capability to move the metal. This frees up the technician to concentrate on the straightening problem.

Simply make the "pulls" a little at a time, relieve the stress, take a measurement. When using a bench, check how close the fixtures are to their corresponding reference points on the unibody, and start the sequence again.

Normally, work from the center section outward. First, correct the length. Then, correct side sway damage. Finally, correct height.

Remember! Approach the straightening operation as though it was going to be done with your bare hands. Determine how the metal should be moved to mold it back into shape if the only tools available were your hands. How many areas could be moved at one time and in which directions? This is the key to effective straightening.

Due to the high-strength and heat-sensitive characteristics of many unibody vehicles, do not attempt to make an alignment or straightening pull in one step. Instead, use a sequence that consists of a pull, hold the pull, more pull, hold, and so on. This will allow more time for working the metal, allow the metal more time to "relax," and allow more time for the process of alignment for clamping, repair or reattachment by welding, and so on.

That is, start the hydraulics moving, slowly and carefully. Watch the movement closely. Is it doing what it is supposed to do? If it is on the right track, keep on going. If not, determine why and make the angle or direction adjustment and try again.

STRAIGHTENING DIRECTIONS

There are a number of setups for "pulling" and "pushing" in different directions. We will review the most common ones.

The "pulling" (straightening) arrangement with the *vector system* is determined by a simple triangle. By changing the shape of the triangle formed by the pulling equipment and chains, you can alter straightening directions.

The setup shown in Figure 21–42 is used for pulling up and out. The ram, the base unit, and the chain form the triangle. As the ram is extended, one side of the triangle becomes longer. This causes the ram to swing to the right because the chain is locked to the ram. As it swings to its new position, the damaged vehicle is pulled upward and over. Tremendous forces can be exerted in a carefully planned direction using this principle.

Figure 21–43 shows a triangular arrangement that will provide more of a straight out pull. Note that the ram is placed at an angle to the right of true vertical. As force is applied, the ram will swing to the right pulling the damaged sections with it. When setting up a pull, make sure the ram is at the proper height. This can be controlled by adding the proper length of tubing onto the ram.

Remember! At no time should pulling continue if the chain between the ram and the anchor goes beyond perpendicular, Figure 21–43. If this occurs, chain overloading could result because of the added stress placed on the anchored end of the chain. To avoid this condition, be sure that the chain lock head is not placed behind the chain anchor.

As higher, longer pulls are needed, the vector system automatically trades power for motion, Figure 21–44. Some systems allow several feet (meters) of chain travel in one continuous pull. The system can easily pull up, out, or down.

In a typical pull setup at frame rail height, the power ram is set so that the angle between the ram and the pulling chain is equal to the angle between the ram and the anchor. With several tons (kilograms) of force and a large amount of chain travel, you can easily make lower, tough, structure pulls.

For an out and down pull, less tubing is needed, Figure 21–45A. Another way to make a down pull is to attach a chain between the vehicle and floor anchors. By pulling on the chain bridge, the vehicle is forced down, Figure 21–45B.

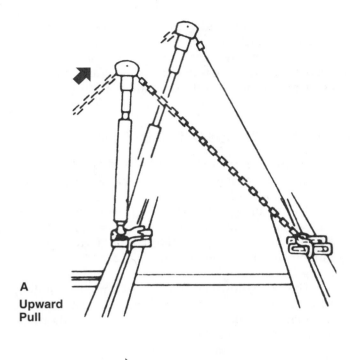

A
Upward
Pull

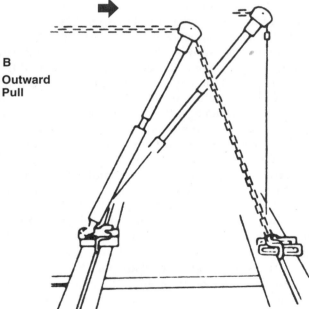

B
Outward
Pull

Figure 21–42 Note how the vector, or angles, of ram and chain alter pulling direction. *(Blackhawk Collision Repair, Inc., Subsidiary of Hein-Werner® Corporation)* (A) Upward pull is produced by this arrangement. (B) Straight outward pull is produced with this configuration.

A horizontal pull on a rail can be accomplished by placing the ram at about a 45-degree angle, Figure 21–45C.

By adding tubing to the ram, a straight outward pull on the cowl can be accomplished, Figure 21–45D.

To pull straight out at the roof line, use the ram with extension tubes as shown in Figure 21–45E.

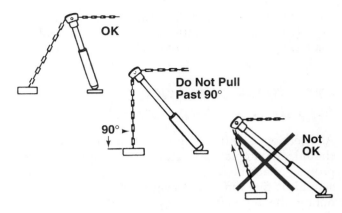

OK

Do Not Pull Past 90°

90°

Not OK

Figure 21–43 When setting up equipment, do not allow the angle of pull to go beyond 90 degrees or a strain can break the pulling chain or other parts. *(Blackhawk Collision Repair, Inc., Subsidiary of Hein-Werner® Corporation)*

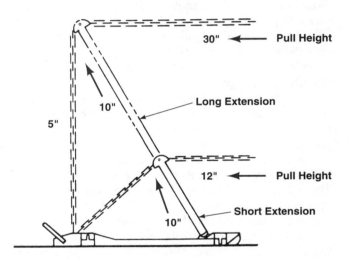

30" ← **Pull Height**

10"

Long Extension

5"

12" ← **Pull Height**

Short Extension

10"

Figure 21–44 Note how changing the length of the ram tube affects pulling direction. *(Blackhawk Collision Repair, Inc., Subsidiary of Hein-Werner® Corporation)*

Upward pulls are very easy to set up, Figure 21–45F. In most cases, the ram is in a vertical position. This pull setup will produce an upward and slightly outward pull.

The same type of setup can be used at roof height by adding extensions to the ram, Figure 21–45G.

Although pushing is not used to the extent it once was in collision damage repair, the capability to push is still important, Figure 21–45H. The vector system provides push capability from any angle around the vehicle by means of a simple triangular setup.

It is also possible to push from underneath the vehicle at whatever angle is needed, Figure 21–46A. This push setup can be used to effectively remove sag at the cowl area, Figure 21–46B.

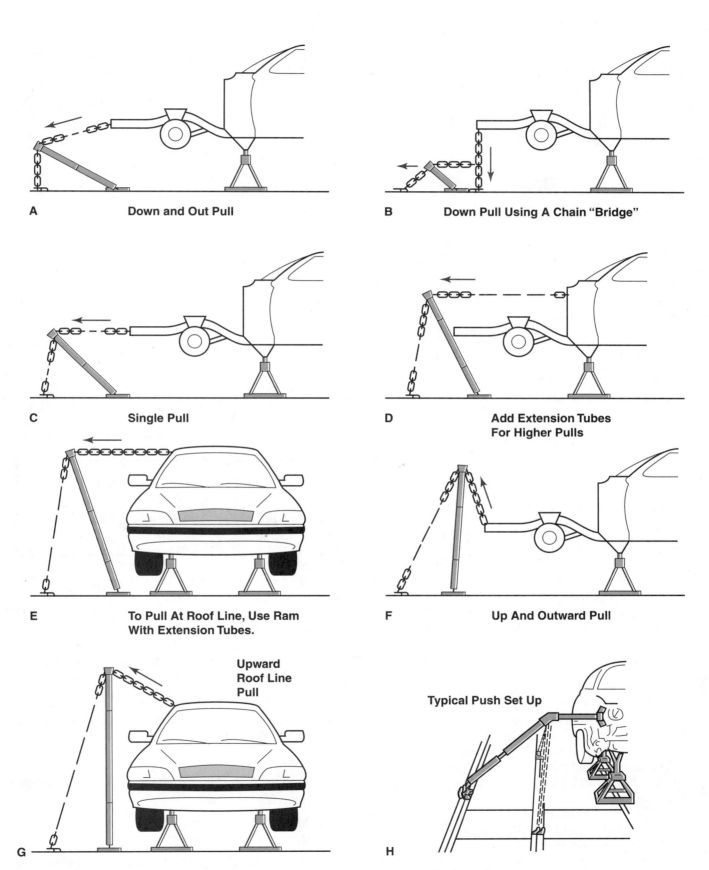

A   Down and Out Pull

B   Down Pull Using A Chain "Bridge"

C   Single Pull

D   Add Extension Tubes For Higher Pulls

E   To Pull At Roof Line, Use Ram With Extension Tubes.

F   Up And Outward Pull

Upward Roof Line Pull

G

Typical Push Set Up

H

Figure 21–45 Study these single-pull setups. (Blackhawk Collision Repair, Inc., Subsidiary of Hein-Werner® Corporation)

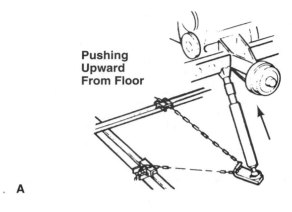

Pushing Upward From Floor

A

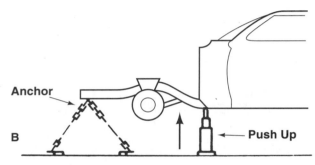

Anchor

B   **Push Up**

Figure 21–46 Note these two setups for pushing. *(Blackhawk Collision Repair, Inc., Subsidiary of Hein-Werner® Corporation)* (A) The ram is being used to push up and sideways on the vehicle. (B) With the front of the frame rails anchored, the ram is being used to remove sag at the cowl area.

OVERPULLING

Overpulling is done by pulling the damage slightly beyond its original dimension. If done in a controlled way, the metal will flex back slightly when tension is released. The unibody/frame reference points will then line up properly. If done too much, overpulling might not be a correctable error. See Figure 21–47.

Overpulling damage results from failing to measure accurately and often. To prevent overpull damage on unibody vehicles, measure the progress when pulling the damaged area.

Remember, you can stretch a piece of string into a straight line, but there is no way to push it straight back. Any damaged metal that is pulled or stretched beyond the critical control dimension is difficult to shrink or compress back. In most instances, the only way the overpulled panel can be repaired is by replacement of body parts. This is a very time-consuming and costly mistake.

When straightening damage, the metal will creep or spring back when the pulling tension is released. A slight overpull may be necessary to obtain the desired dimensions. Use the following straightening sequence when intentionally overpulling:

Overpull

Original

Figure 21–47 Overpulling stretches parts beyond their original dimensions. If done in a controlled manner, it can speed repairs. If done too much, parts will have to be replaced.

1. "Pull" the damage.

2. Stress relieve.

3. Release the pulling tension to allow the metal to return.

4. Measure.

5. Repeat as needed to obtain the desired dimensions.

The key is to "pull" and measure during the entire straightening process. Failure to do so can result in an area being overpulled. An overpull occurs to the point where the metal will not spring back.

> *WARNING! Never use slotted bolt holes to obtain alignment of adjacent panels. This will not correct the problem.*

DANGER! *Because the force of the pull on a unibody is transmitted through the entire body structure, the technician must use extreme caution when straightening. This includes:*

1. Monitor the anchoring locations during the entire straightening operation.

2. Watch for weld seams that are splitting.

3. Listen for popping sounds. These can indicate welds breaking.

4. Clean the jaws of the clamps prior to installation.

5. Remove any grease or undercoating from the attachment points.

6. Make sure that wires, lines, and hoses will not be damaged when pulling.

7. Never stand directly in front of or behind the pulling equipment.

8. Do not overload the straightening equipment. If this happens, release the tension and rethink the "pull."

9. Use safety chains, blankets, or tarps wrapped over pulling chains to minimize backlash if a chain or clamp breaks or slips off.

10. Monitor the pressure gauge on the pulling equipment. Refer to the manufacturer's recommendations for maximum pressure allowed.

ALIGNING FRONT END DAMAGE

First, "pull" the side member on the replacement side in the direction opposite to the impact direction. Then, repair the fender apron and side member on the repair side. Also, repair the front fender apron and side member installation areas on the replacement side.

There are many cases where the entire inner fender apron or side member on the repair side is deflected left or right only. Measure the diagonal dimensions A and B, as shown in Figure 21–48. Then, correct that distance while keeping an eye on the repair condition. The operation can be done efficiently if the fender apron upper reinforcement is pulled at the same time as the side member.

If there is severe damage to the side member on the repair side, separate the front crossmember and radiator upper support and repair them separately. Grip the inside face of the side member. While pulling it forward, pull the broken piece from the inside or push it from the outside. After repairing the bent portion, match up the dimensions to the standard diagonal dimensions.

To repair the other front fender apron and side member area, the main repairs are near the instrument panel and the cowl panel. If the impact was

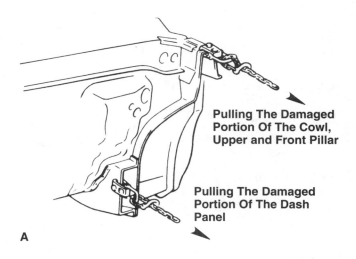
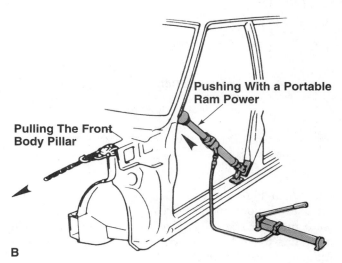

Figure 21–49 With major front damage, you may have to cut away damaged parts so you can pull near the cowl area. (A) With the fender apron cut off, attach clamps next to the cowl and pull. (B) Pulling at the cowl and pushing from behind with a portable power unit may be needed.

severe, the damage will extend into the front body pillar (the door would fit poorly in this case). Simply gripping the front edge of the side member of the fender apron and pulling will not repair this damage, Figure 21–49A. In this case, cut the inner fender apron and side member near the installation area, clamp near the major panel damage, and pull (keep an eye on the door fit conditions).

At the same time that the pillar is being pulled forward, pushing can be done from the interior side with a power ram, Figure 21–49B.

During front unibody aligning, confirm pulls by measuring the dimensions at reference holes. This is often done at the front floor reinforcement and rear of the front fender installation holes.

If the impact to the front side member structure is severe, there is a tendency for it to take the shape

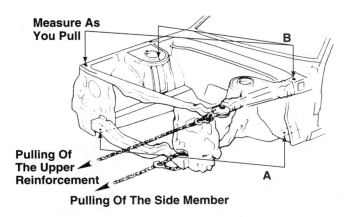

Figure 21–48 As you pull out front damage, measure your progress.

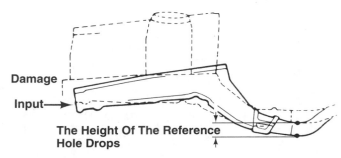

Figure 21–50 During a front collision, this frame rail was pushed back and down.

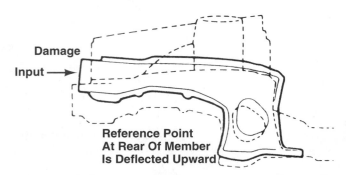

Figure 21–51 During a front collision, this frame rail was pushed back and up.

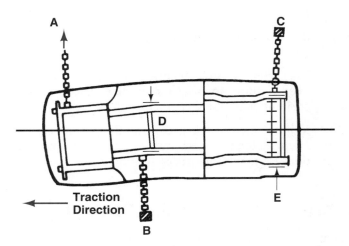

Figure 21–52 Note the setup and "pull" for correcting lateral damage. Blocking devices should be used to link the side rails together.

shown in Figure 21–50. The height of the standard measuring point might be distorted. Further, the front side member often has a reference point in the rear that has a tendency to be deflected upward when damaged. See Figure 21–51.

To correct lateral bending damage of the front, the clamping point receiving the greatest force is point B in Figure 21–52, which must be clamped securely. If point C is not secured, point A cannot be pulled. A blocking device should be used at points D and E. Holding devices should also be placed across the underbody to help transmit pulling force to both sides of the vehicle.

REAR DAMAGE REPAIRS

Usually, the rear bumper is impacted during rear-end collisions. The impact force will usually propagate through the ends of the rear side members or nearby panels. The rear collision will also cause damage to the kick-up area. The wheelhousings may deform causing the entire quarter panel to move forward. If the impact is severe enough, it will have an effect on the roof, door panels, and center body pillar.

Attach clamps or hooks to the rear portion of the rear side member, rear floor pan, or quarter panel rear end portion, Figure 21–53. "Pull" while measuring the

Figure 21–53 Straightening rear damage is not as complex in most instances because you do not have to deal with steering and the engine. *(Hein-Werner® Corporation)*

dimensions of each part of the underbody and door openings. As you "pull," inspect panel fit and part clearances.

Relieve the stress in the quarter panel by pulling on the side member only. If the wheel housing or the roof side inner panel is clamped and pulled along with the rear side member, the clearances with the door panel can be corrected. See Figure 21–54.

With rear structure twist from a front impact, the rear lower preliminary pulling will restore some of the lower alignment points. Moving forward with subsequent "pulls," the alignment and number of anchoring points will, of course, move right along with them.

After pulling, damaged sections requiring replacement can be cut away.

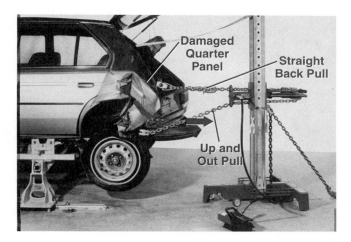

Figure 21–54 Here, a portable unit is being used to straighten a damaged quarter and repair the door opening. After pulling, the quarter panel will be replaced. *(Wedge Clamp International, Inc.)*

STRAIGHTENING SIDE DAMAGE

If there is a severe impact to the center of the rocker panel, the floor pan will deform. The entire body will take on a curved shape, like a "banana." To align this type of damage, use a method similar to straightening a piece of bent wire. The two ends of the body are pulled apart and the side that is caved in is pulled outward. This can often be repaired with three-way pulling, Figure 21–55.

Anchoring unitized body vehicles for side "pulls" can be very difficult due to the limited chain hook-up areas, Figure 21–56. When straightening the side of a vehicle, the center section can be anchored by passing a chain around the pinchweld clamp and hooking to the edge of the bench. Tension/anchoring can be applied by attaching the pull chain to the pinchweld clamp.

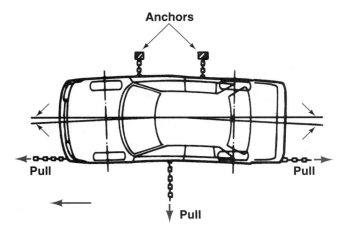

Figure 21–55 For inside damage, you may have to pull in three directions as shown. *(Used with permission from Nissan North America, Inc.)*

Figure 21–56 Multiple pulls are being used to straighten side damage. *(Wedge Clamp International, Inc.)*

The portable beam and knee can be used as a side anchor with either inside or outside contact, Figure 21–57. By attaching the pull chain to the portable beam and knee, it can be used as a pulling attachment.

It is advisable to make an end-to-end stretch pull when pulling outward on the center section of a vehicle. This is shown in Figure 21–58.

If pulling high on the body, tie the vehicle down on the opposite side, Figure 21–59.

STRAIGHTENING SAG

Blocking under the low area and pulling down on the high end will correct sag. The vehicle must also be tied down to the straightening system at the opposite end.

Anchoring the high portion of the vehicle to the straightening system with chains and pushing up at the low spot will also remove the sag, Figure 21–60. When using the pulley and base for the downward pull, the tower pull chain must be in the lowest position.

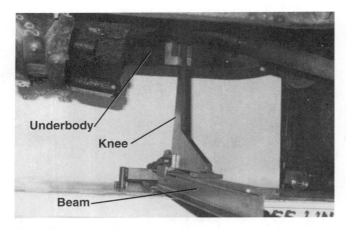

Figure 21-57 A portable beam and knee have been placed under the vehicle to aid pulling. *(Bee Line Co.)*

Figure 21-58 Here, equipment is being used to make an end-to-end stretch pull. *(Bee Line Co.)*

Figure 21-59 When pulling high on a vehicle, make sure the vehicle is anchored securely. *(Chief Automotive Systems, Inc.)*

A

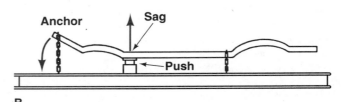

B

Figure 21-60 Sag is a common problem with front impacts. *(Bee Line Co. and Chief Automotive Systems, Inc.)* (A) A hydraulic ram has been placed under the low spot on the vehicle. (B) By anchoring the front section, the ram will push up and remove sag from the cowl area.

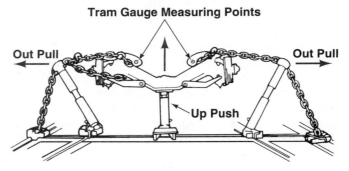

Figure 21-61 Note the typical method for correcting crossmember sag. *(Blackhawk Collision Repair, Inc., Subsidiary of Hein-Werner® Corporation)*

Sag can also occur at the front frame crossmember. The ends of the crossmember will be closer than normal and the center will be too low. This condition can be corrected by using three hydraulic rams and two chains. Look at Figure 21–61.

STRAIGHTENING DIAMOND DAMAGE

To straighten a diamond condition, place a pulling tower or ram on each end of the bench base on opposite sides. Adjust chain height and attach chains to the vehicle as described for end pull corrections. Block or anchor one side of the vehicle to prevent side movement. Activate the pull ram while measuring the results.

STRESS RELIEVING

Stress is defined in metallurgical books as the internal resistance a material offers to being deformed when subjected to a specific load (force). In the collision repair industry, **part stress** can be defined as the internal resistance a material offers to corrective techniques. This resistance or stress can be caused by:

1. Deformation
2. Overheating
3. Improper welding techniques

   Indicators of stress include:

1. Misaligned door, hood, trunk, and roof openings, Figure 21–62
2. Dents and buckles in aprons and rails
3. Misaligned suspension and motor mounts
4. Damaged floor pans and rack and pinion mounts
5. Cracked paint and undercoating

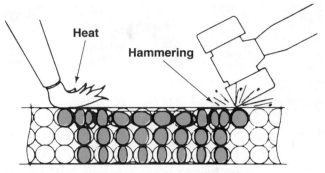

Grain Begins To Relax And Return To Its Original State Relieving Stress.

Figure 21–63 Stress can be relieved with hammering and/or heat.

6. Pulled or broken spot welds
7. Split seams and seam sealer

Stress relieve locked-up metal by hammering and/or heating while pulling, Figure 21–63. First, mark the creases or folds to make them more visible, Figure 21–64. Pull the damaged metal to tension, then loosen it up with hammering and heat if needed. Increase the tension and loosen it again. See Figure 21–65.

If the metal is severely bent, it may be necessary to use a little heat. However, use heat carefully. Some components should NOT be heated. Check with the vehicle manufacturer for heating procedures.

Heat only on the corners and double panels where there is sufficient strength. Use heat carefully and as a means of releasing locked-up metal. Do not use heat as a means to soften up an area.

If the damage requires the use of heat, follow the manufacturer's recommendations. These instructions will be written for the exact metal composition of the vehicle.

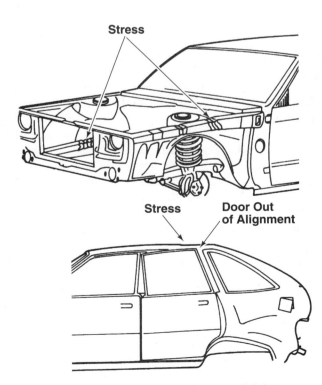

Figure 21–62 These are signs of stress and deformation.

Figure 21–64 Mark the creases before pulling so they are more visible. *(Chief Automotive Systems, Inc.)*

Figure 21–65 The technician is hammering on stress areas while pulling force is applied. *(Chief Automotive Systems, Inc.)*

Never attempt to cool the heated area by using water or compressed air. Allow it to cool naturally. Rapid cooling can cause the metal to become hard and, in some cases, brittle.

The best way to monitor heat applications is with a **heat crayon, thermal melt stick,** or **thermal paint.** Stroke or mark the cold piece with the crayon. When the stated temperature has been reached, the mark will liquefy. Heat crayons are quite precise and far more accurate than watching for specific color change.

NOTE! For more information on heating and welding steel and aluminum, refer to the index.

STRESS CONCENTRATORS

Stress concentrators are designed into unibody vehicles to control and absorb collision forces, minimize structural damage, and increase occupant protection. They result in a localized concentration of stress as a load is applied. See Figure 21–66.

Do not remove designed stress concentrators. Follow the vehicle manufacturer's recommendations for straightening or replacing of parts that have designated stress concentrators.

A quality repair can be achieved only if function, durability, and appearance have been restored. When stress is not removed, the following possibilities can occur:

1. Fatigue can be caused by loading and unloading of the suspension and steering part.

2. In the event of a second similar collision, less force is required to cause the same or greater damage. This could jeopardize the occupants of the vehicle.

3. The vehicle can dimensionally distort causing handling problems.

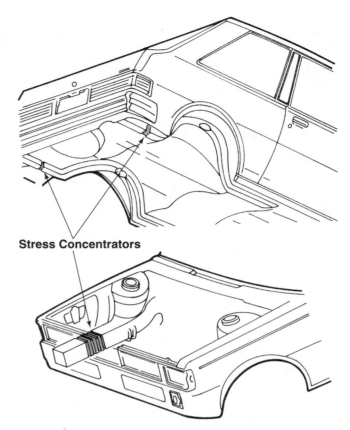

Stress Concentrators

Figure 21–66 Stress concentrators are often designed into a unibody structure.

FINAL STRAIGHTENING/ALIGNMENT CHECKS

Once the repair is completed, including all straightening and welding operations, the alignment procedures are ready for a final check. Final measurements should be made and compared to the unibody/frame dimensions book.

Begin the final checks by slowly walking around the vehicle looking for obvious signs of misalignment.

Review the repair order or estimate to be sure everything was repaired. It is easier to straighten now than to wait until more steps have been completed and then find additional damage.

Among other items that should be carefully inspected are:

1. Check down low at the alignment between the door and rocker sill. This should be a straight gap.

2. Check the general alignment of all the upper body areas. Make sure everything looks as though nothing was ever out of alignment.

3. Open and close the doors and deck lid. Do they feel tight and secure when latched? Make sure they close smoothly and open easily.

4. If applicable, trial install large bolt-on parts, like the fenders and the hood. This will determine if all the bolt holes are aligned or if further pulling is needed. If needed, make final adjustments. Make sure everything aligns properly before removing the vehicle from the straightening equipment.

Summary

- Vehicle straightening involves the use of high-powered hydraulic equipment, mechanical clamps, chains, and a measuring system to bring the frame back to its original shape.
- Straightening equipment includes anchoring equipment, pulling equipment, hydraulic rams, and pulling posts.
- Stress is the internal resistance a material offers to being deformed when subjected to a specific force. It can be caused by deformation, overheating, or improper welding techniques.
- Indicators of stress include misaligned doors, dents in aprons and rails, cracked paint and undercoating, and split seams and seam sealer.
- Once repairs are complete, it is time for a final alignment check. All measurements must be checked against factory specifications to ensure quality work.

Technical Terms

straightening
single pull method
multiple pull method
composite force
straightening equipment
anchoring equipment
pulling equipment
hydraulic rams
pulling posts
anchor pots
anchor rails
anchor clamps
anchor chains
rack straightening
   system
power winch
bench system
bench-rack system
straightening system
   accessories
bench computer
operating manual
portable puller

engine crane
engine stand
engine holder
portable pulling-pushing
   arm
portable hydraulic
   rams
portable power units
pinchweld clamps
wheel clamps
pulling clamps
welded-pull tab
pulling adapters
strut pulling plate
vector system
overpulling
overpulling damage
part stress
stress relieving
heat crayon
thermal melt stick
thermal paint
stress concentrators

Review Questions

1. Which of the following is not an in-floor system?
   a. Anchor-pot system
   b. Modular rail systems
   c. Chainless anchoring system
   d. None of the above

2. Generally, how should you apply straightening force to a damaged vehicle?

3. Portable hydraulic rams have the capability to:
   a. Pull
   b. Push
   c. Spread
   d. All of the above

4. Define the term "anchoring equipment."

5. Define the term "straightening equipment."

6. Why are anchor clamps used instead of hooks when pulling?

7. When should the repair or replacement of severely damaged sections that cannot be restored by the straightening operation take place?

8. How is a power winch used in a collision repair shop?

9. When planning the straightening process, list six rules to remember.

10. List eleven basic safety rules for pulling out damage.

11. What are the four most important points to remember when pulling/straightening?

12. On most unibody vehicles, multiple anchoring is accomplished using four _____ _____.

13. _____ _____ _____ are designed to bolt onto the top of shock towers for pulling.

14. What are stress concentrators?

ASE–Style Review Questions

1. *Technician A* says that you never overpull. *Technician B* says that overpulling is needed since the frame tends to snap back when pulling tension is released. Who is correct?
   a. Technician A
   b. Technician B
   c. Both Technicians A and B
   d. Neither Technician

2. *Technician A* says that it is ideal to have at least four undamaged reference points on the vehicle that can be used to set the vehicle up properly on the straightening equipment. *Technician B* says a straightening system should be used whenever the damage involves the suspension, steering, or power train mounting points. Who is correct?
   a. Technician A
   b. Technician B
   c. Both Technicians A and B
   d. Neither Technician

3. *Technician A* always removes the suspension and driveline completely from a unibody vehicle before putting it on a straightening bench. *Technician B* says that single-pull systems cannot be used on unibody vehicles. Who is correct?
   a. Technician A
   b. Technician B
   c. Both Technicians A and B
   d. Neither Technician

4. When anchoring a unibody vehicle in preparation for pulling, *Technician A* leans toward "over anchoring." *Technician B* sometimes temporarily welds a piece of steel to a section to be pulled. Who is correct?
   a. Technician A
   b. Technician B
   c. Both Technicians A and B
   d. Neither Technician

5. *Technician A* says that the only way that overpull damage can be repaired is by part replacement. *Technician B* says heat can be used to correct an overpull. Who is correct?
   a. Technician A
   b. Technician B
   c. Both Technicians A and B
   d. Neither Technician

6. Which of the following is true?
   a. Cracked paint and undercoating is a sign of stress.
   b. Most of the stress relieving will be "cold work."
   c. The best way to monitor heat applications is with a heat crayon.
   d. High-strength steel (HSS) can be heated.

7. Technicians are planning to straighten a badly damaged vehicle. *Technician A* says to pull the damage in the same sequence as it occurred during the collision. *Technician B* says to straighten a damaged vehicle it should be pulled in the reverse sequence that the damage occurred. Who is correct?

   a. Technician A
   b. Technician B
   c. Both Technicians A and B
   d. Neither Technician

8. *Technician A* says never to stand in line with a pull chain. *Technician B* says to cover pull chains with heavy blankets. Who is correct?
   a. Technician A
   b. Technician B
   c. Both Technicians A and B
   d. Neither Technician

9. *Technician A* anchors a unibody vehicle with chains. *Technician B* uses pinchweld clamps. Who is correct?
   a. Technician A
   b. Technician B
   c. Both Technicians A and B
   d. Neither Technician

10. An inner fender apron is resisting straightening force. A minor buckle is found in the apron. *Technician A* says to use controlled heat and hammer blows to remove the stress from the panel. *Technician B* says to replace the panel. Who is correct?
    a. Technician A
    b. Technician B
    c. Both Technicians A and B
    d. Neither Technician

Activities

1. Visit a collision repair shop. Observe an experienced technician using straightening equipment. Write a report on your visit.

2. Read the operating manuals for different types of straightening equipment. Compare their differences and similarities.

3. Inspect several badly damaged vehicles. Write a report on what repairs would have to be made.

Section 6

Other Operations

Chapter 22
Replacing Structural Parts
and Corrosion Protection

Chapter 23
Interior Repairs

Chapter 24
Mechanical Repairs

Chapter 25
Wheel Alignment

Chapter 26
Electrical Repairs

Chapter 27
Restraint System Repairs

22

Replacing Structural Parts and Corrosion Protection

Objectives

After studying this chapter, you should be able to:

- List the parts of the vehicle that are considered structural.

- List the steps necessary for replacing a part along factory seams.

- Describe how spot welds are separated.

- Explain how new body panels can be positioned on a vehicle.

- List the steps for welding new body panels in place.

- Describe how to install foam panel fillers.

- Section rails, rocker panels, A- and B-pillars, floor pans, and trunk floors.

- Define corrosion and describe the common factors in its formation.

- Identify the principal methods of corrosion protection.

- Choose the correct anticorrosion materials and equipment.

- List common types of seam sealers and explain where each should be used.

Introduction

Most collisions will involve at least some parts replacement. While many parts on a vehicle add to its structural integrity, some parts play a greater role. These parts are known as structural parts. *Structural parts* can be defined as those parts of the vehicle that:

1. Support the weight of the vehicle
2. Absorb collision energy
3. Absorb road shock

The illustration in Figure 22–1 shows parts that are generally considered to be structural. These parts include:

1. Core support/tie bar
2. Front rails
3. Strut towers
4. Rocker panels
5. A-pillar (hinge pillar)
6. B-pillar (center pillar)
7. Rear rails
8. Rear strut towers
9. Suspension crossmembers

Restoring corrosion protection is also a vital aspect of collision repair. Failing to do so can result in major structural failure. Corrosion of structural parts can severely impair the handling and crash worthiness of the vehicle. This damage may not be evident until a collision, but then it will be too late.

NOTE! For complete comprehension of the information in this chapter, you should have already studied Chapter 9, Welding, Heating, and Cutting.

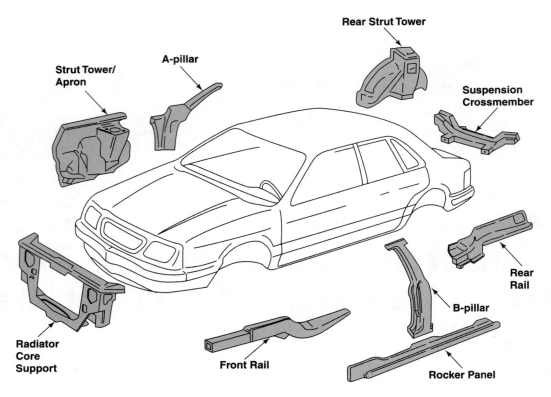

Figure 22-1 Note the typical structural parts of a vehicle.

REPAIR OR REPLACE GUIDELINES

The decision to repair or replace a structural part will be based on the judgment and skill of the technician/appraiser. A simple rule is:

If the part is bent, repair it.
If the part is kinked, replace it.

Whenever possible and practical, the part should be repaired rather than replaced. See Figure 22–2.

Figure 22-2 This badly kinked part would require replacement. It is too severely damaged for straightening.

TYPES OF STRUCTURAL REPLACEMENTS

Structural replacement can take one of two forms: replacement at factory seams and sectioning.

Replacement of panels along *factory seams* (the end or edge of a panel) should be done when practical and economical. The result is almost identical to factory production in both strength and appearance. Replacement of damaged parts along factory seams is a common practice in collision repair. Parts should be replaced along factory seams whenever it is practical and possible.

Sectioning involves cutting the part in a location other than a factory seam. This might or might not be a factory-recommended practice. Special care must be taken. Sectioning a part should be analyzed to make sure it will NOT jeopardize structural integrity.

There are three general types of structural parts:

1. Closed sections, such as rocker panels, front rails, A-, and B-pillars

2. Hat channels, such as rear rails

3. Single layer, flat parts, such as floor pans and trunk floors

Most manufacturers have specific recommendations for parts replacement. Always follow the procedures described in the body repair manual.

When planning to section a structural part:

1. Check the body repair manual for model-specific sectioning procedures.

2. If specific sectioning recommendations do NOT exist, follow the general guidelines presented here. Body repair manuals are available through various publishers and vehicle manufacturers.

Keep in mind that the vehicle must be returned to precollision condition. Failure to restore precollision crushability may affect future air bag deployment, leading to liability exposure.

When choosing a sectioning location, look for a uniform area with enough clearance to perform welding operations.

You should NOT section in or near:

1. Suspension mounting locations

2. Structural part mounting locations

3. Dimensional reference holes

4. Compound shapes

5. Reinforcements (except as noted)

6. Compound structures

7. Collapse/crush zones

8. Engine or drivetrain mounting locations.

DETERMINING SPOT WELD LOCATIONS

To start removal of a structural component, you usually must first locate all factory spot welds. Remove the paint, sealer, or other coatings covering the joint area to find the spot welds. A course scuff wheel mounted in a air tool is often used to remove the paint so you can locate the spot welds on the damaged component.

Another way to do this is to heat the paint film. It is the best choice for loosening the paint film. A coarse wire wheel or brush attached to an air drill or grinder can be used to remove the paint, sealers, and other coatings.

Scrape off thick portions of undercoating or seam sealer before scorching the paint. Do NOT overheat the paint film so that the sheet metal panel begins to turn color. Heat the area only enough to soften the paint and then brush or scrape it off. It is NOT necessary to remove paint from areas where the spot welds are visible through the paint film.

In areas where the spot weld locations are NOT visible after the paint is removed, carefully drive a

chisel between the panels. Doing so will cause an outline of the spot welds to appear.

SEPARATING SPOT WELDS

After the spot welds have been located, there are several methods of separating them. You can use a hole saw, compound drill bit, conventional drill bit, cutoff tool, or spot weld cutting tool. Regardless of which is used, be careful NOT to cut into the lower panel if it is to be used. Also be sure to cut out the spot welds without creating an excessively large hole.

If needed, center punch each spot weld before drilling. This will keep the bit from wandering. See Figure 22–3.

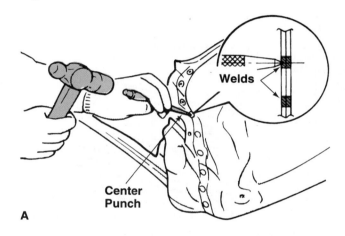

A

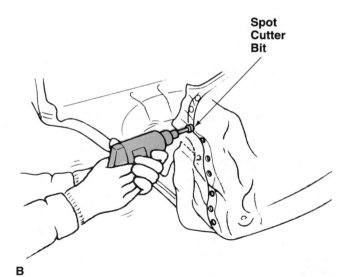

B

Figure 22–3 Spot welds can easily be seen and removed after any paint hiding them is removed. *(Used with permission from Nissan North America, Inc.)* (A) Center punching will help the drill stay in the center of spot welds. (B) Use a drill to cut out the spot weld. Do not drill through the lower panel if it is to be reused.

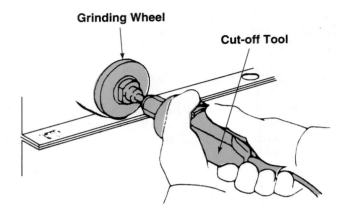

Grinding Wheel

Cut-off Tool

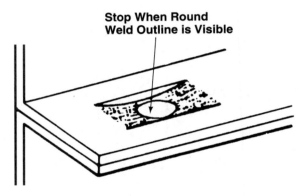

**Stop When Round
Weld Outline is Visible**

Figure 22–4 A cutoff tool provides another way of removing spot welds. (A) Carefully hold the cutoff tool over the spot weld to grind it away. (B) To prevent panel damage, stop grinding when you reach the lower panel.

Make sure the bits are sharp, and use the same pressure on the drill as if you were drilling mild steel. The speed at which the bit turns should be slower than for mild steel. This will keep the heat from affecting the metal.

Figure 22–4 shows how a cutoff tool can be used to remove spot welds.

After the spot welds have been removed, drive a chisel between the panels to separate them. Be careful NOT to cut or bend the undamaged panel. Look at Figure 22–5.

Remember! A chisel should never be used by itself to remove spot welds. This will create excessive damage to adjacent panels.

SEPARATING CONTINUOUS WELDS

In some vehicles, panels are joined by continuous MIG welding. Since the welding bead is long, use a

Figure 22–5 Only after removing spot welds can you use a chisel to separate panels. (Mustang Monthly Magazine)

grinding wheel or cutoff tool to cut through the weld. Be careful NOT to cut into or through the undamaged panels. Hold the cutoff tool at a 45-degree angle to the joint. After grinding through the weld, use an air chisel to separate the panels.

PREPARING VEHICLE FOR NEW PANEL

Always refer to the appropriate body repair manual for the type and placement of welds. The manual will also give other details for the specific vehicle.

After removing the damaged panels, prepare the vehicle for installation of the new panels:

1. Grind off the welding marks, Figure 22–6. Use a wire brush to remove dirt, corrosion, paint, seal-

Figure 22–6 After removing the damaged panel, carefully grind and smooth any small beads left from spot welds. (Mustang Monthly Magazine)

ers, and so on from the joint surfaces. Zinc coatings should NOT be removed.

2. Remove paint and undercoating from the backsides of the panel joining surfaces on parts that will be spot welded during installation.

3. Smooth the mating flanges with a hammer and dolly.

4. Apply weld-through primer to areas where bare metal is exposed.

PREPARING REPLACEMENT PANEL/PART

Since all new parts are coated with primer, it is important that this coating be removed from the flanges to allow the welding current to flow properly. Also, holes for plug welds must be drilled precisely.

To prepare a replacement panel for welding, follow these steps:

1. Use a disc sander to remove the paint from both sides of the spot weld area. Do NOT grind into the panel. Do NOT heat it too much. You do NOT want the panel to turn blue or warp.

   DANGER! *Whenever possible, grind so that sparks fly down and away. Always wear proper eye and hand protection when grinding.*

2. Apply weld-through primer to the welding surfaces where the zinc coating was removed.

3. Drill or punch holes for plug welding. Always refer to the body repair manual for the size of plug weld holes. Always be sure to duplicate the location and number of spot welds used at the factory. See Figure 22–7.

4. Be sure all plug weld holes are of the proper diameter. If a recommended hole size is NOT given in the repair manual, drill ⁵⁄₁₆ inch (8 mm) holes.

5. If the new panel is sectioned to overlap any of the existing panels, rough cut the new panel to size using an air saw or a cutoff wheel. The edges should overlap the portion of the remaining panel by ¾ to 1 inch (19 to 25 mm).

Remember! If the overlap portion is too large, it will make positioning of the panel difficult. If the overlap is too small, structural integrity will suffer.

POSITIONING NEW PANELS

Aligning new panels is a very important step in repairing unibody vehicles. Improperly aligned panels will affect both the appearance and quality of the repaired vehicle.

Use a measuring system to determine the installation position. Remember that the fit of the new and old parts must be within specifications. Whether structural or cosmetic panels are being replaced, proper measurement and fit are critical.

Panel alignment marks can be provided to help position parts before welding. Refer to the manufacturer's published materials for more information on panel alignment marks.

Once you have the part in relative position, use locking pliers or self-taping screws to hold it in place. Use self-tapping screws only in places where there is not sufficient room for locking pliers. You will have to remove and weld each hole made by a self-tapping screw. Measure the position of the panel and adjust its location if needed. See Figure 22–8.

Figure 22–8 Before final welding, parts should be positioned and held in place with locking pliers. Self-tapping screws can also be used in tight places that must be drawn together. Measure the structure as you tack weld parts in place. This will allow you to adjust the position of parts before final welding.

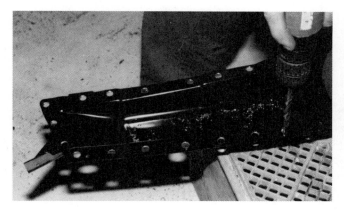

Figure 22–7 The correct number and size holes should be drilled or punched in the replacement parts. Holes should be positioned as explained in factory service literature.

Panel Positioning and Measurement

The vehicle must be properly positioned on the straightening bench before the new panel can be correctly aligned. All straightening must be done before replacing panels. Otherwise, proper alignment of the new panels will be impossible.

As an example of a typical procedure for structural part replacement, we will describe the replacement of a front fender apron, front crossmember, and core support/tie bar.

1. Match the assembly reference marks on the installation areas of the front fender apron and the rail. Secure them with locking pliers. Parts that have no assembly reference marks should be positioned in the same location as the old parts.

2. Align the parts by measuring the distance between reference points. Temporarily install the front crossmember by tack welding one spot. Make any length adjustments by light tapping. Use a hammer against a wood block to adjust the panel without damage, Figure 22–9A.

3. Mark a positioning line at the end of the part that is NOT welded. Use locking pliers to hold the parts together. You can also drill small holes and fasten them together with sheet metal screws. Mark a line on the apron area but do NOT weld the panels together.

4. Use a measuring system to match the height of the new parts to the parts on the opposite side of the vehicle, Figure 22–9B. Support the new parts with a jackstand to make sure that the height does NOT change at all.

5. Measure the lower diagonal and width dimensions, Figure 22–9C. Support the parts with jack stands so that the height does NOT change. Then, adjust the rail as needed to obtain the correct dimensions. Confirm the height dimensions again.

6. Take care to position the front crossmember so that both the left and right ends are uniform.

7. Once the dimensions of the rail match the dimensions found on the dimension chart, secure the part in place. The suspension crossmember can also be installed with fixtures. Use a sufficient number of plug welds to fasten the joining area of the rail to the front crossmember.

8. Make sure that the apron's upper length has NOT changed. Confirm by checking for shifting of the marked line.

9. Measure the diagonal dimensions between the fender rear installation hole and the strut tower hole or fender front installation hole, Figure 22–9D.

10. Measure the width dimension of the strut tower and the front fender bolt hole and fasten them together. If the width dimension does NOT match the body dimension manual, make a small adjustment. Be careful of changes in diagonal dimensions. Temporarily install and fasten the upper core support and the core support, Figure 22–9E.

11. Measure the rail width dimensions. Set the measuring system to the proper measurement and adjust the apron as needed. Lightly fasten the support with locking pliers and tap it softly by hand to move it into place.

12. Measure the diagonal dimensions for the side supports, Figure 22–9F. Be sure these dimensions match.

13. Temporarily install the front fender and inspect it for proper fit with the door and hood. If the clearance is NOT correct, it might be because the fender apron or the rail height is NOT correct.

14. Measure as previously described, and verify the overall dimensions once more before welding.

WELDING NEW PANELS/PARTS

When the position and dimensions of the new panel are satisfactory, it can be permanently welded in place, Figure 22–10. Under ideal circumstances, the general welding procedure for panels should be:

1. Apply weld-through primer to bare metal surfaces.

2. Clamp parts in position.

3. Tack weld parts in position.

4. Remeasure part positions.

5. Plug weld parts in final position.

6. Grind cosmetic weld surfaces.

   Remember that it is very important to duplicate exactly the location and number of original factory spot welds.

SECTIONING

As mentioned, sectioning involves cutting and replacing panels at locations other than factory seams. When body parts need to be replaced, replacing them

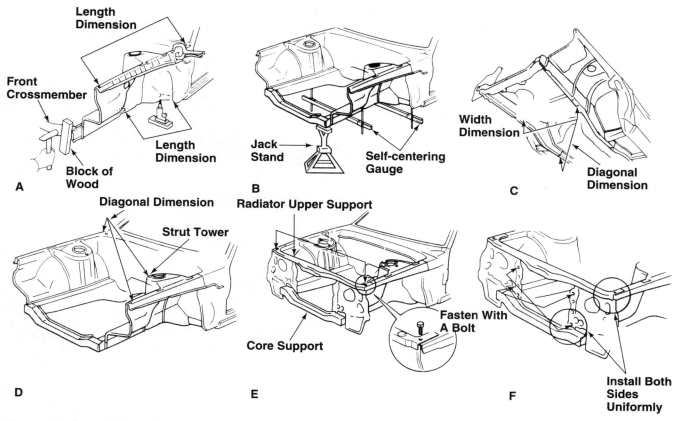

Figure 22–9 Carefully measure and position the panel before welding. (A) Use a wood block and light blows to adjust the panel as needed. (B) Jack stands should be used to support the panel. Use a measuring system to check panel positioning. (C) Measure the lower diagonal and width dimensions. (D) Note the basic method for measuring fender apron dimensions. (E) After the inner component, the apron in this example, is secured, install the next component, core support. F: Measure each component's location before final welding.

at factory seams is the logical first choice. However, this is impractical when many seams have to be separated in undamaged areas. In some repairs, sectioning of parts such as rails, pillars, and rocker panels may be required to make their repair economically feasible. See Figure 22–11.

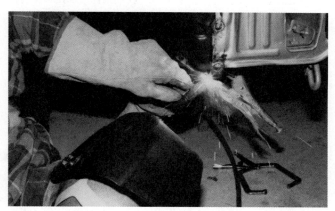

Figure 22–10 Follow all safety rules when welding structural panels in place. Wear welding helmet, welding respirator, and leather gloves.

Remember that sectioning requires precision, as well as strict adherence to recommended procedures. Always check the body repair manual for the vehicle manufacturer's procedure for the specific sectioning location.

As a collision repair technician, never forget that sectioning finally comes down to sound judgment. It is the technician who must assure the quality of the repair through the proper application of tested and proven procedures.

Unibody parts to be sectioned involve these types of construction:

1. Closed sections, such as rocker panels and A- and B-pillars

2. Hat or open U-channels, such as rear rails

3. Single layer or flat parts, such as floor pans and trunk floors

With structural parts, it is best to treat them all like high-strength steel. Then you do not have to identify every piece of high-strength steel. Also, you can use a MIG welder on all repairs.

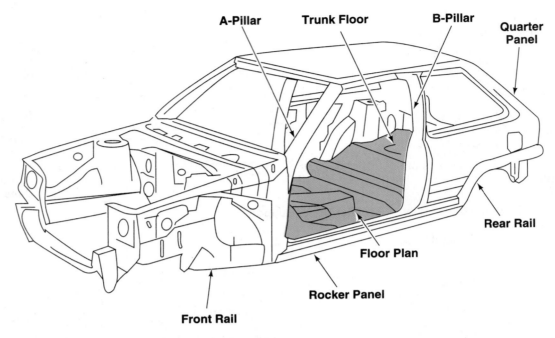

Figure 22-11 Note these common sectioning areas on a vehicle.

Using New Versus Recycled Assemblies

Recycled assemblies are undamaged parts from another damaged vehicle that are used for repairs. The use of recycled assemblies in collision repair makes sense for a number of reasons:

1. Fewer welds need to be made when using recycled assemblies compared to new, separate parts.

2. Less factory corrosion protection is disturbed.

3. Less measuring is required than when welding separate new parts and attaching them to the vehicle.

4. There is an abundance of recycled assemblies available in most areas.

Section Locations

When deciding where to section, look for an area with a uniform cross section. Check the vehicle manufacturer's body repair manuals and bulletins provided by the vehicle manufacturer. Much of this literature provides specific instructions on how and where to section.

Sectioning Joint Types

There are three basic types of sectioning joints. They are the:

1. Lap joint

2. Offset butt joint

3. Butt joint with insert

One of these joints, or a combination of these joints, will be used for all sectioning procedures. The type of joint used for a specific repair will depend upon the location and design of the structural part. See Figure 22-12.

A butt joint with insert is used mainly on closed sections, such as rocker panels, A- and B-pillars, and rails. *Inserts* make it easy to fit and align the joints correctly. They also help make the welding process easier.

Another basic joint is an *offset butt joint* without an insert. This type is also known as a **staggered butt joint**. The staggered butt joint is used on A- and B-pillars and front rails.

The third type is a lap joint, which is used on rear rails, floor pans, trunk floors, and B-pillars.

The configuration and makeup of the part being sectioned might call for a combination of joint types. Sectioning a B-pillar, for instance, might require the use of an offset cut with a butt joint in the outside piece and a lap joint in the inside piece. See Figure 22-13.

Using Lap Joints. Lap joints are used on floor pans, trunk floors, rear rails and other flat or hat channel shaped parts.

When welding lap joints:

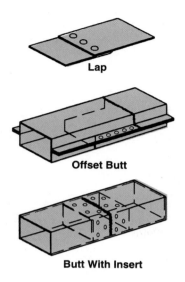

Lap

Offset Butt

Butt With Insert

Figure 22–12 These are the three basic types of joints used for sectioning. *(I-CAR)*

1. Parts should be overlapped ¼ inch to 1 inch (6–25 mm) as specified. See Figure 22–14.

2. Use plug welds in the overlap area. Weld from the top piece down.

3. Close the bottom edge with a continuous weld bead.

4. Seal lap joints in floor pans and trunk floors to prevent water and fumes from entering.

Using Offset Butt Joints. Offset butt joints are used for closed sections such as A- and B-pillars. They are used on B-pillars to avoid disturbing the D-ring anchor. Offset butt joints are also used where three or more pieces join in the construction of a part, making it difficult or impossible to use an insert.

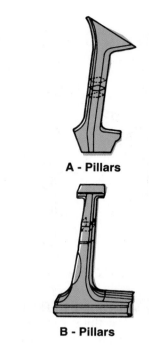

A - Pillars

B - Pillars

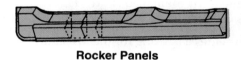

Rocker Panels

Rails

Figure 22–13 A butt joint with an insert produces a very strong repair.

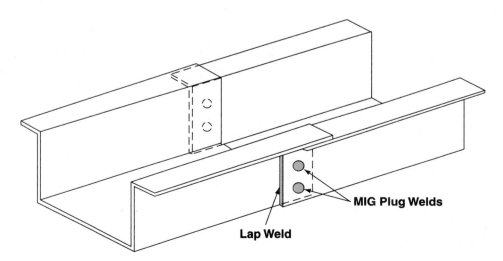

MIG Plug Welds

Lap Weld

Figure 22–14 Lap joints should overlap about an ¼ inch to 1 inch (6–25 mm). Use both plug and continuous welds to join parts. *(I-CAR)*

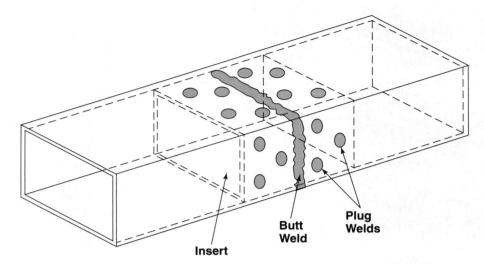

Figure 22–15 Note how a butt joint with an insert is typically made. *(I-CAR)*

Using Butt Joint with Insert. The butt joint with insert is used on closed sections, such as rocker panels, A-pillars, B-pillars, and front rails. Look at Figure 22–15.

The insert is made from a section of either the replacement or damaged part, Figure 22–16. The insert is used for the following reasons:

1. It provides a backing for the MIG butt weld.

2. It keeps burn-through to a minimum.

3. It ensures a completely closed joint.

4. It aligns the parts for the best possible fit.

NOTE! When sectioning front and rear rails, DO NOT put inserts in collapse/crush zones. Strengthening these parts will change the way collision damage is absorbed, possibly endangering the passengers.

Sectioning—Butt Joint Method

The procedures for sectioning with a butt joint with insert are as follows:

1. The length of the insert is twice the width of the part at its widest point.

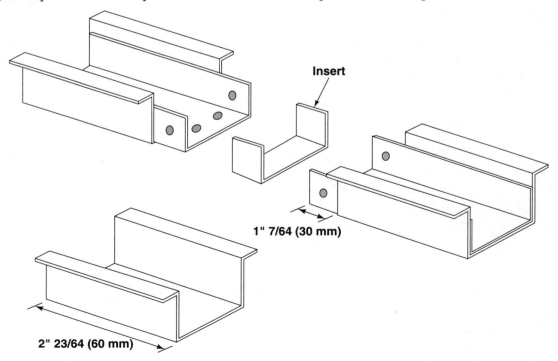

Figure 22–16 An insert can be made by cutting off excess material from the new or old part. *(Used with permission from Nissan North America, Inc.)*

2. Cut insert pieces from either the damaged part or replacement panel. Use a section that closely matches contours.

3. Trim away flanges. Make lengthwise cuts in the insert.

4. Check the fit of the individual pieces of the insert. There should be no gaps between the insert and the inside of the rail or pillar.

5. Drill 5⁄16-inch (8-mm) plug weld holes so that heat affect zones from the butt weld and plug welds do not overlap.

6. Remove burrs from all panels.

7. Remove all paint, undercoating, and seam sealers. Do NOT remove galvanizing.

8. Apply weld-through primer.

9. Make test welds.

10. Fit the insert and secure it in place.

11. Butt the pieces up, leaving a gap the thickness of the panels being welded.

12. Check measurements and fit.

13. Tack weld and measure again.

14. Complete and dress all welds. Make sure the joint is completely closed and the weld has penetrated the insert.

15. Prep for corrosion protection and topcoats.

The most common joints used to section front and rear rails are the butt joint with insert and the lap joint. The butt joint with insert is used most often on front rails. Lap joints are commonly used on rear hat channel rails.

There is an additional combination lap joint–offset butt joint procedure that is used on front rails. This procedure should only be used when tested repair procedures are available for the specific vehicle being repaired.

Making Cut Templates. Parts must be measured and marked prior to making cuts. This can easily be done by making a cutting template:

1. Decide where to section.

2. Cut damaged section off the vehicle.

3. Apply a sturdy, wide tape to the remaining section.

4. Use a razor to cut the tape around various reference holes and the cut line.

5. Remove the tape.

6. Align holes cut in the tape to identical holes on the replacement part.

7. Apply the tape template to the replacement panel.

8. Cut along the tape template cut line.

Heavy paper and masking tape can also be used to create a template.

Sectioning Front Rails—Insert Method

The procedures for front rail sectioning using a butt joint with insert are as follows:

1. Decide where to section.

2. Measure and mark the cuts on both parts.

3. Make straight cuts along cut lines.

> **WARNING!** *Do NOT throw any parts away until the repair is finished.*

4. Make and fit the insert to the rail sections.

5. Drill 5⁄16-inch (8-mm) plug weld holes so that heat affect zones from the butt weld and plug welds do not overlap.

6. Remove burrs from all panels.

7. Remove all paint, undercoating, and seam sealers. Do NOT remove galvanizing.

8. Apply weld-through primer.

9. Make test welds.

10. Fit the insert and secure it in place.

11. Butt the pieces up, leaving a gap the thickness of the panels being welded.

12. Check measurements and fit.

13. Tack weld and measure again.

14. Complete and dress all welds. Make sure the joint is completely closed and that the weld has penetrated the insert.

15. Prep for corrosion protection and topcoats.

Sectioning Rails—Combined Joint Method

Rails are also sectioned using combined lap and offset butt joints. The following procedures and

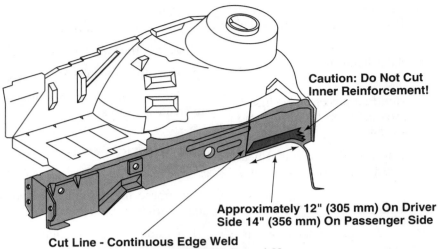

**Caution: Do Not Cut
Inner Reinforcement!**

**Approximately 12" (305 mm) On Driver
Side 14" (356 mm) On Passenger Side**

**Cut Line - Continuous Edge Weld
Overlap: 1/16" (1.5 mm) To 1/4" (6 mm) Max**

Figure 22–17 Since sectioning of the front frame rails is critical to the integrity of the vehicle, use great care. This is a typical example on a unibody vehicle. Always refer to manufacturer's procedures, however. Do not cut inner reinforcements. *(Used by permission from Tech-Cor, Inc.)*

measurements are based on Figure 22–17. Always refer to an industry-accepted and -tested procedure for specifics.

1. If necessary, remove the engine, transmission, and cradle as a unit.

2. Locate and remove factory spot welds that attach the upper rail to the cowl at the base of the windshield.

3. Remove paint, undercoating, and seam sealers from the spot weld areas.

4. Remove the two spot welds that hold the upper rail to the rear outer flange of the strut tower. These spot welds can be accessed through the hole at the rear of the upper rail.

5. Remove spot welds that attach the strut tower to the rail extension panel at the base of the strut tower.

6. Remove paint, undercoating, and seam sealers from inside the engine compartment.

7. Remove sound-deadening material from the wheelhouse.

8. Prepare to section the rail in front of the strut tower center point. Because of the inner reinforcement, this area is ideal for sectioning. See Figure 22–17.

9. Make a staggered cut on the inner and outer lower rail. An electric reciprocating saw with a laminated blade makes the best cuts.

10. Before cutting, remove the two spot welds that attach the reinforcement to the inside of the lower

rail. One is on the wheelhouse side of the rail. The other is on the bottom of the rail. Refer to Figure 22–18.

11. Cut the engine side of the rail near the end of the inner reinforcement. In the example shown in Figure 22–17, this is about 12 inches (305 mm) from the cowl.

12. Cut the wheelhouse side of the rail behind the engine side cut. In the example shown in Figure 22–18, this distance is about 3–5 inches (76–127 mm).

13. Use a staggered butt joint for sectioning the rail. The parts should overlap to aid in alignment and welding. In Figure 22–17, the overlap should be $\frac{1}{16}$–$\frac{1}{4}$ inch (1.5–6 mm). To prepare for the overlap, cut the corners of the original part. In Figure 22–17, this distance should be $\frac{1}{4}$ inch (6 mm).

14. Position the replacement part OVER the original part. This will allow the corrosion protection treatments inside the rail to better penetrate the joint. Any part of the cuts exposed after fitting must be welded shut.

15. Separate the opposite side lower rail extension from the lower rail. First, drill the spot welds that attach the radiator core support and apron extension panel.

16. Next, carefully fold up the apron extension panel to expose the other spot welds that attach the rail extension to the rail. This should loosen the damaged part so it can be removed.

17. Prepare the used assembly. Remove the spot welds and make the lower rail offset cut the same

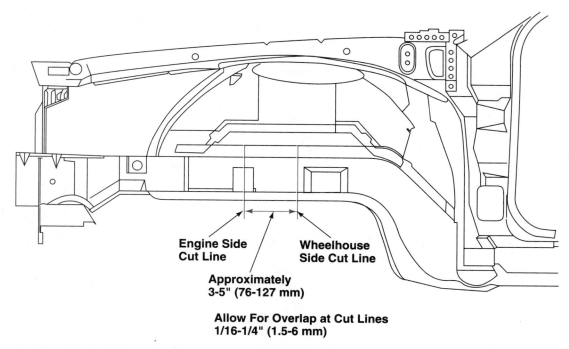

Engine Side Cut Line

Wheelhouse Side Cut Line

Approximately 3-5" (76-127 mm)

Allow For Overlap at Cut Lines 1/16-1/4" (1.5-6 mm)

Figure 22–18 Note how an offset cut is used to section the front rails. *(Used by permission from Tech-Cor, Inc.)*

as on the damaged parts. Be sure to inspect, measure, and if necessary, straighten the used assembly.

NOTE! When cutting the replacement rail, remember to add to the length measurement on the replacement rail for the overlap. In the example shown in Figure 22–17, add 1/16–1/4 inch (1.5–6 mm).

18. Align the replacement assembly and secure it in place. Measure to make sure parts are dimensionally correct.

19. Weld the assembly into position using I-CAR's MIG welding guidelines.

20. Clean the welds carefully to reduce the chance of corrosion.

21. Apply corrosion protection.

This example was for sectioning the driver side rail. Although the dimensions will be different, the passenger side can also be completed the same way.

Be sure to work with the locations of welds and reinforcements, NOT just the dimensions. Measurements given here are approximate. Locations of reinforcements can vary by model year.

Be sure NOT to confuse this procedure with front full body sectioning. This procedure is NOT front full body sectioning. Section closed sections using a butt joint with insert or offset butt joint. In this procedure,

the joint is offset, with the inner and outer rails cut at different points. This produces a staggered butt joint. A slight overlap is used to ensure a tight fit for welding purposes. In the example shown in Figure 22–18, this overlap should be 1/16–1/4 inch (1.5–6 mm). This eliminates the need for inserts.

Areas with inner reinforcements should NOT be sectioned. Cutting in the area of reinforcements can create problems. In the example described here, the rail is carefully cut around the reinforcement. The reinforcement adds a backup. This results in a joint with the best possible features:

1. An offset

2. Tight fit due to the overlap

3. Reinforcement as a bridge between the two parts

Test results indicate that this type of repair performs the same as the original part during a collision.

Sectioning Hat Channels—Rear Rails

Most rear rails are hat channels attached to the trunk floor. To section hat channel rear rails:

1. Decide where to section. Stay forward of all collapse zones.

2. Measure and mark the cuts on both parts.

3. Remove spot welds attaching rail to the floor pan. Drill from the trunk side.

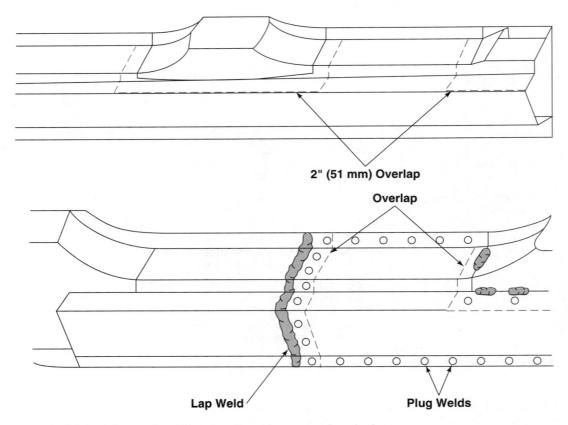

2" (51 mm) Overlap

Overlap

Lap Weld

Plug Welds

Figure 22–19 Note the typical cut and weld locations for replacing a rocker panel.

4. Make straight cuts along cut lines.

5. Drill two to three ⁵⁄₁₆-inch (8-mm) plug weld holes in the overlap area.

6. Remove burrs from all panels.

7. Remove all paint, undercoating, and seam sealers from the floor pan. Do NOT remove galvanizing.

8. Apply weld-through primer.

9. Make test welds.

10. Fit the replacement panel, overlapping the original by ¼ inch to 1 inch (6–25 mm) and secure it in place.

11. Check measurements and fit.

12. Tack weld and measure again.

13. Plug weld the overlap area. Alternate sides to minimize heat buildup.

14. Plug weld through the trunk floor into the rail flanges.

15. Weld along the lap joint.

16. Dress all welds.

17. Prep for corrosion protection and topcoats.

A "window" is cut into the outer rail to access a weld site on the reinforcement. A butt joint with insert repair is made to the rail reinforcement.

The more complicated the structure, the more challenging the repair procedure. Manufacturers are using multiple part designs more often to increase the rigidity of the passenger compartment. This is of increased importance with convertibles.

Sectioning Rocker Panels

When sectioning rocker panels, the two joints most commonly used are a butt joint with insert or a lap joint. In some cases, only the outer panel is replaced or the rocker may be replaced with or without the B-pillar attached. If the B-pillar is attached, then a B-pillar section will have to be made at the same time.

Use a butt weld with insert when installing a recycled rocker panel with B-pillar attached, or when installing a recycled quarter panel. If working on a three-piece design, remove the outer panel first.

Basic guidelines for replacing rocker panels are:

1. Carefully plan the repair before making the first cut.

2. When cutting rocker panels or other structural parts, use a reciprocating saw to get an accurate cut and to reduce the heat affect zone.

3. Use inserts whenever possible.

4. Be sure to completely weld all seams.

5. Use plug welds to secure any inserts used.

To remove and replace the outer rocker panel:

1. Cut around the pillar bases, leaving an overlap area. Stay about 2 inches (51 mm) from the pillars to avoid any reinforcements.

2. Cut outer panel ends. Keep in mind that you will overlap those areas when replacing the rocker panel.

3. Cut or drill out the spot welds along the pinch-weld seams. Separate the panels/components.

4. Cut the replacement panel to allow for a 1-inch (25-mm) overlap of the pillar bases and panel ends, Figure 22–19.

5. Drill ⁵⁄₁₆-inch (8-mm) plug weld holes so that heat affect zones from the butt weld and plug welds do not overlap. Plug welds along the pinchweld should be made following the manufacturer's recommendations. If recommendations are not available, duplicate the number and location of the original welds. Plug welds also have to be made in the overlap areas.

6. Remove burrs from all panels.

7. Remove all paint, undercoating, and seam sealers. Do NOT remove galvanizing.

8. Apply weld-through primer.

9. Make test welds.

10. Check measurements and fit.

11. Tack weld and measure again.

12. Complete and dress all welds. Make sure the joints are completely closed.

13. Prep for corrosion protection and topcoats.

Sectioning Rocker Panels— Insert Method

To section a rocker panel using a butt joint with insert:

1. Plan the cut to be near the middle of the door opening.

2. Cut straight across the rocker panel.

3. Make an insert from either the damaged or replacement part.

4. Check the insert for fit, Figure 22–20.

5. Cut the replacement part to fit.

6. Drill ⁵⁄₁₆-inch (8-mm) plug weld holes on both the vehicle and replacement parts. Stagger the holes to avoid overlapping heat affect zones.

7. Remove burrs from all panels/components.

8. Remove all paint, undercoating, and seam sealers. Do NOT remove galvanizing.

9. Apply weld-through primer.

10. Make a test weld.

11. Fit the insert and secure it in place.

12. Butt the pieces up leaving a gap about the thickness of the panels being welded.

13. Check measurements and fit.

14. Tack weld and measure again.

15. Complete and dress all welds. Make sure the joint is completely closed and that the butt weld has penetrated the insert.

16. Complete other necessary sectioning or welding procedures.

17. Prep for corrosion protection and topcoats.

Sectioning Multi-Part Rocker Panels

The *multiple part rocker* assembly is made up of several pieces of sheet metal with internal reinforcements. Reinforcements make it more difficult to use inserts. It is also more important to plan the exact order the work will be done. This will ensure that all the welding will be done in the right sequence. Also, the various rocker parts will be cut so as to provide maximum reinforcement to one another.

1. Plan the cut in the middle of the door.

2. Cutting points for various parts of rocker assembly should be planned so panels overlap, providing continuity for one another.

3. If more than just the outer rocker is being replaced, you will have to separate the rocker assembly into its individual parts.

4. Work from the inside out, starting with the inner rocker or main floor side member. Then, attach the flat reinforcement, the channel reinforcement, and finally the outer rocker. Check the fit at various points to make sure all is correct.

5. Cut a piece of the rocker to form an insert, providing backing for the weld.

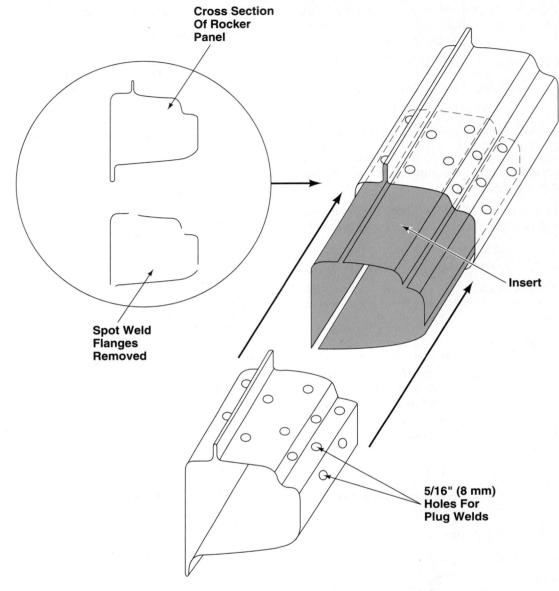

Cross Section Of Rocker Panel

Spot Weld Flanges Removed

Insert

5/16" (8 mm) Holes For Plug Welds

Figure 22–20 An insert can be made and used with a rocker panel replacement. *(I-CAR)*

6. Be sure to weld all butt and lap joints completely closed. Weld in segments to minimize heat distortion. Use weld-through primer.

7. Finish the inside of all cavities and outside weld joints with two-part epoxy primer. Apply anticorrosion compound to enclosed areas.

   NOTE! You may have to treat some of the cavities before final assembly in order to be able to reach all areas.

Sectioning A-Pillars

A-pillars use either two- or multiple-piece construction. They can be sectioned using a butt joint with insert or an offset butt joint.

A butt joint with insert or an offset butt joint can be used on two-piece A-pillars. On multiple-piece A-pillars, the design of the part will guide repair planning. There is often an inner reinforcement at the upper and lower ends, so an offset butt joint may be the only choice. Reinforcement locations vary, so refer to a body repair manual to properly plan the repair. Use an insert if the design of a multiple-piece pillar allows.

Sectioning A-Pillars—Insert Method

Sectioning an A-pillar is similar to sectioning other closed sections. To section an A-pillar using a butt joint with insert:

1. Plan the cut in the middle of the pillar. Stay clear of channel areas for motorized seat belt systems.

2. Use a jig to help locate and guide the cut.

3. Remove the damaged part.

4. Cut the replacement part to fit.

5. Make and fit the insert to A-pillar sections.

6. Drill ⁵⁄₁₆-inch (8-mm) plug weld holes so that heat affect zones from the butt weld and plug welds do not overlap.

7. Remove burrs from all panels/components.

8. Remove all paint, undercoating, and seam sealers. Do NOT remove galvanizing.

9. Apply weld-through primer.

10. Make test welds.

11. Fit the insert and secure it in place.

12. Butt the pieces up leaving a gap about the thickness of the panels being welded.

13. Check measurements and fit.

14. Tack weld and measure again.

15. Complete and dress all welds. Make sure the joint is completely closed and that the weld has penetrated the insert.

16. Prep for corrosion protection and topcoats.

Sectioning A-Pillars—Butt Joint Method

A-pillars can also be sectioned using an offset butt joint. To section an A-pillar using the offset butt joint:

1. Choose cut locations that are between spot welds.

2. Using a jig to guide and locate the cuts, cut through one side of the A-pillar. Cut through the flange but NOT into the flange on the other side.

3. Make a second cut on the opposite side 2–4 inches (50–100 mm) from the first cut. Cut through the

flange but NOT into the flange on the other side.

4. Drill out the spot welds.

5. Cut the replacement part to fit, Figure 22–21.

6. Drill ⁵⁄₁₆-inch (8-mm) plug weld holes so that heat affect zones from the butt weld and plug welds do not overlap.

7. Remove burrs from all panels.

8. Remove all paint, undercoating, and seam sealers. Do NOT remove galvanizing.

9. Apply weld-through primer.

10. Make test welds.

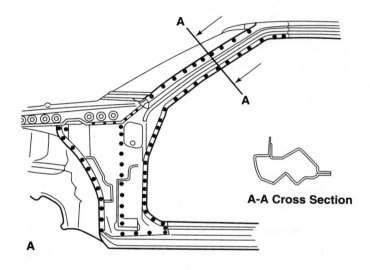

A-A Cross Section

A

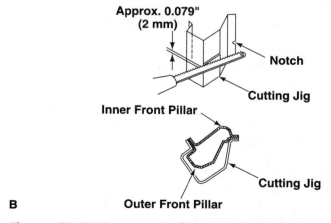

Approx. 0.079" (2 mm)

Notch

Cutting Jig

Inner Front Pillar

Cutting Jig

Outer Front Pillar

B

Figure 22–21 Sectioning A-pillars is similar to sectioning other parts. Check the manufacturer's recommendations for exact procedures. *(Used with permission from Nissan North America, Inc.)* (A) Note a typical location of the area to be cut and of the spot welds that must be removed to replace an A-pillar. (B) A cutting jig can be made and placed over the component to help you make a straight cut through the pillar.

11. Check measurements and fit.

12. Tack weld and measure again.

13. Complete and dress all welds. Make sure the joint is completely closed.

14. Prep for corrosion protection and topcoats.

Replacing Foam Fillers

Some manufacturers place foam inside panels/components. **Foam fillers** are used to add rigidity and strength to structural parts. They also reduce noise and vibrations. Cutting and welding will damage the foam filler. Replacing the foam fillers must be part of the repair procedure.

Some vehicle manufacturers are using urethane foam in A- and B-pillars, and other locations. The manufacturer may or may not consider the foam filler to be structural. The use and location of foam fillers are different from vehicle to vehicle. Follow manufacturer's recommendations for replacing or sectioning foam-filled panels.

Some OEM replacement parts come with the foam already in the part. When the parts come without foam filler, or foam filler needs to be replaced, a product designed specifically for this application must be used to fill the panel.

> *WARNING! Single-part urethane foams made for home use CANNOT be used for replacing automotive foam fillers.*

Cutting and welding will damage the foam. Replacing the foam fillers must be part of the repair procedure. When sectioning foam-filled A-pillars, the foam filler is removed in the repair area. It is then replaced after all welding is completed. See Figures 22–22 and 22–23.

Some manufacturers have specific recommendations for the type of replacement foam filler needed. The repair usually calls for a foam filler that, when cured, doesn't change in volume due to differences in temperature and humidity. There may also be specific requirements for foam density given in ounces per cubic inch (grams per cubic centimeter).

Sectioning B-Pillars

B-pillars can be sectioned using a butt joint with insert or an offset butt and lap joint combination. The B-pil-

lar sectioning procedure can be complicated by:

1. Seat belt D-rings and anchors

2. Reinforcements built in at the rocker panel and roof

3. Motorized seat belt track and mounting points

4. Seat belt retractors and tensioners

When sectioning B-pillars on vehicles equipped with explosive belt tensioners, disconnect and remove the seat belt mechanism. Other restraint systems may require the removal of motors, tracks, cables, and belts.

Care should be taken when reinstalling restraint system parts. Check the service manual for installation procedures. Installation sequence and proper torque can affect the operation of restraint systems.

B-pillars should be sectioned using the butt joint with insert whenever the design of the B-pillar allows it.

Sectioning B-Pillars—Insert Method

To section two-piece B-pillars using the butt joint with insert:

1. Cut below the D-ring anchor point, away from motorized seat belt system tracks. See Figure 22–24.

2. Make a three-sided insert from the damaged or replacement part. A three-sided insert is used because the D-ring reinforcement prevents the use of a full insert.

3. Cut panels so that the replacement panel overlaps inside of the existing part by 3½ inches (90 mm) and the outer panels are butted together.

4. Drill ⁵⁄₁₆ inch (8-mm) plug weld holes so that heat affect zones from the butt, lap, and plug welds don't overlap.

5. Remove burrs from all panels/components.

6. Remove all paint, undercoating, and seam sealers. Do NOT remove galvanizing.

7. Apply weld-through primer.

8. Make test welds.

9. Fit the insert and secure it in place.

10. Butt the pieces up, leaving a gap of about the thickness of the panels being welded.

11. Check measurements and fit.

12. Tack weld and measure again.

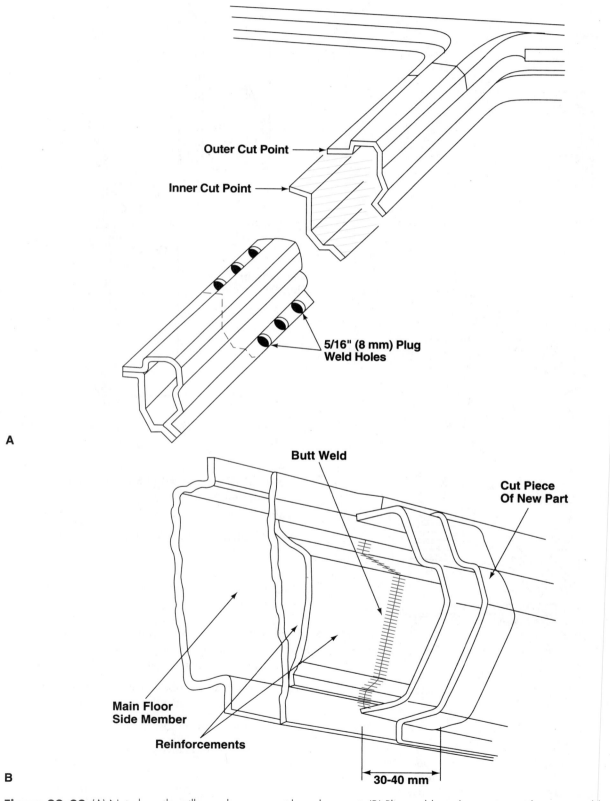

Figure 22–22 (A) Note how the pillar and component have been cut. (B) Plug welds and a continuous butt joint weld will secure the component properly. *(Used with permission from Nissan North America, Inc.)*

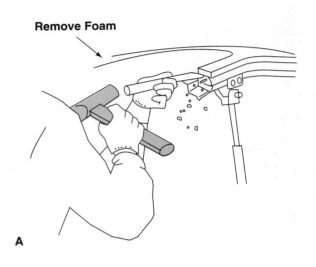

Remove Foam

A

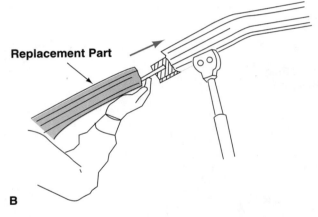

Replacement Part

B

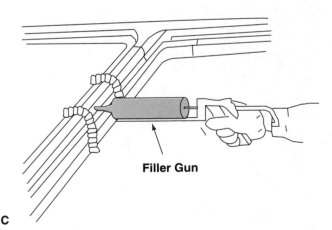

Filler Gun

C

Figure 22–23 Study the basic steps for replacing a part filled with foam. *(Used with permission from Nissan North America, Inc.)* (A) After cutting off the damaged component, remove the foam from the area to be welded. (B) Fit the replacement component with cut on the vehicle. (C) After welding, inject the recommended type of foam into the panel.

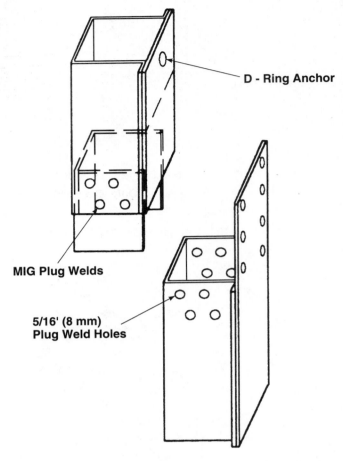

D - Ring Anchor

MIG Plug Welds

5/16' (8 mm)
Plug Weld Holes

Figure 22–24 B-pillars are normally sectioned using a butt joint with an insert. *(I-CAR)*

13. Lap weld the upper edge and complete the plug and butt welds. Dress all welds. Make sure the joints are completely closed.

14. Prep for corrosion protection and topcoats.

Sectioning B-Pillars—Combination Joint Method

B-pillars with complex designs are sectioned using the offset butt and lap joint. To section B-pillars using the offset butt and lap joint, Figure 22–25:

1. Cut the outer panel above the D-ring mount being careful NOT to cut into the inner panel.

2. Cut the inner panel below the D-ring reinforcement.

3. Cut out spot welds along the pinchweld and separate the pieces.

4. Cut the replacement inner panel to overlap the existing inner panel by approximately 3½ inches (90 mm).

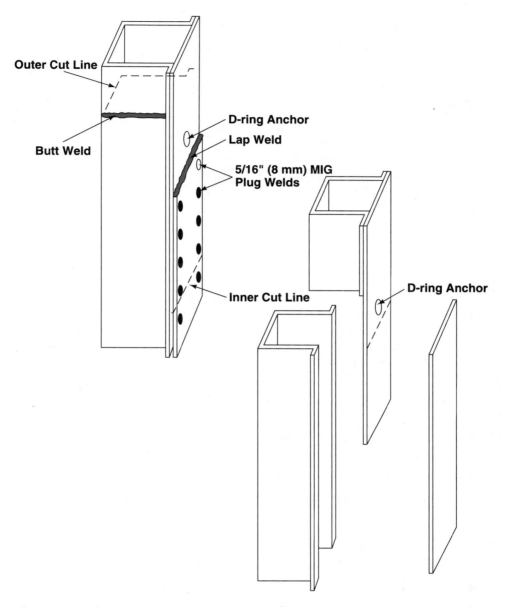

Figure 22–25 B-pillars with complex designs are sectioned using the offset butt and lap joint. *(I-CAR)*

5. Cut the replacement outer panel to butt against the existing outer panel.

6. Drill ⁵⁄₁₆-inch (8-mm) plug weld holes along pinch-weld flanges. Stagger the holes (new inner to new outer) so that heat affect zones do not overlap.

7. Remove burrs from all panels/components.

8. Remove all paint, undercoating, and seam sealers. Do NOT remove galvanizing.

9. Apply weld-through primer.

10. Make test welds.

11. Check measurements and fit.

12. Tack weld and measure again.

13. Lap weld the upper edge of the new inner panel. Complete the plug and butt welds. Dress all welds. Make sure the joints are completely closed.

14. Prep for corrosion protection and topcoats.

Sectioning Floors

Floor pans and trunk floors are sectioned using lap joints. It is critical to completely seal the joints following the sectioning procedure to keep moisture and exhaust gases out of the passenger compartment. Do

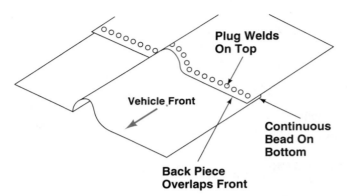

Figure 22–26 Floors are normally sectioned using lap joints. The back part should be placed on top of the front component to help keep out road debris and water. (I-CAR)

NOT cut through any reinforcements or critical areas, such as seat belt anchoring points.

When sectioning floor pans and trunk floors, the rear section always overlaps the front section, Figure 22–26. This allows road splash to stream past the bottom edge of the joint and NOT strike the joint head on.

To section floor pans:

1. Make cuts in areas free of reinforcements, seat, or seat belt anchoring points.

2. Cut the replacement panel to overlap the existing panel by at least 1 inch (25 mm).

3. Drill ⁵⁄₁₆-inch (8-mm) plug weld holes.

4. Remove burrs from all panels/components.

5. Remove all paint, undercoating, and seam sealers. Do NOT remove galvanizing.

6. Apply weld-through primer.

7. Make test welds.

8. Check measurements and fit.

9. Tack weld and measure again.

10. Weld the plug welds from the top down.

11. Weld a continuous bead along the bottom lap joints.

12. Dress the welds and prime.

13. Apply caulk to top side joints and primer and seam sealers to under side joints.

14. Prep for corrosion protection and topcoats.

Sectioning Trunk Floors

The procedure for sectioning trunk floors is the same as that for sectioning floor pans, with some variations:

1. There is generally a suspension crossmember under the trunk floor near the rear suspension. Whenever possible, section the trunk floor above the crossmember rear flange.

2. Plug weld the trunk floor lap joint to the crossmember, welding the plug welds from the top. See Figure 22–27. The bottom, continuous lap weld is not needed because of the strength added by the crossmember.

3. On vehicles that have no crossmember, the lower edge must have a continuous lap weld.

4. In either case, caulk the top side joint, and apply a primer and seam sealer to the under side joint. Topcoat as necessary.

Full Body Sectioning

One of the most drastic repairs that can be performed is full body sectioning. **Full body sectioning** is replacing the entire rear section of a collision damaged vehicle with the rear section of a salvage vehicle. It may be more economical than trying to rebuild the damaged vehicle using new parts. Full body sectioning requires the highest quality workmanship possible.

Jigs are often used to help locate and guide the cut when sectioning. They are used on any sectioning procedure where an offset butt joint is used. Precise cuts are essential when sectioning.

Full body sectioning procedures require sectioning the two A-pillars, two rocker panels, and the floor pan. When this procedure is properly performed, sectioned vehicles have been shown to be as strong and serviceable as an undamaged vehicle.

Full body sectioning is complicated by antilock brake systems. The replacement vehicle must be so equipped, or the ABS system parts must be retrofitted.

Location and function of body computers may change, even within the same production year of a given vehicle. Check the locations of these computers on both the damaged vehicle and the salvage section.

I-CAR recommends that full body sectioning be done as a rear-body section only. I-CAR does NOT recommend front full body sectioning at other locations for the following reasons:

1. VIN would NOT match registration and title.

2. Matching mileage, condition, and accessories would be difficult.

3. Powertrain warranty could be affected.

When performing a full body sectioning, complete disclosure must be made to the vehicle owner. A conference between the owner, insurance appraiser, and

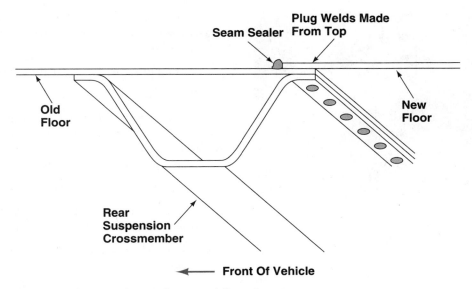

Figure 22-27 Note the common method for installing a trunk floor. *(I-CAR)*

repairer is suggested. Then, all parties will be aware of exactly what will be done to repair the vehicle.

Replacing Adhesives

Some vehicle manufacturers use structural adhesives along certain weld seams. These two-part epoxy adhesives are sometimes called **weld-bond adhesives** because spot welds are placed through the adhesive.

Weld-bond adhesives are used to add strength and rigidity to the vehicle body. They also improve corrosion protection in weld seams. Adhesives also help control noise and vibrations.

Parts most commonly weld-bonded are:

1. A- and B-pillars

2. Rocker panels

3. Roof panels

4. Rear quarter panels

If adhesives are disturbed by repairs, they must be replaced. Follow recommendations in the body repair manual.

Some manufacturers use **structural adhesives** in place of welds. One example is around the wheel openings and sail panel reinforcement. This is a different type of adhesive than the weld-bond adhesive. Check the body repair manual for information on the use of structural adhesives.

CORROSION

Corrosion is a chemical reaction, called **oxidation** or **rust,** formed when three ingredients are present:

1. Oxygen

2. Exposed metal

3. Moisture, Figure 22-28

Oxidation occurs in two steps:

1. Stable metal and oxygen atoms break down into positively and negatively charged particles called ions. See Figure 22-29.

2. These ions are unstable, and combine with each other to form metal oxides, which are more stable.

Figure 22-28 Corrosion is due to moisture, road salt, acid rain, pollution, and paint film damage.

Full Rustout

Figure 22–29 Corrosion, or rust, can be minor or major. This is an example of major corrosion damage that would require part replacement.

A necessary ingredient for oxidation to occur is an electrolyte.

1. Electrolyte is moisture.

2. Moisture is the only part of the oxidation process that repair shops can do anything about.

3. Moisture is controlled by applying coatings to the metal surface to act as a moisture barrier.

4. Coatings do break down with weather exposure.

5. The best coating systems are those that last the longest.

6. Coatings only slow down the process of corrosion. Stopping it permanently is a difficult challenge.

Corrosion Protection

Even with all of the care taken to protect vehicles, corrosion protection breakdown still occurs. The breakdown falls into three general categories:

1. Paint film failure

2. Collision damage

3. Repair process

During a collision, the protective coatings on a vehicle are damaged, Figure 22–30. This occurs not just in the areas of direct impact, but also in the indirect damage zones. Seams pull apart, caulking breaks loose, and paint chips and flakes. Locating the damage and restoring the protection to all affected areas remains a key challenge for the collision repair technician.

Even touching a bare metal surface with bare hands adds corrosion-causing agents. Make sure bare hands do NOT contact bare metal that has been cleaned for refinishing.

When dealing with the zinc coating on a galvanized body structure, there are two important things to remember:

1. If at all possible, do NOT remove it.

2. If it must be removed, replace it. See Figure 22–31.

To be effective, the coating must be applied to a clean surface. The coating must also be tight and unbroken.

When performing collision repairs:

1. Preserve the original coating as much as possible.

2. Use conversion coatings to create a zinc phosphate coating on the steel surface.

3. Use weld-through primers on bare steel in weld areas.

The collision repair shop's interest in corrosion is twofold. First, the technician must be able to repair corrosion damage. Second, the technician must be able to provide treatment that will prevent corrosion from recurring.

When restoring corrosion protection:

1. Use a complete system.

2. Do NOT mix different manufacturers' products.

3. Follow manufacturer's instructions.

4. Do NOT skip steps.

Corrosion Protection Safety

As with other materials used in collision repair, the use of corrosion protection materials requires that you follow safety rules. The most basic rules are:

1. Epoxy systems can create skin irritation, so wear gloves and avoid skin contact.

2. If skin contact has occurred, wash the affected area with soap and hot water. Then apply a skin cream.

3. If adhesive accidentally contacts the eyes, wash immediately with clean water for 15 minutes. Then consult a physician.

4. Spot welding in weld-bond joints can generate gases that can be harmful if inhaled. Be sure to work in a well-ventilated area and wear a respirator.

Anticorrosion Materials

Anticorrosion materials can be divided into four broad categories:

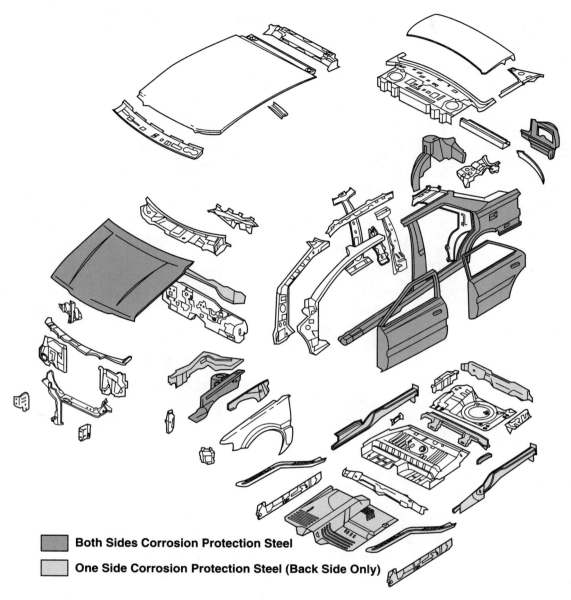

Both Sides Corrosion Protection Steel

One Side Corrosion Protection Steel (Back Side Only)

Figure 22–30 Note the corrosion protection for different panels. *(Copyright Mazda Motor of America, Inc. Used by permission)*

1. ***Anticorrosion compounds*** are either wax- or petroleum-based anticorrosion compounds resistant to chipping and abrasion. They can undercoat, deaden sound, and completely seal the surface. They should be applied to the underbody and inside body panels so that they can penetrate into joints and body crevices to form a pliable, protective film. See Figures 22–32 and 22–33.

2. ***Seam sealers*** prevent the penetration of water, mud, and fumes into panel joints. They serve the important role of preventing corrosion from forming between adjoining surfaces.

3. ***Weld-through primers*** are used between the two pieces of base metal at a weld joint.

4. ***Corrosion Converters*** change ferrous (red) iron oxide to ferric (black/blue) iron oxide. Rust converters may also contain some type of latex emulsion that seals the surface after the conversion is complete. These products offer an interesting alternative for areas that cannot be completely cleaned.

   NOTE! Some manufacturers do not recommend the use of corrosion converters.

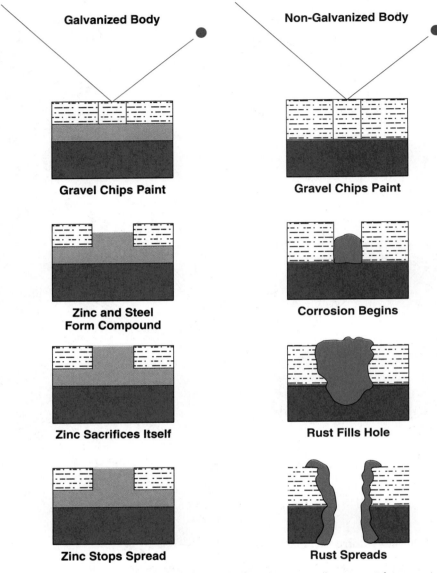

Figure 22–31 Note how corrosion attacks galvanized and nongalvanized parts. *(Volkswagen Of America)*

Applying Corrosion Protection Materials

Care is needed when applying anticorrosion compounds. Keep the material away from parts that conduct heat, electrical parts, labels and identification numbers, and moving parts. Avoid applying corrosion protection materials to:

1. Seat belt retractors and passive restraint guide rails

2. Hidden headlamp assemblies

3. Power window motors and cables

4. Exhaust system

5. Engine and accessories

6. Air filter

7. Air lift shock absorbers

8. Transmission parts

9. Shift linkages

10. Speedometer cables

11. Brake parts

12. Locks, key cylinders, and door latches

13. Power antennas

14. Theft prevention labels

15. Driveshaft

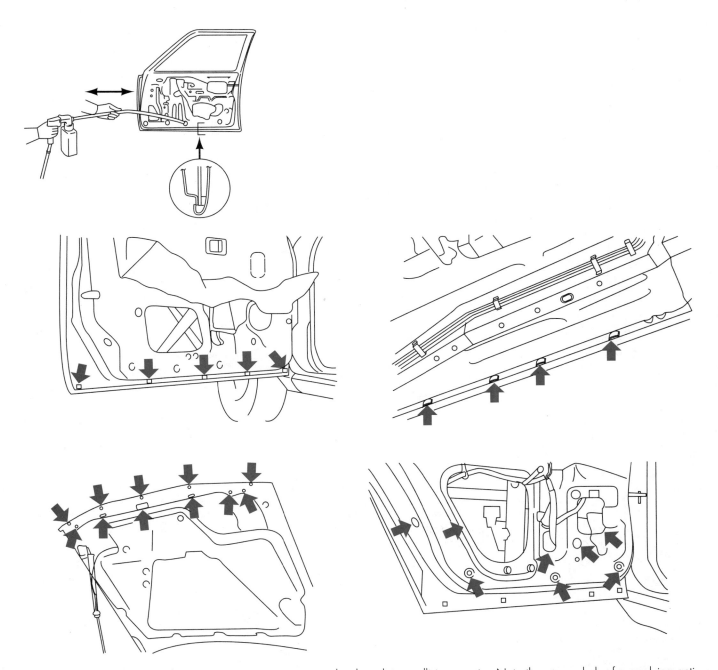

Figure 22–32 Always replace corrosion protection materials when doing collision repairs. Note the access holes for applying anticorrosion materials to the inside of panels. *(Used with permission from Nissan North America, Inc.)*

Typical enclosed and exposed interior and exterior panels are shown in Figure 22–34. The corrosion protection process for exposed joints and seams can be summed up as follows:

1. Thoroughly clean the joint or seam.

2. Apply primers and seam sealers, Figure 22–35.

3. Apply final primer coat(s).

4. Apply a topcoat.

In general, anticorrosion procedures for exposed exterior underbody surfaces are as follows:

1. Clean with a wax and grease remover. If needed, remove any loose sound-deadening materials. They can create moisture pockets for corrosion.

2. Prime with self-etch primer. You can also treat with metal conditioner and conversion coating, and prime with two-part epoxy primer.

3. Apply topcoat and sealer to primed areas or apply anticorrosion compound with a self-etch.

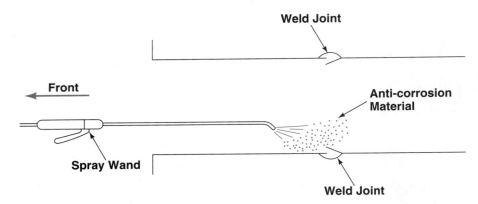

Figure 22–33 Make sure you use a spray wand to replace corrosion-inhibiting material inside panels, especially around welded joints. *(I-CAR)*

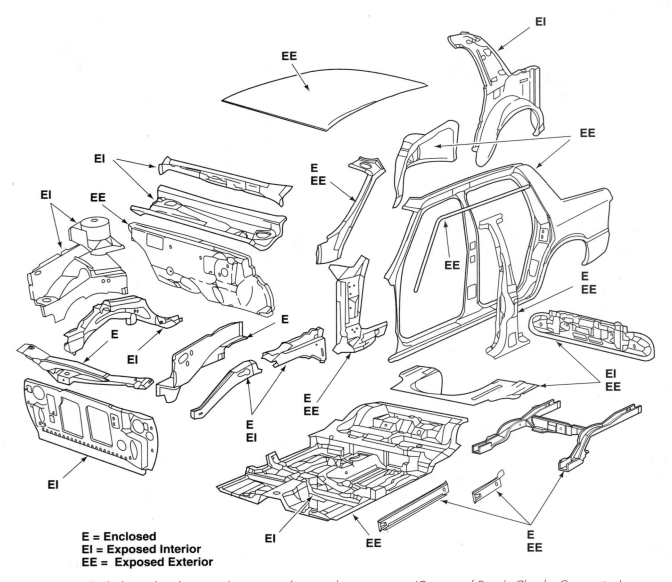

E = Enclosed
EI = Exposed Interior
EE = Exposed Exterior

Figure 22–34 Study the enclosed, exposed interior, and exposed exterior parts. *(Courtesy of DaimlerChrysler Corporation)*

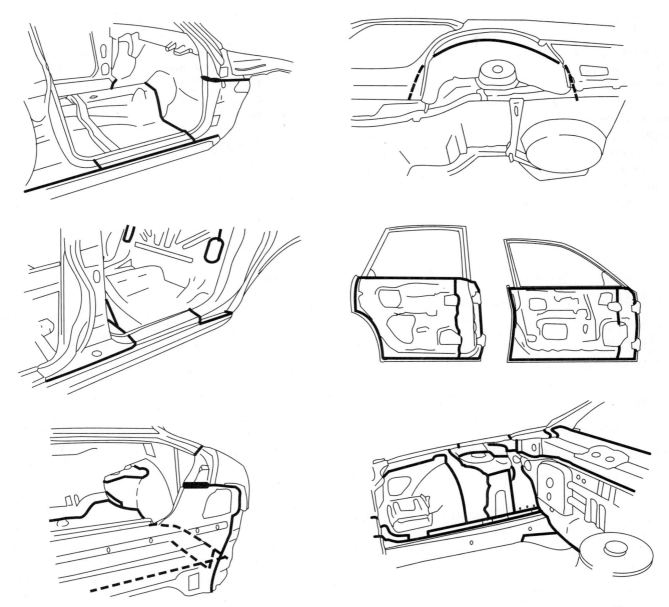

Figure 22-35 Seam sealers must be applied in a continuous bead along the edges of panels. *(Used with permission from Nissan North America, Inc.)*

Summary

- Structural parts are those parts of a vehicle that support the weight of the vehicle, absorb collision energy, and absorb road shock.
- A simple rule regarding repairing versus replacement is:
    If the part is bent, repair it.
    If the part is kinked, replace it.
- Sectioning involves cutting and replacing panels at locations other than at factory seams. Sectioning of parts such as rails, pillars, and rocker panels may be required to make their repair economically feasible.
- Full body sectioning is replacing the entire rear section of a collision damaged vehicle with the rear section of a salvage vehicle. It may be more economical than trying to rebuild a damaged vehicle with new parts.
- Corrosion breakdown falls into three general categories: paint film failure; collision damage; and the repair process.

Technical Terms

structural parts
factory seams
sectioning
panel alignment marks
recycled assemblies
inserts
offset butt joint
multiple part rocker
foam fillers
full body sectioning

jigs
weld-bond adhesives
structural adhesives
corrosion
oxidation
anticorrosion compounds
seam sealers
weld-through primers
corrosion converters

Review Questions

1. List three functions of structural parts.

2. If the part is bent, _____ it. If the part is kinked, _____ it.

3. When should you replace panels along factory seams?

4. What is sectioning?

5. You should not section in or near these eight areas.

6. How do you start removal of a structural panel?

7. List five tools that can be used to remove spot welds.

8. A chisel should never be used by itself to remove spot welds. True or false?

9. All straightening must be done before replacing panels. True or false?

10. List the six general steps for welding panels.

11. These are three basic types of sectioning joints.
    a. Lap joint
    b. Offset butt joint
    c. Butt joint with insert
    d. All of the above
    e. None of the above

12. When cutting panels, Technician A uses a template. Technician B uses a jig. Who is correct?
    a. Technician A
    b. Technician B
    c. Both Technicians
    d. Neither Technician

13. _____ _____ are used to add rigidity and strength to structural parts.

14. Avoid applying corrosion protection materials to these 15 areas or parts.

ASE–Style Review Questions

1. *Technician A* says that the core support tie bar, front rails, strut towers, and rocker panels are structural parts. *Technician B* says that pillars and the suspension crossmember are structural parts. Who is correct?
   a. Technician A
   b. Technician B
   c. Both Technicians A and B
   d. Neither Technician

2. Which of the following locations can be sectioned?
   a. Suspension mounting locations
   b. Structural part mounting locations
   c. Dimensional reference holes
   d. Uniform area with clearance

3. To find spot welds on a damaged panel, *Technician A* uses a coarse scuff wheel mounted in an air tool. *Technician B* heats the panel until the metal glows red hot to remove the paint. Who is correct?
   a. Technician A
   b. Technician B
   c. Both Technicians A and B
   d. Neither Technician

4. When removing spot welds, *Technician A* drills through the top and bottom panel. *Technician B* drills through the top panel only. Who is correct?
   a. Technician A
   b. Technician B
   c. Both Technicians A and B
   d. Neither Technician

5. When preparing a new structural panel for installation, *Technician A* applies weld-through primer to areas where bare metal is exposed. *Technician B* duplicates the location and number of spot welds used at the factory. Who is correct?
   a. Technician A
   b. Technician B
   c. Both Technicians A and B
   d. Neither Technician

6. *Technician A* says to remove the vehicle from the straightening bench before starting installation of new parts. *Technician B* says that further straightening may be needed after installation of new structural parts. Who is correct?
   a. Technician A
   b. Technician B
   c. Both Technicians A and B
   d. Neither Technician

7. When determining where to use new or recycled assemblies for structural repairs, *Technician A* says that less factory corrosion protection is disturbed

when using recycled parts. *Technician B* says that more measuring is required when welding separate new parts and attaching them to the vehicle. Who is correct?
a. Technician A
b. Technician B
c. Both Technicians A and B
d. Neither Technician

8. *Technician A* uses a butt joint with insert rocker panels, A-pillars, B-pillars, and front rails. *Technician B* uses a butt joint with insert on floor pans and hoods. Who is correct?
a. Technician A
b. Technician B
c. Both Technicians A and B
d. Neither Technician

9. A-pillars with multiple piece construction must be sectioned. *Technician A* is going to use a butt joint with insert. *Technician B* is going to use an offset butt joint. Who is correct?
a. Technician A
b. Technician B
c. Both Technicians A and B
d. Neither Technician

10. A floor is going to be sectioned. *Technician A* says to cut the replacement panel to overlap the existing panel by at least 1 inch (25 mm). *Technician B* says to tack weld the panel in place and re-measure before final welding. Who is correct?
a. Technician A
b. Technician B
c. Both Technicians A and B
d. Neither Technician

Activities

1. Inspect a damaged vehicle. List the structural parts that would require replacement.

2. On the same vehicle, summarize the procedures for structural part replacement.

3. Visit a collision repair shop. Ask the owner to allow you to watch a technician replacing a structural part.

23

Interior Repairs

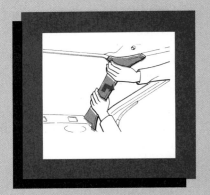

Objectives

After studying this chapter, you should be able to:

- Identify the major parts of a vehicle's interior.

- Remove and replace seats.

- Summarize how to replace seat covers.

- Replace carpeting.

- Service an instrument cluster and other dashboard parts.

- Explain how to replace headliners.

Introduction

During a collision, interior parts can be damaged, Figure 23–1. The dashboard, steering wheel, and other parts can all be damaged. This would include headliners, seats, and other related parts. Interior parts may need to be removed and reinstalled when replacing quarter panels, door panels, roof panels, and other structural parts. For this reason, you should be familiar with procedures for replacing interior parts.

NOTE! Chapter 13 describes the service of doors and window glass, and also explains how to repair air and water leaks. Chapter 27 explains the operation and repair of passive and active restraint systems. Other chapters give information relating to interior repairs; refer to the index as needed.

Figure 23–1 Interior parts can be damaged in a collision. Therefore, you should have basic knowledge of how to replace interior parts. *(Courtesy of General Motors Corporation Service Operations)*

SEAT SERVICE

Seats are often damaged during a collision. They can be damaged by the inertia of the occupant, by side impact intrusion into the passenger compartment, or by stains. You might also have to remove seats for carpet replacement or floor panel repairs.

A **bucket seat** is a single seat for one person. A **bench seat** is a longer seat for several people. Both require similar methods during service.

Seat Removal

Four bolts normally secure the seat to the floor. To remove the front seat hold-down bolts, slide the seat fully backward. This will allow easier access to the front bolts. Then, slide the seat forward to remove the two rear hold-down bolts.

After disconnecting any wiring and other parts attached to the seat, carefully lift it out of the vehicle.

A rear bench seat is often held in position by screws or spring-loaded clips. The screws are normally at the front bottom of the seat. When removed, the seat can be pushed back and lifted up and out.

With spring clips, use your hands to force the seat down and back. This will free the hidden clips and allow you to lift the bench seat out.

Seat Parts

Figure 23–2 shows the parts of a typical front seat:

1. **Seat cushion**—bottom section of the seat, which includes cover, padding, and frame.

2. **Seat back**—rear assembly that includes cover, padding, and metal frame.

3. **Headrest**—padded frame that fits into the top of the seat back.

4. **Headrest guide**—sleeve that accepts headrest post and mounts in the seat back.

5. **Recliner adjuster**—hinge mechanism that allows adjustment of the seat back to different angles.

6. **Seat track**—mechanical slide mechanism that allows seat to be adjusted forward or rearward.

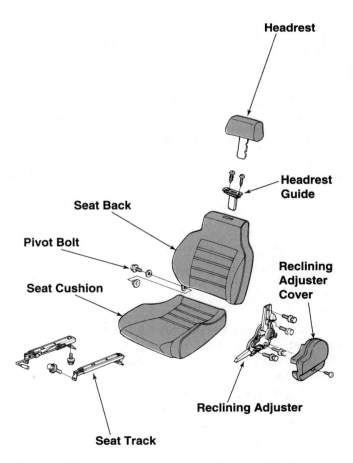

Figure 23–2 Note the major parts of a typical bucket seat. *(© American Honda Motor Co., Inc.)*

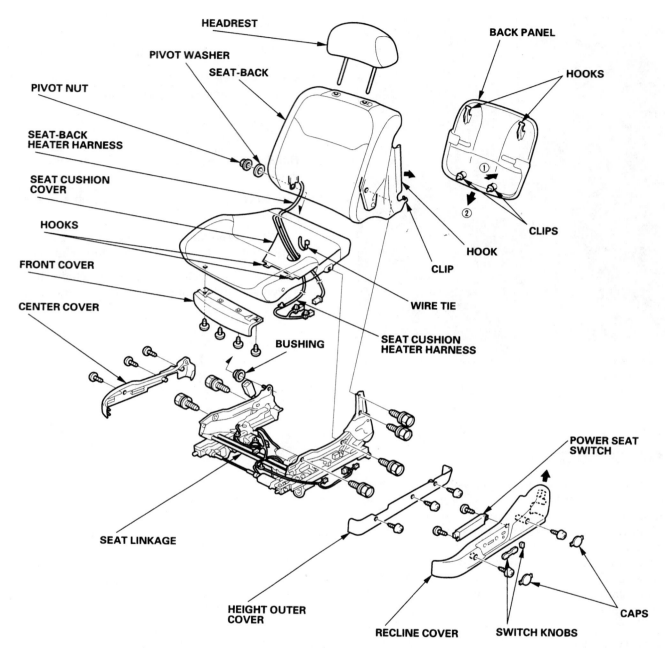

Figure 23–3 Note how the parts of this seat are assembled. This is typical of most modern designs. (© American Honda Motor Co., Inc.)

Seat Cover Service

The seat cover is a cloth, vinyl, or leather cover over the seat assembly. The cover may require replacement when damaged. You must disassemble the seat to replace the covers. See Figures 23–3.

Hog rings and clips normally stretch and hold the seat cover over the seat frame and padding. They are located on the bottom of the seat cushion or rear of the seat back. Remove them and you can lift off the seat cover. The new cover can then be installed in reverse order of removal. Refer to Figure 23–3.

NOTE! When servicing seats, refer to the manufacturer's manual for details. Procedures vary.

CARPETING SERVICE

Carpeting can be stained or torn during a collision. If it cannot be cleaned with a strong carpet cleaner or if torn, the carpeting must be replaced. The major parts of **interior carpeting** are shown in Figure 23–4.

To replace carpeting, you must remove the seats, seat belt anchors, trim pieces, and any other parts

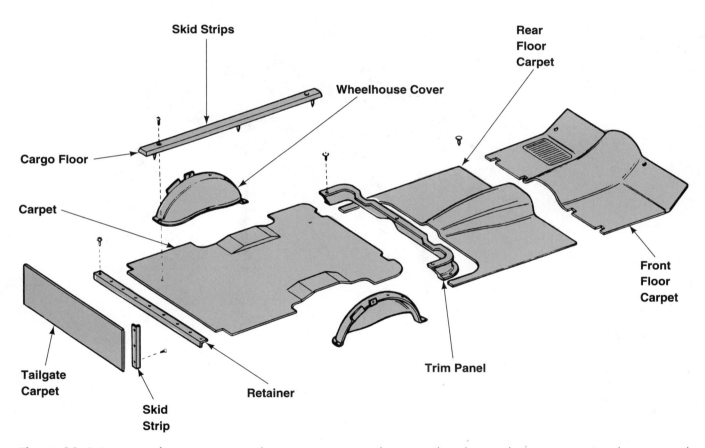

Figure 23–4 Carpeting often comes in several sections. You may not have to replace the complete carpet set. Note the covers and skid strips that help secure carpeting. *(Courtesy of DaimlerChrysler Corporation)*

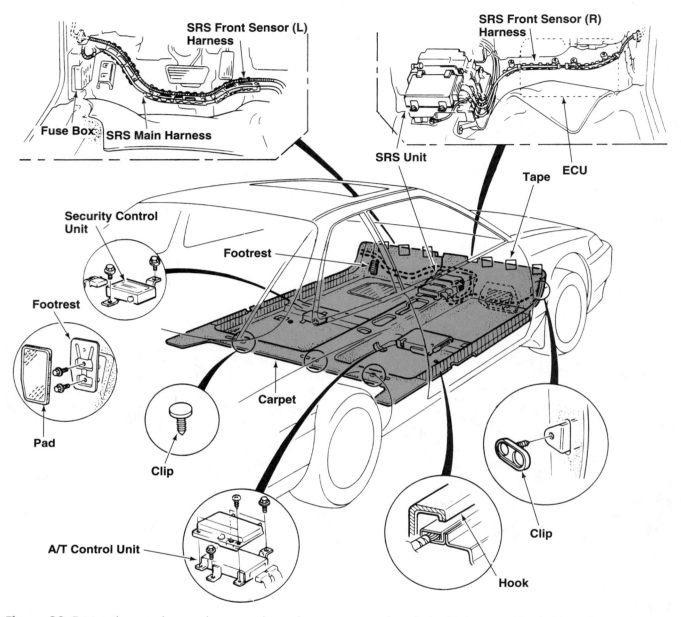

Figure 23–5 Note the parts that must be removed to replace carpeting in this vehicle. (© American Honda Motor Co., Inc.)

mounted over the carpeting. This might include the console, any electronic control units, and wiring harness bolted down to the carpet. Screws and clips hold these parts and the carpet down into place. See Figure 23–5.

After removal, the new carpet is installed in the reverse order of removal. Make sure the new carpet is stretched out smooth and properly centered before installing any fasteners. An adhesive may be required between the carpet and floor in some locations. Refer to the service manual if in doubt.

HEADLINER SERVICE

A **headliner** is a cloth or vinyl cover over the inside of the roof in the passenger compartment. It can be torn or damaged during a collision. Some thick cloth or vinyl covered foam headliners are bonded directly to the roof panel. Others are thin vinyl suspended by metal rods and bonded around the edges of the roof.

To service a headliner, first remove all of the trim pieces around the edges of the roof. Various screws

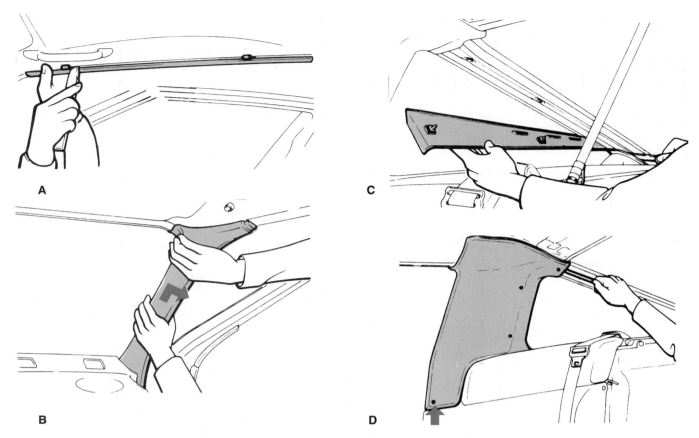

Figure 23–6 These are parts that typically must be removed to replace a headliner. *(Courtesy of DaimlerChrysler Corporation)* (A) Remove the roof side trim. (B) Remove the front pillar trim. (C) Remove the rear pillar trim. (D) Remove the upper quarter trim.

and clips secure the trim pieces. Refer to Figure 23–6. You may also have to remove the sun visors, grab handles, and other parts for headliner service, Figure 23–7.

When installing a foam-backed headliner, be careful not to overbend and kink it. Center it in position. Then, install it in reverse order of removal. Again, refer to the service manual if in doubt.

INSTRUMENT PANEL SERVICE

The *instrument panel* bolts in the front of the passenger compartment on the firewall or cowl. It contains the instrument clusters, radio, glove box, vents, vinyl or leather-covered pad, and other parts.

When the instrument panel is damaged in a collision, it must be removed and replaced. An exploded view of a typical instrument panel is shown in Figure 23–8. Study the relationship of the parts.

Many instrument panel parts can be replaced without unbolting the dash pad. The instrument cluster, vents, and many trim pieces can be removed and replaced with the main part of the dash intact. Vents often snap into place. A thin screwdriver can often be used to release and remove most vents.

Some of the screws and bolts that secure the instrument panel parts can be difficult to find and remove. Some are along the bottom of the dash. Others are on the sides. A few fasteners can be inside openings in the instrument panel. You will have to remove parts to access these fasteners. See Figure 23–9, page 471.

INSTRUMENT CLUSTER SERVICE

An *instrument cluster* contains the speedometer, gauges, indicating lights, and similar parts. It may require service when damaged in a collision or when parts are not working.

To service an instrument cluster, first disconnect the battery. This will prevent the chance of an electrical fire if wires short to ground.

Remove the instrument panel cover. Several screws secure it. Next, remove the screws that hold the cluster to the dash. Pull the cluster out far enough to disconnect the wires and speedometer cable. Then lift the cluster out. Bulbs can be replaced from the rear of the cluster.

To replace gauges or the speedometer, you must disassemble the cluster. To disassemble the instrument cluster, remove the small screws that hold the plastic lens plate over the housing.

> *WARNING! Keep fingers off the inside of the instrument lens. Fingerprints on the inside of the lens can collect dust that cannot be wiped off after you reinstall the cluster.*

With the lens removed, you can then replace gauges and speedometer head. Screws on the rear of the cluster normally hold each unit in place. Again, keep fingerprints off the faces of the gauges and speedometer. They will show up easily after installation.

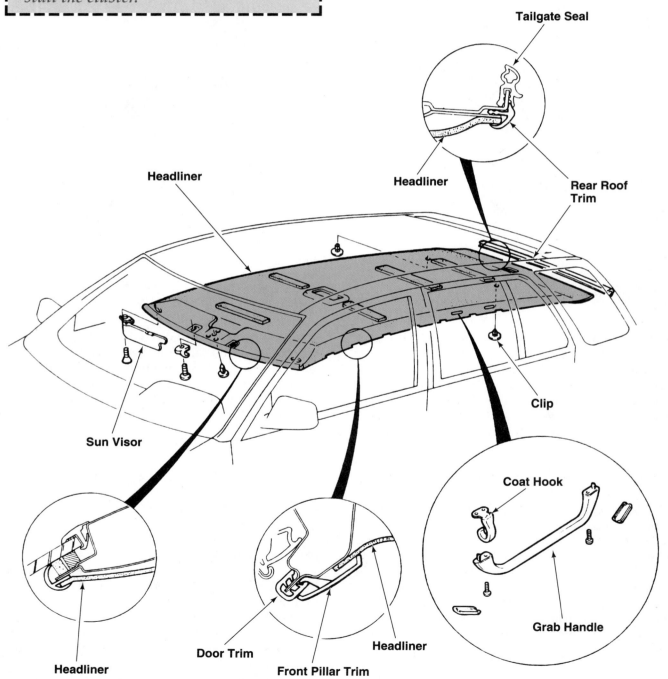

Figure 23–7 This is a thick foam-type headliner. Note the parts that must be serviced to replace the headliner. Also note how the headliner is held secure by trim pieces. (© American Honda Motor Co., Inc.)

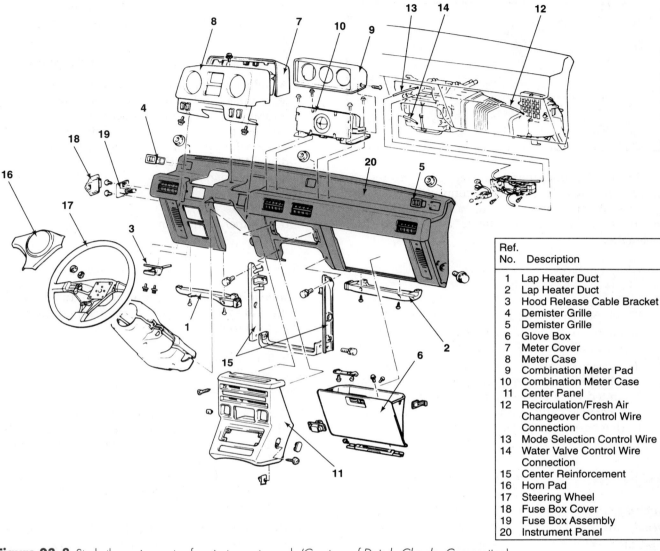

Ref. No.	Description
1	Lap Heater Duct
2	Lap Heater Duct
3	Hood Release Cable Bracket
4	Demister Grille
5	Demister Grille
6	Glove Box
7	Meter Cover
8	Meter Case
9	Combination Meter Pad
10	Combination Meter Case
11	Center Panel
12	Recirculation/Fresh Air Changeover Control Wire Connection
13	Mode Selection Control Wire
14	Water Valve Control Wire Connection
15	Center Reinforcement
16	Horn Pad
17	Steering Wheel
18	Fuse Box Cover
19	Fuse Box Assembly
20	Instrument Panel

Figure 23–8 Study the major parts of an instrument panel. *(Courtesy of DaimlerChrysler Corporation)*

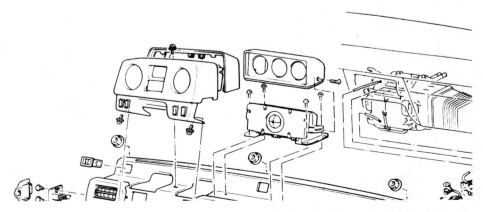

Figure 23–9 Bolts and screws that hold the instrument panel can be hard to find. The service manual will give details about fastener locations. *(Subaru of America and Fuji Industries, Ltd.)*

WARNING! Make sure the speedometer reading of the new or replacement unit is the same as the old one. You are breaking a federal law if you alter a speedometer reading. Check local laws for rules pertaining to reporting speedometer service.

Install the instrument cluster parts in reverse order of removal. Remember to connect all wires and the speedometer cable, if used, to the cluster. Check the operation of all dash lights and gauges after installation.

Summary

- The typical parts of a front seat include the seat cushion, seat back, headrest, headrest guide, recliner adjuster, and seat track.
- To replace carpeting, you must remove the seats, seat belt anchors, trim pieces, and any other parts mounted over the carpeting. This might include the console, any electronic control units, and wiring harness bolted down to the carpet.

Technical Terms

bucket seat	*recliner adjuster*
bench seat	*seat track*
seat cushion	*hog rings*
seat back	*headliner*
headrest	*instrument panel*
headrest guide	*instrument cluster*

Review Questions

1. How do you remove a front bucket seat?

2. List and explain the six basic parts of a front bucket seat.

3. These are used to stretch and hold the seat covers on the seat.
   a. Hog rings
   b. Screws
   c. Bolts
   d. Adhesive

4. To replace carpeting, you must typically remove which parts?

5. A _____ is a cloth or vinyl cover over the inside of the roof in the passenger compartment.

6. List the typical parts of an instrument panel.

7. A vehicle needs to have its instrument cluster lens replaced. Technician A says to keep fingers off the inside of the instrument lens. Technician B says to disconnect wires and the speedometer cable before pulling the cluster too far out of the dash. Who is correct?
   a. Technician A
   b. Technician B
   c. Both Technicians
   d. Neither Technician

ASE–Style Review Questions

1. *Technician A* says that four bolts normally secure the seat to the floor. *Technician B* says the seats are welded to the floor. Who is correct?
   a. Technician A
   b. Technician B
   c. Both Technicians A and B
   d. Neither Technician

2. *Technician A* says that rear bench seats can be held in position by screws. *Technician B* says that spring-loaded clips can hold them. Who is correct?
   a. Technician A
   b. Technician B
   c. Both Technicians A and B
   d. Neither Technician

3. *Technician A* says that seat covers are usually bonded to the cushions. *Technician B* says that hog rings and clips hold the seat covers in place. Who is correct?
   a. Technician A
   b. Technician B
   c. Both Technicians A and B
   d. Neither Technician

4. An instrument cluster is being repaired. *Technician A* says to first disconnect the battery. *Technician B* says to pull the cluster out far enough to disconnect the wires and speedometer cable. Who is correct?
   a. Technician A
   b. Technician B
   c. Both Technicians A and B
   d. Neither Technician

5. *Technician A* says to keep your hands clean when touching the back of an instrument cluster lens. *Technician B* says to keep your fingers off the inside of the instrument lens at all times. Who is correct?

a. Technician A
b. Technician B
c. Both Technicians A and B
d. Neither Technician

6. When installing a new speedometer cluster, *Technician A* says that you are breaking a federal law if you alter a speedometer reading. *Technician B* says to check local laws for rules pertaining to reporting speedometer service. Who is correct?
a. Technician A
b. Technician B
c. Both Technicians A and B
d. Neither Technician

7. *Technician A* says that to replace carpeting the technician must remove the seats, seat belt anchors, trim pieces, and any other parts mounted over the carpeting. *Technician B* says that the carpeting is put in the vehicle last and can be removed with just removing the door trim plates. Who is correct?
a. Technician A
b. Technician B
c. Both Technicians A and B
d. Neither Technician

8. *Technician A* says that seats are seldom damaged in a collision and they never need to be checked. *Technician B* says that the seats should be checked not only for visible damage but they should also be operated through their complete range of movement to assure full operation. Who is correct?

a. Technician A
b. Technician B
c. Both Technicians A and B
d. Neither Technician

9. *Technician A* says that all headliners are bonded directly to the roof. *Technician B* says that some are mounted on metal rods and bonded around the edge of the roof. Who is correct?
a. Technician A
b. Technician B
c. Both Technicians A and B
d. Neither Technician

10. When removing an instrument cluster, *Technician A* says that some of the fasteners may be hard to locate and remove. *Technician B* says that the service manual will give locations of the fasteners. Who is correct?
a. Technician A
b. Technician B
c. Both Technicians A and B
d. Neither Technician

Activities

1. Inspect several vehicles. List any damage or deterioration to the interior. Make a damage report.

2. Remove and replace vehicle seats. Refer to a service manual for specific directions.

24

Mechanical Repairs

Objectives

After studying this chapter, you should be able to:

- Explain the basics of front, rear, and computer-controlled suspension systems.

- Describe the design and operation of parallelogram, rack-and-pinion, power, and four-wheel steering systems.

- Differentiate between drivetrain and suspension problems and those caused by a defective CV-joint.

- Understand how various brake systems work, and describe the procedures for manual and pressure bleeding.

- Perform key cooling and air-conditioning system repairs and maintenance.

- Inspect an exhaust system, and describe the guidelines for working on an emission control system.

Introduction

After a collision, it is often necessary to remove mechanical parts. This would include suspension, steering, drivetrain, and engine parts. See Figure 24–1.

With a front hit, parts right behind the grille are often damaged. The air conditioning condenser, radiator, engine mounts, water pump, engine accessory units, ABS controller, and other mechanical parts can be damaged. With body panels removed, it is much easier to service these parts.

If a vehicle hits a curb or other stationary object with its tires during a collision, tremendous force is transmitted through the wheels and into the steering and suspension systems. Control arms, steering rods, and related parts are often damaged. Other parts on the bottom of the vehicle (engine oil pan, transmission pan, exhaust system, etc.) can also be damaged from the impact.

Today, a collision repair technician should be able to service these assemblies. Some areas of the country require collision repair technicians to be certified to perform mechanical repairs.

Figure 24–1 Mechanical parts are often damaged in a collision. Collision repair technicians should be able to remove and replace most mechanical parts on a vehicle. This technician is replacing rear suspension system parts damaged by a "side hit."

WARNING! When diagnosing and repairing mechanical parts, always refer to the service manual for the specific vehicle. Never attempt to work with mechanical components without the aid of a service manual. It will give the detailed procedures and specifications. See Figure 24–2.

NOTE! Since many mechanical systems are electronically or computer controlled, refer to Chapter 26. This chapter covers these subjects.

FIGURE 24–2 Front impact will often damage the condenser and radiator. Note minor damage to the radiator core. The technician has used masking tape to hold the oil cooler up and out of the way while moving the vehicle and working on it.

POWERTRAIN

The **powertrain/drivetrain** is all of the parts that produce and transfer power to the drive wheels. This includes the engine, transmission or transaxle, drive axle, and other related parts.

Engine

The **engine** provides energy to move the vehicle and power all accessories, Figure 24–3. Most vehicles use gasoline engines while some use diesel engines. The basic parts of a typical internal combustion, piston engine include:

Block—the foundation of the engine; all the other engine parts are either housed in or attached to the block. A **cylinder** is a round hole bored (machined) in the block that guides piston movement.

Piston—transfers the energy of combustion (burning of an air-fuel mixture) to the connecting rod. **Rings** are circular seals installed around the top sides of the piston. They keep combustion pressure and oil from leaking between the piston and cylinder wall (cylinder surface). A **connecting rod** is a link that attaches the piston to the crankshaft.

Crankshaft—changes reciprocating (up and down) motion of the piston and rod into more useful rotary (spinning) motion. Power to turn the driving wheel comes from the rear of the crank and accessories are driven off the front.

Cylinder head—covers and seals the top of the cylinder. It contains valves, rocker arms, and sometimes the camshaft. The **combustion chamber** is a small enclosed area between the top of the piston and bottom of the cylinder head. The burning of the air-fuel mixture occurs in the combustion chamber.

Valves—flow control devices which open to allow air-fuel mixture into and exhaust out of the combustion chamber. Valve springs hold the valves closed when they do not need to be open. They also return the valve train parts to the at-rest position.

Camshaft—controls the operation of the valves. It can be located in the block or the cylinder head. A **lifter** is a cylindrical part that rides on the camshaft lobes and transfers motion to the push rods. The **push rods** are hollow tubes that transfer motion from the lifters to rocker arms. The **rocker arms** are levers that transfer camshaft action from the push rods to the valves.

Flywheel—a heavy metal disc used to help keep the crankshaft turning smoothly. It also connects engine power to the transmission. A larger gear on the outside of the flywheel engages the starting motor when cranking the engine for starting.

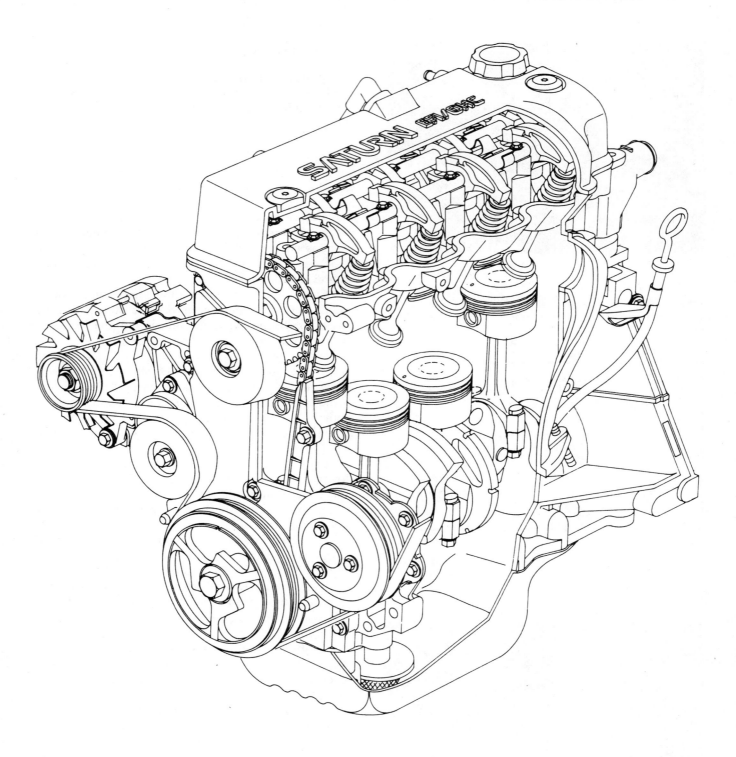

Figure 24-3 Study the basic parts of an engine. The intake valve allows air and fuel to enter the combustion chamber. When the mixture burns, it forces the piston and rod down. This spins the crankshaft to produce usable power for the drivetrain. *(Courtesy of General Motors Corporation, Services Operations)*

ENGINE SUPPORT SYSTEMS

Various **engine support systems** are powered by the engine to protect the engine from damage and to power accessory systems.

Lubrication System

The *lubrication system* forces oil to friction points in the engine. This keeps the moving parts from quickly wearing and failing.

The **oil pump** forces motor oil through passages inside the engine. It can be driven by the crankshaft or by a gear on the camshaft. Oil **galleries** are the passages through the block, heads, and other parts for oil flow through the engine. An **oil pickup** is a tube with a filter screen for drawing oil out of the pan and into the pump. A **pressure relief valve** limits the maximum amount of oil pressure.

The *oil pan* holds a supply of motor oil. Also called the **sump,** it often bolts to the bottom of the engine block. An oil drain plug is provided in the oil pan for draining and changing the engine oil. The *oil filter* traps debris and prevents it from circulating through the engine oil galleries.

Lubrication System Service. In a collision, the oil filter, oil pan, and related parts are sometimes damaged. They can be made of thin metal and can be crushed and ruptured easily.

When starting an engine before or after repairs, check the oil level with the *oil dipstick,* Figure 24–4. Also, always look under the vehicle for oil leakage. If you find an oil leak, shut the engine off immediately. Find and fix the source of the oil leak.

Cooling System

A *cooling system* maintains the correct engine operating temperature. It is often damaged in a collision and must be restored to its precollision condition. Shown in Figure 24–5, the basic parts of a cooling system are:

Antifreeze is used to prevent freeze-up in cold weather and to lubricate moving parts. Antifreeze also prevents engine overheating. A **coolant recovery system** stores an extra supply of coolant for the system.

The *radiator* transfers coolant heat to the outside air. The *radiator pressure cap* prevents the coolant from boiling. A *radiator fan* draws outside air through the radiator to remove heat.

The *water pump* circulates coolant through the inside of the engine, hoses, and radiator. The **water jackets** are passages in the engine for coolant. The *thermostat* regulates coolant flow and system operating temperature.

A *heater system* uses coolant heat and a heater core (small radiator under the dash) to warm the passenger compartment. The **automatic transmission cooler** uses the radiator to reduce automatic transmission fluid temperature.

Cooling System Service. Cooling system problems normally result in engine overheating. In collision repair, problems may be due to a crushed water pump, bent drive pulleys, thermostat housing, or large leak. Always inspect closely for these problems and proper coolant level before starting the engine.

Antifreeze should NOT be reused. Antifreeze contains additives, lubricants, and corrosion inhibitors that break down over time. Antifreeze should be replaced with a 50/50 mixture of antifreeze and water. Always follow antifreeze manufacturer's recommendations. Some warranties will NOT be honored if the antifreeze recommended by the manufacturer is NOT installed.

An *antifreeze tester,* commonly called a *hydrometer,* is used to determine the freeze-up protection of the coolant mixture. Pull a sample of the vehicle's coolant solution into the tester. Then, read the lowest temperature the coolant will withstand without freezing.

Cooling fans are either belt driven or electric. Their operation is critical for proper cooling of both the cooling and air conditioning systems.

When inspecting belt driven fans, inspect the belts for cracks, tears, glazing, and proper tension. Check the fan blades for cracks and the fan clutch for leaks.

When inspecting electric fans, check electrical connections for corrosion and reconnection following repairs. Look for cut, pinched, or burned wires. Make sure the fan comes on when the engine reaches normal operating temperature and when the A/C is turned on.

Radiator caps should be inspected for calcium deposits, which could prevent the cap from operating. They are rated to maintain different pressures and should be replaced according to the pressure recommended by the manufacturer.

A *radiator cap pressure test* is done using a cooling system pressure tester, as illustrated in Figure 24–6. The tester gauge should stop increasing its pressure reading when the cap rating is reached.

The *radiator cap pressure rating* is stamped on the cap. If the cooling system is disassembled during repair or parts are replaced, a pressure test should be performed.

A *cooling system leak test* is performed by installing the tester on the radiator neck. Pump the

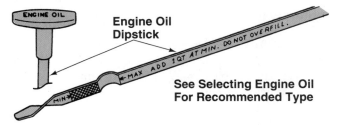

ENGINE OIL

Engine Oil Dipstick

MAX ADD 1QT AT MIN. DO NOT OVERFILL.

MAX ADD 1QT AT MIN.

MIN

See Selecting Engine Oil For Recommended Type

Figure 24–4 The dipstick should show engine oil even with the FULL or MAX mark. *(Courtesy of DaimlerChrysler Corporation)*

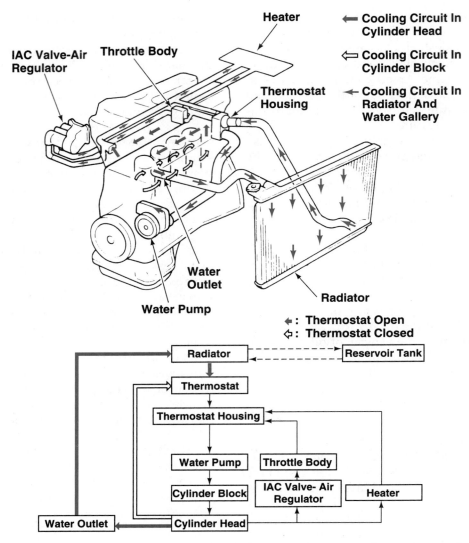

Figure 24–5 Study the basic parts of a cooling system and trace flow. *(Used with permission from Nissan North America, Inc.)*

tester handle until its gauge equals the cap pressure rating. A loss of pressure or coolant means there is a leak, Figure 24–6.

When mounted on the front or side of the engine, the thermostat housing can also be damaged during a collision.

Exhaust System

The **exhaust system** collects and discharges exhaust gases caused by the combustion of the air/fuel mixture within the engine. It also quiets the noise of the running engine.

The **header pipe** is steel tubing that carries exhaust gases from the engine's exhaust manifold to the catalytic converter. The **catalytic converter** is a thermal reactor for burning and chemically changing

exhaust by-products into harmless gases. The **intermediate pipe** is tubing that is sometimes used between the header pipe and catalytic converter or muffler.

A **muffler** is a metal chamber for dampening pressure pulsations to reduce exhaust noise. The **tailpipe** is a tube that carries exhaust gas from the muffler to rear of the vehicle.

Exhaust System Service. The exhaust system can also be damaged during a collision, requiring partial replacement. Its parts may also need removal during major collision repairs.

DANGER! When inspecting or working on the exhaust system, remember that its parts get very hot when the engine is running. Contact with them could cause a severe burn.

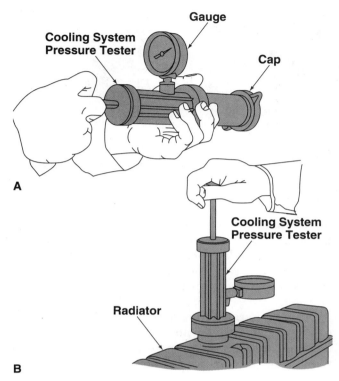

Figure 24–6 A cooling system pressure tester is commonly used to check the condition of the system. *(Courtesy of Daimler-Chrysler Corporation)* (A) With the tester mounted on the radiator, pump in the cap-rated pressure. Then, check for leakage. The system should hold the pressure. (B) Mount the cap on the tester and pump the handle. The cap should release pressure at its rating. Cap pressure rating is printed on the cap.

Because of constant changes in recommended catalytic converter servicing and installation requirements, check with the vehicle manufacturer for the latest data regarding replacement.

To check the exhaust system's condition, grab the tailpipe (when cool). Try to move it up and down and side to side. There should be only slight movement in any direction.

To check further, start the engine (never in a closed shop). Stuff a rag in the tailpipe and feel around every joint for leaks. If one is found, try tightening the clamp. If this does not stop the leak, it must be repaired.

If needed, raise the vehicle. Check the clamps and hangers that fasten the exhaust system to the underbody. Also, jab at all corroded areas in the system with an old screwdriver. If the blade sinks through the metal at any point, that part is badly corroded. You can also tap on parts with a hammer or mallet. A ringing sound indicates that the metal is good. A badly corroded part gives out a dull thud, from thinned metal. If the vehicle is loud, there is usually a large leak or an internally corroded muffler.

Remember! There is only one way to repair faulty exhaust system parts: replace them. It might not be necessary to take off all exhaust system parts. You can usually separate parts and replace them individually.

Fuel System

The vehicle's *fuel system* must carefully feed the correct amount of fuel into the engine. If too much or too little fuel is admitted, the engine will NOT run efficiently.

A *gasoline injection system* has sensors and a computer to control electrically operated fuel valves, called *fuel injectors.* The injectors are located in the intake manifold. When the intake valve opens, the fuel is sprayed into the intake port and pulled into the combustion chamber by air flow.

A *diesel injection system* uses a high pressure, mechanical pump to force fuel directly into the engine's combustion chambers. No spark plugs are needed. The pistons squeeze the intake air which gets heated enough to start the fuel burning.

A **fuel tank** is a container that stores fuel. The *fuel pump* is a mechanical or electric part for forcing fuel to engine. **Fuel lines** are tubing and hoses that route fuel from tank to engine. A *fuel filter* is used for straining out debris in the fuel. The *fuel pressure regulator* is a part that controls the amount of fuel pressure at the fuel injectors.

DANGER! Before disconnecting any part of a fuel system, you may have to relieve fuel pressure. Many fuel injection systems retain pressure even when the engine is not running. If this pressure is not relieved, it will cause fuel to spray in the work area and create a fire hazard.

DRIVETRAIN

The *drivetrain* uses engine power to turn the drive wheels. It includes everything after the engine—the clutch, transmission, drive shaft, drive axles. Drivetrain designs vary. Some cars use a *manual transmission* (hand shifted). Others use an *automatic transmission* (shifts gear automatically using internal oil pressure).

The *transmission* is an assembly with a series of gears for increasing torque to the drive wheels so the car can accelerate properly. It provides high power for acceleration in lower gears and good gas mileage in higher gears. With an automatic transmission, a *torque converter* (fluid coupling) is used in place of a clutch.

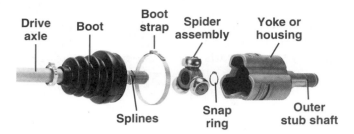

Figure 24–7 Study the parts of a tripod CV-joint. It is sometimes used as the inboard joint. *(Illustration Courtesy of Perfect Circle/Dana Corp.)*

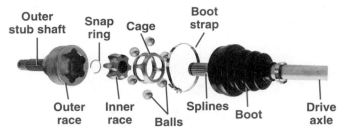

Figure 24–8 Study the parts of a ball and cage or Rzeppa CV-joint. It is used as the outboard, and sometimes the inboard, joint. *(Illustration Courtesy of Perfect Circle/Dana Corp.)*

A ***transaxle*** is a transmission and differential combined into a single housing or case. Both automatic and manual transaxles are available.

A ***clutch*** is a device used to couple and uncouple engine power to a manual transmission or transaxle. It uses a friction disc, pressure plate, flywheel face, and release bearing for activation.

Front-wheel drive vehicles use a transaxle to transfer engine torque to the front drive wheels. **Constant velocity axles** or ***CV-axles*** transfer torque from the transaxle to the wheel hubs. They can be found on rear-wheel drive (RWD), four-wheel drive (4WD), all-wheel drive (AWD) vehicles, and front-wheel drive (FWD) vehicles. Refer to Figures 24–7 and 8.

Front-engine, ***rear-wheel drive*** vehicles use a conventional transmission, drive shaft, and rear axle assembly to transfer power to the rear drive wheels. Look at Figure 24–9.

A ***drive shaft*** is a long tube that transfers power from the transmission to rear axle assembly. It has

universal joints at both ends that allows flexibility of the suspension while maintaining driving force.

The ***rear axle assembly*** is the housing that contains the ring gear, pinion gear, differential assembly, and axles. Rear suspension springs attach to the housing.

A ***differential assembly*** is a unit within the drive axle assembly. It uses gears to allow different amounts of torque (turning force) to be applied to each drive wheel while the vehicle is making a turn.

Drivetrain/Powertrain Service

Begin drivetrain/powertrain inspection by checking the condition of the CV-joint boots. Splits, cracks, tears, punctures, or thin spots require replacement. If the boot appears corroded, this indicates improper greasing or excessive heat. Squeeze-test all boots. If any air escapes, replace the boot. Also replace any boots that are missing.

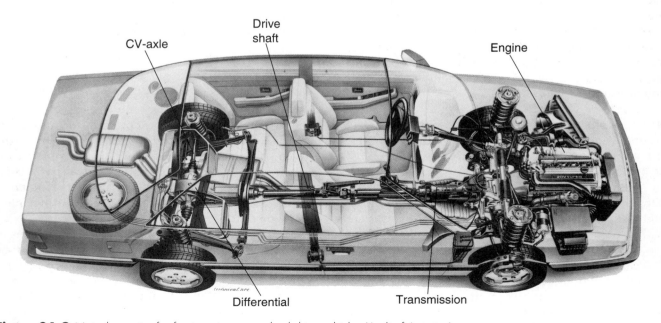

Figure 24–9 Note the parts of a front engine, rear-wheel drive vehicle. *(Audi of America)*

Keep in mind that any discoloring of the housing at the bearing grooves is normal. All CV-joints are heat treated.

If the inner boot appears to be collapsed or deformed, venting it (allowing air to enter) might solve the problem. Place a round-tipped rod between the boot and drive shaft. This equalizes the outside and inside air and allows the boot to return to its normal shape.

Make sure that all boot clamps are tight. Missing or loose clamps should be replaced. If the boot appears loose, slide it back and inspect the grease inside for possible contamination.

A milky or foamy appearance indicates water contamination. A gritty feeling when rubbed between the fingers indicates dirt. In either case, repack the joint with new grease.

On front-wheel drive transaxles with equal length half shafts, inspect the intermediate shaft U-joint, bearing, and support bracket for looseness by rocking the wheel back and forth and watching for any movement.

Various drivetrain and suspension problems can be confused with symptoms produced by a bad CV-joint. The following list of symptoms should help guide the technician to a proper diagnosis.

A vibration that increases with speed is rarely due to CV-joint problems or FWD half shaft imbalance. An out-of-balance tire or wheel, an out-of-round tire or bent rim are the more likely causes. It is possible that a bent half shaft as a result of collision or towing damage could cause a vibration, as could a missing damper weight.

A drive shaft should be checked for signs of contact against the chassis or rubbing. Rubbing can be a symptom of a weak or broken spring or engine mount, or chassis misalignment.

It may be required to remove the drivetrain to make structural repairs. Because modern unibody vehicles tend to have very crowded engine compartments, removal of the drivetrain allows ready access to structural panels.

When servicing parts of a drivetrain in collision repair, always refer to the service manual. It will give the instructions and specs needed to do proper repairs.

Label all wires, hoses, and other parts to help with reassembly, Figure 24–10. This will save time and prevent confusion later.

Sometimes, only engine and transmission removal are needed during collision repair. This might be needed for major frame or unibody straightening or structural part replacement. See Figure 24–11.

When removing a drive shaft, mark its alignment in the rear axle assembly. Some shafts are factory balanced on the vehicle. If you change its orientation, vibration can result.

Proper drive shaft angles must also be maintained to prevent vibration and chatter of the drive shaft and CV-joints. An angle gauge may be needed to check for drive shaft alignment.

Brake System

The **brake system** uses hydraulic pressure to slow or stop wheel rotation with brake pedal application.

The **brake pedal** transfers the driver's foot pressure into the master cylinder. The **master cylinder** develops **hydraulic pressure** (oil pressure) for the system.

Brake lines and **hoses** carry fluid out to the wheel cylinders. The **wheel cylinders** use hydraulic pressure to push the brake pads or shoes outward.

The **brake pads** or **shoes** have a friction lining for rubbing on the brake rotor or drum. The **brake rotors** or **drums** provide heavy metal friction surfaces bolted between the hub and wheel. A **caliper** holds the pis-

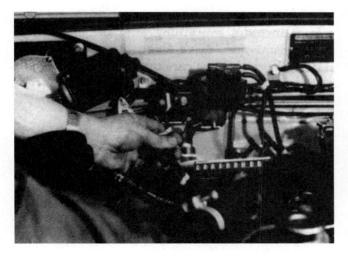

Figure 24–10 To avoid later confusion, label wires and hoses as they are disconnected during service of mechanical-electrical assemblies.

Figure 24–11 Note the construction of this modern four-wheel drive design. The transfer case attaches to the rear of the automatic transmission. The drive shaft extends to the rear axle assembly. A short drive shaft extends along the side of the engine to the front differential. (Mercedes-Benz)

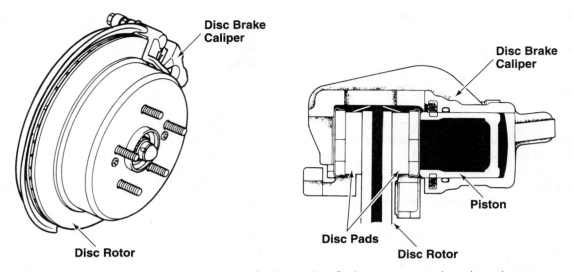

Figure 24–12 With disc brakes, the caliper mounts over the disc. When fluid pressure enters the caliper, the piston is pushed outward. This applies the brake pads to the spinning rotor or disc to slow or stop the vehicle. *(Reprinted with permission)*

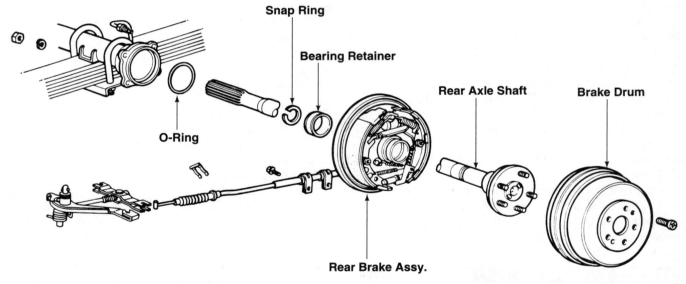

Figure 24–13 With drum brakes, the drum mounts over the brake shoes and wheel cylinders. The wheel cylinders push out on the shoes to apply the brakes. *(Reprinted with permission)*

ton(s) and brake pads on disc brakes. Refer to Figures 24–12 and 13.

The **parking** or **emergency brake** uses a steel cable to physically apply the brake shoes or pads.

Power brakes is a standard hydraulic brake system with a vacuum, hydraulic, or electric assist. A booster unit is added to help apply the master cylinder and brakes.

Brake System Service

When applicable, inspect the brake system for:

1. Kinked or bent brake lines

2. Cut hoses

3. Damaged rotors or drums

4. Backing plate interfering with drum

5. Damage to caliper or mounting area

6. Damage to master cylinder or booster

7. Dash lamp operation if equipped with ABS

Brake lines are seamless steel, reinforced flexible rubber, or nylon. They use two types of flared ends, ISO flare or double flare. Brake lines must be replaced with the same material as used by the factory. Brake lines should be replaced if kinked or severely bent.

When servicing brake lines:

1. Do NOT use copper, aluminum, or rubber hoses.

2. Do NOT repair nylon brake lines.

3. Do NOT use compression fittings.

4. Route in the same locations as the factory.

5. Reinstall all supporting clamps and springs that were removed during repair.

Brake fluid absorbs moisture. Its container should NOT be left open to the atmosphere. Make sure to use the proper type of brake fluid. Vehicles can require DOT 3, 4, or 5. Check the master cylinder label or markings for the type used.

Do NOT reuse fluid drained from system. It may contain contaminants that will damage the system.

Brake system bleeding is done to remove air from the brake fluid. To do this:

1. Clean the bleeder screw.

2. Attach a drain hose to the bleeder screw and submerge the other end in a clear jar partially filled with brake fluid.

3. Bleed one wheel at a time.

4. Start with the wheel closest to, or farthest from, the master cylinder (follow the manufacturer's recommendations).

5. Have a helper pump the brake pedal several times, then hold it with moderate pressure.

6. Slowly open the bleeder screw.

7. Allow fluid/air to flow through it, then close the bleeder screw.

8. Repeat as needed until fluid runs clear without air.

9. Repeat at all wheels, following manufacturer's recommended order, making sure that the master cylinder does not run dry.

10. Refill the master cylinder.

Antilock Brake System Service

Antilock brakes service is similar to conventional brakes. However, electronic parts are added to operate the system. Most ABS brakes have self-diagnosis. The computer will output a trouble code if an electrical-electronic malfunction develops. You can refer to charts in the service manual to see what each number code means. This will tell which part might be at fault.

For example, if a trouble code indicates a problem with one of the wheel speed sensors, make sure it is adjusted properly and undamaged. You may also need to test the sensor and its wiring. See Figure 24–14.

NOTE! For more information on antilock brake system service and trouble codes, refer to Chapter 26.

WARNING! When installing the wheels back onto a vehicle, torque the lug nuts to specs. Use a crisscross pattern as shown in Figure 24–15. The service manual will give lug nut torque values.

STEERING SYSTEMS

The **steering system** transfers steering wheel motion through gears and linkage rods to swivel the front wheels. When you turn the **steering wheel**, a **steering shaft** that extends down through the **steering column** rotates the steering gearbox. The **steering gearbox**, either a worm or rack-and-pinion type, changes the wheel rotation into side movement for turning the wheels. A series of **linkage rods** connect the steering gearbox with the steering knuckles.

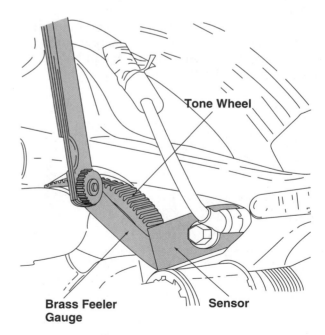

Figure 24–14 When servicing antilock brakes, refer to the service manual for details. Some systems have specifications for clearance between the tone wheel and the wheel speed sensor. Use a brass feeler gauge when adjusting clearance. *(Courtesy of DaimlerChrysler Corporation)*

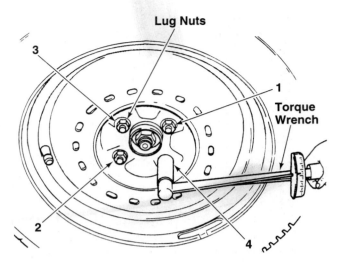

FIGURE 24–15 To avoid warpage of the brake drum, rotor, hub, and wheel, torque the lug nuts to specifications in a criss-cross pattern. *(Courtesy of DaimlerChrysler Corporation)*

Rack-and-Pinion Steering

Rack-and-pinion steering uses a simplified steering system that weighs less and takes up less space than parallelogram steering. For these reasons, it is used on most unibody vehicles. Refer to Figure 24–17.

The rack-and-pinion steering unit takes the place of the idler arm, pitman arm, center link, and gearbox used in parallelogram systems. It consists of a **pinion gear** at the end of the steering shaft, which meshes with a toothed shaft (a bar with a row of teeth cut into one edge) known as the ***rack.*** When the steering wheel is turned, the pinion rotates in a circle and moves the rack sideways—left or right—to turn the wheels.

Two inner ***tie-rod ends*** are threaded onto the end of the rack and are covered by rubber bellows boots. These ***boots*** protect the rack from contamination by dirt, salt, and other road particles. The inner tie-rods are threaded onto the outer tie-rod ends, which connect to the steering arms.

Parallelogram Steering

Parallelogram steering uses two tie-rod assemblies connected to the steering arms to support a long center link, Figure 24–16. The **center link** holds the tie-rods in position. An ***idler arm*** supports the center link on one end. The other end of the center link is attached to the **pitman arm** on the gearbox.

The arrangement forms a parallelogram shape (each end moves equal and parallel to the other end).

On some systems, a **steering damper** is attached to the frame and the steering linkage to help absorb road shocks and wheel shimmy.

Power Steering

The power steering unit is designed to reduce the amount of effort required to turn the steering wheel. It also reduces driver fatigue on long drives and makes it easier to steer the vehicle at slow speeds, particularly during parking.

Power steering can be broken down into two design arrangements: conventional and electronically controlled. In the conventional arrangement, hydraulic power is used to assist the driver. An engine drive ***power steering pump*** forces pressurized oil through the system.

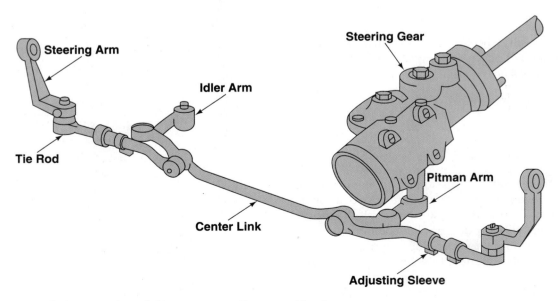

Figure 24–16 Study the parts of parallelogram steering. This is an older design.

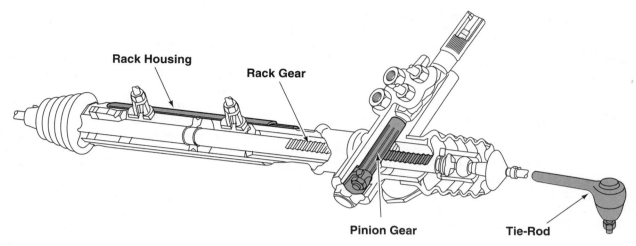

Figure 24–17 Note the parts of common rack-and-pinion steering.

Power steering hoses carry the oil to and from the pump. A **hydraulic piston** on the steering linkage or in the gearbox helps turn the wheels. **Hydraulic valves** control power assist.

With a **computer-assisted steering** system, *sensors* provide feedback for the computer. The *computer* or electronic control unit can then precisely control power assist as variables change. Mechanical steering is still provided with an electrical failure.

With **electronically controlled power steering**, the conventional power steering parts are replaced with electronic controls and an electric motor. The electric motor is mounted in the rack assembly. A DC motor armature with a hollow shaft is used to allow passage of the rack through it. The outboard housing and rack are designed so that the rotary motion of the armature can be transferred to linear movement of the rack.

NOTE! For more information on electronic and computer systems, refer to Chapter 26.

Collapsible Steering Columns

Collapsible steering columns crush under force to absorb impact. This reduces injuries to the driver. Steering column collapse is managed in two parts, upper and lower. The lower section of the column is linked to two or more universal joints. It uses universal joints that allow the section to fold.

The upper section of the steering column uses collapsible steel mesh or shaft and tube. The shaft is locked to the tube and designed to give under pressure. The shaft can be locked to the tube using plastic inserts or steel balls in a plastic retainer.

Collapsible columns need to be measured following a collision. This measurement is compared to dimensions given in the service manual. If the steering column is shorter than the specification given in the service manual, replace the steering column assembly. Never try to repair it.

> *WARNING! When working on a collapsible steering column, do not hammer on or bump column parts. With the column removed from the mounts, it is extremely susceptible to impact damage. See Figure 24–18.*

A slight impact on the column end can collapse the steering shaft or loosen the plastic inserts that maintain column rigidity. When removing the

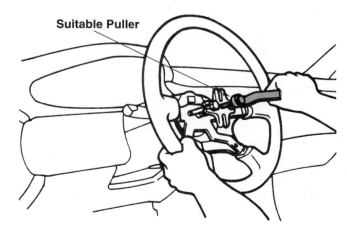

Figure 24–18 A wheel puller is needed to remove the steering wheel. Never hammer on a steering column or damage could result. *(Copyright Mazda Motor of America, Inc. Used by permission)*

steering wheel, use a puller. Do not hammer or pound on any components during removal.

NOTE! Air bags are explained in Chapter 27, Restraint System Repairs.

Steering System Service

To check a rack-and-pinion system, begin by raising the vehicle and taking the weight off the front suspension. Visually inspect the steering system for physical damage. Check the boots for leaks, and inspect the tie-rods. Examine the mounting points for any distortion. Inspect the tie-rod ends. Grab the tie-rod near the tire and try pushing it up and down. Any looseness indicates damage or wear.

Check the inner tie-rod socket by squeezing the bellows until the socket can be felt. With your other hand, push and pull on the tire. Looseness in the socket indicates damage or wear. With both hands, try to swivel a tire right and left while watching for play. If excessive movement is noted, wear or damage is likely. Observe the rack-and-pinion at the same time. Any movement might indicate a problem.

Remember that rack-and-pinion units must be mounted on a level plane. Misalignment of the rack-and-pinion will cause changes in the steering geometry during jounce/rebound. This condition cannot be corrected by changing the length of the tie-rods.

Here are some steering service tips that should be kept in mind:

1. Torque all steering system fasteners to specs with a torque wrench. Refer to the manual for an exact value.

2. Always install new cotter pins. Never reuse old cotter pins. If one were to fall out, a fatal accident could result.

3. Never try to straighten any steering system parts. Always replace bent or damaged parts.

4. Protect a power steering system from dirt and moisture at all times. If the system must be open, be sure to plug or tie off all openings with plastic.

5. Always replace the fluid lost with the recommended type.

6. Check the power steering hose routing when reassembling the system. Always route and hang hoses identical to the factory installation. Avoid contact with other parts. Watch for rubbing against moving parts.

7. Use proper tools when servicing a steering system. If you do not use the right tool, parts can be damaged.

SUSPENSION SYSTEMS

The *suspension system* allows the tires and wheels to move up and down with road surface irregularities. Its major parts are shown in Figure 24–19.

Control arms mount on the frame to swivel up and down. *Ball joints* on the out end of the control arms allow the steering knuckles to swivel and turn. The *steering knuckles* hold the wheel bearings and wheels. The hubs mount on the wheel bearings to hold the wheels or rims. The **wheels** hold the tires. **Springs** support the weight of the car and allow suspension flexing. *Shock absorbers* are dampening devices that absorb spring oscillations (bouncing) to smooth the vehicle's ride quality. They may be gas, oil, or air filled.

In unibody construction, the body provides the critical mounting positions for the suspension and steering. For instance, the inner fender skirts in conventional body-over-frame construction prevent dirt and road splash from entering the engine compartment. In unibody construction, they are called strut towers and inner fender aprons and have the additional task of providing the upper mounting controls for the strut suspension system.

Front Suspension

There are two main types of unibody front suspension systems: the short/long arm (SLA) independent and the strut. The strut, as shown in Figure 24–19, is by far the most commonly used, but independent front systems are making a comeback. The main reason for the return of the modified short/long arm suspension is that it allows a lower hood profile, thus offering improved aerodynamics.

The upper (short) and lower (long) control arms on the traditional independent front suspension function primarily as locators, fixing the position of the system and its components relative to the vehicle. They are attached to the frame with bushings that permit the wheel assemblies to move up and down separately in response to irregularities in the road surface. The outer ends are connected to the wheel assembly with ball joints inserted through each arm into the steering knuckle.

Strut suspension is dramatically different in appearance from the traditional independent front suspension. The MacPherson strut assembly typically includes the spring, upper suspension locator, and shock absorber. It is mounted vertically between the top arm of the steering knuckle and the inner fender panel. Although there are several strut systems in use, all variations operate in basically the same manner.

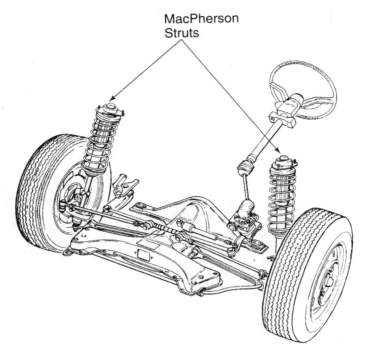

MacPherson
Struts

Figure 24–19 Study the major parts of a strut suspension. *(Courtesy of McQuay-Norris)*

Many unibody vehicles have lower control arms attached directly to the front rails. Others have lower control arms attached to suspension subframes or crossmembers that are bolted directly to the front rails.

The suspension system mounting locations in the unibody structural panels must be able to withstand the load transferred by vertical movement. In the case of strut suspensions, the vertical load is transferred directly from the upper strut mount to the top panel.

Rear Suspension

There are three principal types of rear suspensions: live axle, semi-independent or dead axle, and independent. The introduction of independent rear suspensions was brought about by concerns for improved traction and ride and prompted the introduction of independent front suspensions. If the wheels can move separately on the road and energy forces can be maintained, traction and ride will be improved.

Other Suspension Systems

Trailing arm rear suspension systems provide little, if any, independent movement of the rear wheels. They are commonly used on front drive vehicles, Figure 24–20. Also called a ***dead axle,*** a solid rear axle holds the two wheel assemblies.

Air spring suspension uses air springs to support the vehicle load. The air springs replace or supplement the conventional spring suspension. Air spring suspension provides for automatic front and rear load leveling. It will increase and decrease air pressure in the springs to raise, lower, or level the vehicle as needed.

Air spring suspension parts include a computer, air springs, air compressor, height sensors, and air lines, Figure 24–21. The sensors send signals to the computer about the ride height and attitude of the vehicle. The computer can then react to road and driving conditions. It sends out control signals to the compressor and valves to increase or decrease pressure in each air spring as needed.

Suspension System Service

Always inspect the suspension system for signs of damage from the collision. Damaged or worn suspension parts should be replaced. When worn or damaged, these parts can upset the settings of the entire suspension, steering, and drive line systems.

Loose, worn, or broken attaching parts will allow the rear wheels to shift. This will cause premature tire wear as well as short U-joint life. A metallic jingling sound when driving over small bumps or unusual tracking indicates the need for service.

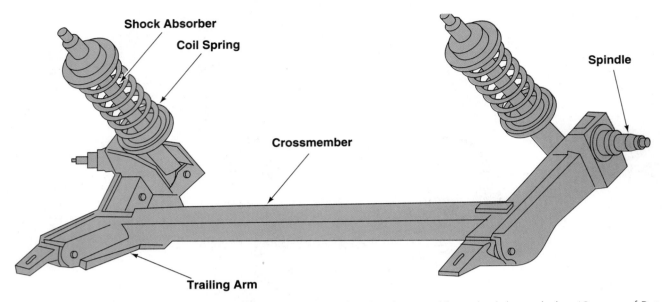

Figure 24–20 Trailing arm suspension, or dead axle, is sometimes found at the rear of front-wheel drive vehicles. *(Courtesy of DaimlerChrysler Corporation)*

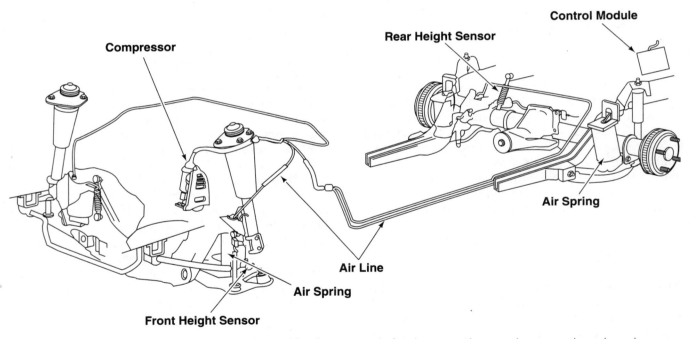

Figure 24–21 Air spring suspension uses synthetic rubber bags instead of coil springs. This provides a smoother ride and an easy way to allow the computer to adjust the suspension system.

WARNING! The manufacturer has designed certain suspension, steering, and alignment features into the unibody vehicle. Do not try to straighten or bend any of the pieces that make up the suspension system.

When servicing suspension system parts, refer to the service manual for details. It will give procedures and specifications for the exact part. Always use a torque wrench to tighten suspension system fasteners to specifications.

AIR CONDITIONING SYSTEMS

An *air conditioning system* is designed to cool the passenger compartment. System designs vary. For

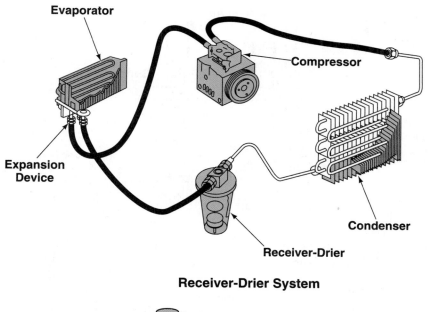

Receiver-Drier System

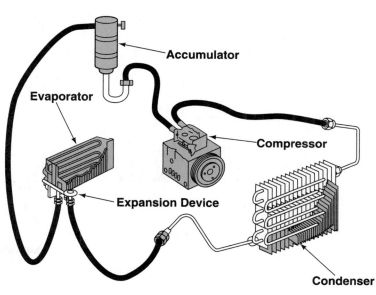

Accumulator System

Figure 24–22 Study the difference between two common air conditioning systems. *(Courtesy of Mitchell International, Inc.)*

example, some air conditioning systems use an accumulator, while others use a receiver-drier. Look at Figure 24–22.

Receiver-driers and **accumulators** serve the same basic purposes. They use a desiccant bag to remove moisture from the system. Their difference is the location. The accumulator is between the evaporator and compressor. The receiver-drier is between the condenser and expansion device. They act as storage tanks.

Air conditioning systems are divided into two sides, high and low, Figure 24–23. The dividing points are the compressor and the expansion device.

The **A/C high-side** contains high pressure/high temperature refrigerant. Its hoses feel hot to the touch.

High-side hoses are generally smaller in diameter than the low-side.

The **A/C low-side** contains low pressure/low temperature refrigerant. Its hoses feel cold to the touch. Low-side hoses are generally larger in diameter than the high-side.

Air Conditioning System Service

Discharging an air conditioning system removes refrigerant from the system and must always be done before parts are removed. Some compressors use a special back seating service valve that allows the

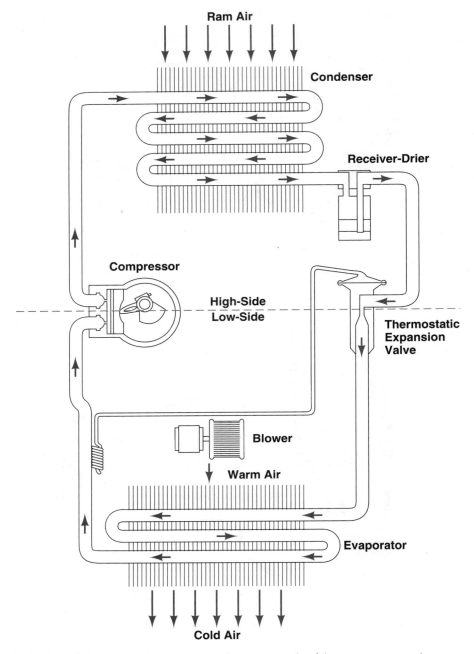

Figure 24–23 The high-side includes the condenser, receiver-drier, output side of the compressor, and connecting lines. The low-side includes the evaporator, expansion valve outlet, and inlet side of the compressor. The evaporator mounts under the dash of the vehicle. *(Courtesy of Mitchell International, Inc.)*

compressor to be removed without completely discharging the system.

When removing refrigerant from the system, remember to use a recovery system, Figure 24–24.

A *recovery system* will capture the used refrigerant and keep it from contaminating the atmosphere. Most will also filter the refrigerant for reuse. Since equipment varies, refer to the user's manual for detailed procedures.

Manufacturers' receiver-drier and accumulator replacement recommendations vary. Generally, if the system has been open for several days, the receiver-drier should be replaced.

Evacuating an air conditioning system is done to remove air and moisture from the system. It must be done anytime air has entered the system. Evacuating is done by vacuum lowering the boiling point of the moisture, converting it to vapor (steam) and removing it.

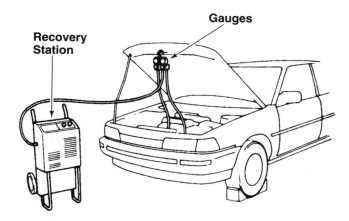

Figure 24–24 A recovery machine is needed to capture used refrigerant, which can then be recycled. This protects our atmosphere from damage.

Before *charging* (filling) the system with refrigerant, determine the amount and type of refrigerant used. This information is found in the service manual, on the label on the radiator support, or on the compressor. Do NOT mix different types of refrigerants. Charging can be done with a gauge set or with a charging station, Figure 24–25.

Purging uses refrigerant to push air and dirt out of the hoses. It prevents air and other contaminants from being pushed into the A/C system. Always purge the gauge hoses before charging.

Refrigerant oil lubricates moving parts in the A/C system. Use only refrigerant oil. Do NOT use any other type of oil. Make sure to use the type recommended for the system being serviced. For example, some oils are designed to be used only with specific types of refrigerant. Using a different type can result in damage to the compressor and seals and other parts.

General rules are to add the amount of oil that was removed during discharge. There are adapters available to use refrigerant pressure to add oil during recharging.

Too much oil can cause reduced cooling. The oil takes up space normally used by the refrigerant. It can also damage the compressor and seals. Too little oil can cause poor lubrication of the system, premature compressor wear, and poor system performance.

An open can of refrigerant oil can collect dirt and moisture. Adding contaminated refrigerant oil to the system can cause corrosion, which can result in the failure of the compressor and other parts.

There are several ways to find refrigerant leaks:

1. Electronic leak detector

2. Refrigerant cans with dye

3. Soap and water solution in a spray bottle

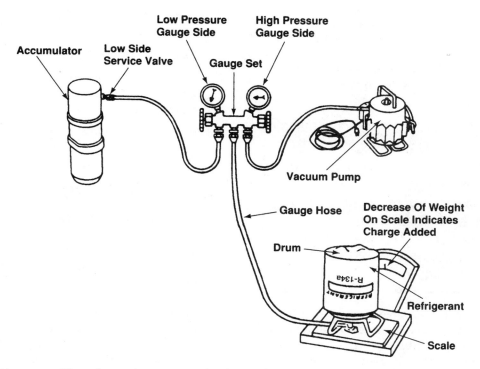

Figure 24–25 Charging or filling of an A/C system can be done with a gauge set or charging station. The correct amount and type of refrigerant must be metered into the system. *(Courtesy of General Motors Corporation, Service Operations)*

A sight glass is used to check the amount of refrigerant in the system. It is located on the receiver-drier or in line. When viewing the sight glass:

1. Clear = completely full or completely empty
2. Oil streaks = no refrigerant
3. Foam or constant bubbles = low refrigerant charge
4. Clouded = desiccant being circulated through system

Electronic leak detectors are battery-operated instruments that use an audio sound to alert you to the presence of a gas leak. They are designed to detect different types of gases.

When checking for refrigerant leaks, always check along the bottom of the hoses, fittings, seals, and other possible leakage points. This is because the refrigerant is heavier than air and is easier to detect below these parts.

Due to their possible depleting effect on the ozone layer, chlorofluorocarbon (CFC) and hydrochlorofluorocarbon (HCFC) are being phased out. This includes R-12 and other refrigerants used in air conditioning and refrigeration systems. **R-12,** also called CFC-12 or dichlorofluoromethane, is an older refrigerant used in vehicles manufactured before 1992.

Environmental agencies call for a gradual phase-out of most ozone-depleting substances. They state that R-12 systems can be serviced using recovered and recycled refrigerant. After filtering, this R-12 can then be used again in another vehicle. This reuse is designed to extend the supply of refrigerant.

R-134a, also called HFC-134A, is the refrigerant used in vehicles produced after 1992; it is the present replacement for R-12. It is less harmful to the ozone layer. New vehicles are being designed to run on this new refrigerant. The compressor and other parts are designed to be used with R-134a.

WARNING! R-134a is NOT compatible with R-12. Also, R-134a oils are NOT compatible with R-12 oils. They require separate service equipment. To avoid a mistake, R-134a uses metric quick-connect service ports. The high-side port is larger, so the same charging hoses cannot be used.

Mixing R-12 and R-134a, even in trace amounts, can be fatal to a system. This mistake can cause damage to seals, bearings, compressor reed valves, and pistons. Mixing refrigerants can also cause desiccants used in R-12 systems to break down and form harmful acids when used with R-134a.

Refrigerant Handling Tips

Although most A/C repairs and service work is done in specialty airconditioning shops, there are a number of pointers that the collision repair technician must keep in mind.

DANGER! Wear hand and face protection when working on an air conditioning system. When refrigerant escapes from the system or the supply tank, it can cause severe frostbite burns.

Most A/C failures are caused by moisture entering the system. Moisture interacting with refrigerant causes sludge and hydrochloric acid to form. This will attack the delicate parts of the compressor. The resulting acid will also eat away at the aluminum components of the evaporator and condenser.

The release of R-12 into the atmosphere is prohibited by current environmental regulations. Never vent the refrigerant into open air. Use a recovery/recycling machine.

When removing or opening up the A/C unit, seal all openings. This can be done with synthetic rubber, tight fitting caps, plugs, or plastic wraps. Use sturdy rubber bands or wire ties to hold plastic wraps in place securely.

If an A/C system has been open to the atmosphere more than a few hours, do the following:

1. Change the oil.
2. Flush each component separately with nitrogen gas before charging.
3. Replace the receiver-drier or accumulator.
4. During evacuation, hold the system at high vacuum for a minimum of 30 minutes to pull out air and moisture.
5. Recharge without leaking refrigerant.

EMISSION CONTROL SYSTEMS

Emission control systems are used to prevent potentially toxic chemicals from entering the atmosphere. The most common of these are the exhaust gas recirculation (EGR) valve, catalytic converter, air injection, and positive crankcase ventilation (PCV) systems. See Figure 24–26.

The **exhaust gas recirculation (EGR)** valve opens to allow engine vacuum to siphon exhaust into the intake manifold. The EGR valve consists of a poppet and a vacuum actuated diaphragm. When ported vacuum is applied to the diaphragm, it lifts the poppet off its seat. Intake vacuum then siphons exhaust

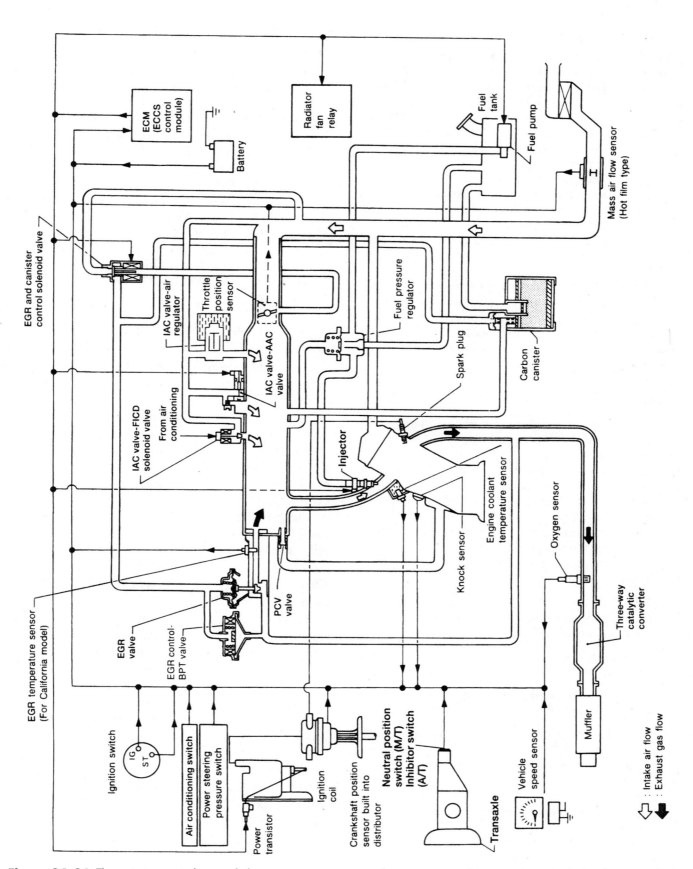

Figure 24-26 The emission control system helps prevent toxic emissions from entering earth's atmosphere. Study this diagram, which shows several emission control systems. *(Used by permission from Nissan North America, Inc.)*

into the engine. The exhaust entering the combustion chambers lowers peak combustion temperatures. This reduces oxides of nitrogen pollution.

The *catalytic converter,* mounted in the exhaust system, plays a major role in emission control. The catalytic converter works as a gas reactor. Its catalytic function is to speed up the heat-producing chemical reaction between the exhaust gas and oxygen to reduce air pollutants in the exhaust.

The *fuel evaporative system* pulls fumes from the gas tank and other fuel system parts into a charcoal canister. The **charcoal canister** absorbs and stores vaporized fuel. When the engine is started, these vapors are drawn into the engine and burned. This prevents this source of pollution from entering our atmosphere.

The *positive crankcase ventilation system,* abbreviated *PCV,* channels engine crankcase blow-by gases into the engine intake manifold. They are then drawn into the engine and burned. This prevents crankcase fumes from entering the atmosphere.

Emission Control System Service

Many times emission control systems are damaged in a collision. They must be serviced as part of the repair. The Clean Air Act, which is a federal law, makes the collision repair technician responsible for the emission control system. The law requires technicians to restore emission control systems to their original design. It also prescribes penalties for shops and technicians who alter emission control systems or fail to restore them to proper working condition.

The following guidelines must be strictly adhered to when working on emission control systems:

1. Damaged parts must be replaced with good parts. Eliminating damaged parts to avoid replacing them is against the law.

2. Using parts that prevent proper operation of the emission control system is also against the law.

3. Proper repairs to the emission control system must be made to manufacturer's specifications.

4. All replacement parts for emission control systems must satisfy the original design requirements of the manufacturer.

Summary

- The powertrain/drivetrain is all of the parts that produce and transfer power to the drive wheels. This includes the engine, transmission or transaxle, drive axle, and other related parts.

- Three vehicle steering systems are parallelogram, rack-and-pinion, and power.

- The suspension system allows the tires and wheels to move up and down with road surface irregularities. The types of suspension systems are front, rear, and computer-controlled.

- Refrigerant leaks in an air conditioning system may be found by using an electronic leak detector, refrigerant cans with dye, or soap and water solution in a spray bottle.

- Emission control systems are used to prevent potentially toxic chemicals from entering the atmosphere. The most common of these systems are the exhaust gas recirculation (EGR) valve, catalytic converter, air injection, and positive crankcase ventilation (PCV) systems.

Technical Terms

powertrain/drivetrain	*radiator cap pressure test*
engine	*radiator cap pressure*
block	*rating*
cylinder	*cooling system leak test*
piston	*exhaust system*
combustion	*header pipe*
connection rod	*catalytic converter*
crankshaft	*muffler*
cylinder head	*fuel system*
combustion chamber	*gasoline injection system*
valves	*fuel injectors*
valve springs	*diesel injection system*
camshaft	*fuel pump*
lifter	*fuel filter*
push rod	*fuel pressure regulator*
rocker arms	*manual transmission*
flywheel	*automatic transmission*
lubrication system	*transmission*
oil pan	*torque converter*
oil drain plug	*transaxle*
oil filter	*clutch*
oil dipstick	*front-wheel drive*
cooling system	*CV-axles*
antifreeze	*rear-wheel drive*
radiator	*drive shaft*
radiator pressure cap	*universal joints*
radiator fan	*rear axle assembly*
water pump	*differential assembly*
thermostat	*brake system*
heater system	*master cylinder*
antifreeze tester	*brake lines*
hydrometer	*brake pads*

brake shoes
brake rotors
brake drums
caliper
power brakes
brake system bleeding
steering system
steering column
steering gearbox
parallelogram steering
center link
idler arm
pitman arm
rack-and-pinion steering
rack
tie-rod ends
power steering pump
power steering hoses
sensors
computer
collapsible steering
   column
suspension system
control arms
ball joints

steering knuckles
shock absorbers
dead axle
air spring suspension
air conditioning system
receiver-driers
accumulators
A/C high-side
A/C low-side
discharging
recovery system
evacuating
charging
purging
electronic leak detector
R-12
R-134a
emission control systems
exhaust gas recirculation
   (EGR)
catalytic converter
fuel evaporative system
positive crankcase
   ventilation system
PCV

Review Questions

1. This would NOT be part of an engine.
   a. Block
   b. Control arm
   c. Connecting rod
   d. Cylinder head

2. An _____ _____, commonly called a _____, is needed to determine the freeze-up protection of the coolant mixture.

3. How do you test a cooling system for leakage?

4. Parallelogram steering is an older system being phased out by rack-and-pinion steering. True or false?

5. List seven steering service tips.

6. These mount on the out end of the control arms to allow the steering knuckles to swivel and turn.
   a. Bushings
   b. Coil springs
   c. Shocks
   d. Ball joints

7. Do not try to straighten or bend any of the pieces that make up the suspension system. True or false?

8. Explain the difference between the A/C high- and low-sides.

9. Why must you use a refrigerant recovery system?

10. Why should you evacuate an air conditioning system?

11. What are three ways of finding refrigerant leaks?

12. Which emission control subsystem is responsible for channeling blow-by gases into the fuel intake area?
    a. Engine control
    b. Positive crankcase ventilation
    c. Evaporative
    d. Exhaust gas recirculation

13. What is the engine's temperature control?
    a. Thermostat
    b. Radiator cap
    c. Cooling fan
    d. Coolant

14. When checking for refrigerant leaks, always check along the _____ of the hoses, fitting, seals, and other possible leakage points.

15. Why must you wear hand and face protection when working on an air conditioning system?

16. List four guidelines for working on emission control systems.

ASE–Style Review Questions

1. *Technician A* says that antifreeze contains additives, lubricants, and corrosion inhibitors that break down over time. *Technician B* says that antifreeze should be replaced with a 50/50 mixture of antifreeze and water. Who is correct?
   a. Technician A
   b. Technician B
   c. Both Technicians A and B
   d. Neither Technician

2. Upon examining a collapsed steering column, *Technician A* decides to attempt a repair of the collapsed portion. *Technician B* says the collapsed portion must be replaced. Who is correct?
   a. Technician A
   b. Technician B
   c. Both Technicians A and B
   d. Neither Technician

3. Which of the following is a replacement for R-12?
   a. R-13a
   b. X-29a
   c. R-22
   d. R-134a

4. When evacuating an A/C system that has been open to the atmosphere for an extended period, *Technician A* holds the system at high vacuum for 30 minutes. *Technician B* replaced the receiver-drier. Who is correct?
   a. Technician A
   b. Technician B
   c. Both Technicians A and B
   d. Neither Technician

5. Which of the following parts would not normally be damaged during a frontal collision?
   a. Radiator
   b. Condenser
   c. Fuel tank
   d. Water pump

6. A brake line is found to be damaged following a collision. *Technician A* says the line should be repaired with a spliced-in section. *Technician B* says that a new copper line must be formed. Who is correct?
   a. Technician A
   b. Technician B
   c. Both Technicians A and B
   d. Neither Technician

7. *Technician A* says that many fuel injection systems retain pressure when the engine is not running. *Technician B* says that there is never fuel pressure when the engine is not running. Who is correct?
   a. Technician A
   b. Technician B
   c. Both Technicians A and B
   d. Neither Technician

8. *Technician A* says to begin drivetrain inspection by checking the condition of the CV-joint boots. *Technician B* says that splits, cracks, tears, punctures, or thin spots in CV-joint boots require replacement. Who is correct?
   a. Technician A
   b. Technician B
   c. Both Technicians A and B
   d. Neither Technician

9. A vehicle suffers from a vibration that increases with speed. *Technician A* says that this symptom is normally due to CV-joint problems. *Technician B* says that an out-of-balance tire or wheel, an out-of-round tire, or a bent rim are the more likely cause. Who is correct?
   a. Technician A
   b. Technician B
   c. Both Technicians A and B
   d. Neither Technician

10. Which of the following parts should be marked for alignment before removal from the rear axle assembly?
    a. Drive shaft
    b. CV-joint
    c. Axle shaft
    d. Brake housing

Activities

1. Inspect a vehicle with collision damage. Use the methods in this chapter to find mechanical damage. Make a report of your findings.

2. Refer to manufacturer's service manuals. Read about the troubleshooting and repair of mechanical systems.

25

Wheel Alignment

Objectives

After studying this chapter, you should be able to:

- Define the term "wheel alignment."

- Inspect tires, steering, and suspension systems before alignment.

- Check and adjust caster, camber, and toe.

- Describe toe-out on turns, steering axis inclination, and tracking.

- Summarize alignment equipment variations.

Introduction

In collision repair, **wheel alignment** involves positioning the vehicle's tires so that they roll properly over the road surface. Wheel alignment is essential to safety, handling, fuel economy, and tire life. See Figure 25–1.

Following collision repair, a vehicle may require an alignment if:

1. There is damage to any steering and suspension parts.
2. There is damage to any steering or suspension mounting locations.
3. There was engine cradle damage or a position change.
4. Suspension or steering parts were removed for access to body parts.
5. There was damage to major structural components.

Remember! Wheel alignment is done to fine-tune unibody/frame adjustments. The job of the collision repair technician is to make sure that everything can be "fine-tuned" and the wheels can be aligned properly.

Figure 25–1 Wheel alignment involves final adjustment of the steering and suspension systems to make the tires roll properly over the road surface. If structural repairs were made properly, this is a relatively simple task using modern equipment. *(Hunter Engineering Company)*

WHEEL ALIGNMENT BASICS

Improper wheel alignment will cause rapid tire wear and poor handling. Basic wheel alignment angles include:

1. Caster
2. Camber
3. Toe
4. Thrust line (tracking)
5. Steering axis inclination
6. Turning radius
7. Included angle

Caster

Caster is the angle of the steering axis of a wheel from true vertical, as viewed from the side of the vehicle. It is a directional stability adjustment. Caster is measured in degrees.

Caster has little effect on tire wear. Caster affects where the tires touch the road compared to an imaginary centerline drawn through the spindle support. Caster is the first angle adjusted during an alignment.

Positive caster tilts the tops of the steering knuckles toward the rear of the vehicle, Figure 25–2. It aids in keeping the vehicle's wheels traveling in a straight line. The wheels resist turning and tend to return to the straight-ahead position.

Negative caster is just the opposite, Figure 25–2. It tilts the tops of the steering knuckles toward the

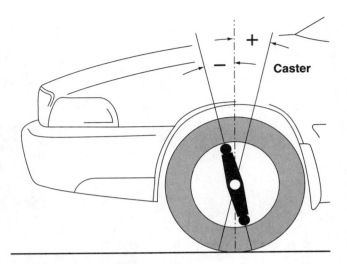

Figure 25–2 Note the difference between positive and negative caster.

front of the vehicle. Negative caster makes the wheels easier to turn. However, it produces less directional stability. The wheels tend to follow imperfections in the road surface.

Caster is designed to provide steering stability. The caster angle for all wheels should be almost equal. Unequal caster angles will cause the vehicle to steer toward the side with less caster. Too much negative caster can cause the vehicle to have "sensitive" steering at high speeds. The vehicle might wander as a result of too much negative caster. See Figure 25–3.

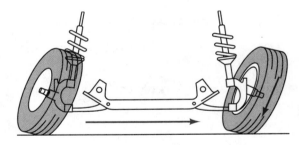

Left Turn, Spindle Moves Down

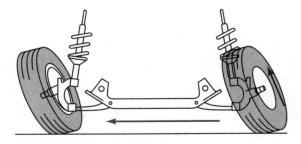

Right Turn, Spindle Moves Up

Figure 25–3 With positive caster, note how the spindle moves down with a left turn and then up with a right turn. *(Hunter Engineering Company)*

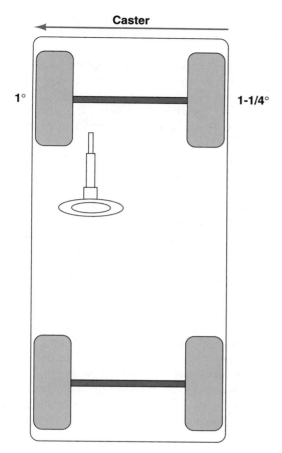

Figure 25–4 The vehicle will pull to the side with the least amount of caster. This can be used to counteract road crown, which produces steering wheel pull. *(I-CAR)*

Several factors can adversely affect caster. The most common problem is worn or loose strut rod bushings and control arm bushings. Caster adjustments are not provided on some strut suspension systems. Where they are provided, they can be made at the top or bottom mount of a strut suspension.

Caster is measured in degrees from true vertical. Specifications for caster are given in degrees positive or negative. Typically, more positive caster is used with power steering. More negative caster is used with manual steering to reduce steering effort. Also, a vehicle pulls to the side with the least amount of caster, Figure 25–4.

Camber

Camber is the angle represented by the tilt of the wheels inward or outward from vehicle centerline when viewed from the front of the vehicle. It assures that all of the tire tread contacts the road surface. Camber is measured in degrees. It is usually the second angle adjusted during a wheel alignment. See Figure 25–5.

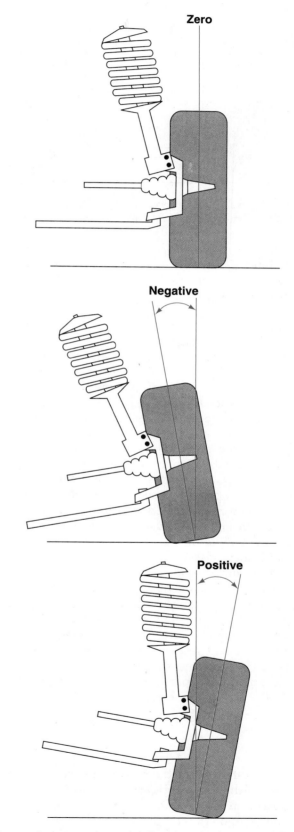

Figure 25–5 Camber is the tilt of the wheel in or out from the centerline when viewed from the front of the vehicle. *(I-CAR)*

Camber Roll

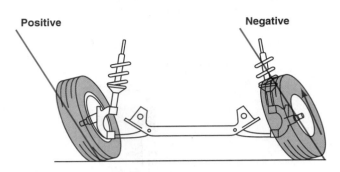

Positive   Negative

Figure 25-6 Camber changes in turns. Therefore, its adjustment affects cornering stability. *(Hunter Engineering Company)*

Camber is usually set equally for all wheels. Equal camber means each wheel is tilted outward or inward the same amount.

Positive camber has the top of the wheel tilted out, when viewed from the front. The outer edge of the tire tread contacts the road.

Negative camber has the top of the wheel tilted inward when viewed from the front. The inner tire tread contacts the road surface more. Note how camber changes when turning, Figure 25–6.

Camber is controlled by the control arms and their pivots. It is affected by worn or loose ball joints, control arm bushings, and wheel bearings. Anything that changes chassis height will also affect camber.

Camber is adjustable on most vehicles. Some manufacturers prefer to include a camber adjustment at the spindle assembly. Camber adjustments are also provided on some strut suspension systems at the top mounting position of the strut. Remember that camber adjustment also changes steering axis inclination (SAI) or the included angle.

Very little adjustment will be required if the strut tower and lower control arm positions are in their proper place. If you find serious camber error and suspension mounts have not been damaged, it is an indication of bent suspension parts. In this case, diagnostic angle and dimensional checks should be made to the suspension parts. Damaged parts must be replaced.

Toe

Toe is the difference in the distance between the front and rear of the left- and right-hand wheels. Toe can be measured in inches or millimeters or degrees, depending upon the equipment used.

Toe adjustment is critical to tire wear. If properly adjusted, toe makes the wheels roll in the same direction. If toe is NOT correct, the misaligned wheels will scuff or drag the tires sideways, causing rapid tire wear.

Remember! **Excessive toe** (in or out) will cause a sawtooth edge on the tire tread from dragging the tire sideways.

Toe-in results when the front of the wheels are set closer than the rear. The wheels point in at the front. Toe-out is the just the opposite. It has the front of the wheels farther apart than at the rear. The wheels point out at the front.

Toe is a very critical tire wearing angle. Wheels that do not track straight ahead have to drag as they travel forward.

Rear-wheel drive vehicles are often adjusted to have toe-in at the front wheels. Toe-in is needed to compensate for tire rolling resistance, play in the steering system, and suspension system action. The tires tend to toe-out while driving. By setting the wheels for a small toe-in of about 1⁄16 inch (1.5 mm), the tires will roll straight ahead over the road surface.

Front-wheel drive vehicles need to have their front wheels set for a slight toe-out. The front wheels pull and propel the vehicle. As a result, they are forced forward by drivetrain torque. This tries to make the wheels point inward while driving. Front-wheel drive toe-out of 1⁄16 inch (1.5 mm) is typical.

Steering Axis Inclination

Steering axis inclination (SAI) is the angle between true vertical and a line through the upper and lower pivot points, as viewed from the front. See Figure 25–7.

On collision damaged vehicles, SAI can help in diagnosing misalignment of a vehicle structure. For example, SAI can be used with structural measurements to diagnose:

1. Strut tower misalignment

2. Shifted engine cradle or crossmember

3. Control arm mounting location damage

4. Misaligned frame or body structure

SAI will generally indicate a damaged structure, but it may also indicate parts damage. For example, if a lower control arm is bent, the lower ball joint location will be changed. The angle between the steering axis and true vertical will also change.

SAI is an engineering angle designed to project the weight of the vehicle to the road surface for stability. SAI also helps the steering system return to straight-ahead after a turn. SAI should be checked because it can help locate other problems. Incorrect

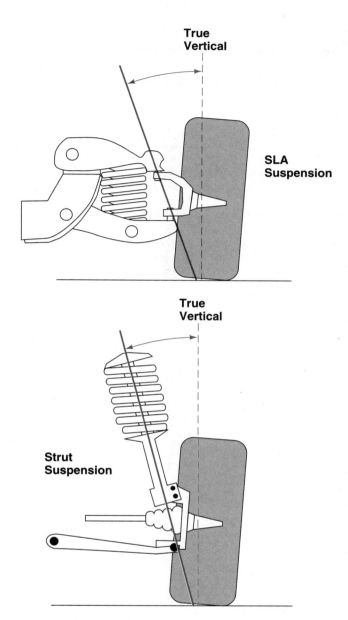

Figure 25–7 Steering axis inclination (SAI) is a diagnostic angle used to indicate structural misalignment. *(I-CAR)*

SAI generally indicates structural misalignment. See the chart in Figure 25–8.

If the vehicle has 0 (zero) SAI, the upper and lower ball joints (or strut pivot points) would be located directly over one another. Problems associated with this simple relationship include tire scrub in turns, lack of control, and increased effort during turn recovery.

If the SAI is tilted, a triangle is formed between ball joints and spindle. An arc is then formed when turning. There is a high point at straight-ahead posi-tion and a drop downward when turning to each side. This motion travels through the control arms to the springs and finally to the weight of the vehicle. The forces generated in a turn are actually trying to lift the vehicle. SAI offsets the lifting forces and helps to pull the tires back to straight-ahead when the turn is finished.

When the SAI angle includes the camber angle, the sum of the two is called the **included angle.** Problems that occur with this built-in angle can be due to a bent spindle, a bent upper or lower control arm or strut, or a badly worn or damaged ball joint. These damaged components must be replaced before doing the alignment. Doing an alignment without checking all the angles can result in error.

Turning Radius

Turning radius, called *toe-out on turns,* is the dif-ferent amount each wheel moves during a turn. It is built into the steering geometry. Turning radius is designed to allow the inside wheel to turn a few degrees more than the outside wheel during a turn. This makes the inside wheel turn in a smaller circle. See Figure 25–9.

As an example, during a left turn, the left wheel will turn approximately 20 degrees while the right wheel will turn approximately 18 degrees. A bent steer-ing arm will affect turning radius. See Figure 25–10.

Thrust Line/Centerline (Tracking)

Tracking is the parallel alignment of the rear wheels, the vehicle centerline, and the front wheels. If a vehi-cle is tracking properly, its rear wheels will follow the centerline of the vehicle when moving straight ahead. Tracking problems require the driver to turn the steer-ing wheel to have the vehicle travel in a straight line. See Figure 25–11.

To check tracking, wet an area of roadway, and drive the vehicle through it. Then, continue driving out of the water, far enough to make measurable tracks. The rear-wheel tracks should be an equal dis-tance from the front-wheel tracks on both sides. Incor-rect tracking is often caused by worn springs, offset rear axle, or rear toe problems.

You must make sure the vehicle runs straight down the road. The rear tires must track (follow) directly behind the front tires when the steering wheel is straight. The geometric centerline of the vehicle should parallel the road direction. This will be the case when rear toe or the rear axle is parallel to the vehicle's geometric centerline in the straight-ahead position.

SAI DIAGNOSTIC REFERENCE CHART				
Suspension System	SAI	Camber	Included Angle	Probable Cause
Short/	Correct	Less	Less	Bent Knuckle
Long Arm	Less	Greater	Correct	Bent Lower Control Arm
Suspension	Greater	Less	Corrrect	Bent Upper Control Arm
System	Less	Greater	Greater	Bent Knuckle
MacPherson	Correct	Less	Less	Bent Knuckle and/or Bent Strut
Strut	Correct	Greater	Greater	Bent Knuckle and/or Bent Strut
Suspension	Less	Greater	Correct	Bent Control Arm or Strut Tower (out at top)
	Greater	Less	Correct	Strut Tower (in at top)
	Greater	Greater	Greater	Strut Tower (in at top) and spindle and/or Bent Strut
	Less	Greater	Greater	Bent Control Arm or Strut Tower (out at top) Plus Bent Knuckle and/or Bent Strut
	Less	Less	Less	Strut Tower (out at top) and Knuckle and/or Strut Bent or Bent Control Arm

Figure 25–8 This chart shows which parts might cause SAI, camber, and included angle problems.

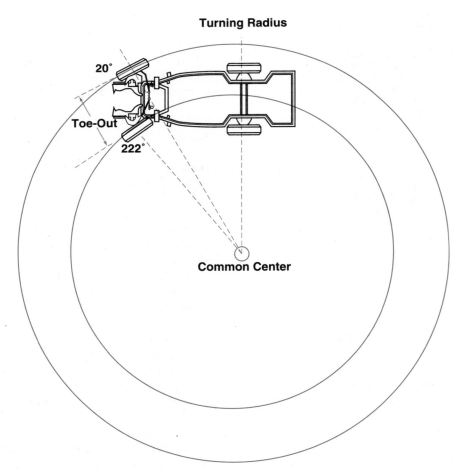

Figure 25–9 In turns, the inner wheel must turn sharper than the outer one. This is termed toe-out on turns, or turning radius. *(Hunter Engineering Company)*

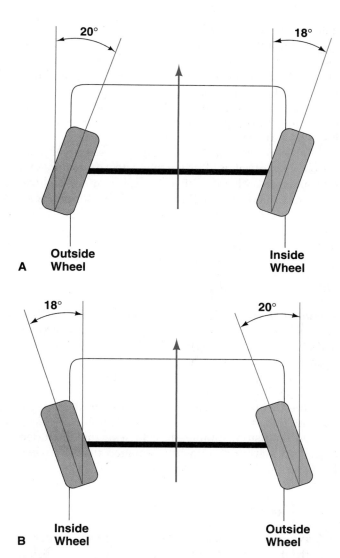

A — Outside Wheel — Inside Wheel

B — Inside Wheel — Outside Wheel

Figure 25–10 To check or read turning radius, the wheels are mounted on turning radius gauges that read in degrees. Refer to the service manual for exact specifications for the vehicle. If toe-out on turns is not correct, replace bent or damaged parts. *(I-CAR)* (A) Turn the steering wheel until the left wheel reads 20 degrees. Then read the other gauge. It should be about two degrees less. (B) Turn the steering wheel the other way until the right wheel reads 20 degrees. The other wheel should now read about 18 degrees.

NOTE! Offset axles resulting from a collision will not be in line with the centerline. You can set toe correctly, but alignment will be incorrect if the axle has been pushed forward or rearward during the collision.

Thrust Angle

Thrust angle is the angle between the thrust line and vehicle centerline, Figure 25–12. It should be zero if it aligns with the vehicle centerline.

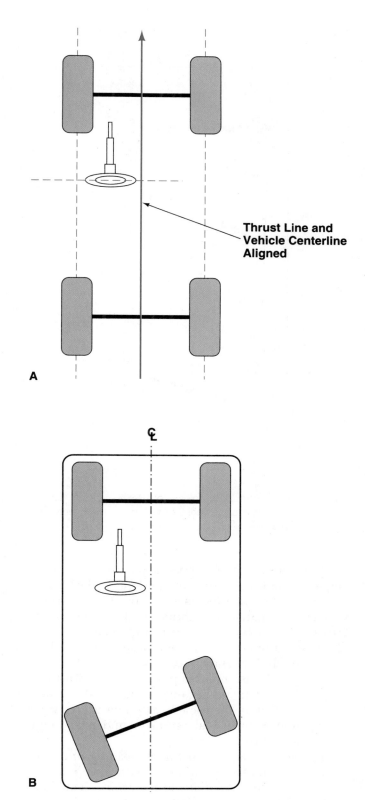

Thrust Line and Vehicle Centerline Aligned

A

B

Figure 25–11 Tracking checks whether the rear and front wheels are following each other. *(Hunter Engineering Company)* (A) Proper tracking has the rear wheels following in the tracks of the front wheels. (B) Improper tracking has the rear wheels turned so they do not follow the front wheels. This is nicknamed "dog tracking."

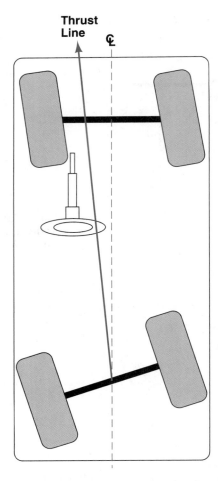

Figure 25–12 Tracking problems require the driver to turn the steering wheel to make the vehicle travel in a straight line. This alters the thrust line of the vehicle. (Hunter Engineering Company)

With most manufacturers, a **positive (+) thrust angle** results if the thrust line projects to the RIGHT of the vehicle centerline as seen from the top of the vehicle. A **negative (−) thrust angle** results if the thrust line projects to the LEFT of the vehicle centerline as seen from the top of the vehicle, Figure 25–13.

Some wheel alignment equipment manufacturers consider the thrust angle as positive if the thrust line projects to the left and negative if the thrust line projects to the right of the vehicle centerline as seen from the top of the vehicle. Make sure you understand how the terms are used.

If a vehicle has a thrust angle reading of 0–0.3°, the vehicle will generally not have a tracking problem.

If rear toe is not parallel to the vehicle centerline, a "thrust" direction to the left or right will be created. The vehicle will tend to travel in the direction of the thrust line, rather than straight ahead.

To correct this problem, begin by setting individual rear toe equal in reference to the geometric centerline. Four-wheel alignment machines check individ-

ual toe on each wheel. Once the rear wheels are in alignment with the geometric centerline, set the individual front toe in reference to the thrust angle. This assures that the steering wheel will be straight ahead for straight-ahead travel.

If you set the front toe to the vehicle geometric centerline, ignoring the rear toe angle, a cocked steering wheel can result.

Remember! Collision damage causing rear axle set-back or offset will adversely affect tracking. The rear axle, as with four-wheel steering, affects vehicle direction. Make sure the axles are in the correct position during an alignment. Modern four-wheel alignment equipment will automatically check axle positions and alignment.

Included Angle

Included angle is the sum of both camber and SAI angles. It is calculated by adding positive camber to the SAI angle. Included angle is calculated by subtracting negative camber from the SAI angle. It should generally be within ½ degree from side to side, Figure 25–14.

As an example, if camber is 2∞ positive and SAI is 12∞, the included angle is 14∞. If camber is 2∞ negative and SAI is 12∞, the included angle is 10∞.

On collision damaged vehicles, the included angle can help in diagnosing damaged parts between the upper and lower pivot points.

Wheelbase

Wheelbase is the measurement between the center of the front and rear wheel hubs. During collision repairs, individual wheelbase can be used as a diagnostic measurement. See Figure 25–15.

Wheelbase is measured to determine the forward and rearward position of each wheel. It is measured from identical known good points on the body structure or frame. Wheelbase measurements can be used to identify caster or axle positioning problems at each wheel.

Wheelbase is generally acceptable if the difference is within ⅛ inch (3 mm) from side to side.

Different individual wheelbases from side to side:

1. Are generally created by frame or chassis misalignment following a collision;

2. May be caused by the movement of an engine cradle;

3. Will change caster by moving the lower ball joint or pivot point forward or rearward in relation to the upper pivot point; and

4. May be designed in by the vehicle maker on some vehicles.

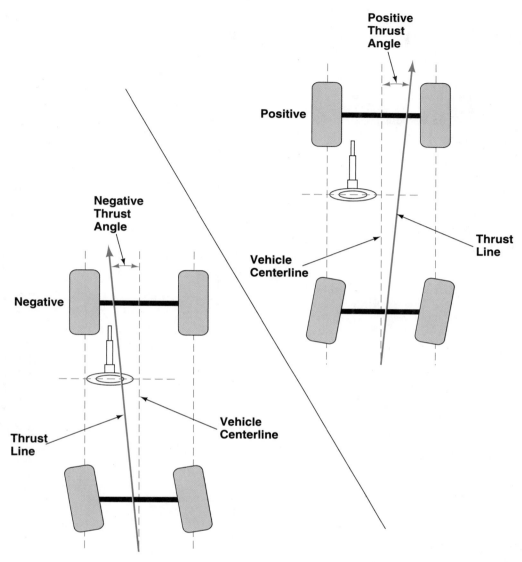

Figure 25–13 Study the difference between positive and negative thrust angles. *(Hunter Engineering Company)*

Set-Back

Set-back is when one wheel is moved back. It creates an unequal wheelbase between the left and right sides of a vehicle. Set-back may be caused by an impact to a front wheel assembly that moves a lower control arm, engine cradle, or radius rod backwards. It may be designed into some suspension systems.

Set-back should be set the same from side to side unless designed into the suspension system. Specifications are generally not listed in a specifications manual, but if it is designed into the suspension system, it can be identified by wheelbase specifications that are listed different from side-to-side.

Excessive set-back may cause a vehicle to pull to the side where set-back exists. Pulling may also occur during braking.

PREALIGNMENT CHECKS

Before making adjustments, conduct the following prealignment checks:

1. VISUALLY INSPECT everything visible while the vehicle sitting on the shop floor: steering wheel effort, play in steering, and power steering fluid. This also includes checking for uneven tire wear and mismatched tire sizes or types. Look for normal wear as well as the results of collision damage. Look for towing damage as well.

Reading tires involves inspecting tire tread wear and diagnosing the cause. Improper camber and toe show up as specific tread patterns.

2. MEASURE RIDE HEIGHT. The vehicle is designed to ride at a specific height, sometimes referred to as ***ride*** or ***curb height,*** Figure 25–16. Ride

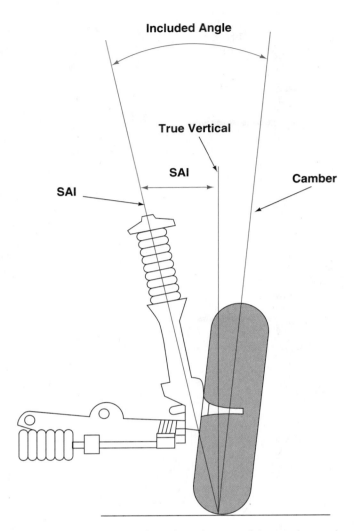

Figure 25–14 Included angle is the sum of the camber and SAI angles. It is calculated by adding positive camber to the SAI angle. (I-CAR)

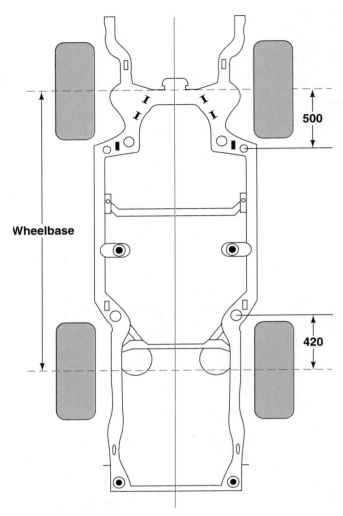

Figure 25–15 During wheel alignment, you may have to check the wheelbase to find the reason for improper wheel alignment. Set-back of one or more wheels may have resulted from collision. (Courtesy of Mitchell International, Inc.)

height specs are published in the service manuals and some alignment spec books.

Ride height is measured from the shop floor to specific points on the vehicle. Ride height should be correct before performing a wheel alignment. Various conditions that affect ride height and alignment include the following.

Vehicle load affects the weight on the front suspension and affects alignment angles. A heavy load in the trunk takes weight off the front suspension and changes caster and camber angles. Coil and leaf springs should be checked and replaced if broken or sagged beyond tolerance. Torsion bars should be checked and adjusted if ride height is incorrect. Spring mounting locations, air spring suspensions, system faults, and shock absorbers can also affect ride height.

If the vehicle leans to one side or seems to be lower on one side than on the other, something is wrong. To

isolate the problem, place the center of the main cross-member in the front of the vehicle on jack (safety) stands. Raise the front of the vehicle, and look at the rear of the vehicle. If the rear looks level, then the problem is in the front suspension, on the side of the vehicle that shows the lean. If the rear is not level, then the problem is with the low side of the rear suspension.

3. INSPECT THE UNDERSIDE of the vehicle for problems. With the vehicle raised, inspect all steering components such as the control arm bushings, upper strut mounts, pitman arm, idler arm, center link, tie-rod ends, ball joints (if equipped) for looseness, popping sounds, binding, and broken boots. Damaged parts must be repaired before adjusting alignment angles.

4. ROAD TEST the vehicle. Begin the alignment with a road test. While driving the vehicle, check to see that the steering wheel is straight. Feel for vibration in the steering wheel as well as in the floor or

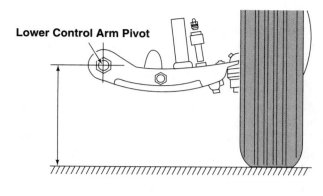

Lower Control Arm Pivot

Front

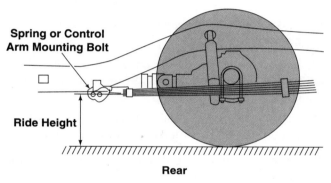

**Spring or Control
Arm Mounting Bolt**

Ride Height

Rear

Figure 25–16 Curb height, or ride height, can affect wheel alignment. The vehicle should be at curb (empty) weight when measuring. Measure from specified points on the vehicle to a level shop floor. If curb height is not correct, springs may have weakened or suspension parts may have been damaged. *(FMC Automotive Equipment Division)*

seats. Notice any pulling or abnormal handling problems, such as hard steering, tire squeal while cornering, or mechanical pops or clunks. This helps find problems that must be corrected before proceeding with the alignment.

Make sure the steering wheel is properly centered during your road test. If not, you will have to adjust the tie-rod ends to center it with the wheel straight ahead.

DIAGNOSTIC CHECKS

The analysis of ride and handling complaints involves the consideration of diagnostic angles. Suspension system parts must be considered as moving parts in vehicle operation; diagnostic checks evaluate the parts as they move.

When a vehicle has a steering problem, the first diagnostic check should be a visual inspection of the entire vehicle for anything obvious: bent wheels, misalignment of the cradle, and so on. If there is nothing

obviously wrong with the vehicle, make a series of diagnostic checks without disassembling the vehicle.

One of the most useful checks that can be made with a minimum of equipment is a jounce-rebound check. Look at Figure 25–17.

Jounce is the motion caused by a wheel going over a bump and compressing the spring. During jounce the wheel moves up toward the chassis. Jounce can be simulated by sitting on the bumper and pushing down on the vehicle. The vehicle should jounce equally on both sides.

Rebound is the motion caused by a wheel going into a dip or returning from a jounce and extending the spring. During rebound, the wheel moves down away from the chassis. Rebound can be simulated by

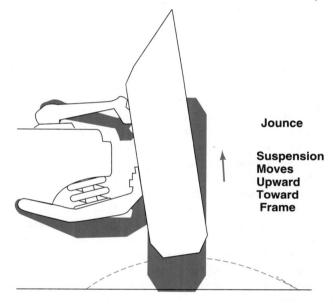

Jounce

**Suspension
Moves
Upward
Toward
Frame**

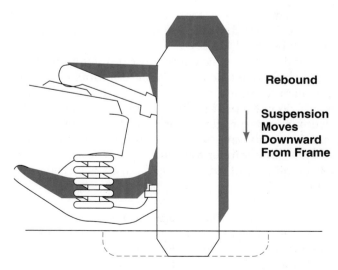

Rebound

**Suspension
Moves
Downward
From Frame**

Figure 25–17 Note the difference between jounce and rebound. *(Hunter Engineering Company)*

lifting up on the bumper. The vehicle should lift equally on both sides.

A *jounce-rebound check* will determine if there is misalignment in the rack-and-pinion gear. For a quick check, unlock the steering wheel and see if it moves during jounce and/or rebound. For a more careful check, use a pointer and a piece of chalk. Use the chalk to make a reference mark on the tire tread and place the pointer on the same line as the chalk mark.

Jounce and rebound the suspension system a few times while someone watches the chalk mark and the pointer. If the mark on the wheel moves unequally in and out on both sides of the vehicle, chances are there is a steering arm or gear out of alignment. If the mark does not move or moves equally in and out on both sides of the vehicle, the steering arm and gear are probably all right.

The next diagnostic check is for turning radius. This check evaluates the proper relationship of the two front wheels as they are turned through a steering arc. To measure turning radius, one wheel is turned a given amount on a turn plate or protractor. The amount of rotation of the opposite wheel is measured with a similar turn plate or protractor. The process is then repeated with the opposite wheel. The results are compared right to left to determine if the two front wheels are rotating through the same arc.

On a typical turning radius check, the left front wheel is turned out 20 degrees and the right wheel rotation is measured. The readings will usually not match with the right wheel. The right wheel should turn in the same amount or about 2 degrees less. The difference accounts for the turning radius difference between the inside and outside wheels during cornering.

The process is repeated with the right wheel. It is turned 20 degrees out. The movement of the left wheel is measured on the protractor or turn plate. The left wheel should turn in the same amount or about 2 degrees less.

By design, a vehicle might use a different turning radius from one side to the other. If in doubt, refer to the manufacturer's specifications. If these measurements do not repeat within 2 degrees, damage to the steering arms or gear is indicated. Turning radius measurements are especially useful in determining whether an improper toe condition is caused by poor wheel alignment or damaged suspension components.

NOTE! Radial tires cause problems when they have defective belts, unusual wear patterns, uneven air pressure, or are mismatched. These tire problems can cause the technician to misdiagnose steering and alignment problems.

Bent wheels are common after a collision. Sometimes, wheel damage can be so small it is difficult to detect visually. A dial indicator can be used to measure actual wheel runout, Figure 25–18.

Loose tie-rod ends are another common problem. They can wear quickly and cause play in the steering. Never try to align wheels if there are loose, worn tie-rod ends, Figure 25–19.

Remember! Before making any adjustment affecting caster, camber, or toe-in, the following checks should be made first:

1. Make sure the vehicle is sitting on a level surface (side to side and front to rear).

2. Rotate the tires if needed. Check the tires for similar tread design, depth, and construction.

3. Make sure all tires are inflated to recommended pressure.

4. Inspect for worn or bent parts and replace. Much of this should be checked during unibody/frame correction.

5. Check and adjust wheel bearings if necessary. Spin tires. Check for looseness or unusual noises.

6. Check for unbalanced loading (proper chassis height). This should be checked after unibody/frame correction. Remove objects from trunk and passenger compartment.

7. Check for loose ball joints, tie-rod ends, steering relay rods, control arms, and stabilizer bar attachments.

8. Check for runout of wheels and tires.

9. Check for defective shock absorbers.

10. Check the condition and type of equipment being used. Follow the manufacturer's instructions for using the equipment.

WHEEL ALIGNMENT PROCEDURES

After a thorough inspection and replacement of any damaged parts, wheel alignment adjustment is next. The main purpose of wheel alignment is to allow the wheels to roll without scuffing, dragging, or slipping on the road.

The adjustment order—caster, camber, toe—is recommended regardless of the make of vehicle or its type of suspension. Methods of adjustment vary. Refer to the manufacturer's service manual for details.

**Front and Rear Wheel
Axial Runout**

    Standard:
        Aluminum Wheel: 0–0.3 mm (0–0.01 in)

**Front and Rear Wheel
Radial Runout**

    Standard:
        Aluminum Wheel: 0–0.3 mm (0–0.01 in)

Figure 25–18 Bent wheels are a common problem in collision repair. A dial indicator will let you accurately check axial and radial runout of the wheels. (© *American Honda Motor Co., Inc.*)

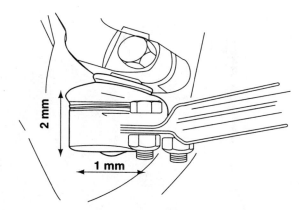

Figure 25–19 Always check the tie-rod ends for wear before adjusting wheel alignment. Play in the joint must be within specifications. (*Saab Cars USA, Inc.*)

A typical alignment procedure for a wheel alignment is as follows:

1. Obtain the manufacturer's specifications for the vehicle's wheel alignment checks and adjustments. Specs will vary depending upon the design of the vehicle.

2. Mount the alignment equipment on the vehicle, following the equipment manufacturer's instructions. Equipment designs and operating procedures vary. Therefore, you must refer to the materials published by the equipment maker for proper use. See Figure 25–20.

3. Check steering and axis inclination.

4. Check and adjust camber. If the equipment shows camber to be within specs, no adjustment is necessary. If it is not within specs, you must change camber the right amount and direction. In Figure 25–21, note typical methods of adjusting camber.

5. Check and adjust caster. If the car tends to pull to one side or if equipment readings are off, change caster. Again, see Figure 25–21 for adjustment methods.

6. Check turning radius. Replace parts if needed.

7. Check and adjust toe LAST. See Figure 25–22. Toe is the last adjustment made in an alignment. It is made at the tie-rods. Toe adjustments on a rack-and-pinion unit must be made evenly on both sides of the rack-and-pinion gear. If they are not made evenly, the rack assembly will be off-center in the gear. This can cause a pull due to the steering wheel being off-center. This pull condition is especially common with power-assisted rack-and-pinion gears. The steering assembly must be centered before these adjustments are made.

8. Document your readings and adjustment values.

Figure 25–20 Alignment equipment basically uses a turning radius gauge (degree wheel) and a caster-camber gauge (level) to measure alignment angles. Tires are centered on top of the turning radius gauges. A wheel adapter holds the caster-camber gauge. *(Hunter Engineering Company)*

ALIGNMENT EQUIPMENT VARIATIONS

Note that most modern wheel alignment equipment is computer controlled, Figure 25–23. It will actually guide you through the alignment process, giving step-by-step instructions. Some equipment will even warn you if any alignment angle is not acceptable. This greatly simplifies the wheel alignment process.

Today most vehicles use a four-wheel alignment. Approximately 80 percent of today's vehicles require front- and rear-wheel alignment. This is due to the growing number of vehicles with independent rear suspensions. Collision repair of any magnitude usually requires at least a four-wheel alignment check.

ROAD TEST AFTER ALIGNMENT

After making your wheel alignment adjustments, road test the vehicle. Make sure the vehicle does not pull to one side of the road, vibrate, or exhibit other troubles.

See Figures 25–24 and 25. They are troubleshooting charts for the steering and suspension systems.

Summary

- Wheel alignment involves positioning the vehicle's tires so that they roll properly over the road surface. Wheel alignment is essential to safety, handling, fuel economy, and tire life.
- Caster is the angle of the steering axis of a wheel from true vertical, as viewed from the side of the vehicle. It is a directional stability adjustment.
- Camber is the angle represented by the tilt of the wheels inward or outward when viewed from the front of the vehicle. It assures that all of the tire tread contacts the road surface. It is usually the second angle adjusted during a wheel alignment.
- Toe is the difference in the distance between the front and rear of the left- and right-hand wheels. It is critical to tire wear.

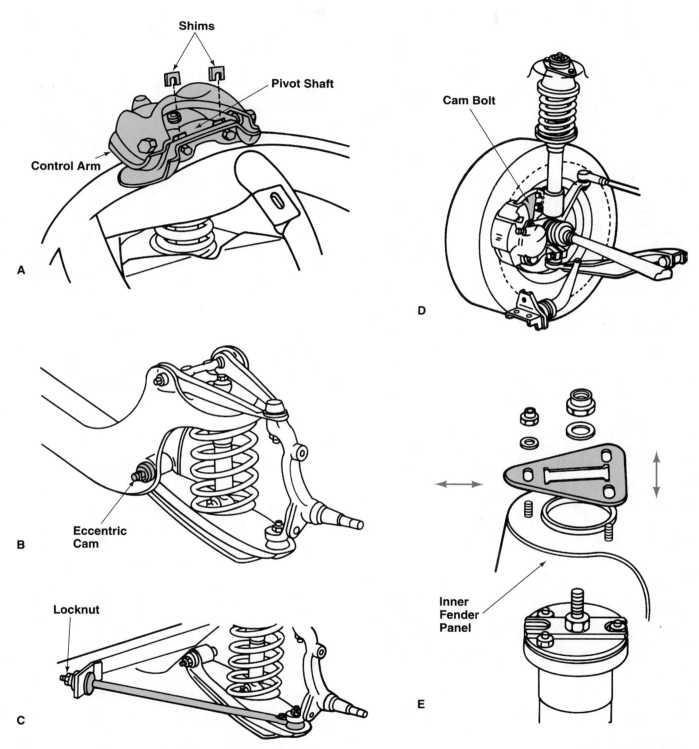

Figure 25–21 Study various ways of adjusting caster, camber, SAI, and included angle. *(Illustrations courtesy of Perfect Circle/Dana Corp.)* (A) Shims are often provided for upper-lower control arm suspension systems. Different thickness shims in front and rear will change caster. Inserting the same size shims only changes camber, SAI, and included angle. (B) Turning the eccentric cam bolt will pivot the control arm in and out for SAI, included angle, and camber-caster adjustment. (C) Shortening or lengthening the strut rod will pivot the outer end of the lower control arm to alter caster. (D) Cam bolts on the lower end of the strut will pivot the steering knuckle as needed for camber, SAI, and included angle adjustment. (E) Many strut suspensions have slotted holes for adjustment of caster, camber, SAI, and included angle. If they are not slotted, aftermarket kits are available to allow for adjustment. *(continued on next page)*

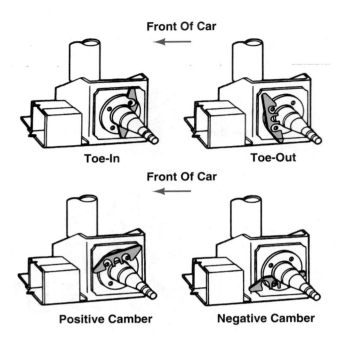

Front Of Car

Toe-In   Toe-Out

Front Of Car

Positive Camber   Negative Camber

Figure 25–21 *(Continued)* (F) Different thickness shims can be placed between the rear knuckle and the strut to adjust alignment.

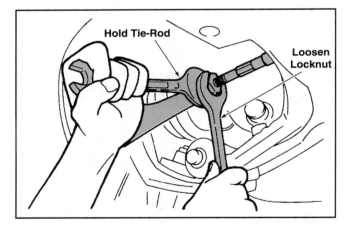

Hold Tie-Rod

Loosen Locknut

Figure 25–22 Toe is adjusted by shortening or lengthening tie-rod ends. If the steering wheel is centered, change the length of the right and left sides equally. This will keep the steering wheel centered. If the steering wheel is not centered, change the tie-rods differently to center the wheel. *(Copyright Mazda Motor of America, Inc. Used by permission)*

Technical Terms

wheel alignment	*toe*
caster	*toe-in*
positive caster	*toe-out*
negative caster	*steering axis inclination*
camber	*(SAI)*
positive camber	*turning radius*
negative camber	*toe-out on turns*

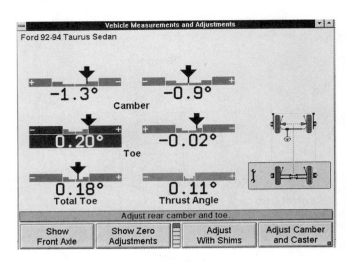

Figure 25–23 Today's computerized alignment equipment will guide you through the alignment procedure. This unit uses arrows and scales to show readings of alignment angles all at once. It also lets you know which readings are within specifications and which are not. *(Hunter Engineering Company)*

tracking	*ride height*
thrust angle	*curb height*
included angle	*jounce*
wheelbase	*rebound*
set-back	*jounce-rebound*
reading tires	*check*

Review Questions

1. List four reasons a vehicle would require an alignment after collision repair.

2. Give six basic wheel alignment angles.

3. _____ _____ tilts the tops of the steering knuckles toward the rear of the vehicle, when viewed from the side of the vehicle.

4. _____ _____ has the top of the wheel tilted out, when viewed from the front. The outer edge of the _____ _____ contacts the road.

5. Camber is adjustable on most vehicles. True or false?

6. Explain the importance of proper toe adjustment.

7. Technician A says that rear-wheel drive vehicles are often adjusted to have toe-in at the front wheels. Technician B says that front-wheel drive vehicles need to have their front wheels set for a slight toe-out. Who is correct?
   a. Technician A
   b. Technician B
   c. Both Technicians
   d. Neither Technician

SUSPENSION PROBLEM DIAGNOSIS

Check	Noise	Instability	Pulls to One Side	Excessive Steering Play	Hard Steering	Shimmy
Tires/Wheels	Road or tire noise	Low or uneven air pressure; radials mixed with belted bias ply tires	Low or uneven air pressure; mismatched tire sizes	Low or uneven air pressure	Low or uneven air pressure	Wheel out of balance or uneven tire wear or overworn tires; radials mixed with belted bias ply tires
Shock Dampers (Struts/ Absorbers)	Loose or worn mounts or bushings	Loose or worn mounts or bushings; worn or damaged struts or shock absorbers	Loose or worn mounts or bushings	—	Loose or worn mounts or bushings on strut assemblies	Worn or damaged struts or shock absorbers
Strut Rods	Loose or worn mounts or bushings	Loose or worn mounts or bushings	Loose or worn mounts or bushings	—	—	Loose or worn mounts or bushings
Springs	Worn or damaged	Worn or damaged	Worn or damaged, especially rear	—	Worn or damaged	—
Control Arms	Steering knuckle control arm stop; worn or damaged mounts or bushings	Worn or damaged mounts or bushings	Worn or damaged mounts or bushings	—	Worn or damaged mounts or bushings	Worn or damaged mounts or bushings
Steering System	Component wear or damage	Component wear or damage	Component wear or damage	Component wear or damage	Component wear or damage	Component wear or damage
Alignment	—	Front and rear, especially caster	Front, camber and caster	Front	Front, especially caster	Front, especially caster
Wheel Bearings	On turns or speed changes: front-wheel bearings	Loose or worn (front and rear)	Loose or worn (front and rear)	Loose or worn (front)	—	Loose or worn (front and rear)
Brake System	—	—	On braking	—	On braking	—
Other	Clunk on speed changes: trans-axle; click on turns: CV joints; ball joint lubrication	—	—	—	Ball joint lubrication	Loose or worn friction ball joints

Figure 25–24 Study this chart that shows suspension system problems that can affect alignment.

STEERING PROBLEM DIAGNOSIS						
Check	**Problem**					
	Noise	**Instability**	**Pull to One Side**	**Excessive Steering Play**	**Hard Steering**	**Shimmy**
Tires/Wheels	Road or tire noise	Low or uneven tire pressure; radial tire lead	Low or uneven tire pressure; radial tire lead	Low or uneven tire pressure	Low or uneven tire pressure	Unbalanced wheel; uneven tire wear; over-worn tires
Tie-Rods	Squeal in turns: worn ends	—	Incorrect toe: tie-rod length	Worn ends	Worn ends	Worn ends
Mounts or Bushings	Parallelogram steering: steering gear mounting bolts, linkage connections; rack-and-pinion steering: rack mounts	Idler arm bushing	—	Parallelogram steering: steering gear mounting bolts, linkage connections; rack-and-pinion steering: rack mounts	Parallelogram steering: steering gear mounting bolts, linkage connections; rack-and-pinion steering: rack mounts	Parallelogram steering: steering gear mounting bolts, linkage connections; rack-and-pinion steering: rack mounts
Steering Linkage Components	Bent or damaged steering rack	Incorrect center link or rack height	Incorrect center link or rack height	Worn idler arm, center link, or pitman arm studs; worn or damaged rack	Idler arm binding	Worn idler arm, center link, or pitman arm studs
Steering Gear	Improper yoke adjustment on rack-and-pinion steering	—	—	Improper yoke adjustment on rack-and-pinion steering; worn steering gear or incorrect gear adjustment on parallelogram steering; loose or worn steering shaft coupling	Parallelogram steering: low steering gear lubricant, incorrect adjustment; rack-and-pinion: bent rack, improper yoke adjustment	—
Power Steering	—	—	—	—	Fluid leaks; loose, worn, or glazed steering belt; weak pump; low fluid level	—
Alignment	—	—	Unequal caster or camber	—	Excessive positive caster, excessive scrub radius (incorrect camber and/or SAI)	Incorrect caster

Figure 25–25 Study this chart for diagnosing steering system problems.

8. SAI can be used with structural measurements to diagnose:
   a. Strut tower misalignment
   b. Control arm mounting location damage
   c. Misaligned frame or unibody structure
   d. Camber change
   e. Toe-out condition

9. Define the term "tracking."

10. This is the measurement between the center of the front and rear wheel hubs.
    a. Scrub
    b. Tracking
    c. Toe
    d. Wheelbase

11. How and why do you read tires?

12. How do you measure vehicle ride height?

13. In your own words, how do you inspect the underside of a vehicle before a wheel alignment?

14. Summarize the eight major steps for a wheel alignment.

ASE–Style Review Questions

1. Following a collision, a vehicle would NOT require an alignment if:
   a. there is damage to any steering and suspension parts.
   b. there is damage to any steering or suspension mounting locations.
   c. there is engine cradle damage or a position change.
   d. there is damage to the deck lid.

2. Which of the following is a directional stability adjustment?
   a. Toe
   b. Caster
   c. Camber
   d. Steering axis inclination

3. A customer complains of rapid tire wear. Which of the following is the most common cause of this problem?
   a. Toe
   b. Caster
   c. Camber
   d. Steering axis inclination

4. Which of the following should the front wheels of front-wheel drive vehicles be set for?
   a. a slight negative camber
   b. a slight positive camber
   c. a slight toe-in
   d. a slight toe-out

5. *Technician A* says to check ride height before a wheel alignment. *Technician B* says to perform a road test before a wheel alignment. Who is correct?
   a. Technician A
   b. Technician B
   c. Both Technicians A and B
   d. Neither Technician

6. Which of the following is the normal sequence for a wheel alignment?
   a. toe, caster, camber
   b. camber, caster, toe
   c. caster, inclination, toe
   d. caster, camber, toe

7. At which of the following locations is toe adjusted?
   a. tie rod ends
   b. top of strut towers
   c. ball joints
   d. steering knuckle

8. At which of the following locations is camber normally adjusted on most vehicles?
   a. tie rod ends
   b. top of strut towers
   c. ball joints
   d. steering knuckle

9. A vehicle suffers from tire wear on the inner edge of the tire. *Technician A* says that the problem is caster. *Technician B* says the problem is due to camber maladjustment. Who is correct?
   a. Technician A
   b. Technician B
   c. Both Technicians A and B
   d. Neither Technician

10. A vehicle steering wheel has a slight pull to the left. *Technician A* says to adjust caster. *Technician B* says to adjust toe. Who is correct?
    a. Technician A
    b. Technician B
    c. Both Technicians A and B
    d. Neither Technician

Activities

1. Visibly inspect the alignment and tire wear on several cars. Try to find one with obvious misalignment. Write a report of your findings.

2. Refer to the operating manual for alignment equipment. Read through the specific procedures for the alignment equipment. Summarize any unique procedure and prepare to discuss it with the class.

26

Electrical Repairs

Objectives

After studying this chapter, you should be able to:

- Define the terms voltage, current, and resistance.

- Use various kinds of electrical test instruments.

- Find electrical problems.

- Explain the operation of automotive electrical-electronic systems.

- Describe the operation of computer systems.

- Use a vehicle's self-diagnosis capabilities.

- Use scanners to find electrical-electronic problems.

Introduction

Electrical systems, such as the ignition, charging, starting, lighting, and computer systems, perform needed functions for a vehicle. It is almost impossible to work on any section of a vehicle without handling some type of electrical part. See Figure 26–1.

Electrical system repair is an essential aspect of repairing a collision-damaged vehicle. Electrical repairs are becoming more common due to the increased use of computer-controlled systems. These repairs can be made complicated by the many computer circuits running throughout the body structure.

This chapter will summarize the operation and repair of electrical-electronic systems.

> **Note!** In this text, the term "electrical" refers to non-solid-state components, like relays, solenoids, circuit breakers. "Electronic" refers to solid-state components, like diodes, transistors, and computers.

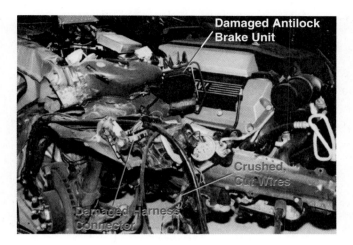

Figure 26–1 During a collision, the wires near the point of impact can be cut or crushed and damaged. As a technician, you should be able to locate and correct electrical problems.

ELECTRICAL TERMINOLOGY

Various vehicle systems are controlled by a series of electrical controls and devices. To understand electricity, you must become familiar with three electrical terms: current, voltage, and resistance.

Current is the movement of electricity (electrons) through a wire or circuit. It is measured in amperes or amps using an ammeter. The common electrical symbol for current is "A" or "I."

Voltage is the pressure that pushes the electricity through the wire or circuit. The **power source** generates the voltage that causes current flow. Voltage is measured in volts using a voltmeter. The symbol for voltage is "E" or "V." See Figure 26–2.

Resistance is a restriction or obstacle to current flow. It tries to stop the current caused by the applied voltage. Circuit or part resistance is measured in ohms using an ohmmeter. The symbol for resistance is R or Ω.

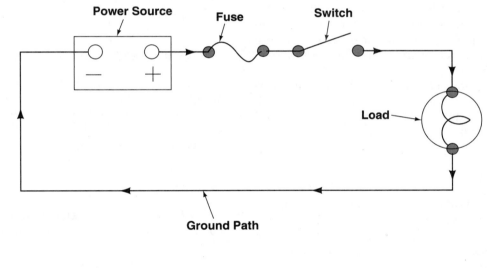

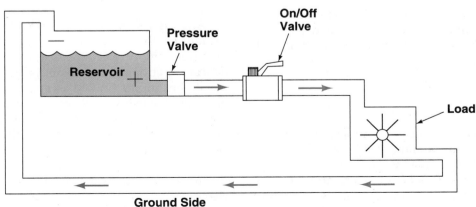

Figure 26–2 Note this comparison of electric and hydraulic circuits. Both use the same principles. *(I-CAR)* A: An electric circuit has a battery as its power source, a switch to control current, and a light bulb as load. B: A hydraulic circuit has a reservoir, a valve, and a hydraulic motor as its basic parts.

A **conductor** carries current to the parts of a circuit. **"Hot wires"** connect the battery positive to the components of each circuit. **Insulation** stops current flow and keeps the current in the metal wire conductor. The body structure provides the **ground** conductor back to the battery negative cable.

Electric Circuits

An **electric circuit** contains a power source, conductors, and a load. Some resistance is designed into a circuit in the form of a load. The **load** is the part of a circuit that converts electrical energy into another form of energy (light, movement, heat, etc.), Figure 26–2. Other parts are added to this simple circuit to protect it from damage and to do more tasks.

A **series circuit** has only one conductor path or leg for current through the circuit. Current must flow through the wires and components one after the other. If any part of the circuit is **opened** (disconnected), all of the series circuit stops working.

A **parallel circuit** has two or more legs or paths for current. Current can flow through either leg independently. One path can be **closed** (electrically connected) and the other opened, and the closed path will still operate.

A **series-parallel circuit** has both series and parallel branches in it. It has characteristics of both circuit types. All the circuits in a vehicle can be classified and tested using the rules of these three circuits.

Ohm's Law

Ohm's Law is a math formula for calculating an unknown electrical value (amps, volts, or ohms) when two values are known. If you know two values, you can mathematically calculate the third unknown value.

For instance, if you know that a circuit has 12.6 volts and 2 amps, what is the circuit resistance? Plug your two known values into Ohm's Law and you can find out (12.6 divided by 2 equals 6.3 amps). See Figure 26–3.

Magnetism

Magnetism involves the study of how electric fields act upon ferrous (iron-containing) objects. Many electrical-electronic parts use magnetism.

A **flux** or **magnetic field** is present around permanent magnets and current-carrying wires. This invisible energy is commonly used to move metal parts. An **electromagnet** is a set of **windings** or wires wrapped around an iron core. When current flows through the windings, a powerful magnetic field is

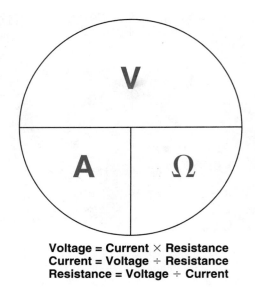

Voltage = Current × Resistance
Current = Voltage ÷ Resistance
Resistance = Voltage ÷ Current

Figure 26–3 Ohm's Law is a simple formula for calculating electrical values in a circuit. (I-CAR)

produced. Electric motors, solenoids, relays, and other parts use this principle.

DIAGNOSTIC EQUIPMENT

Locating an electrical fault is not possible without using **diagnostic tools** (meters, test lights, jumper wires, etc.). Keep in mind that today's delicate electronic systems can be damaged if the wrong methods and equipment are used.

Multimeters

A **multimeter** is a voltmeter, ohmmeter, and ammeter combined into one case. Also called a **VOM** (volt-ohm-ammeter), it can be used to measure actual electrical values for comparison to known good values. Refer to Figure 26–4.

The **digital multimeter (DVOM)** has a number readout for the test value. This type is recommended by auto manufacturers because it will NOT damage delicate electronic components. A high-impedance (10 mega-ohm input) DVOM is recommended to avoid damaging sensitive components. Digital readouts give the precise measurement needed for proper diagnosis. DVOMs are used in conjunction with the vehicle's service manual.

An **analog multimeter (AVOM)** has a pointer needle that moves across the face of a scale when making electrical measurements. Use of an AVOM can damage sensitive electronic components. It should be used only when testing all-electrical, NOT electronic, circuits. AVOMs help show a fluctuating or changing reading, such as from an intermittent (changing) problem.

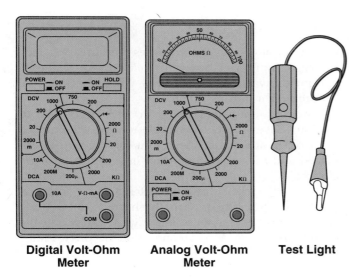

Digital Volt-Ohm Meter   **Analog Volt-Ohm Meter**   **Test Light**

Figure 26–4 These are basic electrical testing tools. Each has its advantages and disadvantages. *(Courtesy of DaimlerChrysler Corporation)*

Test Light, Jumper Wires

An externally powered ***test light*** is often used to determine if there is current flowing through the circuit. One lead of the test lamp is connected to a good ground. The other lead connects to a point in the circuit. If the lamp lights, current is present at that point. See Figure 26–5.

Whenever a technician has an electrical system problem that does not directly concern the computer, a test light is handy.

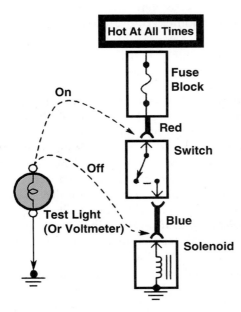

Figure 26–5 A test light provides a fast way of probing for voltage. Make sure you are using a high-impedance instrument when testing computer circuits. *(Courtesy of General Motors Corporation, Service Operations)*

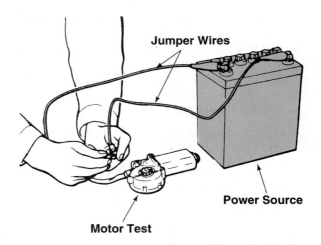

Figure 26–6 Jumper wires can be used to bypass resistive parts or to connect power directly to parts to check their operation.

Connect the light to the voltage source and to ground. If the test light does not glow, there is an open circuit somewhere. If the light is on, but the part does not work, the part is probably bad.

> *WARNING! Test lights should NOT be used randomly, only when specified in the service manual procedures. They are NOT to be used on solid-state circuits. They can damage electronic parts if their impedance (internal resistance) is too low.*

Jumper wires are used to temporarily bypass circuits or components for testing. They consist of a length of wire with an alligator clip at each end. They can be used to test circuit breakers, relays, lights, and other components, Figure 26–6.

Using Multimeters

When measuring resistance, always disconnect the circuit from the power source. The multimeter must never be connected to a circuit in which current is flowing. Doing so can damage the meter.

Use the service manual to determine the normal resistance of the part being checked. For example, a computer system sensor may have a normal resistance reading of 45–55 ohms. You would select the "200" range on the multimeter. This is because the resistance reading of 44–55 ohms will be 200 ohms or less.

Always refer to the multimeter owner's manual before using the multimeter for diagnostic purposes. You must make sure that you understand how to use the multimeter before performing diagnostics. All multimeters are fairly similar, but there can be differences between makes.

To measure resistance:

1. Connect the multimeter test leads to opposite ends of the circuit or wire being tested.

2. Set the range selector switch on the highest range position. Turn the multimeter on.

3. Reduce the range setting until the meter shows a reading near the middle of its scale. Some DVOMs have an **auto ranging** function that adjusts the settings automatically.

Checking Continuity

A circuit remains closed and operational when it has **continuity** or a continuous conductor path. Typically, the service manual will ask for a continuity check of a wire or part. The continuity check is done to see if the electrical circuit has a complete path without any opens.

1. Connect the multimeter test leads to the opposite ends of the wire or part being tested.

2. Set the range selector switch on the highest resistance range position.

3. Read the meter. An **infinite reading** (maximum resistance) shows an open. A zero reading shows continuity.

Measuring Voltage

The multimeter allows you to select either alternating current voltage (ACV) or direct current voltage (DCV). The ACV is selected when measuring alternating current voltage. **AC current** is the type of current that is found in your home wiring.

DCV is selected when measuring direct current voltage. **DC current** is what is normally measured in the automobile. Some signals from sensors, however, can be AC.

When using the multimeter to measure voltage, the selection of the range scale is very important. If the voltage reading will be 12–14 volts, select the 20 V range on the multimeter. This range is selected because the voltage reading will be less than 20 volts.

Measuring Current

Current or amperage is sometimes measured to check the consumption of power by a load. For example, current draw is often measured when checking the condition of a starting motor.

Modern ammeters have an **inductive pickup** that slips over the wire or cable to measure current. With older ammeters, the circuit had to be disconnected and the meter connected in series to measure current.

A high current draw indicates a low resistance, like from a dragging or partially shorted motor. A low current draw means there is a high resistance in the circuit, like from a bad connection or dirty motor brushes.

Checking for Shorts

Many times adjoining wires within a wiring harness get pinched together or severed in a collision. A multimeter check will detect a severed wire.

When checking for a short between two adjoining wires in a harness:

1. Set the multimeter range selector switch the same as for a continuity check.

2. Connect the multimeter test leads to the opposite ends of the adjoining wires.

3. Because the wires are insulated, there should be a reading of almost infinity on the multimeter. If the multimeter reads zero the wires are shorted to ground or another ground wire.

WIRING DIAGRAMS

To determine and isolate electrical problems, it is often necessary to trace through the electrical circuit using a **wiring diagram,** Figure 26–7.

Abbreviations are used on wiring diagrams so that more information can be given. The service manual will have a chart explaining each abbreviation, number, and symbol on the diagram. The abbreviations for one auto maker are in Figure 26–8.

Electrical symbols are graphic representations of electrical-electronic components. Symbol charts can be found at the front of the wiring diagram section of the service manual. One is given in Figure 26–9.

The symbol used to identify a part is either a universal symbol or an auto manufacturer's symbol. Since manufacturers use different symbol designs for some parts, follow the symbol chart for the specific wiring diagram being used.

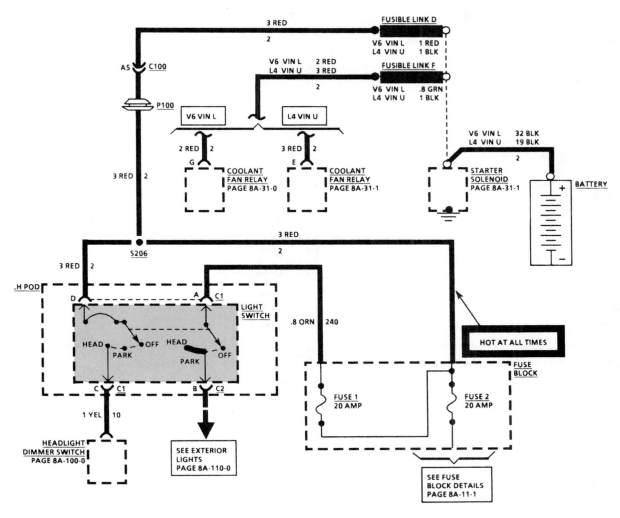

Figure 26–7 A wiring diagram is like a "road map" of the circuit being tested. It shows how wires lead to different locations or parts in the circuit. It also gives color codes, abbreviations, and other information for troubleshooting. *(Courtesy of General Motors Corporation, Service Operations)*

Most wires on wiring diagrams will be identified by their insulation color. **Wire color coding** allows you to find a specific wire in a harness or in a connector, Figure 26–10. The color code abbreviation chart can also be found at the front of the wiring diagram section of the service manual. Remember, different auto makers use different color coding abbreviations on their wiring diagrams.

Many wiring diagrams found in the service manuals also have circuits numbered. **Circuit numbering** is used to specify exactly which part of the circuit the service manual is referring to.

Remember! A wiring diagram is more like a "book" than a "picture." You can not understand a wiring diagram just by glancing at it. Like a book, you must read the diagram carefully all the way through for a complete understanding.

A **wiring harness** has several wires enclosed in a protective covering. A vehicle has several wiring harnesses, usually named after their location in the vehicle. The service manual will give illustrations with code numbers for locating parts and connections, Figure 26–11.

The service manual may also give a **part location diagram** for finding electrical parts (harnesses, sensors, switches, computers, etc.). One is given in Figure 26–12.

ELECTRIC COMPONENTS

A **switch** is used to turn a circuit on or off manually (by hand). When the switch is CLOSED (on), the circuit is complete (fully connected) and will operate.

ABBREVIATIONS

Abbreviation	Meaning of Abbreviation	Abbreviation	Meaning of Abbreviation
A	Ampere (S)	ICM	Ignition control module
ABS	Anti-lock brake system	IG	Ignition
ASM	Assembly	kW	kilowatt
AC	Alternating current	LH	Left hand
A/C	Air conditioning	M/T	Manual transmission
ACC	Accessories	OD	Over drive
A/T	Automatic transmission	OPT	Option
C/B	Circuit breaker	RH	Right hand
CSD	Cold start device	RR	Rear
EI	Electronic ignition	RWAL	Rear wheel anti-lock brake system
EBCM	Electronic brake control module	ST	Start
ECGI	Electronic control gasoline injection	STD	Standard
ECM	Engine control module	SW	Switch
EGR	Exhaust gas recirculation	SWB	Short wheel base
EFE	Early fuel evaporation	TCM	Transmission control module
4A/T	4-speed automatic transmission	3A/T	3-speed automatic transmission
4X4	Four-wheel drive	V	Volt
FL	Fusible link	VSV	Vacuum switching valve
FRT	Front	W	Watt (S)
H/L	Headlight	WOT	Wide open throttle

Figure 26–8 The service manual will give a chart such as this one explaining abbreviations. *(Reprinted with permission by American Isuzu Motors Inc.)*

Legend Of Symbols Used On Wiring Diagrams

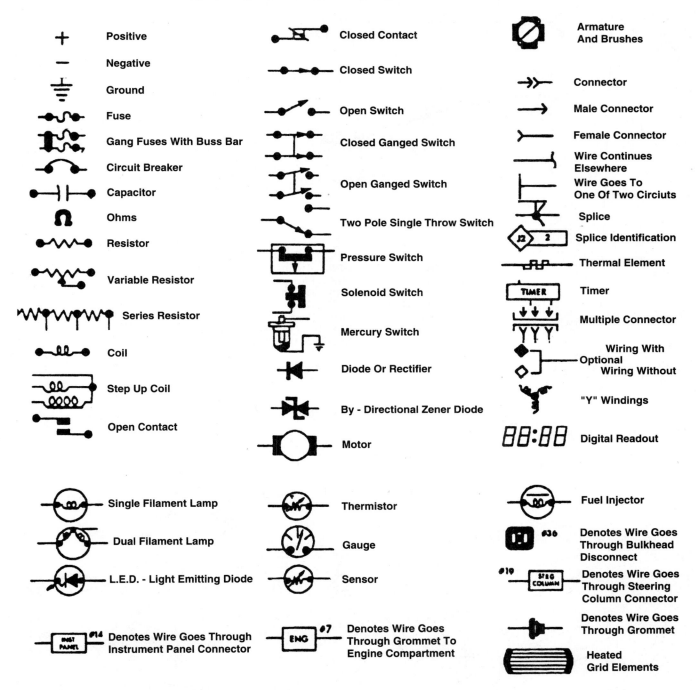

Figure 26–9 Study the symbols typically used on wiring diagrams. *(Courtesy of DaimlerChrysler Corporation)*

Various types can be found in today's vehicles, Figure 26–13 on page 556. A bad switch will often be open in both the closed and open positions.

An *inertia switch* is designed to shut the electric fuel pump off after a collision. Shown in Figure 26–14 on page 556, it must be reset for the engine to restart after a collision. A button on the side of the inertia switch must be pressed to reclose the fuel pump circuit.

A *solenoid* is an electromagnet with a movable core or plunger. When energized, the plunger is pulled into the magnetic field to produce motion. Solenoids are used in many applications: door locks,

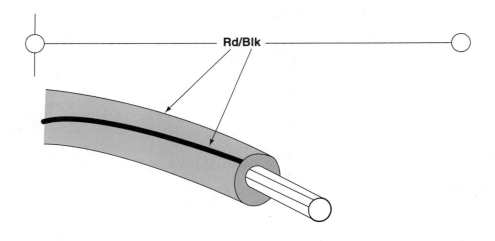

WHT.....White	ORN.....Orange
YEL......Yellow	PNK.....Pink
BLK......Black	BRN.....Brown
BLU......Blue	GRY.....Gray
GRN.....Green	PUR.....Purple
RED......Red	LT BLU.....Light Blue
	LT GRN.....Light Green

Figure 26–10 Wire color codes are abbreviated on wiring diagrams. Study these common abbreviations. (© American Honda Motor Co., Inc.)

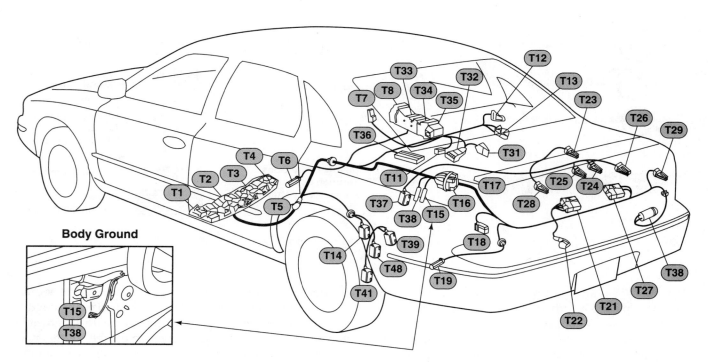

Figure 26–11 Wire harness illustrations give the location of connectors, grounds, and components. Charts in the manual explain the numbers and letters denoting harness configuration. (Used with permission from Nissan North America, Inc.)

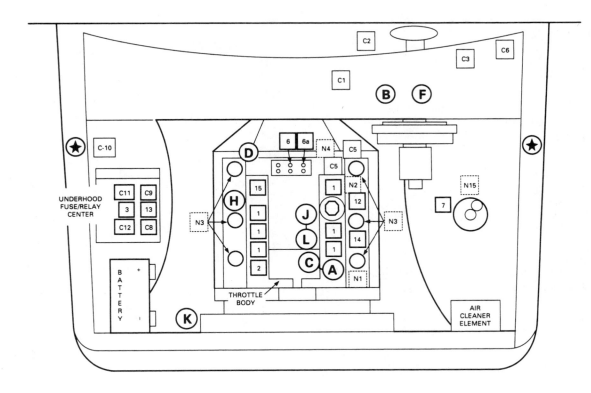

COMPUTER HARNESS

C1   **Engine control module (ECM)** *
C2   **DLC diagnostic connector**
C3   **"Check Engine" Malfunction indicator lamp**
C5   **ECM harness ground**
C6   **Fuse panel**
C8   **ECM main relay**
C9   **Fuel pump fuse 15A**
C10   **Injector resistor**
C11   **Oxygen 10A heater fuse**
C12   **30A ECM main fusible link**

ECM CONTROLLED COMPONENTS

1   **Fuel injector**
2   **Idle air control**
3   **Fuel pump relay**
6   **Ignition control module ignition control**
6a   **Ignition coils**
7   **Knock sensor Module under charcoal canister**
12   **Exhaust Gas Recirculation (EGR) VSV**

13   **Air conditioning relay**
14   **Evaporate emission canister purge VSV**
15   **Induction air control plate system VSV**

ECM INFORMATION SENSORS

A   **Manifold Absolute Pressure**
B   **Heated oxygen sensor**
C   **Throttle position sensor**
D   **Engine coolant temperature**
F   **Vehicle speed sensor**
H   **Crank angle sensor**
J   **Knock sensor (under intake assembly)**
K   **Power steering pressure switch (in-line)**
L   **Intake air temperature**

⊚ **EGR VALVE**

✪ **CHASSIS GROUNDS**

EMISSION COMPONENTS (NOT ECM CONTROLLED)

N1   **Crankcase vent valve (PCV)**
N2   **Exhaust gas recirculation valve back pressure transducer**
N3   **Spark plugs**
N4   **Fuel rail test fitting (for fuel pressure test)**
N15   **Fuel vapor canister**

*   **The ECM is located behind the console in the lower dash.**

Figure 26–12 Component location diagrams show where sensors and other parts are located on the vehicle. *(Reprinted with permission by American Isuzu Motors Inc.)*

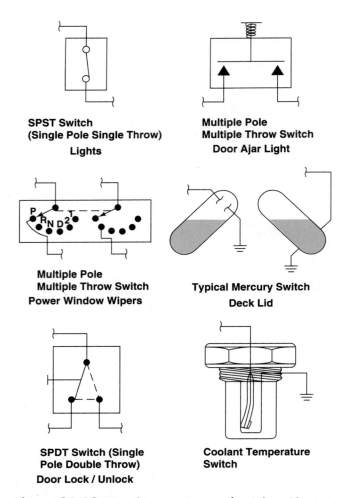

**SPST Switch
(Single Pole Single Throw)
Lights**

**Multiple Pole
Multiple Throw Switch
Door Ajar Light**

**Multiple Pole
Multiple Throw Switch
Power Window Wipers**

**Typical Mercury Switch
Deck Lid**

**SPDT Switch (Single
Pole Double Throw)
Door Lock / Unlock**

**Coolant Temperature
Switch**

Figure 26–13 Note the various types of switches. *(Courtesy of DaimlerChrysler Corporation)*

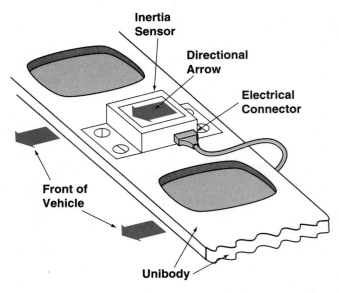

Figure 26–14 The inertia switch is often found in the trunk. A button on the side of the switch must be pressed to reset the switch so the fuel pump will operate.

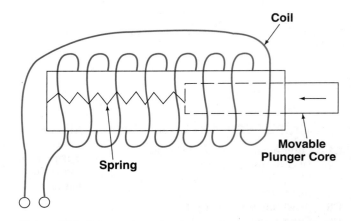

Figure 26–15 A solenoid is the winding wrapped around a movable core. Current flow produces a magnetic field that pulls the core into the winding. *(I-CAR)*

engine idle speed, emission control systems, etc. See Figure 26–15. A bad solenoid can develop winding opens, shorts, or high resistance problems.

A *relay* is a remote control switch. A small switch can be used to energize the relay. The relay coil then acts upon a movable arm to close the relay contacts. This allows a small switch to control high current going to a load. A relay is commonly used with electric motors since they draw heavy current, Figure 26–16. The service manual will often give relay locations for the specific make and model vehicle, Figure 26–17.

A bad relay will often have burned points that prevent current flow to the load. It can also develop coil opens and shorts that keep the points from closing.

Motors use permanent and electromagnets to convert electrical energy into a rotation motion for doing work. Some examples are the electric starting motor for the engine and stepper motors for computer control of parts.

Faulty motors can have worn bushings, brushes that decrease efficiency. They can also have winding shorts and opens that prevent motor operation.

CIRCUIT PROTECTION DEVICES

Circuit protection devices (see Figure 26–18) prevent excess current from burning wires and components. With an overload or short, too much current tries to flow. Without a fuse or breaker, the wiring in the circuit would heat up. The insulation would melt and a fire could result.

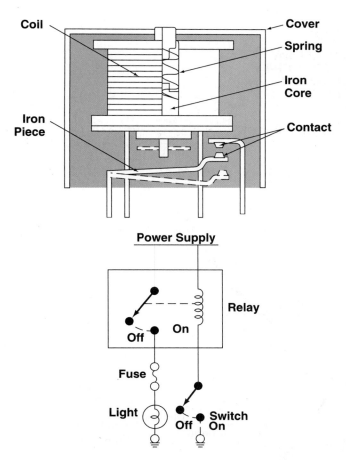

Figure 26–16 A relay is a coil mounted next to contact points. A small amount of current through the coil attracts and closes the points. Then, a large load current flows through the contacts. *(Courtesy of Mitsubishi Motor Sales of America, Inc.)*

FUSES

Fuses burn in half with excess current to protect a circuit from further damage. They are normally wired between the power source and the rest of the circuit. There are three types of fuses in automotive use: cartridge, blade, and ceramic. Refer to Figure 26–19.

The **cartridge fuse** is found on most older domestic vehicles and a few imports. It is composed of a strip of metal enclosed in a glass or transparent plastic tube. To check the fuse, look for a break in the internal wire or metal strip. Discoloration of the glass cover or glue bubbling around the metal end caps is an indication of overheating.

Late-model domestic vehicles and many imports use **blade** or **spade fuses**. To check the fuse, pull it from the fuse panel and look at the element through the transparent plastic housing. Look for internal breaks and discoloration.

The **ceramic fuse** is used on many European imports. The core is a ceramic insulator with a con-

ductive metal strip along one side. To check this fuse, look for a break in the contact strip on the outside of the fuse.

All fuse types can be checked with a circuit tester or multimeter. A **blown fuse** will have infinite resistance.

Fuse ratings are the current at which the fuse will blow. Fuse ratings are often printed on the fuse. Always replace a fuse with one of the same amp rating. If you do NOT, part damage can result from excess current flow.

A **fuse box** holds the various circuit fuses, breakers, and flasher units for the turn and emergency lights. It is often under the instrument panel, behind a panel in the foot well, or in the engine compartment.

> **WARNING!** *Never permanently bypass a fuse or circuit breaker with a jumper wire. Do it only for test purposes.*

FUSE LINKS

Fuse links or fusible links are smaller-diameter wire spliced into the larger circuit wiring for over-current protection. Fuse links are normally found in the engine compartment near the battery. They are often installed in the positive battery lead that powers the ignition switch and other circuits that are live with the key off. Look at Figure 26–20.

Fuse link wire is covered with a special insulation that bubbles when it overheats. This indicates that the fuse link has melted. If the insulation appears good, pull lightly on the wire. If the fuse link stretches, the wire has burned in half. When it is hard to determine if the fuse link is burned out, perform a continuity check.

> **WARNING!** *Do NOT mistake a resistor wire for a fuse link. The resistor wire is generally longer and is clearly marked "Resistor—do NOT cut or splice."*

When replacing fuse links, first cut the protected wire where it is connected to the fuse link. Then, solder a new fuse link of the same rating in place. Since the insulation on manufacturers' fuse links is flameproof, never fabricate a fuse link from ordinary wire because the insulation may NOT be flameproof.

A number of new electrical systems use maxi-fuses in place of traditional fuse links. Maxi-fuses look

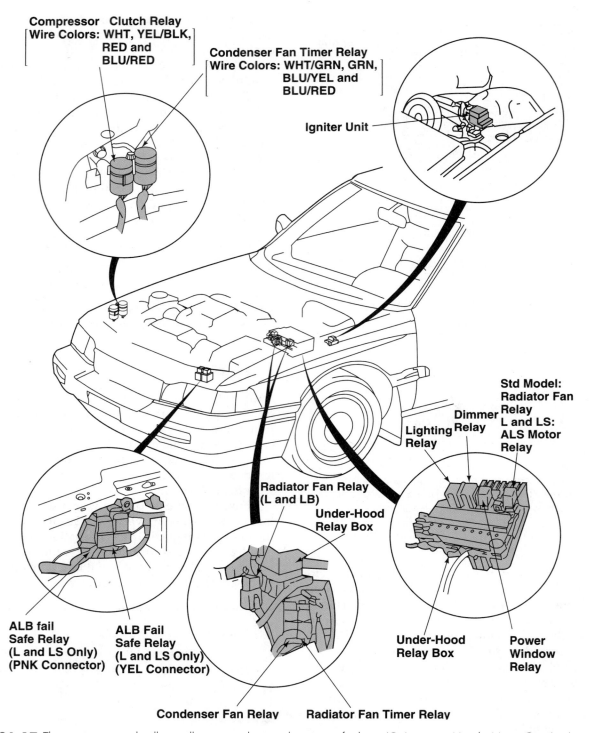

Compressor Clutch Relay
[Wire Colors: WHT, YEL/BLK, RED and BLU/RED]

Condenser Fan Timer Relay
[Wire Colors: WHT/GRN, GRN, BLU/YEL and BLU/RED]

Igniter Unit

Std Model:
Radiator Fan Relay
L and LS:
ALS Motor Relay

Dimmer Relay

Lighting Relay

Radiator Fan Relay (L and LB)

Under-Hood Relay Box

ALB fail Safe Relay (L and LS Only) (PNK Connector)

ALB Fail Safe Relay (L and LS Only) (YEL Connector)

Under-Hood Relay Box

Power Window Relay

Condenser Fan Relay

Radiator Fan Timer Relay

Figure 26–17 The service manual will give illustrations showing locations of relays. (© American Honda Motor Co., Inc.)

and operate like two-prong blade fuses, except they are much larger and can handle more current. Maxi-fuses are located in their own underhood fuse block.

Maxi-fuses are easier to inspect and replace than fuse links. To check a maxi-fuse, look at the fuse ele-ment through the transparent colored plastic side housing. If there is a break in the element, the maxi-fuse has blown. To replace it, pull it from its fuse box or panel. Always replace a blown maxi-fuse with a new one having the same ampere rating.

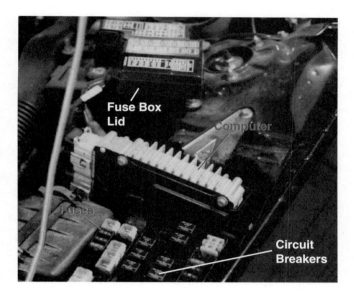

Figure 26–18 Fuses and circuit breakers are normally located in engine compartment, under dash, or under rear seat cushion. Each fuse or circuit breaker will be identified or labeled on plastic lid or cover.

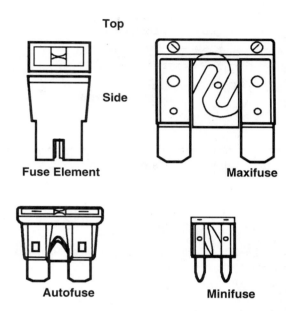

Figure 26–19 Fuses protect circuits from over-current damage. *(Courtesy of General Motors Corporation, Service Operations)*

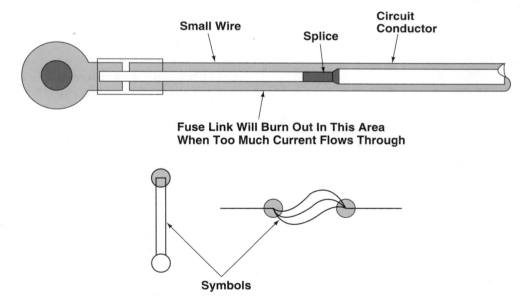

Figure 26–20 Fuse links are often located in the engine compartment right after the battery. They have a small wire that will burn out to protect the circuit from shorts. *(Courtesy of DaimlerChrysler Corporation)*

CIRCUIT BREAKERS

Circuit breakers heat up, and open with excess current to protect the circuit, Figure 26–21. They do NOT suffer internal damage like a fuse. Many circuits are protected by circuit breakers. They can be fuse panel mounted or in-line. Like fuses, they are rated in amperes.

Each circuit breaker conducts current through an arm made of two types of metal bonded together (known as a **bimetal arm**). If the arm starts to carry too much current, it heats up. As one metal expands farther than the other, the arm bends, opening the contacts and breaking the current flow.

In the **cycling breaker,** the bimetal arm will begin to cool once the current to it is stopped. Once it returns to its original shape, the contacts close and power is restored. If the current is still too high, this cycle of breaking the circuit will be repeated.

Cycling circuit breakers are generally used in circuits that are prone to occasional overloads. These

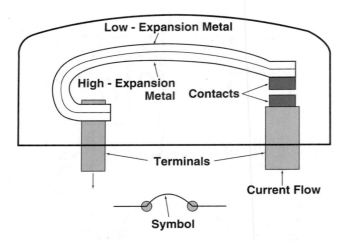

Figure 26–21 A circuit breaker uses a bimetal arm to open a circuit with too much current.

include power windows, in which a jammed or sticking window can overwork the motor. In this situation, a circuit breaker prevents the motor from burning out.

Noncycling breakers open and must be manually reset to close the circuit. In automotive work, two types of noncycling circuit breakers are used. One is reset by removing the power from the circuit. The other type is reset by depressing a reset button.

> *WARNING! When replacing fuses and circuit breakers, install one with the same amp rating. A higher rated unit could cause an electrical fire.*

ELECTRICAL-ELECTRONIC SYSTEMS

The modern vehicle has numerous electrical-electronic systems, each designed to do a specific task. Some of these are discussed below.

Ignition System

The *ignition system* produces an electric arc (spark) in a gasoline engine to cause the fuel to burn. It must fire the spark plugs at the right time. The resulting combustion pressure forces the pistons down to spin the crankshaft.

The *ignition coil* is a step-up transformer that produces high voltage (30,000 volts or more) needed to make current jump the spark plug gap. A switching device is either contact points or an electronic circuit that causes the ignition coil to discharge its electrical

energy. Either a distributor or the computer controls the operations of the ignition coil.

Spark plug wires are high-tension wires that carry coil voltage to each spark plug. The *spark plugs* use ignition coil high voltage to ignite the fuel mixture in the engine's combustion chambers.

Electronic Fuel Injection

Electronic fuel injection uses solenoid type injectors (fuel valves) and computer control to meter gasoline into the engine. An in-tank electric pump pulls fuel out of the gas tank. A second, in-line fuel pump then pushes fuel up to the engine.

A **pressure regulator** on the *fuel rail* (large fuel log or line at engine) limits maximum system pressure.

Starting and Charging Systems

The *starting system* has a large electric motor that turns the engine flywheel. This spins or "cranks" the crankshaft until the engine starts and runs on its own power. See Figure 26–22.

The *ignition switch* in the steering column is used to connect battery voltage to a starter solenoid or relay. Other ignition switch terminals are connected to other electrical circuits. A *starter solenoid,* when energized, connects the battery and starting motor.

The *starting motor* is a large DC motor for rotating the engine flywheel. It normally bolts to the rear, lower, side of an engine. A few are mounted inside the engine, under the intake manifold. The *flywheel ring gear* meshes with the starter-mounted gear while cranking.

The *charging system* recharges the battery and supplies electrical energy when the engine is running, Figure 26–23. An alternator or belt-driven DC generator produces this electricity.

A *voltage regulator,* usually mounted in the alternator, controls alternator output. *Charging system voltage* is typically 13 to 15 volts.

To quickly check the condition of a charging system, connect a voltmeter across the battery. With the engine off, you will read battery voltage. It should be above 12.5 volts. If it is not, the battery needs charging or is defective. When you start the engine with all electrical accessories ON (lights, radio, etc.), the voltage must stay above battery voltage. If it does NOT, there is something wrong with the charging system.

Engine No-Start Problem. Remember! If an engine cranks but fails to start, check for "spark" and "fuel." Both are needed for an engine to operate. See Figure 26–24.

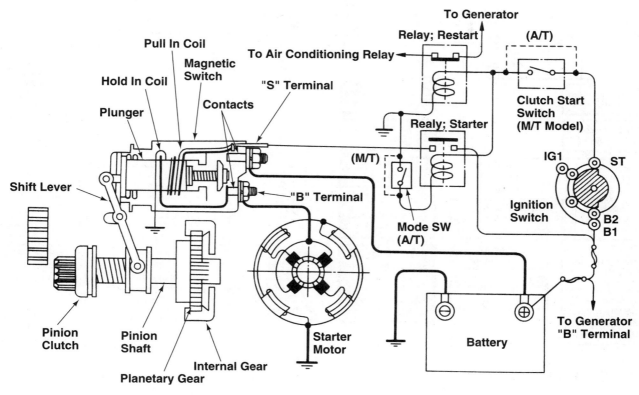

Figure 26–22 Study this typical starting system. Trace the wires from the battery, through the ignition switch, relay, solenoid, and motor. *(Reprinted with permission by American Isuzu Motors Inc.)*

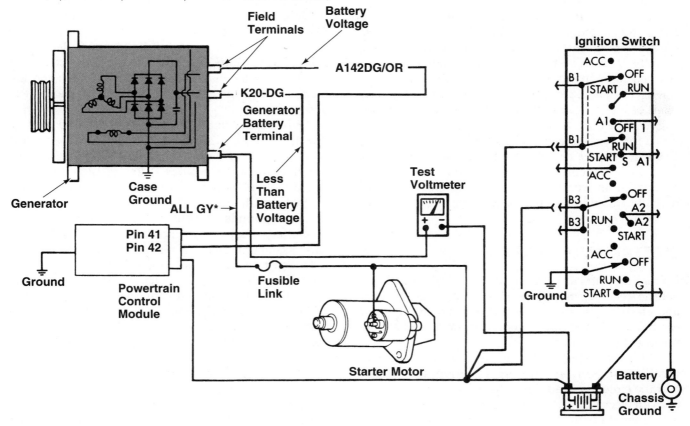

Figure 26–23 Study this charging system circuit. *(Courtesy of DaimlerChrysler Corporation)*

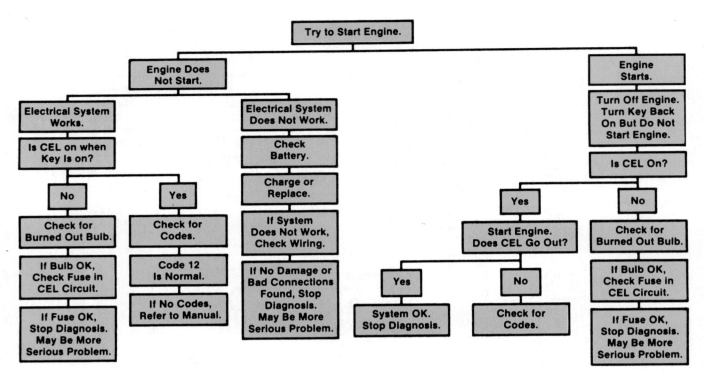

Figure 26–24 Read through this typical troubleshooting chart.

To CHECK FOR SPARK, pull off one of the spark plug wires. Install an old spark plug into the wire and lay the spark plug on the engine ground. When you crank the engine, a bright spark should jump across the spark plug gap. If it does not, something is wrong with the ignition system (blown fuse, damaged wires, crushed components, etc.).

If you have spark, CHECK FOR FUEL. This can often be done by installing a pressure gauge on the engine's fuel rail. A special test fitting is usually provided for a pressure gauge. With the engine cranking or the key on, the gauge should read within specs. If it does not, something is keeping the electric fuel pump(s) from working normally. Check for a clogged fuel filter, blow pump fuse, or wiring problem.

ELECTRICAL PROBLEMS

Common electrical problems in a collision-damaged vehicle generally result from impact damage to wires and electrical-electronic components. These problems can be classified as opens, shorts, grounds, and abnormal resistance values.

An *open circuit* is an unwanted break in an electrical circuit. With an open circuit, current flow ceases and the circuit is dead. This effect can occur in a wire or in an electrical part, such as a light bulb filament.

A *short circuit* is an unwanted wire-to-wire connection in an electrical circuit, Figure 26–25. A short occurs when the insulation is worn between two adjacent wires and the wires contact each other. Shorts often occur when wires are damaged or pinched due to collision damage.

A short can also result to ground, termed a *grounded circuit.* A grounded circuit condition occurs when the insulation on a wire is worn and the metal in the wire touches the metal of the vehicle body or frame.

When this occurs, the current is allowed to flow directly to ground without flowing to the electrical parts in the circuit. In other words, the current takes the **path of least resistance.** A low resistance or short to ground will blow the fuse or trip the circuit breaker for that circuit.

Bad grounds prevent current from returning to the battery. They are the most common cause of electrical system problems. If just the ground for one part is faulty, it will affect only the one circuit. If a major ground is faulty, many circuits will fail all at once. A major ground is one that connects the battery to the engine block or body. Sometimes a short is intermittent—it shorts momentarily when the vehicle bounces heavily or jars. A flickering dash light is an example of this.

An **abnormal resistance** is due to a bad connection or partial short. When tested, the circuit or component will show a value that is not within specs. For example, a high resistance in a light circuit will make the bulbs burn dimly. This is because the abnormally

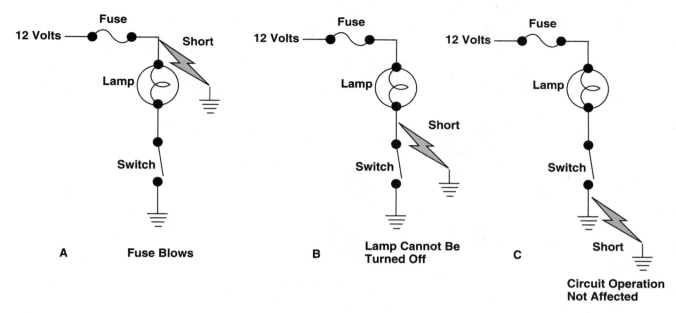

Figure 26–25 The location of a short will have different effects on the circuit. *(I-CAR)* (A) A short at this location would blow fuse. (B) A short here would bypass the switch and the bulb would always glow, even with the switch open. (C) A short at ground would not affect the circuit.

high circuit resistance is not allowing enough current to flow through the bulbs.

Diagnostic Charts

Diagnostic charts give possible causes for electrical problems and symptoms. They are important tools for collision repair technicians and insurance appraisers. A general electrical diagnostic chart is in Figure 26–26.

Most auto manufacturers use some type of diagnostic flow chart to aid technicians in the repair of electrical/electronically controlled systems. Diagnostic flow charts provide a systematic approach to troubleshooting and repair. They are found in service manuals and are given by vehicle make and model.

During the diagnosis, start at the top of the chart and follow the sequence down. The chart will point to the next area to move to after each check is performed. Work through the entire chart, step by step. Perform the repairs or parts replacements as indicated.

BATTERIES

A *battery* stores electrical energy chemically. It provides current for the starting system. It also provides current to all other electrical systems when the engine is not running.

Because of the potential electrical problems caused by a collision, always disconnect both battery cables right away. This will prevent the chance of an electrical fire while working.

When removing or disconnecting the battery, the ignition must be off to prevent voltage spikes. *Voltage spikes* are voltage surges that can destroy many microcircuits in today's electronic systems.

It is a good idea to remove the battery completely from the vehicle before doing any collision repair work. Once the battery has been placed on a bench, it should be checked for a cracked case and similar damage. In many cases with frontal collisions, the battery may need to be replaced.

A charge indicator that shows battery state of charge is often built into the top of the battery. It shows different colors to indicate battery condition.

A voltmeter or a specialized **battery tester** will also check battery condition (specific gravity or voltage). A good, fully charged battery will show 12.5 to 12.6 **open circuit volts** (nothing turned on). See Figure 26–27.

Charging a Battery

The procedure for charging a battery with a battery charger is as follows:

1. Disconnect the negative battery cable.

2. Check the battery casing for damage.

3. Check the water level and add water if necessary.

4. Loosen the vent caps, if the battery is so equipped.

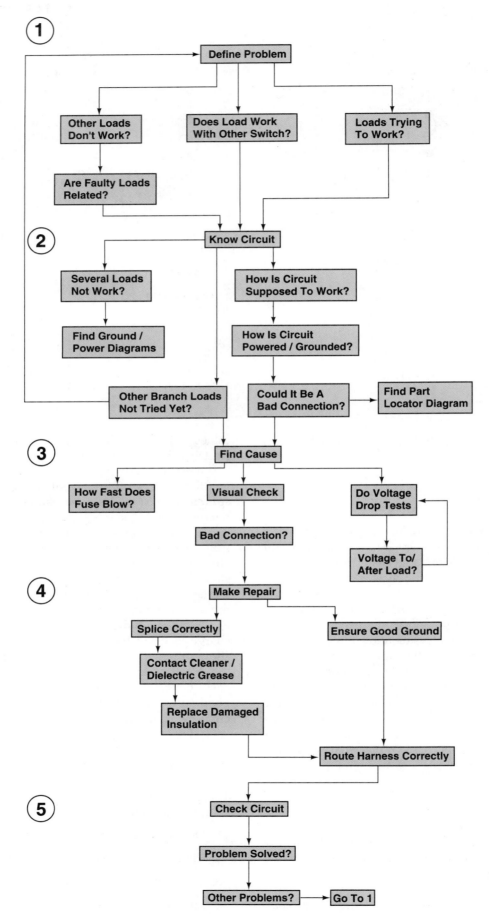

Figure 26–26 Read through this chart showing a basic troubleshooting method. *(I-CAR)*

Figure 26-27 An accurate voltmeter will check state-of-charge of the battery. A fully charged battery should have 12.6 volts.

5. Attach the charger clamps to the battery. Connect the red cable to the battery positive. Connect the black cable to negative.

6. Set the charger to the recommended settings.

7. Turn on the charger.

8. When charging is completed, turn off the charger.

9. Remove the charger clamps, negative first.

10. Replace the vent caps (if applicable).

11. Attach the negative battery cable.

Jump Starting

Although it is a common practice in some collision repair shops, avoid jump starting whenever possible. The discharged battery can explode. This is true of both the vehicle being started and the vehicle providing the jump.

Jumper cables are used to connect two batteries together when one is "dead" (discharged). Connect the jumpers positive to positive and negative to negative. Connect the last jumper to a negative ground away from the battery. This will prevent any spark near the battery, Figure 26–28.

Special care is also necessary when charging or jump starting to avoid damaging computer circuits. Make sure everything is turned off and do not connect the jumper cables backwards.

When jump starting cannot be avoided, use the following procedure:

1. Inspect the battery casing for damage.

2. Position the two vehicles so the jumper cables will reach. Make sure the vehicles are NOT touching.

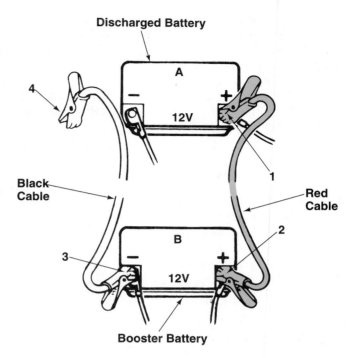

Figure 26-28 Jumper cables must be connected carefully to prevent sparks or part damage. Connect positive to positive and negative to negative in the proper sequence. Note how the last connection is away from the battery on any good engine ground. *(Volkswagen of America)*

3. On the vehicle to be started, turn off all switches and lamps.

4. Connect one end of the jumper cable to the positive terminal of the discharged battery.

5. Connect the other end of the jumper cable to the positive terminal of the booster battery.

6. Connect one end of the other cable to the negative terminal of the booster battery.

7. Connect the other end to an engine bolt head, A compressor, alternator mounting bracket, or other ground. Do NOT connect it to the negative battery terminal of the discharged battery.

8. Make sure that the jumper cables are NOT in the way of moving engine parts.

9. Start the engine in the vehicle with the booster battery. Let it run a minute, then start the other vehicle.

10. Leave all switches off. Reduce engine speed to idle on both vehicles to prevent damage to the electrical system.

11. Disconnect all cables in reverse order, starting with the ground connection on the vehicle with the discharged battery.

Before reinstalling the battery after body work, clean its terminals. Use a battery post tool to remove a thin layer of oxidized metal from the terminals and battery cable ends. This will reduce resistance at the battery connections.

When installing a battery, make sure the ignition is off. Reconnect the positive battery cable and then the negative battery cable.

Battery Safety

DANGER! Batteries produce hydrogen gas when discharging or being charged. This gas, if ignited, can make a battery explode! Acid and battery parts can fly into the shop. Charge only in a well-ventilated area.

1. Keep batteries away from welding operations, open flames, sparks, or other heat sources.

2. Do NOT charge batteries with cracked cases.

3. Do NOT smoke near batteries.

4. Batteries will charge faster if they are warm.

5. Do NOT charge a frozen battery.

Electrolyte in the battery is a mixture of water and sulfuric acid. When handling batteries, keep in mind that the electrolyte contains powerful sulfuric acid.

Wear eye protection whenever handling or charging the battery. Do NOT tilt the battery when carrying it. Be careful NOT to let electrolyte get on clothing, skin, or painted surfaces.

If the battery acid does spill, neutralize with baking soda. Immediately wash the skin or eyes with water and seek medical attention.

LIGHTING AND OTHER ELECTRIC CIRCUITS

The **lighting system** feeds electricity to light bulbs throughout the vehicle. A few are the headlight, turn light, stop light, backup light, emergency flasher, and interior light circuits.

Automotive lighting systems have become increasingly more sophisticated. Headlights and tail lights have grown into multiple-light systems. Indicator lights on the instrument panel commonly warn of failure or improper operation of the charging system, seat belts, brake system, parking brakes, door latches, computer systems, and other items.

When lights fail to function, check the bulb first. If the bulb is good, trace for an open feeding current to the dead bulb.

Other electrical circuits (horn, power windows, etc.) use the same principles just discussed. If needed, refer to the service manual for diagrams of each circuit. Circuit designs vary from vehicle to vehicle.

REPAIRING WIRING AND CONNECTIONS

Some damaged wiring can be repaired if correct procedures are followed, Figure 26–29. Be sure to follow the manufacturer's recommendations for wire repair.

When servicing electrical wiring, remember the following rules:

1. Never tug on the connectors. Use the special tools designed to separate the connectors. This will minimize the possibility of having electrical problems caused by intermittent contact.

2. Route wiring in the same location as the OEM.

3. Protect the electrical connectors from moisture and corrosion by using dielectric grease.

4. Use the same size and type of wiring for repairs. This is especially critical with sensor circuits that are affected by the slightest resistance change.

5. Air bag systems require specific repair procedures and materials.

SOLDER REPAIR

Soldering uses moderate heat and solder to join wires or other parts. All copper wire joints should be soldered, if possible.

A pencil-type soldering iron is recommended for this procedure rather than a soldering gun, Figure 26–30. This is because guns have heavy coils that can create induced current in nearby components and damage them.

Be sure to use *rosin-core solder,* NOT acid-core, when soldering electrical connections. Acid fluxes create corrosion and can damage electronic components. In addition, dielectric grease is frequently used to protect connections from corrosion. When using this product, which has no acid content, follow the manufacturer's instructions and be sure NOT to cut into the moisture-sealed connector joints with test equipment probes.

Use the following procedure to make a solder connection:

1. Remove all traces of dirt and corrosion.

2. Strip the insulation back about ½ inch (13 mm) on each wire.

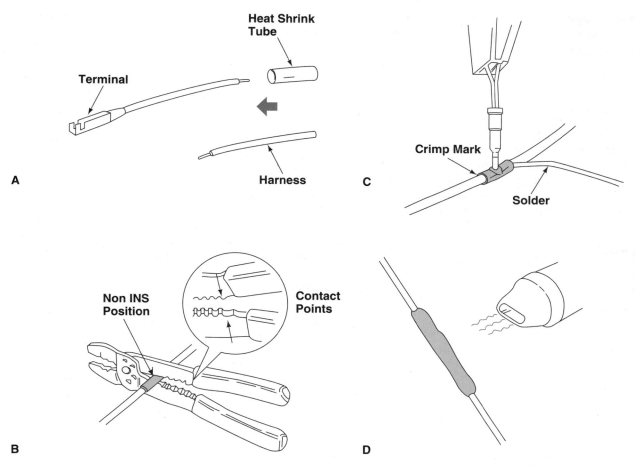

Figure 26–29 Note use of a connector to splice wire. (A) Slide a heat shrink tube over the wire. (B) Crimp the connector down over the wire ends. (C) Solder the connector to the wires to ensure good electrical connection. (D) Use a heat gun to shrink the plastic tube around the repair.

3. Wind the wires tightly together.

4. Heat the wire joint, NOT the solder. Touch the solder to the joint. Allow it to flow into the joint. Do NOT use too much solder.

5. Allow the joint to cool.

6. Gently pull on both ends of the wire to make sure the repair is secure.

When soldering in a restricted area, under the instrument panel for example, hold the soldering iron above the joint. The solder will flow down toward the heat source and joint.

A **shrink tube** can also be used over the joint to further protect the connection from corrosion. When heated, the tube shrinks in size and provides a coating over the repair much like the original insulation.

Liquid electrical tape can also be applied over the joint with a brush. This will help make the electrical connection waterproof.

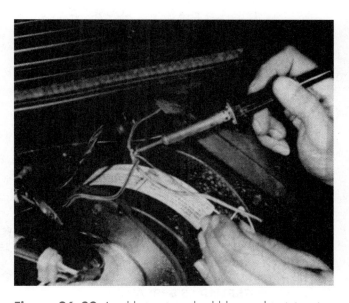

Figure 26–30 A soldering iron should be used to join wires and prevent problems later.

Electrical tape is made of flexible, thin plastic. It can be used around wire splices when the connection will NOT be exposed to moisture.

Using Splice Sleeves

Splice sleeves are compressed down over wires to connect them. Some manufacturers use aluminum in their wiring harnesses. Since aluminum cannot be soldered, splice sleeves are good for this task. Avoid using splice sleeves when you can solder wires together. They can loosen with prolonged vibration.

To install a splice sleeve, proceed as follows:

1. Remove all traces of dirt and corrosion.

2. Strip the insulation back about ⅜ inch (9 mm) on each wire.

3. Match the splice sleeve to the wire diameter.

4. Insert the wire into the splice sleeve until the wire meets the stop in the center of the sleeve.

5. Select the proper sized jaw on the crimp tool for the splice sleeve being used.

6. Crimp both ends of the splice sleeve.

7. Heat the splice sleeve with a heat gun until the sealer oozes out of the ends.

8. Allow it to cool.

9. Gently pull on both ends of the wire to make sure the repair is secure.

ELECTRICAL CONNECTOR SERVICE

An automotive electrical connector includes two plastic, snap-together fittings. They allow several wires to connect together securely. Various connector designs are used on vehicles and each requires a different method for disconnection.

If needed, inspect electrical connectors when trying to find opens. They can be torn apart or damaged in the collision, Figure 26–31. Figure 26–32 shows several ways to disconnect connectors.

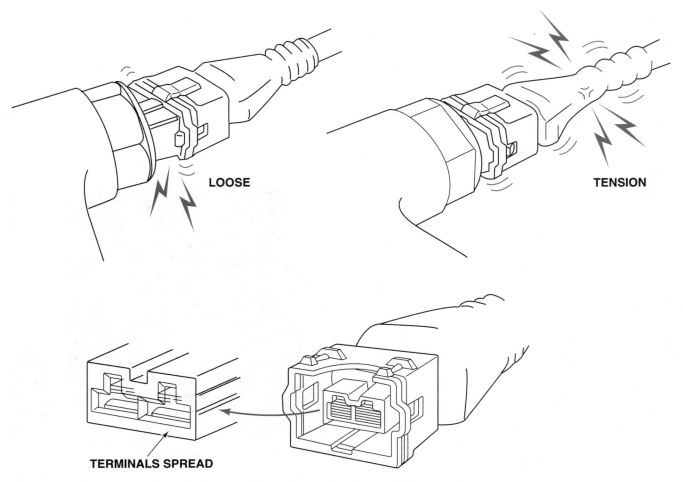

LOOSE

TENSION

TERMINALS SPREAD

Figure 26–31 Inspect connectors for these problems. They can be damaged during a collision.

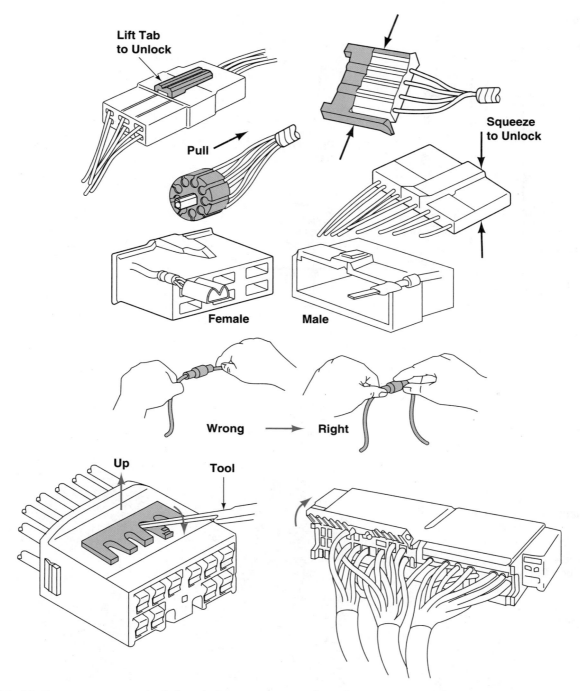

Figure 26–32 There are various methods for unlocking connectors. Refer to the service manual if in doubt.

COMPUTER SYSTEMS

Almost all vehicle systems are now controlled/monitored by computer. This would include the fuel, ignition, charging, suspension and brake, climate control, and other systems.

As shown in Figure 26–33, a basic *computer system* consists of:

1. Sensors (input devices)

2. Actuators (output devices)

3. Computer (electronic control unit)

The *sensors* are devices that convert a condition (temperature, pressure, part movement, etc.) into an electrical signal. They send electrical input signals back to the computer, Figure 26–34. Once the computer analyzes the sensor data, it produces a preprogrammed output that is sent to system actuators.

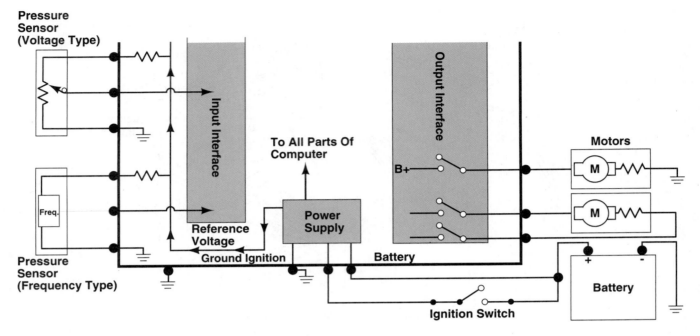

Figure 26–33 Study this basic computer system circuit. *(I-CAR)*

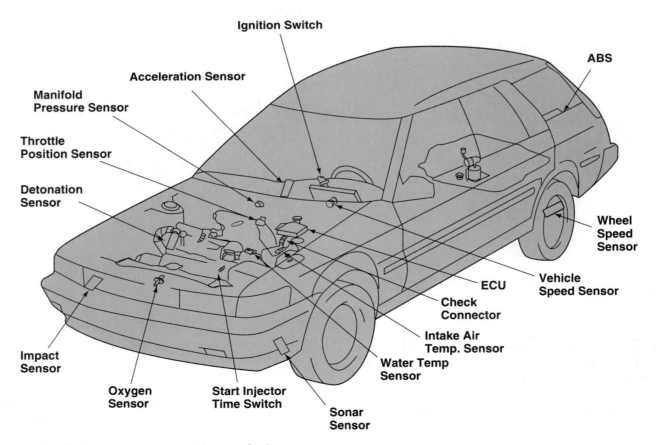

Figure 26–34 Numerous sensors provide inputs for the computer.

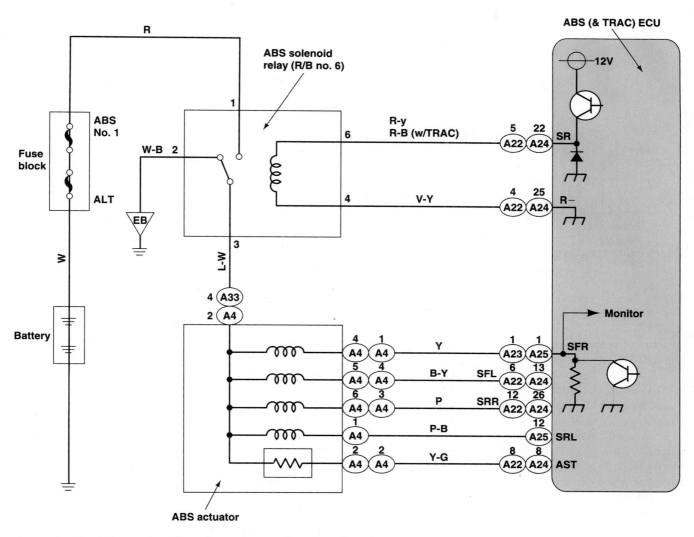

Figure 26–35 Relays, solenoids, and motors respond to outputs from the computer system.

Actuators are devices (solenoids or servo motors, for example) that move when responding to electrical signals from the computer. In this way, a computer system can REACT to sensor inputs and then ACT upon these conditions by operating the motors or solenoids. Refer to Figure 26–35.

The **computer** is a complex electronic circuit that produces a known electrical output after analyzing electrical inputs. Today's vehicles can have one or more computers that monitor and control the operation of electrical systems. See Figure 26–36.

SCANNING COMPUTER PROBLEMS

A **scan tool** is now the "most important tool" of the automobile technician. It provides the fastest way to use on-board diagnostics to find electrical-electronic problems in computer control systems. A scan tool "talks" to the vehicle's computers and can tell you if it detects any problems in the vehicle.

A scanner will convert computer or electronic control module (EMC) data into an alphanumeric display explaining the problem and service procedures in plain English.

Scan Tool Cartridges

Modern scan tools are removable cartridges which are small computer circuits that hold information about the specific vehicle. When the cartridge is installed into the scanner, the scanner has all the information needed to help analyze problems stored in the on-board computer system.

The cartridge normally plugs into the bottom of the scanner. Be careful not to force it into place and damage the small metal terminals. Also avoid

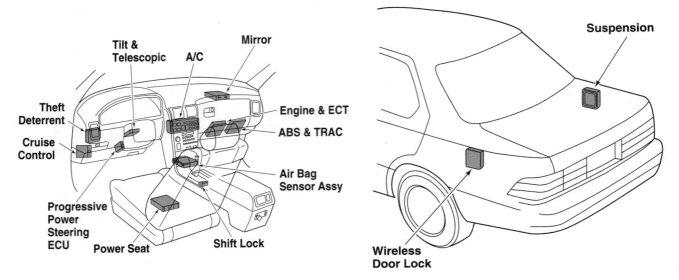

Figure 26–36 Computers can be found in various locations in the vehicle.

touching the terminals. Static electricity could "blow" the delicate electronics in the unit.

A vehicle "make" cartridge provides data for each vehicle manufacturer (GM, Ford, Chrysler, non-American). New cartridges must be purchased as on-board diagnostic systems are improved.

Some scanners also come with a troubleshooting cartridge that gives extra information on how to verify the source of different problems. This is a handy resource when diagnosing hard-to-find problems.

Advanced scanners will hold two cartridges, one for the vehicle make and the troubleshooting cartridge. This scanner allows you to select different installed cartridges and even prompts you on the correct cable adapter and other essential information

Scan Tool Control

The scan tool's menu allows you to scan different test functions: trouble codes and data, functional tests, custom setup, road test, and troubleshooter.

Codes and data allows you to retrieve stored trouble codes and electrical operating values from the on-board computer. This is the most commonly used function.

Functional tests allows you to perform specialized tests of the computer, emission systems, ignition system, sensor circuits, fuel injectors, and other components.

Road test allows you to take an instantaneous "snapshot" of the data while driving the vehicle. When an intermittent problem occurs, activate the scanner and it will store operating values for later evaluation.

Troubleshooter will guide you through diagnosis of various problems. It will give hints about common problems, list tests that should be performed, and give other useful information.

Most scanners come equipped with a printer port. They have a jack for connecting to an external printer, to an oscilloscope, or to a personal computer. You can print a hard copy showing the data stream electrical values for reference or use the scope or PC to further analyze your readings.

Connecting the Scan Tool

If a malfunction indicator light (MIL) in the dash glows or if you cannot find the source of a performance problem, connect the scan tool to the diagnostic connector on the vehicle. See Figure 26–37.

The diagnostic connector allows the scanner to be connected directly to the computer wiring harness on the vehicle. The scan tool can then automatically convert trouble code numbers into a display that gives information for finding problems quickly. This eliminates the need to refer to a service manual to find out what the trouble code number means.

Early diagnostic connectors came in various configurations. An adapter was often needed so the scanner connector would fit the vehicle's connector. Adapters are labeled for easy location for each make of vehicle.

With modern On-board Diagnostics II (OBD II), the standardized connector is a 16-pin personal computer–type connector. The scan tool cable should slide easily into the vehicle's diagnostic connector. If it does not, something is wrong. Never force the two together

Figure 26–37 A scan tool, when connected to the vehicle's diagnostic connector, will allow you to retrieve trouble codes, perform sensor and actuator tests, and quickly analyze the condition of most computer system circuits. *(Courtesy of Snap-on Tools Corporation)*

or you could damage the pins or terminals in the connectors. The most common locations for the diagnostic connector include the following:

- Under the right side of the dash and visible (OBD II standard location)

- Under the left side of the dash or near the fuse box

- Near the inner fender panel in the engine compartment

- On the side of the fuse box

- Under the center console

- Near the firewall in the engine compartment

If it is not powered through the diagnostic connector, connect the scan to battery power. If you use the cigarette lighter receptacle, make sure you have a good connection. To use the tester in the engine compartment, utilize the battery adapter with alligator clips.

Using the Scanner

Most manufacturers recommend that you start and fully warm the engine before scanning. However, if the problem symptoms occur only when the engine is cold, you may not want to warm the engine.

Turn off all accessories that could trip false trouble codes. Even opening the door while scanning could confuse the scanner. Lower the window so that you can start the engine without opening the car door.

Next, prepare the tool for the make of vehicle to be analyzed. Refer to Figure 26–37. The scanner will usually need you to input specific characters from the vehicle identification number. This will let the scanner know which engine, transmission, and other equipment are installed on the vehicle. The VIN plate is on the dash. You must input specific numbers and letters from the VIN. Scroll through the letters and press Y, or yes, when the character is correct or press the correct keypad character. Some vehicles will automatically download the VIN information into the scan tool for you. Next, select what information you would like the scan tool to give you. Engine-off diagnosis is done by triggering on-board diagnosis with the ignition key in the ON position but without the engine running. This pulls stored trouble codes out of the computer memory chips. This check is usually performed before the engine-running diagnostic, especially, when the MIL in the dash is glowing.

When in the Engine-off/Key-on mode for more than 30 minutes, connect a battery charger to the vehicle. This will prevent extended current draw from draining the battery. False trouble codes could result from a partially "dead" battery.

Most technicians look for stored trouble codes first using stored diagnostic information. This is a quick way to go right to any hard (continuous) failure so it can be repaired.

A general rule is to correct the cause of the LOWER NUMBER CODE FIRST. Sometimes fixing the lowest code will clear other codes because of component interaction.

If a problem exists in any circuit, the scanner will not only give you a trouble code number, but it will also give you a brief summary of what the number represents. Always remember that code does NOT mean that a certain component is bad. It simply indicates the possible problem location or circuit. Multimeter testing of the component and circuit is still needed to verify the exact source of trouble. To prevent electronic control unit damage, always use a high-impedance, digital multimeter when testing.

Engine-running diagnosis has the engine warmed to full operating temperature and idling to check electrical values. It checks the condition of the sensors, actuators, computer, and wiring while they are operating under normal conditions of engine heat, vibration, and pressure.

Follow the scanner instructions to start this test. For example, the scanner instructions will typically say to run the engine for two minutes at 2,000 rpms. This will assure that the engine and its components are fully warm and in closed loop mode. The scan tool will usually tell you whether the system is closed or open loop, whether the system is lean or rich, and other information related to system operation.

The troubleshooter cartridge might also give tips for finding a difficult-to-locate problem. If you have a code for a rich mixture, it might tell you to check the coolant temperature sensor, to check for high fuel pressure, or for leaking injectors. All could cause and trip a rich mixture trouble code. You might want to measure fuel pressure or do an injector leak down test.

The troubleshooter cartridge might even tell you to disconnect a questionable sensor and compare the readings with the circuit open. With the sensor disconnected, the scanner should read within the prescribed values for infinite resistance in the circuit. This will isolate the problem to either the wiring and ECM or the sensor quickly. If the reading does not change with the wiring disconnected, you know you have an open in the wiring. The troubleshooter cartridge might also inform you of an improved replacement sensor. A wiggle test is done by moving wires and harness connectors while scanning to find soft (intermittent) failures. If wiggling a wire produces a new trouble code, check that electrical connection more closely.

Switch, Actuator, and Snapshot Diagnostic

A switch diagnostic involves activating various switches while using a scanner. The scanner will tell you which switch to move and will monitor its operation. The scanner will quickly indicate if the switch is working normally. Switch test capabilities vary with make and model year.

An actuator diagnostic uses the scanner and vehicle computer to energize specific output devices with the engine off. This will let you find out if the actuators are working. Actuator diagnostic tests might do any of the following:

- Fire the ignition coil
- Open and close the injectors
- Cycle the idle speed motor or solenoid
- Operate other output devices

You can then watch or listen to make sure the actuators are working.

Most actuator tests will make a specific or unique sound. A servo motor will make a whining sound. An injector will make a swishing sound. A solenoid, a click sound. If you do not hear an actuator fire, you may need to test the input voltage to the actuator or the actuator itself.

Not all vehicle manufacturers provide switch and actuator tests. Refer to the service manual or scanner operating manual for details.

With a modern scan tool, you can also scan for problems while driving the vehicle to simulate conditions when the trouble happens. A road test might help find intermittent problems that don't trigger hard (continuous) trouble codes. For example, if the car only acts up when driving at a specific speed during cold start-up, you can scan under these same conditions. You can then take a "snapshot" or "freeze-frame" right when the symptom occurs. The stored values can be evaluated when the vehicle is back at the shop.

You would need to use logical deduction, scan values, specification values, and an understanding of system operation to find problems that do not set a code or ones that cause false codes.

Erasing Trouble Codes

Erasing trouble codes, also termed clearing diagnostic codes, removes the stored codes from computer memory. There are various methods used to erase trouble codes from the computer:

- Use the scanner tool to remove stored diagnostic codes from the on-board computer. This is the best way to remove old codes after repairs.
- Disconnect the battery ground strap or cable. This will also erase digital clock memory and other memories, however.
- Unplug the fuse to computer or ECU.
- Codes will erase automatically after 30–50 engine starts.

After erasing trouble codes, you might want to again energize on-board diagnosis. If no trouble codes are then displayed, you have corrected the problem.

PROTECTING ELECTRONIC SYSTEMS

The last thing a technician wants to do when a vehicle comes into the shop for collision repair is create problems. This is especially true when it comes to electrical systems and electronic components. There are proper ways to protect automotive electrical systems and electronic components during storage and repair.

1. Disconnect the battery cables (negative cable first) before doing any kind of welding. To avoid the possibility of explosion, completely remove the battery when welding under the hood or on the front end. Also, make sure the ground connection is clean and tight. Position the ground clamp as close as possible to the work area to avoid current seeking its own ground.

2. Whenever disconnecting or removing the battery on a computer-controlled vehicle, remember that the memory for radio station selection, seat posi-

tion, climate control setting, and any other "driver programmable" options are erased. When delivering the vehicle, advise the customer that he or she must reprogram these settings. Or better yet, record them before beginning and reprogram them yourself before delivery.

3. Static electricity can cause problems. Avoid it by grounding yourself before handling new displays. One way is to touch a good ground with one hand before handling the display with the other hand. Another way is to use a grounding strap that attaches to the wrist and then to the vehicle.

4. Avoid touching bare metal contacts. Oils from the skin can cause corrosion and poor contacts.

5. Be careful about the placement of welding cables. Keep the electrical path as short as possible by placing the ground clamp near the point of welding. Also, do NOT let the welding cables run close to electronic displays or computers.

6. Take care when handling electronic displays and gauges. Never press on the gauge face because this could damage it.

7. If a fault code indicates a problem with the oxygen sensor, extra caution is required. The oxygen sensor wire carries a very low voltage and must be isolated from other wires. If it is NOT, nearby wires could add more induced voltage. This gives false data to the computer and can result in a driveability problem. Some manufacturers use a foam sleeve around the oxygen sensor wire to keep it separate and insulated from other wires.

8. The sensor wires that connect to the computer should never be rerouted. The resulting problem might be impossible to find. When replacing sensor wiring, always check the service manual and follow the routing instructions.

9. Remove any computer that could be affected by welding, hammering, grinding, sanding, or metal straightening. Be sure to protect the removed computer by wrapping it in a plastic bag to shield it from moisture and dust.

10. Be careful NOT to damage wiring when welding, hammering, or grinding.

11. Be careful NOT to damage connectors and terminals when removing electronic components. Some may require special tools to remove them.

12. Always route wiring in the same place it was originally. If you do NOT, electronic crossover from the current-carrying wires can affect the sensing and control circuits. Reuse or replace all electrical shielding for the same reason.

Summary

- Current is the movement of electricity (electrons) through a wire or circuit. It is measured in amps.
- Voltage is the pressure that pushes the electricity through the wire or circuit. It is measured in volts.
- Resistance is a restriction or obstacle to current flow. It is measured in ohms.
- Some diagnostic equipment used in finding electrical faults includes the multimeter, test light, and jumper wires.
- The modern vehicle has numerous electrical-electronic systems, including the ignition, starting, and charging systems.
- Wires are often severed and electrical and electronic components damaged during a collision. A scan tool will help you quickly find the location of circuit faults. It will display which circuits or components are not operating within specifications.

Technical Terms

electrical system	jumper wires
current	continuity
voltage	infinite reading
resistance	AC current
conductor	DC current
hot wires	inductive pickup
insulation	wiring diagram
ground	electrical symbols
electric circuit	wire color coding
load	wiring harness
series circuit	part location diagram
opened	switch
parallel circuit	inertia switch
closed	solenoid
series-parallel circuit	relay
Ohm's Law	motors
magnetism	fuses
flux	cartridge fuse
magnetic field	blown fuse
electromagnet	fuse ratings
windings	fuse box
diagnostic tools	fuse link
multimeter	fusible links
VOM	circuit breaker
digital multimeter	ignition system
(DVOM)	ignition coil
analog multimeter	spark plug wires
(AVOM)	spark plugs
test light	fuel rail

starting system
ignition switch
starting solenoid
starting motor
flywheel ring gear
charging system
voltage regulator
charging system voltage
open circuit
short circuit
grounded circuit
diagnostic charts
battery

voltage spikes
jumper cables
electrolyte
soldering
rosin-core solder
liquid electrical tape
electrical tape
computer system
sensors
actuators
computer
scan tool

Review Questions

1. Define the terms "current," "voltage," and "resistance."

2. A _____ _____ has only one conductor path or leg for current through the circuit.

3. A _____ _____ has two or more legs or paths for current.

4. If a circuit has 12.6 volts applied and 4.2 ohms, how much current would flow through the circuit?

5. What is a multimeter?

6. A lighting circuit has an intermittent problem. The bulbs go on and off. Technician A says to use a DVOM to find the trouble. Technician B says to use an AVOM. Who is correct?
   a. Technician A
   b. Technician B
   c. Both Technicians
   d. Neither Technician

7. Why should you NOT use test lights on electronic circuits?

8. Explain the difference between AC and DC electricity.

9. The service manual may also give a _____ _____ _____ for finding electrical components (harnesses, sensors, switches, computers, etc.).

10. How does a fuse operate?

11. With the engine running, how much voltage should be across the battery terminals?
   a. 12 to 15 volts
   b. 18 volts
   c. 12.6 volts
   d. 13 to 15 volts

12. If an engine cranks but fails to start, what should you do?

13. How do you use jumper wires?

14. Be sure to use _____ _____, NOT _____, when soldering electrical connections. Acid fluxes create corrosion and can damage electronic components.

15. In your own words, explain the basic parts and operation of a computer system.

16. How do you use a scan tool?

17. Computers can be located:
   a. Under the dash
   b. Under the seats
   c. Under the hood
   d. Behind kick panels
   e. All of the above
   f. None of the above

ASE–Style Review Questions

1. *Technician A* says that voltage is like electrical pressure that pushes current through a circuit. *Technician B* says that voltage is the flow of electrons through a circuit. Who is correct?
   a. Technician A
   b. Technician B
   c. Both Technicians A and B
   d. Neither Technician

2. Which of the following circuits will go completely "dead" if a load device burns open?
   a. Parallel circuit
   b. Series circuit
   c. Series-parallel circuit
   d. Hydraulic circuit

3. If a circuit is shorted to 0.1 ohm, how much current will try to flow through the circuit of 12.6 volts is applied?
   a. 0.126 amps
   b. 1.26 amps
   c. 12.6 amps
   d. 126 amps

4. A properly charged automotive battery should show how much open circuit voltage?
   a. 12 volts
   b. 12.1 volts
   c. 12.6 volts
   d. 12.8 volts

5. *Technician A* says to connect jumper cables positive-to-positive and negative-to-negative. *Technician B* says to connect the last jumper cable to a ground away from the battery. Who is correct?

a. Technician A
b. Technician B
c. Both Technicians A and B
d. Neither Technician

6. *Technician A* says that batteries can explode. *Technician B* says that battery acid can cause injury. Who is correct?
   a. Technician A
   b. Technician B
   c. Both Technicians A and B
   d. Neither Technician

7. A malfunction indicator light in the dash stays on after collision repair work is complete. *Technician A* says to connect a scan tool to the vehicle's diagnostic connector. *Technician B* says it is better to connect a multimeter to the correct pins on the diagnostic connector for troubleshooting. Who is correct?
   a. Technician A
   b. Technician B
   c. Both Technicians A and B
   d. Neither Technician

8. A trouble code for an oxygen sensor circuit has been retrieved. *Technician A* says to replace the oxygen sensor. *Technician B* says to test the oxygen sensor and its circuit. Who is correct?
   a. Technician A
   b. Technician B
   c. Both Technicians A and B
   d. Neither Technician

9. After repairs, a vehicle's engine fails to idle normally and the scan tool finds a trouble code for the idle air motor. *Technician A* says to test the idle air motor with a multimeter. *Technician B* says to check for an opening in the wiring harness going to the idle air motor. Who is correct?
   a. Technician A
   b. Technician B
   c. Both Technicians A and B
   d. Neither Technician

10. *Technician A* says that the diagnostic connector on OBD II vehicles should be visible on the firewall of the vehicle. *Technician B* says that various scan tool connector configurations may be required. Who is correct?
    a. Technician A
    b. Technician B
    c. Both Technicians A and B
    d. Neither Technician

Activities

1. Read the operating manual for a multimeter. Write a report on its use and safety rules.

2. Use a voltmeter to check the action of a charging system. How much voltage is the system producing across the battery? Discuss your findings with the class.

27

Restraint System Repairs

Objectives

After studying this chapter, you should be able to:

- Explain the difference between an active and passive restraint system.

- Describe the parts that make up a seat belt system and their functions.

- Learn how to service seat belts.

- Become familiar with the different types of child car seats.

- Describe the operation of air bag systems.

- List the basic steps for troubleshooting air bag systems.

- Repair air bag systems safely.

Introduction

A *restraint system* is designed to help hold passengers in their seats and prevent them from being injured during a collision. All new vehicles come equipped with some form of restraint system. See Figure 27–1.

This chapter will summarize the operation and repair of seat belt and air bag systems. This is important because these systems can be damaged or deployed during a collision. Since they are crucial to safety, you must know how to service a restraint system properly.

Now that all new vehicles come equipped with air bags, collision repair personnel must be well versed in the operation and service of these important safety systems.

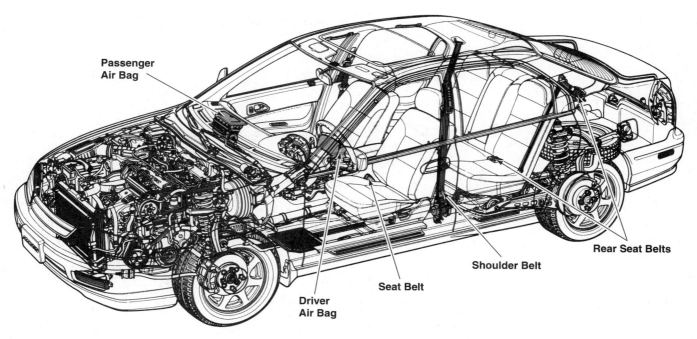

Figure 27–1 Restraint systems are now common. As a collision repair technician, you must know service rules for seat belt and air bag systems. (© American Honda Motor Co., Inc.)

In the figure, the following parts are labeled: Passenger Air Bag, Driver Air Bag, Seat Belt, Shoulder Belt, Rear Seat Belts.

ACTIVE AND PASSIVE RESTRAINTS

An *active restraint system* is one that the occupants must make an effort to use. For example, in most vehicles the seat belts must be fastened for crash protection. Conventional, manually operated seat belts would be classified as an active restraint system.

A *passive restraint system* is one that operates automatically. No action is required to make it functional. Two types are automatic seat belts and air bags.

SEAT BELT SYSTEMS

Seat belts are strong nylon straps with special ends attached for securing people in their seats. *Lap belts* are the seat belts that extend across a person's lap. *Shoulder belts* extend over a person's chest and shoulder, Figure 27–2. A *seat belt buckle mechanism* allows you to put the seat belt on and take it off.

Seat belt anchors allow one end of the belts to be bolted to the body structure. Look at Figure 27–3. A *belt retractor* is used to remove slack from the belts so they fit snugly. Various mechanisms are used in belt retractors. One is shown in Figure 27–4, page 582.

A *seat belt reminder system* uses sensors and a warning system to remind the driver to fasten his or her seat belt. On active systems, the driver's side front seat belt uses a 4- to 8-second fasten seat belt reminder light and sound signal. This is designed to

remind you if the lap and shoulder belts are NOT fastened when the ignition is turned on. If the driver's seat belt is not buckled, the reminder light and sound signal will automatically shut off after a few seconds.

On the passive system, the belt warning light will glow for a few seconds. An audible signal will sound if the driver's lap and shoulder belt is NOT buckled. The system will also signal if the ignition is on and the driver's door is open or if a system failure occurs.

The active belt system consists of a single continuous length of webbing. The webbing is routed from the anchor (at the rocker panel), through a self-locking latch plate (at the buckle), around the guide assembly (at the top of the center pillar), and into a single retractor in the lower area of the center pillar.

The passive system for coupes and late model sedans differs from the active in that two retractors are used. One is provided for the seat belt and a second for the shoulder belt.

Seat Belt Service

When servicing or replacing lap and shoulder belts, remember:

1. Do NOT intermix types of seat belts on front or rear seats.

2. Keep sharp edges and damaging objects away from belts.

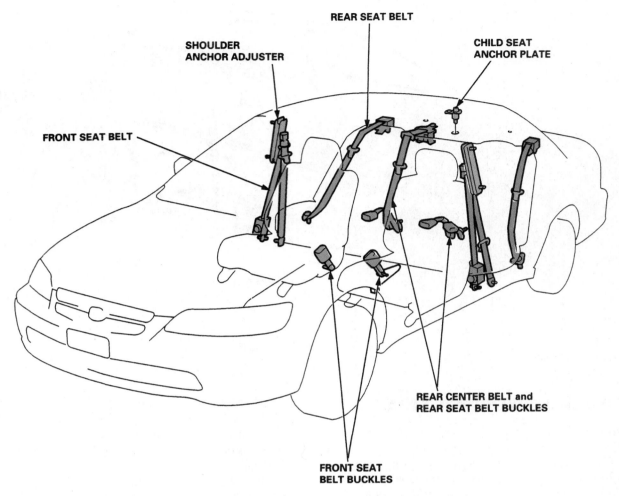

Figure 27–2 Note the major parts of seat belts. *(Used with permission from Nissan North America, Inc.)*

3. Avoid bending or damaging any portion of the belt buckle or latch plate.

4. Do NOT attempt repairs on lap or shoulder belt retractor mechanisms or lap belt retractor covers. Replace with new replacement parts.

5. Tighten all seat and shoulder belt anchor bolts as specified in the service manual. Use a torque wrench.

A visual and functional inspection of the belts themselves is very important to assure maximum protection for vehicle occupants. The following inspection checklist provides a typical, detailed seat belt inspection.

1. Check for twisted webbing due to improper alignment when connecting the buckle.

2. Fully extend the webbing from the retractor. Inspect the webbing and replace with a new assembly if the following conditions are noted.

In situations where the manufacturers do NOT have specific recommendations, perform a visual inspection of the system. Check seat belt webbing for:

1. Twists

2. Cuts or damage

3. Broken or pulled threads

4. Cut loops

5. Color fading or stains

Any of these defects may cause the seat belt webbing to be weakened and possibly fail during a collision.

To inspect seat belt buckles, do the following:

1. Insert the tongue of the seat belt into the buckle until a click is heard. Pull back on the webbing quickly to assure that the buckle is latched properly.

2. Replace the seat belt assembly if the buckle will NOT latch.

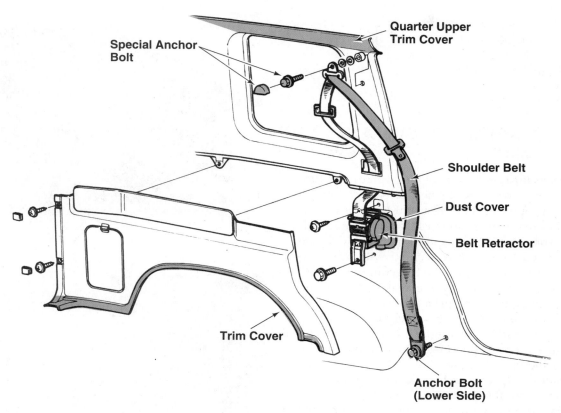

Figure 27-3 Always check the action of seat belt retractors after a collision. They must operate normally. *(Reprinted with permission by American Isuzu Motors Inc.)*

3. Depress the button on the buckle to release the belt. The belt should release with normal finger pressure.

4. Replace the seat belt assembly if the buckle cover is cracked or the push button is loose. Also replace the unit if the pressure required to release the buckle is too great.

To inspect a belt retractor assembly, remember to:

1. Grasp the seat belt webbing. While pulling from the retractor, give the belt a fast jerk. The belt should lock up.

2. Drive the vehicle in an open area away from other vehicles. Drive at about 5 to 15 mph (8 to 24 km/h). Quickly apply the foot brake. The belt should lock up.

3. If the retractor does NOT lock up under these conditions, remove and replace the seat belt assembly.

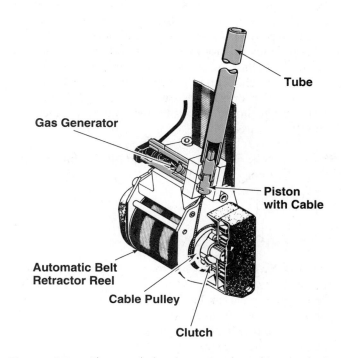

Figure 27-4 This seat belt tensioner works with an air bag system. It fires a gas charge to help pull seat belts tight around a person. It must be replaced after a collision. *(Illustration provided courtesy of Mercedes Benz-North America, Inc.)*

To inspect seat belt anchors, you should check the seat belt anchorage for signs of movement or deformation. Replace if necessary. Position the replacement anchor exactly the same as in the original installation.

CHILD RESTRAINT SYSTEMS AND AIR BAG SAFETY

A *child restraint seat* is a small removable appliance available from most automobile dealers or through various retail outlets. These restraint seats are available in either rearward-facing or forward-facing models, and in some cases may be used in either way. The seat is intended to be tightly secured with the vehicle's shoulder or lap belt and held secure with a special locking clip supplied by the seat manufacturer. The seat should never be used without the special security clip or placed in the vehicle in a direction other than for which it is designed. Many late-model vehicles are equipped with integral seats built into the rear seat by the manufacturer.

Most late-model cars and trucks have air bags for added protection for front seat passengers. When used in conjunction with seat belts, these air bags have been very effective in reducing serious injuries to older children and adults. However, they have been known to cause serious injury to infants and small children sitting in the front seat when the air bag is deployed. This is also true of older children and adults if they are not properly restrained with lap and shoulder belts. The safest procedure to practice is to always seat children in the rear seat; if necessary, they should be properly secured in a portable restraint seat, which should be positioned so that it is facing in the correct direction.

DANGER! *Children should NEVER be seated in the front.*

AIR BAG SYSTEMS

An *air bag system* automatically deploys a large nylon bag during severe collisions, Figure 27–5. One or more air bags can be used to help protect the driver and passengers from injury.

The *driver's side air bag* is mounted inside the steering wheel center pad. It is activated during a frontal collision but may not deploy from side collisions.

The *passenger side air bag* is mounted behind a small door in the side of the instrument panel. This air bag also deploys during frontal impacts. The angle of

A

B

Figure 27–5 (A) Driver and passenger air bags deploy during frontal collisions to keep occupants from hitting the steering wheel, instrument panel, and windshield. *(Courtesy of General Motors Corporation Service Operations)* (B) Side air bags deploy to protect occupants from hitting door, side glass, or other objects. *(I-CAR)*

impact and design of the system determines when the driver's and passenger's air bags deploy. Many systems operate with a frontal impact of within about 30 degrees of the vehicle's centerline.

Side impact air bags deploy from the door panels or from the side of the front seats. They may not deploy during a frontal impact. They are becoming more common and are used by several auto manufacturers. During a side impact from another vehicle, injury usually results when a passenger impacts the side window glass. Side impact air bags help protect people from this type of injury.

Rear seat air bags fit into the rear cushion of the front seats. They inflate to protect the passengers in the rear seat from injury in a frontal collision. They are not very common but can be found in a few expensive luxury vehicles.

While the location and design of the air bag system varies from manufacturer to manufacturer, all air bag systems have similar parts, Figure 27–6. These include:

1. Air bag module—inflator mechanism and nylon bag that expand to protect driver or passenger during collision.

2. Air bag system sensors—inertia sensors that signal computer of collision.

3. Control unit—computer that operates system and detects faults.

4. Wiring harness—wiring and connectors that link system parts.

5. Dash indicator lamp—dash bulb that warns of system problem.

DANGER! *The air bag system is designed to be used with seat belts.*

Air Bag Module

The **air bag module** is composed of the nylon bag and an igniter-inflator mechanism enclosed in a metal-plastic housing. All air bag module components are packaged in a single container mounted in the center of the steering wheel pad or dash, Figure 27–7. This entire assembly must be serviced as one unit when repair of the air bag system is required.

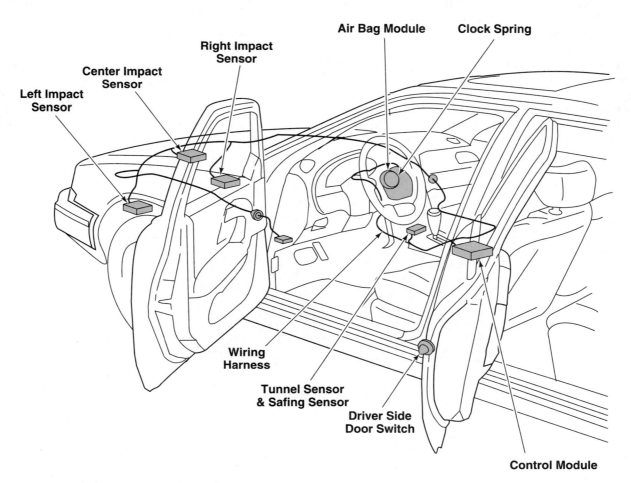

Figure 27–6 Study the major parts of an air bag system.

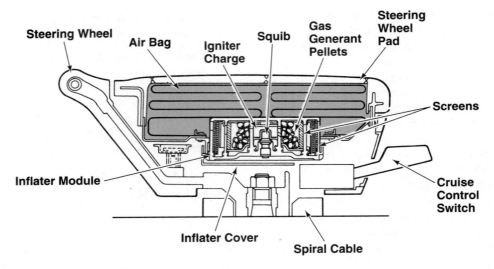

Figure 27-7 The driver air bag module mounts in the steering wheel. Note the parts.

The **air bag** is a strong nylon bag attached to the metal frame of the module. Vent holes in the bag allow for rapid deflation after deployment.

The **air bag igniter** produces a small spark when an electrical signal is sent from the control unit. When the electrical current is applied, the igniter produces an electric arc across two small pins. This spark ignites an igniter charge. This charge rapidly burns and causes the gas generating pellets to burn and produce rapid gas expansion. See Figure 27-8.

Once ignited, the **propellant charge** is progressive, burning sodium azide, which converts to nitrogen gas as it burns. Heat causes the chemicals in the

unit to produce a large amount of nitrogen gas. The nitrogen gas then fills the air bag in a fraction of a second. As the air bag inflates, the steering wheel cover is forced to split open allowing full inflation.

The occupant is then protected by the gas-filled bag instead of being restrained by the seat belts alone. In addition, there is some facial protection against flying objects.

Almost as soon as the bag is filled, the gas is cooled and vented. This deflates the bag right after the collision energy is absorbed. Vents allow rapid deflation. This prevents the driver or passenger from being pinned in the vehicle and also allows normal vision right after deployment.

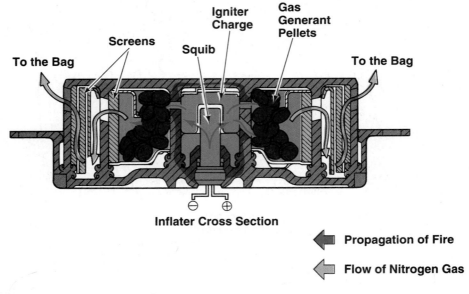

Figure 27-8 Study the parts of an inflator module. A squib spark fires the igniter. The igniter makes the gas generating pellets burn, causing gas expansion into the air bag.

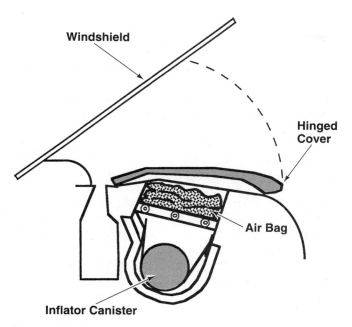

Figure 27–9 The passenger side air bag works in same way, but is under a small door in the instrument panel. (I-CAR)

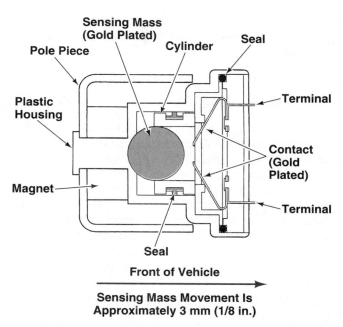

Figure 27–10 This inertia sensor uses a metal ball held by a magnet. Collision force throws the ball forward and into the contacts. This signals the computer of a collision. (I-CAR)

Passenger side air bags are very similar in design to the driver's unit. The actual capacity of gas required to inflate the bag is much greater. The bag must span the extra distance between the occupant and the dashboard at the passenger seating location. See Figure 27–9.

Another part relating to an air bag is the knee bolster. A **knee bolster** cushions the driver's knees from impact and helps prevent the driver from sliding under the air bag during a collision. It is located underneath the steering column and behind the steering column trim.

Air Bag Sensors

Two or more sensors are used in air bag systems, impact sensors and arming sensors.

Impact sensors are the first sensors to detect a collision because they are mounted at the front of the vehicle. Impact sensors are usually located in the engine compartment, while the safing sensor is usually located in the passenger compartment.

The **safing** or **arming sensor** ensures that the particular collision is severe enough to require that the air bag be deployed.

Both impact and arming sensors are inertia sensors. **Inertia sensors** detect a rapid deceleration to produce an electrical signal. Some air bag sensors use a small metal ball held in place by a permanent magnet, Figure 27–10. The sensor ball is thrown forward by the inertia of the collision. It then touches two electrical terminals that closes the sensor circuit to the computer.

Another air bag sensor design uses a weight attached to a coil spring. During impact, the weight is thrown forward. This overcomes spring tension, to close the sensor contact. This also closes the circuit to signal the computer of a possible collision. Both the impact sensor and a safing sensor must close at the same time for air bag inflation. They work together to provide a fail-safe system to prevent accidental air bag deployment. When both an impact sensor and a safing sensor close, the diagnostic control module sends a signal to the igniter, which starts a chemical reaction to inflate the bag.

It is important to remember that the tandem action of at least one main sensor and a safing sensor will activate the system.

Seat cushion sensors detect the weight of a person sitting in the passenger seat. If no one is sitting in the passenger seat, the air bag system may not deploy the passenger air bag. This saves the considerable cost of having to replace the bag without its having protected someone.

Air Bag Controller

The **air bag controller** analyzes inputs from the sensor to determine if bag deployment is needed. If at least one impact sensor and the arm sensor are closed, it sends current to the air bag module. This fires the air bag. The electronic control unit also provides failure data and trouble codes for use in servicing various aspects of most systems.

Air Bag System Servicing

Before servicing a vehicle equipped with an air bag, the system must be **disarmed** (all sources of electricity for the igniter disconnected). Procedures for disarming the air bag system vary.

Manufacturers may specify removal of the system fuse or disconnection of the module. Always refer to the service manual for exact procedures for disarming the system. This will help prevent electrical system damage and accidental deployment of the new air bag. Always disconnect the negative battery cable.

> *WARNING! Air bag systems may be equipped with an energy reserve module that allows the air bag to deploy in the event of a power failure. It must be removed from the system or allowed to discharge for a period of time ranging from a few seconds to 30 minutes after disconnecting the battery.*

DANGER! Even with the battery disconnected, an energy reserve module can deploy the air bag. If you are working near the bag, serious injury could result.

After air bag deployment, use a shop vacuum to clean the passenger compartment. Residual powder, which is an eye and skin irritant, can be present. Vacuum the instrument panel, vents, seats, carpet, and other surfaces contaminated with this powder. See Figure 27–11.

Air bag system parts replacement after a deployment will vary. Check for specific manufacturer recommendations on parts replacement. See Figure 27–12.

When replacing air bag system sensors, double-check that the system is disarmed before removing any sensor. The service manual will give sensor locations. Make sure you have the correct replacement sensor. During installation, check that the **sensor arrow** (directional arrow stamped on the sensor) is facing forward. If a sensor is installed backwards, the air bag will NOT deploy during the next collision.

To remove the deployed air bag, remove the small screws from the rear of the steering wheel. You can then lift out the module and disconnect its wires.

Wear safety glasses and a respirator while removing the deployed bag. This will protect you from the residual powder.

Inspect all parts for damage. Parts that have visible damage should be replaced. This would include the steering wheel, steering column, and related parts. Damage to the electrical wiring may also require wiring harness replacement.

Obtain the correct replacement parts from the manufacturer. Also refer to the service manual for exact procedures. System designs vary.

DANGER! When carrying a live (undeployed) air bag module, be sure the bag and trim cover are pointed away from your body. This will help reduce the chances of serious injury if the bag accidentally inflates. When laying a module down on a work surface, make sure the bag and trim cover are face up to minimize a "launch effect" of the module if the bag suddenly inflates.

Do NOT carry any system parts by the wire harness or pigtails. Follow the manufacturer's policies if any part is dropped or shows visible signs of damage. Follow the manufacturer's policies for replacing parts following a deployment.

Do NOT attempt to repair any parts unless specified by the manufacturer. Do NOT apply any electrical power to any part unless specified by the manufacturer.

The air bag module must be replaced following a deployment. Often the coil or **clock spring,** which is the electrical connection between the steering column and the air bag module, will also have to be replaced. See Figure 27–13.

Remember! There is no need to fear working with air bag equipped vehicles, but they must be treated with respect, Figure 27–14. Common sense can go a long way when working on these systems. By following some simple rules, air bags can be safely serviced.

1. Always have the service manual on hand when working with an air bag equipped vehicle.

2. When servicing a vehicle that has an undeployed air bag, follow the manufacturer's instructions for disarming the system. You should also have the system disarmed when installing a new air bag.

3. Wear rubber gloves and eye protection when servicing the air bag following a deployment. In case of skin or eye irritation, wash thoroughly with water and seek medical attention.

4. Disarm the air bag system prior to performing any welding operations.

5. Keep arms out of the steering wheel spokes when working on an air bag. It can shatter bones if acci-

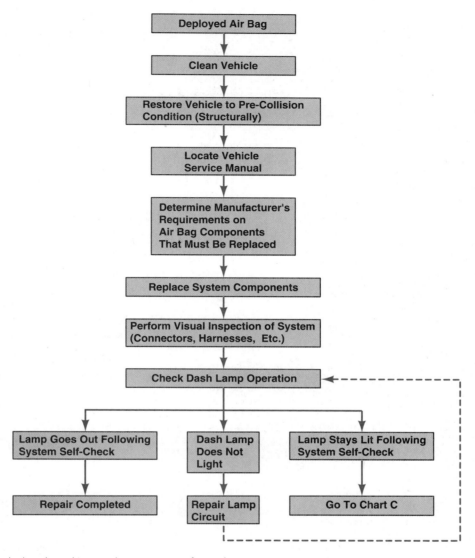

Figure 27-11 Study this chart showing the major steps for air bag system service. *(I-CAR)*

Figure 27-12 After air bag deployment, the passenger compartment should be vacuumed clean. Wear eye protection and a respirator before removing deployed air bags. Note that passenger air bag deployment often breaks the windshield.

dentally deployed. Also keep your head to one side of the bag during installation.

6. Follow the manufacturer's guidelines on force drying paint on vehicles equipped with an air bag.

7. Air bag disposal procedures vary depending on whether the air bag has been deployed or not.

8. If the air bag module is defective or the vehicle is to be scrapped, the air bag should be manually deployed using the procedures described in the service manual. Do NOT dispose of an undeployed air bag.

9. On air bag modules that cannot be manually deployed, the disposal procedure is to ship it back to the manufacturer using the packaging that the replacement module came in. By using the

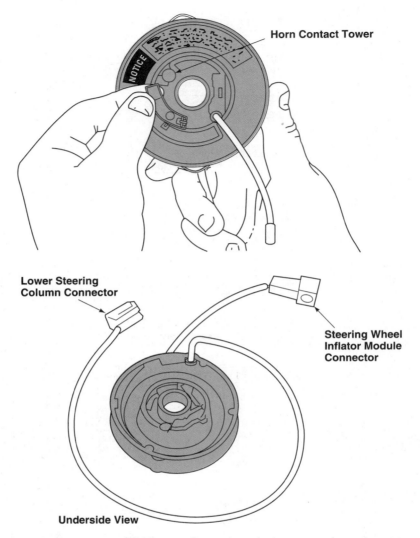

Horn Contact Tower

Lower Steering Column Connector

Steering Wheel Inflator Module Connector

Underside View

Figures 27–13 The clock spring may require replacement after air bag deployment. Refer to the service manual directions. *(Courtesy of DaimlerChrysler Corporation)*

Figure 27–14 When installing new air bag parts, always follow the service manual instructions. Here a technician is installing a new side air bag on the door. Stay to one side and keep your head away when installing air bags.

replacement part's packaging, all of the needed warning labels are already on the package.

The air bag system performs a self-check every time the ignition is turned to the ON position. During the self-check the air bag dash lamp indicator will light steady or blink. When the self-check is completed, the lamp should go OFF. If the lamp stays lit, there is a system fault present.

Make sure a final sweep is made for codes or collision information using the approved scan tool. Carefully recheck the wire and harness routing before releasing the car.

A final inspection of the repair should include checking to make sure the sensors are firmly fastened to their mounting fixtures, with the arrows on them facing forward. Be certain all the fuses are correctly rated and replaced.

Summary

- An active restraint system is one that the occupants must make an effort to use, such as manually operated seat belts. A passive restraint system operates automatically, such as automatic seat belts and air bags.
- A child seat may be rear facing, front facing, or a combination. It is often secured with a lap and shoulder seat belt.
- Most air bag systems have these components: air bag module, air bag system sensors, control unit, wiring harness, and dash indicator lamp.

Technical Terms

restraint system	rear seat air bag
active restraint system	air bag module
passive restraint system	air bag
seat belt	air bag igniter
lap belts	propellant charge
shoulder belt	knee bolster
seat belt buckle	impact sensor
mechanism	safing sensor
seat belt anchor	arming sensor
belt retractor	inertia sensor
child restraint seat	seat cushion sensor
air bag system	air bag controller
driver's side air bag	disarmed
passenger side air bag	energy reserve module
side impact air bag	clock spring

Review Questions

1. An active restraint system is one that the occupants must make an effort to use. True or false?

2. Seat belt _____ allow one end of the belts to be bolted to the body structure.

3. How does a seat belt reminder system operate?

4. What five things must you check when inspecting seat belt webbing?

5. You should avoid placing children in the front seat. True or false?

6. List and explain the five major parts of an air bag system.

7. The _____ _____ _____ is composed of the nylon bag and an igniter-inflator mechanism enclosed in a metal-plastic housing.

8. How does gas cause an air bag to inflate?

9. Explain the operation of air bag system sensors.

10. Before servicing a vehicle equipped with an air bag, the system must be _____. Even with the battery disconnected, the _____ _____ can fire the air bag.

11. How do you handle a live air bag module?

12. List nine rules for working with air bag systems.

ASE–Style Review Questions

1. *Technician A* says you can intermix types of seat belts on front or rear seats. *Technician B* says to tighten all seats and shoulder belt anchor bolts as specified in the service manual. Who is correct?
   a. Technician A
   b. Technician B
   c. Both Technicians A and B
   d. Neither Technician

2. *Technician A* says that you should NOT bleach or dye seat belt webbing. *Technician B* says to clean seat belts with a mild soap and water solution. Who is correct?
   a. Technician A
   b. Technician B
   c. Both Technicians A and B
   d. Neither Technician

3. *Technician A* says that child seats should only be used in the rear seat of a vehicle. *Technician B* says that child seats should be in the front seat. Who is correct?
   a. Technician A
   b. Technician B
   c. Both Technicians A and B
   d. Neither Technician

4. *Technician A* says that to disarm an air bag system you must usually remove the system fuse or disconnect the air bag module. *Technician B* says to always refer to the service manual for exact procedures for disarming the system. Who is correct?
   a. Technician A
   b. Technician B
   c. Both Technicians A and B
   d. Neither Technician

5. *Technician A* says that disconnecting the battery will disarm an air bag system. *Technician B* says to wait at least one minute after battery disconnection before working on the undeployed air bag system. Who is correct?
   a. Technician A
   b. Technician B
   c. Both Technicians A and B
   d. Neither Technician

6. After air bag deployment, *Technician A* says to use a shop vacuum to clean the passenger compartment. *Technician B* says to blow the compartment out with a blowgun. Who is correct?
   a. Technician A
   b. Technician B
   c. Both Technicians A and B
   d. Neither Technician

7. When installing impact sensors, *Technician A* says you can install the sensors in any direction. *Technician B* says to check that the sensor arrow is facing forward. Who is correct?
   a. Technician A
   b. Technician B
   c. Both Technicians A and B
   d. Neither Technician

8. *Technician A* says that when carrying a live (undeployed) air bag module, be sure the bag and trim cover are pointed away from your body. *Technician B* says that when laying an air bag down on a work surface, make sure the bag and trim cover face up to minimize a "launch effect" of the module if the bag suddenly inflates. Who is correct?
   a. Technician A
   b. Technician B
   c. Both Technicians A and B
   d. Neither Technician

9. *Technician A* Says that an undeployed air bag should be deployed before disposal. *Technician B* says you can dispose of undeployed air bags like any other discarded part. Who is correct?
   a. Technician A
   b. Technician B
   c. Both Technicians A and B
   d. Neither Technician

10. Which of the following should NOT be done following servicing of an air bag system?
    a. Make a final sweep for trouble.
    b. Watch for indicator light in dash.
    c. Tap on impact sensor with a hammer.
    d. Check wire and harness routing.

Activities

1. Inspect a vehicle's restraint systems. List the types of restraints installed in the vehicle. Does it have shoulder harnesses and air bag(s)? List your findings.

2. Refer to a repair manual for a specific make and model vehicle. Summarize the procedures for service after air bag deployment.

Section 7

Estimating and Entrepreneurship

Chapter 28
Estimating Repair Costs

Chapter 29
Entrepreneurship and Job Success

28

Estimating Repair Costs

Objectives

After studying this chapter, you should be able to:

- Explain how damage repair estimates are determined.

- Identify and explain the most common abbreviations used in collision estimating guides.

- Describe the methods of determining the repairability of a damaged vehicle.

- Explain the difference between flat-rate labor time and overlap labor time when making a cost estimation.

- Make a rough estimate of the time required to refinish a given collision repair job.

- Explain the difference between direct and indirect damage and locate both types.

- Apply a systematic approach to damage analysis.

- Perform steering and suspension quick checks.

- Identify the key operating features of manual and computerized estimating systems.

- Compare manual and computerized estimating.

- Explain how electronic photos and a computer can streamline the estimating process.

Introduction

An *estimate,* also called a *damage report* or *appraisal,* calculates the cost of parts, materials, and labor for repairing a collision damaged vehicle. Developed by the estimator, it is a written or printed summary of the repairs needed. The estimate is used by the customer, insurance company, shop management, and technician. See Figure 28–1.

Estimates must be accurately written. Repair costs are a major consideration for both the owner and the collision repair shop. Therefore, the profit margins for a collision repair shop depend heavily on the accuracy of estimates. Insurance companies can also write their own estimates.

This chapter will help you understand how both manual and electronic estimates are prepared. As you will learn, computerization has streamlined all aspects of collision repair shop operations. The computer-written estimate drives and integrates other aspects of the collision repair facility. Once initial vehicle and customer data is entered into the computer, everything from consulting estimating guides and vehicle dimension manuals to billing can be done electronically.

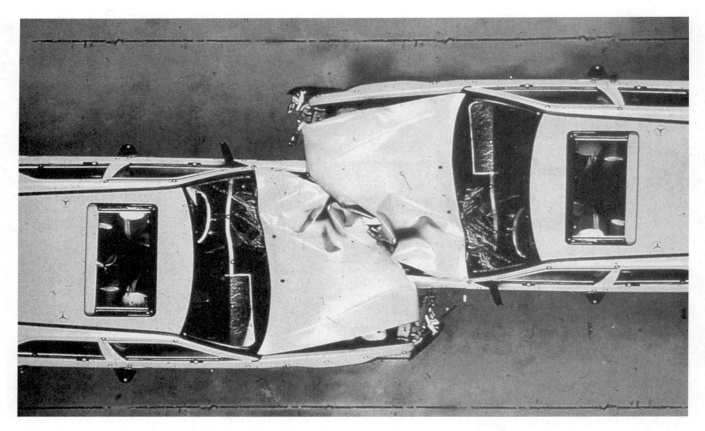

Figure 28–1 Estimating collision repair costs can be a challenge. You must evaluate many variables to determine what must be done to repair the vehicle. *(Used by permission from Tech-Cor Inc.)*

DAMAGE ANALYSIS

For an estimate to clearly establish a true cost of repairs, a thorough damage analysis must be performed on the vehicle. **Damage analysis** involves locating all damage using a systematic series of inspections, measurements, and tests. This allows repairs to be done right the first time and it also prevents cost overruns. See Figure 28–2.

Before starting damage analysis, you should:

1. If possible, discuss the collision with the owner or driver of the vehicle to obtain information that may help during damage analysis.

2. Identify the vehicle completely. Include VIN, year, make, model, engine, and optional equipment, Figure 28–3.

3. List mileage.

4. Identify and note all precollision damage.

5. Check wheels and tires, including the spare. Damaged wheels may provide clues about the collision.

6. Confirm the point of impact and analyze how the damage has traveled.

DIRECT AND INDIRECT DAMAGE

One of the reasons unibody vehicles are a challenge to repair is because of the way the body reacts to collision forces. There are two types of damage that must be identified. They are direct (primary) damage and indirect (secondary) damage.

Direct damage occurs in the area of immediate impact as a direct result of the vehicle striking an

Figure 28–2 Besides obvious damage, there is often hidden damage under the vehicle structure that must be found. *(Used by permission from Tech-Cor Inc.)*

Figure 28–3 The estimator must have a thorough knowledge of vehicle construction and repair methods to develop an accurate estimate of repair costs. *(Used by permission from Tech-Cor Inc.)*

object. Direct damage is usually easy to locate and analyze. See Figure 28–4.

Indirect damage is caused by the shock of collision forces traveling through the body and inertial forces acting upon the rest of the unibody. Indirect damage can be more difficult to completely identify and analyze. It may be found anywhere on the vehicle, Figure 28–5.

Before doing anything else, take time to carefully perform an overall visual inspection and try to determine the direction of impact and the areas of indirect damage.

VEHICLE INSPECTION

The direction of impact will affect the parts damaged. If the vehicle was hit in the front, side, or rear, you will know to check specific areas and parts for damage. See Figure 28–6.

You should use a damage analysis checklist to make sure nothing is overlooked, Figure 28–7.

Inspect the entire vehicle for damage, looking for:

1. Alignment of doors, Figure 28–8.

2. Alignment of the hood and deck lid.

3. Gaps between panels.

4. How the doors, hood, and deck lid open and close, Figure 28–9.

5. Ripples in the roof, fenders, or quarter panels away from the direct impact.

6. Cracked or stressed paint.

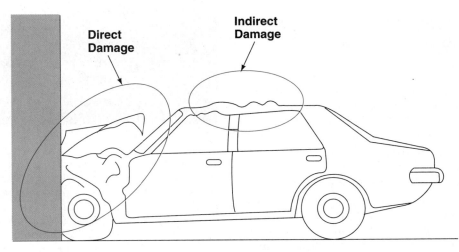

Figure 28–4 Direct or primary damage occurs at the point of impact. Indirect or secondary damage occurs elsewhere as forces travel through the vehicle structure. *(Used with permission from Nissan North America, Inc.)*

Figure 28–5 Note that the indirect damage at the rear of this vehicle was caused by a frontal collision. *(Used by permission from Tech-Cor Inc.)*

7. Cracked seam sealers.

8. Cracked or broken glass.

9. Smooth operation of windows.

10. Damage to interior (instrument panel, seats, seat belts, etc.), deployed air bag, stained carpet, and other problems, Figure 28–10.

11. Indications of previous damage.

12. Remove parts if needed to analyze hidden damage, Figure 28–11.

13. If it was a frontal collision, inspect parts in the engine compartment for damage.

14. Check under the vehicle for fluid leaks, which are signs of mechanical damage.

You should also:

1. Note the angle of impact(s).

2. Mark areas of direct damage.

3. Note the indicators of indirect damage.

4. Note areas of preexisting damage.

Always raise a badly damaged vehicle off the floor so that a good visual inspection can be made of all underbody and drivetrain parts. In some unibody vehicles, it might even be necessary to remove the drivetrain and suspension parts to make a thorough damage inspection.

In other words, the estimator must give the entire vehicle a thorough inspection—top to bottom and front to rear—and take nothing for granted. If something is missed on the original estimate, it often becomes difficult to reopen it for further negotiations later. On the other hand, some estimates do include a hidden damage clause that permits added charges to

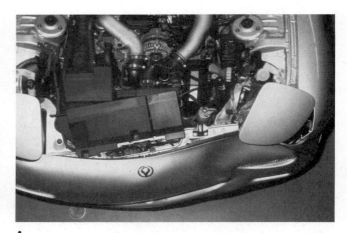

A

B

C

Figure 28–6 The direction of impact tells the estimator much about the resulting damage. (A) A frontal impact will often damage the bumper, grille, hood, fenders, aprons, frame rails, radiator, and sometimes engine parts. *(Used by permission from Tech-Cor Inc.)* (B) Side impact will often damage the door, quarter panel, pillars, roof, and rocker panel. *(Used by permission from Tech-Cor Inc.)* (C) Rear impact will often damage the rear bumper, back panel, valance, tail lights, trunk floor, quarter panels, wheel house, and rear rails.

DAMAGE ANALYSIS CHECKLIST

1. ❑ **STRUCTURAL MEMBERS**
   - ❑ Front Rails
   - ❑ Upper Rails
   - ❑ Tie bars, Upper and Lower (Core Support)
   - ❑ Crossmembers
   - ❑ Side Panels
   - ❑ Apron Assembly
   - ❑ Cowl Panel
   - ❑ A-Pillars
   - ❑ B-Pillars
   - ❑ Rocker Panel
   - ❑ Floor Pan
   - ❑ Rear Rails
   - ❑ Rear Crossmembers
   - ❑ Rear Floor Pan
2. ❑ **FRONT BUMPER**
   - ❑ Front Cover
   - ❑ Reinforcement
   - ❑ Energy Absorber
   - ❑ Molding
   - ❑ Valance
3. ❑ **GRILLE**
   - ❑ Grille
   - ❑ Emblem
   - ❑ Support
4. ❑ **FRONT LAMP**
   - ❑ Lamps
   - ❑ Assemblies
   - ❑ Trim/Door/Moldings
5. ❑ **COOLING SYSTEM**
   - ❑ Radiator
   - ❑ Recovery Tank
   - ❑ Fan
   - ❑ Fan Motor
   - ❑ Shroud
   - ❑ Hoses
   - ❑ Pump/Pulley/Belt
6. ❑ **ENGINE**
   - ❑ Crankshaft Pulley
   - ❑ Engine Mounts
   - ❑ Mount Brackets
   - ❑ Transmission Mounts
   - ❑ Oil Pan
   - ❑ Pumps (Fuel, Water, Power Steering)
   - ❑ Valve Cover(s)

   Exhaust
   - ❑ Catalytic Converter
   - ❑ Exhaust Manifold(s)
   - ❑ Pipe
   - ❑ Muffler

Alternator
   - ❑ Assembly and Mounts
   - ❑ Pulley/Belt

Electrical
   - ❑ Main Computer Module
   - ❑ Ignition Coil
   - ❑ Distributor Cap
   - ❑ Wires

Emission System
   - ❑ Air Pump
   - ❑ Vapor Canister

Air Conditioner
   - ❑ Condenser
   - ❑ Compressor
   - ❑ Pulley/Belt
   - ❑ Drier
   - ❑ Evaporator Core
   - ❑ Blower Motor
   - ❑ A/C Lines/Hoses

7. ❑ **HOOD**
   - ❑ Panel
   - ❑ Hinges
   - ❑ Latch
   - ❑ Latch Release
   - ❑ Moldings/Trim
8. ❑ **FRONT FENDER**
   - ❑ Fender
   - ❑ Wheelhouse
   - ❑ Moldings/Trim
   - ❑ Side Marker Lamps
9. ❑ **WHEELS/TIRES**
   - ❑ Wheel Covers
   - ❑ Wheel
   - ❑ Tires
   - ❑ Spare Tire
10. ❑ **FRONT SUSPENSION**
   - ❑ ABS Dash Indicator Lamp Brakes- Rotors/Calipers/Drums
   - ❑ Bearing/Hub Assemblies
   - ❑ Splash Shields
   - ❑ Brake Lines
   - ❑ Strut/Shock Assemblies
   - ❑ Strut Mounts
   - ❑ Springs
   - ❑ Control Arms & Strut Rods
11. ❑ **FRONT DRIVE AXLE**
   - ❑ Shaft Assembly
   - ❑ Axle Shaft
   - ❑ CV (Constant Velocity) Joints
   - ❑ Boots

Figure 28–7 A damage analysis checklist, like this one, is handy during visual inspection. Read through the checklist. *(I-CAR)*

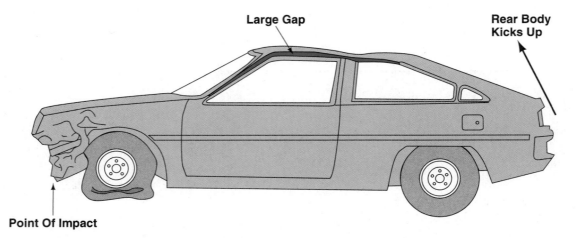

Figure 28-8 When inspecting damage, look for gaps between panels away from the point of impact. This can help you find indirect damage. *(I-CAR)*

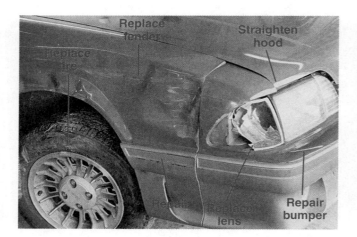

Figure 28-9 Note the damage on this vehicle.

Figure 28-11 Sometimes it is necessary to remove parts to find all the damage. *(Used by permission from Tech-Cor Inc.)*

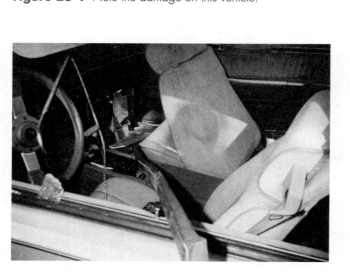

Figure 28-10 Interior damage can vary from light to major. Check for deployed air bag and damaged instrument panel, seats, seat belts, carpet, and other parts that could add to repair costs.

the original estimate if hidden damages are discovered later.

Taking photographs of the vehicle to keep with the documentation is also a good idea. Photos can be useful if there are any questions about a repair. Videotaping the vehicle before and after repairs can also be helpful.

Modern computerized estimating systems can utilize electronic photos and digitize video footage for finalizing the estimate when at the office computer.

INTERIOR INSPECTION

The interior of the vehicle must also be inspected. Check for damage caused by collision forces, unrestrained passengers, or cargo. Inspect glass and mirrors. Check door handles and door locks for proper operation. Refer again to Figure 28-10.

Inspect alignment of the glove box door. Misalignment may indicate instrument panel damage. Check interior controls for proper operation. Also inspect the console and shift lever.

Check seats and the restraint system for damage. Check seat belt buckles, webbing, and anchoring points. Check for deployed air bags and operation of motorized seat belts.

DAMAGE QUICK CHECKS

Damage quick checks can be done to analyze problems with the body structure, steering, and suspension systems. They can assist in determining if further steering system or suspension measurements are necessary. Quick checks include:

1. Steering wheel center check

2. Jounce/rebound steering gear check

3. Strut rotation check

4. Run-out check

5. Strut position check

A *steering wheel center check* involves making sure the steering wheel has not been moved off center due to part damage. It is done as follows:

1. Turn the steering wheel all of the way through its travel, lock to lock.

2. Count the number of turns from one lock position to the other.

3. Divide this number by two, and turn the steering wheel back that distance. The steering wheel should now be centered. Mark the center top of the steering wheel with a piece of tape.

4. Check the position of the steering wheel.

5. Check the position of the front tires, Figure 28–12A.

To analyze the steering wheel position:

1. If the steering wheel looks centered and the front tires point straight ahead, steering gear, steering column, and steering arms are probably NOT damaged.

2. If the steering wheel is more than slightly off center and the tires are straight, steering gear damage is possible.

3. If the steering wheel is centered and the tires are NOT pointed straight ahead, steering arm damage is possible.

The *jounce/rebound steering gear check* involves pushing down on the front or rear of the

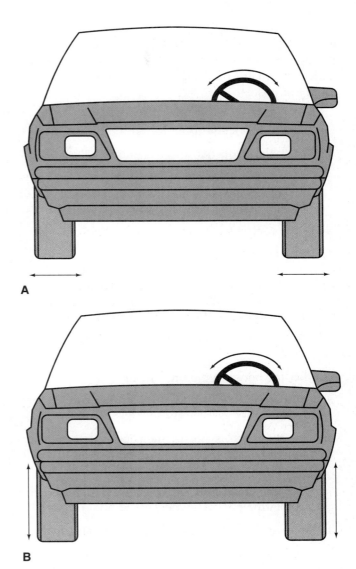

Figure 28–12 Note two quick checks to find steering and suspension damage. (I-CAR) (A) Center the steering wheel between full right and left positions. Then check to see if the tires are facing straight ahead. If either is not facing straight ahead, damage is indicated. (B) While another technician rotates the steering wheel one way and then the other, watch the front of body. If it rises and falls slightly, mechanical damage is indicated.

vehicle to load the suspension, and then allowing the vehicle to bounce back up. This check can help to identify damage to the steering gear, column, or tie rods.

1. Locate the top center of the steering wheel rim.

2. Mark the center top of the steering wheel with a piece of tape.

3. At the front of the vehicle, perform several jounce/rebound sequences while observing the steering wheel.

If the steering wheel rotates back and forth during the jounce/rebound sequences, then steering system damage is possible, Figure 28–12B. This check can indicate damage to the rack-and-pinion, which can cause bump steer.

To do a **strut rotation check** to find damage:

1. Loosen strut shaft upper lock nut ½ to 1 full turn.

2. Rotate the strut shaft with a wrench.

3. While rotating the strut shaft, use the fender lip as a reference and look for any noticeable change in camber. Movement can indicate a bent strut shaft.

4. When completed, retighten the lock nut.

If damage is suspected, inspect the strut carefully for signs of impact, bent housing or other obvious damage. This check will NOT work with struts that don't have an accessible strut shaft.

To do a **strut position check** to analyze damage:

1. Turn the wheels until they are pointed straight ahead.

2. Measure between the tire and strut.

3. Compare to the distance on the other side, Figure 28–13.

4. If a difference is noted, then look for a bent or damaged strut, bent wheel, damaged drum or rotor, or damaged steering knuckle.

A **wheel run-out check** will show if there is damage to a rotating part of a wheel assembly. To do this:

1. Remove any trim rings, wheel covers, and so on.

2. Raise the wheel assembly off the ground.

3. Using a safety stand or other steady rest, place a pointer (screwdriver, pen or pencil, etc.) to within ⅛ inch (3 mm) of the first step of the wheel (where the tire bead is seated).

4. Spin the wheel and watch the distance between the wheel and the pointer. See Figure 28–14.

5. If more than a visible variance exists, remove that wheel, replace with a known good wheel, and repeat the check.

6. If the variance is still present, then a rotating part of the wheel assembly is damaged.

7. When spinning the tire, listen for noise that might indicate wheel bearing damage.

DIMENSION MANUALS

When analyzing damage for making an estimate, you may need to use unibody/frame measuring equipment to determine the extent of the damage. See Figure 28–15.

After taking measurements on the vehicle, compare them to the body dimension manual. The manual will give illustrations of known good distances from specific body/frame points. By comparing your measurements to these known good measurements, you can determine the extent and direction of damage, Figure 28–16.

Severe collisions from any direction often cause the frame or unibody to distort. At one time the estimator could check this damage with the naked eye and would be fairly close in giving an estimate. But with the unibody, the "eyeballing" technique is sometimes NOT enough to detect misalignment. It is far better to use measuring gauges or devices to check for misalignment. Unless the estimator is thoroughly skilled in the use of such tools, it is wise to consult with an experienced technician to determine the time necessary to get the vehicle back into proper alignment. Refer to Chapter 20 for more information on this subject.

COLLISION ESTIMATING AND REFERENCE GUIDES

Collision estimating and reference guides help with filling out the estimate, Figure 28–17. Whether in manual or electronic form, estimating guides contain:

1. Illustrated parts breakdowns

2. Part names and numbers

3. Labor times

4. Refinish times

5. Part prices

6. Other miscellaneous information

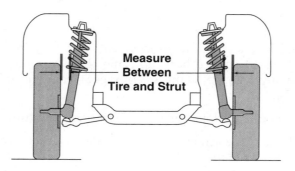

Figure 28–13 Measure between each tire and strut. Different measurements indicate damage. *(Chief Automotive Systems, Inc.)*

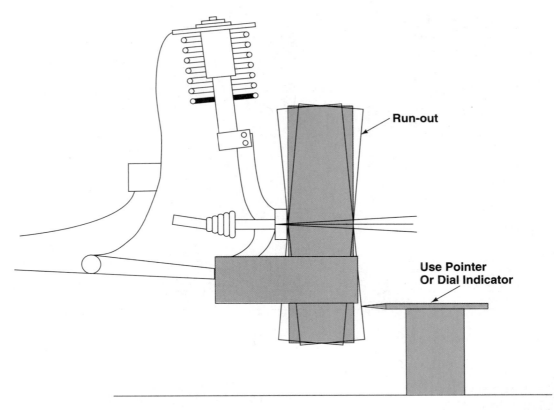

Run-out

Use Pointer
Or Dial Indicator

Figure 28–14 With the tire off the ground, spin the tire. If it wobbles, the wheel is bent. A dial indicator is needed for more accurate measurements of run-out. *(I-CAR)*

Collision estimating guides can be used as a reference for pricing parts. However, never use them to determine the absolute price. Usually, these guides will list the name of the part, the year of the vehicle it will fit, its part number, the estimated time required to replace it, and the current price. The current prices are the factory-suggested list prices. Parts that have been discontinued are usually listed. The price that appears in the guide is the last available one at the time of printing.

Each collision estimating guide will have procedure ("P") pages. These procedure pages provide important information such as:

1. Arrangement of material

2. Explanation of symbols used

3. Definitions of terms used

4. How to read and use the parts illustrations

5. Procedure explanations for labor and refinishing, including which operations are included and which are NOT

6. How discontinued parts information is displayed

7. How interchangeable part information is displayed

8. Additions to labor times

9. Labor times for overlap items

10. How to identify structural operations

11. How to identify mechanical operations

Figure 28–15 When in doubt about the extent of structural damage, use a measuring system. It will accurately show damage. *(Wedge Clamp International, Inc.)*

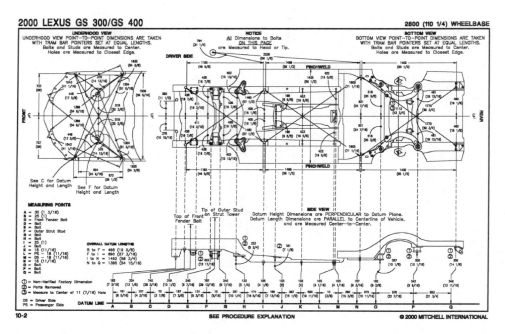

Figure 28–16 Dimension drawings give measurement points and specifications to help analyze damage. (Courtesy of Mitchell International, Inc.)

Figure 28–17 Collision estimating guides give information to help calculate the cost of repairs. Component names, component costs, labor times, and labor rates are given. (Courtesy of Mitchell International, Inc.)

DAMAGE REPORT TERMINOLOGY

Anyone working with collision estimating guides must be familiar with the terms and abbreviations used, Figure 28–18. A few abbreviation examples are:

R & I—to remove and install.

R & R means to remove and replace.

Overhaul—to remove an assembly from the vehicle, disassemble, clean, inspect, replace parts as needed, reassemble, install, and adjust (except wheel and suspension alignment).

Included operations—are those that can be performed individually, but are also part of another operation.

Overlap—occurs when replacement of one part duplicates some labor operation required to replace an adjacent part.

Refer to the procedure pages to identify all terms used.

PART COSTS

When should you repair or replace body parts? This question must be answered almost daily in collision repair shops. Actually, the decision often is a judgment call.

To help you decide whether to repair or replace panels, remember this statement.

If it is bent, repair it!
If it is kinked, replace it!

As explained in earlier chapters, a part is kinked if it is bent more than 90 degrees. A bend is damage of less than 90 degrees.

There are several other factors to consider: the type of surface, location of damage, and extent of damage. The following are some useful guidelines:

1. If the damage is in a flat surface, it will straighten more easily than a sharp fold or buckle along a corner. Kinks along sharply formed edges almost always require replacement.

2. If the damage is located near the end of a rail, with the crush zone unaffected, replacement is NOT as critical as when the damage is in the crush zone or further into the rail. A kink in the crush zone calls for replacement.

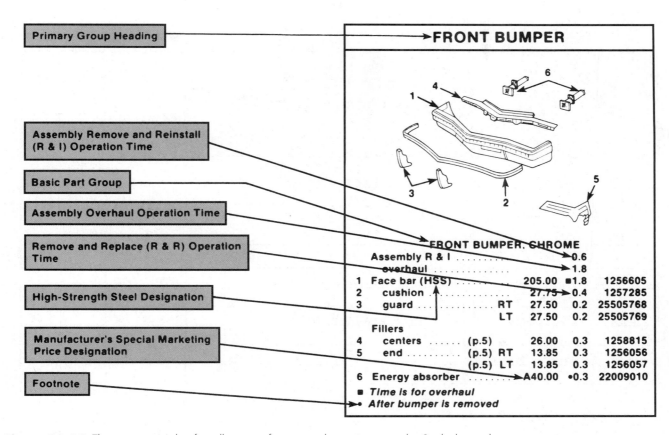

Figure 28-18 This is an example of an illustration from a crash estimating guide. Study the explanations.

3. If the damage is located near the engine or steering mounting areas, repeated stress loading can create fatigue failure. Kinks at these mounting points require replacement rather than repair.

4. Tightly folded metal that is "locked" due to severe work hardening requires replacement.

Part costs are also affected by whether the parts are new or recycled/salvaged. New parts might require more time to install than good recycled/salvaged parts. They will usually cost more too.

Recycled/salvaged parts will already have corrosion protection and sound deadening, where new parts might not. However, recycled/salvaged parts might have minor damage that must be repaired. All of this must be considered when writing the estimate. As can be seen, there are many variables that can affect the repair/replace decision.

WRITING ESTIMATES

An estimate or **damage report** must include basic information about the vehicle. A **manually written estimate** is done longhand by filling in information on a printed form, Figure 28–19. An **electronic/com-** *puter written estimate* is done using a personal computer and printer. You should understand how both are done.

A damage report or estimate includes:

1. Customer information, including name, address, and phone number.

2. Vehicle information—year, make, model, body type, license number, odometer reading, VIN, paint code, and trim colors.

3. Options—remote mirrors, power windows, power door locks, antilock brakes (ABS), air conditioning, and so on.

4. Date the vehicle was received and promised completion date.

5. It may be helpful to make sketches of the damaged areas of the vehicle. Many damage report forms provide a place to do this. With computer estimating systems, electronic photos or videos can also be helpful.

6. List whether new or recycled/salvaged parts will be used during the repair.

7. Summarize cost of parts and labor, Figure 28–20.

B & J
Collision Estimating Services

ESTIMATE OF REPAIRS
№ 002128

SHEET NO. _1_ of _1_ SHEETS

NAME	ADDRESS	PHONE		DATE
KAREN Miller	1143 RAILROAD ST. Cressona, PA. 17929	HOME 395-2719 BUS. 623 7347		12-17-99

YEAR	MAKE	MODEL	LICENSE NO.	MILEAGE	SERIAL/ V.I. NO.
1999	ford	P.U.	MAE 917	14864	1FTDF15YSGNA69994

INSURANCE COMPANY	TYPE OF INSURANCE	ADJUSTER	PHONE	CAR LOCATED AT
Amerisure	COL			mf6. 4/99

PARTS NECESSARY AND ESTIMATE OF LABOR REQUIRED	PAINT COST ESTIMATE		PARTS COST ESTIMATE		LABOR COST ESTIMATE	
① FRONT FACE BAR Chrome NO gds OR PADS			205	82		.5
① " STONE deflecTOR	1	0	40	50		.5
① Left HEADLAMP door (with aRgeNT GRiLL)			52	32		2
① " " Shield			3	20		
① " fRoNT fender	3	1	133	00	1	6
① " " " APRON			60	47	1	0
② wheels 15" 55.60 ea.			111	20		6
② stems and Balance			2	50		5
② HUB CAPS			43	46	—	
REPAIR RADIATOR Support			—		2	0
ALign FRONT End			—		1	5
① Left door TRim Panel			73	40		
STRiPe Left fRoNT fender			15	00		.5
LABOR 25.0 HRS at 23.00			575	00		
(noTe may Be fRoNT suspension damage)						
PAiNT maT.			89	10		
UNderCOAT			15	00		
TOTALS			1,419	.97		

INSURED PAYS $_____ INS. CO. PAYS $_____ R. O. NO._____	GRAND TOTAL	1,419.97
INS. CHECK PAYABLE TO _____		
The above is an estimate, based on our inspection, and does not cover additional parts or labor which may be required after the work has been opened up. Occasionally, after work has started, worn, broken or damaged parts are discovered which are not evident on first inspection. Quotations on parts and labor are current and subject to change. Not responsible for any delays caused by un-availability of parts or delays in parts shipment by supplier or transporter.	TOWING & STORAGE	
ESTIMATOR _____		
	TAX	85.20
AUTHORIZATION FOR REPAIRS. You are hereby authorized to make the above specified repairs to the car described herein.	TOTAL OF ESTIMATE	$ 1,505.17
SIGNED X DATE 19		

Figure 28–19 Here is an example of a handwritten estimate. Read through it carefully.

TIME/DOLLAR CONVERSION TABLE

Dollar Per Hour Rates

Time	.50	$1.00	$10.00	$15.00	$20.00	$25.00	$30.00	$35.00	$40.00	$45.00	$50.00
0.6	.30	.60	6.00	9.00	12.00	15.00	18.00	21.00	24.00	27.00	30.00
0.7	.35	.70	7.00	10.50	14.00	17.50	21.00	24.50	28.00	31.50	35.00
0.8	.40	.80	8.00	12.00	16.00	20.00	24.00	28.00	32.00	36.00	40.00
0.9	.45	.90	9.00	13.50	18.00	22.50	27.00	31.50	36.00	40.50	45.00
1.0	.50	1.00	10.00	15.00	20.00	25.00	30.00	35.00	40.00	45.00	50.00
1.1	.55	1.10	11.00	16.50	22.00	27.50	33.00	38.50	44.00	49.50	55.00
1.2	.60	1.20	12.00	18.00	24.00	30.00	36.00	42.00	48.00	54.00	60.00
1.3	.65	1.30	13.00	19.50	26.00	32.50	39.00	45.50	52.00	58.50	65.00
1.4	.70	1.40	14.00	21.00	28.00	35.00	42.00	49.00	56.00	63.00	70.00
1.5	.75	1.50	15.00	22.50	30.00	37.50	45.00	52.50	60.00	67.50	75.00
1.6	.80	1.60	16.00	24.00	32.00	40.00	48.00	56.00	64.00	72.00	80.00
1.7	.85	1.70	17.00	25.50	34.00	42.50	51.00	59.50	68.00	76.50	85.00
1.8	.90	1.80	18.00	27.00	36.00	45.00	54.00	63.00	72.00	81.00	90.00
1.9	.95	1.90	19.00	28.50	38.00	47.50	57.00	66.50	76.00	85.50	95.00
2.0	1.00	2.00	20.00	30.00	40.00	50.00	60.00	70.00	80.00	90.00	100.00
2.1	1.05	2.10	21.00	31.50	42.00	52.50	63.00	73.50	84.00	94.50	105.00
2.2	1.10	2.20	22.00	33.00	44.00	55.00	66.00	77.00	88.00	99.00	110.00
2.3	1.15	2.30	23.00	34.50	46.00	57.50	69.00	80.50	92.00	103.50	115.00
2.4	1.20	2.40	24.00	36.00	48.00	60.00	72.00	84.00	96.00	108.00	120.00
2.5	1.25	2.50	25.00	37.50	50.00	62.50	75.00	87.50	100.00	112.50	125.00

Figure 28–20 The chart can be used to convert time from an estimating guide into actual labor charge per hour.

At least three copies of the written estimate should be made. One is kept by the shop, one is given to the insurance company, and the other is given to the customer. An estimate is a **firm bid** for a given period of time—usually for 30 days. The reason for a given time period is obvious: part prices change and damaged parts can deteriorate.

The estimate is also considered the authorization to complete the repair work as listed, but ONLY when it is agreed upon and signed by the owner or by the owner and insurance company appraiser. The estimate explains the legal conditions under which the repair work is accepted by the collision repair shop. It protects the shop against the possibility of undetected damage that might be revealed later as repairs progress.

Many insurance policies contain a **deductible clause** which means that the owner is responsible for the first given amount of the estimate (usually $100 to $500). The remaining cost is paid by the insurance company. In such cases, both the customer and insurance company should authorize the estimate.

Another important function of the estimate is that it serves as a basis for writing the work order or operational plan. The work order is usually prepared from the damage appraisal of the estimator (using the written estimate) and a visual inspection by the shop supervisor and/or a technician.

The **work order** outlines the procedures that should be taken to put the vehicle back in top condition. It is also a valuable tool to both the estimator and the shop foreman, since it lists the actual times necessary to do the job.

Following are additional factors to be considered when writing an estimate. Such added time is usually negotiated between the estimator and the insurance company or customer.

1. Time for the setup of the vehicle on frame straightening equipment and damage diagnosis

2. Time for pushing, pulling, cutting, and so on to remove collision-damaged parts

3. Time to straighten or align related parts

4. Time to remove undercoating, tar, grease, and similar materials

5. Time to repair corrosion damage to adjacent parts

6. Time for the free-up of corroded or frozen parts

7. Time for drilling for ornamentation or mounting holes

8. Time to repair damaged replacement parts prior to installation

9. Time to check suspension and steering alignment/toe-in

10. Time for removal of shattered glass

11. Time to rebuild, recondition, and install after-market parts, NOT including refinishing time

12. Time for application of sound-deadening material, undercoating, caulking, and painting of the inner areas

13. Time to restore corrosion protection.

14. Time to R & I main computer module when necessary in repair operations

15. Time to R & I wheel or hub cap locks

16. Time to replace accessories such as trailer hitches, sun roofs, and fender flares

ESTIMATING SEQUENCE

When estimating any type of damage, a logical sequence must be followed. Before making a written estimate, the estimator should visually inspect the entire vehicle, paying special attention to damaged subassemblies and parts that are mounted to (or part of) a damaged part.

The estimator must consider the same points as the technician does before making any decision on repair work. Most estimators start from the outside of the vehicle and work inward, listing everything on paper—by vehicle section—that is bent, broken, crushed, or missing.

For example, if the front grille and some of the related parts are damaged, list the needed repairs or replacements as follows:

1. Front Grille ....Replace

2. Opening Panel ....Straighten and replace

3. Deflector (or Valance Panel).........................Replace

4. Headlight Door..Replace

5. Grille Opening Panel....................................Refinish

Notice that the parts are listed in a definite sequence according to factory disassembly operations or exploded views as provided in service manuals or collision estimating guides.

FLAT RATE OPERATIONS

The *flat rate* is a preset amount of time and money charged for a specific repair operation. As a rule of thumb, repair costs should never exceed replacement costs. If repairs and straightening will NOT produce a quality job, then the estimate should list the required new parts. Sheet metal parts usually offer the most opportunities for repair and straightening. As a result,

sheet metal repair, replacement, and panel refinishing generally account for the largest number of estimate dollars.

To reduce parts costs, many insurance appraisers and some customers might want to use recycled/salvaged parts. Recycled/salvaged parts are removed from damaged vehicles by **recyclers** (salvage yards). There is always a chance that such parts have been previously damaged and repaired. For this reason, it is important to carefully inspect all recycled/salvaged parts before they are installed to be sure that they are in usable condition.

Each damaged vehicle poses different problems that must be solved to arrive at a final decision concerning repair versus replacement. The most difficult questions arise when a vehicle is involved in a major collision.

If the frame rails received extensive damage, for instance, the estimator might want to consider listing their replacement rather than the more costly job of installing a complete new frame assembly. Another option is to list the replacement of a front side rail section or a complete front frame section. These solutions might be more practical (from a time and cost approach) than making repairs with frame straightening equipment.

On the other hand, estimating damage to the unibody shell is often a much more difficult job. Knowing what would be needed to restore the unibody shell to its original dimensions and configuration requires a good deal of experience. If unibody shell distortion is suspected, it can be checked by careful measurement prior to listing any necessary repair of straightening procedures. Generally, a clue to minor unibody distortion (twist, sag, side sway) is if any cracking of stationary glass is noted, yet there is no visible sign of a direct impact with any object.

LABOR COSTS

The procedure pages of crash estimating guides usually provide an explanation of what the labor time does and does NOT include. For example, replacing a panel or fender includes removing, installing, and aligning. It does NOT include the installation of molding or antennas, refinishing, pinstriping, or application of decals. Also NOT considered are corroded or inaccessible bolts, undercoating, and alignment or straightening of damaged adjacent parts.

The labor time reported in crash estimating guides is to be used as a guide only. The times are principally based on data reported by vehicle manufacturers or manual publishers who have arrived at them by repeated performance of each operation under normal shop conditions. An explanation is listed in the guide of the established requirements for the average technician, working under average conditions and following the procedures outlined in the service manuals.

The times reported apply only to standard models listed in identification sections of the crash estimating guide. They do NOT apply to vehicles with equipment other than that supplied by the vehicle manufacturer as standard or regular production options. If other equipment is used, the time might have to be adjusted higher.

All labor times given in crash estimating guides include the time necessary to insure proper fit of the new part. The times are based on new, undamaged parts installed as an individual operation. Additional time has NOT been added to compensate for collision damage to the vehicle. Removal and replacement of exchanged or used parts are also NOT considered. If additional aligning or repairs must be made, such factors should be considered when making an estimate.

When jobs overlap, reductions in the labor times must be considered. This occurs when replacement of one part duplicates some labor operations required to replace an adjacent part. For example, when replacing a quarter panel and rear body panel on the same vehicle, the area where the two parts join is considered overlap. Where a labor overlap condition exists, less time is required to replace adjoining parts collectively than is required when they are replaced individually.

Another labor cost reduction is noted as included operations. These are jobs that can be performed individually but are also part of another operation.

As an example, when replacing a door, the suggested time would include the replacement of all parts attached to the door, except for the ornamentation. It would be impossible to replace the door without transferring these parts. Consequently, the time involved in transferring them would be included operations and should be disregarded because the times for the individual items are already included in the door replacement time.

Flat rate manuals list a labor time plus a materials allowance; they do NOT, however, include the dollar value of the materials required. Some states have consumer protection laws containing very specific requirements that prohibit generically charging for shop supplies and miscellaneous parts without noting each item on the customer's invoice. Other states prohibit charging sales taxes on any part and materials that are not specifically used on a given vehicle. The estimator should, therefore, become familiar with state and local consumer protection laws before writing any estimates. This will reduce the possibility of violating any regulations.

REFINISHING TIME

Making a correct estimate of the amount of time required to refinish panels, doors, hoods, and so forth is a vital part of an estimator's job. Although the wide range of materials and conditions sometimes makes it difficult to arrive at a precise refinishing cost, there are a number of generally approved concepts that will help an estimator arrive at a fair judgment of the amount of materials that will be needed for the job.

Flat rate manuals list a labor time plus a materials allowance (in dollars) for many different refinishing operations. Independently published crash estimating guides list the labor times, but NOT the dollar value of the materials required.

In most crash estimating guides on the market today, the time required for refinishing is shown in the parentheses adjacent to the part name. The refinishing time generally includes:

1. Cleaning/light sanding

2. Masking adjacent panels

3. Priming/scuff sanding

4. Final sanding/cleaning

5. Mixing paint/filling spray equipment

6. Applying color coat

7. Removing masking

8. Buffing or compounding (if required)

9. Cleaning equipment

Refinishing time generally does NOT include:

1. Cost of paints or materials

2. Matching and/or tinting

3. Grinding, filling, and smoothing welded seams

4. Blending into adjacent panels

5. Removal of protective coatings

6. Custom painting

7. Undercoating

8. Anticorrosion materials

9. Sound-deadening materials

10. Refinishing the underside of hood or deck lids

11. Covering the entire vehicle prior to refinishing

12. Additional time to produce custom, non-OEM finishes

The refinishing times given in most estimates are for one color on new replacement parts—outer surfaces only. Additions to refinishing times are usually made for the following operations:

1. Refinishing the underside of the hood

2. Refinishing the underside of the trunk lid

3. Edging the new part

4. Two-tone operations

5. Stone chip (protective material)

6. Clearcoat (basecoat-clearcoat) after deduction for overlap

TOTAL LABOR COSTS

Once all the repairs and labor times have been entered on the estimate form, refer to the estimating guide's time/dollar conversion table. It is used to convert labor time into dollars to fit local labor or operating rates per hour.

When establishing labor rates, the shop overhead (including such items as rent, management and supervision, supplies, and depreciation on equipment) must be determined. Then the actual labor cost of all employees and the profit required to keep the business operating must be added to the shop overhead to obtain a dollar rate for repairs. This rate is usually figured on an hourly basis. Keep in mind that labor times shown in estimating guides are listed in tenths of an hour.

TOTALING THE ESTIMATE

Once all the columns of the estimating form—parts, labor, and refinishing cost—are filled out, they can be added together for a subtotal. To this figure, add any extra charges such as wrecker and towing fees, storage fees, and state and local taxes. These figures, added to the subtotal, will give the grand total of the estimate.

While more and more large collision repair shops are doing repair jobs such as wheel alignment, corrosion protection, and tire placement, smaller operations still "farm-out" or "sublet" these jobs to specialty shops.

When the work is done, the specialty shop bills the collision repair shop for the work. Generally, this is done at a rate less than the normal retail labor cost. In this way, the collision repair shop can charge the normal retail cost and still make a small profit.

Shops that sublet work usually have a "sublet" column on the estimate where the retail labor cost is marked. This figure is added to the rest to obtain the grand total of the estimate.

If the customer wants extra work performed (damage that occurred prior to collision), this should be noted as **customer-requested,** or C/R, repairs. As a general rule, the insurance company will NOT pay for customer-requested repairs. Often a separate estimate must be made in such cases.

COMPUTERIZED ESTIMATING

Computerized estimating systems using a ***PC*** (personal computer) may provide more accurate and consistent damage reports. The use of computers makes dealing with thousands of parts on hundreds of vehicles more manageable. See Figure 28–21.

Computerized estimating systems store the collision estimating guide information in a computer. This eliminates a lot of time spent looking up parts and labor times, and manually entering and totaling them on a form. The computer prepares a damage report, while still allowing the possibility of a manual override when necessary. See Figure 28–22.

Computers have revolutionized the art of estimating by making enormous amounts of data readily accessible.

The ***estimating system*** has all of the data needed at the shop office, usually on CD-ROMs. Modern computer estimating systems store electronic databases containing up-to-date part prices and labor times that can be retrieved from the shop's computer system.

Figure 28–21 Computer estimating is much more efficient than trying to figure repair costs longhand. *(Courtesy of Mitchell International, Inc.)*

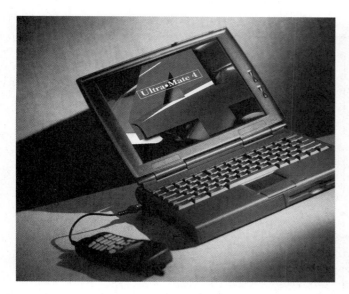

Figure 28-22 This laptop computer can be taken out to the vehicle during damage analysis. It can also be used in the office to finalize the damage report. *(Courtesy of Mitchell International, Inc.)*

With the use of these sophisticated programs, estimators can now use the stored data to make virtually error-free estimates rapidly at a very low cost.

Computer Database

Once information has been gathered at the vehicle, the estimator will take this data into the office. The operator can then use the speed and convenience of a computer and estimating program to work up the estimate. An ***estimating program*** is software (computer instructions) that will automatically help find parts needed and labor rates, and calculate total cost of the repairs, Figure 28-23. The estimating program is often

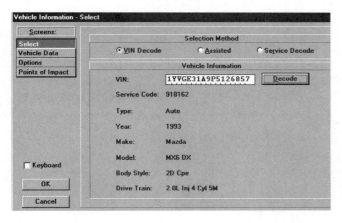

Figure 28-23 An estimating program is designed to help the estimator quickly and accurately identify vehicles and calculate repair costs. *(Courtesy of Mitchell International, Inc.)*

on computer ***floppy disks*** (removable magnetic disks of data) or on a CD-ROM (optical data on disk).

To use the program, you must load the data into the PC's memory. This will then let you pull up images from an electronic dimension manual, Figure 28-24. Modern systems will retrieve exploded views of assemblies in electronic form, Figure 28-25.

Many programs automatically deduct for overlap. The program notes operations as "Incl." if they are part of a larger operation. The program will also identify judgment times or user-entered information.

The estimating program can access a huge database (computer file) of information. The computer ***database*** includes part numbers, part illustrations, labor times, labor rates, and other data for filling out the estimate. Quick searches can be done to move through this database quickly and efficiently.

The most common and modern way of storing an estimating database is the CD-ROM. One or two ***CD-ROMs*** or compact discs can hold all of the crash estimating guides and dimension manuals, as well as other information for every make and model vehicle. The compact disc database contains the same basic information that is in the collision estimating guide. See Figure 28-26.

Labor time information for a given operation is taken from the same sources as in the collision estimating guides.

A CD-ROM allows access to a huge amount of data, which can be pulled into the PC's memory for manipulation.

Advanced computer estimating systems will even allow you to input photos of the damaged vehicle into the computer, Figure 28-27.

An ***electronic camera*** can be used to take photos of the vehicle's damage and store them as digital data. The camera can be connected to the computer to download these ***digital images*** (pictures stored as computer data) into memory or onto the computer's internal hard drive.

The estimator can look at these photos on the computer screen or monitor while finalizing the estimate. The electronic images can also be sent to the insurance adjuster for evaluating the vehicle's damage. See Figure 28-28.

You can also select from either new or recycled/salvaged parts and the program will help you enter the difference in costs and labor to install either type part. You can quickly compare the installation of like kind and quality used parts with new parts. The difference in purchase price and labor times can be quickly analyzed.

If the headlights are damaged, for example, the program will remind you to include aiming headlights or give an exploded view to make sure all damaged parts or part variations are noted.

ENGINE COMPARTMENT

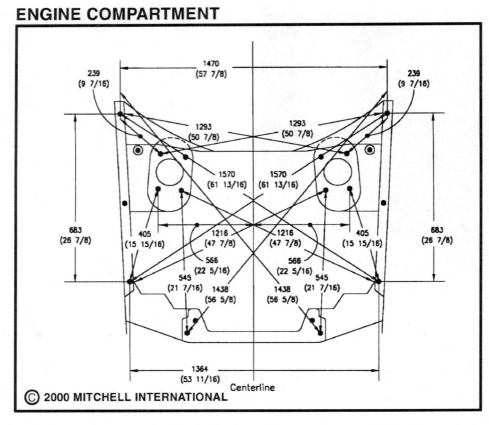

© 2000 MITCHELL INTERNATIONAL

Figure 28–24 Computer estimating systems often have a database of dimension manuals for faster retrieval of specifications. *(Courtesy of Mitchell International, Inc.)*

When estimating painting/refinishing labor and materials, the program will let you select from solid color, base-clearcoat, tri-coat, two-tone paint, and other variables to accurately estimate painting costs. It will add costs for priming, tinting, blending, tape stripes, liquid mask, and other variables that will affect repair costs.

Mechanical repairs to the suspension, steering, or other parts can also be calculated. The estimating program will guide you through selecting parts and labor for correcting mechanical damage: wheel alignment, bleeding brakes, and similar tasks.

Most programs will help you calculate frame/unibody repair time. You can select from a variety of straightening operations, which have general flat rate times.

Once all the information about the repair has been entered into the computer, the estimating program will automatically total all parts and labor, calculate taxes, and add other costs for a grand total.

The estimating program may also interface with other shop programs for payroll, maintenance, business reporting, document scanning, database storage, and shop management.

A ***printer*** can be used to make a hard copy (printed image on paper) for the customer, insurance adjusters, and shop personnel. You can quickly print vehicle dimension drawings, part views, or the estimate. Look at Figure 28–29.

When working with any estimating system, either manual or computer, those working with the damage report must understand:

1. Where vehicle options are listed, so these can be double-checked for accuracy.

2. Which labor rates are shown for a given operation and how to identify them.

3. What times and amounts are judgment items, and how these were generated.

4. Which times and amounts have been overridden.

5. Which times are included, and which are not.

Only by understanding all of these can a judgment be made as to on whether the damage report is accurate. Each system will produce a unique look.

FRONT CLIPS

In some situations, such as a severely damaged front end, an estimator might want to consider the pur-

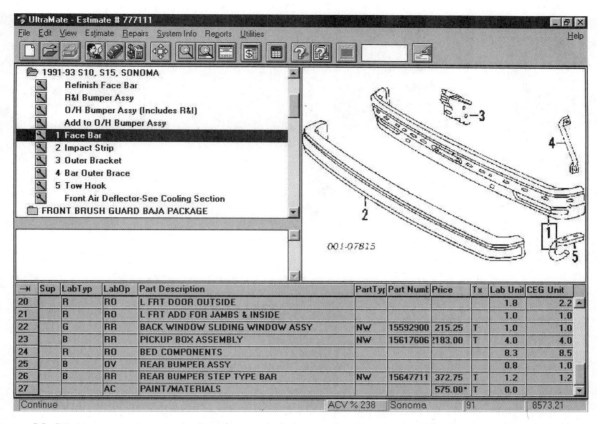

→	Sup	LabTyp	LabOp	Part Description	PartTyp	Part Numt	Price	Tx	Lab Unit	CEG Unit
20		R	RO	L FRT DOOR OUTSIDE					1.8	2.2
21		R	RO	L FRT ADD FOR JAMBS & INSIDE					1.0	1.0
22		G	RR	BACK WINDOW SLIDING WINDOW ASSY	NW	15592900	215.25	T	1.0	1.0
23		B	RR	PICKUP BOX ASSEMBLY	NW	15617606	2183.00	T	4.0	4.0
24		R	RO	BED COMPONENTS					8.3	8.5
25		B	OV	REAR BUMPER ASSY					0.8	1.0
26		B	RR	REAR BUMPER STEP TYPE BAR	NW	15647711	372.75	T	1.2	1.2
27			AC	PAINT/MATERIALS			575.00*	T	0.0	

Figure 28–25 Here is a computer display of an exploded view of a bumper assembly. The estimator can quickly see all the parts and select those that require replacement or repair. *(Courtesy of Mitchell International, Inc.)*

Figure 28–26 CD-ROM provides a vast database for a computer estimating system. One CD-ROM can hold a whole set of books in electronic form. *(Courtesy of Mitchell International, Inc.)*

Figure 28–27 This advanced computer estimating system will let you download pictures taken with an electronic camera or a video camera. The estimator can refer to these images when working on the estimate. *(Courtesy of Mitchell International, Inc.)*

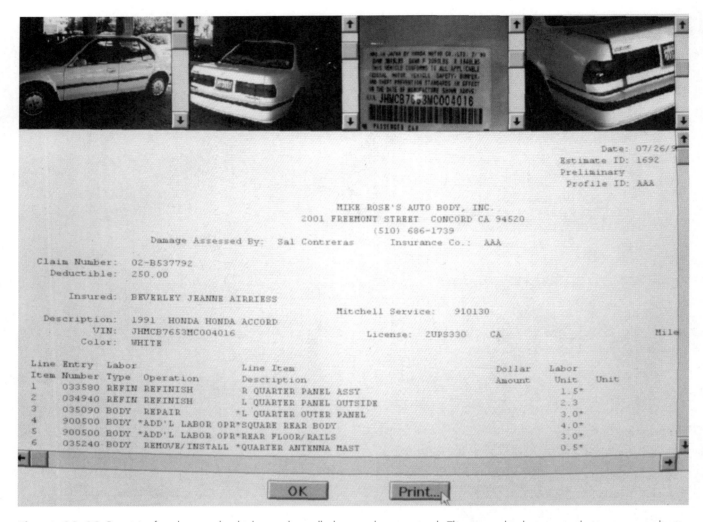

Figure 28-28 Pictures of a damaged vehicle can be pulled up and reexamined. They can also be sent to the insurance adjuster, who can then view vehicle damage without leaving the office. *(Courtesy of Mitchell International, Inc.)*

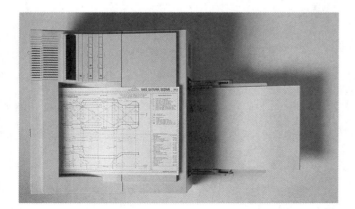

Figure 28-29 Once the computer estimate is done, it can be printed out on a laser printer. At least three copies should be generated. *(Courtesy of Mitchell International, Inc.)*

chase and installation of a recycled/salvaged front end assembly. In the body repair trade this is referred to as a *"front clip."* This assembly generally includes the front bumper and supports, grilles and baffles, and radiator and its supporting members, the hood and its hinges (or hood torsion bars), the front fenders and skirts, all front lights, the wiring, and all other related parts. Often, this method of making collision repairs to the front end will decrease the total time the vehicle is in the shop.

SECTIONING

Another operation that may be flat-rated in the crash estimating guides is sectioning. **Sectioning** involves cutting off the damaged part and welding on a new one. Often times it is less expensive and less time consuming to section a damaged panel or component and weld part of a new panel in place than to replace the entire panel. As discussed earlier, rails and other parts may be sectioned. Sectioning disturbs less OEM corrosion protection and spot welds. In some cases, it is the best repair procedure to choose. However, the crash estimating guides do NOT always provide a flat rate time for sectioning.

Remember! Sectioning of a vehicle may or may not be recommended by the vehicle manufacturer. This procedure should only be performed when a qualified and knowledgeable technician has determined that the operation does not jeopardize the integrity of the vehicle.

TOTAL LOSS

A *total loss* occurs when the cost of the repairs would exceed the value of the vehicle. The insurance company or customer would normally NOT want the vehicle repaired. Instead, the insurance company will write a check to cover the cost of a replacement vehicle. An equivalent year, make, and model vehicle is then purchased by the customer.

The insurance company usually determines if a vehicle is a total loss. The company will evaluate the estimate and market prices for comparable vehicles when making this decision. Older vehicles are written up as a total loss more often than late model vehicles. This is because of their low replacement cost.

The totaled vehicle will usually be auctioned or sold to a salvage yard or recycler. The recycler will then disassemble the vehicle and sell its parts for a profit.

Summary

- An estimate, also called a damage report or appraisal, calculates the cost of parts, materials, and labor for repairing a collision damaged vehicle.
- Direct damage occurs in the immediate area of the collision. Indirect damage is caused by the shock of collision forces traveling through the vehicle and inertial forces acting upon the rest of the unibody.
- Refer to literature such as dimension manuals and collision estimating guides to aid in the estimation process.
- Estimates may be manually written, or electronically (computer) written. Computerized estimating uses a personal computer and printer.

Technical Terms

estimate
damage report
appraisal
damage analysis
direct damage
indirect damage
steering wheel center
  check
jounce/rebound steering
  gear check
strut rotation check
strut position check
wheel run-out check
collision estimating and
  reference guides
R & I
R & R
overhaul
included operations
overlap

manually written
  estimate
electronic/computer
  written estimate
firm bid
deductible clause
work order
flat rate
computerized estimating
PC
downloaded
estimating program
floppy disks
database
CD-ROM
electronic camera
digital images
printer
front clip
sectioning
total loss

Review Questions

1. How many copies of the written estimate should be made?
   a. One
   b. Two
   c. Three
   d. Four

2. An _____, also called a _____ _____, or _____, calculates the cost of parts, materials, and labor for fixing the collision damaged vehicle.

3. List six things you should do before starting damage analysis.

4. When inspecting damaged unibody vehicle, Technician A looks for damage only near the area of impact. Technician B looks for damage both in the area of impact and some distance away. Who is correct?
   a. Technician A
   b. Technician B
   c. Both A and B
   d. Neither A nor B

5. What is the smallest increment in which labor times are listed in crash estimating guides?
   a. Hours
   b. Half hours
   c. Quarter hours

d. Tenths of an hour

6. Explain the difference between direct and indirect damage.

7. When you inspect the entire vehicle for damage, what 14 problems must you look for?

8. How do you analyze the center position of a steering wheel?

9. This check will find a bent wheel.
   a. Strut rotation check
   b. Strut position check
   c. Run-out check
   d. Overrun check

10. List five things included in collision estimating guides.

11. _____ _____ are those that can be performed individually, but are also part of another operation.

12. _____ occurs when replacement of one part duplicates some labor operation required to replace an adjacent part.

13. Give the statement or rule for determining whether to repair or replace a part.

14. A damage report or estimate includes these seven items.

15. The most common and modern way of storing an estimating database is the:
    a. CD-ROM
    b. Floppy disk
    c. Magnetic tape
    d. None of the above

ASE–Style Review Questions

1. *Technician A* says you should discuss the collision with the owner or driver of the vehicle to obtain information that may help during damage analysis. *Technician B* says you should identify the vehicle completely, including VIN, year, make, model, engine, and optional equipment. Who is correct?
   a. Technician A
   b. Technician B
   c. Both Technicians A and B
   d. Neither Technician

2. During a jounce/rebound check of a vehicle, the steering wheel rotates back and forth. Which of the following does this indicate?
   a. No damage

b. Brake system damage
c. Steering damage
d. Engine misalignment

3. When checking for damage, the technician spins each tire while listening for noise. If noise is heard, which of the following parts might be damaged?
   a. Strut
   b. Tie rod
   c. Wheel bearing
   d. Wheel rim

4. *Technician A* says that when analyzing damage for an estimate, you may need to use unibody/frame measuring equipment. *Technician B* says that only a tape measure is needed during estimating. Who is correct?
   a. Technician A
   b. Technician B
   c. Both Technicians A and B
   d. Neither Technician

5. Estimating guides contain all of the following EXCEPT:
   a. illustrated parts breakdowns
   b. part names and numbers
   c. labor times
   d. paint mixing codes

6. Which of the following refers to tasks that can be performed individually, but are also part of another operation?
   a. Included operations
   b. Estimated operations
   c. System operations
   d. Flat rate operations

7. *Technician A* says that if damage is located near the end of a rail, with the crush zone unaffected, replacement is always necessary. *Technician B* says that a kink in the crush zone can normally be repaired. Who is correct?
   a. Technician A
   b. Technician B
   c. Both Technicians A and B
   d. Neither Technician

8. *Technician A* says that new parts might require more time to install than good recycled parts. *Technician B* says that new parts usually cost more than recycled parts. Who is correct?
   a. Technician A
   b. Technician B
   c. Both Technicians A and B
   d. Neither Technician

9. If the labor time for a repair is 4.7 hours and the labor rate is $50 per hour, which of the following is the labor charge?

a. $10
b. $23.50
c. $235
d. $2,350

10. *Technician A* says that most estimating programs will NOT automatically deduct for overlap. *Technician B* says that estimating programs normally deduct for overlap. Who is correct?
a. Technician A
b. Technician B
c. Both Technicians A and B
d. Neither Technician

Activities

1. Manually write an estimate for a vehicle with minor damage. Use manuals to find the cost of parts and labor. Calculate the total repair costs for the damage.

2. Visit a shop with a computerized estimating system. Ask the shop owner if you can observe the system in use.

29

Entrepreneurship and Job Success

Objectives

After studying this chapter, you should be able to:

- Describe desirable traits of collision repair technicians.

- Summarize job responsibilities of industry personnel.

- Define the term "entrepreneur."

- Advance into other automotive related professions.

Introduction

According to economic predictions, there will be a strong demand for collision repair technicians in the future.

Our nation is now, and always will be, a "nation on wheels." You have selected an excellent area of study—collision repair technology. With this knowledge base, you should be able to earn a good living for as long as there are vehicles on the road.

Job openings are common as people move up into other positions and retire. Always read newspapers to stay aware of job openings. Some may pay better or offer better **benefits** (insurance, vacation time, holiday pay, etc.) than your present job.

COLLISION REPAIR TECHNICIAN

A **collision repair technician** is a highly skilled professional capable of doing a wide variety of tasks. This person must have a working knowledge of mechanics, electronics, measurement, welding, cutting, straightening, painting, and so on.

Collision repair technicians earn good money and are in constant demand. After finishing your training and gaining practical experience, you should be proud of becoming a collision repair technician.

Specialized technicians concentrate their study in one area. Some shops have specialized paint technicians, sheet metal technicians, masking technicians, straightening technicians, and so on. This allows each person to be more efficient at their own skill area; more vehicles are repaired and greater profit is made.

RELATED PROFESSIONS

Related professions are jobs that require an understanding of auto body repair, Figure 29–1. There are many related professions that you can advance into

Figure 29–2 Proving yourself as a dependable collision repair technician will help you advance into other positions. This is a vehicle assembly plant with numerous types of job offerings. *(Courtesy of DaimlerChrysler Corporation)*

after gaining a working knowledge of collision repair. By understanding body repair and furthering your education, you may be able to get a job with:

1. Insurance company—adjuster, agent.

2. Materials manufacturer—paint, filler, or other company.

3. Parts suppliers—new, used, and remanufactured parts companies.

4. Auto manufacturer—factory representative, designer, engineer, Figures 29–2 and 29–3.

5. Instructor—teach collision repair classes.

ENTREPRENEURSHIP

An *entrepreneur* is a person who starts his or her own business. This business might be a repair shop,

Figure 29–1 Understanding collision repair technology and terminology may help you get a job in a related profession, like this insurance adjuster position. *(Used by permission from Tech-Cor Inc.)*

Figure 29–3 These engineers are analyzing design aspects of a new vehicle. Knowledge of collision repair would be useful information when designing and testing vehicles. *(Courtesy of General Motors Corporation, Service Operations).*

materials supply house, parts warehouse, or similar endeavor. You might want to consider being an entrepreneur someday.

If you start your own business you would need to understand bookkeeping, payroll, and state and local laws controlling the industry.

After mastering the knowledge needed to do collision repair work, you will become more marketable. These skills of a collision repair technician can help when applying for other jobs or starting a business.

COOPERATIVE TRAINING PROGRAMS

Some schools offer **cooperative education programs** that allow you to have a job and get school credit at the same time. If such opportunities are available, you could be hired to work and train in a real collision repair shop. Some "co-op" programs pay a small wage while you learn.

During cooperative training, you might be a helper technician. You would work under and be further trained by an experienced professional. This is an excellent way of gaining practical, hands-on experience. And remember, experience is the best teacher.

If you are still in school, check with your instructor and guidance counselor about cooperative education programs in your area.

KEEP LEARNING!

Try to learn something job-related every day. Always learn from your mistakes. Read technical magazines and other publications. While working, try to think of more efficient ways of working. Consider new tools and techniques. This attitude will help you become more productive.

USE A SYSTEMATIC APPROACH

Develop a more **systematic approach** for doing your job. A systematic approach involves organizing and using a logical sequence of steps to accomplish a task or job. A systematic approach will result in your selecting more efficient ways of working.
Ask yourself these kinds of questions:

1. Did I take all needed tools to the vehicle?

2. Is there a better tool for a specific task?

3. Have I been reading the manufacturer's instructions (paints, plastic filler, equipment, and so on)?

4. Is my body in the right position to protect myself and feel comfortable?

WORK TRAITS

Work traits are those things, aside from skills, that make you a good or bad employee. Work traits often determine whether you keep a good job or get fired. The most important work traits are:

1. **Reliability** means being at work on time, being at work every day, and doing the job right. This is the most important job trait. Without it, you will have a difficult time keeping a job.

For example, if you often miss work, you will affect everyone in the shop. If the vehicle has been promised on Friday and you are not there, either someone else has to do your work, or the vehicle may not get done on time. Then the customer will be unhappy and will never return to your shop again. If this continues, the "word of mouth" will ruin the reputation of the shop and profits will decline.

2. **Social skills** are important so that other workers and customers like you. Many times you will need the help of another worker to complete a difficult, two-person task. If you are not liked by others, you will find many tasks almost impossible.

A good repair shop will have a "team of workers" who help each other succeed and prosper. They will exchange information, help each other with small tasks, and enjoy working together. You spend much of your life on the job; so why not enjoy it?

3. **Productivity** is a measure of how much work you get done. A highly productive technician will turn out a large amount of work. This will result in higher pay for the technician and more profits for the shop.

You want to balance productivity and quality. If you try to cut corners and get too much done too quickly, quality will often suffer. You must use proper work habits, common sense, and hard work to be productive.

4. **Professionalism** is a broad trait the includes everything from being able to follow orders to pride in workmanship. A professional does everything "by the book." He or she never cuts corners (for instance, leaving out the hard-to-reach bolt) trying to get the repair finished. Always think, dress, and act like a professional.

A professional:

1. Is customer oriented

2. Is up-to-date on vehicle developments

3. Keeps up with advancements in the repair industry

4. Pays attention to detail

5. Ensures that his/her work is up-to-specs

6. Participates in trade associations

Summary

- A collision repair technician is a highly skilled professional capable of a wide variety of tasks. This person must have a working knowledge of mechanics, electronics, measurement, welding, cutting, straightening, painting, and so on.

- Specialized technicians concentrate their study in one area such as painting, sheet metal, or masking. This allows a person to be more efficient in a chosen skill; more vehicles are repaired and greater profit is made.

- An entrepreneur is a person who starts his or her own business. This business may be a collision repair shop, materials supply house, parts warehouse, or similar endeavor.

- Work traits are those things other than skills that rate you as an employee. The most important work traits are reliability, social skills, productivity, and professionalism.

Technical Terms

benefits
related professions
entrepreneur
cooperative education
  programs
systematic approach

work traits
reliability
social skills
productivity
professionalism

Review Questions

1. According to economic predictions, there will be a strong demand for collision repair technicians in the future. True or false?

2. What are job benefits?

3. What are five areas of employment related to collision repair technology?

4. An _____ is a person who starts his or her own business.

5. Some schools offer this type of program which allows you to have a job and get school credit at the same time.
   a. Work release
   b. Work-Ed

c. Cooperative education
d. Continuing education

6. In your own words, describe the need for reliability, social skills, productivity, and professionalism when on the job.

ASE–Style Review Questions

1. *Technician A* says that according to economic predictions, there will be a strong demand for collision repair technicians in the future. *Technician B* says that there will be no new job growth in the collision repair market. Who is correct?
   a. Technician A
   b. Technician B
   c. Both Technicians A and B
   d. Neither Technician

2. *Technician A* says that a collision repair technician needs little or no training and the tasks could be done by anyone. *Technician B* says that a collision repair technician is a highly skilled professional capable of doing a wide variety of tasks. Who is correct?
   a. Technician A
   b. Technician B
   c. Both Technicians A and B
   d. Neither Technician

3. *Technician A* says that all collision technicians do all jobs in the shop. *Technician B* says that some technicians prefer to specialize and concentrate their studies in one area of collision repair. Who is correct?
   a. Technician A
   b. Technician B
   c. Both Technicians A and B
   d. Neither Technician

4. *Technician A* says that an entrepreneur is a person that specializes in frame repair. *Technician B* says that entrepreneurs are people who start their own businesses. Who is correct?
   a. Technician A
   b. Technician B
   c. Both Technicians A and B
   d. Neither Technician

5. *Technician A* says that learning new techniques is a lifelong process that requires continued training. *Technician B* says that when school is over the learning process stops. Who is correct?
   a. Technician A
   b. Technician B
   c. Both Technicians A and B
   d. Neither Technician

6. *Technician A* says that a collision worker who follows a systematic approach to repairs is more efficient. *Technician B* says that as long as all the work gets done no system is needed. Who is correct?
   a. Technician A
   b. Technician B
   c. Both Technicians A and B
   d. Neither Technician

7. Which of the following is NOT a work trait?
   a. Reliability
   b. Social skills
   c. Balance
   d. Productivity

8. *Technician A* says that "professionalism" is a broad trait that includes the ability to follow orders and pride in workmanship. *Technician B* says that "productivity" is a measure of how much work you get done. Who is correct?
   a. Technician A
   b. Technician B
   c. Both Technicians A and B
   d. Neither Technician

9. Which of the following is NOT part of being a professional?
   a. Being customer oriented
   b. Paying attention to detail
   c. Operating a computer
   d. Ensuring that work performed is up to specs

10. *Technician A* says cooperative training is two or more instructors cooperating with each other to provide instruction. *Technician B* says that cooperative education is a program in which a student has a job and gets school credit for it. Who is correct?
    a. Technician A
    b. Technician B
    c. Both Technicians A and B
    d. Neither Technician

Activities

1. Visit a small collision repair related business—parts house, parts supply store, repair shop, and so on. Ask the owner how he or she started the business. Write and give a class report.

2. Visit the library and obtain published materials on starting your own business. Create a wall display describing what it takes to start a business.

GLOSSARY

*A-Pillars Either of the front pillars that support the windshield, the front of the roof, and the door hinges. Also called windshield pillar or A-post.

A/C Low Side One side of the air conditioning system. It contains low pressure/low temperature refrigerant.

Abrasive Blasting The process of using air pressure, a blasting gun, and an abrasive to remove old paint.

*ABS Antilock brake system. A system that prevents wheel lock-up during braking.

AC Current The electrical current found in home wiring.

Accelerator A fast evaporating thinner or reducer for speeding the drying time.

Accidents Unplanned events that hurt people, break tools, damage vehicles, or have other adverse effects on the business and its employees.

*Accumulator A device, located in the low-pressure side of an air conditioning system, that removes moisture and prevents liquid refrigerant and lubricant from entering the compressor.

Acetylene Pressure Pressure leaving acetylene pressure regulator of oxyacetylene welding outfit. Typically ranges from 1 to 12 psi (kPa).

Acid Core Solder Used for doing non-electrical repairs, as on radiators.

Acid Rain Atmospheric moisture, containing acids (usually from industrial sources) that can damage a vehicle's finish.

Acorn Nut A nut closed on one end for appearance and to keep water and debris off the threads.

Acrylic Resin Adhesive A two-part somewhat flexible adhesive (liquid and powder) used on small surface areas. It is waterproof and not affected by gasoline or water.

Active Gas A type of shielding gas that combines with the weld to contribute to weld quality.

Active Restraint System A restraint system where the occupants must make an effort to use it, i.e., fastening the seat belt.

Adhesion Bonding Uses oxyacetylene to melt a filler metal onto the work piece for joining.

**Adhesion Promoter A material applied to a surface, before the application of either adhesive or paint, to strengthen the bond.

*Adhesive Any substance that is capable of bonding other substances together by surface attachment. In an auto glass replacement context, it is a high-strength polyurethane material unless otherwise specified.

Adhesive-Bonded Door Skin A door skin that is bonded, not welded, to the door frame.

Adhesive Bonded Parts Metal or plastic parts that are held together by the use of a high strength epoxy or special glue.

Adhesive Release Agent A chemical that dissolves most types of adhesives.

Adhesive Remover A chemical used to clean off the remaining vinyl glue after removal of a stripe or decal.

Aftermarket Repair Manuals Books published by publishing companies rather than the manufacturer. They are not as detailed but can give enough information needed for most repairs.

Agitator Paint Cup A small air motor in the spray gun moves a mixing paddle up and down keeping the metallic flakes mixed and evenly distributed in the cup.

Air Bag Module An assembly composed of the nylon bag and an igniter-inflator mechanism enclosed in a metal-plastic housing.

Air Bag Controller This analyzes inputs from the sensor to determine if bag deployment is needed.

Air Conditioning Gauges Used to measure operating pressures in a vehicle's air conditioning system.

Air Conversion Unit Acts as in-line regulators to convert high pressure main line air pressure to low pressure, high volume air. They use existing shop air from the compressor.

Air Couplings These allow you to quickly install and remove air hoses and air tools.

Air Cutoff Tool A tool that uses a small abrasive wheel to rapidly cut or grind metal.

Air Dams A part that improves aerodynamics by restricting air passing under the vehicle body, thus reducing turbulence and resistance to airflow.

Air Drills Uses shop air pressure to spin a drill bit.

Air File A long, thin air sander for working large flat surfaces on panels.

Air Filter Drain Valve Allows you to remove trapped water and oil from the filter.

Air Hammer Uses back and forth hammer blows to drive a cutter or driver into the work piece.

Air Hose Leak Test A method used to find body leaks. Place a soapy water solution on possible leakage points and use low pressure air on the inside of the vehicle. Bubbles will form wherever there is a leak.

Air Hoses Sizes Given as inside diameters of passage through hose.

---

* Definitions of terms preceded by an asterisk (*) are supplied courtesy of I-CAR.

Air Line Filter-Drier Used in the air line to remove moisture and debris from the air flowing through the air lines. It is designed to trap and hold water and oil that passes out of the compressor air pump.

Air Nibbler A tool used like tin snips to cut sheet metal. It has an air-powered snipping blade that moves up and down.

Air Polisher Used to smooth and shine painted surfaces by spinning a soft buffing pad.

Air Ratchet A wrench, like the hand wrench, it has a special ability to work in hard-to-reach places.

Air Replacement System Exhausts the air from spray booths continuously during the spraying process.

Air Saw It uses a reciprocating (back and forth) action to move a hacksaw type of blade for cutting parts.

Air Spring Suspension A type of suspension that uses air springs, which use the compressibility of air inside a rubber bladder, to support the vehicle load.

Air Tool Lubrication A few drops of special air tool oil, lightweight motor oil, or a mixture of oil and automatic transmission fluid should be used on all air tools.

Air Vent Hole It allows atmospheric pressure to enter the cup of the spray gun.

Air-Supplied Respirator A half-mask, full face piece, hood, or helmet, to which clean, breathable air is supplied through a hose from a separate air source. It provides protection from isocyanate paint vapors and mists, as well as from hazardous solvent vapors.

Airless Plastic Welding A plastic welding process in which an electric heating element is used to melt a plastic welding rod with no external air supply.

ALDL See Assembly Line Diagnostic Link.

Aligning Punch A long and tapered tool used to align body panels and other parts.

Alignment Rack Used to measure and adjust the steering and suspension of a vehicle after repairs.. Also known as Front End Rack.

***Alkali** Any caustic compound having a pH greater than 7.

All-Wheel Drive Two differentials are used to power all four drive wheels.

Allen Wrench A hex, hexagon, or six-sided wrench. It will install or remove set screws.

***Ambient Temperature** The temperature of the air in the work area.

***American Welding Society (AWS)** An organization that sets standards for all types of welding.

Analog Multimeter (AVOM) A multimeter with a pointer needle that moves across the face of a scale when making electrical measurements.

Anchor Clamps A device bolted to specific points on the vehicle to allow the attachment of anchor chains. They distribute pulling force to prevent metal tearing.

Anchor Pots Small steel cups imbedded in shop floor for anchoring vehicle when pulling out major damage.

Anchor Rails Steel beams in floor for holding vehicle secure while straightening.

Anchoring Equipment It holds the vehicle stationary while pulling and measuring.

Angle Measurement Divides a circle into 360 or more parts called degrees.

Antichip Coating Also called gravel guard, chip guard, or vinyl coating, it is a rubberized material used along a vehicle's lower panels.

***Anticorrosion Compounds** Wax- or petroleum-based coatings used to prevent corrosion. Usually applied inside closed sections of structural members.

***Antiflutter Material** An adhesive used between the inner and outer panels of a hood, deck lid, or roof panel to control vibration.

Antifreeze A liquid used to prevent freeze-up in cold weather and to lubricate moving parts. It also prevents engine overheating.

Antilacerative Glass Glass with one or more additional layers of plastic affixed to the inside of the glass. It gives added protection against shattering and cuts during impact.

***Antiroll Bar** A transverse torsion bar, attached to the suspension system on the underside of the vehicle, to reduce roll.

***Antisway Bar** A suspension part that limits the body lateral movement. Also called track-control rod, or track bar.

***Antitheft Label** An identifying label on certain factory-installed and replacement parts. Factory-installed parts contain the VIN; replacement parts contain "R" and "DOT."

Application Stroke A side-to-side movement of the spray gun to distribute the paint mist properly on the vehicle.

***Argon** An inert gas used for shielding during welding.

Arm Sets A resistance spot welding part that holds welding tips and conducts current to tips.

Arming Sensor A sensor that ensures that the particular collision is severe enough to require that the air bag be deployed.

ASE Tests Multiple-choice questions pertaining to the service and repair of a vehicle.

Assembly Line Diagnostic Link (ALDL) A connector provided for reading fault or trouble codes. It can be located in the passenger or engine compartment.

Automatic Transmission A transmission that shifts through gears automatically using a hydraulic system.

AWS See American Welding Society.

***B-Pillars** Either of the side pillars, near the middle of the vehicle, that support the roof. Also called B-posts.

Backing Strip Also referred to as an insert, it is made of the same metal as the base metal and can be placed behind the weld. It helps proper fit-up and supports the weld.

***Backlite** The glass installed in the rear window of a vehicle.

***Ball Joint** A mechanical joint in which a ball moves freely within a socket. Used primarily to connect steering knuckles to the control arms. Also used to connect tie rods to the steering arms.

Ball Peen Hammer A flat face hammer for general striking and a round peen end for shaping sheet metal, rivet heads, or other objects.

Banding Coat The depositing of enough paint on edges or corners of surfaces.

Bare Metal Treatment A process that prepares the metal for primer.

Base Gauges   First two body gauges hung near the center of the vehicle, usually at torque box area when measuring major damage.

Base Material   The material to be welded, usually metal or plastic.

Basecoat Patch   A small area on the vehicle's surface without clearcoat that enables the painter to check for color match with tri-coat colors.

Battery   A device that stores electrical energy chemically. It provides current for the starting system and current to all other electrical systems when the engine is not running.

Battery Charger   Converts 120 volts AC into 13 to 15 volts DC for recharging drained batteries.

*Bead Height   The height of a weld bead above the surrounding base metal.

*Bead Width   The width of the weld bead, measured at its base.

Bay   A work area for one vehicle. Also termed a repair stall.

Beam Splitters   Also referred to as laser guides, they are capable of reflecting the laser beam to additional targets.

Bench Computer   Assists in the straightening-measuring process.

Bench Grinder   A stationary, electric grinder mounted on a work bench or pedestal stand. It is used to sharpen chisels, punches, and to shape other tool tips.

Bench Seat   A longer seat for several people.

Bench System   Generally a portable or stationary steel table for straightening severe vehicle damage.

Benefits   An employment term referring to insurance, vacation time, holiday pay, etc.

Bent   A term that describes a part where the change in the shape of the part between the damaged and undamaged areas is smooth and continuous, or straightening the part by pulling restores its shape to pre-accident condition without areas of permanent deformation.

*Bevel   The edge of a surface that has been tapered to remove a 90° angle. Also called a V-groove.

*Bezel   A metal or plastic part, used as trim around lamps, knobs, or accessories.

Blasters   Air powered tools for forcing sand, plastic beads, or another material onto surfaces for paint removal.

Bleeder Spray Gun   It allows air to pass through the gun at all times, even when the trigger is not pulled back. It does not have an air control valve. The trigger only controls fluid needle movement.

Block Sanding   A simple back and forth scrubbing action with the sandpaper mounted on a rubber blocking tool.

Body Center Marks   Locations stamped into sheet metal in both upper and lower body areas of vehicle. They can save time when taking measurements to determine extent and direction of major.

Body Clips   Specially shaped retainers for holding trim and other body pieces requiring little strength. The clip often fits into the back of the trim piece and through the body panel.

Body File   A type of file that has very large, open teeth for cutting plastic body filler. They often have a separate blade that fits into a handle.

*Body Fillers   Compounds used to build up and level low areas that cannot be brought back to their original contour by straightening.

Body Jack   Also known as porta-powers, they can be used with frame/panel straighteners or by themselves. It produces a powerful pushing action.

Body Nut   A flange on the nut that helps distribute the clamping force of the thin body panel or trim piece to prevent warpage.

*Body Repair Manual   A model-specific repair manual published by the vehicle manufacturer.

Body Shop Office   An office containing business equipment and where paper work is handled.

Body Shop Tools   Specialized tools designed for working with body parts.

BOF   Body-over-frame vehicle construction.

Bolt   A metal shaft with a head on one end and threads on the other.

Bolt Diameter   Sometimes termed bolt size, it is measured around the outside of the threads.

Bolt Grade Markings   Lines or numbers on the top of the head to identify bolt hardness and strength.

Bolt Head Size   It is the distance measured across the flats of the bolt head.

Bolt Head   The part of a bolt that is used to torque or tighten the bolt.

Bolt Length   Measures the length of the bolt from the end of the threads to the bottom of the bolt head.

Bolt Strength   Indicates the amount of torque or tightening force that should be applied.

Bolt Thread Pitch   It is a measurement of thread coarseness.

Bolt Torque   A measurement of the turning force applied when installing a fastener.

Bookkeeper   A person who performs the accounting duties, i.e., invoices, checks, bills, banking, taxes.

Boots   They protect the rack in the rack-and-pinion steering unit from contamination by dirt, salt, and other road particles.

Box End Wrench   A wrench with closed ends that surround the bolt or nut head.

*Bowing   The rounding, widthwise, of a seat-belt webbing, which inhibits its travel.

Brake System   It uses hydraulic pressure to slow or stop wheel rotation with brake pedal application.

Brake System Bleeding   A process that removes air from the brake fluid.

Breaker Bar   Also called a flex handle, it provides the most powerful way of turning bolts and nuts.

*Breakout Box   An electrical junction box containing test points, which allows the signals in a circuit to be monitored while the system is operating.

Bucket Seat   A single seat for one person.

Bump Caps   Protective head gear worn when working beneath the hood or under the vehicle.

*Bump Steer   A change in steering angle, as the front wheels move over bumps, caused by misalignment of the steering linkage or rack-and-pinion assembly.

*Bumper Cover   A flexible plastic cover that surrounds a bumper structure. Sometimes part of a fascia.

Bumper Shock Absorbers   A part used to absorb some of the impact of a collision and reduce damage.

Bumpers   A part designed to withstand minor impact without damage.

Bumping Hammer A hammer used to bump out large dents.

Burn Mark A mark on the back of a weld that indicates a good weld penetration.

Burn Test A test which involves using a flame and the resulting smoke to determine the type of plastic. It is no longer recommended.

*****Burn-Through** A hole caused by excessive heat buildup which melts the base metal, allowing it to fall out of the weld joint.

Burnback In MIG welding this occurs when the electrode wire melting rate is faster than the wire speed.

C-Clamp A screw attached to a curved frame. It will hold objects on a work surface or drill press while working.

*****C-Pillars** Either of two pillars that support the rear door latch and the roof. Also called C-posts.

Cabinet Blaster A stationary enclosure equipped with a sand or bead blast tool.

Cable Hood Release Consists of hood release handle, hood release cable, hood latch and hood striker.

*****Camber** The inward or outward tilt of a tire at the top, measured in degrees from true vertical, when viewed from the front or rear of the vehicle. An outward tilt, away from the vehicle centerline, is positive camber, and an inward tilt is negative.

*****Canister Purge Valve** A valve in an emission-control system that allows hydrocarbons to flow from the charcoal canister to the engine intake manifold.

Cap Protective head gear to keep hair safe and clean when sanding, grinding, and doing similar jobs.

Cap Screw A term that describes a high strength bolt.

Capillary Attraction In brazing, it is the natural flow of the filler material into the joint.

Captive Spray Gun System Individual guns that are set up and used for applying each different type of material.

Carbon Monoxide (CO) An odorless, invisible gas that comes from vehicle exhaust.

Carburizing Flame Also called a reducing flame, it is obtained by mixing slightly more acetylene than oxygen.

Card Masking Involves using a simple masking pattern to produce a custom paint effect.

Cartridge Filter Respirators A rubber face piece that conforms to the face and forms an airtight seal. Used to protect against vapors and spray mists of nonactivated enamels, lacquers, and other nonisocyanate materials.

Cartridge Gun A device used to apply adhesive or sealer to body panels.

*****Caster** The angle of the steering axis, measured in degrees from true vertical, when viewed from the side of the vehicle. Caster is positive if the steering axis is tilted rearward, and negative if it is tilted forward.

*****Catalytic Converter** An emission control device in the exhaust system which uses a platinum-iridium catalyst to convert carbon monoxide gas and hydrocarbon particles to carbon dioxide gas and water.

CD-ROM A compact disc that stores information.

*****Center Bolt** A special bolt used to hold the plys of a spring together and center the axle on the spring.

Center Link It supports the tie-rods in position in parallelogram steering.

Center Orifice Located on the spray gun air cap at the fluid nozzle. It creates a vacuum for the discharge of the paint.

Center Punch A pointed tool used to start a drilled hole or mark parts. The indentation will keep a drill bit from wandering out of place.

Center Section Also called midsection, it typically includes the body parts that form the passenger compartment.

Certified Crash Tests A test performed using a real vehicle and sensor-equipped dummies that show how much impact the people would suffer during a collision.

*****cfh** Cubic feet per hour. A unit for the flow rate of a gas.

Chain Wrench A wrench used to grasp and turn round objects. It is used to remove stuck or damaged engine oil filters or to adjust exhaust systems.

*****Channel** A piece of U-shaped metal that is attached along the bottom edge of movable glass and connects the glass to the regulator.

Channellock® Pliers Pliers that have several settings for grasping different size objects; also known as rib joint pliers.

Charging A term used to describe filling the air conditioning system with refrigerant.

Charging Station A machine for automatically recharging an air conditioning system with refrigerant.

Charging System It recharges the battery and supplies electrical energy when the engine is running.

Chisel A tool used for some cutting operations such as shearing off rivet heads or separating sheet metal parts.

Chuck Movable jaws that close down and hold the drill bit.

Circular Sander It spins its pad and sandpaper around and around in a circular motion.

Classroom An area provided for lectures, demonstrations and meetings of shop personnel.

*****Clay Product** A mild abrasive specifically made for removing overspray.

*****Clips** Devices which hold decorative chrome to the vehicle body, or hold moldings, etc.

*****Clock Spring** A coil that supplies deployment current to the airbag module from the steering wheel.

Clock Spring The electrical connection between the steering column and the air bag module.

Closed A term that refers to an electrically-connected electrical circuit.

Closed Coat A type of grit on sandpaper and discs, the resin completely covers the grit. This bonds the grit to the paper or disc more securely.

*****Closure Panel** A movable, exterior body panel, such as a hood, door, deck lid, etc.

*****CNG** Compressed natural gas.

*****CO2** The chemical symbol for carbon dioxide. A gas used for shielding during welding.

*****Cold Solder Joint** A solder joint that appears sound, but is electrically defective.

Cold Weather Solvent Designed to speed solvent evaporation, so the paint flows and dries in a reasonable amount of time.

Collapsible Steering Columns They crush under force to absorb impact, thus reducing injuries.

Collision Estimating Guide   A manual or computer software program that gives information to help calculate the cost of repairs.

Collision Estimating Manuals   Guides that give information for calculating the cost of repairs.

Collision Repair   Auto body repair, restoring a damaged vehicle back to its original condition, structurally and cosmetically.

Collision Repair Technician   A highly skilled professional who repairs damage from an accident.

Color Blindness   A physical condition that makes it difficult for a person's eyes to see colors accurately.

Color Coded   Plastic welding rods are identified by color.

Color Corrected Lights   Used in the spray booth to make color show up properly.

Color Directory   A publication containing color chips and other paint-related information for most makes and models of vehicles.

Color Matching Manuals   Guides that contain information needed for finishing panels, so that the repair has the same appearance as the old finish.

Color Matching   The steps needed to make the new finish match the existing finish on the vehicle.

Color Spectrum   By passing light through a prism, light is broken down into its separate colors. White light contains all colors or the full spectrum.

Color Tree   A guide used to locate colors three dimensionally when matching colors.

Colorcoat   See topcoat.

Combination Wrench   A wrench with an open end and a box end.

Combustibles   Substances capable of igniting and burning such as paints, thinners, reducers, gasoline, dirty rags.

Compact Car   The smallest body classification, also known as economy car. It normally uses a small 4-cylinder engine, is very lightweight, and gets the highest gas mileage.

Component   A part, refers to the smallest units on a vehicle.

Composite Force   The force of all pulls combined.

Composite Plastics   Also called "hybrids," they are blends of different plastics and other ingredients designed to achieve specific performance characteristics.

*Compound Shape   A structural shape where the opposite sides are not parallel over its length, or a surface that curves in two or more directions over its area.

Compressive Strength   The property of a material to resist being crushed.

Compressor Drain Valve   Located on the bottom of the compressor storage tank, it allows you to drain out water.

Compressor Oil Plugs   Provided for filling and changing air pump oil on a compressor.

Compressor Pressure Valve   It prevents too much pressure from entering the compressor tank.

Computer   A complex electronic circuit that produces a known electrical output after analyzing electrical inputs.

Computer Estimating   Preparing an estimate via computer. This streamlines the process by automating the written estimate.

Computer Self-Diagnostics   The computer can detect its own problems.

Computer Written Estimate   A damage report that is completed by using a personal computer and printer.

Computerized Measuring Systems   A system that uses data from the measuring system for fast entry and checking of vehicle dimensions.

Computerized Paint Matching Systems   Software that uses data from the spectrophotometer to help in how to match the paint color.

*Condenser   The part of an air-conditioning system where the refrigerant is cooled, changing it from a high-pressure gas to a high-pressure liquid.

Conductor   It carries current to the parts of a circuit.

Cone Concept   As the force of impact penetrates the unibody vehicle, it folds and collapses, and the impact is absorbed by an ever-increasing area of the unibody. Visualize the point of impact as a cone.

Continuity   A continuous conductor path.

Continuous Weld   A single weld bead along a joint.

*Control Arm   A suspension part, usually pivoted at each end, that keeps a wheel or axle in the proper position.

*Control Point   Any of several identifiable points that are very accurately located during a vehicle's original assembly.

*Conversion Coating   The part of a treatment system that modifies the metal surface to improve corrosion protection and paint adhesion.

*Convoluted Area   A series of ridges stamped into a structural part, such as a rail, allowing crushing or bending as a means of absorbing the energy of a collision.

*Coolant   A mixture of glycol and water, circulated through a cooling system to move heat from the engine to the radiator.

Cooling System Leak Test   A test performed by installing the tester on the radiator neck. A loss of pressure or coolant means there is a leak.

Cooperative Education Programs   A curriculum offered by a school that allows you to have a job and get school credits at the same time.

*Corona   The visible glow of the base metal at the weld site caused by the concentration of heat during GMA (MIG) welding.

*Corrosion   The chemical reaction of air, moisture, or corrosive materials on a metal surface. Also called rust or oxidation.

*Corrosive Chemical   An acidic or alkaline substance that, when in contact with unprotected areas of the body, will cause injury.

Cosmetic Filler   Typically a two-part epoxy or polyester filler used to cover up minor imperfections on reinforced plastics.

*Cosmetic Surface   A surface that is finished or decorated to improve its appearance. Includes paint, glass, upholstery, etc.

Cotter Pin   A pin that helps prevent bolts and nuts from loosening or it fits through pin to hold parts together. They are also used as stops and holders on shafts and rods.

Cowl   The body parts at the rear of the front section, right in front of the windshield.

Cream Hardeners   Used to cure body fillers.

Crescent Wrench   Also referred to as an adjustable wrench, it has movable jaws to fit different head sizes.

CRI Color Rendering Index, it is the index for measuring how close a lamp in indoor lighting is to actual daylight.

*Critical Temperature The maximum temperature allowed by a vehicle maker to prevent loss of strength in metal, when heating to relieve stress.

*Crossmember Any of the transverse structural or stiffening members of a vehicle frame or underbody structure.

Curb Height Also called ride height, each of the front and rear wheels must carry the same amount of weight, thus allowing the vehicle to ride at a specific height.

*Cure The process of drying or hardening of a material.

*Cure Time The time required for a chemical or material to dry or set at a given temperature and humidity. Cure time varies with the type of material used and the thickness of the application.

Current Adjuster A circuit relay that is activated by a series coil in the charging circuit.

Current The movement of electricity (electrons) through a wire or circuit.

Custom Body Panels Aftermarket parts that alter the appearance of the vehicle.

Custom Masks Involves drawing a design on thin posterboard, then cutting it out, taping it onto the vehicle and spray painting it for a custom pattern in the paint.

Custom Painting Using multiple colors, metal flake paints, multi-layer masking, and special spraying techniques to produce a personalized look.

Custom-Mixed Colors Paint colors that are mixed to order at the paint supply distributor.

*Cutting Torch A common term for an oxyacetylene torch with a special tip used to cut metal. Can also be a plasma arc cutter.

CV-Axle Constant Velocity Axle, they transfer torque from the transaxle to the wheel hubs.

*CV-Joint Constant velocity joint. A type of universal joint that allows the output shaft to rotate at the same instantaneous speed as the input shaft. Used in halfshaft assemblies for driving wheels that are sprung independently of the differential.

Cyanoacrylates (CAs) One-part, fast-curing adhesives used to help repair rigid and flexible plastics.

Cylinder Head Covers and seals the top of the cylinder.

*D-Pillar Either of two pillars that support the roof and are located rearward of the C-pillars, as in station wagons and vans. Also called D-posts.

*D-Ring The seat-belt mounting point located on the B-pillar or the door surround panel.

Damage Analysis Involves locating all damage, using a systematic series of inspections, measurements, and tests.

Damage Appraisal See Estimating.

*Dash Lamp Any of the dash-mounted warning lamps that are used to display diagnostic trouble codes.

Database A computer file. An estimating database would include part numbers, part illustrations, labor times, labor rates, and other data for filling out the estimate.

Dead Blow Hammer A hammer that has a metal face filled with lead shot (balls) to prevent rebounding. It will NOT bound back up after striking.

Decimal A number value that divides the inch into parts. It is used when precision is important.

Decimal Conversion Chart It allows you to quickly change from fractions, to decimals, to millimeters.

Dedicated Bench A strong, flat work surface to which fixtures are attached.

Deformation Refers to the new, undesired bent shape the metal takes after an impact or collision.

*Delamination The failure of the bond between layers, as when windshield glass separates from the inner layer of vinyl, or when paint peels from the substrate beneath it.

Dent Puller A slide hammer with a threaded tip, a hood tip, or a suction cup.

Destructive Testing A test that forces the weld apart to measure weld strength. Force is applied until the metal or weld breaks.

Diagnosis Charts Charts that give logical steps for finding the source of problems.

Diagonal Cutting Pliers Pliers that have cutting jaws that will clip off wires flush with a surface; they are also known as side cut pliers.

Dial-Up Estimating System A system that relies on data stored in a mainframe computer at a remote location. The data is usually accessed via a telephone modem.

Die A tool used for cutting threads on the outside of bolts or studs.

Differential Assembly A unit within the drive axle assembly. It uses gears to allow different amounts of torque to be applied to each drive while the vehicle is making a turn.

Digital Images Pictures stored as computer data.

Digital Multimeter (DVOM) A multimeter that gives a digital readout for the test value.

Dinging Hammer A hammer used for final shaping.

Disarmed All sources of electricity for air bag igniter are disconnected.

Disc Adhesive A product used to apply nonstick sandpaper to the backing pad of a sander.

Disc Adhesive A special non-hardening glue that comes in a tube. It can be placed on the pad to adhere the paper to the sander.

Disc-Type Grinder The most commonly used portable air grinder. It is operated like the single-action disc sander.

Discharging The removal of refrigerant from the air conditioning system.

*Dogtracking A condition caused by excessive thrust angle where, as the vehicle's wheels move straight ahead, the body appears to be moving at an angle.

Dollies Also referred to as dolly blocks. Used like a small, portable anvil; generally used on the backside of a panel being struck with a hammer.

Door Alignment Tool A tool used to slightly bend and adjust sprung doors and hatches.

Door Frame The main steel frame of the door.

Door Glass Channel It serves as a guide for the glass to move up and down. It is a U-shaped channel lined with a low friction material.

Door Glass It must allow good visibility out of the door.

Door Intrusion Beam A part that is welded or bolted to the metal support brackets on the door frame to increase door strength.

Door Latch It engages the door striker on the vehicle body to hold the door closed.

Door Lock Assembly It usually consists of the outside door handle, linkage rods, the door lock mechanism, and the door latch.

Door Regulator A gear and arm mechanism for moving the glass.

Door Skin The outer panel over the door frame. It can be made of steel, aluminum, fiberglass, or plastic.

Door Striker Wrench A special socket designed to fit over or inside door striker posts. It will allow you to turn and loosen the post for adjustment or placement.

Door Striker It holds the door closed.

Door Trim Panel An attractive cover over the inner door frame.

Double Masking Using two layers of masking paper to prevent bleed-through or finish-dulling from solvents.

Dowel Pin Small rod or pin that helps hold or align parts.

Downdraft Spray Booth Forces air from the ceiling down through exhaust vents in the floor.

Download A computer term meaning to enter data.

*Drag Link A link found in some steering systems that connects the pitman arm to the steering arm. Also called center link.

Draw-Down Bar A machine bar that evenly distributes 1 mil of unreduced paint to a card or sheet of plastic. It is used to check how much hiding the paint will provide.

*Dressing A Weld Grinding or sanding a weld bead or nugget to reduce the height, or make it flush with the base metal.

Drills Used to make accurately sized holes in metal and plastic parts. Both air and electric drills are available.

Drive Shaft A long tube that transfers power from the transmission to rear axle assembly.

Dry Metallic Paint Spray The aluminum flakes are trapped at various angles near the surface of the paint film making the paint appear lighter and more silver.

Dry Sanding A term describing sanding done with coarser non-waterproof sandpaper, without using water.

Dry Sandpaper Designed to be used without water.

*Dry-Set To trial fit a glass assembly before applying adhesive.

*Dry Spray Appearance A paint defect where the paint droplets flash before they flow together, resulting in a dull, overspray appearance.

Dual Action Sander A sander that moves in two directions at the same time. It produces a much smoother surface finish.

Duct Tape A thick tape with a plastic body. It is sometimes used to protect parts from damage when grinding, sanding, or blasting.

Dulled Finish This occurs where the spray pattern overlaps an already sprayed but dry area when painting.

Dust Coat A light, dry coating that is similar to a mist coat.

Dust Respirator A filter that fits over your nose and mouth to block small airborne particles while sanding or grinding.

Dustless Sanding System A system designed to draw airborne dust into a storage container, much like a vacuum cleaner, by using a blower or air pump.

Duty Cycle The number of times the welder can safely operate at a given amperage level continuously over 10 minutes.

*DVOM Digital volt-ohmmeter. A high-impedance instrument used to test electrical systems.

*Dye Penetrant A colored fluid that penetrates microscopic cracks, making them visible to the naked eye.

*E-Coat Electro-deposition primer applied to metal parts, during vehicle assembly and replacement part manufacture, to prevent corrosion.

Earmuffs Provides protection to your eardrums from damaging noise levels in the metalworking areas.

Earplugs Provides protection to your eardrums from damaging noise levels in the metalworking areas.

Edge Masking Taping over panel edges and body lines prior to machine buffing or polishing to protect the paint from burn-through.

*Egg-Crate Absorber A type of energy absorber that relies on the collapse of plastic ribs to absorb energy.

EGR Exhaust Gas Recirculation. It is a valve that opens to allow engine vacuum to siphon exhaust into the intake manifold.

Elastic Deformation The ability of metal to stretch and return to its original shape.

Electric Adhesive Cutter A device that plugs into 120 volts to produce a vibrating action to assist in cutting the adhesive around glass.

Electrical Fires Results when excess current causes wiring to overheat, melt and burn.

Electrical Repairs Repairing electrical components such as wiring, sensors.

Electrical Symbols Graphic representations of electrical-electronic components.

Electrical Systems Anything powered by the battery, i.e. ignition, lights, radio, heat, air conditioning.

Electrode Tips A resistance spot welding part that allows current flow through spot weld.

Electromagnet A set of windings.

Electronic Camera A camera used to take photos of the vehicle's damage and store them as digital data.

Electronic Leak Detector A device that uses a signal generator tool and signal detector to find body leaks.

*Electronic Memory The storage capability of a programmable electronic device, such as a security system, memory seat, mirror, or radio.

Electrostatic Spray Gun Electrically charges the paint particles at the gun to attract them to the vehicle body.

Emblem Adhesive Designed to hold hard plastic and metal parts. It is also referred to as plastic adhesive.

Emergency Telephone Numbers A list of telephone numbers clearly posted next to the shop's telephone including a doctor, hospital, fire and police.

Encapsulated Glass See modular glass.

Energy Reserve Module It allows the air bag to deploy in the event of a power failure.

Engine A machine that provides energy to move the vehicle and power all accessories.

*Engine Cradle A sub-frame bolted to the vehicle frame or underbody, that supports the engine. The engine cradle may also support the transmission and the suspension system. Sometimes called the powertrain cradle.

Engine Crane   A device that can be formed on most racks or benches to raise and remove an engine.

Engine Holder   A device used to support the engine-transmission assembly when the cradle must be unbolted or removed.

English Measuring System   Fractions and decimals are used for giving number values.

English-Metric Conversion Chart   It allows you to convert from English to metric or from metric to English values.

EPA (Environmental Protection Agency)   A government agency responsible for environmental protection.

*Epoxy Primer   A primer that uses a thermoset resin to increase corrosion protection and adhesion, with minimal shrinkage.

*ESD Strap   Electrostatic discharge strap. A wristband, with an electrical contact point and ground wire, which discharges static electricity from the technician's body. Usually contains a high resistance in the ground lead to protect the technician from electrical shock.

Estimate   Also called a damage report or appraisal, it calculates the cost of parts, materials and labor for repairing a collision-damaged vehicle.

Estimating Program   Software that will automatically help find parts needed, labor rates, and calculate total for the repairs.

*Etch   The process of chemically roughing a metal surface.

Exhaust System   It collects and discharges exhaust gases caused by the combustion of the air/fuel mixture within the engine.

Explosions   Air pressure waves that result from extremely rapid burning.

*Evaporator   The part of an air conditioning system where the refrigerant changes from a low-pressure liquid to a low-pressure gas by absorbing heat from the air in the vehicle's passenger compartment.

*Express Power Window   A power window that lowers or raises nonstop when the switch is momentarily pressed.

Extensions   A part that fits between the socket and its drive handle. They allow you to reach in and install the socket when surrounded by obstructions.

Factory Packaged Colors   Ready mixed paints.

*Factory Seams   The edges and seams where panels were joined during vehicle assembly. Also called factory joints.

Fastened Parts   Parts held together with various fasteners (bolts, nuts, clips, etc.); fenders, hood, and grill.

Fault Code   A code representing a specific circuit or part with a problem.

*Featheredge   A tapered edge of a repair area, or the process of making a taper.

Fender Cover   Material placed over the vehicle body to protect the paint from chips and scratches while working.

Fender Washer   A washer with a very large outside diameter for the size hole. They provide better holding power on thin metal and plastic parts.

Fiber-Reinforced Plastics (FPR)   Plastic panels reinforced with fiber.

Fiberglass Body Filler   A plastic filler with fiberglass material added. It is used for rust repair or where strength is important.

Fiberglass Cloth   Made by weaving the fiberglass strands into a stitched pattern.

Fiberglass Mat   A series of long fiberglass strands irregularly distributed to form a patch. It is used to strengthen and form a shape for the resin liquid.

Fiberglass Resin   Another form of plastic body repair material. It is a thick resin liquid.

File   A category of hand tools used to remove burrs, sharp edges, and to do other smoothing tasks.

Filler Filing   Using a coarse "cheese grater" or body file to rough shape the semi-hard filler. You will knock off the high spots and rough edges.

Filler Material   Material from a wire or rod that is added to the weld joint.

Filler Mixing Board   A clean, nonporous surface for mixing filler and hardener.

Filler Over-Catalyzation   A condition caused by using too much hardener for the amount of filler.

Filler Oversanding   A condition that results when the filled in area is below the desired level, which makes it necessary to apply more filler.

Filler Undersanding   A condition that results when the filled area is higher or thicker than the surrounding panel.

*Filler Rod   A plastic rod used in a plastic weld to form a bond and fill a damaged area. Also, a metal rod when welding metal.

*Filler Strip   A strip inserted into a rubber gasket after the glass is installed, forcing the gasket against the glass to form a seal and improve the grip. Sometimes called locking bead or spline.

Filler Under-Catalyzation   A condition caused by not using enough hardener in the filler.

*Filler Wire   The wire used to add metal to a weld puddle, to create a bead or nugget.

*Fillet Weld   A weld where the bead fills the junction between two pieces of base metal joined at an angle, as in a tee or lap joint.

Final Detailing   Term used to describe the location and correction of any defection that may cause customer complaints. The steps are sanding and filing; compounding; machine glazing; and hand glazing.

Fine Line Masking Tape   A very thin, smooth surface plastic masking tape. Also termed flush masking tape.

Finishing Area   An area where the body is painted. Also known as the Paint Booth Area.

Finishing Hammer   A hammer used to achieve the final sheet metal contour.

Fire Extinguisher   An instrument designed to quickly smother a fire. There are several types available for putting out different kinds of fires.

Fire   Rapid oxidation of flammable material producing high temperatures.

Firewall   The panel dividing the front section and center, passenger compartment section. Sometimes termed dash panel or front bulkhead.

First Aid Kit   This includes many medical items such as sterile gauze, bandages, scissors, bandaids, and antiseptics that are needed to treat minor shop injuries.

Fisheye Eliminator   A paint additive that helps smooth the paint when small craters or holes in the paint film are a problem.

Fit Test A test performed prior to using a respirator. It checks for respirator air leaks, making both negative and positive pressure checks.

*Flame Treatment The light scorching of a plastic repair area with a torch, to strengthen the bond of the adhesive to be applied.

*Flange A bend or offset formed along the edge of a panel, or around a hole in a panel.

*Flare A tapered expansion of an opening. A flare is often used when connecting joints in tubing.

*Flash The first stage of drying where some of the solvents evaporate. The surface dulls from a high gloss to a normal gloss.

*Flash Corrosion Corrosion that forms when water dries on bare steel.

Flat Rate A preset amount of time and money charged for a specific repair operation.

Flat Washer A washer used to increase the clamping surface area. They prevent the smaller bolt head from pulling through the sheet metal or plastic.

Flat Welding The pieces are parallel with the bench or shop floor.

Flattener An agent added to paint to lower gloss or shine.

*Flex Agent An additive that increases the flexibility of paint for use on plastic parts.

*Flexible Filler A material used to fill and level repair areas on plastic parts.

Flop Also termed flip-flop, it refers to a change in color hue when viewing from head-on and then from the side.

Floppy Disks Removable magnetic disks of data used with a computer.

Flow Meter A device that measures the movement of air, gas, or a liquid past a given point.

Fluid Control Knob A part of the spray gun; it changes the distance the fluid needle moves away from its seat in the nozzle when the trigger is pulled. This controls paint flow.

Fluid Control Valve On a spray gun, it can be turned to adjust the amount of paint or other material emitted.

Fluid Needle A part of the spray gun; it open or shuts off the flow of paint.

Fluid Needle Valve It seats in the fluid tip on a spray gun to prevent flow or can be pulled back to allow flow.

Fluid Nozzle A part of the spray gun; it forms an internal seat for the fluid needle.

Fluorescent Light Artificial light that has more violets and reds than daylight.

Fluorine Clearcoat Used to improve resistance to UV radiation; protect from weathering, chalking, oxidation, and industrial fallout; improve luster of the finish; and reduce required maintenance, such as waxing and polishing.

*Flux-Cored Wire A type of welding wire having a central core of flux that creates a shielding gas when heated.

*Foam Fillers Plastic foam used in body cavities, primarily to stop the transmission of sound.

Forced Drying Curing fresh paint by using special heat lamps or other equipment to speed the process.

Four-Wheel Drive A system that uses a transfer case to send power to two differentials and all wheels. The transfer can be engaged and disengaged to select two- or four-wheel drive.

Fraction A number value that divides the inch into parts, It is used when precision is not critical.

Frame Horn The very front of the frame rails where the bumper attaches.

Frame Rack Accessories The many chains, special tools, and other parts needed to perform the many different frame straightening operations.

*Frame Rails The fore-and-aft members of a frame or underbody structure.

Frame Straightening Accessories The many chains, special tools, and other parts needed to perform the many different frame straightening operations.

Framed Doors A type of door design, it surrounds the sides and top of the door glass with a metal frame, helping keep the window glass aligned.

Front End Rack Used to measure and adjust the steering and suspension of a vehicle after repairs. Also known as Alignment Rack.

Front Fender Aprons The inner panels that surround the wheels and tires to keep out road debris.

Front Fenders Part that extends from the front doors to the front bumper. They cover the front suspension and inner aprons.

Front Section Also called nose section, it includes everything between the front bumper and the firewall.

Fuel Evaporative System It pulls fumes from the gas tank and other fuel system parts into a charcoal canister.

Full Body Sectioning Replacing the entire rear section of a collision-damaged vehicle with the rear section of a salvage vehicle.

Full Cutout Method A method of windshield replacement used when the original adhesive is defective or requires complete removal.

Full Wet Coat A heavy, glossy coat applied over the tack coat. It makes the paint lay down smooth and shiny.

Full-Face Shield Full face protection used when there is danger from flying particles, as when grinding.

Full-Size Car The largest body classification. It is large, heavy, and often uses a high performance V-8 engine.

Funnel Instruments used when pouring fluid from a container into a small opening.

Fuse A circuit protection device.

Fuse Box This holds the various circuit fuses, breakers, and flasher units.

*Fusible Link A short length of wire, smaller in diameter (lower current capacity) than the rest of the circuit, spliced into a primary circuit to act as a time-delay fuse.

*Fusion The melting and flowing together of two pieces of metal during welding.

Fuse Rating The current at which the fuse will blow.

Fusion Welding Joining different pieces of metal together by melting and fusing them into each other.

*Galvanic Corrosion Corrosion caused by contact between dissimilar metals when moisture is present.

*Galvanizing A protective zinc coating applied during the production of steel.

Gas Flow Rate A measurement of how fast gas flows over the weld puddle in MIG welding.

*Gas Metal Arc Welding (GMAW)   A welding process in which the electrode filler wire is fed continuously into the weld puddle while the puddle is protected by the shielding gas. Also called metal inert gas (MIG) welding.

Gasoline   A highly flammable petroleum or crude oil-based liquid that vaporizes and burns rapidly.

Gauge Measuring Systems   A method that uses sliding metal rods or bars, and adjustable pointers with rules scaled to measure body dimensions.

Get Ready   A thorough clean-up before returning the vehicle to the customer.

Glazing Putty   A material made for filling small holes or sand scratches.

Goggles   Eye protection suitable when handling chemicals.

Graduated Pail   Used to measure liquid materials when mixing.

Grain Structure   In a piece of steel, it determines how much it can be bent or shaped.

Gravity Feed Spray Guns   The cup is mounted on top of the spray gun head so the material flows down without external air pressure.

Grease Gun   A device used to lubricate high friction points on a vehicle's steering and suspension systems.

Grinders   Used for fast removal of material. They are often used to smooth metal joints after welding and to remove paint and primer.

Grinding Discs   Round, very coarse abrasives used for initial removal of paint, plastic, and metal (weld joints).

*Grit   A numerical rating of the coarseness of sandpaper or sanding discs.

Grit Numbering System   Denotes how coarse or fine the abrasive is.

Grit Sizes   A term that refers to the coarseness of sandpaper.

Ground   Electrical path back to the battery negative cable through wires, cables or body structure.

Gun Angle   Consists of two different angles; the angle of the gun to the work place, and the angle of direction of travel.

Gun Arching   Uneven film thickness, paint runs, excessive overspray, dry spray and orange peel caused by not moving the spray gun parallel with the surface.

Gun Body   The part of the spray gun that holds the parts that meter air and liquid. It also holds the spray pattern adjustment valve, fluid control valve, air cap, fluid tip, trigger, and related parts.

Gun Heeling   Uneven film thickness, excessive overspray, dry spray and orange peel caused by allowing the spray gun to tilt.

Hacksaw   A tool sometimes used to cut metal parts.

*Halfshaft   A rotating shaft assembly using two CV-joints to transmit power from a differential to a wheel that is sprung independently of the differential. Commonly used in front-wheel-drive vehicles and some rear-wheel-drive vehicles.

Halo Effect   An unwanted shiny ring or halo that appears around a pearl or mica paint repair. It is caused by the paint being wetter in the middle and drier near the outer edges of the repair.

Hammer-Off-Dolly   A method used to straighten metal just before the final stage of straightening.

Hammer-On-Dolly   A method used to stretch metal and to smooth small, shallow dents and bulges.

Hand Blaster   A small, portable tool for blasting parts and panels on the vehicle.

Hand Cleaner   Used to wash hands.

Hand Compound   Compound designed to be applied by hand on a rag or cloth.

Hand Compounds   An oil based rubbing compound that provides lubrication; best on small or blended areas.

Hand Glazes   Used for final smoothing and shining of the paint.

Hand Scraper   A hand tool that will easily remove gaskets and softened paint (where a paint stripping chemical has been used). They are used on flat surfaces.

Hand Tools   General tools such as wrenches, screwdrivers, pliers and other common tools.

Handscraper   A hand tool that will easily remove gaskets and softened paint (paint stripping chemical used). They are used on flat surfaces.

Hard Brazing   Brazing with materials that melt at temperatures above 900_F (486_C).

*Hardener   An additive that causes a chemical reaction to cure paint, body filler, adhesives, etc.

Hardtop   A body design that does not have a center pillar to support the roof, but must be reinforced to provide extra strength. It is available in 2- and 4-door versions.

Hardtop Doors   A type of door design, the glass extends up out of the door without a frame around it. The glass itself must seal against the weatherstripping in the door opening.

Hatchback   A body design that has a large third door at the back. This design is commonly found on small compact cars, allowing for more rear storage space.

Head-On View   Compares the test panel to the vehicle straight on, or perpendicular.

Headlight Aimers   Equipment used to adjust the direction of the vehicle headlights.

Headlight Aiming Screw   A screw with a special plastic adapter mounted on it. The adapter fits into the headlight assembly.

Headrest Guide   The sleeve that accepts the headrest post and mounts in seat back.

Headrest   The padded frame that fits into the top of the seat back.

*Heat-Affected Zone   A metal area that has been heated beyond its critical temperature, causing it to lose strength.

Heat Crayons   Used to determine the temperature of the aluminum or other metal being heated. They will melt at a specific temperature and warn you to prevent overheating.

Heat Setting   Determines the length of the arc when MIG welding. Also called voltage.

*Heat Sink   A piece of metal that protects wiring, or other sensitive parts, by absorbing excessive heat. Clamp-on heat sinks are often used during repair operations such as soldering.

Heat Sink Compound   A paste that can be applied to parts during welding to absorb heat and prevent warpage.

Helicoil®   Also referred to as a thread insert, it can be used to repair badly damaged internal threads.

*Heat-Treatable Alloy   Any substance whose strength will not be affected by the application of a controlled amount of heat. Usually refers to metals, such as alloys of aluminum, or iron (steel).

Helper   An apprentice type worker that learns by helping experienced technicians.

*HEPA Filter   A high-efficiency particulate air filter. A type of air filter that traps at least 99.97% of all particles that are 0.3 micron or larger in size.

*Hg   The chemical symbol for mercury. The height of a column of mercury, expressed in inches, is a measure of pressure or vacuum. One inch Hg equals 0.491 psi.

High Efficiency Spray Gun   Designed to reduce VOC emissions by only using 10 psi air pressure at the gun head. Also called HVLP for high volume, low pressure or LVLP for low volume, low pressure.

High Spot   Also called a bump, it is an area that sticks up higher than the surrounding surface.

High-Solids Paint   The non-liquid contents of paint needed to reduce air pollution or emissions when painting.

High-Strength Steel (HSS)   Sheet metal that is stronger than low-carbon or mild steel because of heat treatment.

Hinged Parts   Parts that swing up and open such as doors, hoods, and decklids.

Hog Rings   A device that stretches and holds the seat cover over the seat frame and padding.

Hood   The hinged panel for accessing the engine compartment (front-engine vehicle) or trunk area (rear-engine vehicle).

Hood Adjustments   Adjustments made at the hinges, at the adjustable stops, and at the hood latch.

Hood Hinge Spring Tool   A hooked tool that should be used to stretch the spring off and on.

Hood Hinge Spacers   A part that helps adjust the hood.

Hood Hinges   A part that allows the hood to open and close while staying in alignment.

Hood Latch Mechanism   This keeps the hood closed and releases the hood when activated.

Hood Latch Adjustment   This controls how well the hood striker engages the latch mechanism.

Hood Latch   Part of the cable hood release, it has metal arms that grasp and hold the hood striker.

Hood Release Cable   Part of the cable hood release, it is a steel cable that slides inside a plastic housing. One end fastens to the release handle and the other to the hood latch.

Hood Release Handle   Part of the cable hood release, it can be pulled to slide a cable running out to the hood latch.

Hood Stop Adjustment   A device that controls the height of the front of the hood.

Hood Striker   Part of the cable hood release, it bolts to the hood and engages the hood latch when closed.

*Horizontal Datum Plane   A horizontal plane, located at or below a vehicle's underbody, that serves as a reference for height measurements.

Horizontal Welding   The pieces are turned side-ways. Gravity tends to pull the puddle into the bottom piece.

Hose Clamp   A clamp used to hold radiator hoses, heater hoses and other hoses onto their fittings.

Hot Air Plastic Welding   Uses a tool with an electric heating element to produce hot air which blows through a nozzle and onto the plastic.

Hot Weather Solvent   A reducer or thinner. It is designed to slow solvent evaporation to prevent problems.

*HSS   High-strength steel. A type of steel which is stronger than mild steel and able to withstand stress up to 70,000 psi.

HVLP   See High Efficiency Spray Gun

Hydraulic   Equipment operated by oil under pressure.

Hydraulic Jacks   A device used to raise the vehicle.

*Hydraulic Modulator   A part of an antilock brake system that controls the application of hydraulic pressure to each wheel cylinder.

*Hydrometer   An instrument used to measure the specific gravity of a liquid, such as battery acid or radiator coolant.

I-CAR   See Inter-Industry Conference on Auto Collision Repair.

Idler Arm   An arm that duplicates the motion of the pitman arm on the opposite side of the vehicle, in order to maintain steering geometry.

Ignition System   This produces an electric arc in a gasoline engine to cause the fuel to burn. It must fire the spark plugs at the right time. The resulting combustion pressure forces the pistons down to spin the crankshaft.

Impact Driver   A screwdriver that is hit with a hammer to rotate tight or stuck screws. The body of the driver can be rotated to change directions for installing or removing.

Impact Sensor   The first sensor to detect a collision; it is mounted at the front of the vehicle.

Improper Coverage   A term that means the paint film thickness is not uniform or too thin. It is caused by not triggering exactly over the edge of the panel or from improper spray overlap.

Improper Spray Overlap   A term that means you are not covering half of the previous pass with the next pass.

In-Line Oiler   An attachment that will automatically meter oil into air lines for air tools.

In-Shop Estimating System   A system that has all of the data needed at the shop office, usually on CD-ROMs.

Incandescent Light   Artificial light that has more yellow, oranges and reds than daylight.

*Included Angle   The sum in degrees of the camber and SAI.

Included Operations   A repair that can be performed individually, but is also part of another repair.

Incorrect Gun Stroke Speed   A term that means you are moving the gun too slowly or quickly.

Indirect Damage   Damage caused by the shock of collision forces traveling through the body and inertial forces acting upon the rest of the unibody.

Inductive Pickup   A device that slips over the wire or cable to measure current.

Inert Gas   A type of shielding gas which protects the weld, but does not combine with the weld.

*Inertia Switch   A switch that shuts off the electrical power to the fuel pump and other circuits during a collision.

Infrared Drying Equipment Uses special bulbs to generate infrared light for fast drying of paint materials.

Inner Wheel Housings Surround the rear wheels.

Inserts Metal backing sometimes installed in weld repairs of closed sections, such as rocker panels, R- and B-pillars and rails.

Inside Calipers A tool designed for measuring inside parts. They are accurate to about 1/64 in.

Instrument Panel The assembly that includes the soft dash pad, instrument cluster, radio, heater and AC controls, vents and similar parts. Also known as dash assembly.

Insulation It stops current flow and keeps the current in the metal wire conductor.

Integral Unit is built as one-piece instead of two or more.

Inter-Industry Conference on Auto Collision Repair An advanced training organization dedicated to promoting high quality practices in the collision repair industry.

Intermediate Car The medium body classification. It can use 4, 6, or 8 cylinder engine, has average weight, and physical dimensions.

Intermix System A full set of paint pigments and solvents that can be mixed at the body shop.

***Intrusion Beam** A beam, inside the door frame, which protects the occupants during a side impact.

***ISO** Internation Standards Organization. A worldwide group that through consensus sets standards for member nations.

***ISO Codes** International codes, usually molded into plastic parts, which identify the type of plastic.

***Isocyanates** Toxic additives used in some refinish materials, to force curing by molecular cross-linking.

***Isolator** A friction type of energy absorber.

ISO Codes "International Symbols" is a way of identifying an unknown plastic. The code is molded into the plastic part.

Jam Nut A thin nut used to help hold larger, conventional nuts in place.

Joint Fit-Up Refers to holding work pieces tightly together, in alignment, to prepare for welding. It is critical to the replacement of body parts.

Jounce/Rebound Steering Gear Check A test that involves pushing down on the front or rear of the vehicle to load the suspension, and then allowing the vehicle to bounce back up. It is used to identify damage to the steering gear, column, or tie rods.

Jumper Cables Cables used to connect two batteries together when one is discharged.

Jumper Wires A device used to temporarily bypass circuits or components for testing.

Jumpsuit The proper clothing to wear in the paint spraying area.

Key A device used by equipment manufacturers to retain parts in alignment.

Key Used to tighten the chuck on the drill.

Kick Panels The small panels between the front pillars and rocker panels.

***Kingpin** The shaft that a steering knuckle pivots around.

***Kink** In the context of kink vs. bend, a type of damage to steel or aluminum where there is a sharp bend of small radius, typically more than 90°, over a short distance. Because of work hardening, a kinked area cannot be straightened without leaving a crack, tear, or permanent deformation.

Kinked A term that describes a part that has a sharp bend or a small radius, usually more than 90 degrees, or after straightening, there is a visible crack or tear in the metal, or there is permanent deformation that cannot be straightened to its pre-accident shape without the use of excessive heat.

Kneading Hardener Squeezing the tube of hardener back and forth with your fingers to mix the material.

***kPa** Kilopascal. A metric unit for measuring pressure. One kPa equals 0.145 psi.

Lace Painting Involves spraying through lace fabric to produce a custom pattern in the paint.

Ladder Frame A frame that has long frame rails with a series of straight crossmembers formed in several locations.

Laminated Plate Glass Two thin sheets of glass with a layer of clear plastic between them. It is used for windshields.

Lap Belt A seat belt that extends across a person's lap.

***Lap Joint** A type of weld joint made by overlapping two pieces of metal and joining them with a bead along one or both edges.

***Lap Shear Strength** The maximum force per unit area that a lap joint can withstand in shear before failure occurs. Commonly expressed in kilopascals (pounds per square inch).

***Lateral Acceleration** A sensor that provides an electrical acceleration signal to a control module.

Leak Detector A tool for finding refrigerant leaks in lines, hoses and other parts of an air conditioning system.

Leather Shoes To prevent falls and foot injuries, thick leather shoes with nonslip soles should be worn.

Left-Hand Threads Nuts and bolts that must be turned counterclockwise to tighten.

***Let-Down Panel** A test panel used to determine the color match and the number of coats required to match a multi-stage finish.

Lid Shock Absorbers Usually a spring-loaded, gas-filled unit that holds the lid open and causes it to close more slowly.

Lid Torsion Rods Spring steel rods used to help lift the weight of the lid.

Lifetime Tool Guarantee A tool that will be replaced or repaired if it ever fails or breaks.

Lift Safety Catch A device on a hydraulic jack. When engaged it ensures that the lift cannot lower while working underneath the vehicle.

Light Body Filler A filler formulated for easy sanding and fast repairs. It is used as a very thin top coat of filler for final leveling.

Light Temperature A rating of lighting in "Kelvin." For painting, a lamp rating or temperature of 6000-7000 Kelvin is recommended.

Line Wrench Also referred to as tubing or flare nut, it has a small split in its jaw to fit over lines and tubing.

Linear Measurement Straight line measurement of distance. Commonly used when evaluating major structural damage after a collision.

Lint-Free Coveralls The proper clothing to wear in the paint spraying area.

Liquid Masking Material A masking system that seals off the entire vehicle to protect undamaged panels and parts from paint overspray.

Lock Cylinders A tumbler mechanism that engages the key so that you can turn the key and disengage the latch.

*Lock Rod The rod that connects a lock cylinder, night-latch, and power lock motor to the latch mechanism.

Locksmith Tool Also termed "slim jim," it is used to open locked doors on cars and trucks.

Long Spray Distance A spray gun distance that causes a greater percentage of the thinner to evaporate, resulting in orange peel or dry spray.

Long-Hair Fiberglass Filler A fiberglass filler with long strands of fiberglass for even more strength.

Low Impedance Refers to low resistance in meters used to test modern electronic circuits.

Low Spot Also called a dent, it is an area that is recessed below the surrounding surface.

*LPG Liquefied petroleum gas.

*Lumbar Support An inflatable pocket built into a seat back that is used to vary the support of the occupant's lower back.

Lumen Rating The measurement of a lamp's brightness.

LVLP See High Efficiency Spray Gun

Machine Compound A compound formulated to be applied with an electric or air polisher.

Machine Compounds A rubbing compound that is water based to disperse the abrasive while using a power buffer.

Machine Grinding A process used to remove old finish from small flat areas and gently curved areas.

Machine Screw A screw that is threaded its full length and is relatively weak.

MAG Metal Active Gas, describes the welding process when the primary gas of the shielding gas is active.

*Magnaflux A method of detecting cracks in steel parts, using an electric current to activate a magnetic dye.

Magnetic Field Also known as flux, it is an invisible energy present around permanent magnets and current carrying wires.

Magnetic Guide Strip Used to help guide the pin striping tool for making straight or slightly curved pin stripes.

Major Repairs Refers to the replacement of large body sections and frame straightening.

Manometer An instrument that indicates when the intake filters in a spray booth are overloaded.

Manual Estimating Preparing a written estimate using crash estimating books and collision damage manuals.

Manual Transmission A transmission that must be hand shifted through each forward gear.

Manually Written Estimate A damage report that is completed in long hand by filling in information on a printed form.

Manufacturer's Instructions Very detailed procedures for the specific product.

Manufacturer's Specifications Measurements given by a manufacturer for proper repair of their vehicles.

Manufacturer's Warnings Procedures given by the manufacturer for the safe use of their product.

Marble Effect A custom paint job made by forcing crumpled plastic against a freshly painted stripe or area.

Masking Coating Also called masking liquid, it is usually a water-based sprayable material for keeping overspray off body parts.

Masking Cover A specially shaped cloth or plastic cover for masking specific parts.

Masking Machine A machine designed to feed out masking paper while applying masking tape to one edge of the paper.

Masking Paper A special paper designed to be used to cover body parts not to be painted.

Masking Paper Dispenser A device that automatically applies tape to one edge of the masking paper as the paper is pulled out.

Masking Plastic Used just like masking paper to cover and protect parts from overspray.

Masking Tape An adhesive used to hold masking paper or plastic into position. It is a high tack, easy-to-tear tape.

Master Cylinder It develops hydraulic pressure for the brake system.

*mb Millibar. A unit of pressure or vacuum. One mb equals 0.0145 psi.

Measurement Gauges Special tools used to check specific frame and body points. They allow you to quickly measure the direction and extent of vehicle damage.

Measurements Number values that help control processes in collision repair.

Measuring System A special machine used to gauge and check the amount of frame and body damage. It compares known good measurements with those on the vehicle being repaired.

Mechanical Repairs Repairing mechanical components such as water pump, radiator, etc.

Mechanical Systems Chassis parts of an automobile such as steering, braking, and suspension systems.

*Media The material that is used, under air pressure, to remove paint from a surface in a process called media blasting. Types of media are ground-up plastic, sand, or other material.

Melt-Flow Plastic Welding The most commonly used airless welding method. It can be utilized for both single-sided and two-sided repairs.

*Metal Conditioner An acid-based material that is part of a metal treatment system. This material etches the metal surface to improve paint adhesion.

Metallic Paint Paint with large reflective pigment flakes added.

Metalworking Area An area where parts are removed, repaired and installed.

Micrometer A device that is used to measure mechanical parts when high precision is important.

MIG Contact Tip Also call tube, it transfers current to the welding wire as the wire travels through.

MIG Nozzle It protects the contact tip and directs the shielding gas flow.

MIG Welding Gun Also called a torch, it delivers wire, current, and shielding gas to the weld site.

MIG-MAG The welding process, when nearly equal active and inert combinations of shielding gases are used.

Mil Gauge Used to measure the thickness of the paint on the vehicle.

*Mild Steel (MS) Steel which can withstand stress up to 30,000 psi. Also called low-carbon steel, it is a sheet metal that has a low level of carbon and is relatively soft and easy to work.

Minivan A smaller version of the box-shaped van.

Minor Repairs Repairs requiring minimum time and effort, i.e. small dents, scratches.

Mixing Board The surface used for mixing the filler and its hardener.

Mixing By Parts For a specific volume of paint or other material, a specific amount of another material must be added.

Mixing Chart A guide that converts a percentage into how many parts of each material must be mixed.

Modular Glass Also known as encapsulated glass, it has a plastic trim molding attached to its edge to fit the contours of the vehicle more closely.

Molded Core A curved body repair part made by applying plastic repair material over a part and then removing the cured material.

Molding Removal Tool A tool used to separate adhesive-held molding or trim.

*Motorized Seat Belt An automatic seat belt that is applied to an occupant and pre-tensioned by a motor-driven mechanism.

Motors A machine that uses permanent and electromagnets to convert electrical energy into a rotation motion for doing work.

*Mounting Spade A bracket that is adhesively bonded to a glass surface for mounting a mirror or other devices.

MRD A mid-engine, rear-wheel drive vehicle. The engine is located right behind the front seat or centrally located.

*MSDS Material Safety Data Sheet. A document which provides safety, handling, storage, and clean-up information for hazardous materials.

Multi-Stage Finishes A type of glamour finish.

Multiple Part Rocker An assembly of several pieces of sheet metal with internal reinforcements.

*Multiple Pull Making a pull in more than one direction, or from more than one location, at the same time.

Multiple Pull Method A method used on major damage, it involves several pulling directions, and steps are needed.

National Institute for Automotive Service Excellence (ASE) Offers a voluntary certification program that is recommended by the major vehicle manufacturers in the United States.

Needlenose Pliers Pliers that have long, thin jaws for reaching in and grasping small parts.

Negative Pressure Test Part of the Fit Test. Performed by placing the palms of both hands over the cartridges and inhaling. A good fit is evident if the face piece collapses onto your face.

*Neoprene A synthetic rubber material that is resistant to acid and fuels.

*Neutral Flame A flame created by balancing the amounts of oxygen and fuel gas, so that the inner and outer cones are at equal heights.

*Neutralize To cause a substance or surface to be neither acidic, nor alkaline.

New Part Primer Coat Replacement panels that are primed by the manufacturer or part supplier to protect the metal against corrosion.

*NGA National Glass Association. A national trade association that represents companies involved in both auto glass installations and architectural glass work.

*NIOSH National Institute for Occupational Safety and Health. A U.S. research group that develops methods for controlling worker exposure to chemical and physical hazards.

*Nitrile Rubber A type of fuel-resistant rubber used for the manufacture of protective gloves.

Nonbleeder Spray Gun A type of spray gun that releases air only when the trigger is pulled back. It has an air control valve. The trigger controls both fluid needle movements and the air control valve.

*Nugget The fused area of a plug weld or spot weld.

Non-Bleeder Spray Gun It releases air only when the trigger is pulled back. They have an air control valve. The trigger controls both fluid needle movements and the air control valve.

Nut A fastener that uses internal (inside) threads and an odd shaped head that often fits a wrench. When tightened onto a bolt, a strong clamping force holds the parts together.

*OEM Original equipment manufacturer. The original maker of a vehicle or equipment.

OEM Finishes Factory paint jobs.

Office Manager The person who oversees various aspects of the business, i.e., letters, estimates.

Offset Butt Joint A basic joint without an insert. It is also known as a staggered butt joint.

Oil Dipstick A device used to check the oil level.

Oil Drain Plug It is located in the oil pan for draining and changing the engine oil.

Oil Filter It traps debris and prevents it from circulating through the engine oil galleries.

Oil Pan It holds an extra supply of motor oil. Also known as the sump, it bolts to the bottom of the engine block.

½ Inch Impact An impact wrench with a $^1/2$-inch size drive head. It is shaped like a pistol with a hand grip hanging down. It is frequently used to service wheel lug nuts.

One-Part Putty Spot putty applied directly out of tube or container. It takes longer to harden than two-part and can shrink more easily.

Open Coat A type of grit on sandpaper and discs, the resin that bonds the grit to the paper only touches the bottom of the grit.

Open End Wrench A wrench with three-sided jaws on both ends.

Opened A term that refers to a disconnected electrical circuit.

Orifice A hole in the bottom of the viscosity cup.

Outside Calipers   A tool that will make rough external measurements.

Overall Refinishing   Painting the whole vehicle.

Overhead Welding   The piece is turned upside down.

Overlap Strokes   The process of making each spray gun coat cover about half of the previous coat of paint.

Overlay Dry Method   A method of installing stripes or decals that does not involve the use of soapy water. All surfaces are clean and dry.

Overlay Wet Method   A method of installing stripes or decals that involves using soapy water to help simplify the application of larger decals and striping.

Overmasked   A term that means the painter must touch up the part of the vehicle that should have been painted.

Overpulling Damage   Results from failing to measure accurately and often when pulling on unibody vehicles.

*Overspray   Sprayed material that falls outside the intended spray area.

Overspray Leak   The term that describes the results of not sealing the masking paper or plastic and getting overspray on unwanted surfaces.

Oxalic Acid   A mild acid used to remove industrial fallout stains from painted surfaces.

*Oxidation   The residue that forms on metal, or the dull layer which forms on the surface of paint. Oxidation occurs when the base material combines with oxygen.

Oxidizing Flame   It is obtained by mixing slightly more oxygen than acetylene.

Oxygen Pressure   Pressure leaving oxygen pressure regulator of oxyacetylene welding outfit. Typically ranges from 5 to 100 psi (kPa).

Pad Cleaning Tool   A metal star wheel and handle that will clean dried polishing compound out of the pad.

Paint   A term that generally refers to the visible topcoat.

Paint Adhesion Check   A test to see if the old paint is adhering properly.

Paint Binder   The ingredient in a paint that holds the pigment particles together.

Paint Blending   Tapering the new paint gradually into the old paint.

Paint Booth Area   An area where the body is painted. Also known as the Finishing Area.

Paint Drying Room   A dust-free room that speeds up drying, produces a cleaner job, and increases the volume of refinishing work.

Paint Film Breakdown   A term for checking, cracking, or blistering of paint.

Paint Formula   The percentage of each ingredient that is needed to match an OEM color.

Paint Mixing Stick   Used to help mix paints, solvents, catalysts, and other additives right before spraying.

Paint Mixing Instructions   Directions on how to mix the paint; usually given on the paint can label.

Paint Mixing   Blending the paint and thinner or reducer to the correct thickness or viscosity.

Paint Mixing Room   A safe, clean, well-lit area for a painter to store, mix, and reduce paint materials.

Paint Overlap   The area where one vertical painted area overlaps the new area being painted.

Paint Prep Area   An area where the vehicle is readied for painting or refinishing.

Paint Preparation   Preparing the vehicle for spraying or refinishing.

Paint Reference Charts   Found in service manuals, they give comparable paints manufactured by different companies.

Paint Runs   Excess paint thickness flows down when spraying.

Paint Scuffing   A term used to describe roughing the paint for better paint adhesion.

Paint Sealer   An innercoat between the topcoat and the primer or old finish to prevent bleeding.

Paint Selection   Determining the right paint for the job, so that the new paint matches the old and to assure long life.

Paint Settling   A term that describes when the paint sits idle and pigments, metallic flakes, or mica particles settle to the bottom.

Paint Solvent   The liquid solution that carries the pigment and binder so it can be sprayed.

Paint Stirring Stick   Wooden sticks for mixing the contents after being poured into the spray gun cup or container.

Paint Stripper   A powerful chemical that dissolves paint for fast removal of an old finish.

Paint Surface Chips   A result of mechanical impact damage to the paint film: door dings, damage from road debris, etc.

Paint Surface Protrusion   A particle of paint or other debris sticking out of the paint film after refinishing.

Paint System   All materials (primers, catalysts, paints) are compatible and manufactured by the same company.

Paint Thickness   Measurement of distance between surface of paint and body of vehicle. Viscosity of paint; affected by reduction or thinning.

Paint Viscosity   A term that refers to the paint's thickness or ability to resist flow.

Painted Flames   A custom painting technique often used on "hot rods" or older street rods.

Painter Prep Area   A shop area that provides a clean area for a painter to dress, a clean storage for painting suits, and a place to clean and store personal safety equipment.

Painter's Stretch Hood   Protective head gear for the paint booth.

Painting Environment   The suitable conditions for painting, i.e., cleanliness, temperature/humidity, light, and compressed air/pressure.

Paintless Dent Removal   A method of using picks to remove small dents without having to repaint the panel.

Pan   A floor related component, i.e. front floor pan.

Panel Alignment Marks   Marks that can be provided on new parts to help position them for welding.

Panel Cutters   A tool that will precisely cut sheet metal, leaving a clean, straight edge that can be easily welded.

Panel Replacement   Removing and installing a new panel or body part.

Panel Straightening   Bending or shaping the panel back to its original position, using hand tools and equipment.

Part Gap   Also known as clearance, it is the distance measured between two adjacent parts, hood-to-fender gap for example.

Partial Cutout Method   Refers to using some of the original adhesive during windshield replacement.

Partial Frame   A frame that is a cross between a solid frame and a unibody.

Particle Respirator   See Dust Respirator.

Parts Manager   A person in charge of ordering, receiving and distributing parts.

*Passive Restraint System   Any restraint system that does not require activation by the occupant to be effective, such as an automatic seat belt or supplemental airbag system.

*Pathogen   A disease-causing agent, usually in a medium such as blood.

Pattern Control Knob   A part of the spray gun; when closed, the spray pattern is round. As the knob is opened, the spray becomes more oblong in shape.

PC   See Personal Computer.

PCV   Positive Crankcase Ventilation System. It channels engine crankcase blowby gases into the engine intake manifold.

Pearl Paint   Paint with medium size reflective pigment particles, such as mica, that give it luster or shine that tends to change color with the viewing angle.

Percentage Reduction   Materials must be added in certain proportions or parts.

Perimeter Frame   A frame that has the frame rail near the outside or perimeter of the vehicle.

Personal Computer   Also referred to as a PC, it is used to keep track of business transactions, complete damage reports, and performs electrical tests on vehicles.

*PGM   Poly gel mitigator. A type of energy absorber that uses a gel instead of a fluid to absorb energy.

*pH   Potential of hydrogen ions. A measure of the acidity or alkalinity of a liquid. The numerical range of pH is from 0, the most acidic, to 14, the most alkaline. A neutral liquid has a pH of 7.

Phillips Screwdriver   A hand tool that has two crossing blades for a star-shaped screw head.

Physical Injury   A general category that includes cuts, broken bones, strained backs and similar injuries.

Piano Wire   Fine steel wire that can be used to cut the adhesive around glass.

Pick   A tool used to reach into confined spaces for removing tiny dents or dings.

Picking Hammer   A hammer with a pointed end used to raise small dents from the inside, and a flat end for hammer-and-dolly work to remove high spots and ripples.

Pin   A device used by equipment manufacturers to retain parts in alignment.

Pin Punch   It has a straight shank for use after a starting punch. It will push a shaft completely out of a hole.

Pin Striping Brushes   Brushes with very long soft horse hairs that are used to apply paint in a small, controlled area.

Pin Striping Tool   A serrated roller to deposit a painted pin stripe.

*Pinchweld   A flange extending from the body of a vehicle into the opening for glass parts, usually extending from the side pillars and roof.

Pinchweld Clamps   Attached to the body at the front and rear of the rocker panel pinchwelds; four are normally used.

Pipe Wrench   A type of adjustable wrench for holding and turning round objects. It has sharp jaw teeth that dig into and grasp the part.

Pitch   Spacing between resistance spot welds.

*Pitman Arm   The arm that converts the rotary motion of the steering gear sector shaft to the side-to-side movement of the center link in a parallelogram steering system.

*Plastic Adhesion Promoter   A special primer containing adhesion promoters for refinishing plastics.

*Plastic Cleaner   A cleaner used to remove contaminants from a plastic repair area before starting a repair, and again before applying primers or paint.

Plastic Flame Treating   A torch flame is used to oxidize and chemically prepare the plastic for adhesion.

Plastic Gloves   Used to protect hands from the harmful effects of corrosive liquids, undercoats, and finishes.

Plastic Hammer   A hammer used for making light blows where parts can be easily damaged.

Plastic Memory   A term that refers to plastic that wants to keep or return to its original molded shape.

Plastic Speed Welding   A type of welding that uses a specially designed tip to produce a more uniform weld and at a high rate of speed.

Plastic Stitch-Tamp Welding   A type of welding used primarily on hard plastics, like ABS and nylon, to insure a good base/rod mix.

Plastic Welder   A tool used to heat and melt plastic for repairing or joining plastic parts.

*Plastic Welding   A welding process using heat and a filler material to repair plastic parts.

Plastics   Refers to a wide range of materials synthetically compounded from crude oil, coal, natural gas, and other natural substances.

Pliers   A hand tool used for working with wires, clips and pins. They will grasp and hold parts, like your fingers.

Plotting Color   The process of identifying paint color in a graphic way based on value, hue, and chroma.

*Plug Weld   A weld where two or more pieces of metal are joined by filling a hole in the outer pieces, while penetrating into the underlying pieces.

Pneumatic Windshield Cutter   A device that uses shop air pressure and a vibrating action to help cut the adhesive around glass.

Pocket Scale   A pocket size ruler.

Point-To-Point Measurement   A term that refers to the shortest distance between two points.

Polishing   Using very fine compound to bring the paint surface up to full gloss.

Polishing Compound   Also called machine glaze, it is a fine grit compound designed for machine applications.

Polishing Pad   A cotton or synthetic cloth cover that fits over the polisher's backing plate.

*Polyester Putty   A material used to fill and level minor surface imperfections and low areas.

*Polypropylene Primer   A special primer containing adhesion promoters for refinishing polyolefin plastics.

*Porosity   The presence of holes or voids within a weld bead or nugget.

Pop Rivets   A fastener used to hold two pieces of sheet metal together.

Portable Power Units   Small piston and cylinder assemblies for removing minor damage.

Portable Puller   A hydraulic ram and post mounted on caster wheels.

Positive Pressure Test   Part of the Fit Test. Performed by covering up the exhalation valve and exhaling. A proper fit is evident if the face piece billows out without air escaping from the mask.

Postpainting Operations   Things done after painting such as removing masking tape, reinstalling parts, and cleaning the vehicle.

Power Steering Pump   A pump that forces pressurized oil through the power steering system.

Power Tools   Tools that use air pressure or electrical energy to aid repairs.

Premixed   Material that is ready to spray.

Prep Solvent   A fast drying solvent often used to clean a vehicle. It removes wax, oil, grease, and other debris that could contaminate and ruin the paint job.

Pressure Gauge   A device that reads in pounds per square inch or kilograms per square centimeter.

Pressure Multiplier   It usually applies welder arm pressure.

Pressure Tank Spray Gun   Similar to the pressurized pot, except it uses a much larger storage container for paint materials.

Pressurized Pot   Also referred to as a pressure cup spray gun, it uses air pressure inside the paint cup or tank to force the material out of the gun.

*Primary Damage   Damage, at or near the point of impact, caused by the collision. Also called direct damage.

*Primer   An undercoat or chemical applied to a surface to improve the adhesion, durability, and appearance of a topcoat or the bond of an adhesive.

Primer-Filler   A very thick form of primer-surfacer. It is sometimes used when a very pitted or rough surface must be a filled and smoothed quickly.

*Primer-Sealer   An undercoat used to protect the primer from the solvents in the topcoat.

*Primer-Surfacer   A high-solid type of primer used to level and fill small imperfections in a surface.

Printer   A machine used with a computer to make a hard copy (printed images on paper).

Productivity   A measure of how much work you get done.

Professional   Refers to the attitude, work quality and image that a business and its workers project to customers.

Professionalism   A broad trait that includes everything from being able to follow orders to pride in workmanship.

Progression Shop   An assembly line type of organization with specialists in each area of repair.

Proportional Numbers   Denotes the amount of each material needed.

*psi   Pounds per square inch. A unit of pressure or stress activation by the occupant to be effective, such as an automatic seat belt or supplemental airbag system.

Pull Rod   A tool that has a handle and curved end for light pulling of dents.

Pull Welding   You aim or angle the gun back toward the weld puddle.

Pulling Adaptors   Special straightening equipment accessories that allow you to pull in difficult situations.

Pulling Clamps   They bolt around or onto the vehicle, so a pulling chain can be attached to the vehicle.

Pulling Equipment   It uses hydraulic power to force the body structure or frame back into position.

Pulling Posts   Also known as towers, they are strong steel members used to hold the pulling chains and hydraulic rams.

Push Welding   You aim or angle the gun ahead of the weld puddle.

Push-In Clip   Usually made of plastic, they are used to hold body panels.

*R-12 Refrigerant   A type of refrigerant containing chlorofluorocarbon (CFC).

*R-134a Refrigerant   A type of refrigerant containing hydrochlorofluorocarbon (HFC).

*Race   The inner or outer bearing surface of a ball- or roller-bearing assembly. The balls or rollers support the load by rolling between the surfaces of the races.

Rack   A toothed shaft that is part of the rack-and-pinion steering unit.

*Rack-and-Pinion Assembly   A steering gear system in which the rotary motion of a pinion gear is converted to a side-to-side movement of a rack, the ends of which are attached to the steering arms by tie rods.

Rack Straightening System   Generally a drive-on system with a built-in anchoring and pulling mechanism.

Radiator Cap Pressure Test   A test done using a cooling system pressure tester. The tester gauge should stop increasing its pressure reading when the cap rating is reached.

Radiator Core Support   The framework around the front of the body structure for holding the cooling system radiator and related parts.

Radiator Pressure Cap   Prevents the coolant from boiling.

*Radius Rod   A suspension part that controls the fore-and-aft position of an axle or a wheel. Also called radius arm or strut rod.

*Rail Dust   The dust that falls on vehicles being shipped by rail. Similar to industrial fallout, but containing diesel particulates and powdered iron.

Ratchet   A common type of socket drive handle with a small lever that can be moved for either loosening or tightening bolts and nuts.

Reading Tires   Involves inspecting tire tread wear and diagnosing the cause.

Rear Bulkhead Panel   A panel that separates the passenger compartment from the rear trunk area.

Rear Section   Also known as tail section or rear clip, it consists of the rear quarter panels, trunk or rear floor pan, rear frame rails, trunk or deck lid, rear bumper, and related parts.

Rear Shelf   A thin panel behind the rear seat and in front of the back glass. It often has openings for rear stereo speakers. Also known as package tray.

Rear Spoiler   A part that mounts in the trunk lid to alter the airstream at the rear of the body to increase body down force. This helps rear wheel traction at highway speeds.

*Receiver-Drier A device, located in the high-pressure side of an air conditioning system, that removes moisture and stores liquid refrigerant and lubricant.

Recliner Adjuster A hinge mechanism that allows adjustment of the seat back to different angles.

Recovery Station A machine used to capture the old, used refrigerant so that it does not enter and pollute the atmosphere.

Recovery System It captures the used refrigerant from the air conditioning system and keeps it from contaminating the atmosphere.

Recycled Assemblies Undamaged parts from another damaged vehicle that are used for repairs.

*Redress A Tool To reshape, resharpen, or polish a tool.

*Reference Points Identifiable points on a vehicle that may be used for measurement when a control point is not available.

Refinishing Materials A general term referring to the products used to repaint the vehicle.

*Refractometer An instrument used to measure the specific gravity of a liquid.

*Regulator The mechanism that moves the glass when activated by a switch or crank.

Reinforced Reaction Injection Molded (RRIM) Polyurethane A two-part polyurethane composite plastic. Part A is the isocyanate. Part B contains the reinforced fibers, resins, and a catalyst. The parts are mixed and injected into a mold.

Related Professions Any job or work that has something in common with a specific category of work.

*Relative Humidity The amount of water vapor in the air, expressed as a percentage of the maximum amount the air can hold under current conditions of temperature and pressure. Rain or fog forms when the relative humidity reaches 100%.

*Release Agent A solvent used to soften adhesives or sealants.

Reliability A quality meaning dependable.

Repair Stall A work area for one vehicle. Also termed a bay.

*Repair Taper The sloping sides of a repair area, formed by grinding, sanding, or cutting.

Resistance A restriction or obstacle to current flow.

Restraint System Designed to help hold people in their seats and prevent them from being injured during a collision.

Retarder A slow evaporating thinner or reducer used to retard or slow drying.

*Retrofit A post-production change to improve the performance, or change the function, of a part or system, using parts and procedures that have been tested.

Reveal Moldings Used to cover the adhesive, they are held in place by adhesive grooves in the body, or by clips.

Reverse Masking A masking method that requires you to fold the masking paper back and over the masking tape.

Reverse Polarity Also called "DC Reverse," it is when the electrode is positive and the workpiece is negative.

Ribbon Sealer Sealer that comes in strip form and is applied by hand.

*Ride Height The height measurement, from the surface supporting tires to designated points on the vehicle structure, under conditions specified by the vehicle maker.

Right-Hand Threads Nuts and bolts that must be turned clockwise to tighten.

*Rocker Panels Either of two structural members, located below the doors, that support the door pillars and the floor pan.

*Roof Bow Any of several side-to-side structural arches that support the roof. The bows are attached to the roof rails, between the front and rear header panels.

Roof Panel A large multi-piece panel that fits over the passenger compartment.

Rosin Core Solder A type of solder designed for doing electrical repairs.

RRIM Reinforced Reaction Injection Molded, a two-part poly-urethane composite plastic. Part A is the isocyanate. Part B contains the reinforced fibers, resins, and a catalyst. The parts are mixed and injected into a mold.

Rubber Body Insulators Used between the frame and body to reduce noise and vibration.

Rubber Gloves Used to protect hands from the harmful effects of corrosive liquids, undercoats, and finishes.

Rubber Mallet A solid rubber head that is fairly heavy. It is often used to gently bump sheet metal without damaging the painted finish.

Rubberized Undercoat A synthetic-based rubber material applied as a corrosion or rust preventive layer.

*Run Glider A U-shaped track that guides the travel of movable glass and seals against water and air leaks.

*Runout Any radial or lateral variation in the dimensions of a part such as a wheel, brake rotor, or brake drum, measured as the part is rotated on its side.

*Runs and Sags A paint defect caused by excessive paint flowing unevenly down a surface, causing ridges and lines to form.

RWD A front engine, rear-wheel drive vehicle. The engine is in the front and the drive axle is in the rear.

*SAE Society of Automotive Engineers.

Safety Using proper work habits to prevent personal injury and property damage.

Safety Glasses Eye protection suitable for minor danger.

Safety Program A written shop policy designed to protect the health and welfare of personnel and customers.

Safety Signs Provided to give information that helps to improve shop safety, such as fire exits, fire extinguisher locations, dangerous or flammable chemicals and other information.

*SAI Steering axis inclination. The angle of the steering axis, in degrees from true vertical, when viewed from the front of the vehicle.

*Salvage Part A part, removed from a vehicle being scrapped, that is intended to be used as a replacement part.

Sand Scratches This is a result of final sanding with too coarse a sandpaper, or not allowing materials (usually primer-surfacer or spot putty) to dry fully before painting.

Sander Pad A soft mounting surface for the sand paper.

Sanding Discs Round sandpaper normally used on an air-powered orbital sander.

Sanding Sheets Sandpaper in square sheets that can be cut to fit sanding blocks.

Sandpaper A heavy paper coated with an abrasive grit. It is the most commonly used abrasive in auto body repair.

*Sandscratch Swelling A paint defect caused by solvents being absorbed into sandscratches.

Scale The most basic tool for linear measurement, it is also referred to as a ruler.

Scan Tool A device used to read and convert computer fault codes.

Scanner Cartridge Installed into the scanner for the specific make and model car or truck being tested. It holds the information needed for that vehicle.

Scanner A device used to diagnose or troubleshoot vehicle computer system and wiring problems

Screw A fastener often used to hold nonstructural parts on the vehicle.

Screw-on Slide Hammer Removes deeper dents and creases.

Scuff Pads Tough synthetic pads used to clean and lightly scratch the surface of paints so the new paint will stick.

*Scuff-Sand The process of roughing a surface by light sanding, or rubbing with a scuff pad.

*Sealant Any of various liquids which, when applied to a joint, dry to form an airtight seal.

Sealers Used to prevent water and air leaks between parts. They are flexible, which prevents cracking.

*Seam Sealer A material designed to keep moisture and fumes out of the passenger compartment, and protect seams and joints from corrosion.

Seat Back The rear assembly that includes cover, padding, and metal frame.

Seat Belt Anchor A device that allows one end of the seat belt to be bolted to the body structure.

Seat Cushion The bottom section of a seat which includes cover, padding, and frame.

Seat Track The mechanical slide mechanism that allows the seat to be adjusted forward or rearward.

*Secondary Damage Damage, beyond the point of impact, that resulted from the force of the collision. Also called indirect damage.

*Sectioning A repair made by cutting and removing the damaged portion of a panel, and replacing it with an undamaged part, as opposed to replacing the damaged panel at factory seams.

*Sector Shaft The output shaft of a steering gear box.

Sedan A body design with a center pillar that supports the roof. They come in 2- and 4-door versions.

Self-Darkening Filter Lenses A welding filter lens that instantly turns dark when the arc is struck.

Self-Defrosting Glass It has a conducting grid or invisible layer that carries electric current to heat the glass.

*Self-Etching Primer A primer which contains an etching agent, used in place of metal treatment before priming, to provide corrosion protection.

Self-Locking Nut A nut that produces a friction or force fit when threaded onto a bolt or stud.

Self-Stick Sandpaper Sandpaper with adhesive applied to it during manufacturing.

Self-Tapping Screw A screw that has a pointed tip to help cut new threads in parts.

Service Manual Abbreviations Represents technical terms or words and saves space in the service manual.

Service Ports Small test fittings in the air conditioning system to which the air conditioning gauges are connected.

Set Screw A screw that frequently has an internal drive head for an Allen wrench. It is used to hold parts onto shafts.

*Setting Blocks Small, hard rubber blocks that are set along the bottom pinchweld during a windshield installation to maintain the position of the glass while the adhesive cures.

Shading Coats A progressive application of paint on the boundary of spot repair areas so that a color difference is not noticeable. Also referred to as blend coats.

Shakeproof Lock Washer Washers with teeth or bent lugs that grip both the work and the nut. They are also referred to as teeth lock washers.

Shear Strength A measure of how well a material can withstand forces acting to cut or slice it apart.

Sheet Metal Screw A screw that has pointed or tapered tips. They thread into sheet metal for light holding tasks.

Sheet Metal Gauge Instrument used to measure body or repair panel thickness or gauge size (a number system that denotes the thickness of sheet metal).

Sheet Molded Compounds (SMC) Fiber-reinforced composite plastic panels.

*Shielding Gas A gas or mixture of gases used to protect a weld site from atmospheric contamination.

Shock Absorbers Dampening devices that absorb spring oscillations (bouncing) to smooth the vehicle's ride quality. It is a part of the suspension system.

Shop Air Lines Thick steel pipes that feed out from the tank to several locations in the body shop.

Shop Layout The general organization or arrangement of work areas.

Shop Owner A person who is concerned with and understands all phases of work performed in the shop.

Shop Supervisor A person who oversees personnel and day-to-day shop operations.

*Short-Circuit Transfer The preferred mode of filler transfer during MIG welding, where the filler wire alternately shorts to the weld puddle and then burns free, re-igniting the arc.

Short Spray Distance A spray gun distance that causes the high velocity mist to ripple the wet film.

Shoulder Belt A seat belt that extends over a person's chest and shoulder.

Shrinking Hammer A finishing hammer with a serrated or cross-grooved face. They are used to shrink spots that have been stretched by excessive hammering.

Shrinking Metal Removes strain or tension on damaged, stretched sheet metal.

Side Draft Spray Booth Moves air sideways over the car or truck. An air inlet in one wall pushes fresh air into the booth. A vent on the opposite wall removes the booth air.

Side Orifices Located on the spray gun air cap, they determine the spray pattern by means of air pressure.

Side-Tone View Compares the test panel to the vehicle on a 45–60 degree angle.

Single Pull Method   A method that uses only one pulling chain; it works well with minor damage on one part.

Single-Sided Plastic Welds   A type of weld used when the part cannot be removed from the vehicle.

Siphon Spray Gun   It uses air flow through the gun head to form a suction that pulls paint into the air stream.

*****Skim Coat**   A thin layer of material applied to a surface to cover imperfections.

Skin Hem   Folded door edge.

Skip Welding   It produces a continuous weld by making short welds at different locations to prevent overheating.

*****SLA**   Short-long arm. A type of independent suspension system having upper control arms that are shorter than the lower control arms.

Sliding Calipers   A tool that measures inside, outside, and depth dimensions with high accuracy.

Slip-Joint Pliers   Also known as combination pliers, they will adjust for two sizes.

*****Slip Yoke**   The yoke in a universal joint that is splined, transmitting torque to or from the driveshaft while allowing the effective length of the driveshaft to vary as the driven wheels move through the range of the suspension system.

Slotted Nut   Also known as a castellated nut, it is grooved on top, so that a safety wire or cotter pin can be installed into a hole in the bolt.

Smoke Gun   A device used to find air leaks around doors and windows.

Snap Ring   A non-threaded fastener that installs into a groove machined into a part. They are used to hold parts on shafts.

Snap Ring Pliers   Pliers that have tiny tips for fitting into clips or snap rings. Both external and internal types are available.

Snap-Fit Parts   Parts that use clips or interference fit to hold together.

Social Skills   The ability to get along with co-workers and customers.

Socket Drive Size   The size of the square opening for the drive handle.

Socket Wrench   Cylinder-shaped, box end wrench for rapid turning of bolts and nuts.

*****Sodium Hydroxide**   A caustic chemical (lye) found in small amounts with the talc residue, following an airbag deployment.

Soft Brazing   Brazing with materials that melt at temperatures below 900°F (468°C).

Soldering   A process using moderate heat and solder to join wires or other parts.

Soldering Gun   A tool used to heat and melt solder for making electrical repairs.

*****Solenoid**   A remotely operated device that converts an electrical input into a mechanical output for the purpose of actuating a latch, valve, switch, etc.

*****Solid Axle**   A part of a suspension system in which the wheels are mounted at each end of a rigid beam or axle housing.

*****Solvent**   Any volatile liquid in which a substance can be dissolved, such as cleaning fluids, thinners, and fuels.

Sonotrode   In ultrasonic welding it is the equivalent of an electrode.

Sound Deadening Material   Insulation materials that prevent engine and road noise from entering the passenger compartment.

Sound-Deafening Pads   Often bonded to the inside surface of truck cavities and doors to reduce noise, vibration, and harshness. It is made of plastic or asphalt-based material.

Spark Lighter   A device that produces a spark and ignites an oxyacetylene torch.

Specialty Shop   Repair facility that only works on one part or system of a vehicle.

Speed Handle   A tool that can be rotated to quickly remove or install loose bolts and nuts.

*****Speed Sensor**   A sensor that provides an electrical speed signal to a control module.

Spider Webbing   A technique accomplished by forcing paint through the air brush in a very thin, fibrous type spray.

*****Spindle**   The shaft-like portion of a suspension system that supports the wheel and bearings of an undriven wheel. May also refer to the hub of a driven wheel.

Split Lock Washer   A washer used under nuts to prevent loosening by vibration. The ends of these spring-hardened washers dig into both the nut and the work to prevent rotation.

Spoiled Hardener   Hardener that remains thin and watery after thoroughly kneading it.

Spoons   An instrument used to pry out damage, struck with a hammer to drive out damage, and as a dolly in hard-to-reach areas.

*****Spot Weld**   A type of weld where two pieces of metal are fused together at a spot, using pressure and electric current, with no filler material.

Spot Weld Drill   A specially designed drill for removing spot-welded body panels.

Spot Weld Timer   Adjusts the time that the current passes through the joint.

Spout   A device designed to install into a can for handy pouring.

Spray Booth   Designed to provide a clean, safe, well-lit enclosure for painting.

Spray Gun Air Cap   Works with the air valve to control the spray pattern of the paint. It screws over the front of the gun head.

Spray Gun Angle   A term that refers to whether the gun is tilted up or down or sideways.

Spray Gun Cap   It often fits onto the bottom of the spray gun body to hold the material to be sprayed.

Spray Gun Cup   A cup that fits onto the body of the spray gun and holds the material to be sprayed.

Spray Gun Distance   The distance from the gun tip to the surface being painted.

Spray Gun Lubrication   Placing a small amount of oil on packing and high friction points.

Spray Gun Speed   A term that refers to how fast the gun is moved sideways while painting.

Spray Gun Triggering   Involves stopping the paint spray before you stop moving the gun sideways.

Spray Gun Washer   Enclosed equipment that automatically cleans a spray gun.

*****Spray Pattern**   The pattern made by material sprayed from a stationary spray gun as it strikes a flat surface.

Spray Pattern Test   A process to check the operation of the spray gun on a piece of paper.

Spray-Out Panel   A test panel that checks the paint color and also shows the effects of the painter's technique.

Spraying   The physical application of paint using a spray gun.

Spreader   A tool used to apply filler to low spots in body panels.

Spring Compressor   A tool used to remove and install the small spring used on some door hinges.

Spring Hammering   A method of bumping damage with a hammer and a dinging spoon.

Spring Hangers   These are sometimes formed on the frame to hold the suspension system springs.

Spring Hose Clamp   A clamp made of spring steel with barbs on each end. Squeezing the ends opens and expands the clamp.

Spring Scale   A device that measures pulling force.

Spring-Back   The tendency for metal to return to its original shape deformation.

*Spring Shackle   A coupling that attaches the floating end of a leaf spring to the frame or unibody rail, allowing fore-and-aft movement as the spring flexes.

Squeeze-Type Resistance Spot Welding   A type of welding that uses electric current through the base metal to form a small, round weld between the base metals.

*Stabilizer Bar   A transverse torsion bar, attached to the suspension system on the underside of the vehicle, to reduce roll. Also called anti-roll bar.

Standard Screwdriver   A hand tool that has a tip with a single flat blade for fitting into the slot in the screw.

Starting Punch   Also known as a drift punch, it has a fully tapered shank and will drive pins, shafts and rods partially out of holes.

Starting System   A large electric motor that turns the engine flywheel. This spins or "cranks" the crankshaft until the engine starts and runs on its own power.

Station Wagon   A body design that extends the roof straight back to the rear of the body. A rear hatch or window and tailgate open to allow access to the large storage area.

Stationary Parts   Parts that are permanently welded or adhesive bonded into place like the floor, roof, and quarter panels.

*Steering Angle Sensor   A sensor that provides an electrical steering-angle signal to a control module.

*Steering Damper   A shock absorber-type device used to dampen vibration in a steering system.

*Steering Knuckle   That part of the steering system that supports the wheel spindle or hub, and pivots about the steering axis.

*Steering Ratio   The number of turns of the steering wheel required to steer the front wheels from lock to lock.

Steering System   It transfers steering wheel motion through gears and linkage rods to swivel the front wheels.

Steering Wheel Center Check   A test that involves making sure the steering wheel has not been moved off center to do part damage.

Step-Down Gun   Internally restricts the inlet pressure of 40–100 psi to 10 psi at the air cap. They do NOT require the use of a turbine or air conversion unit. They use existing shop air from the compressor.

Stick Arc Welding   Also known as shielded arc welding, it uses a welding rod coated with a flux for fusion welding.

Stitch Welding   A continuous weld in one location, but with short pauses to prevent overheating.

Straight Polarity   Also called "DC straight" it is when the electrode is negative.

Straightening Equipment   Anchoring equipment, pulling equipment, and other accessories used to apply tremendous force to move the frame or body structure back into alignment.

Straightening System   A hydraulic machine for pulling a badly damaged frame or unibody structure back into proper alignment. Also termed a frame rack.

*Stress Relieving   A process using heat or hammering to relax work-hardened metal.

Stressed Hull Structure   A concept where all force applied is not concentrated in one place, but is dispersed over the surface.

Stretched Metal   Metal that has been forced thinner in thickness and larger in surface area by impact.

*STRSW   Squeeze-type resistance spot welding. A type of welding that uses a machine with two arms to apply pressure and electric current to the spot being welded. Also the machine for making such welds.

Structural Adhesive   An adhesive used to bond parts together, in place of welds.

Structural Filler   Used to fill the larger gaps in the panel structure while maintaining strength. It adds to the structural rigidity of the part.

Structural Parts   Those parts of a vehicle which support the weight of the vehicle, absorb collision energy, and absorb road shock.

Strut Position Check   A test to check the position of the struts, thus determining if there is any damage.

Strut Pulling Plates   A device designed to bolt onto the top of shock towers for pulling.

Strut Rotation Check   A test to see if the strut is damaged.

Strut Tower Gauge   Allows visual alignment of the upper body area. It shows misalignment of the strut tower/upper body parts in relation to the vehicle's centerline plane and datum plane.

Stubby Screwdriver   A screwdriver with a very short shank for use in tight or restricted areas.

Stud Spot Welder   It joins pull rods on the surface of a panel so that you do not have to drill holes. It is the best way to pull dents.

Suction Cup Tool   A large synthetic rubber cup for sticking onto parts. They are often used to hold window glass when holding or moving the glass, or also to pop out large surface area dents in panels.

Sunlight   The standard by which other light sources are measured. It contains the entire visible spectrum of light.

Support Members   Used in high stress areas and often bolted to the bottom of the unibody to hold the engine, transmission and suspension in alignment, and to reduce body flex.

Surface Evaluation   A close inspection of the old paint to determine its condition.

Surface High Spot   An area that shows up when sanding quickly cuts through the guide coat.

Surface Inspection   Examining the old body surface condition before preparing it for new paint.

Surface Low Spot   An area that shows up when sanding will not remove the guide coat.

Surface Roughness   A measurement of paint film or other surface smoothness in a limited area.

Surface Temperature Thermometer   A thermometer used to measure the surface of the vehicle and the material to be used on the vehicle.

Surface Thermometer   When using a paint drying room, it measures panel temperatures to prevent overheating damage.

Surface Waviness   A measurement of an area's general levelness or trueness.

Suspension System   It allows the tires and wheels to move up and down with road surface irregularities.

Swirl Marks   Round or curved lines in the paint caused by buffing.

Systemic Approach   An organized and logical sequence of steps to accomplish a task or job.

Tabulation Chart   A chart used to log reference point measurements during a repair.

*Tack   To wipe very small particles from a surface using a special cloth, called a tack cloth.

Tack Cloth   A rag with coating that holds dust and debris.

Tack Coat   Also known as mist coat, it is a very light, mist coat applied to the surface first. It allows the application of heavier wet coats without sagging or runs.

*Tack Weld   A small, temporary weld placed at intervals along a joint to hold the pieces in alignment.

Tack-Rag   A cloth used to remove any loose dust and other debris on vehicle surfaces.

Tailgate   A door that allows access to the rear of a station wagon.

Tape Rule   Also known as tape measure, it will extend out for making very long measurements.

Taper Pin   A pin with a larger diameter on one end than on the other. It is used to locate and position matching parts or to secure small pulleys and gears to shafts.

*TCS   Traction control system. A system which improves traction during vehicle acceleration and cornering.

*Tempered Glass   A strong, break-resistant type of safety glass that, if broken, shatters into small granular pieces.

*Tensile Strength   The maximum force per unit area that a material can withstand in tension before failure occurs. Commonly expressed in kilopascals (pounds per square inch).

*Tensioner   A device in a seat belt system which automatically tightens the belt around an occupant, when there is sudden deceleration of the vehicle.

Test Light   An instrument used to determine if there is current flowing through the circuit.

*Thermal Paint   A paint that either melts or changes color when a specified temperature is reached.

Thermometer   A device used to measure temperature.

*Thermosetting Plastics   Plastics which are heated during manufacture and molded into a fixed shape. After manufacture, the shape cannot be altered by heat.

Thermostat   Regulates coolant flow and system operating temperature.

Thread Pitch Gauge   A device that measures bolt thread pitch.

Three Body Sections   For simplicity and to help communication in the auto body repair, a vehicle is divided into front, center, and rear sections.

*Three-Dimensional Measuring System   A measuring system that can locate points with the dimensions of length, width, and height, relative to three defined reference planes.

Three-Way Tailgate   On most unibody station wagons, it is the rear gate that can be operated as a tailgate with the glass fully down, or as a door with the glass up or down.

*Thrust Angle   The angle between the centerline of the vehicle and an imaginary line that is parallel to the track of the rear wheels. Ideally, the thrust angle should be zero.

*Tie Rod   A rod attached to the steering arm, that controls the steering angle of the wheel.

Tie-Rod Ends   There are two that thread onto the end of the rack and are covered by rubber bellows boots in the rack-and-pinion steering unit.

Tinsnips   The most common metal-cutting tool.

Tinted Glass   It contains a shaded vinyl material to filter out most of the sun's glare. Both laminated and tempered glass can be tinted.

Tinting   Altering the paint color slightly to better match the new finish with the old finish.

Tire Changer   A machine for quickly dismounting and mounting tires on and off wheels.

Tire Pressure Gauge   An instrument that will accurately measure tire air pressure.

*Toe   The angle that the tires point inward or outward relative to straight ahead, when viewed from above the vehicle.

*Tolerance   The allowable plus-or-minus variation from a specified value.

*Tone Ring   A toothed wheel on a hub or axle shaft, that changes the magnetic field of the speed sensor in proportion to the wheel speed. A tone ring may also be called a tone wheel, toothed wheel, reluctor, or gear pulser.

Tool and Equipment Storage Room   An area for safely keeping specialized tools and equipment.

Tool Box   A place to store and protect tools.

Tool Holders   Refers to clip racks, pouches, or trays that help you organize small tools.

*Topcoat   The final paint material applied to a surface. Several coats of topcoat are applied in some cases.

Torch Shrinking   Uses the heat of an oxyacetylene torch to release tension in the panel.

Torque Boxes   The structural parts of the frame designed to allow some twisting to absorb road shock and collision impact.

Torque Pattern   A tightening sequence that assures that parts are clamped down evenly by several bolts or nuts.

Torque Specifications Tightening values for the specific bolt or nut.

Torque Wrench A tool used to measure tightening or twisting force.

*****Torsion Bar** A steel rod that, when twisted, functions as a spring in a suspension system.

*****Torsion Rod** A type of rod that, when twisted, acts as a spring. Also called torsion spring.

Torsional Strength The property of a material that withstands a twisting force.

Touch-Up Spray Gun A very small gun, ideal for painting small repair areas.

*****Track-Control Bar** A suspension part that limits the body lateral movement. Also called antisway bar, or track bar.

*****Trailing Arm** A suspension part, mounted ahead of a wheel or axle, that controls the fore-and-aft position of the wheel or axle.

Tram Gauge A special body dimension measuring tool. It determines the direction and amount of body misalignment damage.

Transfer Efficiency The percentage of paint a spray gun is capable of depositing on the surface.

Transmission An assembly with a series of gears for increasing torque to the drive wheels so the car can accelerate properly.

Travel Speed When welding, it is the speed you move the gun across the joint.

Tread Depth Gauge An instrument that will quickly measure tire wear.

Trim Pad Tool A tool designed to reach behind interior panels to pop out and remove clips.

Trim Screw A screw that has a washer attached to it. It improves the appearance and helps keep the trim from shifting.

Trunk Floor Panel A stamped steel part that forms the bottom of the rear storage compartment.

Turbine Spray System The unit produces its own pressure and does NOT use shop air.

*****Turning Radius** The difference in degrees between the steering angle of the inside wheel and the steering angle of the outside wheel during a turn.

Twin I-Beam Suspension A combination of a solid axle and independent suspension. Two large I-beams form the lower control arms.

Two-Part Adhesive Systems An adhesive that consists of a base resin and a hardener (catalyst). When mixed together they form an acceptable alternative to welding for many plastic repairs.

Two-Part Putty Spot putty that comes with its own hardener for rapid curing.

Two-Sided Plastic Weld The strongest type of weld because you weld both sides of the part.

Two-Stage Paints Consists of two distinct layers of paint: basecoat and clearcoat.

*****U-Joint** Universal joint. A device that couples rotary motion between shafts that meet at an angle.

*****UHSS** Ultra high-strength steel. A type of steel which is stronger than high-strength steel and is able to withstand stresses of up to 110,000 psi.

UL (Underwriters Laboratories) Nonprofit organization that operates laboratories to investigate materials, devices, products, equipment and construction methods to define any hazards that may affect life and property.

*****Undercoat** Any of several types of paint materials that are applied before the basecoat or colorcoat, such as primers, sealers, etc.

*****Undercut** A weld condition where the base metal is burned away, below the surrounding area.

Undermasked A term describing an area that must be cleaned with solvent or polishing compound to remove overspray.

Unibody A type of construction where body parts are welded and bolted together to from an integral frame. Lighter, thinner, high-strength steel alloys or aluminum alloy is used.

*****Unidirectional Glass Cloth** A fiberglass cloth with the fibers running mostly in one direction.

Universal Joint See U-joint.

*****Urethane Adhesive** Any of several strong polymer adhesives that are used to install auto glass. Urethane adhesives are necessary to meet government standards for windshield retention in most late-model passenger vehicles.

USC The abbreviation for United States Customary, which is the system used to measure bolt head size.

Utility Knife A hand tool with a retractable blade used for cutting or trimming.

Vacuum Gauge A device that reads vacuum, negative pressure or "suction." It reads in inches of mercury or kilograms per square centimeter.

Vacuum Suction Cup A tool that uses a remote power source (separate vacuum pump or air compressor airflow) to produce negative pressure (vacuum) in the cup. It is used to pull out larger, deeper dents than a suction cup.

*****Valance** A decorative exterior panel used to fill a gap, or hide other parts.

Variance Chips Several samples of color used to help match paint color variations.

*****VAT** Volt-amp tester. A voltage and amperage tester used to test starters and alternators under load.

Vehicle Construction How the vehicle was manufactured, there are many variations.

Vehicle Dimension Manual A guide that gives body and frame measurements from undamaged vehicles.

*****Vehicle Identification Number (VIN) Plate** A permanently installed plate, displaying the vehicle identification number, which is viewable through the windshield from outside the vehicle.

Vehicle Left Side This is the steering wheel side on vehicles built for American roads.

Vehicle Preparation The final steps prior to painting, including cleaning, sanding, stripping, masking, priming, and other related tasks.

Ventilated Has openings for air circulation.

Vertical Welding The pieces are turned upright. Gravity tends to pull the puddle down the joint.

*****VIN Notch** A gap in the dark band across the bottom of the windshield that allows the vehicle identification number (VIN) to be read from outside the vehicle.

Vinyl   A soft, flexible, thin plastic material often applied over a foam filler.

Vinyl Adhesive   An adhesive designed to bond a vinyl top to the vehicle body.

Vinyl Decals   A more complex and decorative pressure-sensitive material than tape stripes.

Vinyl Overlay Remover   A chemical that dissolves plastic vinyl to aid its removal.

Vinyl Prep   A chemical solution that opens the pores of the vinyl material, preparing it for color and/or clearcoating.

Vinyl Tape   A tough, durable, decorative, striping material adhered to the vehicle's finish. It has a pressure-sensitive backing that bonds securely to the finish

*Viscosity   The thickness of a liquid that affects its ability to flow.

Viscosity Cup   A device used to measure the thickness or fluidity of the mixed materials, usually paint. It is a small stainless steel cup attached to a handle.

Viscosity Cup Seconds   Number of seconds paint should take to flow out of viscosity cup if thinned or reduced properly.

Vise   A device used to secure or hold parts during hammering, cutting, drilling and pressing operations. It normally bolts onto a workbench.

Vise Caps   Soft lead, wood, or plastic vise jaw covers that will protect a part from marring.

Vise Grip Pliers   Also known as locking pliers, they have jaws that lock into position on parts. The jaws can be adjusted to different widths.

Visual Weld Inspection   The most practical method of inspection on a repaired vehicle because the weld is not taken apart. It involves looking at the weld for flaws, proper size and proper shape.

VOC Tracking Systems   A computerized system that helps monitor and record materials used in order to abide by environmental regulations.

*Volatile Organic Compound (VOC)   A solvent that, when released into the atmosphere, forms ozone.

Voltage   The pressure that pushes the electricity through the wire or circuit.

*Wander   An unwanted drift of a vehicle from straight ahead, caused by excessive wear or misalignment of the steering system. Driver corrections usually result in a side-to-side motion.

Washer   A device used under bolts, nuts, and other parts. They prevent damage to the surfaces of parts and provide better holding power.

*Warning Lamp   Any of several dash-mounted lamps that alert the driver of a problem or dangerous condition, such as check engine, low fuel, door ajar, brakes, etc.

*Wash Primer   A primer which contains an etching agent and leaves little or no surface build. It is used in place of metal treatment, before priming, to provide corrosion protection.

Washup   A thorough cleaning of the vehicle before repairs begin.

Water Hose Test   A method used to find body leaks, water is sprayed onto the possible leakage areas, while someone watches for leaks inside the vehicle.

Water Pump   It circulates coolant through the inside of the engine, hoses, and radiator.

Waterbase Paint   Paint that uses water to carry the pigment.

*Wax and Grease Remover   A cleaner used to remove contaminants from a repair area before starting a repair, and again before applying primers or paint.

Weatherstrip Adhesive   An adhesive designed to hold rubber seals and similar parts in place.

Weatherstripping   A rubber seal that prevents leakage at the joint between the movable part (lid, hatch, door) and the body.

*Webbing-Sensitive Retractor   A type of seat-belt retractor that locks up when a sudden pullout of the webbing occurs.

*Weld-Bond Adhesive   An adhesive used between the weld surfaces of a joint to seal and strengthen the joint.

Weld Cracks   Cracks on the top or inside the weld bead.

Weld Distortion   Uneven weld bead.

Weld Legs   The width and height of the weld bead.

Weld Overlap   Excess weld metal mounted on top and either side of weld bead.

Weld Penetration   Indicated by the height of the exposed surface of the weld on the back side. It is needed to assure maximum weld strength.

Weld Porosity   Holes in the weld.

Weld Spatter   Drops of electrode on and around weld bead.

Weld Throat   This refers to the depth of the triangular cross section of the weld.

*Weld-Through Primer   A corrosion-resistant primer applied, before welding, to mating surfaces which are uncoated, or where a zinc coating has been removed.

Weld Undercut   A groove melted along either side of the weld and left unfilled.

Welded Door Skin   Spot welds hold the skin onto its frame.

Welded Parts   Metal or plastic parts that are permanently joined by melting the material, so that it flows together and bonds when cooled.

Welded-Pull Tab   A small piece of steel plate welded onto the unibody. It is used if you need to pull where there is no place to attach a clamp.

Welder Amperage Rating   It gives the maximum current flow through the weld joint. The thickness of the metal being welded determines the amperage needed.

Weldface   The exposed surface of the weld on the side where you welded.

Welding Filter Lens   Sometimes called filter plate, it is a shaded glass welding helmet insert for protecting your eyes from ultraviolet burns.

Weldroot   The part of the joint where the wire electrode is directed.

*Western Union Splice   A method of twisting two wires together, before soldering, to make a mechanically strong joint.

Wet Edge   The area just painted that is still wet when starting to paint a new adjoining area.

Wet Metallic Paint Spray   The aluminum flakes have sufficient time to settle in the wet paint, making the color appear darker and less silver.

Wet Sandpaper   Sandpaper that can be used with water for flushing away sanding debris that would otherwise clog fine grits.

Wheel Alignment Machine   A machine used to measure wheel alignment angles so they can be adjusted.

Wheel Clamps   Wheel clamps are used to load and hold suspension with wheels and tires removed with vehicle on the rack or bench.

Wheel Cylinders   It uses hydraulic pressure to push the brake pads or shoes outward.

Wheel Puller   A device used to remove steering wheels, engine pulleys, and similar pressed-on parts.

Wheel Run-Out Check   A test that will show if there is damage to a rotating part of a wheel assembly.

White Light   A mixture of various colors of light.

Windings   Wires wrapped around an iron core.

Window Antenna Wire   A wire for radio reception, placed either between the layers of laminated glass (windshield), or printed on the surface of the glass (rear window).

Window Regulator   A gear mechanism that allows you to raise and lower the door glass.

Windshield Gasket   A rubber molding shaped to fit the windshield, body and trim.

Windshield Knife   A hand tool designed to cut through the sealant used on some windshields. It has a sharp cutter blade and two handles.

Wire Color Coding   The wire insulation is colored for easy identification.

Wire Speed   In welding, it is how fast the rollers feed the wire into the weld puddle.

Wire Stick-Out   The distance between the end of the gun's contact tip and the metal being welded.

Wiring Diagram   A drawing of an electrical circuit used to determine and isolate problems.

Wiring Harness   Several wires are enclosed in a protective covering.

Woodruff Key   A key used to prevent hand wheels, gears, cams, and pulleys from turning on their shaft.

Work Hardening   The limit of plastic deformation that causes the metal to become hard where it has been bent.

Work Life   The time when it is still possible to disturb the adhesive and have the adhesive set up for a good bond.

Work Order   It outlines the procedures that should be taken to put the vehicle back in top condition.

Work Traits   Qualities, aside from skills, that make you a good or bad employee, i.e., reliability, productivity, professionalism.

Worm Hose Clamp   A clamp that uses a screw that engages a slotted band. Turning the screw reduces or enlarges clamp diameter.

Wrench Size   The distance across the wrench jaws.

X-Frame   A frame that has two long members crossing over in the middle of the frame rails. Also called a backbone frame.

Yield Point   The amount of force that a piece of metal can resist without tearing or breaking.

Yield Stress   The amount of strain needed to permanently deform a test specimen of sheet metal.

*Zinc Coating   A coating of zinc applied to metal during manufacture to improve corrosion resistance. Also called galvanizing.

Zone Concept   It divides the horizontal surfaces of the vehicle into zones defined by character lines and moldings. It requires refinishing of an entire zone or zones with basecoat, mica intermediate coats, and clearcoats.

APPENDIX A

ABBREVIATIONS USED BY COLLISION REPAIR TECHNICIANS AND ESTIMATORS

The estimator and collision repair technician must be able to communicate verbally as well as in writing. Both in estimates and work procedure reports most estimators use abbreviations. Generally, these abbreviations are the same as those used in crash estimating guides. Some abbreviations are even used verbally. For example, three of the most commonly used abbreviations in a body shop are:

- **R&I: Remove and Reinstall.** The item is removed as an assembly, set aside, and later reinstalled and aligned for a proper fit. This is generally done to gain access to another part. For example, R&I bumper would mean that the bumper assembly would have to be removed to install a new fender or quarter panel.
- **R&R: Remove and Replace.** Remove the old parts, transfer necessary items to new part, replace, and align.
- **O/H: Overhaul.** Remove an assembly from the vehicle, disassemble, clean, inspect, replace parts as needed, then reassemble, install and adjust (except wheel and suspension alignment).

In addition to these abbreviations, the following terms are those accepted by most estimating guides, shop manuals, and estimators. They are the ones used in most written forms.

A	Manufacturer has no list price for the part.
ABS	Anti-lock Brake System
A/C	Air Conditioner
ACRS	Air Cushion Restraint System
adj	adjuster or adjustable
A/F	Air Fuel Ratio
AIR	Air Injector Reactor
alt	alternator
alum	aluminum
amp	ampere
approx.	approximately
assy	assembly
A/T	Automatic Transaxle
AT	Automatic Transmission
auth	authority
auto	automatic
aux	auxiliary
B+	Battery Voltage
bat.	battery
bbl	barrel
bk	back
blwr	blower
bmpr	bumper
brg	bearing
brk	brake
brkt	bracket
Bro	Brougham
btry	battery
btwn	between

B-U	Back-Up
bush	bushing
Calif	California
chnl	channel
c/mbr	cross member
cntr	center
col	column
comp	compressor
compt	compartment
conn.	connector
cond	conditioning or conditioner
cont	control
conv	converter or convertible
cor	corner
cov	cover
Cpe	Coupe
C/R	Customer Requested
crossmbr	cross member
c/shaft	crankshaft
ctl	control
Ctry	Country
Cust	Custom
CV	Constant Velocity
cyl	cylinder
D	Discontinued part
d	drilling operational time
da	dash
dbl	double

def	deflector		ID	Inside Diameter
dehyd	dehydrator		ign	ignition
desc	description		in	inch
dia	diameter		incl	includes
diag	diagonal		INJ	Injection
dist	distributor		in-lb	inch-pounds
div	division		inr	inner
Dlxe	DeLuxe		inst	instrument
DP	Dash Pot		inter	intermediate
dr	door		IP	Instrument Control Panels
ea	each		L	Left
ECU	Electronic Control Unit		LED	Light Emitting Diode
EFI	Electronic Fuel Injection		LF	Left Front
EGR	Exhaust Gas Recirculation		LH	Left Hand
elec	electric		lic	license
emiss	emission		lp	lamp
eng	engine		LR	Left Rear
EP	Exhaust Purging		LS	Left Side
equip	equipment		lwr	lower
evap	evaporator			
exc	except		max	maximum
exh	exhaust		mdl	model
extn	extension		mldg	molding
			MT	Manual Transmission
FED	Federal (All States Except Calif.)		mtd	mounted
flr	floor		mtg	mounting
Fndr	Fender		muff	muffler
Fr & Rr	Front & Rear			
frm	from		NAGS	National Auto Glass Specification
fr or rr	front or rear		neg	negative
ft	foot			
ft-lb	foot-pounds		OD	Outside Diameter
FWD	Front Wheel Drive		OD, O/D	Overdrive
			OEM	Original Equipment Manufacturer
gal	gallon		opng	opening
gen	generator		orna	ornament
GND	Ground		O/S	Oversize
grds	guards		otr	outer
grv	groove			
			p	paint operational time
harn	harness		pass	passenger
H'back	Hatchback		PCV	Positive Crankcase Ventilation
hd	heavy duty		pkg	package
HDC	Heavy Duty Cooling		plr	pillar
hdr	header		pnl	panel
HEI	High Energy Ignition		pos	positive
Hi Per	High Performance		PS	Power Steering
horiz	horizontal		psi	pounds per square inch
H.P.	High Performance		pt	pint
hp	horsepower		Pwr	Power
hsg	housing			
HSLA	High Strength Alloy Steel		qtr	quarter
HSS	High Strength Steel			
HT	Hard Top		R	Right
H'Top	Hard Top		rad	radiator
hyd	hydraulic		R-L	Right or Left
Hydra	Hydramatic		rec	receiver

refl	reflector		TE	Thermactor Emission
reg	regulator		tel	telescope
reinf	reinforcement		trans	transmission
reson	resonator		TWC	Three-Way Catalyst
RH	Right-Hand			
Sed	Sedan		U/D	Underdrive
ser	serial or series		upr	upper
shld	shield		U/S	Undersize
sidembr	side member			
sig	single			
SIR	Supplemental Inflatable Restraint		vent	ventilator
spd	speed		vert	vertical
spec	special		vib	vibration
SRS	Supplemental Restraint System		VIR	Valve-In-Receiver
Sta	Station			
stab	stabilizer			
stat	stationary		w/	with
Std	Standard		w/b	wheel base
stl	steel		WB	Wheelbase
strg	steering		WD	Wheel Drive
Sub	Suburban		Wgn	Wagon
sup	super		w'house	wheelhouse
supt	support		whl	wheel
surr	surround		whlse	wheelhouse
susp	suspension		wndo	window
SW	Station Wagon		w/o	without
			wo/	without
tach	tachometer		wshd	windshield
t & t	tilt & telescope or tilt & travel		w'strip	weatherstrip

APPENDIX B

MEASUREMENT EQUIVALENTS

CONVERSION FACTORS		
Multiply	**By**	**To obtain**
Length		
Millimeters (mm)	0.03937	Inches
	0.1	Centimeters (cm)
Kilometers (km)	0.6214	Miles
	3281	Feet
Inches	25.4	Millimeters (mm)
Miles	1.6093	Kilometers (km)
Area		
Inches^2	645.16	Millimeters^2 (mm^2)
	6.452	Centimeters^2 (cm^2)
Feet^2	0.0929	Meters^2 (m^2)
	144	Inches^2
Volume		
Centimeters^3 (cc)	0.06102	Inches^3
	0.001	Liters (L)
Liters (L)	61.024	Inches^3
	0.2642	Gallons
	1.0567	Quarts
Inches^3	16.387	Centimeters^3 (cc)
Feet^3	1728	Inches^3
	7.48	Gallons
	28.32	Liters (L)
Fluid ounces (oz)	29.57	Milliliters (mL)
Mass		
Gram (g)	0.03527	Ounce
Kilograms (kg)	2.2046	Pounds
	35.274	Ounces
Force		
Ounce	0.278	Newton (N)
Pound	4.448	Newton (N)
Kilogram	9.807	Newton (N)
Torque		
Foot-pounds	1.3558	Newton-meters (Nm)
	0.1383	Kilogram/meter (kg/m)
Inch-pounds	0.11298	Newton-meters (Nm)
	0.0833	Foot-pounds

CONVERSION FACTORS (CONTINUED)

Multiply	By	To obtain
Torque		
Kilogram-meters (Kg/m)	7.23	Foot-pounds
	9.80665	Newton-meters (Nm)
Pressure		
Atmospheres	14.7	Pounds/square inch (psi)
	29.92	Inches of mercury (In. Hg)
Inches of mercury (In. Hg)	0.49116	Pounds/square inch (psi)
	13.1	Inches of water
	3.377	Kilopascals (kPa)
Bars	100	Kilopascals (kPa)
	14.5	Pounds/square inch (psi)
Kilogram/cu² (Kg/cm²)	14.22	Pounds/square inch (psi)
	98.07	Kilopascals (kPa)
Kilopascals (kPa)	0.145	Pounds/square inch (psi)
	0.2961	Inches of mercury (In. Hg)
Pascal	1	Newton/square meter (N/m²)
Fuel Performance		
Miles/gallon	0.4251	Kilometers/liter (km/L)
Velocity		
Miles/hours	1.467	Feet/second
	88	Feet/minute
	1.6093	Kilometers/hour (km/h)
Kilometers/hour	0.27778	Meters/second (m/s)

TORQUE CONVERSIONS
(FOOT-POUNDS, NEWTON-METERS, METER-KILOGRAMS)

ft/lb	0	1	2	3	4	5	6	7	8	9
0	0	1.35	2.70	4.05	5.40	6.75	8.10	9.45	10.8	12.1
10	13.5	14.9	16.2	17.6	18.9	20.3	21.6	22.9	24.3	25.6
20	27.0	28.3	29.7	31.0	32.5	33.7	35.1	36.4	37.8	39.1
30	40.5	41.8	43.2	44.5	45.9	47.2	48.6	49.9	51.3	52.6
40	54.0	55.3	56.7	58.0	59.4	60.7	62.1	63.4	64.8	66.1
50	67.5	68.8	70.2	71.5	72.9	74.2	75.6	76.9	78.3	79.6
60	81.0	82.3	83.7	85.0	86.4	87.7	89.1	90.4	91.8	93.1
70	94.5	95.8	97.2	98.5	99.9	101	102	103	105	106
80	108	109	110	112	113	114	116	117	118	120
90	121	122	124	125	126	128	129	130	132	133
100	135	136	137	139	140	141	143	144	145	147
110	148	149	151	152	153	155	156	157	159	160
120	162	163	164	166	167	168	170	171	172	174
130	175	176	178	179	180	182	183	184	186	187

Note: The following formulas can be used:

$$\text{ft-lb} \times 1.35 = \text{Nm}$$
$$\text{ft-lb} + 7.23 = \text{mkg}$$

For meter-kilograms within 5%, divide Newton-meters by 10 (move the decimal point one place left).
For greater accuracy, divide by 9.81.

DECIMAL AND METRIC EQUIVALENTS

Fractions	Decimal (in.)	Metric (mm)	Fractions	Decimal (in.)	Metric (mm)
1/64	.015625	.397	33/64	.515625	13.097
1/32	.03125	.794	17/32	.53125	13.494
3/64	.046875	1.191	35/64	.546875	13.891
1/16	.0625	1.588	9/16	.5625	14.288
5/64	.078125	1.984	37/64	.578125	14.684
3/32	.09375	2.381	19/32	.59375	15.081
7/64	.109375	2.778	39/64	.609375	15.478
1/8	.125	3.175	5/8	.625	15.875
9/64	.140625	3.572	41/64	.640625	16.272
5/32	.15625	3.969	21/32	.65625	16.669
11/64	.171875	4.366	43/64	.671875	17.066
3/16	.1875	4.763	11/16	.6875	17.463
13/64	.203125	5.159	45/64	.703125	17.859
7/32	.21875	5.556	23/32	.71875	18.256
15/64	.234275	5.953	47/64	.734375	18.653
1/4	.250	6.35	3/4	.750	19.05
17/64	.265625	6.747	49/64	.765625	19.447
9/32	.28125	7.144	25/32	.78125	19.844
19/64	.296875	7.54	51/64	.796875	20.241
5/16	.3125	7.938	13/16	.8125	20.638
21/64	.328125	8.334	53/64	.828125	21.034
11/32	.34375	8.731	27/32	.84375	21.431
23/64	.359375	9.128	55/64	.859375	21.828
3/8	.375	9.525	7/8	.875	22.225
25/64	.390625	9.922	57/64	.890625	22.622
13/32	.40625	10.319	29/32	.90625	23.019
27/64	.421875	10.716	59/64	.921875	23.416
7/16	.4375	11.113	15/16	.9375	23.813
29/64	.453125	11.509	61/64	.953125	24.209
15/32	.46875	11.906	31/32	.96875	24.606
31/64	.484375	12.303	63/64	.984375	25.003
1/2	.500	12.7	1	1.00	25.4

APPENDIX C

TAP AND DRILL BIT DATA

DECIMAL EQUIVALENTS AND TAP DRILL SIZES			DECIMAL EQUIVALENTS AND TAP DRILL SIZES		
Drill Size	Decimal	Tap Size	Drill Size	Decimal	Tap Size
1/64	.0156		27	.1440	
1/32	.0312		26	.1470	
60	.0400		25	.1495	10–24
59	.0410		24	.1520	
58	.0420		23	.1540	
57	.0430		5/32	.1562	
56	.0465		22	.1570	10–30
3/64	.0469	0–80	21	.1590	10–32
55	.0520		20	.1610	
54	.0550	1–56	19	.1660	
53	.0595	1–64, 72	18	.1695	
1/16	.0625		11/64	.1719	
52	.0635		17	.1730	
51	.0670		16	.1770	12–24
50	.0700	2–56, 64	15	.1800	
49	.0730		14	.1820	12–28
48	.0760		13	.1850	12–32
5/64	.0781		3/16	.1875	
47	.0785	3–48	12	.1890	
46	.0810		11	.1910	
45	.0820	3–56, 4–32	10	.1935	
44	.0860	4–36	9	.1960	
43	.0890	4–40	8	.1990	
42	.0935	4–48	7	.2010	1/4–20
3/32	.0937		13/64	.2031	
41	.0960		6	.2040	
40	.0980		5	.2055	
39	.0995		4	.2090	
38	.1015	5–40	3	.2130	1/4–28
37	.1040	5–44	7/32	.2187	
36	.1065	6–32	2	.2210	
7/64	.1093		1	.2280	
35	.1100		A	.2340	
34	.1110	6–36	15/64	.2344	
33	.1130	6–40	B	.2380	
32	.1160		C	.2420	
31	.1200		D	.2460	
1/8	.1250		E, 1/4	.2500	
30	.1285		F	.2570	5/16–18
29	.1360	8–32, 36	G	.2610	
28	.1405	8–40	17/64	.2656	
9/64	.1406		H	.2660	

DECIMAL EQUIVALENTS AND TAP DRILL SIZES		
Drill Size	Decimal	Tap Size
I	.2720	5/16–24
J	.2770	
K	.2810	
9/32	.2812	
L	.2900	
M	.2950	
19/64	.2968	
N	.3020	
5/16	.3125	3/8–16
O	.3160	
P	.3230	
21/64	.3281	
Q	.3320	3/8–24
R	.3390	
11/32	.3437	
S	.3480	
T	.3580	
23/64	.3594	
U	.3680	7/16–14
3/8	.3750	
V	.3770	
W	.3860	
25/64	.3906	7/16–20
X	.3970	
Y	.4040	
13/32	.4062	
Z	.4130	
27/64	.4219	1/2–13
7/16	.4375	
29/64	.4531	1/2–20
15/32	.4687	
31/64	.4844	9/16–12
1/2	.5000	

DECIMAL EQUIVALENTS AND TAP DRILL SIZES		
Drill Size	Decimal	Tap Size
33/64	.5156	9/16–18
17/32	.5312	5/8–11
35/64	.5469	
9/16	.5625	
37/64	.5781	5/8–18
19/32	.5937	11/16–11
39/64	.6094	
5/8	.6250	11/16–16
41/64	.6406	
21/32	.6562	3/4–10
43/64	.6719	
11/16	.6875	3/4–16
45/64	.7031	
23/32	.7187	
47/64	.7344	
3/4	.7500	
49/64	.7656	7/8–9
25/32	.7812	
51/64	.7969	
13/16	.8125	7/8–14
53–64	.8281	
27/32	.8437	
55/54	.8594	
7/8	.8750	1–8
57/64	.8906	
29/32	.9062	
59/64	.9219	
15/16	.9375	1–12, 14
61–64	.9531	
31/32	.9687	
63/64	.9844	
1	1.000	

PIPE THREAD SIZES			
Thread	Drill	Thread	Drill
1/8–27	R	1 1/2–11 1/2	1 47/64
1/4–18	7/16	2–11 1/2	2 7/32
3/8–18	37/64	2 1/2–8	2 5/8
1/2–14	23/32	3–8	3 1/4
3/4–14	59/64	3 1/2–8	3 3/4
1–11 1/2	1 5/32	4–8	4 1/4
1 1/4–11 1/2	1 1/2		

METRIC TAP DRILL SIZES		
Diameter and Pitch	Metric Drill	Inch Drill
5 × .80	4.20	11/64
6 × 1.00	5.00	13/64
7 × 1.00	6.00	15/64
8 × 1.25	6.75	17/64
10 × 1.50	8.50	11/32
12 × 1.75	10.25	13/32

Note: Nominal outside diameter minus the pitch equals the tap drill size for 75% thread contact area.

DECIMAL EQUIVALENTS OF NUMBER SIZE DRILLS

No.	Size of Drill (Inches)	No.	Size of Drill (Inches)	No.	Size of Drill (Inches)	No.	Size of Drill (Inches)
1	.2280	21	.1590	41	.0960	61	.0390
2	.2210	22	.1570	42	.0935	62	.0380
3	.2130	23	.1540	43	.0890	63	.0370
4	.2090	24	.1520	44	.0860	64	.0360
5	.2055	25	.1495	45	.0820	65	.0350
6	.2040	26	.1470	46	.0810	66	.0330
7	.2010	27	.1440	47	.0785	67	.0320
8	.1990	28	.1405	48	.0760	68	.0310
9	.1960	29	.1360	49	.0730	69	.0292
10	.1935	30	.1285	50	.0700	70	.0280
11	.1910	31	.1200	51	.0670	71	.0260
12	.1890	32	.1160	52	.0635	72	.0250
13	.1850	33	.1130	53	.0595	73	.0240
14	.1820	34	.1110	54	.0550	74	.0225
15	.1800	35	.1100	55	.0520	75	.0210
16	.1770	36	.1065	56	.0465	76	.0200
17	.1730	37	.1040	57	.0430	77	.0180
18	.1695	38	.1015	58	.0420	78	.0160
19	.1660	39	.0995	59	.0410	79	.0145
20	.1610	40	.0980	60	.0400	80	.0135

DECIMAL EQUIVALENTS OF LETTER SIZE DRILLS

Letter	Size of Drill (Inches)	Letter	Size of Drill (Inches)
A	.234	N	.302
B	.238	O	.316
C	.242	P	.323
D	.246	Q	.332
E	.250	R	.339
F	.257	S	.348
G	.261	T	.358
H	.266	U	.368
I	.272	V	.377
J	.277	W	.386
K	.281	X	.397
L	.290	Y	.404
M	.295	Z	.413

HELICOIL TAP DRILL SIZES

Inches		Metric	
Thread	Tap Drill Size	Thread	Tap Drill Size
1/4–20	17/64	5 × 0.8	13/64
5/16–18	21/64	6 × 1.0	1/4
3/8–16	25/64	7 × 1.0	9/32
7/16–14	29/64	8 × 1.25	21/64
1/2–13	17/32	10 × 1.50	13/32
		12 × 1.75	31/64

APPENDIX D

REFERENCE TABLES

DRY TORQUE RECOMMENDATIONS						
	Grade 5		Grade 6		Grade 8	
Diameter	Coarse	Fine	Coarse	Fine	Coarse	Fine
1/4	108[a]	120[a]	132[a]	156[a]	17	19
5/16	17	20	23	25	34	37
3/8	31	35	40	45	60	68
7/16	50	55	64	72	96	108
1/2	75	85	98	110	145	165
9/16	110	120	140	160	210	235
5/8	150	170	195	220	290	330

[a]Any torque less than 15-lb is given in in-lb.

LUBRICATED TORQUE RECOMMENDATIONS (REDUCED 33 PERCENT)						
	Grade 5		Grade 6		Grade 8	
Diameter	Coarse	Fine	Coarse	Fine	Coarse	Fine
1/4	72[a]	80[a]	88[a]	104[a]	136[a]	152[a]
5/16	135[a]	160[a]	15	17	22	25
3/8	21	23	26	30	40	45
7/16	33	37	43	48	64	72
1/2	50	57	65	73	96	110
9/16	73	80	93	107	140	157
5/8	100	113	130	147	193	220

[a]Any torque less than 15-lb is given in in-lb.

METRIC DRY TORQUE RECOMMENDATIONS			
Diameter (mm)	Pitch	Grade 8.8	Grade 12.9
6	1.00	84[a]	132[a]
7	1.00	132[a]	20
8	1.25	18	29
8	1.00	20	32
10	1.50	33	58
10	1.25	35	61
10	1.00	38	64
12	1.75	59	100
12	1.50	65	110

[a]Any torque less than 15 ft-lb is given in in.-lb.

METRIC LUBRICATED TORQUE RECOMMENDATIONS (REDUCED 33 PERCENT)			
Diameter (mm)	Pitch	Grade 8.8	Grade 12.9
6	1.00	56[a]	88[a]
7	1.00	88[a]	160[a]
8	1.25	144[a]	19
8	1.00	160[a]	21
10	1.50	22	39
10	1.25	23	41
10	1.00	25	43
12	1.75	39	67
12	1.50	43	73

[a]Any torque less than 15 ft-lb is given in in.-lb.

APPENDIX E

TORQUE VALUES

The following charts give the standard torque values for bolts, nuts, and taperlock studs of SAE Grade 5 or better quality.

GENERAL TIGHTENING TORQUE FOR BOLTS, NUTS, AND TAPERLOCK STUDS			
Thread Diameter		**Standard Torque**	
Inches	**Millimeters**	**ft.-lb.**	**N·m***
Standard Thread 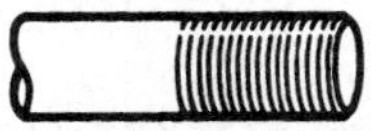		Use these torques for bolts and nuts with standard threads (conversions are approximate).	
1/4	6.35	9 ± 3	12 ± 4
5/16	7.94	18 ± 5	25 ± 7
3/8	9.53	32 ± 5	45 ± 7
7/16	11.11	50 ± 10	70 ± 15
1/2	12.70	75 ± 10	100 ± 15
9/16	14.29	110 ± 15	150 ± 20
5/8	15.88	150 ± 20	200 ± 25
3/4	19.05	265 ± 35	360 ± 50
7/8	22.23	420 ± 60	570 ± 80
1	25.40	640 ± 80	875 ± 100
1-1/8	28.58	800 ± 100	1100 ± 150
1-1/4	31.75	1000 ± 120	1350 ± 175
1-3/8	34.93	1200 ± 150	1600 ± 200
1-1/2	38.10	1500 ± 200	2000 ± 275
		Use these torques for bolts and nuts on hydraulic valve bodies.	
5/16	7.94	13 ± 2	20 ± 3
3/8	9.53	24 ± 2	35 ± 3
7/16	11.11	39 ± 2	50 ± 2
1/2	12.70	60 ± 3	80 ± 4
5/8	15.88	118 ± 4	160 ± 6

GENERAL TIGHTENING TORQUE FOR BOLTS, NUTS, AND TAPERLOCK STUDS (CONTINUED)			
Thread Diameter		**Standard Torque**	
Inches	Millimeters	ft.-lb.	N·m*
Taperlock Stud		Use these torques for studs with taperlock threads.	
1/4	6.35	5 ± 2	7 ± 3
5/16	7.94	10 ± 3	15 ± 5
3/8	9.53	20 ± 3	30 ± 5
7/16	11.11	30 ± 5	40 ± 10
1/2	12.70	40 ± 5	55 ± 10
9/16	14.29	60 ± 10	80 ± 15
5/8	15.88	75 ± 10	100 ± 15
3/4	19.05	110 ± 15	150 ± 20
7/8	22.23	170 ± 20	230 ± 30
1	25.40	260 ± 30	350 ± 40
1-1/8	28.58	320 ± 30	400 ± 40
1-1/4	31.75	400 ± 40	550 ± 50
1-3/8	34.93	480 ± 40	650 ± 50
1-1/2	38.10	550 ± 50	750 ± 70

*1 Newton-meter (N·m) is approximately the same as 0.1 mkg.

The torques shown in the charts that follow are to be used on the part of 37° flared, 45° flared and inverted flared fittings (when used with steel tubing), O-ring plugs, and O-ring fittings.

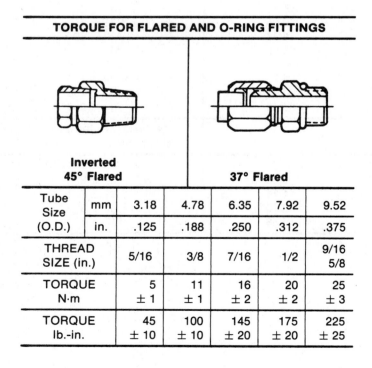

TORQUE FOR FLARED AND O-RING FITTINGS						
Inverted 45° Flared				37° Flared		
Tube Size (O.D.)	mm	3.18	4.78	6.35	7.92	9.52
	in.	.125	.188	.250	.312	.375
THREAD SIZE (in.)		5/16	3/8	7/16	1/2	9/16 5/8
TORQUE N·m		5 ± 1	11 ± 1	16 ± 2	20 ± 2	25 ± 3
TORQUE lb.-in.		45 ± 10	100 ± 10	145 ± 20	175 ± 20	225 ± 25

TORQUE FOR FLARED AND O-RING FITTINGS

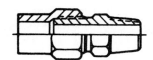

		45° Flared		O-Ring Fitting—Plug			Swivel Nuts		
Tube Size (O.D.)	mm	12.70	15.88	19.05	22.22	25.40	31.75	38.10	50.80
	in.	.500	.625	.750	.875	1.000	1.250	1.500	2.000
THREAD SIZE (in.)		3/4	7/8	11/16	1-3/16 1-1/4	1-5/16	1-5/8	1-7/8	2-1/2
TORQUE N·m		50 ± 5	75 ± 5	100 ± 5	120 ± 5	135 ± 10	180 ± 10	225 ± 10	320 ± 15
TORQUE lb.-ft.		35 ± 4	55 ± 4	75 ± 4	90 ± 4	100 ± 7	135 ± 7	165 ± 7	235 ± 10

INDEX